Springer Monographs in Mathematics

U. Kohlenbach

Applied Proof Theory: Proof Interpretations and Their Use in Mathematics

Springer

Ulrich Kohlenbach
Technische Universität Darmstadt
FB Mathematik
Schloßgartenstr. 7
64289 Darmstadt
Germany
kohlenbach@mathematik.tu-darmstadt.de

ISBN 978-3-540-77532-4 e-ISBN 978-3-540-77533-1

DOI 10.1007/978-3-540-77533-1

Springer Monographs in Mathematics ISSN 1439-7382

Library of Congress Control Number: 2008920614

Mathematics Subject Classification (2000): 03F03, 03F10, 03B30, 41A10, 41A52, 47H9, 47H10

Cover design: WMXDesign GmbH, Heidelberg

Printed on acid-free paper

9 8 7 6 5 4 3 2 1

springer.com

For Gabriele and Katharina

Preface

This book gives an introduction to so-called proof interpretations, more specifically various forms of realizability and functional interpretations, and their use in mathematics. Whereas earlier treatments of these techniques (e.g. [366, 266, 122, 369, 7]) emphasize foundational and logical issues the focus of this book is on applications of the methods to extract new effective information such as computable uniform bounds from given (typically ineffective) proofs. This line of research, which has its roots in G. Kreisel's pioneering work on 'unwinding of proofs' from the 50's, has in more recent years developed into a field of mathematical logic which has been called (suggested by D. Scott) 'proof mining'. The areas where proof mining based on proof interpretations has been applied most systematically are numerical analysis and functional analysis and so the book concentrates on those. There are also some extractions of effective information from proofs (guided by logic) in number theory (G. Kreisel, H. Luckhardt, see e.g. [249, 268, 267, 122]) and algebra (G. Kreisel, C. Delzell, H. Lombardi, T. Coquand and others, see e.g. [252, 84, 77, 74, 76]). However, here mainly methods from structural proof theory such as Herbrand's theorem, ε-substitution and cut-elimination are used and we will refer to the literature for more information on these results.

In this book two kinds of systems play an important role: those with full induction and variants with induction for purely existential formulas (whose central role has been singled out in the context of so-called reverse mathematics, [338]). Further (still weaker) fragments are briefly discussed in comments and referred to in the literature.

Modified realizability (due to G. Kreisel) and functional interpretation (due to K. Gödel) are both first developed in the framework of constructive ('intuitionistic') arithmetic in higher types to which consecutively various non-constructive principles are added.

After this, systems based on ordinary ('classical') logic are studied. It is shown that the combination of Gödel's functional ('Dialectica') interpretation with the so-called negative translation, which embeds certain classical theories into approximately intuitionistic counterparts, can be used to unwind fully non-constructive proofs. Since the main emphasis throughout this book is on ineffective proofs based on ordinary logic and hence on functional interpretation the preceding treatment of modified realizability is largely independent from the rest of this book. However, the study of modified realizability is recommended for a better understanding of the more complicated functional interpretation.

Next, a so-called monotone variant of functional interpretation due to the author is applied which combines functional interpretation with majorizability in the sense of W.A. Howard and allows one to treat the binary König's lemma and related principles such as a strong uniform boundedness principle.

The book presents in detail C. Spector's deep extension of Gödel's functional interpretation to full classical analysis by means of his bar recursive functionals.

As an alternative method to the combination of functional interpretation with negative translation one can – in some circumstances – use instead a combination of modified realizability with negative translation if one inserts the so-called A-translation (due to H. Friedman and A. Dragalin) as an intermediate step. This approach will be briefly discussed as well.

Using suitable standard representations of Polish and compact metric spaces (going back to L.E.J. Brouwer) we develop general metatheorems based on monotone functional interpretation which guarantee the extractability of effective uniform bounds from large classes of proofs in analysis. Moreover, monotone functional interpretation provides an algorithm to carry out such extractions for given proofs.

As an extended case study of the use of these metatheorems and the extraction algorithms a number of concrete proofs in approximation theory (best Chebycheff and L_1-approximation of continuous functions), where this approach has led to new results, are analyzed in great detail

By extending the aforementioned proof interpretations to new formal systems of analysis, where general classes of abstract metric, hyperbolic, CAT(0), normed and uniformly convex spaces are added as new types, very general metatheorems are obtained which guarantee the extractability of strongly uniform bounds which are not only independent from parameters in compact metric spaces but even from parameters in metrically bounded subspaces of such abstract spaces. Further refinement of this approach shows that it actually suffices to suppose certain local boundedness information rather than the boundedness of the whole (sub)space.

Finally, in a second extended case study, these general metatheorems are applied to proofs in metric fixed point theory leading to numerous (even qualitatively) new results which were obtained first by this methodology. These results concern, among other things, the asymptotic behavior of Krasnoselski-Mann iterates of nonexpansive (and more general classes of) functions on hyperbolic spaces. Many more applications in this area are referred to in the literature.

The book is concluded with some speculations about future applications of the methods developed in this book to proofs in the areas of algebraic number theory, partial differential equations, ergodic theory and geometric group theory.

Of course, much work on the general topic of 'computational content of proofs' has been carried out by logical methods other than the ones covered in this book as well as in the context of constructive foundations of mathematics (see e.g. the recent book [349] which, however, only studies logical aspects of calculi and formal systems without mathematical applications). In this book we focus on those methods which have been applied in the past to concrete proofs in different areas of 'core mathematics', have produced new mathematical results in these areas and are likely to be useful in other parts of mathematics as well.

A relevant topic that is beyond the scope of this book is the issue of implementing the techniques developed in suitable programming languages aiming at an automated extraction of algorithms from proofs. Much work in this direction has been done in Munich by the group around H. Schwichtenberg in connection with the MINLOG tool (see e.g. [20, 26, 21]). This work is based on modified realizability and refined versions of the A-translation but, subsequently, also various forms of functional interpretation have been successfully implemented (see e.g. [158]). Due to the enormous difficulties involved in dealing with fully formalized proofs only mathematically rather simple examples have been carried out by such tools yet. As this book focuses on advanced applications to nontrivial proofs in mathematics these implemented tools will be mentioned only briefly with references to the literature.

In addition to standard undergraduate knowledge in mathematics the book only presupposes some familiarity with the basic concepts from elementary recursion theory (up to the Kleene normal form theorem, e.g. [333]) and logic (covered in any introduction to mathematical logic such as [372]). Some previous exposure to constructive ('intuitionistic') logic ([372, 15, 371]) would be helpful but is not required.

Most chapters have exercises and historical comments at the end. These comments, in particular, give detailed references to the relevant literature where the results presented first appeared. In addition to this we give explicit references at key definitions and results themselves in cases where they are neither standard or folklore nor due to the author. Except for a few central results that are joint work with co-authors, we do not label our own results explicitly but refer to the 'historical comments' sections for proper references.

The book's web site

http://www.mathematik.tu-darmstadt.de/∼kohlenbach/prooftheory.html

will list errata and updates.

Darmstadt, *Ulrich Kohlenbach*

December 2007

Acknowledgements

The starting point for this book were some brief BRICS Lecture Notes LS-98-1 for a PhD course on Proof Interpretations which I gave at the University of Aarhus in the spring term 1998. Over many years, these notes were constantly growing incorporating many results from recent research in the area. I am grateful to the participants of my 1998 BRICS PhD course for helpful and clarifying comments and the encouragement to further expand on the course notes.

H. Blasum, E. M. Briseid, G. Ferreira, J. Gaspar, P. Gerhardy, D.-M. Hernest, K.F. Jørgensen, P. Oliva, L. Leuştean and P. Safarik read more recent versions of some chapters of this book and provided numerous corrections and suggestions for improvements. I am also grateful to the participants of my course at the 'Fall school of the Logic seminar of the Mathematical Institute of the Academy of Sciences of the Czech Republic, Pec p. Snezkou, Sept. 14–20, 2003' which was based on early versions of some chapters of this book. The same applies to the participants of a course on this book (together with H. Schwichtenberg) at the summer university of the 'Studienstiftung des deutschen Volkes' of Germany in Salem in August 2006.

Various versions of chapters of this book were used by colleagues for courses and seminars at the Universities of Lisbon, Munich, Oslo and Stanford. I am grateful to Professors J. Avigad, T. Coquand, F. Ferreira, J. Håstad, J. Krajicek, G. Mints, P. Pudlak and H. Schwichtenberg for critical remarks.

I also like to thank Professors S. Feferman, V. F. Hendricks, A. Scedrov, D. Scott and J. Tucker for urging me at various stages to develop my notes into a book.

Above all I am grateful to Professor G. Kreisel for numerous critical comments and invaluable insights on the theme of this book he communicated to me in many letters.

The book was brought into its final shape during the sabbatical term I had from the Technische Universität Darmstadt during March to September 2007. The main part of that sabbatical (March to June 2007) I stayed as guest at the Max-Planck-

Institute for Mathematics in Bonn. I am most grateful to the Max-Planck-Institute and its director, Professor Gerd Faltings, for the hospitality and the generous support provided there.
It has been a pleasure to work with Mrs. Susanne Denskus and Mrs. Ellen Kattner from the Springer-Verlag's professional staff.

Last but not least I like to thank my wife Gabriele and our daughter Katharina for their encouragement, patience and emotional support.

Contents

Common Notations and Terminology

Throughout this book $\mathbb{N}$ denotes the set of natural numbers **including** 0, i.e.

$$\mathbb{N} := \{0,1,2,\ldots\}.$$

$\mathbb{Z},\mathbb{Q},\mathbb{R}$ denote the sets of integers, rational numbers and real numbers, respectively.

$$\mathbb{R}^* := \mathbb{R}\setminus\{0\},\ \ \mathbb{R}_+ := \{x\in\mathbb{R} : x\geq 0\},\ \ \mathbb{R}^*_+ := \{x\in\mathbb{R} : x>0\}.$$

The sets $\mathbb{Q}^*,\mathbb{Q}_+,\mathbb{Q}^*_+$ are defined analogously.

$\mathbb{N}^{\mathbb{N}}$ is the set of all number theoretic functions $f:\mathbb{N}\to\mathbb{N}$ and $2^{\mathbb{N}}$ is the set of all functions $f:\mathbb{N}\to\{0,1\}$. Elements in $\mathbb{N}^{\mathbb{N}}$ and $2^{\mathbb{N}}$ are usually denoted by $f,g,h,\ldots$ but also by $\alpha,\beta,\gamma,\ldots$.

Together with

$$d(f,g) := \begin{cases} 2^{-\min n[f(n)\neq g(n)]}, & \text{if } \exists n\in\mathbb{N}(f(n)\neq g(n)),\\ 0, & \text{otherwise,}\end{cases}$$

$\mathbb{N}^{\mathbb{N}}$ becomes a metric space which is referred to as Baire space. $2^{\mathbb{N}}$ with the induced metric is a compact metric space, called Cantor space.

Many formal systems used in this book are formulated in the language of functionals of finite type. The most important types are

$$0 := \ \text{the type of natural numbers } \mathbb{N}$$

and

$$1 := 0(0) := \ \text{the type of number theoretic functions } f:\mathbb{N}\to\mathbb{N}.$$

E.g. a sentence $\forall f^1\exists g^1\forall n^0(g(n)=f(f(n)))$ expresses

$$\forall f \in \mathbb{N}^{\mathbb{N}} \exists g \in \mathbb{N}^{\mathbb{N}} \forall n \in \mathbb{N}(g(n) = f(f(n))).$$

Instead of '$f \in \mathbb{N}^{\mathbb{N}}$' we also write '$f : \mathbb{N} \to \mathbb{N}$'.

Formulas are usually denoted by A, B, C, etc. The subscript '0', i.e. A_0, B_0, C_0 etc., indicates that the respective formula does not contain any quantifiers, i.e. is quantifier-free.

Occasionally, we use some notation from elementary recursion theory. '$T(e,x,n)$' expresses that e is the number code of a Turing machine whose computation with the input x terminates, where n is the code of that terminating computation. By U we denote the function that extracts from n the result of the computation encoded by n. The e-th partial recursive function, which we denote by $\{e\}$, is the partial function defined by

$$\{e\}(x) := \begin{cases} U(\min n[T(e,x,n)]), \text{ if } \exists n \in \mathbb{N}(T(e,x,n)) \\ \text{undefined, otherwise.} \end{cases}$$

Total recursive functions usually are called computable functions. A subset $A \subseteq \mathbb{N}$ is decidable if its characteristic function is computable.

A well-know fact from elementary recursion theory, which we frequently use, is the undecidability of the so-called special halting set or 'special halting problem'

$$H := \{e \in \mathbb{N} : \exists n \in \mathbb{N}(T(e,e,n)\},$$

i.e. the set of all numbers e such that the Turing machine with number e applied to the input e performs a terminating computation.

A function $\Phi : \mathbb{N}^{\mathbb{N}} \to \mathbb{N}$ is computable if there exists a Turing machine that computes $\Phi(f)$ for any given function $f \in \mathbb{N}^{\mathbb{N}}$ used as an oracle. All these notions extend in the obvious way to several number or function arguments and to n-ary functions f rather than unary functions. As common in functional languages, we often write 'Φf' instead of '$\Phi(f)$'.

In fact, most of our formal systems permit the bijective encoding of finite sequences $(x_0, \ldots, x_{n-1})$ of natural numbers $x_0, \ldots, x_{n-1}$ by numbers $x := \langle x_0, \ldots, x_{n-1} \rangle \in \mathbb{N}$. $lth : \mathbb{N} \to \mathbb{N}$ is the function with $lth(x) = n$ ('length of the sequence encoded by x') and $(\cdot)_{(\cdot)} : \mathbb{N} \times \mathbb{N} \to \mathbb{N}$ is the function satisfying

$$(x)_i = \begin{cases} x_i, \text{ if } i < lth(x) \\ 0, \text{ otherwise.} \end{cases}$$

With $x = \langle x_0, \ldots, x_{n-1} \rangle$ and $y = \langle y_0, \ldots, y_{m-1} \rangle$ the number $x * y$ is the code of the concatenation of the two sequences encoded by x, y, i.e.

$$x * y := \langle x_0, \ldots, x_{n-1}, y_0, \ldots, y_{m-1} \rangle.$$

For $f \in \mathbb{N}^{\mathbb{N}}$ and $n \in \mathbb{N}$, the code $\langle f(0), \ldots, f(n-1) \rangle$ of the sequence of the first n values of f is denoted by $\overline{f}(n)$ or just $\overline{f}n$, where $\overline{f}0$ is the code $\langle\rangle$ of the empty sequence.

If $\mathscr{T}$ is a formal axiomatic system and A a sentence in the language $\mathscr{L}(\mathscr{T})$ of $\mathscr{T}$, then

$$\mathscr{T} \vdash A$$

expresses that A is derivable in $\mathscr{T}$. If $\mathscr{M}$ is a structure that interprets $\mathscr{L}(\mathscr{T})$, then

$$\mathscr{M} \models A$$

expresses that A holds (or is 'valid' or 'true') in $\mathscr{M}$. We then also say that '$\mathscr{M}$ is a model of A'.

Chapter 1
Introduction

Consider a theorem of the following form

$$B :\equiv \forall x \in \mathbb{N} \exists y \in \mathbb{N} \, R(x,y),$$

where $R(x,y)$ is some quantifier-free formula, i.e. a formula not containing any quantifier, in the language of elementary arithmetic (and hence decidable). E.g. $R(x,y)$ may be of the form $p(x,y) = 0$, where p is a polynomial in $\mathbb{N}[x,y]$ or some other elementary computable function.

Suppose that we have a formal proof p of B in some theory $\mathcal{T}$ that proceeds via a lemma having a more complicated logical form

$$A :\equiv \forall u \in \mathbb{N} \exists v \in \mathbb{N} \forall w \in \mathbb{N} \, Q(u,v,w),$$

where Q again is a quantifier-free formula of elementary arithmetic. This means that p has two subproofs p_1 and p_2, where p_1 is a proof of A and p_2 is a proof of $A \to B$, i.e. of

$$\forall u \in \mathbb{N} \exists v \in \mathbb{N} \forall w \in \mathbb{N} \, Q(u,v,w) \to \forall x \in \mathbb{N} \exists y \in \mathbb{N} \, R(x,y),$$

and then proceeds from A and $A \to B$ by the modus ponens rule to derive B.

Even if the proof p of B is ineffective, it is clear that the conclusion B has a computable solution by just forming the following program P for a partial function:

$$P(x) := \begin{cases} \min y \in \mathbb{N} \text{ such that } R(x,y) \text{ holds, if existent} \\ \text{undefined, otherwise.} \end{cases}$$

p establishes that this partial function P actually is total and so is a computable (using the decidability of R) function satisfying

$$\forall x \in \mathbb{N} \, R(x, P(x)).$$

The program P, however, does not use at all how B is proved. Only the truth of B is involved in verifying its termination on any input $x \in \mathbb{N}$. Intuitively, though, one will expect that looking into the computational content of the proof p it should be possible to get more information on P, e.g. some complexity information via a subrecursive bound on the unbounded search involved in P. In order to bring the actual proof p into the play one attempt would be to see whether p_1 might yield an algorithm P_1 satisfying

$$\forall u, w \in \mathbb{N}\, Q(u, P_1(u), w).$$

Then the proof p_2 might contain an algorithm P_2 (no longer operating just on numbers but also on functions $f : \mathbb{N} \to \mathbb{N}$) that transforms any hypothetical function f realizing '$\exists v \in \mathbb{N}$' in A into a realizer for '$\exists y$' in B, i.e.

$$\forall f : \mathbb{N} \to \mathbb{N}(\forall u, w \in \mathbb{N}\, Q(u, f(u), w) \to \forall x \in \mathbb{N}\, R(x, P_2(f, x)).$$

So $P'(x) := P_2(P_1, x)$ is another program realizing the conclusion which might be much more informative as it takes into account how B has been proved. The problem, however, is that an ineffectively proven lemma A in general will not have a computable realizer P_1 :

Let $S(u, v)$ be a quantifier-free decidable predicate on $\mathbb{N} \times \mathbb{N}$ so that $S'(u) :\equiv \exists v \in \mathbb{N}\, S(u, v)$ is undecidable (e.g. we may take the special Halting Problem for Turing machines) and define $Q(u, v, w) :\equiv S(u, v) \vee \neg S(u, w)$. Then

$$A :\equiv \forall u \in \mathbb{N} \exists v \in \mathbb{N} \forall w \in \mathbb{N}\, Q(u, v, w)$$

is provable already in logic but any function P_1 realizing '$\exists v \in \mathbb{N}$' could be used to decide S' since then

$$S'(u) \leftrightarrow S(u, P_1(u)).$$

Another instance of such a lemma (taken from elementary analysis) is the Cauchy property

$$\forall u \in \mathbb{N} \exists v \in \mathbb{N} \forall w \geq v(|a_v - a_w| \leq 2^{-u})$$

for bounded monotone sequences (a_n) in $\mathbb{R}$. By a well-known construction of E. Specker ([342]) there are easily computable such sequences already in $[0, 1] \cap \mathbb{Q}$ without any computable bound on '$\exists v$', i.e. which have no computable Cauchy modulus.

The problem in this example is caused by the fact that the proof of the conclusion B uses a logically more complicated lemma A which in general no longer admits a direct computable witness. One possible strategy to solve this would be to transform the proof p into a direct proof p' which does not use any formulas more complicated than the conclusion. This can be achieved in certain contexts by the process of (partial) cut-elimination which in turn yields so-called Herbrand terms from which a realization of the conclusion can be obtained. However, (partial) cut-elimination is not always available and where it is, the process is of enormous complexity resulting sometimes in a direct proof p' whose length is superexponential in the length of p (see e.g. [344, 296, 305]).

In this book we focus on an alternative strategy which rather than eliminating complicated lemmas interprets them when necessary (as for A) in a more sophisticated way than simply asking for a full witness function: the idea is to interpret all formulas F involved by formulas F^I in such a way that

1) A^I and $(A \to B)^I$ are weak enough to allow for a computational realization which can be extracted from proofs of A and $A \to B$,
2) computational realizations of A^I and $(A \to B)^I$ yield a computational realization of B^I (interpretation of the modus ponens rule) and
3) for $B \equiv \forall x \in \mathbb{N} \exists y \in \mathbb{N} R(x,y)$ a computational realization of B^I provides a program P such that
$$\forall x \in \mathbb{N} R(x, P(x)).$$

To achieve 1) and 2) one has to assign to each formula A in the language $\mathcal{L}(\mathcal{T})$ of $\mathcal{T}$ a new formula A^I (possibly in an extension of the original language) such that

a) all axioms A of $\mathcal{T}$ admit a computational interpretation of A^I (verifiable in some other theory $\mathcal{T}^I$) and
b) all the rules of $\mathcal{T}$ are valid (verifiable in $\mathcal{T}^I$) under the interpretation I.

As a consequence of a) and b), given a proof p of A in $\mathcal{T}$ one can construct a new proof p^I in $\mathcal{T}^I$ of A^I by a simple recursion over p. As the general logical structure of p remains intact, that new proof p^I will not be much longer than p (usually it is at most cubic in p, [159]).

In order to achieve (3) we need that

$$(\forall x \in \mathbb{N} \exists y \in \mathbb{N} R(x,y))^I \equiv \exists f : \mathbb{N} \to \mathbb{N} \forall x \in \mathbb{N} R(x, f(x)),$$

where a computational interpretation of the latter sentence provides a computable f. Of course to be of any use, that f (as well as the computational interpretation of general $\mathcal{T}$-provable sentences) should not just be computable but of certain restricted complexity and carrying additional information which reflects that the conclusion is proved by the restricted means of $\mathcal{T}$ rather than merely being true.

In connection with this (necessarily slightly simplifying) discussion of proof interpretations one should mention, however, that the separation between cut-elimination, normalization and Herbrand theory on the one hand and proof interpretations on the other hand is not as strict as it might appear: often (partial) normalization of the terms extracted by proof interpretations is used and, conversely, proof interpretations can be applied to extract Herbrand terms as well ([118]). However, transforming proofs into functional programs (as is done by functional interpretation) makes it possible to use mathematical properties of these functionals (which often can be established by the use of logical relations without any normalization) such as majorizability, continuity, compactness etc., which cannot be applied directly to proofs.

This book is devoted to the study of proof interpretations and their use to extract new information hidden (both effective data as well as new qualitative uniformities and other strengthenings) in given proofs. Historically, most of these proof interpretations were (just as cut-elimination and its variants) developed for foundational reasons, in particular to give (relative) consistency proofs. E.g. two proof interpretations which play a most important role in this book, Gödel's negative translation and his functional interpretation, were developed to give a consistency proof for first order ('Peano') arithmetic PA. In such contexts, proof interpretations are applied to a hypothetical proof of a contradiction, say '$0 = 1$', or to proofs of universal theorems (so-called 'real' statements in Hilbert's terminology) that can be expressed as quantifier-free open formulas $A_{qf}(a)$: Gödel's interpretation establishes that if $\forall a A_{qf}(a)$ is provable in PA, then $A_{qf}(a)$ can be derived already in a quantifier-free calculus T of primitive recursive functionals in higher types (anticipated already in Hilbert [161]). By a natural 'shift of emphasis' (G. Kreisel) one instead can aim at using such interpretations for interesting proofs of existential theorems to e.g. extract realizers for the existential quantifier as explicit functions of the parameters from the proof. As stressed by Kreisel, (proofs of) universal theorems (which play the key role in Hilbert's consistency program) do not have any impact on the extraction and the complexity of the realizers which shows that Kreisel's emphasis is kind of opposite to the foundational orientation in Hilbert's program. Moreover, for this applied reorientation of proof theory, the fact that the extracted realizer can be verified in a 'constructive' quantifier-free system plays no role as any principles may be used for its verification.
It is precisely this applied reorientation of proof theory this book is about. Certain technical issues related to the original foundational aims of the interpretations (covered extensively in existing literature such as [366]) are bypassed in our presentation while many new aspects relevant for applications appear here for the first time in book format. In particular, the shift of emphasis prompts new variants of the interpretations specially tailored for applications to certain classes of proofs.

Another important practical use of proof interpretations is that they sometimes can be used to generalize a proof p of a theorem A to a proof p^G which might use only a restricted version of the assumptions of A and so prove a more general result A^G. We will see various such generalizations in this book, e.g. some results in fixed point theory obtained previously only for nonexpansive mappings could easily be generalized to other classes such as directionally nonexpansive mappings or the assumption on the existence of a fixed point could be dropped etc. (see chapters 17 and 18). This is achieved by generalizing p^I to a proof $(p^I)^G$ (which often is easy since p^I exhibits the hidden combinatorial core of the proof p) and then looking for a generalized proof p^G whose interpretation $(p^G)^I$ just coincides with $(p^I)^G$ (see also [358] for a discussion of a related phenomenon in ergodic theory):

$$\begin{array}{ccc} p & \xrightarrow{I} & p^I \\ G\downarrow & & \downarrow I^G \\ p^G & \xrightarrow{G^I} & (p^I)^G = (p^G)^I. \end{array}$$

After a discussion of some general aspects of this applied form of proof theory (also called 'Proof Mining') in chapter 2 we start by considering first proofs based on constructive (so-called intuitionistic) logic as initiated by L.E.J. Brouwer and formalized by A. Heyting (chapter 3). Here the law-of-the-excluded-middle schema $A \vee \neg A$ (or – equivalently – the principle $\neg\neg A \to A$) is omitted. For proofs based on intuitionistic logic the problem of the lemma A discussed above does not exist since an intuitionistic proof of any $\forall x \in \mathbb{N} \exists y \in \mathbb{N} F(x,y)$-sentence with F of arbitrary logical complexity yields an algorithm P such that

$$\forall x \in \mathbb{N} F(x, P(x)).$$

One proof interpretation to extract such algorithms, e.g. from proofs in intuitionistic ('Heyting') arithmetic, is the so-called (recursive) realizability technique due to S.C. Kleene which extracts computable solutions. However, in order to obtain solutions of restricted complexity and to be able to deal also with higher objects than such numbers, a variant of this, the so-called modified realizability due to G. Kreisel, is more suitable. We will study this technique in chapter 5. Even in this context we can allow some ineffective lemmas to be used: e.g. let

$$\forall x \in \mathbb{N} \exists y \in \mathbb{N} F(x,y)$$

be constructively proved from an ineffective lemma of the form

$$\forall u \in \mathbb{N} \exists v \leq 1 \forall w \in \mathbb{N} Q(u,v,w)$$

(with Q as above) as assumption. Then modified realizability provides a program $P(f,x)$ such that

$$\forall f : \mathbb{N} \to \{0,1\} \big(\forall u, w\, Q(u, f(u), w) \to \forall x \in \mathbb{N} F(x, P(f,x))\big).$$

Moreover, programs extracted by modified realizability have (for many theories $\mathcal{T}$) the property of being majorizable, which means that another program P^* of the same or even lower complexity can be easily constructed from P such that

$$\forall x^*, x \in \mathbb{N} \forall f^*, f : \mathbb{N} \to \mathbb{N} (x^* \geq x \wedge f^* \; maj \; f \to P^*(f^*, x^*) \geq P(f,x)),$$

where

$$f^* \; maj \; f :\equiv \forall y^*, y \in \mathbb{N} (y^* \geq y \to f^*(y^*) \geq f(y)).$$

Now let 1 denote the constant-1 function, then

$$\forall x \in \mathbb{N} \exists y \leq P'(x) F(x,y),$$

where $P'(x) := P^*(1,x)$ is computable. So we have obtained at least a computable bound. In applications to analysis one, of course, not only needs quantifiers over natural numbers but also over e.g. Polish metric spaces X and compact metric spaces K. The former can be represented by number theoretic functions $g : \mathbb{N} \to \mathbb{N}$ and the latter even in such a way that only functions g with $\forall x \in \mathbb{N}(g(x) \leq M(x))$ for some fixed computable M are needed (chapter 4). For simplicity let us assume that M again is just the constant-1 function and define $B := \{g : \mathbb{N} \to \mathbb{N} : \forall x \in \mathbb{N}(g(x) \leq 1)\}$. By modified realizability combined with majorization (chapters 6 and 9) one can extract from a proof of a sentence

$$\forall x \in \mathbb{N} \forall g \in B \exists y \in \mathbb{N} F(x,g,y)$$

a computable bound $\Phi(x)$ on '$\exists y \in \mathbb{N}$' which is independent from $g \in B$ and hence – if $F'(g) :\equiv F(x,g,y)$ represents a property of some compact metric space K – a bound which is uniform on K. This result, called 'fan rule', also holds if the proof uses certain ineffective principles which do not contribute to majorants of functionals extracted by modified realizability. E.g. this is the case for sentences

$$\forall x \in X \exists y \in K A_{ef}(x,y),$$

where X, K as above and A_{ef} is a so-called $\exists$-free formula, i.e. a formula that does not contain $\exists$ or $\vee$. This is due to the fact that it is possible to directly extract P^* rather than first extracting P and then constructing P^* from P even in cases where a computable P does not exist. The method for doing this is called monotone modified realizability (chapter 7).

The restriction of the underlying logic being intuitionistic is a very strong one. Many more theorems become provable if one allows at least the use of the so-called Markov principle (for numbers) which is not intuitionistically derivable and in one formulation can be stated as

$$\mathrm{M} : \neg \forall x \in \mathbb{N} \neg R(x) \to \exists x \in \mathbb{N} R(x),$$

where $R(x)$ is a decidable formula (possibly with further parameters). It first seems that the only way to give a computational interpretation to M is by using unbounded search as above and, in fact, this is precisely how Kleene realizability interprets this principle with the consequence that no complexity information is obtained and no information from a proof of $\neg \forall x \in \mathbb{N} \neg R(x)$ is used. Modified realizability, on the other hand, does provide subrecursive realizers whose complexity depends on the proof principles used but it does not interpret M (see chapter 5). However, a deep and much more refined interpretation, K. Gödel's so-called functional ('Dialectica') interpretation D, does interpret proofs involving M in a way which avoids the use of unbounded search (chapter 8). In fact, it also validates extensions of M to higher objects f such as functions $f : \mathbb{N} \to \mathbb{N}$ (instead of just numbers $x \in \mathbb{N}$) to which search cannot even be applied. In order to achieve this the interpretation has to interpret throughout the proof negatively occurring universal quantifiers in a positive way. In particular, not only is

$$\exists x \in \mathbb{N} R_1(x) \to \exists y \in \mathbb{N} R_2(y)$$

interpreted by D (as well as by modified realizability) as

$$\exists f : \mathbb{N} \to \mathbb{N} \forall x \in \mathbb{N} \big(R_1(x) \to R_2(f(x))\big)$$

but also

$$\forall x \in \mathbb{N} R_1(x) \to \forall y \in \mathbb{N} R_2(y)$$

is interpreted by D (but not by modified realizability) as

$$\exists g : \mathbb{N} \to \mathbb{N} \forall y \in \mathbb{N} (R_1(g(y)) \to R_2(y)).$$

As before in the case of modified realizability, also functional interpretation can be combined with majorizability (monotone functional interpretation, chapter 9) to extract computable uniform bounds from proofs of $\forall\exists$-sentences of still arbitrary logical complexity which (in addition to M) might use certain ineffective principles (chapters 9 and 15). Now, the latter have to be of a more restricted form, e.g. $\forall x \in X \exists y \in K(G(x,y) = 0)$, where G represents a computable (and $\mathscr{T}$-definable) function $: X \times K \to \mathbb{R}$. This suffices to cover important theorems of classical analysis such as the fact that every function $f \in C[0,1]$ attains its maximum, Brouwer's fixed point theorem, the Cauchy-Peano existence theorem and many more which in logical terminology correspond to the binary ('weak') König's lemma WKL ([338]) and imply the existence of noncomputable real numbers resp. functions $f : \mathbb{N} \to \mathbb{N}$. WKL states that every infinite binary tree has an infinite branch. By a well-known result of S.C. Kleene there are easily decidable such trees which do not have any computable infinite branch.

The most important use of functional interpretation, however, is that by interpreting M one now can even treat proofs based on the full ordinary ('classical') logic (chapter 10) as will be the focus of this book. This is made possible by combining functional interpretation (and monotone functional interpretation) by yet another proof interpretation, namely the aforementioned negative translation which translates systems based on ordinary logic into systems with intuitionistic logic (or at least approximately intuitionistic logic). The first such negative translations were due to K. Gödel and G. Gentzen in 1933 and were refined later e.g. by S. Kuroda and others (chapter 10). As is clear from the counterexample discussed above, it will no longer be possible to extract computable realizers (or even bounds) for theorems of the form

$$A :\equiv \forall u \in \mathbb{N} \exists v \in \mathbb{N} \forall w \in \mathbb{N}\, Q(u,v,w), \ Q \text{ quantifier-free},$$

but this is possible for theorems

$$B :\equiv \forall x \in \mathbb{N} \exists y \in \mathbb{N} R(x,y)$$

with quantifier-free R and even for higher objects such as $x \in X$ in Polish metric spaces X etc. rather than integers. After the use of negative translation one has a constructive proof of

$$B' :\equiv \forall x \in \mathbb{N} \neg\neg\exists y \in \mathbb{N} R(x,y).$$

Whereas modified realizability is not able to recover any positive information from this statement, functional interpretation does so since it interprets **M** which suffices to derive B from B'.

But how is the situation where B is proved from lemmas A and $A \to B$ resolved now? By negative translation it follows that in order to obtain (constructively) B' one only needs the negative translation A' of A, namely

$$A' :\equiv \forall u \in \mathbb{N} \neg\neg\exists v \in \mathbb{N} \forall w \in \mathbb{N} Q(u,v,w).$$

However, the functional interpretation of the latter (in this special case also called 'no-counterexample interpretation', [241]) is

$$\exists \Phi \forall u, g \; Q(u, \Phi(u,g), g(\Phi(u,g)))$$

(here $u \in \mathbb{N}, g : \mathbb{N} \to \mathbb{N}$ and $\Phi : \mathbb{N} \times \mathbb{N}^{\mathbb{N}} \to \mathbb{N}$) which usually does have a simple computable solution Φ. E.g. in the example

$$\forall u \in \mathbb{N} \exists v \in \mathbb{N} \forall w \in \mathbb{N} (S(u,v) \vee \neg S(u,w))$$

discussed above we can just take

$$\Phi(u,g) := \begin{cases} u, \text{ if } \neg S(u,g(u)) \\ g(u), \text{ otherwise.} \end{cases}$$

A majorant satisfying the monotone functional interpretation is even easier to obtain as we can take simply $\Phi^*(u,g) := \max\{u, g(u)\}$. In fact, instead of using the values u and $g(u)$ one can use also $c, g(c)$ for any $c \in \mathbb{N}$, i.e.

$$\Phi_c(u,g) := \begin{cases} c, \text{ if } \neg S(u,g(c)) \\ g(c), \text{ otherwise.} \end{cases}$$

Putting $c := 0$ we see that already $\Phi_0^* := g(0)$ satisfies the monotone functional interpretation in this case. Note also that in contrast to Φ, Φ_0 the majorants Φ^*, Φ_0^* no longer depend on S.

In total, the combination of negative translation and monotone functional interpretation applied to a proof of an implication $A \to B$ extracts a functional

$$\Omega^* : \mathbb{N}^{(\mathbb{N}\times\mathbb{N}^{\mathbb{N}})} \to \mathbb{N}^{\mathbb{N}}$$

that transforms any hypothetical solution $\Phi^* : \mathbb{N}\times\mathbb{N}^{\mathbb{N}} \to \mathbb{N}$ of the monotone functional interpretation of A' into a bound for B. A suitable Φ^* is extracted (again by negative translation and monotone functional interpretation) from a proof of A. As a result one gets

$$\forall u \in \mathbb{N} \exists v \leq \Omega^*(\Phi^*, u) R(u,v).$$

Usually, the level of the function spaces used can be lowered in practice as, typically, Ω^* will be just the application of Φ^* to some concrete function argument(s) (defined in the parameters of the problem).

Not only can one treat proofs in (classical) arithmetic in this way but also substantial fragments of analysis based on noncomputational WKL-related principles (such as the aforementioned analytical principles) thereby obtaining bounds of rather limited complexity. Using monotone functional interpretation one can also show how to eliminate the use of WKL from a large class of proofs. A particular usable extension of WKL is given by a strong uniform boundedness principle (chapter 12) which, although false in the full set theoretic model, can be used in formalizing (and subsequently analyzing) many proofs in analysis yielding bounds of low complexity which in the end can be verified without the use of this principle (which gets eliminated by the monotone interpretation).

An extension of Gödel's functional interpretation to full classical analysis (based on the axiom of dependent choice) was given by C. Spector in 1962. Although the effective realizers in this case (the so-called bar recursive functionals) are of vast complexity they do have the effective majorizability property which allows one to obtain interesting uniformity results as we will use in chapter 17. We will give a detailed account of Spector's result in chapter 11.

The first part of the book which is devoted to the development of the proof interpretations used in the analysis of proofs (chapters 2–14) finishes with a presentation of an (in some cases) alternative to the combination of negative translation and functional interpretation (chapter 14). Here in between negative translation and modified realizability another proof interpretation (called A-translation and due to H. Friedman and A. Dragalin independently) is used to pre-process the result of the negative translation so that modified realizability gives a meaningful result (various refinements combine the use of negative and A-translation into a prima-facie simpler, single step). In general, however, this route tends to require realizers of greater complexity (compared to functional interpretation).

The second part of the book (chapters 15–18) brings the logical theorems on the extractability of effective uniform bounds (developed in the first part) into the form of general metatheorems directly applicable to proofs in analysis. Moreover, several substantial applications to concrete proofs are carried out while further examples are pointed to in the literature. The first set of such metatheorems is formulated in the context of continuous functions on concrete Polish and compact metric spaces such as $C[0,1]$ (with the uniform norm) or $[0,1]^d$ (chapter 15). We show how to extract from proofs effective bounds which are uniform in the sense that they do not depend on parameters from compact metric spaces. The emphasis is on bounds of low complexity and hence on proofs formalizable in fragments of analysis roughly based on analytical principles corresponding to WKL. Various classes of theorems where such uniform bounds carry important information are specified. A particularly interesting class are uniqueness theorems. Here the metatheorems guarantee the extractability of effective rates of so-called strong unicity (also called moduli of uniqueness) which play an important role in various parts of numerical analysis (e.g. in approximation theory). Using such effective moduli one can in fact construct algorithms for the computation of prima-facie ineffectively proven unique existence theorems of the form

$$\forall x \in X \exists! y \in K(G(x,y) =_{\mathbb{R}} 0),$$

where X (resp. K) is a Polish (resp. compact) metric space and $G : X \times K \to \mathbb{R}$ an explicitly given function. Although this is not a $\forall\exists$-sentence but a $\forall\exists\forall$-sentence (due to the universal quantifier hidden in the equality predicate $=_{\mathbb{R}}$ for real numbers) the uniform quantitative information obtainable even from ineffective uniqueness proofs is sufficient to compute y.

In chapter 16 we extract explicit rates of strong unicity for both best Chebycheff approximations as well as best L_1-approximations of functions $C[0,1]$ by polynomials $p \in P_n$ of degree $\leq n$ from the standard uniqueness proofs. In both cases these proofs heavily use noncomputable real numbers by relying on the principle that $f \in C[0,1]$ attains its maximum (which corresponds to WKL, [338]). Nevertheless, we succeed (guided by the proofs of the metatheorems) to extract rates of uniqueness which in both cases have an optimal dependency on the error.

Whereas the applications mentioned so far deal with concrete Polish and compact metric spaces, chapters 17 and 18 consider theorems which hold for general classes of structures such as arbitrary metric, hyperbolic, CAT(0), normed, uniformly convex and inner product spaces which, in particular, are not assumed to be separable or to have a representation by names in $\mathbb{N}^{\mathbb{N}}$. We develop general metatheorems which guarantee the extractability of effective bounds which are even independent from parameters in noncompact (but only metrically bounded) subsets of such structures if the proof does not use any more specific conditions other than the general axioms for these structures. In order to formalize such proofs we add abstract metric, hyperbolic etc. spaces X axiomatically to the systems as a kind of atoms with a new base type for elements of X. So although we cannot quantify over these classes of structures we can refer to them as parameters. A further refinement of this approach – based on a novel form of majorizability relative to a variable reference point $a \in X$ which was first developed in [120] – even allows one to replace the boundedness of the whole space by bounds on local distances between certain relevant terms at hand. Here an a-majorant of an element $x \in X$ of a metric space (X,d) is a number

$n^* \in \mathbb{N}$ such that $n^* \geq d(x,a)$ whereas an a-majorant of a function $f : X \to X$ is a (monotone) function $f^* : \mathbb{N} \to \mathbb{N}$ such that

$$\forall n^* \in \mathbb{N} \forall x \in X (n^* \geq d(x,a) \to f^*(n^*) \geq d(f(x),a)).$$

For higher function spaces this definition is extended in the usual hereditary fashion. The functional interpretation will be parametrized by this reference point $a \in X$ as well which will be fixed only in the end depending on the parameters of the problem at hand.

While in the case of compact metric spaces and continuous functions as treated in chapter 15 the (ineffective) existence of uniform bounds follows from standard compactness arguments so that it is the concrete effective information which is relevant, the extended metatheorems in chapter 17 allow one to derive uniformity results which even ineffectively are nontrivial. In fact, a number of new such results in metric fixed point theory were obtained in this way (see chapter 18). Since here this new qualitative improvement of theorems is most prominent we work in a context of full analysis (with bar recursive bounds) in order to make the results as widely applicable as possible. However, things scale down to fragments as well and in the concrete applications relatively simple bounds are obtained due to the very limited use of analytic principles. Chapter 18 gives a detailed treatment of some applications which deal with the asymptotic behavior of so-called Krasnoselski-Mann iterations of nonexpansive mappings of hyperbolic spaces. We also briefly mention a recent application due to Avigad-Gerhardy-Towsner in ergodic theory. Many other applications are referred to in the literature. The book is concluded with some brief comments about potential future uses of the metatheorems established in chapters 15 and 17 concerning ergodic theory and its connection to combinatorics, uniqueness results for hyperbolic conservation laws, the reduction of Fermat's last theorem to the Taniyama-Weil theorem as well as applications to geometric group theory (an area in which metric structures such as CAT(0)-spaces feature prominently).

Chapter 2
Unwinding proofs ('Proof Mining')

2.1 Introductory remark

In this chapter we give – exemplified by a couple of easy but fundamental examples – a kind of tour-de-force through a number of topics which will be developed in detail throughout the rest of this book. The aim of this chapter is to provide the reader with a guiding line by explaining (in more technical terms than in the previous chapter) the main goals we are aiming at in this book and the difficulties one has to address in the course of this. We recommend to read this chapter first at a somewhat informal level skipping maybe some technical details but to revisit it after the study of the material up to chapter 11.

2.2 Informal treatment of ineffective proofs

Proof interpretations of the kind we are going to study in this book are tools to extract constructive (computational) data from given proofs by recursion on the proof. Such data quite often cannot directly be read off from a proof but are hidden behind the use of quantifiers.
G. Kreisel was the first to formulate the program of unwinding proofs under the general question:

'What more do we know if we have proved a theorem by restricted means than if we merely know that it is true?'

The term 'unwinding of proofs' is due to G. Kreisel. More recently, D. Scott suggested to us to use the more catchy slogan 'proof mining' which we find quite appropriate for this area of applied proof theory.

What do we mean by 'constructive data'?

E.g.

1) Realizing terms from a proof of an existential theorem $A \equiv \exists x B(x)$ (closed).
   A weaker requirement is to construct a list of terms $t_1, \ldots, t_n$ which are candidates for A, i.e. such that $B(t_1) \vee \ldots \vee B(t_n)$ holds.
   More general: If $A \equiv \forall x \exists y B(x,y)$, then one can ask for an algorithm p such that $\forall x B(x, p(x))$ holds (or – weaker – for a bounding function b such that

$$\forall x \exists y \leq b(x) B(x,y),$$

   if e.g. y ranges over the natural numbers).
2) weakening of the assumptions used in the proof: e.g. replacing general assumptions by specific instances of them.

What type of information one can expect (in general) depends of course on the structure of the theorem A to be proved and the principles used in its proof.
A first, very rough, division of the structure of a sentence (i.e. a closed formula) A can be made according to the quantifier complexity of A :

From now on $A_0, B_0, C_0, \ldots$ always denote quantifier-free formulas. Sometimes we also write A_{qf}. Instead of a single variable we may have (here and in the following) also a tuple $\underline{x} = x_1, \ldots, x_n$ of variables.

1) A purely universal, i.e. $A \equiv \forall x A_0(x)$, where A_0 is quantifier-free.
Such sentences A, sometimes called complete, don't ask for any witnessing data. So the problem of extracting data is empty here.

2) A purely existential, i.e. $A \equiv \exists x A_0(x)$. We treat this as a special case of

3) $A \equiv \forall x \exists y A_0(x,y)$. Let's consider the case where $x, y \in \mathbb{N}$ and $A_0 \in \mathscr{L}(\mathrm{PA})$ (here PA denotes first order Peano arithmetic which we assume to contain all primitive recursive functions; see chapter 3 for a precise definition). A_0 is decidable (Exercise: $A_0(\underline{x}) \in \mathscr{L}(\mathrm{PA})$, then one can construct a primitive recursive function term t such that $\mathrm{PA} \vdash \forall \underline{x}(t(\underline{x}) = 0 \leftrightarrow A_0(\underline{x}))$) (see proposition 3.8 below) and, therefore, defines a partial recursive function f, namely

$$f(x) := \begin{cases} \min y[A_0(x,y)], \text{ if } \exists y A_0(x,y) \\ \text{undefined, otherwise.} \end{cases}$$

A just says that f is total recursive.

Questions: How to extract a non-trivial program for f (different from simple unbounded search) from a proof of A? What is the complexity and the rate of growth of f if A is proved in a certain theory $\mathscr{T}$?

Theorems expressing that a set $\{y \in \mathbb{N} : A(y)\} \subseteq \mathbb{N}$ is infinite have the form

$$\forall x \in \mathbb{N} \exists y \geq x A(y).$$

Quite often A can be expressed in a quantifier-free way A_0 in PA, so that this falls under the general form $\forall x \exists y B_0(x,y)$, where

$$B_0(x,y) :\equiv (y \geq x \wedge A_0(y)).$$

As an example consider the following

Proposition 2.1. *There are infinitely many prime numbers.*

The predicate $P(x) :\equiv$ 'x is a prime number' can be expressed in a quantifier-free way as a primitive recursive predicate (see e.g. [194, 371]).

Proof 1 (Euclid): Define $a := 1 + \prod_{\substack{p \leq x \\ p \, prime}} p$. a cannot be divided by any prime number $p \leq x$. By the decomposition of every number into prime factors it follows that a contains a prime factor $q \leq a$ with $q > x$. □

From this proof one immediately gets the bound $g(x) := 1 + x! \, (\geq 1 + \prod_{\substack{p \leq x \\ p \, prime}} p)$. By the Stirling formula we obtain

$$g(x) \sim 1 + (2\pi x)^{\frac{1}{2}} \left(\frac{x}{e}\right)^x = 1 + \sqrt{2\pi} \cdot e^{x \ln x - x + \frac{1}{2} \ln x}.$$

In order to obtain from Euclid's proof an upper bound on the $(r+1)$-th prime number p_{r+1} which only depends on r instead of $x \geq p_r$ one can argue as follows: Euclid's proof yields that

$$p_{r+1} \leq p_1 \cdot \ldots \cdot p_r + 1.$$

From this one obtains (exercise) by induction on r that

$$p_r < 2^{2^r} \text{ for all } r \geq 1.$$

Proof 2 (Euler): Suppose that there are only finitely many prime numbers $p_1, \ldots, p_r$ (listed in increasing order, $r \geq 1$). One has

$$\begin{aligned}
\sum_{0 \leq \alpha_1, \ldots, \alpha_r \leq n} \frac{1}{p_1^{\alpha_1} \cdot \ldots \cdot p_r^{\alpha_r}} &= \left(\sum_{i=0}^{n} \frac{1}{p_1^i}\right) \cdot \ldots \cdot \left(\sum_{i=0}^{n} \frac{1}{p_r^i}\right) \\
&< \frac{1}{1 - \frac{1}{p_1}} \cdot \ldots \cdot \frac{1}{1 - \frac{1}{p_r}} = \frac{p_1}{p_1 - 1} \cdot \ldots \cdot \frac{p_r}{p_r - 1} \\
&\leq \frac{2}{1} \cdot \frac{3}{2} \cdot \frac{4}{3} \cdot \ldots \cdot \frac{p_r}{p_r - 1} = p_r
\end{aligned}$$

(note that this holds for all $n \in \mathbb{N}$).
It follows (using the decomposition into prime numbers) that for all $n \in \mathbb{N}$

$$\sum_{i=1}^{n}\frac{1}{i}\leq p_r.$$

But this contradicts the fact that $\sum\limits_{i=1}^{\infty}\frac{1}{i}=\infty$. □

Quantitative analysis of Euler's proof:

We need a quantitative version of '$\sum\limits_{i=1}^{n}\frac{1}{i}\overset{n\to\infty}{\longrightarrow}\infty$', more precisely we need a bound on $\exists n(\sum\limits_{i=1}^{n}\frac{1}{i}>p_r)$. It is known that $\sum\limits_{i=1}^{n}\frac{1}{i}-\ln(n)\searrow C$, where $C\approx 0.5772...$ is the so-called Euler-Mascheroni constant. Hence for $n_r:=\lceil e^{p_r-C}\rceil$ we have $\sum\limits_{i=1}^{n_r}\frac{1}{i}>p_r$ (and this is essentially optimal). From the proof above it follows that for all $n\in\mathbb{N}$

$$\sum_{0\leq\alpha_1,\ldots,\alpha_r\leq n}\frac{1}{p_1^{\alpha_1}\cdot\ldots\cdot p_r^{\alpha_r}}\leq p_r.$$

Hence there must be an i $(1\leq i\leq n_r)$ which contains a prime factor p with $p_r<p\leq i\leq n_r$. So put together

$$\exists p(p\ prime\ \wedge p_r<p\leq\lceil e^{p_r-C}\rceil).$$

Applying this argument to all prime numbers $p_1<\ldots<p_{r_x}\leq x$ we obtain

$$\forall x\exists p(p\ prime\ \wedge x<p\leq\lceil e^{x-C}\rceil).$$

So we can take $g(x):=\lceil e^{x-C}\rceil$ (or an appropriate upper bound of this to make it computable).

Conclusion: Euler's proof yields a bound that is slightly better than the one from Euclid's proof.

Improvement of the analysis: The estimate at the beginning of Euler's proof can be improved (using that $p_i\geq i+1$) to

$$\begin{aligned}&\sum_{0\leq\alpha_1,\ldots,\alpha_r\leq n}\frac{1}{p_1^{\alpha_1}\cdot\ldots\cdot p_r^{\alpha_r}}=\left(\sum_{i=0}^{n}\frac{1}{p_1^i}\right)\cdot\ldots\cdot\left(\sum_{i=0}^{n}\frac{1}{p_r^i}\right)\\&<\frac{1}{1-\frac{1}{p_1}}\cdot\ldots\cdot\frac{1}{1-\frac{1}{p_r}}\leq\frac{1}{1-\frac{1}{2}}\cdot\ldots\cdot\frac{1}{1-\frac{1}{r+1}}\\&=\frac{2}{1}\cdot\frac{3}{2}\cdot\frac{4}{3}\cdot\ldots\cdot\frac{r+1}{r}=r+1.\end{aligned}$$

Analogously to the previous analysis we now get that (for $r\geq 1$) the $r+1$-th prime number p_{r+1} is upper bounded by $g(r):=\lceil e^{r+1-C}\rceil$ which is exponential in r (and no longer in $x\geq p_r$) and constitutes a significant improvement over the double exponential upper bound (in r) from Euclid's proof (e.g. from the former bound one gets the lower bound $\ln x$ for $x\geq 1$ (exercise) for the Euler π-function

$\pi(x) := |\{p : p \text{ prime} \wedge p \le x\}|$ whereas the bound from Euclid's proof only yields $\ln\ln x$ (for $x \ge 2$) as a lower bound (exercise, see also [149]).

Remark 2.2. Euler's proofs uses as a lemma the fact that $\sum\limits_{i=1}^{\infty} \frac{1}{i} = \infty$, i.e.

$$\forall k\, \exists n \, (\sum_{i=1}^{n} \frac{1}{i} \ge k),$$

which itself (just as the conclusion) is of the form $\forall x \exists y A_0(x,y)$. Hence what the analysis of Euler's proof actually provides is a procedure that transforms a rate of divergence of the harmonic series into a bound on a prime number $p \ge x$. In the analysis above we directly applied this to the rate of divergence resulting from

$$\sum_{i=1}^{n} \frac{1}{i} - \ln(n) \searrow C.$$

Proof 3: Let $p_1, \ldots, p_r$ $(r \ge 1)$ be the first r primes and define for $x \ge 1$ $N(x) := \{n \le x : n \ge 1 \wedge n \text{ is not divisible by any prime } p > p_r\}$. We can express $n \in N(x)$ in the form $n = n_1^2 m$ where m is 'squarefree', i.e. is not divisible by a square of any prime.
We have $m = p_1^{b_1} \cdot p_2^{b_2} \cdot \ldots \cdot p_r^{b_r}$, where $b_i \in \{0,1\}$. There are 2^r possible exponents and consequently at most 2^r different values of m. Also, because of $n_1 \le \sqrt{n} \le \sqrt{x}$, there are not more than $\sqrt{x}$ different values of n_1. Hence $|N(x)| \le 2^r\sqrt{x}$. Now if there were only finitely many primes $p_1, \ldots, p_r$, then $|N(x)| = x$ for every x and so $2^r\sqrt{x} \ge x$ for all x which is a contradiction.

From this proof one gets a bound as follows: Let $p_1, \ldots, p_r$ be the first r primes. Define $x := (2^r)^2 + 1 = 2^{2r} + 1$. Then $2^r\sqrt{x} < x$. Hence $\exists n \le x(n$ is divisible by some prime $p > p_r)$ and so $\exists p(p \text{ prime} \wedge p_r < p \le 2^{2r} + 1 = 4^r + 1)$.
So we get again a bound $g(r) := 4^r + 1$ which is exponential in r rather than p_r.

For another proof (in fact a variant of proof 3) see the exercise 1. Still further proofs can be found in [2].

Discussion:

1) All three proofs provide more information than the mere fact that 'there are infinitely many primes' is true. By making their quantitative content explicit one can compare them with respect to their numerical quality.
2) The unwindings of the proofs 1)–3) were straightforward and didn't require any tools from logic as guiding principles. However there are more complicated proofs where the use of proof-theoretic tools turned out to be decisive in practice (see e.g. [122, 267, 204, 205]). The final verification of the data extracted will always be again an ordinary mathematical proof (obtained by a proof-theoretic transformation of the original proof) which does not rely on any logical metatheorems (in contrast to the verification of the general procedure of transformation).

This differs from many model theoretic applications to mathematics where the provability or the truth in some model of the conclusion is established without exhibiting a proof which doesn't rely on model theoretic theorems.

3) Already the a-priori information, provided by a general metatheorem, that e.g. a certain computable bound must be extractable from a given proof which is formalizable in a certain system $\mathcal{T}$ can be an important step in actually finding such a bound even if the latter is carried out by ad hoc methods and doesn't follow closely any proof-theoretic procedure.

Remark 2.3. If A does not have the form $\forall x \exists y A_0(x,y)$ right away it may have so after some logical transformations, e.g.

$$A :\equiv \big(\exists x \forall y A_0(x,y) \to \forall u \exists v B_0(u,v)\big)$$

is logically equivalent to the prenex normal form

$$A^{pr} :\equiv \forall u, x \exists v, y \big(A_0(x,y) \to B_0(u,v)\big)$$

so that the reasoning above applies to the A^{pr}.

4) $A \equiv \exists x \forall y A_0(x,y)$: From a proof of A (even in first order logic without equality $\mathrm{PL}_{-=}$) one cannot (in general) obtain a realization $\forall y A_0(t,y)$ nor a list of candidates such that $\bigvee\limits_{i=1}^{n} \forall y A_0(t_i,y)$ ($t, t_1, \ldots, t_n$ not containing y) holds:

Proposition 2.4. *There exists a logically valid sentence $A \equiv \exists x \forall y A_0(x,y) \in \mathcal{L}(\mathrm{PA})$ in the language of Peano arithmetic* PA *such that there is* **no** *list of closed terms $t_1, \ldots, t_k \in \mathcal{L}(\mathrm{PA})$ such that*

$$\mathrm{PA} \vdash \bigvee_{i=1}^{k} \forall y A_0(t_i,y).$$

Proof: Take $P(x) :\equiv Prov_{\mathrm{PA}}(x, \ulcorner \overline{0} = \overline{1} \urcorner)$ and $A_0(x,y) :\equiv P(x) \vee \neg P(y)$ (here '$Prov_{\mathrm{PA}}(x, \ulcorner \overline{0} = \overline{1} \urcorner)$' expresses primitive recursively 'x is the Gödel number of a PA-proof of $0 = 1$' (see e.g. [194]). Suppose there are closed terms $t_1, \ldots, t_k$ such that

$$(1)\ \mathrm{PA} \vdash \bigvee_{i=1}^{k} \forall y A_0(t_i,y).$$

Within PA each t_i can be computed to a numeral $\overline{n}_i$:

$$(2)\ \mathrm{PA} \vdash t_i = \overline{n}_i \text{ for } 1 \leq i \leq k.$$

By (1) and (2) we have

$$(3)\ \mathrm{PA} \vdash \bigvee_{i=1}^{k} \forall y A_0(\overline{n}_i,y).$$

By the consistency of PA we know that

$$(4)\ \mathbb{N} \models \bigwedge_{i=1}^{k} \neg P(\overline{n}_i).$$

Hence by the numeralwise representability of primitive recursive predicates in PA we have

$$(5)\ \mathrm{PA} \vdash \bigwedge_{i=1}^{k} \neg P(\overline{n}_i).$$

But (3) and (5) imply

$$(6)\ \mathrm{PA} \vdash \forall y \neg Prov_{\mathrm{PA}}(y, \ulcorner \overline{0} = \overline{1} \urcorner),$$

which contradicts Gödel's second incompleteness theorem. □

However, although PA is not able to verify $\bigvee_{i=1}^{k} \forall y A_0(t_i, y)$ for any tuple of terms t_i **we** can (using the consistency of PA) verify this on the meta-level: In fact, for **any** term t, e.g. for 0, we know that $\forall y A_0(t, y)$ is true in $\mathbb{N}$ simply because

$$\mathbb{N} \models \forall y \neg Prov_{\mathrm{PA}}(y, \ulcorner \overline{0} = \overline{1} \urcorner).$$

However, there are other examples where –in general– even this is not possible, e.g. take

$$A_e :\equiv \exists x \forall y \big(T(\overline{e}, \overline{e}, x) \vee \neg T(\overline{e}, \overline{e}, y)\big),$$

where T is the (primitive recursive) Kleene-T-predicate, i.e. $T(x,y,z) :\equiv$ 'the Turing machine with Gödel number x applied to the input y terminates with a computation whose Gödel number is z' (see e.g. [371]).
In general we are not able to determine closed terms $t_1, \ldots, t_k$ such that

$$\mathbb{N} \models \bigvee_{i=1}^{k} \forall y \big(T(\overline{e}, \overline{e}, t_i) \vee \neg T(\overline{e}, \overline{e}, y)\big),$$

since this would allow us to decide whether $\exists x T(\overline{e}, \overline{e}, x)$ or not (simply check whether $\bigvee_{i=1}^{k} T(\overline{e}, \overline{e}, t_i)$ is true or not).

In fact, for

$$A :\equiv \forall x \exists y \forall z \big(T(x,x,y) \vee \neg T(x,x,z)\big)$$

A is provable in PA using only the logical axioms and rules and hence in $\mathrm{PL}_{-=}$, but there is no computable bound g on '$\exists y$', i.e. no computable g such that

$$\forall x \exists y \leq g(x) \forall z \big(T(x,x,y) \vee \neg T(x,x,z)\big)$$

since this would make the (special) halting problem $\{x \in \mathbb{N} : \exists y \in \mathbb{N}(T(x,x,y))\}$ decidable by the then computable function

$$f(x) := \begin{cases} 0, \text{ if } \exists y \le g(x)(T(x,x,y)) \\ 1, \text{ otherwise.} \end{cases}$$

We sometimes make use of the following definition:

Definition 2.5. A formula $A \in \mathcal{L}(\mathrm{PL})$ in prenex normal form is called Π^0_n-formula if it has n-alternating blocks of equal quantifiers starting with a block of universal quantifiers, i.e.

$$\forall \underline{x}_1 \exists \underline{x}_2 \ldots \forall/\exists \underline{x}_n A_0(\underline{x}_1, \ldots, \underline{x}_n),$$

where $\underline{x}_i$ are tuples of variables. If the formula starts with a block of existential quantifiers

$$\exists \underline{x}_1 \forall \underline{x}_2 \ldots \exists/\forall \underline{x}_n A_0(\underline{x}_1, \ldots, \underline{x}_n)$$

it is called Σ^0_n-formula.

Remark 2.6. The upper index '0' only is relevant in theories with higher order quantifiers (over functions and functionals) where then it indicates that all the quantifiers range over the first order ('base type') variables.
Many theories, such as PA, allow the contraction of tuples of variables into single variables.

As we discussed above, infinity statements (for quantifier-free properties) in number theory have the form of Π^0_2-formulas. So an important class of Σ^0_2-formulas are finiteness statements, i.e.

$$\exists x \forall y > x \neg A_0(y,a),$$

where a, y are the only free variables in $A_0(y,a)$. We now may ask for

- a computable bound $h(a)$ on the height of the solutions, i.e.

$$\forall a \forall y > h(a) \neg A_0(y,a)$$

  or
- a computable bound $N(a)$ on the number of solutions, i.e.

$$N(a) \ge |\{y : A_0(y,a)\}|.$$

It is clear that any h also is an upper bound N on the number of solutions but in general the existence of a computable function N does not imply the existence of a computable height function h as the following example (due to [267]) shows: consider again Kleene's T-predicate and define

$$A_0(y,a) :\equiv T(a,a,y).$$

Then clearly $N(a) := 1$ is a bound on the number of solutions but by the undecidability of the special halting problem there is no computable (in a) bound $h(a)$ on y.

In general, for Σ_2^0-finiteness theorems not even a computable (in the parameters) bound N exists ([267]): Let

$$A_0(y,a) :\equiv T(a,a,j_2(y)) \wedge 0 < j_1(y) \leq j_2(y),$$

where we refer to some standard pairing/unpairing functions j, j_1, j_2 (see definition 3.30 below). Here the number of solutions and their maximal size coincide and so neither of them is computable.

Famous finiteness theorems in mathematics are Roth's theorem ([317]) on the number of exceptionally good rational approximations of irrational algebraic numbers and Falting's theorem establishing the Mordell conjecture ([97]). In both cases effective bounds N are known but no computable bounds h.

Roth's theorem says the following

Theorem 2.7 (Roth [317]). An algebraic irrational number α has only finitely many exceptionally good rational approximations, i.e. for $\varepsilon > 0$ there are only finitely many $q \in \mathbb{N}$ such that

$$R(q) :\equiv q > 1 \wedge \exists ! p \in \mathbb{Z} : (p,q) = 1 \wedge |\alpha - pq^{-1}| < q^{-2-\varepsilon}.$$

The first polynomial bound N in the case of Roth's theorem was obtained in Luckhardt [267] by extracting Herbrand terms (see theorem 2.18 and the subsequent discussion below) from a proof of Roth's theorem due to Esnault and Viehweg [94] using certain growth properties of these terms (following general ideas from Kreisel [249]).

Theorem 2.8 (Luckhardt [267]). The following upper bound on $\#\{q : R(q)\}$ holds:

$$\#\{q : R(q)\} < \frac{7}{3}\varepsilon^{-1} \log N_\alpha + 6 \cdot 10^3 \varepsilon^{-5} \log^2 d \cdot \log(50\varepsilon^{-2} \log d),$$

where $N_\alpha < \max(21 \log 2h(\alpha), 2\log(1+|\alpha|))$ and h is the logarithmic absolute homogeneous height.

Independently, a roughly similar bound was obtained in Bombieri-van der Poorten [38] using a more ad hoc strategy.

Two elementary examples of non-constructive proofs in number theory:

Proposition 2.9. $\exists a,b \in \mathbb{R}\big(a,b \textit{ irrational} \wedge a^b \textit{ rational}\big)$.

Proof: Case 1: $\sqrt{2}^{\sqrt{2}}$ is rational. Put $a := b := \sqrt{2}$.
Case 2: $\sqrt{2}^{\sqrt{2}}$ irrational. Put $a := \sqrt{2}^{\sqrt{2}}, b := \sqrt{2}$. □

Remark 2.10. In the example above, the matrix 'a,b irrational $\wedge a^b$ rational' is more complex than Π_1^0 : Using the representation of real number from chapter 4 below 'a,b irrational' is in Π_2^0 and 'a^b rational' is in Σ_2^0.

From this proof we get two candidates for (a,b), namely $(\sqrt{2},\sqrt{2})$ and $(\sqrt{2}^{\sqrt{2}},\sqrt{2})$ but no decision which one satisfies the proposition.

Remark 2.11. 1) From a deep result of Gelfand and Schneider, stating that if a,b are algebraic, $a \neq 0,1$ and b irrational, then a^b is transcendental, it follows that $\sqrt{2}^{\sqrt{2}}$ is transcendental and, therefore, irrational. So it is the pair $(\sqrt{2}^{\sqrt{2}},\sqrt{2})$ which satisfies the proposition.

2) While it requires the Gelfand-Schneider theorem to determine which of the candidates $(\sqrt{2},\sqrt{2})$ and $(\sqrt{2}^{\sqrt{2}},\sqrt{2})$ satisfies the proposition, there is a trivial argument (which we learned from G. Stolzenberg) that provides an explicit solution to proposition 2.9: take $a := \sqrt{2}$ and $b := 2\log_2(3)$. b is irrational since $\log_2(3) = m/n$ for some $m,n \in \mathbb{N}^*$ would imply that $2^m = 3^n$ which is impossible. Clearly, $a^b = 3$ is rational.

Here is another example (communicated by H. Friedman) of a simple non-constructive proof in number theory:

Proposition 2.12. $(e - \pi$ *is irrational*$)$ *or* $(e + \pi$ *is irrational*$)$.

Proof: One easily formalizes the proof of the irrationality of e as given e.g. in [149] in PA. If both $e - \pi$ and $e + \pi$ were rational, then also their sum $2e$ and, therefore, e would be rational which is a contradiction. □

Remark 2.13. In 1996, it was proved by Yu.V. Nesterenko ([285]) that e^π and π are algebraically independent and hence $\pi + e^\pi$ is transcendental whereas for $e + \pi$ and $e - \pi$ individually the question of transcendence is still open.

2.3 Herbrand's theorem and the no-counterexample interpretation

We have seen that already for Σ^0_2, Π^0_3-sentences A (i.e. $A \equiv \exists n \forall m A_0(n,m)$ or $A \equiv \forall k \exists n \forall m A_0(k,n,m)$ where A_0 is recursive) it is not possible in general to compute witnesses resp. bounds. However one can obtain such witness candidates and bounds (and even realizing function(al)s) for a weakened version of A, namely its so-called Herbrand normal form A^H:

Definition 2.14. $A \equiv (\forall y_0)\exists x_1 \forall y_1 \ldots \exists x_n \forall y_n A_0(y_0,x_1,y_1,\ldots,x_n,y_n)$. Then the Herbrand normal form of A is defined as

$$A^H :\equiv (\forall y_0)\exists x_1,\ldots,x_n A_0(y_0,x_1,f_1(x_1),\ldots,x_n,f_n(x_1,\ldots,x_n)),$$

where $f_1,\ldots,f_n$ are new function symbols, called Herbrand index functions.

Remark 2.15. In theories with function variables and function quantifiers we take the Herbrand normal form of A to be

$$A^H :\equiv \forall(y_0), f_1, \ldots, f_n \exists x_1, \ldots, x_n A_0(y_0, x_1, f_1(x_1), \ldots, x_n, f_n(x_1, \ldots, x_n)).$$

In the following PL denotes first order predicate logic with equality.
For prenex sentences A, A and A^H are equivalent with respect to logical validity, i.e.

$$\models A \Leftrightarrow \models A^H$$

(this fact is also expressed by saying that A^H is a validity normal form) but are not logically equivalent since in general

$$\text{PL} \nvdash A^H \rightarrow A.$$

However the converse implication holds

$$\text{PL}_{-=} \vdash A \rightarrow A^H.$$

Remark 2.16. The dual normal form in which the existentially quantified variables in a prenex normal formula are replaced by new function symbols depending on the universally quantified variables from the universal quantifiers to the left is called Skolem normal form and denoted by A^S, i.e. for

$$A :\equiv \forall x_1 \exists y_1 \ldots \forall x_n \exists y_n A_0(x_1, y_1, \ldots, x_n, y_n)$$

$$A^S :\equiv \forall x_1, \ldots, x_n A_0(x_1, f_1(x_1), \ldots, x_n, f_n(x_1, \ldots, x_n)).$$

The function symbols $f_1, \ldots, f_n$ are called Skolem functions.
For prenex sentences A, the Skolem normal form is a satisfiability normal form.
Unfortunately, the terminology differs for different authors. Sometimes the name Skolem normal form is used for what we call Herbrand normal form.

Let $\text{PL}^2_{(-=)}$ denote the extension of $\text{PL}_{(-=)}$ obtained by the addition of n-ary function variables (for every n) and function quantifiers.
Let furthermore AC denote the schema of choice

$$\text{AC: } \forall \underline{x} \exists y A(\underline{x}, y) \rightarrow \exists f \forall \underline{x} A(\underline{x}, f(\underline{x})) \;\; (\underline{x} = x_1, \ldots, x_n),$$

then it is an easy exercise to show that

$$\text{PL}^2_{-=} + \text{AC} \vdash A \leftrightarrow A^H.$$

We now consider again the sentence

$$A \equiv \forall x \exists y \forall z \big(P(x,y) \vee \neg P(x,z)\big),$$

where P is some predicate symbol. In contrast to A, the Herbrand normal form A^H of A

$$A^H \equiv \exists y\big(P(x,y) \vee \neg P(x,g(y))\big)$$

allows an interpretation in form of a list of candidates (uniformly in x,g) for '$\exists y$', namely (x,gx) and also (c,gc) for any constant c does the job since the disjunction

$$A^{H,D} :\equiv \big(P(x,c) \vee \neg P(x,g(c))\big) \vee \big(P(x,g(c)) \vee \neg P(x,g(g(c)))\big)$$

is a tautology.
A tautology remains a tautology if we replace all occurrences of a term s by a variable y: Replace $g(c)$ by y and $g(g(c))$ by z. Then $A^{H,D}$ becomes

$$A^D :\equiv \big(P(x,c) \vee \neg P(x,y)\big) \vee \big(P(x,y) \vee \neg P(x,z)\big),$$

which still is a tautology. From A^D we can derive A by a so-called direct proof (which uses only appropriate quantifier introduction rules, the shift of quantifiers over $\vee$ and contraction):

$$P(x,c) \vee \neg P(x,y) \vee P(x,y) \vee \neg P(x,z)$$

$$\Downarrow (\forall\text{-introduction})$$

$$P(x,c) \vee \neg P(x,y) \vee \forall z\big(P(x,y) \vee \neg P(x,z)\big)$$

$$\Downarrow (\exists\text{-introduction})$$

$$P(x,c) \vee \neg P(x,y) \vee \exists y \forall z\big(P(x,y) \vee \neg P(x,z)\big)$$

$$\Downarrow (\forall\text{-introduction})$$

$$\forall y\big(P(x,c) \vee \neg P(x,y)\big) \vee \exists y \forall z\big(P(x,y) \vee \neg P(x,z)\big)$$

$$\Downarrow (\exists\text{-introduction})$$

$$\exists u \forall y\big(P(x,u) \vee \neg P(x,y)\big) \vee \exists y \forall z\big(P(x,y) \vee \neg P(x,z)\big)$$

$$\Downarrow (\text{contraction})$$

$$\exists y \forall z\big(P(x,y) \vee \neg P(x,z)\big)$$

$$\Downarrow (\forall\text{-introduction})$$

$$\forall x \exists y \forall z\big(P(x,y) \vee \neg P(x,z)\big)$$

Definition 2.17. A formula A in the language of first order predicate logic with equality (PL) is called a quasi-tautology if it is a tautological consequence of instances of $=$-axioms.

Theorem 2.18 (Herbrand's Theorem).
Let $A \equiv \exists x_1 \forall y_1 \ldots \exists x_n \forall y_n A_0(x_1,y_1,\ldots,x_n,y_n)$. Then the following holds:
$\mathrm{PL}_{=} \vdash A$ *iff there are terms $t_{1,1},\ldots,t_{1,k_1},\ldots,t_{n,1},\ldots,t_{n,k_n}$ (built up out of the constants, free variables and function symbols of A and the index functions used for the*

formation of A^H) such that

$$A^{H,D} :\equiv \bigvee_{j_1=1}^{k_1} \ldots \bigvee_{j_n=1}^{k_n} A_0\big(t_{1,j_1}, f_1(t_{1,j_1}), \ldots, t_{n,j_n}, f_n(t_{1,j_1}, \ldots, t_{n,j_n})\big)$$

is a tautology.
The terms $t_{i,j}$ can be extracted constructively from a given PL$_{-=}$*-proof of A and conversely one can construct a* PL$_{-=}$*-proof for A out of a given tautology $A^{H,D}$.*
The theorem holds for PL *if 'tautology' is replaced by 'quasi-tautology'.*

Proof: See e.g. [332]. □

The most difficult part of the proof of Herbrand's theorem is the construction of the Herbrand terms $t_{i,j}$. The reverse direction for PL$_{-=}$ follows similar to the special case treated above: the f_i-terms in $A^{H,D}$ are replaced by new variables (starting from terms of maximal size) yielding an index-function-free Herbrand disjunction A^D. From this A is derived by a direct proof. For PL the reverse direction is more complicated to establish since also instances of equality axioms $\underline{x} = \underline{y} \rightarrow f_i(\underline{x}) = f_i(\underline{y})$ are now allowed in the proof of $A^{H,D}$.

In applications, the Herbrand disjunction A^D without index function has been particular useful (see [249],[267]). Although it is quite complicated to write down the general form of such a disjunction it is easy for Π^0_3-sentences (which is sufficient for many applications in mathematics):

For sentences $A \equiv \forall x \exists y \forall z A_0(x,y,z)$, A^D can always be written in the form

$$A_0(x,t_1,b_1) \vee A_0(x,t_2,b_2) \vee \ldots \vee A_0(x,t_k,b_k),$$

where the b_i are new variables and t_i does not contain any b_j with $i \leq j$ (see [249]).

Herbrand's theorem immediately extends to so-called open theories, i.e. first order theories $\mathcal{T}$ whose non-logical axioms $G_1, \ldots, G_m$ are all purely universal ($G_i \equiv \forall a_i G^i_0(a_i)$), if '(quasi-)tautology' is replaced by 'tautological consequence of instances of equality axioms and the non-logical axioms'.

Proof: Apply Herbrand's theorem for logic to

$$\tilde{A} :\equiv \exists x_1 \forall y_1 \ldots \exists x_n \forall y_n \exists a_1, \ldots, a_m \big(\bigwedge_{i=1}^{m} G^i_0(a_i) \rightarrow A_0(x_1, y_1, \ldots, x_n, y_n)\big).$$

□

Warning: For the extension of Herbrand's theorem to open theories $\mathcal{T}$ it is important that the index function used in defining A^H are new and do not occur in the non-logical axioms. In particular if we have a schema of purely universal axioms then in the statement of Herbrand's theorem this schema is always understood with respect to the original language (without the index functions). Otherwise the reverse direction in Herbrand's theorem in general would fail (see [202] for a discussion of this and related matters thereby pointing out errors in the literature).

In general Herbrand's theorem in the form stated above does not hold for theories which are not open, e.g. it fails for PA.
However there are ways to extend the general idea behind Herbrand's theorem to theories like PA and beyond: in this book we will discuss various forms of Gödel's functional interpretation (chapters 8, 9, 10) and the so-called no-counterexample interpretation (due to G. Kreisel [241, 242], see further below in this chapter and chapter 10). We conclude this chapter by motivating the latter and also indicating its limitations:

Let's consider again the example

$$A \equiv \forall x \exists y \forall z \big(P(x,y) \lor \neg P(x,z)\big).$$

If P is formulated in some theory like PA with decidable prime formulas, e.g. if $P(x,y) \equiv T(x,x,y)$, then we can realize the Herbrand normal form A^H of A instead of using a disjunction also by a computable functional of type level 2 which is defined by cases:

$$\Phi(x,g) := \begin{cases} x \ \text{ if } \neg T(x,x,g(x)) \\ g(x) \ \text{ otherwise.} \end{cases}$$

From this definition it easily follows that

$$\forall x, g \big(T(x,x,\Phi(x,g)) \lor \neg T(x,x,g(\Phi(x,g)))\big).$$

If A is not provable in PL but e.g. in PA we no longer can expect that functionals as simple as Φ above will be sufficient. In addition to the use of definition by cases we also have to allow certain recursive definitions whose complexity depends on the strength of the theory in which A is proved. In this book we will show e.g. in the case of PA (and subsystems) what functionals are needed.

Definition 2.19. Let $A \equiv \exists x_1 \forall y_1 \ldots \exists x_n \forall y_n A_0(x_1, y_1, \ldots, x_n, y_n)$. If a tuple of functionals $\Phi_1, \ldots, \Phi_n$ realizes the Herbrand normal form A^H of A, i.e. if

$$\forall \underline{f} A_0\big(\Phi_1(\underline{f}), f_1(\Phi_1(\underline{f})), \ldots, \Phi_n(\underline{f}), f_n(\Phi_1(\underline{f}), \ldots, \Phi_n(\underline{f}))\big)$$

is true (where $\underline{f} = f_1, \ldots, f_n$), then we say that $\underline{\Phi}(= \Phi_1, \ldots, \Phi_n)$ satisfies the no-counterexample interpretation of A (short: $\underline{\Phi}$ n.c.i. A).
If A starts with a universal quantifier $\forall y_0$ then y_0 is considered as a 0-place index function and Φ_i now depends on y_0 and $\underline{f}$.

Motivation for the name 'no-counterexample interpretation':

Let A be as above. Then $\neg A$ is equivalent to

$$\forall x_1 \exists y_1 \ldots \forall x_n \exists y_n \neg A_0(x_1, y_1, \ldots, x_n, y_n).$$

So a counterexample to A is given by functions $f_1, \ldots, f_n$ such that

$$(+)\ \forall \underline{x} \neg A_0(x_1, f_1(x_1), \ldots, x_n, f_n(x_1, \ldots, x_n))$$

holds. Hence functionals $\underline{\Phi}$ satisfying the n.c.i. of A produce a counterexample to $(+)$ i.e. to the existence of counterexample functions $f_1, \ldots, f_n$.

The no-counterexample interpretation can indeed be realized for many interesting classical theories (in particular Peano arithmetic **PA** for which it was designed by Kreisel) and fragments thereof by certain subrecursive classes of functionals. E.g. we will show in chapter 10 that theorems of the fragment $\mathbf{PA}_1$ of **PA** with the schema of induction restricted to purely existential (Σ^0_1-)formulas always have functionals satisfying the no-counterexample interpretation which are primitive recursive in the sense of Kleene. This was first shown by Parsons ([299]) and will be proved in chapter 10. Full Peano arithmetic requires primitive recursive functionals in higher types in the extended sense of Gödel [133] (see chapter 3 and – again – chapter 10 below).

Definition 2.20. A function $f : \mathbb{N}^k \to \mathbb{N}$ is called primitive recursive if it can be defined by the following schemas:

1) The initial functions $Z(x) = 0$ (Zero), $P^p_i(x_0, \ldots, x_{p-1}) = x_i$ $p \geq 1, i < p$ (Projections), $S(x) = x + 1$ (Successor) are primitive recursive.
2) If $h_0(x_0, \ldots, x_{p-1}), \ldots, h_{l-1}(x_0, \ldots, x_{p-1})$ and $g(y_0, \ldots, y_{l-1})$ are primitive recursive functions, then also

$$f(x_0, \ldots, x_{p-1}) = g(h_0(x_0, \ldots, x_{p-1}), \ldots, h_{l-1}(x_0, \ldots, x_{p-1}))$$

is primitive recursive.
3) If $g(x_0, \ldots, x_{p-1})$ and $h(z, y, x_0, \ldots, x_{p-1})$ are primitive recursive functions, then also f defined by

$$f(0, x_0, \ldots, x_{p-1}) = g(x_0, \ldots, x_{p-1}),$$
$$f(y+1, x_0, \ldots, x_{p-1}) = h(f(y, x_0, \ldots, x_{p-1}), y, x_0, \ldots, x_{p-1})$$

is primitive recursive.

Definition 2.21. A functional F is called primitive recursive (of level or 'type' ≤ 2) in the sense of Kleene if it can be defined by the following schemas ($\underline{x} = x_0, \ldots, x_{p-1}$ is a list of number variables and $\underline{f} = f_0, \ldots, f_{q-1}$ is a list of function variables for any $p, q \geq 1$):

(i) (Projections) $F(\underline{x},\underline{f}) = x_i$ (for $i < p$) and (Zero) $F(\underline{x},\underline{f}) = 0$,
(ii) (Function application) $F(\underline{x},\underline{f}) = f_i(x_{j_0},\ldots,x_{j_{l-1}})$
(for $i < q$ and $j_0,\ldots,j_{l-1} < p$ and f_i of arity l),
(iii) (Successor) $F(\underline{x},\underline{f}) = x_i + 1$ (for $i < p$),
(iv) (Substitution)
$F(\underline{x},\underline{f}) = G(H_0(\underline{x},\underline{f}),\ldots,H_{l-1}(\underline{x},\underline{f}),\lambda y.K_0(y,\underline{x},\underline{f}),\ldots,\lambda y.K_{j-1}(y,\underline{x},\underline{f}))$,
(v) (Primitive recursion)
$F(0,\underline{x},\underline{f}) = G(\underline{x},\underline{f}),\;\; F(y+1,\underline{x},\underline{f}) = H(F(y,\underline{x},\underline{f}),y,\underline{x},\underline{f})$.

Remark 2.22. The class of primitive recursive functionals of level ≤ 2 in the sense of Kleene which do not have any function arguments $\underline{f}$ coincides with the class of primitive recursive functions. Exercise!

We will now demonstrate the no-counterexample on two simple examples (the second of which will play an important role in applications to metric fixed point theory in chapter 18 below):

Example 1: Consider the following proposition which is an immediate consequence of the least number principle for natural numbers (which can formally be proved using Σ^0_1-induction):

$$(+)\; \forall f : \mathbb{N} \to \mathbb{N} \forall k \in \mathbb{N} \exists n \geq k \forall m \geq k (f(n) \leq f(m)).$$

$(+)$ is ineffective in the sense that there is no computable bound $\Phi(f,k)$ on n. In fact, the next two propositions give even stronger results:

Proposition 2.23. *There is no computable functional* $\Phi : \mathbb{N}^{\mathbb{N}} \to \mathbb{N}$ *such that*

$$\forall f : \mathbb{N} \to \mathbb{N} \exists n \leq \Phi(f) \forall m \in \mathbb{N} (f(n) \leq f(m)).$$

Proof: Assume that on the contrary such a computable Φ would exist. Consider the constant-1 function $1 := \lambda k.1$. Since Φ is computable, $\Phi(f)$ only depends on finitely many values of f, i.e. $\Phi : \mathbb{N}^{\mathbb{N}} \to \mathbb{N}$ is continuous w.r.t. the product topology on $\mathbb{N}^{\mathbb{N}}$ and the discrete topology on $\mathbb{N}$. Hence

$$(*)\; \exists l \in \mathbb{N} \forall g : \mathbb{N} \to \mathbb{N} (\forall i \leq l (g(i) = 1) \to \Phi(1) = \Phi(g)).$$

Now define

$$g(i) := \begin{cases} 1, \text{ if } i \leq \max(l, \Phi(1)) \\ 0, \text{ otherwise.} \end{cases}$$

Then $\Phi(g) = \Phi(1)$ by $(*)$, but also $\Phi(g) > \Phi(1)$ since $g(j) = \min\{g(i) : i \in \mathbb{N}\} = 0$ for some $j \leq \Phi(g)$, whereas $g(i) = 1$ for all $i \leq \Phi(1)$. □

Proposition 2.24. *There exists a primitive recursive function* $f_0 : \mathbb{N} \to \mathbb{N}$ *such that there is no computable function* $\Phi : \mathbb{N} \to \mathbb{N}$ *with*

$$\forall k \in \mathbb{N} \exists n \leq \Phi(k) \big(n \geq k \wedge \forall m \geq k (f_0(n) \leq f_0(m))\big).$$

Proof: Let $e \in \mathbb{N}$ be such that

$$\{k \in \mathbb{N} : \{e\}(k) \downarrow\} = \{k \in \mathbb{N} : \exists n \in \mathbb{N}\, T(e,k,n)\}$$

is undecidable, where the primitive recursive Kleene T-predicate satisfies

$$(1)\ \forall k, n_1, n_2 \in \mathbb{N}(T(e,k,n_1) \wedge T(e,k,n_2) \rightarrow n_1 = n_2).$$

Define $f_0(n) := g(j_1(n), j_2(n))$, where

$$g(k,n) := \begin{cases} j(k,0), \text{ if } T(e,k,n) \\ j(k,n+1), \text{ otherwise} \end{cases}$$

and $j : \mathbb{N} \times \mathbb{N} \rightarrow \mathbb{N}$ is any primitive recursive bijection with primitive recursive projections j_1, j_2 (e.g. we may take as j the standard Cantor pairing function, see definition 3.30 below). It is clear that f_0 is primitive recursive. Now suppose that $\Phi : \mathbb{N} \rightarrow \mathbb{N}$ is computable and satisfies

$$(2)\ \forall k \in \mathbb{N} \exists n \leq \Phi(k)\big(n \geq k \wedge \forall m \geq k(f_0(n) \leq f_0(m))\big).$$

Define (primitive recursively in Φ) a function $\tilde{\Phi} : \mathbb{N} \rightarrow \mathbb{N}$ by

$$(3)\ \tilde{\Phi}(0) := 0,\ \tilde{\Phi}(l+1) := \max\{\tilde{\Phi}(l), \Phi(\tilde{\Phi}(l)) + 1\}.$$

By induction on l we show that

$$(4)\ \forall l \in \mathbb{N} \forall m \geq \tilde{\Phi}(l)\,(f_0(m) \geq l):$$

The case $l = 0$ is trivial. $l \mapsto l+1$: By induction hypothesis we have

$$(5)\ \min\{f_0(m) : m \geq \tilde{\Phi}(l)\} \geq l.$$

(2) yields

$$(6)\ \exists n \leq \Phi(\tilde{\Phi}(l))\,\big(n \geq \tilde{\Phi}(l) \wedge f_0(n) = \min\{f_0(m) : m \geq \tilde{\Phi}(l)\}\big).$$

The injectivity of f_0 (which follows using (1)) implies that n is uniquely determined by $f_0(n) = \min\{f_0(m) : m \geq \tilde{\Phi}(l)\}$ and thus (using (3))

$$\forall m \geq \tilde{\Phi}(l+1) = \max\{\tilde{\Phi}(l), \Phi(\tilde{\Phi}(l)) + 1\}\,\big(f_0(m) > \min\{f_0(k) : k \geq \tilde{\Phi}(l)\}.$$

Hence (using (5))

$$\forall m \geq \tilde{\Phi}(l+1)\,\big(f_0(m) \geq l+1\big)$$

which finishes the proof of (4).
Now let $k_0 := \tilde{\Phi}(j(k,0)+1)$. Then by (4)

$$(7)\ \forall m \geq k_0\big(f_0(m) > j(k,0)\big).$$

Hence

$$\{e\}(k)\downarrow\leftrightarrow \exists n\, T(e,k,n) \leftrightarrow \exists m\big(j_1(m)=k \wedge f_0(m)=j(k,0)\big)$$
$$\leftrightarrow \exists m < k_0\big(j_1(m)=k \wedge f_0(m)=j(k,0)\big),$$

where the latter clearly is decidable which is a contradiction. □

In contrast to these negative results we have a primitive recursive (in the sense of Kleene) functional Φ satisfying the no-counterexample interpretation of $(+)$:

Proposition 2.25. *There exists a primitive recursive (in the sense of Kleene) functional Φ such that for all $f,g:\mathbb{N}\to\mathbb{N}$*

$$\forall k\in\mathbb{N}\big(\Phi(f,g,k)\geq k \wedge \big(g(\Phi(f,g,k))\geq k \to f(\Phi(f,g,k))\leq f(g(\Phi(f,g,k)))\big)\big).$$

Proof: We construct an upper bound $\Phi^*(f,g,k)$ for $\Phi(f,g,k)$, i.e.

$$\forall k\in\mathbb{N}\exists n\leq \Phi^*(f,g,k)\,\big(n\geq k\wedge (g(n)\geq k\to f(n)\leq f(g(n)))\big).$$

Φ can then be constructed from Φ^* by primitive recursive bounded search.
Let $f,g:\mathbb{N}\to\mathbb{N}, k\in\mathbb{N}$. We first show that

$$(*)\ \exists i\leq f(k)\,\big(g^{(i)}(k)\geq k\wedge (g^{(i+1)}(k)\geq k\to f(g^{(i)}(k))\leq f(g^{(i+1)}(k)))\big),$$

where $g^{(0)}(k):=k, g^{(i+1)}(k):=g(g^{(i)}(k))$.
Case 1: $\exists i<f(k)\,\big(g^{(i+1)}(k)<k\big)$. Let i_0 be the least such i. Then $g^{(i_0+1)}(k)<k\wedge g^{(i_0)}(k)\geq k$. Hence the claim is satisfied with i_0.
Case 2: $\forall i<f(k)\,\big(g^{(i+1)}(k)\geq k\big)$ and hence $\forall i\leq f(k)\,\big(g^{(i)}(k)\geq k\big)$. Assume that

$$\forall i\leq f(k)\,\big(f(g^{(i)}(k))>f(g^{(i+1)}(k))\big).$$

Then

$$f(g^{(f(k)+1)}(k))<f(k)-f(k)$$

which is a contradiction. So again $(*)$ follows for some $i\leq f(k)$.
Now define

$$\Phi^*(f,g,k):=\max\{g^{(i)}(k): i\leq f(k)\}.$$

□

Example 2: Let $(a_n)_{n\in\mathbb{N}}$ be a nonincreasing sequence of rational numbers in $[0,1]$. Since rational numbers can be coded by natural numbers one can consider (a_n) as a number theoretic function. The order relation $\leq$ and the usual arithmetical operations between rational numbers are primitive recursive in their codes (identifying below '2^{-k}' with its encoding).

Consider the proposition stating that (a_n) is a Cauchy sequence, i.e.

$$(++)\ \forall k\in\mathbb{N}\exists n\in\mathbb{N}\forall m\in\mathbb{N}\ (|a_{n+m}-a_n|<_{\mathbb{Q}} 2^{-k}).$$

By a well-known result of E. Specker [342] even for certain primitive recursive sequences (a_n) (so-called Specker sequences) there is in general no computable bound $f(k)$ on n. However, we have the following:

Proposition 2.26. *There exists a primitive recursive functional in the sense of Kleene satisfying the no-counterexample interpretation of $(++)$. In fact, there exists a primitive recursive Φ such that*

$$\forall k \in \mathbb{N} \forall g \in \mathbb{N}^{\mathbb{N}} \exists n \le \Phi(g,k)\ (|a_{n+g(n)} - a_n| <_{\mathbb{Q}} 2^{-k}).$$

Proof: For $g : \mathbb{N} \to \mathbb{N}$ define $\tilde{g} : \mathbb{N} \to \mathbb{N}$ by $\tilde{g}(n) := n + g(n)$. We first show that

$$(*)\ \forall k \in \mathbb{N} \forall g \in \mathbb{N}^{\mathbb{N}} \exists i \le 2^k (a_{\tilde{g}^{(i)}(0)} - a_{\tilde{g}^{(i+1)}(0)} <_{\mathbb{Q}} 2^{-k}).$$

Assume that on the contrary for some $k \in \mathbb{N}$ and $g \in \mathbb{N}^{\mathbb{N}}$

$$\forall i \le 2^k (a_{\tilde{g}^{(i)}(0)} - a_{\tilde{g}^{(i+1)}(0)} \ge_{\mathbb{Q}} 2^{-k}).$$

Then (using that $\tilde{g}^{(0)}(0) = 0$)

$$a_0 - a_{\tilde{g}^{(2^k+1)}(0)} \ge (2^k + 1) \cdot 2^{-k} > 1$$

which is a contradiction and so finishes the proof of $(*)$.
Since (a_n) is nonincreasing, $(*)$ implies that

$$\forall k \in \mathbb{N} \forall g \in \mathbb{N}^{\mathbb{N}} \exists i \le 2^k (|a_{\tilde{g}^{(i)}(0)} - a_{\tilde{g}^{(i)}(0)+g(\tilde{g}^{(i)}(0))}| <_{\mathbb{Q}} 2^{-k}).$$

We now take $\Phi(g,k) := \tilde{g}^{(2^k)}(0)\ (= \max\{\tilde{g}^{(i)}(0) : i \le 2^k\})$. □

Using the primitive recursive decidability of $<_{\mathbb{Q}}$ one can apply primitive recursive bounded search to get a primitive recursive realizer $\Psi((a_n), g, k)$ for '$\exists n$' from the bound $\Phi(g,k)$ in proposition 2.26. The bound $\Phi(g,k)$ is also valid for sequences (a_n) of real numbers in $[0,1]$. Moreover, using the monotonicity of (a_n) and the proof above we can state the result as follows (where $[n; n+g(n)] := \{i \in \mathbb{N} : n \le i \le n + g(n)\}$):

Proposition 2.27. *Let (a_n) be any nonincreasing sequence in $[0,1]$ then*

$$\forall k \in \mathbb{N} \forall g \in \mathbb{N}^{\mathbb{N}} \exists n \le \Phi(g,k) \forall i,j \in [n; n+g(n)] (|a_i - a_j| <_{\mathbb{R}} 2^{-k}),$$

where

$$\Phi(g,k) := \tilde{g}^{(2^k)}(0) \text{ with } \tilde{g}(n) := n + g(n).$$

Moreover, there exists an $i \le 2^k$ such that n can be taken as $\tilde{g}^{(i)}(0)$.

Note that the bound $\Phi(g,k)$ in proposition 2.27 does not depend on (a_n) at all. Hence, using the fact that only sequence elements a_k for $k \le n + g(n)$ are touched, we obtain the following (explicit version of a) 'finite convergence principle' which recently was considered by T. Tao ([357, 358]):

Corollary 2.28. *For all $k \in \mathbb{N}, g \in \mathbb{N}^{\mathbb{N}}$ there exists an $M \in \mathbb{N}$ such that for all nonincreasing finite sequences $0 \leq a_M \leq \ldots \leq a_0 \leq 1$ of length $M+1$ in $[0,1]$ there exists an $n \in \mathbb{N}$ with*

$$n + g(n) \leq M \wedge \forall i, j \in [n; n+g(n)](|a_i - a_j| <_{\mathbb{R}} 2^{-k}).$$

Moreover, we can compute M as $M := \tilde{g}^{(2^k+1)}(0)$, where $\tilde{g}(n) := n + g(n)$.

Remark 2.29. 1) For nonincreasing sequences in $[0,C]$ for some $C \in \mathbb{N}$ one can take $\Phi(g,k,C) := \tilde{g}^{(C \cdot 2^k)}(0)$ and M as $\tilde{g}^{(C \cdot 2^k+1)}(0)$.
2) The property

$$\forall k \in \mathbb{N} \forall g \in \mathbb{N}^{\mathbb{N}} \exists n \in \mathbb{N} \forall i, j \in [n; n+g(n)]\, (|a_i - a_j| <_{\mathbb{R}} 2^{-k})$$

of a sequence (a_n) of reals, which is nothing else but the Herbrand normal form of the following (equivalent) reformulation of the usual Cauchy property of (a_n) (treating '$\forall i, j(\ldots)$' as a Σ^0_1-formula to which it is equivalent since $<_{\mathbb{R}}$ is Σ^0_1 and the universal quantifiers are bounded)

$$\forall k \in \mathbb{N} \exists n \in \mathbb{N} \forall m \in \mathbb{N} \forall i, j \in [n; n+m]\, (|a_i - a_j| <_{\mathbb{R}} 2^{-k}),$$

is (for given k, g) called 'metastability' in Tao [357] and $[n; n+g(n)]$ a region where (a_n) is 'metastable' with error 2^{-k}.

There is, however, a problem in using the no-counterexample interpretation as a tool to extract such realizing functionals in a modular way i.e. by a recursion over the proof-tree which keeps the basic structure of the proof unchanged (which is of crucial importance for actually analyzing concrete and – in particular – not fully formalized proofs). In fact, Parsons' and Gödel's results where obtained by using a different more complicated interpretation, the so-called Gödel functional ('Dialectica') interpretation ([133]), which we will treat in chapters 8, 9, 10. In contrast to the no-counterexample interpretation, which only refers to functionals of type level 2, functional interpretation uses – even for first order systems like **PA** – functionals of arbitrary finite types to achieve an interpretation which respects the modus ponens. We conclude this chapter by indicating why functionals of type 2 are not sufficient whereas higher types allow one to resolve the problem.

The modus ponens problem:

Consider an instance

$$\frac{A \quad A \to B}{B}$$

of the modus ponens rule where A, B are sentences in $\mathcal{L}(\mathbf{PA})$ of the form

$$A :\equiv \forall x \exists y \forall z A_0(x,y,z) \text{ and } B :\equiv \forall u \exists v B_0(u,v),$$

and A_0, B_0 are quantifier-free and suppose we have functionals satisfying the no-counterexample interpretation of A and $A \to B$. In order to make the latter precise we first have to choose a prenex normal form of $A \to B$, say

$$(A \to B)^{pr} :\equiv \forall u \exists x \forall y \exists z, v \big(A_0(x,y,z) \to B_0(u,v)\big).$$

The no-counterexample interpretation of A and $(A \to B)^{pr}$ asks for functionals realizing the Herbrand normal forms

$$A^H :\equiv \forall x, g \exists y A_0(x,y,g(y))$$

and

$$\big((A \to B)^{pr}\big)^H :\equiv \forall u, f \exists x, z, v \big(A_0(x, f(x), z) \to B_0(u,v)\big)$$

of A and $(A \to B)^{pr}$, i.e. for functionals $\varphi_0, \varphi_1, \varphi_2, \varphi_3$ such that

$$\forall x, g A_0(x, \varphi_0(x,g), g(\varphi_0(x,g)))$$

and

$$\forall u, f \big(A_0(\varphi_1(u,f), f(\varphi_1(u,f)), \varphi_2(u,f)) \to B_0(u, \varphi_3(u,f))\big).$$

In order to solve the modus problem one has to solve (in the parameter u) the following system of equations for solutions x, f, g :

$$\begin{cases} x = \varphi_1(u,f), \\ \varphi_0(x,g) = f(\varphi_1(u,f)), \\ g(\varphi_0(x,g)) = \varphi_2(u,f). \end{cases}$$

However, we will show that no primitive recursive functional – not even in the extended sense of Gödel allows one to solve this system of equations as a functional in $u, \varphi_0, \varphi_1, \varphi_2$. Indeed, the solvability of this system of equation will turn out to correspond to the consistency of the schema of arithmetical comprehension

$$\exists f \forall x \big(f(x) = 0 \leftrightarrow A(x)\big),$$

where $A(x)$ contains only number quantifiers but maybe function parameters (see chapter 11).
The solution requires so-called bar recursion (of type 0) which was introduced by C. Spector [343] and which goes beyond Gödel's primitive recursive functionals. We will discuss this further in chapter 11 below.

Moreover, one can construct concrete sentences A and B of the logical form as above such that A and any prenex normal form of $A \to B$ have primitive recursive functionals in the sense of Kleene satisfying their no-counterexample interpretations but where B has no primitive recursive realizing function (but only one in the extended sense of Gödel's primitive recursion in higher types defined in chapter 3). For A of the form $\forall x \exists y \forall z A_0(x,y,z)$ primitive recursion with equality between functions, i.e. of type 1 (see chapter 3), suffices for B but for more complex formulas A one has to exhaust all finite types to realize B while A and $A \to B$ in general still have no-counterexample interpretations using only primitive recursion of type 0, i.e. in the sense of Kleene (these results are proved in [215] which provides a thorough discussion of the modus ponens problem for the no-counterexample interpretation).

The reason for the weakness of the no-counterexample interpretation is the weakness of the Herbrand normal form F^H of formulas F of complexity $\exists x \forall y \exists z F_0(x,y,z)$ or higher (such as $(A \to B)^{pr}$ above). Then the passage from F^H to F requires AC (though only from numbers to numbers) for $\forall$-formulas (and beyond), which in general are undecidable. I.e. one has to apply F^H to noncomputable index functions to derive F. For A of the form above, AC for the quantifier-free (and hence decidable) formula $\neg A_0$ is enough to prove $A^H \to A$ but already for $(A \to B)^{pr}$ this no longer is the case. In chapter 13 we will show that for **any** theorem A of full Peano arithmetic one can define a logically equivalent sentence $\tilde{A}$ in prenex normal form such that $\tilde{A}^H$ is provable using only quantifier-free induction (see proposition 13.1).

For the time being we confine ourselves with indicating how the above instance of the modus ponens can be treated if one uses an interpretation which doesn't stop at type level 2, namely Gödel's functional interpretation ([133]) which – for classical proofs (where it always is combined with the so-called negative translation) – will be developed in chapter 10:

The functional interpretation of A and $A \to B$:
Whereas we don't change the interpretation of A we use the following transformations of $A \to B$:

$$
\begin{aligned}
&(A \to B) \quad \rightsquigarrow \\
&\big(\forall x, g \exists y A_0(x,y,g(y)) \to \forall u \exists v B_0(u,v)\big) \rightsquigarrow \\
&\big(\exists Y \forall x, g A_0(x, Y(x,g), g(Y(x,g))) \to \forall u \exists v B_0(u,v)\big) \rightsquigarrow \\
&(+)\ \forall u, Y \exists x, g, v \big(A_0(x, Y(x,g), g(Y(x,g))) \to B_0(u,v)\big).
\end{aligned}
$$

Note that only AC applied to quantifier-free formulas (though to objects more complicated than numbers only) is needed to prove the equivalence between $A \to B$ and $(+)$.

We say that the functionals $\Phi_0, \Phi_1, \Phi_2, \Phi_3$ satisfy the functional interpretation of A and $A \to B$ if

$$\forall x, g A_0(x, \Phi_0(x,g), g(\Phi_0(x,g)))$$

and

$$\Phi_1(u,Y), \Phi_2(u,Y), \Phi_3(u,Y) \text{ realize } x, g, v \text{ in } (+).$$

A solution of the modus ponens problem is then given just by putting

$$Y := \Phi_0,\ x := \Phi_1(u, \Phi_0),\ g := \Phi_2(u, \Phi_0)$$

yielding the conclusion

$$\forall u B_0(u, \Phi_3(u, \Phi_0)).$$

So a realizing function for $\forall u \exists v B_0(u,v)$ is simply obtained by applying Φ_3 to Φ_0. Note that Φ_0 already is a functional $\mathbb{N} \times \mathbb{N}^{\mathbb{N}} \to \mathbb{N}$ (i.e. has type level 2 in the sense of chapter 3) and so $\Phi_3 : \mathbb{N} \times \mathbb{N}^{(\mathbb{N} \times \mathbb{N}^{\mathbb{N}})} \to \mathbb{N}$ has type level 3 which goes beyond the realm of the no-counterexample interpretation.

Let us compare further the no-counterexample interpretation and the functional interpretation (combined with negative translation): consider the so-called 'Infinite Pigeonhole Principle' (IPP) stating that for any partition of $\mathbb{N}$ into finitely many subsets at least one of these sets has infinitely many elements: Let $C_n := \{0, \ldots, n\}$.

$$\text{(IPP)}: \forall n \in \mathbb{N} \forall f : \mathbb{N} \to C_n \exists i \leq n \forall k \in \mathbb{N} \exists m \geq k \big(f(m) = i\big).$$

The Herbrand normal form of (IPP) is

$$\text{(IPP)}^H \equiv \forall n \in \mathbb{N} \forall f : \mathbb{N} \to C_n \forall F : C_n \to \mathbb{N} \exists i \leq n \exists m \geq F(i) \big(f(m) = i\big).$$

The no-counterexample interpretation of (IPP) has the following trivial solution:

$$M(n,f,F) := \max\{F(i) : i \leq n\} \text{ and } I(n,f,F) := f(M(n,f,F))$$

are realizers for '$\exists m$' and '$\exists i$' in (IPP)^H. These realizers by no means reflect the true complexity of (IPP) and its potential contribution to the complexity of programs or bounds extractable from proofs based on (IPP). In fact, (IPP) corresponds to the so-called bounded collection principle for universal formulas whose strength is known to be in between induction for Σ^0_2-formulas (called Σ^0_2-IA) and induction for Σ^0_1-formulas (called Σ^0_1-IA and defined in the exercises below). For a detailed study of these principles see e.g. [211] and chapter 13. In particular, as (IPP) implies Σ^0_1-IA it may cause arbitrary primitive recursive growth of functions provably total by the use of (IPP) (this is not in conflict with the trivial solution of the n.c.i of (IPP) but just shows again the failure of n.c.i. to interpret the modus ponens rule without causing a complexity explosion).

The functional interpretation of (the negative translation of) (IPP) (i.e. the ND-interpretation in the sense of chapter 10 of (IPP)) is arrived at in the following way

$$\begin{aligned}
&\text{(IPP)} \rightsquigarrow \\
&\forall n \in \mathbb{N} \forall f : \mathbb{N} \to C_n \exists i \leq n \exists g : \mathbb{N} \to \mathbb{N} \forall k \in \mathbb{N} \big(g(k) \geq k \wedge f(g(k)) = i\big) \rightsquigarrow \\
&\forall n \in \mathbb{N} \forall f : \mathbb{N} \to C_n \forall K : C_n \times \mathbb{N}^{\mathbb{N}} \to \mathbb{N} \exists i \leq n \exists g : \mathbb{N} \to \mathbb{N} \\
&\qquad\qquad \big(g(K(i,g)) \geq K(i,g) \wedge f(g(K(i,g))) = i\big) \equiv: \text{(IPP)}^{ND}.
\end{aligned}$$

The functional interpretation of (IPP) requires functionals $I(n,f,K)$ and $G(n,f,K)$ realizing '$\exists i$' and '$\exists g$'. As follows from the soundness theorem for ND in chapter 10 (theorem 10.7) I and G precisely constitute the computational contribution resulting from the use of (IPP) in a proof. Such functionals can be defined by a complicated (though very restricted form of) primitive recursion of level 1 (see chapter 10) which, however, can be written in a rather short form using a finite version of bar recursion as was shown by P. Oliva [293] (see chapter 11).

It is clear that to derive (IPP) from $(\text{IPP})^{ND}$ one only has to consider computable (in f) functionals K which, therefore, are continuous in g (in the sense of the Baire space topology (see chapter 4). One then can replace g by some finite initial segment of g that can be encoded into a number m (see chapter 3).
Let $[m](i) := g(i)$ for $i < \text{length}(m)$ and $[m](i) := 0$ otherwise. Then we can reformulate $(\text{IPP})^{ND}$ as

$$\forall n \in \mathbb{N} \forall f : \mathbb{N} \to C_n \forall K : C_n \times \mathbb{N}^{\mathbb{N}} \overset{cont.}{\to} \mathbb{N} \exists i \leq n \exists m \in \mathbb{N}$$
$$\big([m](K(i,[m])) \geq K(i,[m]) \wedge f([m](K(i,[m]))) = i\big).$$

One can now define a functional $\Omega(n, f, K)$ which searches for the least code $\langle i, m \rangle$ of a pair (i, m) satisfying

$$i \leq n \wedge \big([m](K(i,[m])) \geq K(i,[m]) \wedge f([m](K(i,[m]))) = i\big).$$

Clearly, Ω is computable in its arguments and hence for continuous K it is continuous in f. In the case at hand this is obvious (even for general K) as f is only evaluated at the argument $[m](K(i,[m]))$. Hence $\Omega(n, \cdot, K)$ is bounded on the whole compact subspace $(C_n)^{\mathbb{N}}$ of $\mathbb{N}^{\mathbb{N}}$. This allows one to conclude the following 'finite' version of (IPP)

$$\forall n \in \mathbb{N} \forall K : C_n \times \mathbb{N}^{\mathbb{N}} \overset{cont.}{\to} \mathbb{N} \exists M \in \mathbb{N} \forall f : C_M \to C_n \exists i \leq n \exists m \leq M$$
$$\big(Image([m]) \subseteq C_M \wedge [m](K(i,[m])) \geq K(i,[m]) \wedge f([m](K(i,[m]))) = i\big).$$

It is possible to represent continuous functionals K by number theoretic functions $\alpha : \mathbb{N} \times \mathbb{N} \to \mathbb{N}$ such that (see also definition 3.58 in chapter 3)

$$\forall i \leq n \forall g \in \mathbb{N}^{\mathbb{N}} \exists m \in \mathbb{N} \, (\alpha_K(i, \overline{g}m) \neq 0)$$

and

$$\forall i \leq n \forall g \in \mathbb{N}^{\mathbb{N}} \, (\alpha_K(i, \overline{g}(\min m\, [\alpha_K(i, \overline{g}m) \neq 0])) - 1 = K(i, g)).$$

Here $\overline{g}m$ encodes the initial segment of g of length m (see definition 3.30). As the last equation shows, K can effectively be recovered from α (this would no longer be the case if we simply had taken $\alpha'(i,m) := K(i,[m])$ as we then have to search for the least k such that $\alpha'(i, \overline{g}l)$ remains constant for all $l \geq k$ which is not effective). Without loss of generality one may assume that α satisfies

$$\forall i \leq n \forall m, k \in \mathbb{N} \, (m \subseteq k \wedge \alpha(m) > 0 \to \alpha(i,m) = \alpha(i,k)),$$

where $m \subseteq k$ expresses that the finite sequence encoded by m is an initial segment of the finite sequence encoded by k (for more information on this see Kohlenbach [223]).

The above 'finite' version of (IPP) is very similar to T. Tao's ([357]) formulation of his 'finite' version of this principle. As mentioned by Tao, the principle is not fully finitizable due to the hidden quantifiers in the continuity assumption (resp. the assumption that a certain finite set-function – corresponding to α' above – is eventually constant in Tao's formulation). This is also the reason why to compute M in this finite version one either needs unbounded search or has to enrich K with a modulus of uniform continuity functional $\omega_K(h)$ on $\{g : g \leq_1 h\}$ (see below for the latter).

Because of this it is not the finite version of (IPP) which is useful in concrete unwindings of proofs involving the principle (IPP) but the (primitive recursive in the sense of Gödel) functionals $I(n,f,K), G(n,f,K)$ realizing $(\text{IPP})^{ND}$ which, combined with the majorization technique developed in chapter 6 below, yield a uniform (and monotone) 'bound' (in the sense of being a majorant of G) $G^*(n,K)$ that no longer depends on f (for I the construction of $I^*(n,K) := I^*(n) := n$ is trivial).

Remark 2.30. G^* being a majorant of G (essentially) means that for all K^* being a majorant of K in sense of

$$\forall n^*, n, g^*, g\,(n^* \geq n \wedge g^* \; maj \; g \to K^*(n^*, g^*) \geq K(n,g))$$

one has

$$\forall n^*, n \forall f : \mathbb{N} \to C_n\,(n^* \geq n \to G^*(n^*, K^*) \; maj \; G(n,f,K)),$$

where for functions g^*, g :

$$g^* \; maj \; g :\equiv \forall n^*, n\,(n^* \geq n \to g^*(n^*) \geq g(n)).$$

So, in particular, for all $n \in \mathbb{N}$ and $f : \mathbb{N} \to C_n$

$$\forall k\,(G^*(n,K^*)(k) \geq G(n,f,K)(k)).$$

Now let $\omega_K(i,h)$ be a modulus of uniform continuity for $K(i,g)$ on $\{g : g \leq h\}$ (where $\leq$ is defined pointwise) for $i \leq n$, i.e.

$$\forall i \leq n \forall g_1, g_2 \leq h\,(\forall k \leq \omega_K(i,h)(g_1(k) = g_2(k)) \to K(i,g_1) = K(i,g_2)).$$

Given ω_K one can easily compute a majorant K^* of K and applying subsequently $\omega_K(i,h)$ to the bound $h := G^*(n,K^*)$ one can construct (in n, K, ω_K) a bound on the 'finite' version of (IPP) above which no longer relies on unbounded search.

The variant of functional interpretation which directly extracts such uniform bounds ('majorants') G^* we call monotone functional interpretation (see chapter 9 and – combined with negative translation – chapter 10). By the soundness theorem of monotone functional interpretation (and the soundness of negative translation) such majorizing terms, as provided by monotone functional interpretation, of principles or lemmas used in a proof are all that is needed in extracting uniform bounds from proofs.

2.4 Exercises, historical comments and suggested further reading

Exercises:

1) Verify the estimate $p_r < 2^{2^r}$ stated in the discussion of Euclid's proof of proposition 2.1.
2) Let $\pi(x)$ be the number of all primes $\leq x$ (for $x \geq 1$). From the estimates we obtained by analyzing proofs 1)-3) of proposition 2.1 derive the following lower bounds on $\pi(x)$:

   a. From Proof 1 (Euclid): $\pi(x) \geq \ln\ln x$ for $x \geq 2$.
   b. From Proof 2 (Euler): $\pi(x) \geq \ln x$ for $x \geq 1$.
   c. From Proof 3: $\pi(x) \geq \frac{\ln x}{2\ln 2}$ for $x \geq 1$.

3) Consider
$\Psi(x) := |\{n \in \mathbb{N} : 1 \leq n \leq x \wedge n \text{ is not divisible by any square number} \neq 1\}|$.
Show that $\Psi(x) \geq x - \sum_{\substack{p \text{ prime} \\ p \leq x}} [\frac{x}{p^2}]$ and use this to show that there are infinitely many primes. Use this proof to obtain an upper bound $g(j)$ for the next prime p_{j+1} as in the 3rd proof of this statement above. Can you improve the bound we obtained from the latter (see Hacks [148])?
4) (Ulrich Berger) Consider the open first order theory $\mathcal{T}$ in the language of first order logic with equality and a constant 0 and two unary function symbols S, f. The only non-logical axiom of $\mathcal{T}$ is $\forall x(S(x) \neq 0)$.

   (i) Prove that $\mathcal{T} \vdash \exists x\big(f(S(f(x))) \neq x\big)$.
   (ii) Construct from the proof finitely many closed terms $s_1,\ldots,s_m$ and $t_1,\ldots,t_n$ such that
   $$\mathrm{PL} \vdash \bigwedge_{i=1}^{m}(S(s_i) \neq 0) \to \bigvee_{j=1}^{n}(f(S(f(t_j))) \neq t_j).$$

5) Prove remark 2.22.
6) ([306, 231]) Let $(a_n),(b_n),(c_n)$ be sequences in $\mathbb{R}_+$ such that $\sum b_n$ and $\sum c_n$ are bounded and
$$\forall n \in \mathbb{N}(a_{n+1} \leq (1+b_n)a_n + c_n).$$
Show that (a_n) is convergent and hence a Cauchy sequence. Construct a primitive recursive functional $\Phi(A,B,C,g,k)$ such that
$$\forall k \in \mathbb{N} \forall g \in \mathbb{N}^{\mathbb{N}} \exists n \leq \Phi(A,B,C,g,k) \forall i,j \in [n;n+g(n)]\ (|a_i - a_j| < 2^{-k})$$
for all $A,B,C \in \mathbb{N}$ be such that

$$a_0 \le A, \sum b_n \le B, \sum c_n \le C.$$

7) Construct primitive recursive functionals $\underline{\Phi}$ which satisfy the n.c.i. of (some prenex normal form of) the second order axiom of Σ^0_1-induction:

$$\Sigma^0_1\text{-IA}: \begin{cases} \forall f\big(\exists y(f(0,y)=0) \wedge \forall x(\exists y(f(x,y)=0) \to \exists y(f(x+1,y)=0)) \\ \qquad \to \forall x \exists y(f(x,y)=0)\big) \end{cases}$$

uniformly as a functional in f and the index functions.
8) Let $(a_n)_{n\in\mathbb{N}}$ be a sequence of non-negative rational numbers. Use Σ^0_1-IA to prove that

$$(+)\ \forall k \exists n \forall m(a_n \le_{\mathbb{Q}} a_m + 2^{-k})$$

and construct a primitive recursive functional satisfying the no-counterexample interpretation of $(+)$ (see also exercise 3 in chapter 4 below).

Historical comments and suggested further reading:

1) More information on the general program of unwinding proofs (proof mining) can be found in [249, 250, 252, 251, 99, 268, 84, 16, 122, 206, 210, 219, 226, 236, 229, 270].
2) For detailed accounts of Herbrand's theorem see [62, 122, 202, 332, 249, 267, 114].
3) More material on the no-counterexample interpretation can be found in [122, 215, 241, 242, 350, 351, 353, 326] as well as chapter 10 below.

In particular, Kohlenbach [215] provides a thorough discussion of the modus ponens problem for the no-counterexample interpretation.

A partially modular approach to Herbrand's theorem via Gödel's functional interpretation (see chapter 8) can be found in Gerhardy-Kohlenbach [118].

A detailed complexity analysis of Herbrand's theorem and the closely related cut elimination theorem is given in Gerhardy's articles [114] and [115].

For early applications of the no-counterexample interpretation as well as the ε-substitution method (originally due to D. Hilbert and W. Ackermann), which is closely related to Herbrand's theorem and on which Kreisel's original treatment of his no-counterexample interpretation is based, to proofs in number theory (e.g. Littlewood's theorem on the sign changes of $\pi(n) - li(n)$) and algebra see Kreisel's original papers on the subject [241, 242] and also [243]. As briefly discussed above, Luckhardt [267] presents (inspired by Kreisel [249]) an important application of Herbrand terms extracted from two proofs of Roth's theorem in diophantine approximation resulting in the first polynomial bounds on the number of solutions (see also Luckhardt [268]). Applications of the ε-substitution method to the solution of Hilbert's 17th problem and subsequent work in this direction are discussed in Delzell [84] (see also Delzell's papers [79, 80, 81, 82, 83] although some do not use proof theory directly). Again this work is inspired by ideas of G. Kreisel going back to the 50's (see e.g. Kreisel [245]). An analysis of two variants of the proof of Furstenberg and Weiss of van der Waerden's theorem by means of cut-elimination and the no-counterexample interpretation, respectively, is given in Girard [122] (see pp. 237–251 and 483–496). Bellin [16] presents an application of the no-counterexample interpretation to Ramsey's theorem.

Applications of cut-elimination to coherence theorems in category theory are given in Mints [276, 278, 279] and Babaev-Solovjov [9].

For other approaches to proof mining not (or only briefly) treated in this book see e.g. the work of Coquand et al. [72, 73] and Berger-Schwichtenberg [21] (see, however, chapter 14). Interesting connections between proof theory and combinatorics can be found in Ketonen-Solovay [181] and – recently – Weiermann [376].

Chapter 3
Intuitionistic and classical arithmetic in all finite types

3.1 Intuitionistic and classical predicate logic

In the following we formulate an axiomatic system for intuitionistic first order predicate logic IL. The particular Hilbert-type axiomatization we choose is due to [133] and specially suited to carry out proof interpretations inductively over the proof tree. Of course, like any Hilbert-type system also this axiomatization is not very convenient for actually carrying out proofs for which a natural deduction style calculus is to be recommended. For a proof of the equivalence of these calculi see [366] (1.1.5–1.1.11).

Intuitionistic first order predicate logic without equality $\mathbf{IL}_{-=}$

I. **The language $\mathscr{L}(\mathbf{IL}_{-=})$ of $\mathbf{IL}_{-=}$:**
As logical constants we use $\wedge, \vee, \rightarrow, \perp$ (absurdity or 'falsum'), $\exists, \forall$. $\mathscr{L}(\mathrm{IL}_{-=})$ contains variables $x, y, z, \ldots$ (which can be free or bound). Furthermore, for any arity $n \geq 0$ we have (possibly empty) denumerable sets of function symbols $f_1, f_2, f_3, \ldots$ and (for $n \geq 1$) predicate symbols $P_1, P_2, P_3, \ldots$.
0-place function symbols are called constants and usually denoted by $c_1, c_2, c_3, \ldots$.

Terms:

(i) Variables and constants are terms.
(ii) If $t_1, \ldots, t_n$ are terms and f is an n-ary function symbol, then $f(t_1, \ldots, t_n)$ is a term.

Terms that do not contain any variables are called closed.

Formulas:

(i) If $t_1, \ldots, t_n$ are terms and P an n-ary predicate symbol, then $P(t_1, \ldots, t_n)$ is a (prime) formula. Moreover, $\perp$ is a (prime) formula.
(ii) If A, B are formulas, then $(A \wedge B)$, $(A \vee B)$ and $(A \rightarrow B)$ are formulas.

(iii) If A is a formula and x a variable, then $(\forall x A)$ and $(\exists x A)$ are formulas.

As usual, formulas which do not contain free variables (i.e. variables occurring not bound by any quantifier) are called closed or sentences.

Abbreviations:
$\neg A :\equiv A \to \bot$, $A \leftrightarrow B :\equiv (A \to B) \wedge (B \to A)$.

Conventions on parentheses: Negation and quantifiers bind stronger than $\vee, \wedge$ which bind stronger than $\to, \leftrightarrow$. Using this convention we can safely drop many parentheses around formulas, e.g. we simply write (dropping also outermost parentheses) $A \wedge B \to \neg C \vee D$ instead of $((A \wedge B) \to ((\neg C) \vee D))$.

II. **Axioms of $\mathrm{IL}_{-=}$:**

(i) $A \vee A \to A$, $A \to A \wedge A$ (axioms of contraction);
(ii) $A \to A \vee B$, $A \wedge B \to A$ (axioms of weakening);
(iii) $A \vee B \to B \vee A$, $A \wedge B \to B \wedge A$ (axioms of permutation);
(iv) $\bot \to A$ (ex falso quodlibet);
(v) $\forall x A \to A[t/x]$, $A[t/x] \to \exists x A$, where t is free for x in A and $A[t/x]$ is the result of replacing every free occurrence of x in A by t (quantifier axioms).

III. **Rules of $\mathrm{IL}_{-=}$:**

(i)
$$\frac{A, A \to B}{B}, \quad \frac{A \to B, B \to C}{A \to C}$$
(modus ponens and syllogism);

(ii)
$$\frac{A \wedge B \to C}{A \to (B \to C)}, \quad \frac{A \to (B \to C)}{A \wedge B \to C}$$
(exportation and importation);

(iii)
$$\frac{A \to B}{C \vee A \to C \vee B} \text{ (expansion);}$$

(iv)
$$\frac{B \to A}{B \to \forall x A}, \quad \frac{A \to B}{\exists x A \to B}, \text{ where } x \text{ is not free in } B$$
(quantifier rules).

Remark 3.1. Most of the time we will use for notational simplicity the slightly imprecise notation '$A(t)$' instead of '$A[t/x]$'.

Classical first order predicate logic without equality $\mathrm{PL}_{-=}$ results from $\mathrm{IL}_{-=}$ by adding the law-of-excluded-middle (LEM) schema

$$A \vee \neg A.$$

The Brouwer-Heyting-Kolmogorov '(BHK)' proof interpretation of the intuitionistic logical constants (our exposition makes use of [318]).

This interpretation is an informal attempt to explain the meaning of the logical constants of $IL_{=}$ in terms of proof constructions. Here 'proof' is understood as 'verification by a construction' and not as a formal proof in some fixed deductive framework like HA below.

(i) There is no proof for $\bot$.
(ii) A proof of $A \wedge B$ is a pair (q, r) of proofs, where q is a proof of A and r is a proof of B.
(iii) A proof of $A \vee B$ is a pair (n, q) consisting of an integer n and a proof q which proves A if $n = 0$ and resp. B if $n \neq 0$.
(iv) A proof p of $A \to B$ is a construction which transforms any hypothetical proof q of A into a proof $p(q)$ of B.
(v) A proof p of $\forall x A(x)$ is a construction which produces for every construction c_d of an element d of the domain a proof $p(c_d)$ of $A(d)$.
(vi) A proof p of $\exists x A(x)$ is a pair (c_d, q), where c_d is the construction of an element d of the domain and q is a proof of $A(d)$.

Discussion: There is one problem with the BHK-interpretation: from a strictly constructive point of view one would like to have a constructive verification of 'p is a proof of A' in case this is true, i.e. one would like to recognize a proof if one sees it. For $(i), (ii), (iii), (vi)$ there is no problem with this requirement. But for the universal statements in $(iv), (v)$ one would need an additional clause as suggested by Kreisel in [247]:

(iv)' A proof p of $A \to B$ is a pair (r, q), where q is a construction which transforms any hypothetical proof s of A into a proof $q(s)$ of B and r is a proof which verifies that q is such a construction.
(v)' A proof p of $\forall x A(x)$ is a pair (r, q) where q is a construction which produces for every construction c_d of an element d of the domain a proof $q(c_d)$ of $A(d)$ and r is a proof of the fact that q is such a construction.

Remark 3.2. There are various ways to formalize the idea behind the BHK-interpretation which give rise to various forms of so-called realizability interpretations. The first version of realizability, the so-called recursive realizability, was introduced by Kleene in [193]. In this book we will focus on a typed variant of Kleene's type-free interpretation which is called 'modified realizability' and is due to Kreisel [244, 246].
Recently S. Artemov has developed a so-called 'logic of proofs' where 'proof' in the BHK-clauses is interpreted as 't is a proof (polynomial) for A' referring to some standard proof (not: provability) predicate e.g. for PA. Using this interpretation he proves a completeness result for intuitionistic propositional logic (see [4]).

Intuitionistic first order predicate logic with equality IL

IL results from $\mathrm{IL}_{-=}$ by adding a special binary predicate symbol $=$ to the language together with the

Equality axioms:

(i) $x = x$, $x = y \to y = x$, $x = y \wedge y = z \to x = z$.
(ii) $x_1 = y_1 \wedge \ldots \wedge x_n = y_n \to f(x_1, \ldots, x_n) = f(y_1, \ldots, y_n)$ for any n-ary function symbol f.
(iii) $x_1 = y_1 \wedge \ldots \wedge x_n = y_n \to \big(P(x_1, \ldots, x_n) \to P(y_1, \ldots, y_n)\big)$ for any n-ary predicate symbol P.

Classical first order predicate logic with equality PL results from IL by adding the law-of-excluded-middle schema

$$\mathrm{LEM}: \ A \vee \neg A.$$

3.2 Intuitionistic ('Heyting') arithmetic HA and Peano arithmetic PA

$\mathscr{L}(\mathrm{HA})$ contains the logical constants of $\mathscr{L}(\mathrm{IL})$, number variables $x, y, z, \ldots$, a constant 0 (zero), a unary function symbol S (successor), function symbols for all primitive recursive functions (more precisely for all derivations of primitive recursive functions).

Axioms and rules of HA:

(i) axioms and rules of IL (based on $\mathscr{L}(\mathrm{HA})$),
(ii) successor axioms:

$$\begin{cases} S(x) \neq 0, \\ S(x) = S(y) \to x = y, \end{cases}$$

(iii) defining equations for the primitive recursive functions,
(iv) axiom schema of complete induction

$$\mathrm{IA}: \ A(0) \wedge \forall x\big(A(x) \to A(S(x))\big) \to \forall x A(x)$$

for every formula $A \in \mathscr{L}(\mathrm{HA})$.

Convention: We often write x' or $x+1$ for $S(x)$.

Remark 3.3. 1) In HA one can prove that $\perp \leftrightarrow 0 = 1$ and so we may identify $\perp$ with $0 = 1$, where $1 := S(0)$. Then the axioms involving falsum, namely $\perp \to A$ and $\neg S(x) = 0$ even become redundant: for $0 = 1 \to A$ it suffices to establish this for all prime formulas $s = t$ which follows by primitive recursion. $S(x) = 0 \to 0 = 1$ is proved similarly (exercise).

2) Instead of the axiom schema IA we could have formulated HA equivalently using the rule of induction

$$\text{IR:} \quad \frac{A(0)\ ,\ A(x) \to A(S(x))}{A(x)}.$$

Exercise!

Lemma 3.4.

$$\text{HA} \vdash \forall x(x = 0 \vee x \neq 0).$$

Proof: Induction on x : If $x = 0$, then $x = 0 \vee x \neq 0$ by weakening.
For $S(x)$ we have $S(x) \neq 0$ and hence $S(x) = 0 \vee S(x) \neq 0$ again by weakening. □

Proposition 3.5. *The following rule of double induction is derivable in* HA*:*

$$\frac{A(x,0)\ ,\ A(0,y)\ ,\ A(x,y) \to A(S(x),S(y))}{A(x,y)}.$$

Proof: We leave the tedious proof as exercise resp. refer to the literature: [371]. □

In the following, let $+, \cdot, \overline{sg}, pd, \dot{-}, |\cdot - \cdot|$ be defined primitive recursively as follows (to bring these informal primitive recursions into the official format of primitive recursion one has to make use of projections to introduce dummy arguments, exercise):

$$x + 0 = x, x + S(y) = S(x + y);$$

$$x \cdot 0 = 0, x \cdot S(y) = x \cdot y + x;$$

$$\overline{sg}(0) = 1, \overline{sg}(S(x)) = 0;$$

$$pd(0) = 0, pd(S(x)) = x;$$

$$x \dot{-} 0 = x, x \dot{-} (S(y)) = pd(x \dot{-} y);$$

$$|x - y| = (x \dot{-} y) + (y \dot{-} x).$$

Remark 3.6. In the presence of the defining axioms for the primitive recursive functions including the predecessor function pd the successor axiom $S(x) = S(y) \to x = y$ actually becomes redundant since

$$S(x) = S(y) \to x = pd(S(x)) = pd(S(y)) = y.$$

Lemma 3.7. HA *proves the following basic facts:*

1) $x + y = 0 \leftrightarrow x = 0 \wedge y = 0.$
2) $x \cdot y = 0 \leftrightarrow x = 0 \vee y = 0.$
3) $\overline{sg}(x) = 0 \leftrightarrow x \neq 0.$
4) $\overline{sg}(x) \cdot y = 0 \leftrightarrow (x = 0 \to y = 0).$
5) $|x - y| = 0 \leftrightarrow x = y.$

Proof: Exercise (use lemma 3.4 and double induction, i.e. proposition 3.5). □

Proposition 3.8. *Let $A_0(\underline{x})$ be a quantifier-free formula of $\mathcal{L}(\mathrm{HA})$ whose free variables are among $\underline{x} = x_1, \ldots, x_n$. Then there is an n-ary primitive recursive function symbol f of* HA *such that*

$$\mathrm{HA} \vdash \forall \underline{x}(f(\underline{x}) = 0 \leftrightarrow A_0(\underline{x})).$$

Proof: Immediate from lemma 3.7 since all prime formulas of $\mathcal{L}(\mathrm{HA})$ are of the form $t = s$. □

Corollary 3.9. *Let A_0 be a quantifier-free formula of $\mathcal{L}(\mathrm{HA})$. Then*

$$\mathrm{HA} \vdash A_0 \vee \neg A_0.$$

In particular, quantifier-free formulas are provably stable, i.e.

$$\mathrm{HA} \vdash \neg\neg A_0 \rightarrow A_0.$$

Proof: Lemma 3.4 and proposition 3.8. □

Classical ('Peano') arithmetic PA results from HA by adding the law-of-excluded-middle schema

$$\mathrm{LEM}: \; A \vee \neg A.$$

3.3 Extensional intuitionistic ('Heyting') and classical ('Peano') arithmetic in all finite types

In chapter 5 we will show, in particular, that a certain subclass $\mathcal{C}$ of the class of all total computable functions suffices to provide witnesses for all HA-provable sentences of the form $\forall x \exists y A(x,y)$. However, to describe this class we have to go beyond the primitive recursively defined functions contained in HA as HA proves $\forall x \exists y A(x,y)$-sentences such that for no primitive recursive function f, $\forall x A(x, f(x))$ is true over $\mathbb{N}$. E.g. HA can prove the totality of the so-called Ackermann function (defined in the exercises to this chapter) which is not primitive recursive (for the latter see e.g. [341]). In order to describe algorithmically a sufficiently rich (and actually optimal, though we will not prove this) such class of functions we need to consider so-called functionals of higher type defined by a generalized form of primitive recursion. Even in fragments we consider where the recursion to define functionals of higher types is restricted to yield only ordinarily primitive recursive functions $f : \mathbb{N} \rightarrow \mathbb{N}$ (or even functions of much lower complexity) to enrich the language with variables and quantifiers for functionals of all finite types is important to carry out the proof interpretations we are mainly interested in. To some extent higher types (though usually quite low) are also needed to formalize proofs in analysis.

The set $\mathbf{T}$ of all finite types (over $\mathbb{N}$) is generated inductively by the clauses

$$(i)\; 0 \in \mathbf{T}, \; (ii)\; \rho, \tau \in \mathbf{T} \Rightarrow \tau(\rho) \in \mathbf{T}.$$

The type 0 is the type natural numbers. Objects of type $\tau(\rho)$ are functions which map objects of type ρ to objects of type τ.

Remark 3.10. Some authors write $(\rho)\tau$, $(\tau\rho)$ or $(\rho \to \tau)$ instead of $\tau(\rho)$. The notation $(\rho \to \tau)$ has the benefit of indicating directly the formation of a function space. Because of this we will also use it occasionally. Moreover, it visualizes the so-called Curry-Howard correspondence (or isomorphism, see [349] for a comprehensive treatment) between formulas (of the implicative fragment of intuitionistic propositional logic) and types as well as between proofs and terms (see below). The drawback is that complicated types get much longer to write than in our notation.

We often omit brackets which are uniquely determined and write e.g. $0(00)$ instead of $0(0(0))$.
One easily notices that every type $\rho \neq 0$ can uniquely be written as $\rho = 0(\rho_k)\dots(\rho_1)$ for suitable k and types $\rho_1,\dots,\rho_k$. We usually use $\rho = 0\rho_k\dots\rho_1$ as shorthand for this if it is clear to which types $\rho_1,\dots,\rho_k$ we refer so that there is no danger of confusion.

The set $\mathbf{P} \subset \mathbf{T}$ of pure types is defined by

$$(i)\ 0 \in \mathbf{P},\ (ii)\ \rho \in \mathbf{P} \Rightarrow 0(\rho) \in \mathbf{P}.$$

Pure types are often denoted by natural numbers:

$$0(n) := n+1 \text{ (e.g. } 00 = 1, 0(00) = 2).$$

The type level or degree $deg(\rho)$ of a type ρ is defined as

$$deg(0) := 0,\ deg(\tau(\rho)) := \max(deg(\tau), deg(\rho)+1)$$

(note that for pure types ρ, $deg(\rho)$ is just the number which denotes ρ).

Objects of type ρ with $deg(\rho) > 1$ are usually called functionals.
We sometimes write '$\tau \leq n$' instead of '$deg(\tau) \leq n$'.

The language $\mathcal{L}$(E-HA$^\omega$) of E-HA$^\omega$ is based on a many-sorted version $\mathrm{IL}^\omega_{=}$ of $\mathrm{IL}_{=}$ which contains variables $x^\rho, y^\rho, z^\rho, \dots$ and quantifiers $\forall x^\rho$, $\exists y^\rho$ for every type ρ. As constants E-HA$^\omega$ contains 0^0 (zero), S^{00} (successor), $\Pi^{\rho\tau\rho}_{\rho,\tau}$ (projector), $\Sigma_{\delta,\rho,\tau}$ (combinator of type $\tau\delta(\rho\delta)(\tau\rho\delta)$) and (simultaneous) recursor constants $\underline{R}_{\underline{\rho}} = (R_1)_{\underline{\rho}},\dots,(R_k)_{\underline{\rho}}$, where R_i has type $\rho_i(\rho_k 0\underline{\rho}')\dots(\rho_1 0\underline{\rho}')\underline{\rho}'0$ for all $\delta,\rho,\tau,\underline{\rho}$ $(= (\rho_1)\dots(\rho_k))$ in $\mathbf{T}$. Here we use the notation $\underline{\rho}' := (\rho_k)\dots(\rho_1)$.
Furthermore $\mathcal{L}$(E-HA$^\omega$) contains a binary predicate symbol $=_0$ for equality between objects of type 0.
Sometimes we write $t \in \rho$ to express that t is of type ρ.

Terms of E-HA$^\omega$ are built up by

(i) constants c^ρ and variables x^ρ of type ρ are terms of type ρ
(ii) if $t^{\tau\rho}$ is a term of type $\tau\rho$ and s^ρ is a term of type ρ, then $t(s)$ is a term of type τ.

Again, we often simply write $ts_1\ldots s_k$ – or $t(s_1,\ldots,s_k)$ – instead of $t(s_1)\ldots(s_k)$. Some authors write (ts) instead of $t(s)$.

Formulas of E-HA$^\omega$ are built up by

(i) prime formulas (also called 'atomic formulas') $s =_0 t$ are formulas (where s^0, t^0 are terms of type 0);
(ii) if A, B are formulas, then also $(A \wedge B)$, $(A \vee B)$ and $(A \to B)$ are formulas;
(iii) if A is a formula and x^ρ a variable of type ρ, then also $(\forall x^\rho A)$ and $(\exists x^\rho A)$ are formulas.

We adopt the same conventions on parentheses as before.

Abbreviations:

1) Higher type equations $s =_\rho t$ between terms s,t of type $\rho = 0\rho_k\ldots\rho_1$ (where $k \geq 1$) are abbreviations for

$$\forall y_1^{\rho_1},\ldots,y_k^{\rho_k}\big(sy_1\ldots y_k =_0 ty_1\ldots y_k\big),$$

where $y_1,\ldots,y_k$ are variables which don't occur in s,t.
2) As before: $\neg A :\equiv A \to \bot$, where $\bot :\equiv (0 =_0 1)$; $A \leftrightarrow B :\equiv (A \to B) \wedge (B \to A)$.

Axioms and rules of E-HA$^\omega$

(i) all axioms and rules of $\mathrm{IL}^\omega_{=}$;
(ii) equality axioms for $=_0$:
$x =_0 x$, $x =_0 y \to y =_0 x$, $x =_0 y \wedge y =_0 z \to x =_0 z$;
(iii) higher type extensionality:

$$E_\rho : \forall z^\rho, x_1^{\rho_1}, y_1^{\rho_1},\ldots,x_k^{\rho_k}, y_k^{\rho_k}\Big(\bigwedge_{i=1}^{k}(x_i =_{\rho_i} y_i) \to z\underline{x} =_0 z\underline{y}\Big),$$

where $\rho = 0\rho_k\ldots\rho_1$;
(iv) successor axioms;
(v) induction schema

$$\mathrm{IA}:\ A(0) \wedge \forall x^0\big(A(x) \to A(Sx)\big) \to \forall x^0 A(x),$$

where $A(x^0)$ is an arbitrary formula of E-HA$^\omega$;
(vi) axioms for $\Pi_{\rho,\tau}, \Sigma_{\delta,\rho,\tau}$ and $\underline{R}_{\underline{\rho}}$:

$$(\Pi) : \Pi_{\rho,\tau}x^\rho y^\tau =_\rho x^\rho,$$

$$(\Sigma) : \Sigma_{\delta,\rho,\tau}xyz =_\tau xz(yz) \quad (x^{\tau\rho\delta}, y^{\rho\delta}, z^\delta),$$

$$(\underline{R}) : \begin{cases} (R_i)_{\underline{\rho}}0\underline{y}\underline{z} =_{\rho_i} y_i \\ (R_i)_{\underline{\rho}}(Sx^0)\underline{y}\underline{z} =_{\rho_i} z_i(\underline{R}_{\underline{\rho}}x\underline{y}\underline{z})x \ \text{ for } i = 1,\ldots,k, \end{cases}$$

where $\underline{\rho} = \rho_1, \ldots, \rho_k$, $\underline{y} = y_1, \ldots, y_k$, $\underline{z} = z_1, \ldots, z_k$ with y_i of type ρ_i and z_i of type $\rho_i 0 \underline{\rho}'$.

Remark 3.11. 1) As a many-sorted system, the fact that the sorts $\tau\rho$ and ρ, τ are connected via the term formation rule stating that with $t^{\tau\rho}, s^{\rho}$ also $t(s)$ is a term (of type τ) strictly speaking needs to be expressed via application symbols $App_{\rho,\tau}$, where then $App_{\rho,\tau}(t^{\tau\rho}, s^{\rho})$ stands for $t(s)$. We suppress this cumbersome notation and simply write $t(s)$. However, we have to keep in mind that in order to specify a model for E-HA$^{\omega}$ (see section 3.6 below) we also have to give interpretations to $App_{\rho,\tau}$ (usually this will be the obvious set-theoretic application).

2) The reflexivity, symmetry and transitivity of the defined higher type equalities $=_{\rho}$ are derivable from the corresponding axioms for $=_0$. Using the extensionality axioms one can prove

$$x =_{\rho} y \wedge A(x) \to A(y)$$

by induction on the complexity of A (for the case of prime formulas one first proves by induction on the terms that $x =_{\rho} y \to r[x/z^{\rho}] =_{\tau} r[y/z^{\rho}]$) (exercise).

3) Instead of the axioms E_{ρ} for all types ρ we could have used equivalently $E_{\rho,\tau}$ for all $\rho, \tau \in \mathbf{T}$, where

$$E_{\rho,\tau} : \ \forall z^{\tau\rho}, x^{\rho}, y^{\rho} \, (x =_{\rho} y \to zx =_{\tau} zy)$$

(exercise).

Definition 3.12. Later on we will need also a 'weakly extensional' variant WE-HA$^{\omega}$ of E-HA$^{\omega}$, where the extensionality axioms E_{ρ} are weakened to a quantifier-free rule of extensionality

$$\text{QF-ER:} \quad \frac{A_0 \to s =_{\rho} t}{A_0 \to r[s/x^{\rho}] =_{\tau} r[t/x^{\rho}]},$$

where A_0 is quantifier-free and $s^{\rho}, t^{\rho}, r^{\tau}$ are terms of WE-HA$^{\omega}$ ($\rho, \tau \in \mathbf{T}$ arbitrary).

Remark 3.13. 1) The special case of QF-ER with $\tau = 0$ already implies the general case.

2) Note that QF-ER allows one (by taking $A_0 :\equiv x =_0 y$) to derive full extensionality for equality of type 0, i.e.

$$\forall x^0, y^0 \left(x =_0 y \to r[x/z^0] =_{\tau} r[y/z^0]\right).$$

Also, QF-ER suffices to prove the following rule

$$\frac{A_0 \to s =_{\rho} t}{A_0 \to (B[s/x^{\rho}] \to B[t/x^{\rho}])},$$

where B is an arbitrary formula and s, t are free for x^{ρ} in B.

Warning: We will prove later (chapter 9) that WE-HA$^{\omega}$ does not satisfy the deduction theorem.

Remark 3.14. 1) One can show in WE-HA$^\omega$ that the simultaneous primitive recursors $\underline{R}_{\underline{\rho}}$ can in fact be reduced to the single recursors R_ρ (i.e. the case $k = 1$) using appropriate embeddings of tuples of types in a suitable common higher type and tuple codings of functionals. For details see [366](1.6.17). Nevertheless, we prefer to include simultaneous primitive recursion as a primitive concept since it is used in this form in the soundness proofs of our proof interpretations in chapters 5 and 8. Moreover, we will later (chapter 17) extend our framework to new types where such a reduction does not seem to be possible anymore (unless one introduces product types explicitly which would be another alternative).
2) Occasionally, we will denote the set of all closed terms of WE-HA$^\omega$ by T and use T_n to denote the subset of closed terms involving only recursors R_ρ with $deg(\rho) \leq n$.

In the following FV(t) (FV(A)) denotes the set of all free variables of t (A).

WE-HA$^\omega$ allows the definition of λ-abstraction in the following sense:

Lemma 3.15. *For every term $t[x^\rho]^\tau$ (here x refers to all occurrences of x in t) one can construct in* WE-HA$^\omega$ *a term $(\lambda x^\rho.t[x])$ of type $\tau\rho$ (with $FV(\lambda x^\rho.t[x]) = FV(t[x]) \setminus \{x\}$) such that*

$$\text{WE-HA}^\omega \vdash (\lambda x^\rho.t[x])(s^\rho) =_\tau t[s/x].$$

In contexts where is no danger of ambiguity, we omit the outer parentheses around $(\lambda x.t[x])$.

Proof: Define

$$\lambda x.x := \Sigma\Pi\Pi,$$
$$\lambda x.t := \Pi t, \text{ if } x \notin \text{FV}(t),$$
$$\lambda x.(ts) := \Sigma(\lambda x.t)(\lambda x.s), \text{ if } x \in \text{FV}(ts)$$

(here Π, Σ of suitable types). □

Notation: Instead of $\lambda x_1 \ldots \lambda x_k.t$ we often write $\lambda x_1, \ldots, x_k.t$.

Remark 3.16. It is easy to see that using R_0 and lemma 3.15 one can define all primitive recursive functions so that HA can be viewed as a subsystem of WE-HA$^\omega$. For details see [366](1.6.9). Note that the use of R_ρ in [366](1.6.9) can be replaced by R_0 if T_{ψ_1} and T_{ψ_2} are replaced by $T_{\Psi_1}(x_1, \ldots, x_n)$ and $\lambda u^0, z^0.T_{\Psi_2}(u, z, x_1, \ldots, x_n)$ respectively.

Proposition 3.17. *Let $A_0(\underline{x})$ be a quantifier-free formula of $\mathcal{L}$(*WE-HA$^\omega$*) whose free variables are contained among $\underline{x}$. Then one can construct a closed term t such that*

$$\text{WE-HA}^\omega \vdash \forall \underline{x}(t\underline{x} =_0 0 \leftrightarrow A_0(\underline{x})).$$

Proof: Analogously to the proof of proposition 3.8 using lemma 3.15 and the previous remark. □

Corollary 3.18. *Let A_0 be a quantifier-free formula of $\mathscr{L}(\text{WE-HA}^\omega)$. Then*

$$\text{WE-HA}^\omega \vdash A_0 \vee \neg A_0.$$

In particular, quantifier-free formulas are provably stable, i.e.

$$\text{WE-HA}^\omega \vdash \neg\neg A_0 \to A_0.$$

Proposition 3.19. *For each type $\rho \in \mathbf{T}$ there exists a closed term t (using only the recursor R_0 of type 0) such that*

$$\text{WE-HA}^\omega \vdash \forall x^0, y_1^\rho, y_2^\rho([x =_0 0 \to txy_1y_2 =_\rho y_1] \wedge [x \neq_0 0 \to txy_1y_2 =_\rho y_2]).$$

Proof: Define
$$\chi(x^0, y_1^0, y_2^0) := R_0(x, y_1, \lambda n^0, m^0.y_2).$$
Then
$$\chi(0, y_1, y_2) =_0 y_1 \text{ and } \chi(S(x), y_1, y_2) =_0 y_2.$$
Hence, since $x \neq_0 0 \to x =_0 S(pd(x))$,

$$x \neq_0 0 \to \chi(x, y_1, y_2) =_0 y_2.$$

Let $\rho = 0\rho_k \ldots \rho_1$ and $\underline{v}^{\underline{\rho}} := v_1^{\rho_1}, \ldots, v_k^{\rho_k}$ be distinct variables. Now define

$$t := \lambda x^0, y_1^\rho, y_2^\rho, \underline{v}^{\underline{\rho}}.\chi(x, y_1\underline{v}, y_2\underline{v}).$$

□

Definition 3.20. IL^ω is $\text{IL}^\omega_{-=}$ together with $=_0$ and the equality axioms for $=_0$ and QF-ER. $\text{PL}^\omega_{(-=)}$ (WE-PA^ω, E-PA^ω) is the extension of $\text{IL}^\omega_{(-=)}$ (WE-HA^ω, E-HA^ω) obtained by adding the law-of-excluded-middle schema LEM (i.e. $A \vee \neg A$ for **arbitrary** formulas A).

The set-theoretic functionals denoted by the closed terms of E-HA^ω (see the model $\mathscr{S}^\omega$ defined in section 3.6 below) are called the 'Gödel primitive recursive functionals of finite type'. They were introduced first in [133] (but see also [161]). In the exercises to this chapter we show that the Gödel primitive recursive functionals of type degree 1 form a larger class than the ordinary primitive recursive functions. In fact, the former class coincides with the provably recursive functions of Peano arithmetic PA: that this class contains the provably recursive functions of PA follows from Gödel's ([133]) functional interpretation (see chapter 9). The other inclusion follows from work of Parsons ([299]) and others.
The combinators Π and Σ are due already to [323]. The correspondence between these combinators and the usual Hilbert-style axiomatization of the implicative fragment of intuitionistic propositional logic given by (the modus ponens rule and) the schemata

$$A \to (B \to A)$$

$$(A \to (B \to C)) \to ((A \to B) \to (A \to C))$$

as well as the correspondence between typed λ-terms and the natural deduction style formalization of that fragment are known as the 'Curry-Howard'-isomorphism (see [166, 349]).

We now consider various fragments of full arithmetic in all types which will be used later. Most of the techniques used in this book apply to all of these systems down to $\text{G}_2\text{A}^\omega$ (to be defined below) whose provably recursive functions are bounded by polynomials. This shows that the proof interpretations on which the applications to proof mining in analysis discussed in chapters further below are based do not produce any non-polynomial growth of numerical bounds by themselves. So if a given proof implicitly contains a bound say of polynomial growth, then the unwinding process will produce such a bound. However, we will usually only sketch how the results proved in this book for (W)E-HA$^\omega$, (W)E-PA$^\omega$ can be adapted to these fragments and refer to the literature for more details. This is in order to avoid to have to deal with too many formal systems (a general malaise in the area of proof theory) but also for the following important reason: weak formal systems are only needed to state **a-priori** that a certain proof allows one to extract data of certain low complexity because it can be formalized in such a weak system. For actual proof mining where such data are explicitly extracted one will instead work in a stronger framework to get an easier formalization of the proof. If the proof indeed contains numerical data of low complexity the proof mining procedure will produce such if the procedure is faithful. It is only to show the latter point why it is of relevance to verify that in principle all techniques used in the procedure applied can be adapted to work for such weak systems as well.

3.4 Fragments of (W)E-HA$^\omega$ and (W)E-PA$^\omega$

Let $\widehat{\text{(W)E-PA}}^{\omega}\!\upharpoonright$ ($\widehat{\text{(W)E-HA}}^{\omega}\!\upharpoonright$) be the fragment of (W)E-PA$^\omega$ ((W)E-HA$^\omega$) where we only have the recursor R_0 for type-0-recursion and the induction schema is restricted to the schema of quantifier-free induction

$$\text{QF-IA} : A_0(0) \wedge \forall x^0 \big(A_0(x) \to A_0(S(x))\big) \to \forall x^0 A_0(x),$$

where A_0 is quantifier-free and may contain parameters of arbitrary types.

The set-theoretic functionals denoted by the closed terms of $\widehat{\text{E-HA}}^{\omega}\!\upharpoonright$ are called the 'Kleene primitive recursive functionals of finite type'. They were (for pure types) first introduced in [195] under the name of S1-S8 computable functionals. In contrast to the Gödel primitive recursive functionals, the Kleene primitive recursive functionals of type degree 1 are just the ordinary primitive recursive functions (see

e.g. [7] for a proof of this).
The systems $\widehat{\text{(W)E-HA}}^\omega\!\upharpoonright$ were introduced in [98] (see also [299]). Proposition 3.17, corollary 3.18 and proposition 3.19 hold analogously for $\widehat{\text{WE-HA}}^\omega\!\upharpoonright$ instead of WE-HA$^\omega$.

In the presence of the quantifier-free axiom of choice schema for numbers

$$\text{QF-AC}^{0,0}: \ \forall x^0 \exists y^0 A_0(x,y) \to \exists f^1 \forall x^0 A_0(x,f(x)) \ (A_0 \text{ quantifier-free}),$$

the schema of induction for Σ^0_1-formulas

$$\Sigma^0_1\text{-IA}: \ \exists y^0 A_0(0,y) \wedge \forall x^0 \big(\exists y A_0(x,y) \to \exists y A_0(S(x),y)\big) \to \forall x \exists y A_0(x,y),$$

where A_0 is quantifier-free (with parameters of arbitrary types), becomes derivable from QF-IA:

Proposition 3.21. $\widehat{\text{WE-PA}}^\omega\!\upharpoonright + \text{QF-AC}^{0,0} \vdash \Sigma^0_1$ -IA.

Proof: Assume $\exists y_0 A_0(0,y_0)$ and $\forall x, y_1 \exists y_2 \big(A_0(x,y_1) \to A_0(S(x),y_2)\big)$. By QF-AC^{0,0} we get (using that x, y_1 can be coded into a single variable, see lemma 3.30 below)

$$\exists f \forall x, y_1 \big(A_0(x,y_1) \to A_0(S(x), f(x,y_1))\big).$$

Define

$$\begin{cases} \Phi(0,y,f) :=_0 y \\ \Phi(S(x),y,f) :=_0 f(x, \Phi(x,y,f)) \end{cases}$$

(note that this can be done by R_0). Then by QF-IA one easily shows that

$$\forall x A_0(x, \Phi(x,y_0,f))$$

for y_0 such that $A_0(0,y_0)$ and, therefore, $\forall x \exists y A_0(x,y)$. □

Whereas inspection of the proof above shows that the result also holds for the intuitionistic system $\widehat{\text{WE-HA}}^\omega\!\upharpoonright + \text{QF-AC}^{0,0}$, the next result requires classical logic: QF-AC^{0,0} also allows one to prove the schema of Δ^0_1-comprehension

$$\Delta^0_1\text{-CA}: \ \forall x^0 \big(\exists y^0 A_0(x,y) \leftrightarrow \forall y^0 B_0(x,y)\big) \to \exists f^1 \forall x^0 \big(f(x) = 0 \leftrightarrow \exists y^0 A_0(x,y)\big),$$

where again parameters in all types are allowed (A_0, B_0 quantifier-free).

Proposition 3.22. $\widehat{\text{WE-PA}}^\omega\!\upharpoonright + \text{QF-AC}^{0,0} \vdash \Delta^0_1$ -CA.

Proof: Exercise! □

As the proof of lemma 3.15 shows we still can define λ-abstraction in $\widehat{\text{(W)E-PA}}^\omega\!\upharpoonright$. Using λ-abstraction and R_0 one can define recursors $\widehat{R}_\rho$ such that

$$\begin{cases} \widehat{R}_\rho 0 y z \underline{w} =_0 y \underline{w} \\ \widehat{R}_\rho (Sx) y z \underline{w} =_0 z(\widehat{R}_\rho x y z \underline{w}) x \underline{w}, \end{cases}$$

where y is of type $\rho = 0\rho_k \ldots \rho_1$, $\underline{w} = w_1^{\rho_1} \ldots w_k^{\rho_k}$ and the type of z is $\rho 00$ (exercise).

The crucial difference between $\widehat{R}_\rho$ and the much stronger R_ρ is that in the case of the former $\widehat{R}_\rho xyz\underline{w}$ may only be used with the fixed set of parameters $\underline{w}$ in the recursion step while in the case of R_ρ we can use the whole functional $R_\rho xyz$.

3.5 Fragments corresponding to the Grzegorczyk hierarchy

We now define a hierarchy of systems $\mathrm{G}_n\mathrm{A}^\omega$ corresponding (w.r.t. the definable and provably total functions) to n-th level of the so-called Grzegorczyk hierarchy [147]. Following Ritchie [314] we base our definition on the n-th branch of the Ackermann function [1]:

Definition 3.23. Let $n \in \mathbb{N}$. We define (by recursion on n from the outside) the n-th branch of the Ackermann function $A_n : \mathbb{N} \times \mathbb{N} \to \mathbb{N}$ by

$$A_0(x,y) := y' \quad \text{(Here and in the following } x' \text{ denotes the successor } Sx \text{ of } x\text{)},$$

$$A_{n+1}(x,0) := \begin{cases} x, \text{ if } n = 0 \\ 0, \text{ if } n = 1 \\ 1, \text{ if } n \geq 2, \end{cases}$$

$$A_{n+1}(x,y') := A_n(x, A_{n+1}(x,y))$$

Remark 3.24. 1) $A_1(x,y) = x + y$, $A_2(x,y) = x \cdot y$, $A_3(x,y) = x^y$,
$A_4(x,y) = x^{x^{\cdot^{\cdot^{x}}}}$ (y times).
2) For each fixed $n \in \mathbb{N}$ the function A_n is primitive recursive. However, as first shown by Ackermann [1], the diagonal function $A(x) := A_x(x,x)$ is no longer primitive recursive (see e.g. [341]).

The **intuitionistic Grzegorczyk arithmetic $\mathbf{G}_n\mathbf{A}_i^\omega$ of level n in all finite types** and its **classical variant $_n\mathbf{A}^\omega$** :

The language $\mathcal{L}(\mathrm{G}_n\mathrm{A}_i^\omega)$ of $\mathrm{G}_n\mathrm{A}_i^\omega$ is the extension of $\mathcal{L}(\mathrm{IL}^\omega_{=})$ resulting from the addition of the constant 0^0, the projectors $\Pi_{\rho,\tau}$, the combinators $\Sigma_{\delta,\rho,\tau}$, function constants S^{00} (successor), $\max_0^{000}, \min_0^{000}, A_0^{000}, \ldots, A_n^{000}$ and functional constants $\Phi_1^{001}, \ldots, \Phi_n^{001}$, $\mu_b^{00(000)}$ (bounded μ-operator), $\tilde{R}_0$ (bounded recursion of type 0) of type $01(000)00$. Furthermore we have a predicate symbol $=_0$ for equality at type 0 and a predicate symbol $<_0$ for objects of type 0.

In addition to the axioms and rules of $\mathrm{IL}^{\omega}_{=}$ the theory $\mathrm{G}_n\mathrm{A}^{\omega}_i$ contains the following (with $x \le_0 y$ being defined as $x < y \vee x = y$):

1) the Π, Σ-axioms (as in the case of E-HA$^{\omega}$).
2) the equality axioms for $=_0$:
   $x =_0 x$, $x =_0 y \to y =_0 x$, $x =_0 y \wedge y =_0 z \to x =_0 z$.
   $x_1 =_0 x_2 \wedge y_1 =_0 y_2 \wedge x_1 < y_1 \to x_2 < y_2$.
3) $<_0$-axioms: $\neg x <_0 0$, $x <_0 Sy \leftrightarrow x <_0 y \vee x =_0 y$, $x <_0 y \vee x =_0 y \vee y <_0 x$.
4) S-axioms: $Sx =_0 Sy \to x =_0 y$, $\neg 0 =_0 Sx$.
5) (max) : $\max_0(x,y) \ge_0 x$, $\max_0(x,y) \ge_0 y$, $\max_0(x,y) =_0 x \vee \max_0(x,y) =_0 y$.
6) (min) : $\min_0(x,y) \le_0 x$, $\min_0(x,y) \le_0 y$, $\min_0(x,y) =_0 x \vee \min_0(x,y) =_0 y$.
7) The defining recursion equations for $A_0, \ldots, A_n$ from the definition 3.23 above.
8) Defining recursion equations for $\Phi_1, \ldots, \Phi_n$:

$$\begin{cases} \Phi_1 f0 =_0 f0 \\ \Phi_1 fx' =_0 \max_0(fx', \Phi_1 fx) \end{cases}$$

and

$$\begin{cases} \Phi_i f0 =_0 f0 \\ \Phi_i fx' =_0 A_{i-1}(fx', \Phi_i fx), \quad \text{for } i \ge 2. \end{cases}$$

9)

$$(\mu_b) : \begin{cases} y \le_0 x \wedge f^{000}xy =_0 0 \to fx(\mu_b fx) =_0 0, \\ y <_0 \mu_b fx \to fxy \ne_0 0, \\ \mu_b fx =_0 0 \vee (fx(\mu_b fx) =_0 0 \wedge \mu_b fx \le_0 x). \end{cases}$$

10) Defining recursion equations for $\tilde{R}_0$ (bounded recursion of type 0):

$$\begin{cases} \tilde{R}_0 0yzv =_0 y \\ \tilde{R}_0 x'yzv =_0 \min_0(z(\tilde{R}_0 xyzv)x, vx). \end{cases}$$

11) All $\mathbb{N}, \mathbb{N}^{\mathbb{N}}, \mathbb{N}^{(\mathbb{N}^{\mathbb{N}})}$-true purely universal sentences $\forall \underline{x} A_0(\underline{x})$, where $\underline{x}$ is a tuple of variables whose types have a degree ≤ 2, where B^A denotes the set of all set-theoretic functions $f : A \to B$.
12) The quantifier-free extensionality rule QF-ER.

$\mathrm{G}_n\mathrm{A}^{\omega}$ is the variant of $\mathrm{G}_n\mathrm{A}^{\omega}_i$ with the law-of-excluded-middle schema $A \vee \neg A$ added. Analogously, we define $\mathrm{G}_\infty\mathrm{A}^{\omega}$
If we add $(E) = \bigcup_\rho \{(E_\rho)\}$ to $\mathrm{G}_n\mathrm{A}^{\omega}$,$\mathrm{G}_n\mathrm{A}^{\omega}_i$ we obtain theories which are denoted by E-$\mathrm{G}_n\mathrm{A}^{\omega}$, E-$\mathrm{G}_n\mathrm{A}^{\omega}_i$. $\mathrm{G}_n\mathrm{R}^{\omega}$ denotes the set of all closed terms of $\mathrm{G}_n\mathrm{A}^{\omega}$.
(E)-$\mathrm{G}_\infty\mathrm{A}^{\omega}_{(i)} := \bigcup_{n \ge 1} \{$ (E)-$\mathrm{G}_n\mathrm{A}^{\omega}_{(i)}\}$ and $\mathrm{G}_\infty\mathrm{R}^{\omega} := \bigcup_{n\ge 1} \{\mathrm{G}_n\mathrm{R}^{\omega}\}$.

Remark 3.25. 1) Our axioms contain w.r.t. $0, S, +, \cdot, <$ what is called Robinson's system Q which specifies (numeralwise) the meaning of $0, S, +, \cdot, <$ when interpreted in the standard model $\mathbb{N}$ and hence – when augmented by the schema of full induction – results in a system containing a version of Peano arithmetic (see e.g. [363] (I.9) for more information on this). Also the meaning of the other constants is – when interpreted in the full type-structure over $\mathbb{N}$ (to be defined below) – uniquely determined by the axioms.

2) The functionals Φ_1, Φ_2 and Φ_3 have the following meaning: $\Phi_1 fx = \max(f0, f1, \ldots, fx)$, $\Phi_2 fx = \sum_{y=0}^{x} fy$, $\Phi_3 fx = \prod_{y=0}^{x} fy$. In general, for $i \geq 2$, $\Phi_i fx$ is the iteration of the $(i$-$1)$-th branch A_{i-1} of the Ackermann function on the f-values $f0, \ldots, fx$.

3) The axioms on μ_b formalize that

$$\mu_b fx := \begin{cases} \min y \leq_0 x(fxy =_0 0), \text{ if such an } y \leq x \text{ exists}, \\ 0, \text{ otherwise}. \end{cases}$$

4) As in the case of $\widehat{\text{WE-HA}}^{\omega}\!\upharpoonright$ one can define λ-abstraction in $\mathrm{G}_n\mathrm{A}^{\omega}_i$ as well as bounded recursors $\tilde{R}_\rho$ that satisfy

$$\begin{cases} \tilde{R}_\rho 0yzv\underline{w} =_0 y\underline{w} \\ \tilde{R}_\rho x'yzv\underline{w} =_0 \min_0(z(\tilde{R}_\rho xyzv\underline{w})x\underline{w}, vx\underline{w}), \end{cases}$$

where $\rho = 0\rho_k \ldots \rho_1$ and $\underline{w} = w_1^{\rho_1} \ldots w_k^{\rho_k}$.

5) Our definition of $\mathrm{G}_n\mathrm{A}^{\omega}$ contains some redundancies (which however we want to keep for greater flexibility of our language): E.g. Φ_i $(i > 1)$ can be defined from $A_i, \tilde{R}_0, \min_0$ and Φ_1: With $f^M := \lambda x.\Phi_1 fx$ one can show that $\Phi_i fx \leq \Phi_i f^M x \leq A_i(f^M(x) + 1, x + 1)$. Hence Φ_i can be defined by $\tilde{R}_0$ using $A_i(f^M(x) + 1, x + 1)$ as boundary function v.

6) The axiom of quantifier-free induction

$$(*)\ \forall f^1, x^0 \big(f0 =_0 0 \wedge \forall y < x(fy =_0 0 \to fy' =_0 0) \to fx =_0 0\big)$$

can be expressed as a universal sentence $\forall f^1, x^0 A_0$ by prop. 3.28 below and thus is an axiom of $\mathrm{G}_n\mathrm{A}^{\omega}_i$. $(*)$ implies every instance (with parameters of arbitrary type) of the schema of quantifier-free induction in the form (equivalent to QF-IA)

$$\forall x^0 \big(A_0(0) \wedge \forall y < x(A_0(y) \to A_0(y')) \to A_0(x)\big)$$

since again by prop. 3.28 there exists a term t such that $tx =_0 0 \leftrightarrow A_0(x)$ for quantifier-free A_0. Now apply $(*)$ to $f := t$.

7) Our reason for including all true universal sentences as axioms in 11) is as follows: it is an old observation made by G. Kreisel that proofs of universal lemmas have no impact on the extraction of programs or bounds from proofs. In fact, all the extraction techniques developed in this book allow one to treat universal

lemmas (and most of the time even more general classes of lemmas) as axioms. For the systems (W)E-PA$^\omega$ etc. we make this explicit in our main theorems on program and bound extractions rather than adding them beforehand as axioms already to the definition of (W)E-PA$^\omega$ since, occasionally, we state 'foundational' corollaries on the proof-theoretic strength of systems, consistency proofs, conservation theorems over PA or PRA etc. that would be spoiled by this. However, for systems as weak as say $G_2A^\omega|\backslash$ such foundational issues are less relevant and to verify some basic properties of functions such as $\max(x,y)$, $|x-y|$ and codings etc. would be very tedious to carry out using just QF-IA. From $n \geq 3$ on, usually the standard proofs of these facts known from primitive recursive arithmetic PRA go through even in the variant of G_nA^ω with the universal axioms 11) replaced by the schema QF-IA and we make free use of this in chapter 13. The reason for our restriction of the types in the universal axioms we add is that in some places in this book we deal with principles which are valid only in the type structure $\mathcal{M}^\omega$ of the so-called majorizable functionals due to [27] (which will be discussed further below) but not in the full type structure $\mathcal{S}^\omega$ of all set-theoretic functionals. Since both type structures coincide up to type 1 and for the type 2 the inclusion $\mathcal{M}_2^\omega \subset \mathcal{S}_2^\omega$ holds, the implication $\mathcal{S}^\omega \models \forall x^\rho A_0 \Rightarrow \mathcal{M}^\omega \models \forall x^\rho A_0$ is obvious **if** $deg(\rho) \leq 2$.

In the following we make free use of the fact that universal lemmas are included as axioms if true.

Already in $G_1A_i^\omega$ we can, using $\tilde{R}_0$ and trivial bounding functions, define the following functions:

Definition 3.26. 1) $prd(0) =_0 0,\ prd(x') =_0 x$ (predecessor),

2) $\begin{cases} sg(0) =_0 0, \\ sg(x') =_0 1, \end{cases}$ and $\begin{cases} \overline{sg}(0) =_0 1, \\ \overline{sg}(x') =_0 0. \end{cases}$

3) $\begin{cases} x \dot{-} 0 =_0 x \\ x \dot{-} y' =_0 prd(x \dot{-} y), \end{cases}$

4) $|x-y| =_0 \max(x \dot{-} y, y \dot{-} x)$ (symmetrical difference).

For the rest of this section we usually omit the type in $=_0$ and simply write $=$.

Remark 3.27. The following basic properties of the functions defined above are all purely universal and so (because of the universal axioms 11) trivially provable already in $G_1A_i^\omega$:
$sg(x) = 0 \leftrightarrow x = 0$, $\overline{sg}(x) = 0 \leftrightarrow x \neq 0$, $sg(x) \leq 1$, $\overline{sg}(x) \leq 1$, $prd(x) \leq x$, $|x-y| = 0 \leftrightarrow x = y$, $x = 0 \vee x = S(prd(x))$, $\max(x,y) = 0 \leftrightarrow x = 0 \wedge y = 0$, $\min(x,y) = 0 \leftrightarrow x = 0 \vee y = 0$, $\max(x,y) = y \leftrightarrow x \leq y$, $x < y \leftrightarrow Sx \dot{-} y = 0$.

Proposition 3.28. *Let $n \geq 1$. For each formula $A \in \mathcal{L}(G_nA^\omega)$ which contains no quantifiers except for bounded quantifiers of type 0 one can construct a closed term t_A in G_nA^ω such that*

$$\mathrm{G}_n\mathrm{A}_i^\omega \vdash \forall x_1^{\rho_1},\ldots,x_k^{\rho_k}\big(t_A x_1\ldots x_k =_0 0 \leftrightarrow A(x_1,\ldots,x_k)\big),$$

where $x_1,\ldots,x_k$ contain all the free variables of A.

Proof: Induction on the logical structure of A using the remark above. Bounded quantifiers are captured by μ_b:

$$\mathrm{G}_n\mathrm{A}_i^\omega \vdash \exists y \leq_0 x A_0(x,y,\underline{a}) \overset{(\mu_b)}{\leftrightarrow} A_0(x,\mu_b(\lambda x,y.t_{A_0}xy\underline{a},x),\underline{a}).$$

Similarly for the bounded universal quantifier. □

Proposition 3.29. *Let $n \geq 1$ and $A_0(\underline{x}) \in \mathscr{L}(\mathrm{G}_n\mathrm{A}^\omega)$ be a quantifier-free formula, where $\underline{x} = x_1^{\rho_1}\ldots x_k^{\rho_k}$ contain all the free variables of A_0, and $t_1^{0\rho_k\ldots\rho_1}, t_2^{0\rho_k\ldots\rho_1}$ are closed terms of $\mathrm{G}_n\mathrm{A}^\omega$. Then there exists a closed term $\Phi^{0\rho_k\cdots\rho_1}$ in $\mathrm{G}_n\mathrm{A}^\omega$ such that*

$$\mathrm{G}_n\mathrm{A}_i^\omega \vdash \forall \underline{x}\left(\Phi\underline{x} =_0 \begin{cases} t_1\underline{x}, \text{ if } A_0(\underline{x}) \\ t_2\underline{x}, \text{ if } \neg A_0(\underline{x}).\end{cases}\right)$$

Proof: Define

$$\Phi := \lambda \underline{x}.\tilde{R}_0(t_{A_0}\underline{x})(t_1\underline{x})(\lambda y^0,z^0.t_2\underline{x})(\lambda y^0.t_2\underline{x}),$$

where t_{A_0} is as in the previous proposition. One easily verifies that Φ does the job. □

We now show how to encode pairs, tuples and finite sequences of numbers:

Definition 3.30 (and lemma). For **n ≥ 2** we can define the well-known surjective Cantor pairing function j with its projections in $\mathrm{G}_n\mathrm{R}^\omega$:

$$j(x^0,y^0) := \begin{cases} \min u \leq_0 (x+y)^2+3x+y[2u =_0 (x+y)^2+3x+y] \text{ if existent} \\ 0^0, \text{ otherwise,}\end{cases}$$

$$j_1 z := \min x \leq_0 z[\exists y \leq z(j(x,y)=z)],$$

$$j_2 z := \min y \leq_0 z[\exists x \leq z(j(x,y)=z)].$$

Using j, j_1, j_2 we can define a coding of k-tuples for every **fixed** number k by

$$\nu^1(x_0) := x_0,\ \nu^2(x_0,x_1) := j(x_0,x_1),\ \nu^{k+1}(x_0,\ldots,x_k) := j(x_0,\nu^k(x_1,\ldots,x_k)),$$

$$\nu_1^1(x) := x \text{ and (for } k>1)\ \nu_i^k(x) := \begin{cases} j_1\circ(j_2)^{i-1}(x), \text{ if } 1 \leq i < k \\ (j_2)^{k-1}(x), \text{ if } 1 < i = k.\end{cases}$$

Indeed, these functions satisfy the following properties of a surjective tuple coding:

$$\nu_i^k(\nu^k(x_1,\ldots,x_k)) = x_i \quad (1 \leq i \leq k)$$

and

$$\nu^k(\nu_1^k(x), \dots, \nu_k^k(x)) = x.$$

Following essentially [366] we now extend this tuple coding to a coding of finite sequences:

$$\langle\rangle := 0, \ \langle x_0, \dots, x_k\rangle := S(\nu^2(k, \nu^{k+1}(x_0, \dots, x_k))).$$

Using $\tilde{R}_0$ we define functions $lth, \Pi(x,y) \in \mathrm{G}_n\mathrm{R}^\omega$ such that for every fixed k

$$lth(\langle\rangle) = 0, \quad lth(\langle x_0, \dots, x_k\rangle) = k+1$$

and (for $x = \langle x_0, \dots, x_m\rangle$)

$$\Pi(x,y) = \begin{cases} x_y, \text{ if } y \leq m \\ 0^0, \text{ otherwise.} \end{cases}$$

Define

$$lth(x) := \begin{cases} 0^0, \text{ if } x =_0 0 \\ j_1(x \dot{-} 1) + 1, \text{ otherwise,} \end{cases}$$

$$\Pi(x,y) := \begin{cases} 0^0, \text{ if } lth(x) \leq y \\ j_1 \circ (j_2)^{y+1}(x \dot{-} 1), \text{ if } 0 \leq y < lth(x) \dot{-} 1 \\ (j_2)^{lth(x)}(x \dot{-} 1), \text{ if } lth(x) > 0 \wedge y = lth(x) \dot{-} 1 \end{cases}$$

To improve the readability we normally write $(x)_y$ instead of $\Pi(x,y)$.
That $\Pi(x,y)$ is definable even in $\mathrm{G}_2\mathrm{R}^\omega$, follows from the fact that the iteration $\varphi xy = (j_2)^y(x)$ of j_2 is definable in $\mathrm{G}_2\mathrm{R}^\omega$ since $\varphi xy \leq x$ for all x, y so that we can use $\lambda y.x$ as bounding function.

We need $\mathrm{G}_3\mathrm{R}^\omega$ to define a coding of initial segments of **variable** length of a function f. Indeed, there is a functional $\Phi_{\langle\rangle} \in \mathrm{G}_3\mathrm{R}^\omega$ such that $\Phi_{\langle\rangle} fx = \langle f0, \dots, f(x \dot{-} 1)\rangle$. Of course we cannot write $\langle f0, \dots, f(x \dot{-} 1)\rangle$ for variable x. However the meaning of $\Phi_{\langle\rangle} fx$ can be expressed via $(\Phi_{\langle\rangle} fx)_y = fy$ for all $y < x$ (and $= 0$ for $y \geq x$).
To achieve this, we first define

$$\begin{cases} \tilde{f}0 = f0 \\ \tilde{f}x' = j(fx', \tilde{f}x) \end{cases}$$

in $\mathrm{G}_3\mathrm{R}^\omega$: One easily verifies (using $j(x,x) \leq 4x^2$) that $\tilde{f}x \leq 4^{3^x}\left(f^M x\right)^{2^x}$ for all x. Hence the definition of $\tilde{f}$ can be carried out by $\tilde{R}_0$ using as our bounding function

$$\lambda x. j\left(fx', 4^{3^x}\left(f^M x\right)^{2^x}\right) \in \mathrm{G}_3\mathrm{R}^\omega.$$

With $\tilde{j}(x,y) := j(y,x)$ we see that $\tilde{f}x$ means $\tilde{j}(\ldots\tilde{j}(\tilde{j}(f0,f1),f2)\ldots fx)$. Hence $\widehat{f}x := (\lambda y.f(x \dot{-} y))x$ has the meaning $j(f0,\ldots j(f(x-2),j(f(x-1),fx))\ldots)$. Finally, we are now in the position to define $\Phi_{\langle\rangle} \in \mathrm{G}_3\mathrm{R}^\omega$:

$$\Phi_{\langle\rangle} fx := \begin{cases} 0^0, \text{ if } x = 0 \\ \widehat{(f_x)}x + 1, \text{ otherwise,} \end{cases}$$

where

$$f_x y := \begin{cases} x, \text{ if } y = 0 \\ f(y \dot{-} 1), \text{ otherwise.} \end{cases}$$

Again for better readability, we usually write $\overline{f}x$ instead of $\Phi_{\langle\rangle} fx$.

Next we define a function $*$ in $\mathrm{G}_3\mathrm{R}^\omega$ by

$$n * m := \Phi_{\langle\rangle}(fnm)(lth(n) + lth(m)),$$

where

$$(fnm)(k) := \begin{cases} (n)_k, \text{ if } k < lth(n) \\ (m)_{k \dot{-} lthn}, \text{otherwise.} \end{cases}$$

Then

$$\langle x_0,\ldots,x_k\rangle * \langle y_0,\ldots,y_m\rangle = \langle x_0,\ldots,x_k,y_0,\ldots,y_m\rangle.$$

Note that $\Phi_{\langle\rangle}$ and $*$ are not definable in $\mathrm{G}_2\mathrm{R}^\omega$ since their definitions involve an iteration of the polynomial j.

Remark 3.31. 1) For detailed information on this as well as various other codings see [341] and also [101] (where j is called 'Cauchy's pairing function').
2) One easily shows that $(x+y)^2 + 3x + y$ is always even so that the case 'otherwise' in the definition of j never occurs and $2j(x,y) = (x+y)^2 + 3x + y$ for all x,y.

Definition 3.32. For arbitrary $\rho \in \mathbf{T}$ we define the relation $x_1 \geq_\rho x_2$ between functionals x_1,x_2 of type ρ by induction on ρ:

$$\begin{cases} x_1 \geq_0 x_2 \text{ is defined already,} \\ x_1 \geq_{\tau\rho} x_2 :\equiv \forall y^\rho (x_1 y \geq_\tau x_2 y). \end{cases}$$

$x_1 \leq_\rho x_2 :\equiv x_2 \geq_\rho x_1$.

Lemma 3.33. *Let $\rho = \tau\rho_k\ldots\rho_1$. Then*

$$\mathrm{G}_1\mathrm{A}_i^\omega \vdash x_1 \geq_\rho x_2 \leftrightarrow \forall y_1^{\rho_1},\ldots,y_k^{\rho_k}(x_1\underline{y} \geq_\tau x_2\underline{y}).$$

We now for the first time make use of an important structural property of the closed terms of all our systems. This property of 'majorizability' which is due to W.A.

Howard will play key roles (in different ways) in the proofs of numerous results in this book. In the present chapter it is used to prove results on the growth of the definable functionals of G_nA^ω. We will focus on the cases $n = 1,2,3$ as only those (actually only $n = 2,3$) are of practical interest (the whole hierarchy is treated in [207]).

Definition 3.34 (W.A. Howard [163]). We define the relation $x^*\ maj_\rho\ x$ (x^* majorizes x) between functionals of type ρ by induction on ρ:

$$\begin{cases} x^*\ maj_0\ x :\equiv x^* \geq_0 x, \\ x^*\ maj_{\tau\rho}\ x :\equiv \forall y^*, y(y^*\ maj_\rho\ y \to x^*y^*\ maj_\tau\ xy). \end{cases}$$

Lemma 3.35. $G_1A_i^\omega$ *proves:*

(i) $\tilde{x}^* =_\rho x^* \wedge \tilde{x} =_\rho x \wedge x^*\ maj_\rho\ x \to \tilde{x}^*\ maj_\rho\ \tilde{x}$.
(ii) $x^*\ maj_\rho\ x \wedge x \geq_\rho y \to x^*\ maj_\rho\ y$.
(iii) *For* $\rho = \tau\rho_k \ldots \rho_1$:

$$x^*\ maj_\rho\ x \leftrightarrow \forall y_1^*, y_1, \ldots, y_k^*, y_k \Big(\bigwedge_{i=1}^{k} (y_i^*\ maj_{\rho_i}\ y_i) \to x^*\underline{y}^*\ maj_\tau\ x\underline{y} \Big).$$

Proof: Induction on the type respectively on k. □

Remark 3.36. 1) The relation maj_ρ is a kind of hereditary form of $\geq_\rho$ (combined with monotonicity) and is (in contrast to $\geq_\rho$) a so-called logical relation in the sense of G. Plotkin which implies a nice behavior w.r.t. substitution (see lemma 3.35 (iii)). Because of this, results on the majorization of complex terms can be established directly by induction on the term structure without any use of normalization.
2) The previous lemma can be proved also in $\widehat{\text{WE-HA}}^\omega\!\upharpoonright$ since only the transitivity of $\leq_0$ is used (which can be proved by QF-IA) but no general universal axioms 11).

Next we need some basic properties of the constants in $G_nA_i^\omega$:

Lemma 3.37. *Provably in* $G_nA_i^\omega$ *(if applicable) the following holds:*

1) $\Pi\ maj\ \Pi$ *and* $\Sigma\ maj\ \Sigma$.
2) $0\ maj\ 0$, $S\ maj\ S$, $A_i\ maj\ A_i$ $(i \leq 2)$, $A_j^* := \lambda x^0, y^0.\max(A_j(x,y),1)\ maj\ A_j$ $(j \geq 3)$.
3) $\min\ maj\ \min$, $\max\ maj\ \max$.
4) $\Phi_1^* := \lambda f^1, x^0.f(x)\ maj\ \Phi_1$, $\Phi_2^* := \lambda f^1, x^0.f(x)\cdot(x+1)\ maj\ \Phi_2$, $\Phi_j^* := \lambda f^1, x^0.A_j(f(x)+1, x+1)\ maj\ \Phi_j$ $(j \geq 3)$.
5) $\mu_b^* := \lambda f, x.x\ maj\ \mu_b$.
6) $\tilde{R}_0^* := \lambda x, y, z, v.\max_0(y, v(prd(x)))\ maj\ \tilde{R}_0$.

Proof: Exercise! □

Definition 3.38. 1) $G_nR^\omega_- \subset G_nR^\omega$ denotes the subset of all closed terms in G_nR^ω that are built up from $\Pi_{\rho,\tau}, \Sigma_{\delta,\rho,\tau}, A_0, \dots, A_n, 0^0, S, prd, \min_0$ and $\max_0$ only (i.e. which do not contain occurrences of $\Phi_1, \dots, \Phi_n, \tilde{R}_0$ or μ_b).
2) $G_nR^\omega_-[\Phi_1]$ is the set of all closed terms built up from $G_nR^\omega_-$ plus Φ_1.

Proposition 3.39. *Let $n \ge 1$. To each term $t^\rho \in G_nR^\omega$ one can construct by direct induction on the structure of t a term $t^{*\rho} \in G_nR^\omega_-$ such that*

$$G_nA_i^\omega \vdash t^* \; maj_\rho \; t.$$

Proof: 1. Replace every occurrence of $\tilde{R}_0$ in t by $\tilde{R}_0^*$, which is built up from Π, Σ (which are used for defining the λ–operator) and the monotone functions $\max_0$ and prd.
2. Replace all occurrences of $A_3, \dots, A_n$ in t by $A_3^*, \dots, A_n^*$.
3. Replace all occurrences of $\Phi_1, \dots, \Phi_n, \mu_b$ in t by $\Phi_1^*, \dots, \Phi_n^*, \mu_b^*$.
Let t^* be the term which results from t after having carried out 1.–3. By construction, $t^* \in G_nR^\omega_-$. Moreover, t^* is constructed by replacing every constant c in t by a closed term s_c^* such that $s_c^* \; maj \; c$ (lemma 3.37). Since t is built up from constants only this implies (using lemma 3.35) $t^* \; maj \; t$. □

Corollary to the proof:
One can even achieve that the majorizing term t^* does not contain $S, prd, \max_0$ or $\min_0$ (though this in general will give a less good bound if we use t^* as a bound for t^0): this follows using $\lambda x^0.x \; maj_1 \; prd$ and $A_1 \; maj \; \max_0, \min_0$ and – as majorants for $\tilde{R}_0$ and A_j $(j \ge 3)$ – $\tilde{R}^* := \lambda x, y, z, v.(y + v(x))$ and $A_j^*(x, y) := A_j(x+1, y) + 1$, where '$x + 1$' can be replaced by $A_0 0 x$.

The majorizing term t^* constructed in proposition 3.39 has a much simpler form than t as it does not contain any functional of type degree > 1 except for the projectors and combinators Π and Σ. We know show that if t^* has type $(\le)2$, then we can rewrite the term $t^* x^1$ even in the form $\widehat{t}[x]$, where $\widehat{t}[x]$ no longer contains any projector or combinator (but does contain x^1) :

Proposition 3.40. *Let $n \ge 1$ and $\rho = 0\rho_k \dots \rho_1$ with $deg(\rho_i) \le 1$ for $i = 1, \dots, k$ (i.e. $deg(\rho) \le 2$). Moreover, let $t^\rho \in G_nR^\omega_-$. Then one can construct (by 'logical' normalization, i.e. by carrying out all possible Π, Σ-reductions) a term $\widehat{t}[x_1^{\rho_1}, \dots, x_k^{\rho_k}]$ such that*

1) $\widehat{t}[x_1, \dots, x_k]$ contains at most $x_1 \dots, x_k$ as free variables,
2) $\widehat{t}[x_1, \dots, x_k]$ is built up only from $x_1, \dots, x_k, A_0, \dots, A_n, S^1, 0^0, prd, \min_0, \max_0$,
3) $G_nA_i^\omega \vdash \forall x_1^{\rho_1}, \dots, x_k^{\rho_k} (\widehat{t}[x_1, \dots, x_k] =_0 t x_1 \dots x_k)$.

Proof: We perform reductions $\Pi st \rightsquigarrow s$ and $\Sigma str \rightsquigarrow sr(tr)$ inside of $tx_1 \dots x_k$ as long as no further such reduction is possible and denote the resulting term by $\widehat{t}[x_1, \dots, x_k]$. The well-known strong normalization theorem for typed combinatory logic (see e.g. [370] (theorem 1.2.18) or [349] (theorem 5.3.6)) ensures that this situation will always occur after a finite number of reduction steps. Since $\Pi xy = x$ and $\Sigma xyz =$

$xz(yz)$ are axioms of $G_nA_i^\omega$ the quantifier-free rule of extensionality yields (even without appeal to the universal axioms 11)

$$G_nA_i^\omega \vdash \forall x_1^{\rho_1},\dots,x_k^{\rho_k}\,(\widehat{t}[x_1,\dots,x_k] =_0 tx_1\dots x_k).$$

We now prove that $\widehat{t}[x_1,\dots,x_k]$ does not contain any longer any combinator Π, Σ : Suppose that on the contrary $\widehat{t}[x_1,\dots,x_k]$ contains an occurrence of Σ (resp. Π). Then Σ (Π) must occur in the form $\Sigma, \Sigma t_1$ or $\Sigma t_1 t_2$ ($\Pi, \Pi t_1$) but not in the form $\Sigma t_1 t_2 t_3$ (resp. $\Pi t_1 t_2$) since in the latter case we could have carried out the reduction $\Sigma t_1 t_2 t_3 \rightsquigarrow t_1 t_3(t_2 t_3)$ (resp. $\Pi t_1 t_2 \rightsquigarrow t_1$) contradicting the construction of $\widehat{t}$. All the terms $s = \Sigma, \Sigma t_1, \Sigma t_1 t_2, \Pi, \Pi t_1$ have a type whose degree is ≥ 1. Hence s can occur in $\widehat{t}$ only in the form $r(s)$, where $r = \Sigma, \Sigma t_4, \Sigma t_4 t_5, \Pi$ or Πt_4 since these terms are the only reduced ones requiring an argument of type ≥ 1 that can be built up from $x_1^{\rho_1},\dots,x_k^{\rho_k},\Sigma,\Pi,A_i,S^1,0^0$ and $\max_0$ (here we use that $deg(\rho_i) \leq 1$). We notice that the cases $r = \Sigma t_4 t_5$ and $r = \Pi t_4$ cannot occur since otherwise r(s) would allow a reduction of Σ resp. Π. Hence $r(s)$ is again a Π,Σ-term having a type of degree ≥ 1 and, therefore, has to occur within a term r' for which the same reasoning as for r applies and so on. Since $\widehat{t}$ is finite this process has to stop which gives a contradiction. □

Remark 3.41. Proposition 3.40 gets false if $\deg(\rho) = 3$: Define $\rho := 0(0(000))$ and $t^\rho := \lambda x^{0(000)}.x(\Pi_{0,0})$. Then $tx =_0 x(\Pi_{0,0})$ contains Π but no Π-reduction applies.

Corollary 3.42. *Let $n \geq 1$, ρ be as in proposition 3.40 and $t^\rho \in G_nR^\omega$. Then one can construct (by majorization and subsequent 'logical' normalization) a term $t^*[x_1^{\rho_1},\dots,x_k^{\rho_k}]$ such that*

1) *$t^*[x_1,\dots,x_k]$ contains at most $x_1\dots,x_k$ as free variables,*
2) *$t^*[x_1,\dots,x_k]$ is built up only from $x_1,\dots,x_k,0^0,A_0,\dots,A_n$.*
3) *$G_nA_i^\omega \vdash \lambda x_1,\dots,x_k.t^*[x_1,\dots,x_k]$ maj t, i.e.*

$$\forall x_1^*,x_1,\dots,x_k^*,x_k\Big(\bigwedge_{i=1}^{k}(x_i^* \; maj_{\rho_i}\, x_i) \rightarrow t^*[x_1^*,\dots,x_k^*] \geq_0 tx_1\dots x_k.$$

Proof: The corollary follows from propositions 3.39 (and the corollary to its proof) and 3.40 together with lemma 3.35. □

We are now in the position to estimate the growth of the functions definable by terms in G_1R^ω, G_2R^ω and G_3R^ω. Note that we only need majorization and logical normalization for this.

Proposition 3.43. *The growth of the functions defined by closed terms t^1 of G_nR^ω ($n = 1,2,3$) can be calibrated as follows:*

$$\begin{cases} t^1 \in \mathrm{G_1R}^\omega \Rightarrow \exists c_1, c_2 \in \mathbb{N}: \ \mathrm{G_1A}_i^\omega \vdash \forall x^0 (tx \le_0 c_1 x + c_2) \ \textit{(linear growth)}, \\ t^1 \in \mathrm{G_2R}^\omega \Rightarrow \exists k, c_1, c_2 \in \mathbb{N}: \ \mathrm{G_2A}_i^\omega \vdash \forall x^0 (tx \le_0 c_1 x^k + c_2) \\ \qquad \textit{(polynomial growth)}, \\ t^1 \in \mathrm{G_3R}^\omega \Rightarrow \exists k, c \in \mathbb{N}: \ \mathrm{G_3A}_i^\omega \vdash \forall x^0 (tx \le_0 2_k^{cx}), \ \textit{where } 2_0^a = a, 2_{k'}^a = 2^{2_k^a} \\ \qquad \textit{(finitely iterated exponential growth)}. \end{cases}$$

The result can also be extended to tuples of number variables: for t^ρ with $\rho = 0\underbrace{(0)\ldots(0)}_{m-times}$ we have

$$\begin{cases} t^\rho \in \mathrm{G_1R}^\omega \Rightarrow \exists c_1, \ldots, c_{m+1} \in \mathbb{N}: \ \mathrm{G_1A}_i^\omega \vdash \forall x_1^0, \ldots, x_m^0 (t\underline{x} \le_0 \sum_{i=1}^m c_i x_i + c_{m+1}), \\ t^\rho \in \mathrm{G_2R}^\omega \Rightarrow \exists p \in \mathbb{N}[x_1, \ldots, x_m]: \ \mathrm{G_2A}_i^\omega \vdash \forall \underline{x} (t\underline{x} \le_0 p\underline{x}), \\ t^\rho \in \mathrm{G_3R}^\omega \Rightarrow \exists k, c_1, \ldots, c_m \in \mathbb{N}: \ \mathrm{G_3A}_i^\omega \vdash \forall \underline{x} (t\underline{x} \le_0 2_k^{c_1 x_1 + \ldots + c_m x_m}). \end{cases}$$

The constants $c_i, k \in \mathbb{N}$ as well as the coefficients of p can be effectively computed from a given closed term t by majorization and normalization.

Proof: By corollary 3.42 one can in all three cases construct a term $\widehat{t}[\underline{x}]$ built up from $\underline{x}^0, 0^0, A_0, \ldots, A_n$ $(n = 1,2,3)$ such that $\widehat{t}[\underline{x}] \ge_0 t\underline{x}$ for all $\underline{x}$. For the particular cases this yields the following:
$n = 1$: Consider a term $t^\rho \in \mathrm{G_1R}^\omega$, where $\rho = 0\underbrace{(0)\ldots(0)}_{m}$. $\widehat{t}[x_1^0, \ldots, x_m^0]$ is built up from $x_1^0, \ldots, x_m^0, 0^0, A_0$ and A_1 only. Both $A_0(x_1, x_2) = 0 \cdot x_1 + 1 \cdot x_2 + 1$ and $A_1(x_1, x_2) = 1 \cdot x_1 + 1 \cdot x_2 + 0$ are functions having the form $c_1 x_1 + c_2 x_2 + c_3$ or – more generally – $c_1 x_1 + \ldots + c_k x_k + c_{k+1}$. Since substitution of such functions again yields a function which can be written in this form it follows that $\widehat{t}[x_1, \ldots, x_m] = c_1 x_1 + \ldots + c_m x_m + c_{m+1}$ for suitable constants $c_1, \ldots, c_{m+1}$.
$n = 2$: For $t^\rho \in \mathrm{G_2R}^\omega$, $\widehat{t}[x_1, \ldots, x_m]$ is built up from $x_1^0, \ldots, x_m^0, 0^0, A_0, A_1, A_2$. Since A_0, A_1 and A_2 are polynomials (in two variables) and substitution of polynomials in several variables yields a function which can be written again as a polynomial, it is clear that $\widehat{t}[x_1, \ldots, x_m] = p(x_1, \ldots, x_m)$ for a suitable polynomial in $\mathbb{N}[x_1, \ldots, x_m]$. In the case $m = 1$, $p(x)$ can be bounded by $c_1 x^k + c_2$ for suitable numbers c_1, c_2.
$n = 3$: Let $t^\rho \in \mathrm{G_3R}^\omega$. For $\tilde{A}_3(x,y) := A_3(\max_0(x,2), \max_0(y,2))$ the following holds

$$\tilde{A}_3 \ maj \ A_0, A_1, A_2, A_3.$$

Replace in $\widehat{t}[x_1, \ldots, x_m]$ all occurrences of A_i (with $i \le 3$) by $\tilde{A}_3$ and denote the resulting term by $\tilde{t}[x_1, \ldots, x_m]$. Then $\tilde{t}[x_1, \ldots, x_m]$ can be bounded by y_k, where $y_0 := y$, $y_{k'} := y^{y_k}$ and $y := \max(x_1, \ldots, x_m, 2)$. and hence $\forall \underline{x}(2_{\tilde{k}}^{\underline{x}} \ge t\underline{x})$ for a suitable $\tilde{k} \ge k$, where $2_0^{\underline{x}} := x_1 + \ldots + x_m$ and $2_{k'}^{\underline{x}} := 2^{2_k^{\underline{x}}}$. □

In concrete extractions of bounds from given proofs one will use, of course, all kinds of auxiliary functions to get a sharper estimate than the above perspicuous, but crude, calibration in terms of $0, S, +, \cdot, \exp$ only.

We now show how that calibration extends to functionals of type ρ with $deg(\rho) \leq 2$. To keep the notational complexity low we only formulate things for $\rho = 1(1)$. Before we can state the result we need the following

Definition 3.44. A functional $\Phi^{1(1)}$ is called linear (polynomial, elementary recursive resp.) if it can be written as a term $\tilde{t}[f,x]$ which is built up only out of $x, f, 0^0, S^1, +$ ($x, f, 0^0, S^1, +, \cdot$ resp. $x, f, 0^0, S^1, +, \cdot, (\cdot)^{(\cdot)}$).

Remark 3.45. If $\Phi^{1(1)}$ is linear (polynomial, elementary recursive) and f is a linear (polynomial, elementary recursive) function, then $\lambda x.\Phi f x$ again is a linear (polynomial, elementary recursive) function.

Proposition 3.46. *Let $t^{1(1)}$ be a closed term of $\mathrm{G}_1\mathrm{R}^\omega$. Then there exists a linear functional Φ given by some term $\tilde{t}[f,x]$ as above such that*

$$\mathrm{G}_1\mathrm{A}_i^\omega \vdash \forall x^0, f^1 \left(\tilde{t}[f^M, x] \geq t f x\right),$$

where $f^M := \Phi_1 f$, i.e. $f^M(x) := \max\{f(0), \ldots, f(x)\}$.
Moreover, f^M can be replaced by h^M for any h satisfying $h \geq_1 f$.
Analogously for $t^{1(1)} \in \mathrm{G}_2\mathrm{R}^\omega$ (resp. $t^{1(1)} \in \mathrm{G}_3\mathrm{R}^\omega$), where then Φ is a polynomial (resp. elementary recursive) functional and $\mathrm{G}_1\mathrm{A}_i^\omega$ is replaced by $\mathrm{G}_2\mathrm{A}_i^\omega$ (resp. $\mathrm{G}_3\mathrm{A}_i^\omega$).

Proof: The proposition follows from corollary 3.42 and the fact that

$$\mathrm{G}_1\mathrm{A}_i^\omega \vdash h \geq_1 f \to h^M \; maj_1 \; f.$$

□

So, in particular, if $t^{1(1)} \in \mathrm{G}_2\mathrm{R}^\omega$ and f is bounded by a polynomial $p \in \mathbb{N}[x]$, then tf again is bounded by some polynomial $r \in \mathbb{N}[x]$ (note that p always is monotone so that $p^M = p$). In fact, this holds even in a certain uniform sense:

Proposition 3.47. *Let $t^{1(1)} \in \mathrm{G}_2\mathrm{R}^\omega$. Then one can construct a polynomial $q \in \mathbb{N}[x]$ such that*

$$\begin{cases} \text{For every polynomial } p \in \mathbb{N}[x] \\ \text{one can construct a polynomial } r \in \mathbb{N}[x] \text{ such that} \\ \forall f^1 \left(f \leq_1 p \to \forall x^0 (t f x \leq_0 r(x))\right) \text{ and } deg(r) \leq q(deg(p)) \end{cases}$$

The result also holds in the case where t has tuples $f_1^1, \ldots, f_k^1, x_1^0, \ldots, x_l^0$ of arguments with $f_1, \ldots, f_k \leq_1 p$ and $r \in \mathbb{N}[x_1, \ldots, x_l]$.

Proof: Let $p \in \mathbb{N}[x]$ and $\tilde{t}[f,x]$ be constructed to tf according to proposition 3.46. Then $\tilde{t}[p,x] \geq_0 t f x$ for all $f \leq_1 p$ and $\tilde{t}[p,x]$ is built up from $x, 0^0, A_0, A_1$ and p only.

As in the proof of proposition 3.43 one concludes that $\tilde{t}[p,x]$ can be written as a polynomial r in x. The existence of the polynomial q bounding the degree of r in the degree of p follows from the fact that the degree of a polynomial $p_1 \in \mathbb{N}[x_1,\ldots,x_m]$ obtained by substitution of a polynomial p_2 for one variable in a polynomial p_3 is $\leq \deg(p_2)\cdot\deg(p_3)$ and that $\deg(p_2+p_3), \deg(p_2\cdot p_3) \leq \deg(p_2)+\deg(p_3)$. □

The analysis of the growth of the functionals definable in $\mathrm{G}_n\mathrm{A}^\omega$ made use for the first time of the important notion of majorizability from [163]. In connection with the definition of a model for E-PA$^\omega$ (and its extension by bar recursion, see chapter 11) a variant – called 'strong majorization' and denoted by 's-maj' – was introduced semantically in [27] (see the next section). This notion has the following syntactic counterpart:

Definition 3.48 (Bezem [27]). We define the relation x^* *s-maj*$_\rho$ x (x^* strongly majorizes x) between functionals of type ρ by induction on ρ:

$$\begin{cases} x^* \ s\text{-}maj_0\ x :\equiv x^* \geq_0 x, \\ x^* \ s\text{-}maj_{\tau\rho}\ x :\equiv \forall y^*, y(y^* \ s\text{-}maj_\rho\ y \to x^*y^* \ s\text{-}maj_\tau\ x^*y, xy). \end{cases}$$

The properties basic properties of *s-maj* are

Lemma 3.49. $\mathrm{G}_1\mathrm{A}_i^\omega$ *(as well as* $\widehat{\text{WE-HA}}^\omega\!\restriction$*) proves:*

(i) $\tilde{x}^* =_\rho x^* \wedge \tilde{x} =_\rho x \wedge x^* \ s\text{-}maj_\rho\ x \to \tilde{x}^* \ s\text{-}maj_\rho\ \tilde{x}$.
(ii) $x^* \ s\text{-}maj_\rho\ x \to x^* \ s\text{-}maj_\rho\ x^*$.
(iii) $x_1 \ s\text{-}maj_\rho\ x_2 \wedge x_2 \ s\text{-}maj_\rho\ x_3 \to x_1 \ s\text{-}maj_\rho\ x_3$.
(iv) $x^* \ s\text{-}maj_\rho\ x \wedge x \geq_\rho y \to x^* \ s\text{-}maj_\rho\ y$.
(v) *For* $\rho = \tau\rho_k\ldots\rho_1$:

$$x^* \ s\text{-}maj_\rho\ x \leftrightarrow \forall y_1^*, y_1, \ldots, y_k^*, y_k \Big(\bigwedge_{i=1}^{k} (y_i^* \ s\text{-}maj_{\rho_i}\ y_i) \to x^*\underline{y}^* \ s\text{-}maj_\tau\ x^*\underline{y}, x\underline{y} \Big).$$

Proof: (i)–(iv) follow by induction on ρ (where we use (ii) in the proof of (iii)). (v) follows by induction on k using again (ii). □

Analogously to proposition 3.39 one proves:

Proposition 3.50. *For all* $n \geq 1$ *the following holds: To each term* $t^\rho \in \mathrm{G}_n\mathrm{R}^\omega$ *one can construct by induction on the structure of* t *(without normalization) a term* $t^{*\rho} \in \mathrm{G}_n\mathrm{R}^\omega_-$ *such that*

$$\mathrm{G}_n\mathrm{A}_i^\omega \vdash t^* \ s\text{-}maj_\rho\ t.$$

Proof: The same term as constructed in the proof of proposition 3.39 also works for *s-maj* (exercise). □

3.6 Models of E-PA$^\omega$

The full set-theoretic model:

We define the type-structure $\mathcal{S}^\omega$ of **all** set-theoretic functionals as follows:

$$\begin{cases} S_0 := \mathbb{N} \\ S_{\tau\rho} := \big\{ \text{all set-theoretic functionals } \varphi : S_\rho \to S_\tau \big\} \\ \mathcal{S}^\omega := \langle S_\rho \rangle_{\rho \in \mathbf{T}}. \end{cases}$$

The following proposition is obvious:

Proposition 3.51. *$\mathcal{S}^\omega$ is a model of* E-PA$^\omega$.

The model of all sequentially continuous functionals: The type-structure $\mathcal{C}^\omega$ of all (sequentially) continuous functionals was introduced in [321]. We first need some preparatory definitions:

Definition 3.52 (Kuratowski [258]). Let X be a set together with a relation of convergence '$\rightarrow$' between sequences (p_n) of X and elements $p \in X$. As usual we write '$p_n \rightarrow p$' instead of '$\rightarrow ((p_n), p)$'.
$(X, \rightarrow)$ is called a 'limit space' (short: 'L-space') if the following axioms are satisfied:

1) $p_n \rightarrow p$ implies that for every subsequence $(p_{k_n})\quad (k_1 < k_2 < \ldots)$ of (p_n) also $p_{k_n} \rightarrow p$;
2) if $p_n = p$ for almost all $n \in \mathbb{N}$, then $p_n \rightarrow p$;
3) if not $p_n \rightarrow p$, then there exists a sequence $k_1 < k_2 < \ldots$ such that no subsequence of (p_{k_n}) converges to p;
4) if $p_n \rightarrow p$ and $p_n \rightarrow q$, then $p = q$.

Definition 3.53. Let $(X, \rightarrow_X)$ and $(Y, \rightarrow_Y)$ be two L-spaces. A function $f : X \rightarrow Y$ is called continuous if $f(p_n) \rightarrow_Y f(p)$ whenever $p_n \rightarrow_X p$. The set of all continuous functions from X to Y is denoted by $\mathcal{C}(X,Y)$.

On $\mathcal{C}(X,Y)$ one can define the following relation of convergence

$$f_n \rightarrow f :\equiv \forall (p_n), p \big(p_n \rightarrow_X p \Rightarrow f_n(p_n) \rightarrow_Y f(p)\big).$$

Lemma 3.54. *$(\mathcal{C}(X,Y), \rightarrow)$ is again an* L*-space.*

Proof: Exercise! □

Definition 3.55 (Scarpellini [321]). The type-structure of sequentially continuous functionals is defined as follows:

$$\begin{cases} C_0 := \mathbb{N},\ p_n \to_0 p :\equiv \exists k \forall m > k(p_m = p); \\ C_{\tau\rho} := \mathscr{C}(C_\rho, C_\tau), \\ f_n \to_{\tau\rho} f :\equiv \forall (p_n) \in C_\rho^{\mathbb{N}}, p \in C_\rho \big(p_n \to_\rho p \Rightarrow f_n(p_n) \to_\tau f(p)\big). \\ \mathscr{C}^\omega := \langle C_\rho \rangle_{\rho \in \mathbf{T}}. \end{cases}$$

Remark 3.56. Since $(C_0, \to_0)$ clearly is an L-space, lemma 3.54 implies that $(C_\rho, \to_\rho)$ is an L-space for any $\rho \in \mathbf{T}$.

Proposition 3.57 (Scarpellini [321]). *$\mathscr{C}^\omega$ is a model of* E-PA$^\omega$.

Proof: For $0^0, \Pi, \Sigma, S^1$ this is rather straightforward. For the recursors one shows by induction on n that the functionals $\underline{R}n$ belong to $\mathscr{C}$ for all n. It then follows immediately that also the recursors $\underline{R}$ themselves belong to $\mathscr{C}$. For details see [321]. □

The model of all extensional hereditarily continuous functionals: A different type structure of continuous functionals, the so-called extensional hereditarily continuous functionals ECF$^\omega$, is based on a notion of 'continuous functional' due to [196] and [244]. As proved in [169], ECF$^\omega$ is in fact isomorphic to $\mathscr{C}^\omega$. The main proof-theoretic use of ECF$^\omega$ is due to the fact that the functionals in ECF$^\omega$ are represented by number theoretic functions $\alpha \in \mathbb{N}^\mathbb{N}$, their so-called associates, for which an equivalence relation $=_\rho$ is defined for each type so that the equivalence classes correspond to the functionals being represented. This makes it possible for formalize certain semantic arguments in systems with number and function quantifiers such as the so-called elementary intuitionistic analysis EL of Kreisel and Troelstra (see [366], which in turn is based on [197], for a comprehensive treatment of all this). We sketch here only the definition of ECF$^\omega$: The definition is based on two versions of so-called partial continuous function application (in the following $\alpha, \beta, \gamma, \ldots$ range over unary number theoretic functions and $x, y, z, \ldots$ over natural numbers):

Definition 3.58. We define $\alpha|\beta$ and $\alpha(\beta)$ by

$$(\alpha|\beta)(x) \simeq y :\equiv \alpha(\langle x \rangle * \overline{\beta}(\mu z[\alpha(\langle x \rangle * \overline{\beta}z) \neq 0])) \dot{-} 1 \simeq y,$$
$$\alpha(\beta) \simeq y :\equiv \alpha(\overline{\beta}(\mu z[\alpha(\overline{\beta}z) \neq 0])) \dot{-} 1 \simeq y.$$

$(\alpha|\beta)(x)$ and $\alpha(\beta)$ are partial recursive functionals in x, α, β resp. α, β. Thus, using some basic notation from ordinary recursion theory, there are codes $n_0, n_1 \in \mathbb{N}$ for corresponding oracle Turing machines such that

$$\{n_0\}(x, \alpha, \beta) \simeq y \leftrightarrow (\alpha|\beta)(x) \simeq y,$$
$$\{n_1\}(\alpha, \beta) \simeq y \leftrightarrow \alpha(\beta) \simeq y.$$

$\alpha|\beta \simeq \gamma$ is defined as $\forall x \in \mathbb{N}((\alpha|\beta)(x) \simeq \gamma(x))$.

Definition 3.59 (Kleene [195], Kreisel [244], Troelstra [366]). The type structure ECF$^\omega$ of all extensional hereditarily continuous functionals of finite type is defined by simultaneously declaring ECF_ρ and an extensional equality $=_\rho$:

$$\mathrm{ECF}_0 := \mathbb{N},\ \ x =_0 y :\equiv x,y \in \mathbb{N} \wedge x = y;$$

$$\mathrm{ECF}_{0(0)} := \mathbb{N}^{\mathbb{N}},\ \ \alpha =_1 \beta :\equiv \alpha,\beta \in \mathbb{N}^{\mathbb{N}} \wedge \forall x \in \mathbb{N}(\alpha(x) = \beta(x));$$

$$\mathrm{ECF}_{0(\rho)} :=$$

$$\{\alpha \in \mathbb{N}^{\mathbb{N}} : \forall \beta \in \mathrm{ECF}_\rho \exists x \in \mathbb{N}(\alpha(\beta) \simeq x) \wedge \forall \beta_1,\beta_2(\beta_1 =_\rho \beta_2 \to \alpha(\beta_1) \simeq \alpha(\beta_2))\},$$

$$\alpha_1 =_{0(\rho)} \alpha_2 :\equiv \alpha_1,\alpha_2 \in \mathrm{ECF}_{0(\rho)} \wedge \forall \beta \in \mathrm{ECF}_\rho(\alpha_1(\beta) \simeq \alpha_2(\beta)) \text{ for } \rho \neq 0;$$

$$\mathrm{ECF}_{\tau(0)} := \{\alpha \in \mathbb{N}^{\mathbb{N}} : \forall x \in \mathbb{N} \exists \gamma \in \mathrm{ECF}_\tau(\alpha|\lambda y.x \simeq \gamma)\},$$

$$\alpha_1 =_{\tau(0)} \alpha_2 :\equiv \alpha_1,\alpha_2 \in \mathrm{ECF}_{\tau(0)} \wedge \forall x \in \mathbb{N}(\alpha_1|\lambda y.x =_\tau \alpha_2|\lambda y.x) \text{ for } \tau \neq 0;$$

$$\mathrm{ECF}_{\tau(\rho)} :=$$

$$\{\alpha \in \mathbb{N}^{\mathbb{N}} : \forall \beta \in \mathrm{ECF}_\rho \exists \gamma \in \mathrm{ECF}_\tau(\alpha|\beta \simeq \gamma) \wedge \forall \beta_1,\beta_2(\beta_1 =_\rho \beta_2 \to \alpha|\beta_1 =_\tau \alpha|\beta_2)\},$$

$$\alpha_1 =_{\tau(\rho)} \alpha_2 :\equiv \alpha_1,\alpha_2 \in \mathrm{ECF}_{\tau(\rho)} \wedge \forall \beta \in \mathrm{ECF}_\rho(\alpha_1|\beta =_\tau \alpha_2|\beta) \text{ for } \rho,\tau \neq 0.$$

$$\mathrm{ECF}^\omega = \langle \mathrm{ECF}_\rho \rangle_{\rho \in \mathbf{T}}.$$

So in ECF^ω, the application operation $App_{\tau,\rho}$ between (representatives of) functionals in $\mathrm{ECF}_{\tau(\rho)}$ and ECF_ρ is not the set-theoretic application but defined depending on ρ,τ as follows

$$App_{0,0}(\alpha,x) := \alpha(x),$$
$$App_{0,\rho}(\alpha,\beta) := \alpha(\beta) \text{ for } \rho \neq 0,$$
$$App_{\tau,0}(\alpha,x) := \alpha|\lambda y.x \text{ for } \tau \neq 0,$$
$$App_{\tau,\rho}(\alpha,\beta) := \alpha|\beta \text{ for } \rho,\tau \neq 0.$$

The following proposition is shown in [366]:

Proposition 3.60. ECF^ω *is a model of* E-PA$^\omega$. *Moreover, as interpretations* $[t]_{\mathrm{ECF}^\omega}$ *for the closed terms* t^ρ $(\rho \neq 0)$ *of* E-PA$^\omega$ *one can take computable functions* $\alpha \in \mathrm{ECF}_\rho \cap REC$.

The model of all strongly majorizable functionals: The following type structure $\mathcal{M}^\omega$ of all strongly majorizable functionals was constructed in [27] making use of the variant *s-maj* of Howard's majorization relation *maj*.

Definition 3.61 (Bezem [27]). The type structure $\mathcal{M}^\omega$ of all hereditarily strongly majorizable set-theoretic functionals of finite type is defined as

$$\begin{cases} M_0 := \mathbb{N},\ \ n\ s\text{-}maj_0\ m :\equiv n \geq m \wedge n,m \in \mathbb{N}, \\ x^*\ s\text{-}maj_{\tau(\rho)}\ x :\equiv x^*,x \in M_\tau^{M_\rho} \wedge \forall y^*,y \in M_\rho(y^*\ s\text{-}maj_\rho\ y \to x^*y^*\ s\text{-}maj_\tau\ x^*y, xy), \\ M_{\tau(\rho)} := \left\{x \in M_\tau^{M_\rho} : \exists x^* \in M_\tau^{M_\rho}(x^*\ s\text{-}maj_{\tau(\rho)}\ x)\right\} \quad (\rho,\tau \in \mathbf{T}) \end{cases}$$

(Here $M_\tau^{M_\rho}$ denotes the set of all total set-theoretic mappings from M_ρ into M_τ). $\mathcal{M}^\omega := \langle M_\rho \rangle_{\rho \in \mathbf{T}}$.

Remark 3.62. An easy induction on ρ shows that for $x^*, x \in M_\rho$ the relation $s\text{-}maj_\rho$ as defined in definition 3.61 coincides with the interpretation of the corresponding syntactic relation from definition 3.48 in the model $\mathcal{M}^\omega$, i.e.

$$\forall x^*, x \in M_\rho (x^* \; s\text{-}maj_\rho \, x \leftrightarrow x^* [s\text{-}maj_\rho]_{\mathcal{M}^\omega} x).$$

The relation defined in definition 3.61, however, also applies to functionals which prima facie are only in say $M_\tau^{M_\rho}$ and is used to determine whether such a functional actual is in $M_{\tau(\rho)}$. In the following lemmas we always refer to the relation from definition 3.61.

Lemma 3.63. 1) $x^* \; s\text{-}maj_\rho \, x \rightarrow \; x \in M_\rho \wedge x^* \; s\text{-}maj_\rho \, x^* \rightarrow x^*, x \in M_\rho$.
2) *For all* $\rho = \tau\rho_k \ldots \rho_1 \;\; (k \geq 1)$ *and all* $x^*, x : M_{\rho_1} \rightarrow (M_{\rho_2} \rightarrow \ldots \rightarrow M_\tau) \ldots)$ *the following holds*

$$x^* \; s\text{-}maj_\rho \, x \leftrightarrow$$
$$\forall y_1^*, y_1, \ldots, y_k^*, y_k \Big(\bigwedge_{i=1}^{k} y_i^* \; s\text{-}maj_{\rho_i} \, y_i \rightarrow x^* y_1^* \ldots y_k^* \; s\text{-}maj_\tau \, x^* y_1 \ldots y_k, x y_1 \ldots y_k \Big).$$

Proof: 1) is proved by induction on ρ. 2) follows by induction on k using 1). □

Remark 3.64. The implication '$\leftarrow$' in the second claim of this lemma is used often to establish that a functional $x : M_{\rho_1} \rightarrow (M_{\rho_2} \rightarrow \ldots \rightarrow M_\tau) \ldots)$ actually is in $M_{\tau\rho_k \ldots \rho_1}$.

Definition 3.65. Let $x \in M_\rho^{M_0}$, where $\rho = 0\rho_k \ldots \rho_1$. Then we define

$$x^M(n) := \lambda \underline{v}. \max\{x i \underline{v} \; : \; i \leq n\},$$

where $\underline{v} = v_1^{\rho_1}, \ldots, v_k^{\rho_k}$.

Lemma 3.66. *Let* $x, \widehat{x} \in M_\rho^{M_0}$ *be such that*

$$\forall n \in \mathbb{N} \, (\widehat{x} n \; s\text{-}maj_\rho \, x n).$$

Then

$$\widehat{x}^M \; s\text{-}maj_{\rho 0} \, x^M, x$$

and hence $\widehat{x}^M, x^M, x \in M_{\rho 0}$.
In particular, this implies that $(\cdot)^M \; s\text{-}maj_{\rho 0(\rho 0)} \, (\cdot)^M \in M_{\rho 0(\rho 0)}$.
As a special case it follows that

$$x_1^* \; s\text{-}maj_\rho \, x_1 \wedge x_2^* \; s\text{-}maj_\rho \, x_2 \rightarrow \max_\rho(x_1^*, x_2^*) \; s\text{-}maj_\rho \; \max_\rho(x_1, x_2), x_1, x_2,$$

where $\max_\rho(x_1, x_2) := \lambda \underline{v}. \max_0(x_1 \underline{v}, x_2 \underline{v})$.

Proof: Let $\rho = 0\rho_k \ldots \rho_1$ and $y_1^*, y_1, \ldots, y_k^*, y_k$ be such that $\bigwedge_{i=1}^{k} (y_i^* \text{ s-maj}_{\rho_i}\ y_i)$. One easily shows by induction on n (using lemma 3.63.1) that

$$\forall n \forall m \le n(\widehat{x}^M n\underline{y}^* \ge_0 \widehat{x}^M m\underline{y}, x^M m\underline{y}, xm\underline{y})$$

which by lemma 3.63.2) yields the claim. □

Remark 3.67. Note that $(\cdot)^M$ is definable as a closed term of type $\rho 0(\rho 0)$ already in $\mathrm{G_1A}^\omega$.

Corollary 3.68. $M_\rho^{M_0} = M_{\rho 0}$ *for each* $\rho \in \mathbf{T}$.

Proof: $x \in M_\rho^{M_0}$ implies that for each $n \in \mathbb{N}$ there exists an $x_n^* \in M_\rho$ with

$$x_n^* \ s\text{-}maj_\rho\ x(n).$$

Using $\mathrm{AC}^{0,\rho}$ on the meta-level we obtain a sequence $x^* \in M_\rho^{M_0}$ with

$$\forall n \in \mathbb{N}\ (x^* n\ s\text{-}maj_\rho\ xn).$$

Lemma 3.66 now yields that $(x^*)^M\ s\text{-}maj_{\rho 0}\ x \in M_{\rho 0}$. □

Proposition 3.69 (Bezem [27]). $\mathcal{M}^\omega$ *is a model of* E-PA$^\omega$.

Proof: Using lemma 3.63.2) it is immediate that $0, S, \Pi, \Sigma$ all majorize themselves and hence belong to $\mathcal{M}^\omega$. For the recursors $\underline{R}_{\underline{\rho}}$ one shows by induction on n that for all $\underline{y}^*, \underline{y} \in M_{\underline{\rho}}$ with $\underline{y}^*\ s\text{-}maj_{\underline{\rho}}\ \underline{y}$, i.e. $\bigwedge_{i=1}^{k} (y_i^*\ s\text{-}maj_{\rho_i}\ y_i)$ and $\underline{z}^*, \underline{z} \in M_{\underline{\rho}0\underline{\rho}}$ with $\underline{z}^*\ s\text{-}maj_{\underline{\rho}0\underline{\rho}}\ \underline{z}$ we have

$$(*)\ \forall n \in \mathbb{N}\ (\underline{R}_{\underline{\rho}} n\underline{y}^*\underline{z}^*\ s\text{-}maj_{\underline{\rho}}\ \underline{R}_{\underline{\rho}} n\underline{y}\underline{z}).$$

Let $n = 0$: Then $\underline{R}_{\underline{\rho}} 0\underline{y}^*\underline{z}^* = \underline{y}^*\ s\text{-}maj_{\underline{\rho}}\ \underline{y} = \underline{R}_{\underline{\rho}} 0\underline{y}\underline{z}$.

$n \mapsto n+1$: $\underline{R}_{\underline{\rho}}(n+1)\underline{y}^*\underline{z}^* = \underline{z}^*(\underline{R}_{\underline{\rho}} n\underline{y}^*\underline{z}^*)n \overset{3.63, I.H.}{s\text{-}maj_{\underline{\rho}}} \underline{z}(\underline{R}_{\underline{\rho}} n\underline{y}\underline{z})n = \underline{R}_{\underline{\rho}}(n+1)\underline{y}\underline{z}$.
This finishes the proof of $(*)$. Lemma 3.63.2 yields that

$$\forall n \in \mathbb{N}(\underline{R}_{\underline{\rho}} n\ s\text{-}maj\ \underline{R}_{\underline{\rho}} n).$$

By lemma 3.66 it now follows that

$$(R_i)_{\underline{\rho}}^* := (R_i)_{\underline{\rho}}^M\ s\text{-}maj\ (R_i)_{\underline{\rho}} \in M_{\rho_i(\underline{\rho}0\underline{\rho})\underline{\rho}0}.$$

□.

The next proposition shows that the three models $\mathcal{S}^\omega, \mathcal{C}^\omega$ and $\mathcal{M}^\omega$ start to differ from type 2 on:

Proposition 3.70. 1) $S_1 = C_1 = \mathrm{ECF}_1 = M_1 = \mathbb{N}^{\mathbb{N}}$.
2) $C_2 \subset M_2 \subset S_2$, *where both '$\subset$' are strict.*

Proof: 1) By definition we have that $S_0 = C_0 = \mathrm{ECF}_0 = M_0 = \mathbb{N}$ and $\mathrm{ECF}_1 = \mathbb{N}^{\mathbb{N}}$. It is trivial to observe that every $f \in \mathbb{N}^{\mathbb{N}}$ is continuous in the sense of $\mathscr{C}^\omega$. Since $\forall n \in \mathbb{N}\ (f(n) \geq f(n))$ we get by lemma 3.66 that f^M *s-maj*$_1$ $f \in M_1$.

2) We first show the inclusions. Since $M_2 \subseteq M_0^{M_1} \stackrel{1)}{=} S_0^{S_1} = S_2$ the second inclusion is trivial. Now let $x \in C_2$. Then (by 1) $x \in M_0^{M_1}$. We have to construct a majorant $x^* \in M_0^{M_1}$: $x \in C_2$ implies that x is a continuous function $\mathbb{N}^{\mathbb{N}} \to \mathbb{N}$ between the Baire space $\mathbb{N}^{\mathbb{N}}$ (with its usual metric) and $\mathbb{N}$ with the discrete metric. Now let $f \in \mathbb{N}^{\mathbb{N}}$. Then $K_f := \{g \in \mathbb{N}^{\mathbb{N}} : \forall n \in \mathbb{N}\ (g(n) \leq f(n))\}$ is a compact subspace of the Baire space. Hence x is uniformly continuous on K_f and so the following definition is well-defined: $x^*(f) := \max\{x(g) : g \in K_f\}$. We claim that x^* *s-maj*$_2$ x. Let f *s-maj*$_1$ g. Then $f(n) \geq g(n)$ for all $n \in \mathbb{N}$. Hence $g \in K_f$ and so $x^*(f) \geq x(g)$. Moreover, $x^*(f) \geq x^*(g)$ since $K_g \subseteq K_f$.
To show that $S_2 \not\subseteq M_2$ consider $x \in S_2$ defined as follows

$$x(f) := \begin{cases} n, & \text{for the least } n \text{ such that } f(n) = 0 \text{ if existent} \\ 0, & \text{otherwise.} \end{cases}$$

Suppose x^* *s-maj*$_2$ x for some $x^* \in S_2$. Consider the function $1_n(k) := 1$ for $k < n$ and $1_n(k) := 0$ for $k \geq n$ and let 1^1 be the constant-1 function. Then 1 *s-maj*$_1$ 1_n and hence $x^*(1^1) \geq x(1_n)$ for all $n \in \mathbb{N}$, but $x(1_n) = n$ which is a contradiction.
To show that $M_2 \not\subseteq C_2$ consider $x \in M_2$ defined as follows

$$x(f) := \begin{cases} 1, & \text{if } \exists n \in \mathbb{N}(f(n) = 0) \\ 0, & \text{otherwise.} \end{cases}$$

Clearly, $x \in M_2$ since x is majorized by the constant-1 functional. But also clearly $x \notin C_2$. $\square$

Let $\mathscr{T}^\omega$ be a model of E-PA$^\omega$ and t a closed term of E-PA$^\omega$. Then we use $[t]_{\mathscr{T}^\omega}$ to denote the interpretation of t in $\mathscr{T}^\omega$. If t has a type of degree ≤ 2, then – intuitively – it is clear (using proposition 3.70.1) that $[t]_{\mathscr{S}^\omega} = [t]_{\mathscr{C}^\omega} = [t]_{\mathscr{M}^\omega}$. The following proposition confirms this:

Proposition 3.71. *Let t^2 be a closed term of* E-PA$^\omega$ *of type* 2
Then $[t]_{\mathscr{S}^\omega}(f) = [t]_{\mathrm{ECF}^\omega}(f) = [t]_{\mathscr{C}^\omega}(f) = [t]_{\mathscr{M}^\omega}(f)$ for all $f \in \mathbb{N}^{\mathbb{N}}$, where in the case of ECF$^\omega$ *'$\cdot(\cdot)$' refers to the partial continuous function application.*
This result also holds for closed terms t^ρ whose type ρ is of degree ≤ 2.

Proof: We give the details for $[t]_{\mathscr{S}^\omega}(f) = [t]_{\mathscr{C}^\omega}(f)$ (the claims $[t]_{\mathscr{S}^\omega}(f) = [t]_{\mathscr{M}^\omega}(f)$ and $[t]_{\mathscr{S}^\omega}(f) = [t]_{\mathrm{ECF}^\omega}(f)$ are proved analogously). We define a so-called logical relation (first studied systematically by G.Plotkin) $\approx_\rho$ by induction on ρ as follows:

$$x \approx_0 y :\equiv x, y \in S_0 = C_0 \wedge x = y,$$
$$x \approx_{\tau\rho} y :\equiv x \in S_{\tau\rho} \wedge y \in C_{\tau\rho} \wedge \forall u \in S_\rho, v \in C_\rho (u \approx_\rho v \to xu \approx_\tau yv).$$

One easily shows for all constants c^ρ of E-PA$^\omega$ that

$$[c]_{\mathscr{S}^\omega} \approx_\rho [c]_{\mathscr{C}^\omega}.$$

This immediately implies that for all closed terms t

$$[t]_{\mathscr{S}^\omega} \approx_\rho [t]_{\mathscr{C}^\omega}.$$

Now let $deg(\rho) \leq 2$. It is clear that it suffices to consider the case $\rho = 2$. By proposition 3.70.1) we have that

$$x \approx_1 y \leftrightarrow x, y \in S_1 = M_1 = \mathbb{N}^{\mathbb{N}} \wedge \forall n \in \mathbb{N} (x(n) = y(n)).$$

In particular, $x \approx_1 x$ for all $x \in \mathbb{N}^{\mathbb{N}}$. Hence $[t]_{\mathscr{S}^\omega} \approx_2 [t]_{\mathscr{C}^\omega}$ implies that

$$\forall x \in \mathbb{N}^{\mathbb{N}} ([t]_{\mathscr{S}^\omega}(x) = [t]_{\mathscr{C}^\omega}(x)).$$

$\square$

Remark 3.72. For $\mathscr{C}^\omega, \mathscr{M}^\omega, \mathscr{S}^\omega$, where we still have at least inclusions for the type 2, the proof above can even be used to show that for closed terms t^3 of E-PA$^\omega$ of type 3 we have

$$\forall x \in C_2 \subset M_2 \subset S_2 ([t]_{\mathscr{S}^\omega}(x) = [t]_{\mathscr{C}^\omega}(x)) = [t]_{\mathscr{M}^\omega}(x))$$

and

$$\forall x \in M_2 \subset S_2 ([t]_{\mathscr{S}^\omega}(x) = [t]_{\mathscr{M}^\omega}(x)).$$

Proof: Exercise!

3.7 Exercises, historical comments and suggested further reading

Exercises:

1) Convince yourself that the axioms and rules of IL$_{=}$ are sound under the BHK interpretation.
2) Convince yourself that $\neg\neg A \to A$ in general is not valid under the BHK interpretation.
3) Show that IA is derivable from IR. Compare the complexity of the induction formula $\tilde{A}(x)$ of the IR-instance needed to prove an IA-instance with induction formula $A(x)$ with that of $A(x)$.
4) ([161]) It is known that the function $\alpha : \mathbb{N} \times \mathbb{N} \to \mathbb{N}$ defined by the equations

$$(*)\begin{cases}\alpha(0,y)=y'\\ \alpha(x',0)=\alpha(x,1)\\ \alpha(x',y')=\alpha(x,\alpha(x',y))\end{cases}$$

is not primitive recursive (in the sense of Kleene). In fact, α is a variant due to R. Peter of the well-known Ackermann function.

Show that α is definable in WE-HA$^\omega$ by a closed term $t^{0(0)(0)}$ using R_1 (i.e. WE-HA$^\omega$ proves the equations $(*)$ for t).

5) Prove the statements in remark 3.11.2) and 3.11.3).
6) Prove the statement in remark 3.14.
7) Prove proposition 3.22.
8) The so-called bounded collection principle CP is given by

$$\text{CP: } \forall a^0(\forall x <_0 a \exists y^0 A(x,y) \to \exists z^0 \forall x <_0 a \exists y <_0 z A(x,y)),$$

where A is an arbitrary formula with arbitrary parameters allowed (including a) but with z not free in A.

a. Prove WE-HA$^\omega \vdash$ CP.
b. Proved $\widehat{\text{WE-HA}}^\omega\restriction$+QF-AC$^{0,0} \vdash \Sigma^0_1$-CP, where Σ^0_1-CP is the restriction of CP to Σ^0_1-formulas (with arbitrary parameters of arbitrary types).
Show the same result for G$_n$A$^\omega_i$ instead of $\widehat{\text{WE-HA}}^\omega\restriction$.

9) Prove the claim at the end of section 3.4.
10) Show that WE-PA$^\omega$+QF-AC^{0,1} proves the following: if Φ^2 restricted to $\{f : f \leq_1 h\}$ is uniformly continuous (for any h), then Φ is pointwise continuous everywhere, i.e.

$$\forall\Phi^2\Big(\forall h^1\exists n^0\forall f,g \leq_1 h(\overline{f}n = \overline{g}n \to \Phi f = \Phi g) \to \\ \forall f^1\exists n^0\forall g^1(\overline{f}n = \overline{g}n \to \Phi f = \Phi g)\Big).$$

Analogously for $\Phi^{1(1)}$.

11) Show that G$_3$A$^\omega$+(IPP) proves the following finite form of Σ^0_1-comprehension

$$\forall n\exists j\forall i < n\big((j)_i = 0 \leftrightarrow \exists y A_0(i,y)\big),$$

where A_0 is quantifier-free and may contain arbitrary parameters. Hint: Let $\nu(x)$ be the code of the binary sequence of length n satisfying

$$(\nu(x))_i = 0 \leftrightarrow \exists y \leq x A_0(i,y)$$

for all $i < n$. Now define the coloring $f(x) := \nu(x)$ of $\mathbb{N}$ by $\leq \overline{1}n$-colors.

12) Use the previous exercise to show that

$$\text{G}_3\text{A}^\omega\text{+(IPP)} \vdash \Sigma^0_1\text{-IA}.$$

13) Prove lemma 3.37 (Hint: see [207] where this lemma is proved for s-maj instead of maj. With minor modifications the proof applies to maj as well).
14) Prove proposition 3.50.
15) Prove lemma 3.54.
16) Fill in the details of the proof of proposition 3.57.
17) Define the functional

$$\Phi(z^2, x^1, y^1) := \min n^0[x(n) =_0 y(n) \rightarrow zx =_0 zy].$$

Show that $\Phi \in \mathscr{C}^\omega$ and $\Phi \in \mathscr{S}^\omega$ but $\Phi \notin \mathscr{M}^\omega$ ([163]).

18) Show that in ECF^ω the so-called fan functional

$$\Phi_{\mathrm{FAN}}(x^2) := \min n\ [\forall y_1, y_2 \in B(\forall i < n(y_1 i = y_2 i) \rightarrow xy_1 = xy_2)],\ \text{where}$$

$B := \{x \in \mathbb{N}^{\mathbb{N}} : \forall n \in \mathbb{N}(xn \leq 1)\}$, exists and even has a recursive associate α (such functionals are called recursively countable; as shown by Tait (unpublished) and Gandy/Hyland [111] Φ_{FAN} is not Kleene computable, in the sense of his schemata S1-S9 [195], over ECF^ω).

19) For x^1 define

$$\overline{(x, n^0)}(k^0) := \begin{cases} xk, & \text{if } k < n \\ 0^0, & \text{otherwise} \end{cases}$$

and

$$\mu_1(x^2, y^1) := \begin{cases} \min n^0[x(\overline{y,n}) =_0 xy], & \text{if } \exists n(x(\overline{y,n}) =_0 xy) \\ \text{undefined}, & \text{otherwise.} \end{cases}$$

Show that $\mu_1 \in \mathscr{C}^\omega$, but $\mu_1 \notin \mathscr{M}^\omega$ and $\mu_1 \notin \mathscr{S}^\omega$.

20) Define

$$\mu_1^l(x^2, y^1) := \begin{cases} \min n^0[x(\overline{y,n}) =_0 xy], & \text{if } \exists n(x(\overline{y,n}) =_0 xy) \\ 0^0 \text{ otherwise.} \end{cases}$$

Show that $\mu_1 \in \mathscr{S}^\omega$ and $\mu_1 \in \mathscr{C}^\omega$, but $\mu_1 \notin \mathscr{M}^\omega$.

21) Define

$$\mu_2(x^2, y^1) := \begin{cases} \min n^0[x(\overline{y,n}) < n], & \text{if } \exists n(x(\overline{y,n}) < n) \\ \text{undefined}, & \text{otherwise.} \end{cases}$$

Show that $\mu_2 \in \mathscr{M}^\omega$ and $\mu_2 \in \mathscr{C}^\omega$, but $\mu_2 \notin \mathscr{S}^\omega$

22) Show that $\mathscr{C}^\omega \models \mathrm{QF\text{-}AC}^{1,0}$, where

$$\mathrm{QF\text{-}AC}^{1,0} : \forall f^1 \exists x^0 A_0(f, x) \rightarrow \exists F^2 \forall f^1 A_0(f, F(f))$$

with A_0 quantifier-free.

23) Show that $\mathscr{M}^\omega \not\models \mathrm{QF\text{-}AC}^{1,0}$ ([201]).

24) Show the following 'weak continuity' property of $\mathcal{M}^\omega$ ([22]):

$$\mathcal{M}^\omega \models \forall y^{0(\rho 0)}, x^{\rho 0} \exists n^0 \forall \tilde{x}^{\rho 0} \big(\overline{x,n} =_{\rho 0} \overline{\tilde{x},n} \to y\tilde{x} \leq n \big).$$

Use this to give an alternative proof of the previous exercise.

25) Prove the claim in remark 3.72.

Historical comments and suggested further reading:
For an introduction to intuitionistic logic and Heyting arithmetic as well as constructivism in general we refer to van Dalen [372] (chapter 5) and the first volume of Troelstra-van Dalen [371]. Most of the material on HA and its finite type extensions is taken from Troelstra [366]. Primitive recursion in higher types (in the sense of the recursors R_ρ was first considered in Hilbert [161] and Gödel [133]. The fragments $\widehat{\text{(W)E-PA}}^\omega\!\restriction$ are studied in Feferman [98]. For pure types the functionals definable in these fragments correspond to the primitive recursive functionals defined by the schemata S1-S8 of Kleene [195]. The fragments $\text{G}_n\text{A}^\omega$ were introduced and studied first in Kohlenbach [207] from which most material of the corresponding section is taken. See Troelstra [365] for further information on the BHK-interpretation. A thorough treatment of the Curry-Howard isomorphism can be found in Sørensen-Urzyczyn [349]. For more information on (W)E-HA$^\omega$ and its numerous variants see Troelstra [366] and Troelstra-van Dalen [371].
The model $\mathcal{C}^\omega$ of sequentially continuous functionals was defined first in Scarpellini [321] (with further refinements in Scarpellini [322]) as a model of bar recursion (see chapter 11). The model ECF$^\omega$ of all extensional hereditarily continuous functionals is due (independently) to Kleene [196] and Kreisel [244]. For uses of ECF$^\omega$ in the proof theory of systems of intuitionistic analysis see again Troelstra [366]. The precise relationship between ECF$^\omega$ and $\mathcal{C}^\omega$ was clarified in Hyland [169]. For a comprehensive treatment of the recursion theory for ECF$^\omega$ see Normann [287]. The model $\mathcal{M}^\omega$ of all hereditarily strongly majorizable functionals is due to Bezem [27] (see also chapter 11).

Chapter 4
Representation of Polish metric spaces

4.1 Representation of real numbers

In this chapter we develop the so-called standard representation of complete separable metric spaces X, so-called 'Polish' metric spaces. Via this representation, elements of X are represented by number theoretic functions, i.e. objects f of type-1. Moreover, we will show that the representation can be arranged in such a way that **every** function f^1 represents a unique element in X. In general, an element in X will have many representatives f. On the representatives we define an equivalence relation $f_1 =_X f_2 :\equiv (f_1, f_2$ represent the same X-element). Instead of having explicitly to introduce elements of X as equivalence classes of representatives, we can use the representatives themselves and then state that the function or predicate in question respects the equivalence relations. E.g. a function $F : X \to Y$ between two Polish spaces represented in this way is just a functional $\Phi^{1(1)}$ satisfying

$$\forall f_1^1, f_2^1 (f_1 =_X f_2 \to \Phi(f_1) =_Y \Phi(f_2)).$$

For **compact** metric spaces K we can arrange that the elements $x \in K$ are represented already by functions $f \leq_1 M$ which are bounded by a fixed function M depending on K only. This representation goes back to L.E.J. Brouwer (see also the historical comments at the end of this chapter). Again we can achieve that **every** $f \leq_1 M$ represents a unique element in K.

The main benefit of this representation is that quantification of the kind $\forall/\exists x \in X$ and $\forall/\exists x \in K$ can be expressed in our typed systems just as quantification $\forall/\exists f^1$ and $\forall/\exists f \leq_1 M$ without introducing any further quantifiers which we would have to do if not every function f could be viewed as representing an element in X. In the latter case '$\forall x \in X(\ldots)$' would translate into

$$\forall f^1 (\text{if } f \text{ represents an element of } X \to \ldots)$$

with further quantifiers added by the implicative premise.

Our treatment (taken from [204] (section 3)) is based on ideas and constructions

from [15, 57] and [371] and – for a comprehensive treatment of a general theory of representations – [377]).
We develop this representation in some detail since, for the application of our proof-theoretical extraction of effective data, the logical form of the representation is crucial. As we will in this book mainly use the systems (W)E-PA$^\omega$, $\widehat{\text{(W)E-PA}}^\omega\!\upharpoonright$ and their intuitionistic variants, we carry out the representation only for $\widehat{\text{WE-HA}}^\omega\!\upharpoonright$ but not for weaker fragments. With minor changes the representation can also be done already in $G_3A_i^\omega$ but for $G_2A_i^\omega$ things have to be changed significantly (see [209] for a treatment of reals and some function spaces in such a weak setting).

Real numbers in $\widehat{\text{WE-HA}}^\omega\!\upharpoonright$

As common in computable ([377]) and constructive analysis ([32]) as well as complexity theory on the reals ([199]) and also reverse mathematics ([338]), real numbers are represented as Cauchy sequences of rational numbers with fixed given Cauchy modulus. It will be convenient to use 2^{-n} as Cauchy modulus. In order to carry out this representation we first have to define the ordered field $(\mathbb{Q},+,\cdot,0,1,<)$ of rational numbers within $\widehat{\text{WE-HA}}^\omega\!\upharpoonright$:
Rational numbers are represented as codes $j(n,m)$ of pairs (n,m) of natural numbers (i.e. type-0 objects): $j(n,m)$ represents the rational number $\frac{\frac{n}{2}}{m+1}$ if n is even, and the negative rational number $-\frac{\frac{n+1}{2}}{m+1}$ otherwise. Here we use the surjective Cantor pairing function j from definition 3.30. So each natural number can be conceived as code of a uniquely determined rational number. On the representatives of the rational numbers, i.e. on $\mathbb{N}$, we define an equivalence relation $n_1 =_{\mathbb{Q}} n_2$ which expresses that n_1 and n_2 represent the same rational number:

$$n_1 =_{\mathbb{Q}} n_2 :\equiv \frac{\frac{j_1 n_1}{2}}{j_2 n_1 + 1} = \frac{\frac{j_1 n_2}{2}}{j_2 n_2 + 1}$$

in the case where both $j_1 n_1$ and $j_1 n_2$ are even and analogous in the remaining cases. For $b,d > 0$ here $\frac{a}{b} = \frac{c}{d}$ is defined to hold iff $ad =_0 cb$ when $bd > 0$.
In order to express the statement that n represents the rational r, we write $n =_{\mathbb{Q}} \langle r \rangle$ or simply $n = \langle r \rangle$. Of course $\langle \cdot \rangle$ is not a function of r since r possesses infinitely many representatives. Rational numbers are, strictly, speaking equivalence classes on $\mathbb{N}$ w.r.t. $=_{\mathbb{Q}}$. By using only their representatives and $=_{\mathbb{Q}}$ one can avoid formally introducing the set $\mathbb{Q}$ of all these equivalence classes. Alternatively one could always select a canonical representative by stipulating

$$c^1(n) :=_0 \min m \leq_0 n[n =_{\mathbb{Q}} m]$$

which is the code of the maximally simplified fraction representing the rational number encoded by n. Clearly $c(n) =_{\mathbb{Q}} n$ and $n =_{\mathbb{Q}} m \to c(n) =_0 c(m)$. However, the use of general representatives is convenient and anyway will be unavoidable for an effective representation of real numbers.

On $\mathbb{N}$ one can easily define primitive recursive operations $+_{\mathbb{Q}}, \cdot_{\mathbb{Q}}$ and predicates $<_{\mathbb{Q}}$, $\leq_{\mathbb{Q}}$ (in the sense of Kleene) such that e.g. $\langle r_1 \rangle +_{\mathbb{Q}} \langle r_2 \rangle =_{\mathbb{Q}} \langle r_3 \rangle$ iff $r_1 + r_2 = r_3$ for the rational numbers r_1, r_2, r_3 which are represented by $\langle r_1 \rangle, \langle r_2 \rangle, \langle r_3 \rangle$ (analogous for $\cdot_{\mathbb{Q}}, <_{\mathbb{Q}}, \leq_{\mathbb{Q}}$). In view of our representation of $\mathbb{Q}$ we have the following embedding of $\mathbb{N}$ into $\mathbb{Q}$:

$$n \mapsto \langle n \rangle := j(2n, 0); 0_{\mathbb{Q}} := \langle 0 \rangle, 1_{\mathbb{Q}} := \langle 1 \rangle.$$

It can easily be verified (within $\widehat{\text{WE-HA}}^{\omega}\upharpoonright$) that $(\mathbb{N}, +_{\mathbb{Q}}, \cdot_{\mathbb{Q}}, 0_{\mathbb{Q}}, 1_{\mathbb{Q}}, <_{\mathbb{Q}})$ is an ordered field (which represents $(\mathbb{Q}, +, \cdot, 0, 1, <)$ in $\widehat{\text{WE-HA}}^{\omega}\upharpoonright$).

Using the above representation, each function $f : \mathbb{N} \to \mathbb{N}$ (i.e. each functional of type 1) can be conceived of as an infinite sequence of codes of rationals and, therefore, as a representative of an infinite sequence of rationals.

We now officially represent real numbers by functions f^1 satisfying

$$(*)\ \forall n \big(|fn -_{\mathbb{Q}} f(n+1)|_{\mathbb{Q}} <_{\mathbb{Q}} \langle 2^{-n-1} \rangle \big).$$

$(*)$ implies that for all n, k, m with $k > m \geq n$:

$$|fm -_{\mathbb{Q}} fk|_{\mathbb{Q}} \leq_{\mathbb{Q}} \sum_{i=m}^{k-1} |fi -_{\mathbb{Q}} f(i+1)|_{\mathbb{Q}} \leq_{\mathbb{Q}} \sum_{i=n}^{\infty} |fi -_{\mathbb{Q}} f(i+1)|_{\mathbb{Q}} < \langle 2^{-n} \rangle,$$

i.e. each f which satisfies $(*)$ in fact represents a Cauchy sequence of rationals with Cauchy modulus 2^{-n}. Condition $(*)$ is particularly convenient to define a construction which guarantees that each function f represents some real number: define the following functional $f \mapsto \widehat{f}$ which is primitive recursive (in the sense of [194] and hence can be carried out in $\widehat{\text{WE-HA}}^{\omega}$):

$$(**)\ \widehat{f}n := \begin{cases} fn \text{ if } \forall k < n \big(|fk -_{\mathbb{Q}} f(k+1)|_{\mathbb{Q}} <_{\mathbb{Q}} \langle 2^{-k-1} \rangle \big), \\ fk \text{ for the least } k < n \text{ with } |fk -_{\mathbb{Q}} f(k+1)|_{\mathbb{Q}} \geq_{\mathbb{Q}} \langle 2^{-k-1} \rangle \text{ otherwise.} \end{cases}$$

$\widehat{f}$ always satisfies $(*)$. If $(*)$ is already valid for f then $\forall n (fn =_0 \widehat{f}n)$. Thus each function f^1 codes a uniquely determined real number, namely the real number which is given by the Cauchy sequence coded by $\widehat{f}$. In the other direction, if f represents a Cauchy sequence of rationals with modulus 2^{-n}, then $gn := f(n+1)$ satisfies $(*)$ and, therefore, represents the real number, given by f, in our sense. Hence nothing is lost by our restriction of sequences satisfying $(*)$ and the construction $\widehat{f}$ makes it possible to reduce quantification over $\mathbb{R}$ to $\forall f^1$ resp. $\exists f^1$ without adding further quantifiers. Likewise for the operations on $\mathbb{R}$ below, we do not have to assume that f_1, f_2 represent real numbers but directly can formulate the operations on the level of $\widehat{f_1}, \widehat{f_2}$.

Remark 4.1. The construction $f \mapsto \widehat{f}$ can even be carried out in $G_3A_i^{\omega}$ and for a representation of real numbers based on the rate of convergence $1/k$ instead of 2^{-k} even in $G_2A_i^{\omega}$ (see e.g. [207]).

On the representatives (in the sense above) of real numbers (i.e. on the functionals of type 1) f_1, f_2 we define an equivalence relation $=_{\mathbb{R}}$ by

$$f_1 =_{\mathbb{R}} f_2 :\equiv \forall n(|\widehat{f_1}(n+1) -_{\mathbb{Q}} \widehat{f_2}(n+1)|_{\mathbb{Q}} <_{\mathbb{Q}} \langle 2^{-n}\rangle).$$

$f_1 =_{\mathbb{R}} f_2$ holds iff f_1 and f_2 represent the same real number (w.r.t. the usual identity relation on the reals).
This time there is no continuous (and hence no computable) way of selecting a canonical representative of a real number given an arbitrary representative of that real, i.e. there is no continuous (in the sense of Baire space) $c^{1(1)}$ with $c(f) =_{\mathbb{R}} f$ and $f_1 =_{\mathbb{R}} f_2 \to c(f_1) =_1 c(f_2)$ for all f^1, f_1^1, f_2^1 (exercise).
In contrast to $=_{\mathbb{Q}}$, the relation $=_{\mathbb{R}}$ is not decidable but in Π^0_1.

$$f_1 <_{\mathbb{R}} f_2 :\equiv \exists n(\widehat{f_2}(n+1) -_{\mathbb{Q}} \widehat{f_1}(n+1) \geq_{\mathbb{Q}} \langle 2^{-n}\rangle) \in \Sigma^0_1,$$

$$f_1 \leq_{\mathbb{R}} f_2 :\equiv \neg(f_2 <_{\mathbb{R}} f_1) \in \Pi^0_1.$$

It is not difficult to define functionals $+_{\mathbb{R}}, -_{\mathbb{R}}, \cdot_{\mathbb{R}}$ etc. on our codes of real numbers, which represent the elementary operations $+,-,\cdot$ etc. on $\mathbb{R}$: For example, define $f_1 +_{\mathbb{R}} f_2$ by

$$(f_1 +_{\mathbb{R}} f_2)(k) := \widehat{f_1}(k+1) +_{\mathbb{Q}} \widehat{f_2}(k+1).$$

By applying $\widehat{f_i}$ to $k+1$ instead of k it is ensured that $f_1 +_{\mathbb{R}} f_2$ has the right speed of convergence to pass $(**)$ unchanged. So it is clear that $f_1 +_{\mathbb{R}} f_2 =_{\mathbb{R}} f_3$ holds iff $x_1 + x_2 = x_3$ for the real numbers x_1, x_2, x_3 which are represented by f_1, f_2, f_3. $+_{\mathbb{R}}$ is a functional of type $1(1)(1)$. $-_{\mathbb{R}}$ is defined analogously.
If $n = \langle r\rangle$ codes the rational number r, then $\lambda k.n$ represents r as a real number. $0_{\mathbb{R}} := \lambda k.0_{\mathbb{Q}}, 1_{\mathbb{R}} := \lambda k.1_{\mathbb{Q}}$.
$\mathbb{R}$ denotes the set of all equivalence classes on the set of functions f w.r.t. $=_{\mathbb{R}}$. As in the case of $\mathbb{Q}$, we use $\mathbb{R}$ only informally and deal exclusively with the representatives and the operations defined on them. $(\mathbb{N}^{\mathbb{N}}, +_{\mathbb{R}}, \cdot_{\mathbb{R}}, 0_{\mathbb{R}}, 1_{\mathbb{R}}, <_{\mathbb{R}})$ is an Archimedean ordered field (provable in $\widehat{\text{WE-HA}}^{\omega}\!\upharpoonright$), which represents $(\mathbb{R}, +, \cdot, 0, 1, <)$ in $\widehat{\text{WE-HA}}^{\omega}\!\upharpoonright$.
One easily verifies the following fact which expresses that $\widehat{f}(k)$ is a rational 2^{-k}-approximation of f:

Lemma 4.2. $\widehat{\text{WE-HA}}^{\omega}\!\upharpoonright \vdash \forall k(|f -_{\mathbb{R}} \lambda n.\widehat{f}(k)|_{\mathbb{R}} <_{\mathbb{R}} \langle 2^{-k}\rangle)$.

Each functional $\Phi^{1(0)}$ can be conceived of as an infinite sequence of codes of real numbers and, therefore, as a representative of a sequence of real numbers. We have the following Cauchy completeness:

Lemma 4.3. $\widehat{\text{WE-HA}}^{\omega}\!\upharpoonright$ *proves that*

$$\forall \Phi^{1(0)}(\forall n; m,k \geq n(|\Phi m -_{\mathbb{R}} \Phi k|_{\mathbb{R}} \leq_{\mathbb{R}} \langle 2^{-n}\rangle) \to \exists f^1 \forall n(|\Phi n -_{\mathbb{R}} f|_{\mathbb{R}} \leq_{\mathbb{R}} \langle 2^{-n}\rangle)).$$

In fact, f can be defined as

$$fk := \widehat{\Phi(k+3)}(k+3).$$

4.2 Representation of complete separable metric ('Polish') spaces

Complete separable metric spaces, which we call: CSM-spaces or 'Polish' (metric) spaces, are represented as completions $(\widehat{X},\widehat{d})$ of countable metric spaces (X,d). Such countable spaces are given as follows: Assume that the elements of X are coded by natural numbers and that f_X is a (prim. rec.) enumeration of the set $\langle X\rangle$ of all these codes. (X,d) can now be represented by a pseudo metric d_X (more precisely: a functional of type $1(0)(0)$, which represents a pseudo metric) on $\mathbb{N}$ such that $d_X(n,m) =_{\mathbb{R}} \langle d(x,y)\rangle$, where $f_X n$ and $f_X m$ are codes of $x,y \in X$ and $\langle d(x,y)\rangle$ is a representative of the real number $d(x,y)$. In general, d_X will not be a metric on $\mathbb{N}$, i.e. $d_X(n,m) =_{\mathbb{R}} 0_{\mathbb{R}} \not\to n =_0 m$.
In the following we assume that $d_X(n,m) =_1 \widehat{d_X(n,m)}$ (for otherwise we simply could define $d'_X(n,m) := \widehat{d_X(n,m)}$). The completion $(\widehat{X},\widehat{d})$ of (X,d) is now represented as the completion of $(\mathbb{N},d_X)$: An element of this completion is given by a function h satisfying

$$(*)\ \forall n\big(d_X(hn,h(n+1)) <_{\mathbb{R}} \langle 7\cdot 2^{-n-1}\rangle\big).$$

$(*)$ implies that $(hn)_{n\in\mathbb{N}}$ is a Cauchy sequence in $(\mathbb{N},d_X)$ with modulus 2^{-n+3}. As for the representation of $\mathbb{R}$ above, we want to have that **each** function h represents a (uniquely determined) element of the completion. If we would try to define $\widehat{h}$ in the same way as we defined $\widehat{f}$ used in the representation of $\mathbb{R}$ above, then this operation would not be computable in h since $<_{\mathbb{R}}$ is (in contrast to $<_{\mathbb{Q}}$) not decidable. In order to overcome this difficulty we first modify $(*)$ to

$$(**)\ \forall n\big([d_X(hn,h(n+1))](n+1) <_{\mathbb{Q}} \langle 6\cdot 2^{-n-1}\rangle\big).$$

Now $\widehat{h}$ can be defined as a functional in h in $\widehat{\text{WE-HA}}^{\omega}\!\upharpoonright$ by

$$\widehat{h}(n) := \begin{cases} h(n) \text{ if } \forall k < n\big([d_X(hk,h(k+1))](k+1) <_{\mathbb{Q}} \langle 6\cdot 2^{-k-1}\rangle\big),\\ h(k) \text{ for } \min k < n : [d_X(hk,h(k+1))](k+1) \geq_{\mathbb{Q}} \langle 6\cdot 2^{-k-1}\rangle,\\ \qquad\qquad\qquad\qquad \text{otherwise.}\end{cases}$$

In the following we always refer to this definition of $\widehat{h}$ unless we deal with real numbers. $\widehat{h}$ fulfills $(**)$ and, therefore, $(*)$ for all h. If $(**)$ is already valid for h, then $\forall n(hn =_0 \widehat{h}n)$. Hence each h may be thought of as being a representative of a (uniquely determined) element of the completion of $(\mathbb{N},d_X)$, namely of that element, which is represented by $\widehat{h}$. In the other direction, each representative of a Cauchy sequence in $(\mathbb{N},d_X)$ with Cauchy modulus 2^{-n} fulfills $(**)$ and, therefore, also is a representative in our sense (of the element given by this Cauchy sequence). Using the construction $\widehat{h}$ we now can extend the pseudo metric d_X to a pseudo metric $\widehat{d_X}$ on $\mathbb{N}^{\mathbb{N}}$:

$$\widehat{d_X}^{1(1)(1)}(h_1,h_2)(n) :=_0 [d_X(\widehat{h_1}(n+5),\widehat{h_2}(n+5))](n+5).$$

By the summand $+5$, the 'right' rate of convergence of $\widehat{d_X}(h_1,h_2)$ is ensured:

$$\forall k\big(|\widehat{d_X}(h_1,h_2)(k) -_{\mathbb{Q}} \widehat{d_X}(h_1,h_2)(k+1)| <_{\mathbb{Q}} \langle 2^{-k-1}\rangle\big).$$

Furthermore we have

Lemma 4.4. $\widehat{\text{WE-HA}}^{\omega}\!\upharpoonright \vdash \forall k\big(\widehat{d_X}(f,\lambda m.\widehat{f}(k)) <_{\mathbb{R}} \langle 2^{-k+3}\rangle\big)$.

$(\mathbb{N}^{\mathbb{N}},\widehat{d_X})$ is the completion of $(\mathbb{N},d_X)$:

Lemma 4.5. $\widehat{\text{WE-HA}}^{\omega}\!\upharpoonright \vdash \forall \Phi^{10}\Big(\forall n; m,k \geq n\big(\widehat{d_X}(\Phi m,\Phi k) \leq_{\mathbb{R}} \langle 2^{-n}\rangle\big)$
$\rightarrow \exists f^1 \forall n\big(\widehat{d_X}(\Phi n,f) \leq_{\mathbb{R}} \langle 2^{-n}\rangle\big)\Big)$.
In fact, f can be defined as: $fk := \widehat{\Phi(k+5)}(k+5)$.

$(\mathbb{N},d_X)$ is canonically embedded into $(\mathbb{N}^{\mathbb{N}},\widehat{d_X})$ by $i:\mathbb{N}\rightarrow\mathbb{N}^{\mathbb{N}}$, $i(n) := \lambda k^0.n$. $d_X(n,m) =_{\mathbb{R}} \widehat{d_X}(i(n),i(m))$ for all m,n.

Definition 4.6. $(\mathbb{N}^{\mathbb{N}},\widehat{d_X})$ is called **standard representation** of $(\widehat{X},\widehat{d})$.

We now define an equivalence relation $=_{\widehat{X}}$ on the set $\mathbb{N}^{\mathbb{N}}$ by

$$h_1 =_{\widehat{X}} h_2 :\equiv \big(\widehat{d_X}(h_1,h_2) =_{\mathbb{R}} 0_{\mathbb{R}}\big).$$

$\widehat{d_X}$ induces a metric on the equivalence classes with respect to $=_{\widehat{X}}$. This metric space is isometric to $(\widehat{X},\widehat{d})$, which justifies the expression 'standard representation of $(\widehat{X},\widehat{d})$'.

Remark 4.7. Later in this book we call CSM-spaces most of the time 'Polish spaces' though this is a minor deviation from standard terminology according to which CSM-spaces are called 'Polish metric spaces' whereas a Polish space officially is a separable completely metrizable topological space (the metric will not be uniquely determined). In our case the space always comes with a metric w.r.t. which it is a complete separable metric space.

Example 4.8. 1) $\mathbb{R}^n$ resp. $[0,1]^n$ endowed with the Euclidean metric
$\widehat{d_E}\big((x_1,\dots,x_n),(y_1,\dots,y_n)\big) := \big(\Sigma_{i=1}^n |x_i-y_i|^2\big)^{\frac{1}{2}}$ is the completion of the space $\mathbb{Q}^n$ resp. $[0,1]^n\cap\mathbb{Q}^n$ with the metric
$d_E\big((r_1,\dots,r_n),(\tilde{r}_1,\dots,\tilde{r}_n)\big) := \big(\Sigma_{i=1}^n |r_i-\tilde{r}_i|^2\big)^{\frac{1}{2}}$ on $\mathbb{Q}^n$ resp. $[0,1]^n\cap\mathbb{Q}^n$.

2) The space $C[0,1]$ of all continuous functions $f:[0,1]\to\mathbb{R}$ together with the maximum metric $\widehat{d_\infty}$ is the completion of (A,d_∞) where A is the set of all finite tuples of rational numbers and $d_\infty\big((r_0,\ldots,r_m),(\tilde{r}_0,\ldots,\tilde{r}_n)\big):=$ $\sup\limits_{x\in[0,1]}\big|(r_mx^m+\ldots+r_1x+r_0)-(\tilde{r}_nx^n+\ldots+\tilde{r}_1x+\tilde{r}_0)\big|$.

   $C[0,1]$ is represented in $\widehat{\text{WE-HA}}^\omega\restriction$ as the standard representation of the completion $(\widehat{A},\widehat{d_\infty})$ of (A,d_∞). This representation of $f\in C[0,1]$ is effectively equivalent to the representation of f as a uniformly continuous function which is given with a modulus of uniform continuity $\omega_f:\mathbb{N}\to\mathbb{N}$:

$$\forall n\in\mathbb{N},x,y\in[0,1]\big(|x-y|<2^{-\omega_f(n)}\to|f(x)-f(y)|<2^{-n}\big).$$

   Then it suffices to consider f restricted to the rational numbers in $[0,1]$ as f on $[0,1]$ can be recovered from this by ω_f (for more details see the end of section 4.3). Clearly, the representation above can be converted into the latter using that one can explicitly write down a modulus of uniform continuity on $[0,1]$ for a polynomial (in chapter 16 we will give a particularly efficient such modulus using the so-called Markov inequality). In the other direction one uses the effective Weierstraß approximation theorem which for functions given with a modulus of uniform continuity can be proved in e.g. (RCA_0 and hence in) $\widehat{\text{E-PA}}^\omega\restriction+\text{QF-AC}^{0,0}$ as shown in [338]. If f is just given as a pointwise continuous function in the sense of the representation used in [338] then one needs the binary König's lemma (discussed in chapters 9 and 10 below) to convert this into a representation of f in our sense (see [338] and also [223]).

3) Let L_p denote (as usual) the space of (equivalence classes w.r.t. identity except on null sets) functions $f:[0,1]\to\mathbb{R}$ such that $|f|^p$ is integrable on $[0,1]$ ($1\le p<\infty$): On the set A of 2) we define the metric

   $d_p\big((r_0,\ldots,r_m),(\tilde{r}_0,\ldots,\tilde{r}_n)\big):=\big(\int_0^1|(r_mx^m+\ldots+r_0)-(\tilde{r}_nx^n+\ldots+\tilde{r}_0)|^pdx\big)^{\frac{1}{p}}$.

   L_p is represented in $\widehat{\text{WE-HA}}^\omega\restriction$ as the standard representation of the completion $(\widehat{A},\widehat{d_p})$ of (A,d_p).

4) Let l_p be the space of all sequences (x_n) in $\mathbb{R}$ such that $\sum\limits_{i=0}^\infty|x_i|^p$ converges ($1\le p<\infty$). On the set A from 2) we define a metric by

   $d_{l_p}((r_0,\ldots,r_m),(\tilde{r}_0,\ldots,\tilde{r}_n)):=\Big(\sum\limits_{i=0}^{\max(m,n)}|r_i-\tilde{r}_i|^p\Big)^{\frac{1}{p}}$, where $r_i:=0$ (resp. $\tilde{r}_i=0$) for $i>m$ (resp. $i>n$).

   l_p is represented in $\widehat{\text{WE-HA}}^\omega\restriction$ as the standard representation of the completion $(\widehat{A},\widehat{d_{l_p}})$ of (A,d_{l_p}). Over e.g. $\widehat{\text{E-PA}}^\omega\restriction+\text{QF-AC}^{0,0}$ this representation is in canonical 1-1 correspondence with the space of sequences (x_n) in $\mathbb{R}$ for which $\sum\limits_{i=0}^\infty|x_n|^p$ is convergent, which can (by the monotonicity of the sequence of partial sums) be written as (referring to the representation of $\mathbb{R}$)

$$\exists x^1 \forall n^0 \exists m^0 (x -_{\mathbb{R}} \sum_{i=0}^{m} |x_n|^p <_{\mathbb{R}} 2^{-n}).$$

Since $<_{\mathbb{R}} \in \Sigma^0_1$ one can use QF-AC^{0,0} to obtain a rate of convergence. From this on it is clear how to rewrite (a representation of) (x_n) into a representative in the sense above (the converse direction is an easy exercise). Note that if one expresses the convergence of the sum in the classically equivalent way by stating the boundedness of the sum, then one needs AC^{0,0} for Π^0_1-formulas (which amounts to arithmetical comprehension, see chapter 11) to conclude the existence of a rate of convergence.

Definition 4.9. A CSM-space $(\widehat{X},\widehat{d})$ is called WE-HA$^\omega$-definable ($\widehat{\text{WE-HA}}^\omega\upharpoonright$-definable), if it possesses a standard representation $(\mathbb{N}^{\mathbb{N}},\widehat{d_X})$ where $\widehat{d}_X^{1(1)(1)}$ (i.e. actually $d_X^{1(0)(0)}$) is given by a closed term of WE-HA$^\omega$ (resp. of $\widehat{\text{WE-HA}}^\omega\upharpoonright$) and $\widehat{d_X}$ represents provably in WE-HA$^\omega$ (resp. $\widehat{\text{WE-HA}}^\omega\upharpoonright$) a pseudo-metric on $\mathbb{N}^{\mathbb{N}}$.

Remark 4.10. The metric spaces 1)–4) above are all $\widehat{\text{WE-HA}}^\omega\upharpoonright$-definable and, furthermore, the definitions of these spaces in definition 4.9 are classically equivalent to the usual mathematical definitions (though not necessarily provably so in $\widehat{\text{WE-PA}}^\omega\upharpoonright$ or WE-PA$^\omega$). In the case of $C[0,1]$ we will use the representation of f with a given modulus of uniform continuity (mentioned above) when we carry out the extraction of moduli of uniqueness from concrete mathematical proofs in approximation theory (see chapter 16). The enriching of data by such a modulus ω_f is more convenient in practice than the presentation of f as a Cauchy sequence of polynomials having rational coefficients, since such a sequence is in general quite complicated to construct, whereas a modulus ω_f can often easily be written down. Thus, standard representations of CSM-spaces are only used for proving the general metatheorems. For the unwinding of concrete proofs, the most useful definition of these objects is used.

Definition 4.11. Let $(\widehat{X},\widehat{d})$ be a CSM-space. Then for each $x_0 \in \widehat{X}$, $r \in \mathbb{R}^+$, the open ball with radius r and center x_0 is defined by $B(x_0,r) := \left\{x \in \widehat{X} : \widehat{d}(x,x_0) < r\right\}$.

Lemma 4.12. *Let* $(\widehat{X},\widehat{d})$ *be a* CSM-*space.*

1) *Assume that* h_{x_0}, h_y *represent* $x_0, y \in \widehat{X}$ *in* $(\mathbb{N}^{\mathbb{N}},\widehat{d_X})$. *Then the following implication holds*
$$\overline{h}_{x_0}(m+1) = \overline{h}_y(m+1) \to y \in B(x_0, 2^{-m+4}).$$

2) *To each* $x_0 \in \widehat{X}$, $m \in \mathbb{N}$ *there exists an* $n \in \mathbb{N}$ *such that for all* $y \in B(x_0, 2^{-m})$ *there exists a representative* h_y *in* $(\mathbb{N}^{\mathbb{N}},\widehat{d_X})$ *with* $\overline{h}_y m =_0 n$.
Moreover, for a given representative h_{x_0} *of* x_0, n *can be taken as*
$n := <\widehat{h}_{x_0}(3),\ldots,\widehat{h}_{x_0}(m+2)>$.

Proof:

1) By lemma 4.4 and the definition of the embedding i the following holds:

$$\widehat{d_X}\left(h_{x_0}, i(\widehat{h}_{x_0}(m))\right) <_{\mathbb{R}} \langle 2^{-m+3}\rangle \text{ and } \widehat{d_X}\left(h_y, i(\widehat{h}_y(m))\right) <_{\mathbb{R}} \langle 2^{-m+3}\rangle.$$

Since the assumption $\overline{h}_{x_0}(m+1) = \overline{h}_y(m+1)$ implies that $\widehat{h}_{x_0}(m) = \widehat{h}_y(m)$ we obtain

$$\widehat{d_X}(h_{x_0}, h_y) \leq_{\mathbb{R}} \widehat{d_X}\left(h_{x_0}, i(\widehat{h}_{x_0}(m))\right) +_{\mathbb{R}} \widehat{d_X}\left(i(\widehat{h}_{x_0}(m)), i(\widehat{h}_y(m))\right)$$
$$+_{\mathbb{R}} \widehat{d_X}\left(i(\widehat{h}_y(m)), h_y\right) <_{\mathbb{R}} \langle 2^{-m+3} + 2^{-m+3}\rangle = \langle 2^{-m+4}\rangle.$$

2) Let h_{x_0}, h_y be representatives of x_0, y in $(\mathbb{N}^{\mathbb{N}}, \widehat{d_X})$ and assume that $m \geq 1$ (for $m = 0$ we may take $n := \langle\rangle =_0 \overline{h}_y 0)$. Then

$$n := < \widehat{h}_{x_0}(3), \ldots, \widehat{h}_{x_0}(m+2) > \text{ and } \tilde{h}_y(k) := \begin{cases} \widehat{h}_{x_0}(k+3) \text{ if } k < m, \\ \widehat{h}_y(k+3) \text{ if } k \geq m, \end{cases}$$

fulfill the lemma: $\overline{\tilde{h}}_y m =_0 n$. Furthermore, $\tilde{h}_y$ represents y, i.e. $\widehat{d_X}(\tilde{h}_y, h_y)) =_{\mathbb{R}} 0_{\mathbb{R}}$: It is sufficient to show that $\widehat{\tilde{h}}_y =_1 \tilde{h}_y$. i.e.

$$(*)\ \forall k \Big(d_X\left(\tilde{h}_y(k), \tilde{h}_y(k+1)\right)(k+1) <_{\mathbb{Q}} \langle 6 \cdot 2^{-k-1}\rangle \Big).$$

The only problematic case is when $k = m - 1$. In this case we, however, have (using tacitly the embedding of $\mathbb{Q}$ into $\mathbb{R}$ on the level of the representation):

$$d_X\left(\tilde{h}_y(m-1), \tilde{h}_y(m)\right)(m) =_{\mathbb{Q}} d_X\left(\widehat{h}_{x_0}(m+2), \widehat{h}_y(m+3)\right)(m)$$
$$\overset{\text{L.4.2}}{\leq_{\mathbb{R}}} d_X\left(\widehat{h}_{x_0}(m+2), \widehat{h}_y(m+3)\right) + 2^{-m}$$
$$\overset{\text{L.4.4}}{\leq_{\mathbb{R}}} \widehat{d_X}(h_{x_0}, h_y) + 2^{-m+1} + 2^{-m} + 2^{-m}$$
$$\overset{y \in B(x_0, 2^{-m})}{\leq_{\mathbb{R}}} +2^{-m} + 2^{-m+1} + 2^{-m} + 2^{-m} =_{\mathbb{R}} 5 \cdot 2^{-m}.$$

□

Remark 4.13. Lemma 4.12.1 in particular implies that each CSM-space $(\widehat{X}, \widehat{d})$ is the uniformly continuous image of the Baire space $(\mathbb{N}^{\mathbb{N}}, d)$, where

$$d(f,g) := \begin{cases} 2^{-\min n[f(n) \neq g(n)]}, \text{ if } \exists n (f(n) \neq g(n)) \\ 0, \text{ otherwise}, \end{cases}$$

namely the image under the function $\Phi(f^1) :=$ the unique element of $\widehat{X}$ represented by $\widehat{f}$.

The proof of lemma 4.12 can be carried out in $\widehat{\text{WE-HA}^\omega}\upharpoonright$, i.e.

Corollary 4.14. *For each CSM-space $(\widehat{X},\widehat{d})$ given in standard representation $(\mathbb{N}^{\mathbb{N}},\widehat{d}_X)$, $\widehat{\text{WE-HA}}^{\omega}\upharpoonright\backslash$ proves*

1) $\forall h_0^1, h^1, m^0 \big(\overline{h}_0(m+1) =_0 \overline{h}(m+1) \to \widehat{d}_X(h_0,h) <_{\mathbb{R}} \langle 2^{-m+4}\rangle\big)$ *and*
2) $\forall h_0^1, m^0 \exists n^0 \forall h^1 \big(\widehat{d}_X(h_0,h) <_{\mathbb{R}} \langle 2^{-m}\rangle \to \exists \tilde{h}\big(\widehat{d}_X(h,\tilde{h}) =_{\mathbb{R}} 0_{\mathbb{R}} \wedge \overline{\tilde{h}}m =_0 n\big)\big)$.
   Moreover, n can be taken as $n :=< \widehat{h}_{x_0}(3),\ldots,\widehat{h}_{x_0}(m+2) >$.

Definition 4.15. Let $\widehat{X}$ and $\widehat{Y}$ be CSM-spaces. A closed term $\Phi^{1(1)}$ of WE-HA$^{\omega}$ ($\widehat{\text{WE-HA}}^{\omega}\upharpoonright\backslash$) represents provably in WE-HA$^{\omega}$ ($\widehat{\text{WE-HA}}^{\omega}\upharpoonright\backslash$) a function $\widehat{X} \to \widehat{Y}$ if

$$\text{WE-HA}^{\omega}\ (\widehat{\text{WE-HA}}^{\omega}\upharpoonright\backslash) \vdash \forall f_1^1, f_2^1 \big(f_1 =_{\widehat{X}} f_2 \to \Phi f_1 =_{\widehat{Y}} \Phi f_2\big).$$

Remark 4.16. 1) Definition 4.15 is justified by the following fact: If Φ represents a function $\widehat{X} \to \widehat{Y}$ in the sense of definition 4.15, then Φ induces (provable in WE-HA$^{\omega}$ or in $\widehat{\text{WE-HA}}^{\omega}\upharpoonright\backslash$) a function on the equivalence classes on $\mathbb{N}^{\mathbb{N}}$ w.r.t. $=_{\widehat{X}}$ and $=_{\widehat{Y}}$. Modulo an isometry between $\widehat{X},\widehat{Y}$ and their standard representations this function in turn induces a function $\widehat{X} \to \widehat{Y}$.
2) Using negative translation (developed in chapter 10) one can show that

$$\text{WE-PA}^{\omega} \vdash \forall f_1^1, f_2^1 \big(f_1 =_{\widehat{X}} f_2 \to \Phi f_1 =_{\widehat{Y}} \Phi f_2\big)$$

implies

$$\text{WE-HA}^{\omega} \vdash \forall f_1^1, f_2^1 \big(f_1 =_{\widehat{X}} f_2 \to \Phi f_1 =_{\widehat{Y}} \Phi f_2\big).$$

This also holds for $\widehat{\text{WE-PA}}^{\omega}\upharpoonright\backslash$ and $\widehat{\text{WE-HA}}^{\omega}\upharpoonright\backslash$.

Definition 4.17. A function $F : \widehat{X} \to \widehat{Y}$ is called WE-HA$^{\omega}$-definable ($\widehat{\text{WE-HA}}^{\omega}\upharpoonright\backslash$-definable) if a closed term $\Phi_F^{1(1)}$ in WE-HA$^{\omega}$ ($\widehat{\text{WE-HA}}^{\omega}\upharpoonright\backslash$) exists such that

1) Φ_F represents provably in WE-HA$^{\omega}$ ($\widehat{\text{WE-HA}}^{\omega}\upharpoonright\backslash$) a function $\widehat{X} \to \widehat{Y}$ and
2) the function $\widehat{X} \to \widehat{Y}$ represented by Φ_F coincides with F, i.e. if $\delta(f^1)$ and $\delta'(f^1)$ denote the uniquely determined elements in $\widehat{X}$ and $\widehat{Y}$, respectively, represented by f (w.r.t. the above standard representation), then

$$\forall f^1 \big(\delta'(\Phi_F(f)) = F(\delta(f))\big).$$

This means that the following diagram commutes:

$$\begin{array}{ccc} \mathbb{N}^{\mathbb{N}} & \stackrel{\Phi_F}{\longrightarrow} & \mathbb{N}^{\mathbb{N}} \\ \delta\downarrow & & \downarrow\delta' \\ \widehat{X} & \stackrel{F}{\longrightarrow} & \widehat{Y}. \end{array}$$

In a similar way one could treat also partial functions F where then Φ_F might be partial as well. However, partiality of functions corresponds to hidden data (witnesses for membership in the domain) which we have to make explicit for the extraction of computational information from proofs as this information might depend on these hidden data as inputs. E.g. a function $f : \mathbb{R}^*_+ \to \mathbb{R}$ we treat as a total function $f' : \mathbb{N} \times \mathbb{R} \to \mathbb{R}$ between the Polish spaces $\mathbb{N} \times \mathbb{R}$ and $\mathbb{R}$, namely as $f'(n,x) := f(\max\{2^{-n},x\})$, rather than as a partial function on $\mathbb{R}$. I.e. we make the function total by enriching the input: a strictly positive real number is a pair of a real number and a witness n of its positivity.

Remarks and conventions: The function F in definition 4.17 may be given in set–theoretical terms which are not expressible within WE-HA$^\omega$. In particular, 4.17.2) may be unprovable in WE-HA$^\omega$ or E-PA$^\omega$+QF-AC etc. In the following, if we say that a certain sentence involving $\widehat{X},\widehat{Y},F$ holds provably in WE-HA$^\omega$, we always mean that the corresponding statement expressed in terms of the standard representations of $\widehat{X},\widehat{Y}$ and Φ_F is provable in WE-HA$^\omega$. We always consider $\widehat{X},\widehat{Y}$ as given with fixed standard representations $(\mathbb{N}^\mathbb{N},\widehat{d_X})$, $(\mathbb{N}^\mathbb{N},\widehat{d_Y})$ and F as represented by a fixed functional Φ_F. Assume e.g. that $\widehat{X}$ and $\widehat{Y}$ are WE-HA$^\omega$-definable CSM-spaces and that $F : \widehat{X} \times \widehat{Y} \to \mathbb{R}$ is a WE-HA$^\omega$-definable function. Then the sentence $\forall x \in \widehat{X} \exists y \in \widehat{Y} \big(F(x,y) = 0\big)$ is represented in $\mathscr{L}$(WE-HA$^\omega$) by

$$(*)\ \forall x^1 \exists y^1 \big(\Phi_F(x,y) =_\mathbb{R} 0_\mathbb{R}\big)$$

with Φ_F a closed term of WE-HA$^\omega$. $(*)$ has the logical form

$$(**)\ \forall x^1 \exists y^1 \forall k^0 A_0(x,y,k),$$

where $A_0 \in \mathscr{L}$(WE-HA$^\omega$) is quantifier-free.
If $\mathscr{T}$ is a theory in the language of WE-HA$^\omega$, then

$$\mathscr{T} \vdash \forall x \in \widehat{X} \exists y \in \widehat{Y} \big(F(x,y) = 0\big)$$

stands for

$$\mathscr{T} \vdash \forall x^1 \exists y^1 \forall k^0 A_0(x,y,k).$$

Proposition 4.18. *If $F : \widehat{X} \to \widehat{Y}$ is a* WE-HA$^\omega$*-definable ($\widehat{\text{WE-HA}}^\omega{\restriction}\backslash$-definable) function, then F possesses provably in* WE-HA$^\omega$ *($\widehat{\text{WE-HA}}^\omega{\restriction}\backslash$) a modulus*

$$\omega_F^{001} \in \text{WE-HA}^\omega\ (\widehat{\text{WE-HA}}^\omega{\restriction}\backslash)$$

of pointwise continuity, i.e.

$$\text{WE-HA}^\omega \vdash \forall f_0^1, f^1, n^0 \big(\widehat{d_X}(f_0,f) <_\mathbb{R} \langle 2^{-\omega_F f_0 n}\rangle \to \widehat{d_Y}(\Phi_F f_0, \Phi_F f) <_\mathbb{R} \langle 2^{-n}\rangle\big).$$

(Analogous for $\widehat{\text{WE-HA}}^\omega{\restriction}\backslash$).
Note that ω_f is in general not extensional in f_0 with respect to $=_{\widehat{X}}$.

Proof: The proposition follows from corollary 4.14 and the fact that each closed term of type 1(1) in WE-HA$^\omega$ (resp. $\widehat{\text{WE-HA}}^\omega\!\upharpoonright$) possesses a modulus of pointwise continuity in WE-HA$^\omega$ (resp. $\widehat{\text{WE-HA}}^\omega\!\upharpoonright$) (provable in WE-HA$^\omega$ resp. $\widehat{\text{WE-HA}}^\omega\!\upharpoonright$; see [366] (3.7.8) and [200] (3.22.1)). □

Complete separable **normed** spaces are represented in $\widehat{\text{WE-HA}}^\omega\!\upharpoonright$ as special metric spaces with the metric $d(x,y) := \|x-y\|$. In addition to the metric also the vector space operations have to be represented in WE-HA$^\omega$ or $\widehat{\text{WE-HA}}^\omega\!\upharpoonright$. The spaces in the examples above are also $\widehat{\text{WE-HA}}^\omega\!\upharpoonright$-definable as normed spaces (over $\mathbb{R}$).

4.3 Special representation of compact metric spaces

Compact spaces usually are defined via the Heine-Borel covering property. However, this property is unprovable in WE-PA$^\omega$+QF-AC^{1,0} (also with QF-AC in all types as discussed in chapter 10) even for the most common compact spaces as e.g. $[0,1] \subset \mathbb{R}$ or the Cantor space 2^ω as it requires the so-called binary König's lemma WKL to be discussed in chapters 9–10 and 12 (see also [338]). For metric spaces, however, there are two additional (classically) equivalent definitions of compactness, namely sequential compactness (every sequence contains a convergent subsequence), and total boundedness (together with completeness). While sequential compactness needs even stronger ineffective tools to be provable e.g. for $[0,1]$, namely arithmetical comprehension as discussed in chapter 11 below (and actually is equivalent to this principle, see [338]), the total boundedness of the usual compact metric spaces can be established fully elementary. Following Brouwer's treatment of compact metric spaces in intuitionistic mathematics as well as the practice in computable analysis we, therefore, adopt this definition as the basis of our representation of compact spaces (which itself goes back to Brouwer).

Definition 4.19. A metric space $(\widehat{X},\widehat{d})$ is called **compact** if it is complete and totally bounded, i.e. if there exist functions $x_{(\cdot,\cdot)} : \mathbb{N}\times\mathbb{N} \to \widehat{X},\ \alpha : \mathbb{N}\to\mathbb{N}$, which yield a finite sequence $x_{k,0},\ldots,x_{k,\alpha(k)}$ in $\widehat{X}$ for each $k\in\mathbb{N}$ such that $\forall x\in\widehat{X}, k\exists i \leq \alpha(k)\big(\widehat{d}(x,x_{k,i}) < 2^{-k}\big)$.

Remark 4.20. 1) Each compact metric space is separable: the countable set $X := \bigcup_{k\in\mathbb{N}}\left\{x_{k,0},\ldots,x_{k,\alpha(k)}\right\}$ is dense in $\widehat{X}$.
2) As mentioned above, in constructive mathematics the various classically equivalent definitions of compactness need to be distinguished. Compact metric spaces in our sense are often called 'CTB-spaces', where 'CTB' stands for 'complete totally bounded'.

Since by remark 4.20 compact metric spaces are special CSM-spaces, we could use the standard representation for CSM-spaces described above. However, we need a

special form of this representation, which has the property that all representatives are bounded by some function M, i.e. are elements of $\{x \in \mathbb{N}^{\mathbb{N}} : \forall n(x(n) \leq M(n))\}$, and every element of this set represents a uniquely determined point in the compact space:
Let $(\widehat{X},\widehat{d})$ be a compact metric space with a 2^{-k}-net $x_{k,0},\dots,x_{k,\alpha(k)}$ in $\widehat{X}$ for each $k \in \mathbb{N}$. We code $x_{k,i}$ by $j(k,i)$ (if n is not of the form $j(k,i)$ with $i \leq \alpha(k)$, then n codes the element $x_{0,\alpha(0)}$). Now we consider the standard representation $(\mathbb{N}^{\mathbb{N}},\widehat{d_X})$ of $(\widehat{X},\widehat{d})$ w.r.t. $X := \bigcup_{k\in\mathbb{N}}\{x_{k,0},\dots,x_{k,\alpha(k)}\}$ using this coding. For each $f \in \mathbb{N}^{\mathbb{N}}$ there exists (primitive recursively in f, d_X and α) a function $g_f \leq_1 M$ such that f and g_f represent the same element in $\widehat{X}$, i.e. $\widehat{d_X}(f,g_f) =_{\mathbb{R}} 0_{\mathbb{R}}$, where $M(k) := \max\{j(k+2,0),\dots,j(k+2,\alpha(k+2))\}$:
Define $\tilde{f}(k) := \widehat{f}(k+1)$, then $\forall k(d_X(\tilde{f}(k),\tilde{f}(k+1)) <_{\mathbb{R}} \langle 7\cdot 2^{-k-2}\rangle)$. For every $k \in \mathbb{N}$ exists (primitive recursively in f, d_X and α) an $i_k \leq M(k)$ such that $d_X(i_k,\tilde{f}(k)) <_{\mathbb{R}} \langle 2^{-k-1}\rangle$: Define i_k as the least $i \leq M(k)$ such that

$$(*)\ d_X(i,\tilde{f}k)(k+3) <_{\mathbb{Q}} \langle 2^{-k-2}\rangle + \langle 2^{-k-3}\rangle$$

(such an $i \leq M(k)$ always exists since, by M–definition, there exists an $i \leq M(k)$ with $d_X(i,\tilde{f}k) <_{\mathbb{R}} \langle 2^{-k-2}\rangle$).
The construction of i_k implies

$$d_X(i,\tilde{f}k) <_{\mathbb{R}} \langle 2^{-k-2}\rangle + \langle 2^{-k-3}\rangle + \langle 2^{-k-3}\rangle = \langle 2^{-k-1}\rangle.$$

Hence

$$\begin{aligned} d_X(i_k,i_{k+1}) &\leq_{\mathbb{R}} d_X(i_k,\tilde{f}k) + d_X(\tilde{f}k,\tilde{f}(k+1)) + d_X(\tilde{f}(k+1),i_{k+1}) \\ &<_{\mathbb{R}} \langle 2^{-k-1}\rangle + \langle 7\cdot 2^{-k-2}\rangle + \langle 2^{-k-2}\rangle = \langle 5\cdot 2^{-k-1}\rangle \end{aligned}$$

and, therefore,

$$d_X(i_k,i_{k+1})(k+1) <_{\mathbb{Q}} \langle 5\cdot 2^{-k-1}\rangle + \langle 2^{-k-1}\rangle = \langle 6\cdot 2^{-k-1}\rangle.$$

Now define $g_f(k) := i_k$. The above shows that $\widehat{g_f} =_1 g_f$. It follows that g_f represents the same element of $\widehat{X}$ as f and $\forall k \in \mathbb{N}(g_f(k) \leq M(k))$.
$\{f \in \mathbb{N}^{\mathbb{N}} | f \leq_1 M\}$ together with the restriction of $\widehat{d_X}$ on $\{\dots\}$, which we also denote by $\widehat{d_X}$, represents the compact metric space $(\widehat{X},\widehat{d})$.

Remark 4.21. 1) Inspection of the construction $f \mapsto g_f$ shows that it can be defined already by a closed term of G_3A^ω in f, d_X, α.
2) Note that $\overline{f_1}(n+1) =_0 \overline{f_2}(n+1) \to \overline{g}_{f_1}(n) =_0 \overline{g}_{f_2}(n)$, i.e. the construction $f \mapsto g_f$ is a uniformly continuous selfmapping of the Baire space.

Definition 4.22. $\left(\{f \in \mathbb{N}^{\mathbb{N}} : f \leq_1 M\},\widehat{d_X}\right)$ is called standard representation of the compact metric space $(\widehat{X},\widehat{d})$. A compact metric space $(\widehat{X},\widehat{d})$ is called WE-HA$^\omega$

(resp. $\widehat{\text{WE-HA}}^{\omega}\upharpoonright\backslash$)-definable if it possesses a standard representation $\left(\{f \in \mathbb{N}^{\mathbb{N}} : f \leq_1 M\}, \widehat{d_X}\right)$ where $M^1, \widehat{d}_X^{1(1)(1)}$ (i.e. M^1, d_X^{100}) are closed terms of WE-HA$^{\omega}$ (resp. $\widehat{\text{WE-HA}}^{\omega}\upharpoonright\backslash$), and if d_X represents provably in WE-HA$^{\omega}$ (resp. $\widehat{\text{WE-HA}}^{\omega}\upharpoonright\backslash$) a pseudo-metric on $\langle X\rangle$ such that

$$\text{WE-HA}^{\omega} \vdash \forall i \in \langle X\rangle, k \exists j \leq M(k)\big(d_X(i,j) < 2^{-k-2}\big).$$

Examples:

1) $[0,1]^n (\subset \mathbb{R}^n)$ is a $\widehat{\text{WE-HA}}^{\omega}\upharpoonright\backslash$-definable compact metric space (w.r.t. the Euclidean metric d_E). Exercise!
2) Let K, C be positive rational numbers such that $2K > (n-1)C$. Then $M := \big\{(x_1,\ldots,x_n) \in \mathbb{R}^n : |x_i| \leq K \wedge |x_i - x_j| \geq C \ (1 \leq i < j \leq n)\big\}$
   (endowed with the Euclidean metric) is a $\widehat{\text{WE-HA}}^{\omega}\upharpoonright\backslash$-definable compact metric space (exercise, see also the end of this chapter for a much easier treatment of this space).

Remarks and conventions: Let $\widehat{X}$ be a CSM-space, K a compact metric space and $F : \widehat{X} \times K \to \mathbb{R}$ a WE-HA$^{\omega}$ (resp. $\widehat{\text{WE-HA}}^{\omega}\upharpoonright\backslash$)-definable function on the WE-HA$^{\omega}$ (resp. $\widehat{\text{WE-HA}}^{\omega}\upharpoonright\backslash$)-definable spaces $\widehat{X}, K$. Then a sentence

$$\forall x \in \widehat{X} \forall y \in K\big(F(x,y) > 0\big)$$

is represented in WE-HA$^{\omega}$ ($\widehat{\text{WE-HA}}^{\omega}\upharpoonright\backslash$) by

$$(+)\ \forall x^1 \forall y \leq_1 M\big(\Phi_F(x,y) >_{\mathbb{R}} 0_{\mathbb{R}}\big)$$

with $M, \Phi_F \in$ WE-HA$^{\omega}$ (resp. $\widehat{\text{WE-HA}}^{\omega}\upharpoonright\backslash$). $(+)$ can be written equivalently as

$$(++)\ \forall x^1 \forall y \leq_1 M \exists k^0 (|\Phi_F(x,y)|_{\mathbb{R}} >_{\mathbb{R}} \langle 2^{-k}\rangle),$$

where $\exists k^0(|\Phi_F(x,y)|_{\mathbb{R}} >_{\mathbb{R}} \langle 2^{-k}\rangle) \in \mathcal{L}(\text{WE-HA}^{\omega})$ ($\in \mathcal{L}(\widehat{\text{WE-HA}}^{\omega}\upharpoonright\backslash)$ is a Σ^0_1-formula. We usually rewrite sentences $(+)$ in the form $(++)$ since we are interested in extracting bounds on '$\exists k^0$' from proofs of $(+)$.

Similarly, a sentence

$$\forall x \in \widehat{X} \exists y \in K\big(F(x,y) = 0\big)$$

is represented in WE-HA$^{\omega}$ ($\widehat{\text{WE-HA}}^{\omega}\upharpoonright\backslash$) by

$$(*)\ \forall x^1 \exists y \leq_1 M\big(\Phi_F(x,y) =_{\mathbb{R}} 0_{\mathbb{R}}\big)$$

with $M, \Phi_F \in$ WE-HA$^{\omega}$ (resp. $\widehat{\text{WE-HA}}^{\omega}\upharpoonright\backslash$). $(*)$ can be rewritten as

$$(**)\ \forall x^1 \exists y \leq_1 M \forall k^0 (|\Phi_F(x,y)|_{\mathbb{R}} \leq_{\mathbb{R}} \langle 2^{-k}\rangle),$$

where $\forall k^0(|\Phi_F(x,y)|_{\mathbb{R}} \leq_{\mathbb{R}} \langle 2^{-k}\rangle) \in \mathcal{L}(\text{WE-HA}^\omega)$ $(\in \mathcal{L}(\widehat{\text{WE-HA}}^\omega\restriction))$ is a Π^0_1-formula. Again, we usually rewrite $(*)$ as $(**)$ since we are interested in extracting bound on '$\forall k^0$' from proofs which use $(*)$ as an assumption.

Proposition 4.23. *Let $F : \widehat{X} \times K \to \widehat{Y}$ be a* WE-HA$^\omega$*-definable function, where K is a* WE-HA$^\omega$*-definable compact metric space K and $\widehat{X}, \widehat{Y}$ are* WE-HA$^\omega$*-definable CSM-spaces. Then $F(x,\cdot) : K \to \widehat{Y}$ possesses (uniformly in a representative of x as parameter) provably in* WE-HA$^\omega$ *a modulus $\omega_F^{1(1)}$ of uniform continuity, i.e. $\omega_F^{1(1)}$ is a closed term such that*

$$\text{WE-HA}^\omega \vdash$$
$$\forall x^1 \forall f_1^1, f_2^1, n^0 \big(\widehat{d}_K(f_1,f_2) <_{\mathbb{R}} \langle 2^{-\omega_F(x,n)}\rangle \to \widehat{d}_Y(\Phi_F(x,f_1), \Phi_F(x,f_2)) <_{\mathbb{R}} \langle 2^{-n}\rangle\big).$$

An analogous result holds for $\widehat{\text{WE-HA}}^\omega\restriction$.

Proof: Let F be given by a closed term $\Phi_F^{1(1)(1)}$ of WE-HA$^\omega$. By proposition 9.10 (which will be proved in chapter 9 below) we can construct a closed term $\omega_{\Phi_F}^{1(1)}$ such that for x^1, n^0 and $k := \omega_{\Phi_F}^{1(1)}(x,n)$

$$\forall g_1, g_2 \leq M(\overline{g}_1(k) =_0 \overline{g}_2(k) \to (\overline{\Phi_F(x,g_1)})(n) =_0 (\overline{\Phi_F(x,g_2)})(n)),$$

i.e. $\lambda n.\omega_{\Phi_F}(x,n)$ is a modulus of uniform continuity for $\Phi_F(x,\cdot)$ on $\{g : g \leq_1 M\}$. Here M^1 is the fixed function from the standard representation of K as a compact metric space.

Now let x^1, f_1^1, f_2^1 be given and define

$$k := \omega_F(x,n) := \omega_{\Phi_F}(x,n+2)+1.$$

Assume that

$$\widehat{d}_K(f_1,f_2) <_{\mathbb{R}} \langle 2^{-\omega_F(x,n)}\rangle.$$

By corollary 4.14 there exist functions $\tilde{f}_1, \tilde{f}_2$ with

$$\bigwedge_{i=1}^{2} (\tilde{f}_i =_K f_i) \wedge \overline{\tilde{f}}_1(k) =_0 \overline{\tilde{f}}_2(k).$$

By the construction $f \mapsto g_f$ from the standard representation of compact metric spaces K and remark 4.21.2 applied to $\tilde{f}_1, \tilde{f}_2$ there exist $g_1, g_2 \leq_1 M$ such that

$$g_1 =_K f_1 \wedge g_2 =_K f_2 \wedge \overline{g}_1(k-1) =_0 \overline{g}_2(k-1).$$

By the definition of k and the $\widehat{\cdot}$–construction we get

$$(\widehat{\Phi_F(x,g_1)})(n+1) =_0 (\widehat{\Phi_F(x,g_2)})(n+1)$$

and so (using lemma 4.2)

$$\widehat{d_Y}(\Phi_F(x,g_1),\Phi_F(x,g_2)) <_{\mathbb{R}} \langle 2^{-n}\rangle.$$

Using the extensionality of $\Phi_F(x,\cdot)$ w.r.t. $=_K$ and $=_{\widehat{Y}}$ yields

$$\bigwedge_{i=1}^{2}(\Phi_F(x,g_i) =_{\widehat{Y}} \Phi_F(x,f_i)).$$

Hence

$$\widehat{d_Y}(\Phi_F(x,f_1),\Phi_F(x,f_2)) <_{\mathbb{R}} \langle 2^{-n}\rangle.$$

□

Notational convention: For the rest of this book we will usually simply write '2^{-k}' instead of its rational or real number code '$\langle 2^{-k}\rangle$' when it will be clear from the context that we refer to the representation. Also we will usually simply write X instead of $(\widehat{X},\widehat{d})$.

For the special compact space $[0,1]$ the representation of compact metric spaces above can be somewhat simplified. As this case will play a particularly important role in chapters further below we will present this case here:

Definition 4.24.

$$\tilde{x}(n) := j(2k_0, 2^{n+2}-1), \text{ where } k_0 = \max k \leq 2^{n+2}[\frac{k}{2^{n+2}} \leq_{\mathbb{Q}} \widehat{x}(n+2)]$$

($k_0 := 0$ if no such k exists).
Note that $\lambda x^1.\tilde{x}$ can easily be defined by a closed term in $\widehat{\text{WE-HA}}^{\omega}{\restriction}\backslash$.

One easily verifies the following

Lemma 4.25. Provably in $\widehat{\text{WE-HA}}^{\omega}{\restriction}\backslash$, for all x^1:

1) $0_{\mathbb{R}} \leq_{\mathbb{R}} x \leq_{\mathbb{R}} 1_{\mathbb{R}} \to \tilde{x} =_{\mathbb{R}} x$,
2) $0_{\mathbb{R}} \leq_{\mathbb{R}} \tilde{x} \leq_{\mathbb{R}} 1_{\mathbb{R}}$,
3) $\tilde{x} =_{\mathbb{R}} \tilde{\tilde{x}}$,
4) $\tilde{x} \leq_1 N := \lambda n.j(2^{n+3}, 2^{n+2}-1)$,
5) $x >_{\mathbb{R}} 1_{\mathbb{R}} \to \tilde{x} =_{\mathbb{R}} 1_{\mathbb{R}}$, $x <_{\mathbb{R}} 0_{\mathbb{R}} \to \tilde{x} =_{\mathbb{R}} 0_{\mathbb{R}}$.
6) $x =_{\mathbb{R}} y \to \tilde{x} =_{\mathbb{R}} \tilde{y}$.

Using the construction $x \mapsto \tilde{x}$ from definition 4.24 we also can represent $[0,1]^d$ for every fixed number d by a bounded set $\{x^1 : x \leq_1 N_d\}$ of functions, where N_d : $\nu^d(N,\ldots,N)$ (with ν^d as in definition 3.30 and N as in definition 4.24):

$x(\leq N_d)$ represents the vector in $[0,1]^d$ which is represented by $(\widetilde{(\nu_1^d x)},\ldots,\widetilde{(\nu_d^d x)})$. If (in the other direction) $x_1,\ldots,x_d$ represent real numbers $r_1,\ldots,r_d \in [0,1]$, then $x := \nu^d(\tilde{x}_1,\ldots,\tilde{x}_d) \leq_1 \nu^d(N,\ldots,N)$ represents $(r_1,\ldots,r_d) \in [0,1]^d$ in this sense.

Remark 4.26. For $a,b\in\mathbb{R}$ with $a\leq_{\mathbb{R}} b$, quantification $\forall x\in[a,b]\,A(x)$ (respectively $\exists x\in[a,b]\,A(x)$) reduces to quantification over $[0,1]$ (and, therefore, –modulo our representation– over $\{x: x\leq_1 N\}$) by $\forall\lambda\in[0,1]\,A((1-\lambda)a+\lambda b)$ and analogously for $\exists x$. This transformation immediately generalizes to $[a_1,b_1]\times\cdots\times[a_d,b_d]$ using $\lambda_1,\ldots,\lambda_d$.

Since an explicit representation of the intervals $[-m,m]$ for $m\in\mathbb{N}$ is of particular importance, we just indicate how to generalize directly definition 4.24 and 4.25 from $[0,1]$ to $[-m,m]$ (dropping for notational simplicity the dependence of $\tilde{x}$ on m):

$$\tilde{x}(n):=\begin{cases} j(2k_0,2^{n+2}-1), \text{ where } k_0=\max k\leq m2^{n+2}[\frac{k}{2^{n+2}}\leq_{\mathbb{Q}}\widehat{x}(n+2)], \\ \qquad\qquad\text{if } \widehat{x}(n+2)\geq_{\mathbb{Q}} 0_{\mathbb{Q}}, \\ j(2k_0+1,2^{n+2}-1), \text{ where } k_0=\min k\leq m2^{n+2}[-\frac{k}{2^{n+2}}\leq_{\mathbb{Q}}\widehat{x}(n+2)], \\ \qquad\qquad\text{if existent and } =0 \text{ otherwise,if } \widehat{x}(n+2)<_{\mathbb{Q}} 0_{\mathbb{Q}}. \end{cases}$$

As in lemma 4.25 it follows that

1) $|x|_{\mathbb{R}}\leq_{\mathbb{R}}(m)_{\mathbb{R}}\to\tilde{x}=_{\mathbb{R}}x$,
2) $|\tilde{x}|_{\mathbb{R}}\leq_{\mathbb{R}}(m)_{\mathbb{R}}$,
3) $\tilde{x}=_{\mathbb{R}}\tilde{\tilde{x}}$,
4) $\tilde{x}\leq_1 N_m:=\lambda n.j(m2^{n+3}+1,2^{n+2}-1)$,
5) $x>_{\mathbb{R}}(m)_{\mathbb{R}}\to\tilde{x}=_{\mathbb{R}}(m)_{\mathbb{R}},\ x<_{\mathbb{R}}(-m)_{\mathbb{R}}\to\tilde{x}=_{\mathbb{R}}(-m)_{\mathbb{R}}$.

For most applications it suffices to have instead of the standard representation of compact metric spaces (K,d) a representation by functions $f\in\mathbb{N}^{\mathbb{N}}$ which are bounded by some primitive recursive function M in the sense that $f\leq_1 M$ and which satisfy a Π^0_1-condition $A(f)\equiv\forall m^0 A_{qf}(f_x,m)$ on the level of the representation. I.e. every element $x\in K$ has a representation by a function $f_x\leq_1 M$ which satisfies $A(f_x)$ and each function $f\leq_1 M$ satisfying $A(f)$ represents a unique element of K. This suffices to ensure that for properties B of elements of K which can be expressed on the level of the representatives by a $\exists$-formula $B_1(f,l)\equiv\exists n^0 B_{qf}(f,l,n)$ (e.g. $\exists l^0(|\Phi_F(f)|_{\mathbb{R}}>_{\mathbb{R}}2^{-l})$ as in the remarks after definition 4.22) statements of the kind

$$\forall x\in K\exists l\in\mathbb{N}\,B(x)$$

have the form

$$\forall f\leq_1 M(\forall m^0 A_{qf}(f,m)\to\exists l^0,n^0 B_{qf}(f,l,n))$$

and hence are equivalent to

$$\forall f\leq_1 M\exists m^0,l^0,n^0(A_{qf}(f,m)\to B_{qf}(f,l,n)),$$

i.e. result in the same logical form (+) as in the case of spaces given in standard representation (see the remarks after definition 4.22, similarly for statements of the form $(*)$). Such a weaker form of the representation is often much easier to achieve and also applies to many closed subspaces of compact spaces which might not have a constructive standard representation as compact spaces on their own. E.g. the second of our examples above has an immediate representation in this weaker sense since the conditions '$|x_i -_{\mathbb{R}} x_j|_{\mathbb{R}} \geq_{\mathbb{R}} C$' are purely universal. As another example we give such a representation of the compact subset $K_{\omega,m} \subset C[0,1]$ of all continuous functions $f : [0,1] \to [-m,m]$ which have ω^1 as a modulus of uniform continuity in the sense of

$$\forall x,y \in [0,1], n^0(|x-y| < 2^{-\omega(n)} \to |f(x) - f(y)| \leq_{\mathbb{R}} 2^{-n}).$$

Define a primitive recursive function by

$$q(n) := \begin{cases} \min l \leq_0 n[l =_{\mathbb{Q}} n], \text{ if } 0 \leq_{\mathbb{Q}} n \leq_{\mathbb{Q}} 1 \\ 0^0, \text{ otherwise.} \end{cases}$$

Every rational number $\in [0,1] \cap \mathbb{Q}$ has a unique code by a number $\in q(\mathbb{N})$ and $\forall n^0(q(q(n)) =_0 q(n))$. Also every such number codes an element of $\in [0,1] \cap \mathbb{Q}$. We may conceive every number n as a representative of a rational number $\in [0,1] \cap \mathbb{Q}$, namely of the rational coded by $q(n)$. Now a function in $K_{\omega,m}$ can be represented via (the encoding of) its restriction to the rational numbers (from which is can uniquely be recovered via ω), i.e. as a function $f^{1(0)} \leq \lambda k.N_m$ (with N_m as above) satisfying

$$A_{\omega,m}(f) :\equiv \forall m^0, k^0, n^0(|q(m) -_{\mathbb{Q}} q(k)| < 2^{-\omega(n)} \to |\widetilde{f(q(m))} -_{\mathbb{R}} \widetilde{f(q(k))}| \leq_{\mathbb{R}} 2^{-n}),$$

where $A_{\omega,m}$ is Π^0_1. It is clear that $f^{1(0)}$ can be encoded in a function g^1 via $g(n) := f(j_1(n), j_2(n))$.

When dealing with $C[0,1]$ we will mainly refer to this representation of elements of $C[0,1]$ as pairs $(f^{1(0)}, \omega^1)$ where f represents the restriction of the function to the rational numbers in $[0,1]$ and ω is a modulus of uniform continuity of f. In these data one can easily write down closed terms of E-HA$^\omega$ (and fragments thereof) which compute e.g. the Riemann integral $\int_0^1 f(x)dx$ or the uniform norm $\|f\|_\infty := \sup_{x\in[0,1]} |f(x)|$ (exercise, see also [209]).

4.4 Fragments, exercises, historical comments and suggested further reading

Fragments: The representation of $\mathbb{R}^d$, $[0,1]^d$ and other Polish spaces and compact metric spaces developed in this chapter can without major changes be carried out in G$_3$A$^\omega_{(i)}$ already. For G$_2$A$^\omega_{(i)}$ some major modifications are necessary. E.g. the rate of convergence 2^{-k} must be replace by $1/(k+1)$. Moreover, the representation of $C[0,1]$ as Cauchy sequences of polynomials with rational coefficients can no longer be used directly since we do not have the exponential function. However, one can use a representation based on polygons instead with the same effect (see [209] for details on all this. For the simplified representation mentioned at the end of this chapter not even this is necessary. In $(f^{1(0)}, \omega^1)$ one can define $\int_0^1 f(x)dx$ and $\sup_{x\in[0,1]} f(x)$ even by closed terms of $G_2A^{\omega}_{(i)}$ (see again [209]).

Exercises:

1) Show that there is no continuous (in the sense of Baire space) functional $c^{1(1)}$ with $c(f) =_{\mathbb{R}} f$ and $f_1 =_{\mathbb{R}} f_2 \to c(f_1) =_1 c(f_2)$ for all f^1, f_1^1, f_2^1.
2) Fill in the details in the proofs of some lemmas in this chapter which we omitted.
3) Show that $\widehat{\text{WE-PA}}^{\omega}\!\upharpoonright + \Sigma^0_1$-IA (and hence – by proposition 3.21 – $\widehat{\text{WE-PA}}^{\omega}\!\upharpoonright +$ QF-AC^{0,0}) proves

$$\forall a^1_{(\cdot)}\ (\forall n(a_n \geq_{\mathbb{R}} 0) \to \forall k \exists n \forall m\ (a_n \leq_{\mathbb{R}} a_m + 2^{-k})).$$

4) Construct closed terms $\Phi_{\sup_{[0,1]}}$ and Φ_I in $\widehat{\text{WE-HA}}^{\omega}\!\upharpoonright$ which compute $\sup\limits_{x\in[0,1]} f(x)$ and $\int_0^1 f(x)dx$, respectively, in the data $(f^{1(0)}, \omega^1)$.

Historical comments and suggested further reading: The presentation in this chapter follows closely section 3 of Kohlenbach [204] which in turn is influenced by section 20 of chapter 1 from Beeson [15]. The standard representation of compact metric spaces (i.e. complete totally bounded metric spaces) is already due to Brouwer and for general Polish (i.e. complete separable) metric spaces can be found e.g. in Troelstra [364] (see Troelstra-van Dalen [371], chapter 7, for a more recent account). The special form of our representation of $\widehat{X}$ by which **every** element of $\mathbb{N}^{\mathbb{N}}$ represents a member of $\widehat{X}$ is due to Beeson [15]. Standard representations of Polish and compact metric spaces are also frequently used in computable analysis, e.g. in the so-called type-2 approach to computable analysis (see e.g. Weihrauch [377]) as well as in the area of reverse mathematics (see e.g. Simpson [338]). For a comprehensive treatment of the Baire and Cantor space as well as general Polish spaces see Moschovakis [282].

Chapter 5
Modified realizability

5.1 The soundness and program extraction theorems

In the following we make use of the notation

$$\underline{y}\underline{x} := y_1\underline{x}, \ldots, y_n\underline{x},$$

where $\underline{y} = y_1, \ldots, y_n$ and $\underline{x} = x_1, \ldots, x_k$ are tuples of functionals of suitable types and $y_i\underline{x} := y_i x_1 \ldots x_k$.

Definition 5.1 (modified realizability, Kreisel [244, 246]). For each formula A of $\mathscr{L}$(E-HA$^\omega$) we define a formula $\underline{x}$ *mr* A (in words: '$\underline{x}$ modified realizes A') of $\mathscr{L}$(E-HA$^\omega$) whose free variables are contained in that of A and $\underline{x}$, where $\underline{x}$ is a – possibly empty – tuple of variables which do not occur free in A. The length of $\underline{x}$ and the types of these variables are determined by the **logical** structure of A, since the definition of $\underline{x}$ *mr* A proceeds by induction over the logical structure of A:

(i) $\underline{x}\ mr\ A :\equiv A$ with the empty tuple $\underline{x}$, if A is a prime formula.
(ii) $\underline{x}, \underline{y}\ mr\ (A \wedge B) :\equiv \underline{x}\ mr\ A \wedge \underline{y}\ mr\ B$.
(iii) $z^0, \underline{x}, \underline{y}\ mr\ (A \vee B) :\equiv [(z =_0 0 \to \underline{x}\ mr\ A) \wedge (z \neq_0 0 \to \underline{y}\ mr\ B)]$.
(iv) $\underline{y}\ mr\ (A \to B) :\equiv \forall \underline{x}(\underline{x}\ mr\ A \to \underline{y}\underline{x}\ mr\ B)$.
(v) $\underline{x}\ mr\ (\forall y^\rho A(y)) :\equiv \forall y^\rho (\underline{x}y\ mr\ A(y))$.
(vi) $z^\rho, \underline{x}\ mr\ (\exists y^\rho A(y)) :\equiv \underline{x}\ mr\ A(z)$.

Definition 5.2. 1) A formula $A \in \mathscr{L}$(E-HA$^\omega$) is called $\exists$-free if it is built up from prime formulas by means of $\wedge, \to$ and $\forall$ only.
2) A formula $A \in \mathscr{L}$(E-HA$^\omega$) is called negative if it is built up from negated prime formulas by means of $\wedge, \to$ and $\forall$ only.

Remark 5.3. From corollary 3.18 we recall that in WE-HA$^\omega$ all prime formulas P are decidable and, therefore, $\neg\neg P \leftrightarrow P$ is provable in WE-HA$^\omega$. Hence every $\exists$-free formula is equivalent to a negative formula in WE-HA$^\omega$.

Remark 5.4. 1) For $\exists$-free formulas A we have $(\underline{x}\ mr\ A) \equiv A$ with $\underline{x}$ being the empty tuple.
2) $(\underline{x}\ mr\ A)$ is always an $\exists$-free formula.

We will also need a variant 'modified realizability with truth' *mrt* of *mr*:

Definition 5.5. $\underline{x}\ mrt\ A$ is defined analogously to $\underline{x}\ mr\ A$ except that clause (iv) is replaced by

$$(iv)'\ \underline{y}\ mrt\ (A \to B) :\equiv \forall \underline{x}(\underline{x}\ mrt\ A \to \underline{y}\underline{x}\ mrt\ B) \wedge (A \to B).$$

The name 'modified realizability with truth' is motivated by the following

Lemma 5.6. WE-HA$^\omega \vdash (\underline{x}\ mrt\ A) \to A$, *for every formula* A.

Proof: Straightforward. □

Proposition 5.7. WE-HA$^\omega \vdash (\underline{x}\ mrt\ \neg A) \leftrightarrow \neg A$ *for every formula* A*, where* $\underline{x}$ *is the empty tuple.*

Proof: Exercise! □

The schema of choice AC:$= \bigcup\limits_{\rho,\tau \in \mathbf{T}} \{\mathrm{AC}^{\rho,\tau}\}$ is given by

$$\mathrm{AC}^{\rho,\tau}: \ \forall x^\rho \exists y^\tau A(x,y) \to \exists Y^{\tau\rho} \forall x^\rho A(x, Yx),$$

where A is an arbitrary formula of E-HA$^\omega$.

The independence-of-premise-schema $\mathrm{IP}^\omega_{ef} := \bigcup\limits_{\rho \in \mathbf{T}} \{\mathrm{IP}^\rho_{ef}\}$ for $\exists$-free formulas is given by

$$\mathrm{IP}^\rho_{ef}: \ \big(A \to \exists x^\rho B(x)\big) \to \exists x^\rho \big(A \to B(x)\big),$$

where A is $\exists$-free and doesn't contain x free.

In the following we use $\mathcal{O}^\rho := \lambda x_1^{\rho_1}, \ldots, x_k^{\rho_k}.0^0$ for $\rho = 0(\rho_k)\ldots(\rho_1)$.

Theorem 5.8 (soundness of *mr***, Troelstra [366]).** *Let* A *be an arbitrary formula in* $\mathcal{L}$(E-HA$^\omega$) *and* Δ_{ef} *be an arbitrary set of* $\exists$*-free sentences. Then the following rule holds*

$$\text{E-HA}^\omega + \mathrm{AC} + \mathrm{IP}^\omega_{ef} + \Delta_{ef} \vdash A \ \Rightarrow \ \text{E-HA}^\omega + \Delta_{ef} \vdash \underline{t}\ mr\ A,$$

where $\underline{t}$ *is a suitable tuple of terms of* E-HA$^\omega$ *with* $FV(\underline{t}) \subseteq FV(A)$ *which can be extracted from a given proof of* A.

Remark 5.9. Let in $A(\underline{a})$ the tuple $\underline{a}$ denote all the free variables in A. Then the statement that there are terms $\underline{t}[\underline{a}]$ containing at most the free variables $\underline{a}$ such that $\underline{t}[\underline{a}]\ mr\ A(\underline{a})$ can equivalently be stated as: there are closed terms $\underline{s}$ such that $\underline{s}\underline{a}\ mr\ A(\underline{a})$. Just take $\underline{s} := \lambda \underline{a}.\underline{t}[\underline{a}]$.

Proof of theorem 5.8: Induction on the length of the derivation of A: We first treat the logical axioms and rules and then complete the proof by considering the non-logical axioms.

1) Logical axioms. $A \vee A \to A$: If $z^0, \underline{x}, \underline{y}$ *mr* $(A \vee A)$, then $\underline{t}z^0\underline{x}\,\underline{y}$ *mr* A, where $\underline{t}$ is such that $\underline{t}z^0\underline{x}\,\underline{y} := \begin{cases} \underline{x}, & \text{if } z = 0 \\ \underline{y}, & \text{if } z \neq 0 \end{cases}$

(t_i can easily be defined using R_0: exercise!). Hence $\underline{t}$ *mr* $(A \vee A \to A)$.

$A \to A \wedge A$ is realized by $\lambda \underline{x}.[\underline{x}, \underline{x}]$.

$A \to A \vee B$: Let $\underline{x}$ *mr* A, then $(0, \underline{x}, \underline{\mathcal{O}})$ *mr* $(A \vee B)$ and hence $\lambda \underline{x}.[0, \underline{x}, \underline{\mathcal{O}}]$ *mr* $(A \to A \vee B)$ (here $\underline{\mathcal{O}}$ is a suitable tuple $\mathcal{O}_1^{\rho_1}, \ldots, \mathcal{O}_k^{\rho_k}$ with suitable types ρ_i so that $\underline{\mathcal{O}}$ *mr* B is syntactically correct).

$A \wedge B \to A$: if $(\underline{x}, \underline{y})$ *mr* $A \wedge B$, then $\underline{x}$ *mr* A. Hence $\lambda \underline{x}, \underline{y}.\underline{x}$ *mr* $(A \wedge B \to A)$.

$\perp \to A$ is realized by $\underline{\mathcal{O}}$ where $\underline{\mathcal{O}}$ is such that $\underline{\mathcal{O}}$ *mr* A is syntactically correct. Then $\underline{\mathcal{O}}$ *mr* $(\perp \to A)$.

$A \vee B \to B \vee A$ is realized by $\lambda z^0, \underline{x}, \underline{y}.[\overline{sg}(z), \underline{y}, \underline{x}]$, where $\overline{sg}(z) := \begin{cases} 0^0, & \text{if } z \neq 0 \\ 1^0, & \text{otherwise.} \end{cases}$

$A \wedge B \to B \wedge A$ is realized by $\lambda \underline{y}, \underline{x}.[\underline{x}, \underline{y}]$.

$\forall x^\rho A(x) \to A(t^\rho)$: Let $\underline{y}$ *mr* $\forall x A(x)$. Then $\underline{y}(t)$ *mr* $A(t)$. Hence $\lambda \underline{y}.\underline{y}(t)$ *mr* $(\forall x A(x) \to A(t))$.

$A(t^\rho) \to \exists x^\rho A(x)$ is realized by $\lambda \underline{y}.[t, \underline{y}]$, where $\underline{y}$ is a tuple of variables such that $\underline{y}$ *mr* $A(t)$ is well-formed.

2) Rules. $\frac{A,\, A \to B}{B}$: Assume $\underline{t}$ *mr* A and $\underline{s}$ *mr* $(A \to B)$. Let $\underline{r}$ be the terms which result from $\underline{s}\,\underline{t}$ by replacing all free variables $\underline{a}$ which occur in A but not in B by $\underline{\mathcal{O}}$. Then $\underline{r}$ *mr* B.

$\frac{A \to B,\, B \to C}{A \to C}$: $\underline{s}$ *mr* $(A \to B)$, $\underline{t}$ *mr* $(B \to C)$. If $\underline{x}$ *mr* A, then $\underline{s}\underline{x}$ *mr* B and hence $\underline{t}(\underline{s}\underline{x})$ *mr* C. Thus $\lambda \underline{x}.\underline{t}(\underline{s}\underline{x})$ *mr* $(A \to C)$ (if necessary replace free variables which don't occur in $A \to C$ by $\mathcal{O}$).

$\frac{A \wedge B \to C}{A \to (B \to C)}$ and $\frac{A \to (B \to C)}{A \wedge B \to C}$ are trivially satisfied: use the terms from the premise for the conclusion.

$\frac{A \to B}{C \vee A \to C \vee B}$: Assume $\underline{t}$ *mr* $(A \to B)$ and $z^0, \underline{x}, \underline{y}$ *mr* $(C \vee A)$. Then either $z = 0$, $\underline{x}$ *mr* C or $z \neq 0$, $\underline{y}$ *mr* A. In the second case we have $\underline{t}\,\underline{y}$ *mr* B. Hence $\lambda z^0, \underline{x}, \underline{y}.[z^0, \underline{x}, \underline{t}\,\underline{y}]$ *mr* $(C \vee A \to C \vee B)$.

$\frac{B \to A(x^\rho)}{B \to \forall x^\rho A(x)}$: Assume $\underline{t}[x]$ *mr* $(B \to A(x))$ and $\underline{z}$ *mr* B. Then $\lambda x.(\underline{t}[x]\underline{z})$ *mr* $\forall x A(x)$ and, therefore, $\lambda \underline{z}, x.(\underline{t}[x]\underline{z})$ *mr* $(B \to \forall x A(x))$.

$\frac{A(x^\rho) \to B}{\exists x^\rho A(x) \to B}$: Assume $\underline{t}[x]$ *mr* $(A(x) \to B)$ and $x, \underline{z}$ *mr* $\exists x A(x)$. Then $\underline{z}$ *mr* $A(x)$ and, therefore, $\underline{t}[x]\underline{z}$ *mr* B. Thus $\lambda x, \underline{z}.(\underline{t}[x]\underline{z})$ *mr* $(\exists x A(x) \to B)$.

3) Axioms for $=_0, S, \Pi, \Sigma, R$ **and** E_ρ**:** These axioms are all $\exists$-free and, therefore, realized by themselves.

4) The induction schema: Let $\underline{x}$ *mr* $A(0)$ and $\underline{y}$ *mr* $\forall z^0(A(z) \to A(z+1))$. Define $\underline{t}$ by simultaneous primitive recursion such that

$$\begin{cases} \underline{t}\,\underline{x}\,\underline{y}0 = \underline{x} \\ \underline{t}\,\underline{x}\,\underline{y}(z+1) = \underline{y}z(\underline{t}\,\underline{x}\,\underline{y}z). \end{cases}$$

By induction on z^0 one shows that $\underline{t}\,\underline{x}\,\underline{y}z$ *mr* $A(z)$ and hence $\underline{t}\,\underline{x}\,\underline{y}$ *mr* $\forall z A(z)$.
5) The interpretations for AC and $\mathrm{IP}^{\omega}_{ef}$ are trivial (note that AC and $\mathrm{IP}^{\omega}_{ef}$ are not needed to verify their mr-interpretation).
6) The sentences in Δ_{ef} are (by remark 5.4.1) realized by themselves with the empty tuple of realizers. □

Remark 5.10. At various places in the soundness proof we could have used an arbitrary term of suitable type and, for definiteness, chose $\mathcal{O}$. In applications it can sometimes be crucial to use a variable instead which will be fixed only later depending on the parameters of the situation at hand.

Remark 5.11. Theorem 5.8 (and its proof) also extends immediately to the case where the language $\mathcal{L}(\text{E-HA}^{\omega})$ is extended by new constants c^{ρ} of any type ρ which may occur in the axioms Δ_{ef}. Then the terms $\underline{t}$ are built-up out of the original constants, the new constants and the free variables of A.

Theorem 5.12 (Characterization theorem for *mr*, Troelstra [366]). *Let A be an arbitrary formula of $\mathcal{L}(\text{E-HA}^{\omega})$. Then*

$$\text{E-HA}^{\omega} + \text{AC} + \text{IP}^{\omega}_{ef} \vdash A \leftrightarrow \exists \underline{x}(\underline{x}\ mr\ A).$$

Proof: Easy induction on the logical structure of A. □

The soundness theorem together with the characterization theorem for modified realizability allows us to derive the following

Theorem 5.13 (Main theorem on program extraction by mr).
Let $\forall x^{\rho} \exists y^{\tau} A(x,y)$ be a sentence of $\mathcal{L}(\text{E-HA}^{\omega})$ with arbitrary types ρ, τ and Δ_{ef} be an arbitrary set of $\exists$-free sentences. Then the following rule holds

$$\begin{aligned} &\text{E-HA}^{\omega} + \text{AC} + \text{IP}^{\omega}_{ef} + \Delta_{ef} \vdash \forall x^{\rho} \exists y^{\tau} A(x,y) \ \Rightarrow \\ &\text{E-HA}^{\omega} + \text{AC} + \text{IP}^{\omega}_{ef} + \Delta_{ef} \vdash \forall x^{\rho} A(x,tx), \end{aligned}$$

where t is a closed term of E-HA$^{\omega}$ *which is extracted from a given proof of the premise by modified realizability.*
In particular, the conclusion yields that

$$\mathcal{S}^{\omega} \models \forall x^{\rho} A(x,tx),$$

if $\mathcal{S}^{\omega} \models \Delta_{ef}$.
The theorem also applies to the situation with tuples $\underline{x}, \underline{y}$ of variables (of any types).

Proof: From the premise we get by the soundness theorem a tuple of closed terms $t_1, \ldots, t_n$ such that

$$\text{E-HA}^\omega + \Delta_{ef} \vdash \underline{t} \; mr \; \forall x \exists y A(x,y),$$

i.e.

$$\text{E-HA}^\omega + \Delta_{ef} \vdash t_2,\ldots,t_n \; mr \; \forall x A(x, t_1 x).$$

Together with the characterization theorem this yields the conclusion of the theorem with $t := t_1$. □

The (proof of the) previous theorem shows in particular that subproofs of $\exists$-free lemmas used in a given proof do not need to be analyzed at all for the extraction of programs by modified realizability. The significance of this is further expressed by the following proposition which implies that theorem 5.13 also holds for arbitrary negated sentences $\neg A$ included into Δ_{ef} and with the independence-of-premise schema $\text{IP}^\omega_\neg = \bigcup\limits_{\rho \in \mathbf{T}} \{\text{IP}^\rho_\neg\}$ for negated formulas

$$\text{IP}^\rho_\neg : \; \big(\neg A \to \exists x^\rho B(x)\big) \to \exists x^\rho \big(\neg A \to B(x)\big)$$

instead of IP^ω_{ef}.

Proposition 5.14. *Let A be an arbitrary formula of $\mathcal{L}$(E-HA$^\omega$). Then one can construct an $\exists$-free formula B such that*

$$\text{E-HA}^\omega + \text{AC} + \text{IP}^\omega_{ef} \vdash \neg A \leftrightarrow B.$$

Proof: By the characterization theorem we have

$$\text{E-HA}^\omega + \text{AC} + \text{IP}^\omega_{ef} \vdash \neg A \leftrightarrow \forall \underline{y}((\underline{y} \; mr \; A) \to \bot),$$

where '$\forall \underline{y}(\underline{y} \; mrA) \to \bot)$' is $\exists$-free as – by remark 5.4.2) – '$\underline{y} \; mr \; A$' is $\exists$-free. □

Remark 5.15. For HA instead of E-HA$^\omega$+AC+IP$^\omega_{ef}$ the above result is not correct: as shown in Troelstra [366] (3.8.2) the negated sentence

$$\neg \forall x (\neg\neg \exists y T(x,x,y) \to \exists y T(x,x,y))$$

is not equivalent to any $\exists$-free sentence in $\mathcal{L}$(HA) over HA.

Conversely (as mentioned already in remark 5.3), every $\exists$-free formula A is already in E-HA$^\omega$ provably equivalent to $\neg\neg A$ and hence to a negated formula (here one uses that the prime formulas of E-HA$^\omega$ are decidable).

Corollary 5.16. 1) *Over* E-HA$^\omega$+AC, *the principles* IP$^\omega_{ef}$ *and* IP$^\omega_\neg$ *are equivalent.*
2) *In theorem 5.13 we can also allow arbitrary negated sentences in Δ_{ef}.*

A class of formulas that is more general than the $\exists$-free formulas are the Harrop formulas:

Definition 5.17. A formula in $\mathcal{L}$(E-HA)$^\omega$ is called a Harrop formula if it belongs to the following inductively defined class of formulas:

1) Prime formulas are Harrop formulas.

2) With A, B also $A \wedge B$ and $\forall x A$ are Harrop formulas.
3) If B is a Harrop formula, then also $A \rightarrow B$ is a Harrop formula (where A is arbitrary).

Clearly any negated formula $\neg A \equiv A \rightarrow \bot$ is a Harrop formula. Over E-HA$^\omega$ every Harrop formula A is equivalent to a negated formula, namely to $\neg\neg A$:

Lemma 5.18. *For any Harrop formula A one has*

$$\text{E-HA}^\omega \vdash A \leftrightarrow \neg\neg A.$$

Proof: Induction on A, using the stability of prime formulas. We leave the details as an exercise. □

From corollary 5.16 and lemma 5.18 it follows that we may also include Harrop formulas in our set of axioms Δ_{ef}.

Definition 5.19 (Troelstra [366]). The subset Γ_1 of formulas $\in \mathcal{L}(\text{E-HA}^\omega)$ is defined inductively by

1) Prime formulas are in Γ_1 (note that in our theories quantifier-free formulas can be written as prime formulas $s =_0 t$, see proposition 3.17).
2) $A, B \in \Gamma_1 \Rightarrow A \wedge B, A \vee B, \forall x A(x), \exists x A(x) \in \Gamma_1$.
3) If A is $\exists$-free and $B \in \Gamma_1$, then $(\exists \underline{x} A \rightarrow B) \in \Gamma_1$.

Lemma 5.20. *For $A \in \Gamma_1$ we have*

$$\text{E-HA}^\omega \vdash (\underline{x} \; mr \; A) \rightarrow A.$$

Proof: Straightforward induction on the generation of Γ_1. □

Corollary 5.21. E-HA$^\omega$+ AC + IP$^\omega_{ef}$ *is conservative over* E-HA$^\omega$ *with respect to formulas $A \in \Gamma_1$.*
In particular E-HA$^\omega$+ AC + IP$^\omega_{ef}$ *is consistent relative to* E-HA$^\omega$ *since* $(0 = 1) \in \Gamma_1$.

Remark 5.22. One can show by much more complicated methods that E-HA$^\omega$+ AC is conservative over HA. For a 'neutral' version of E-HA$^\omega$ without extensionality this is due to [138, 139, 140] (see also [277, 312]). The extension to E-HA$^\omega$ is due to [14].
In contrast to corollaries 5.21 and 5.24 (below), this result does not relativize to subsystems with restricted induction (see [214]).

In the following, the notation $\mathcal{T} \overset{+}{\_} A$ indicates that we consider both cases: $\mathcal{T}$ and $\mathcal{T} + A$.

Theorem 5.23 (soundness of ***mrt*****, Troelstra [369]).** *Let Λ be an arbitrary set of negated formulas $\neg B$ of $\mathcal{L}$(E-HA$^\omega$) and* H$^\omega$:= E-HA$^\omega \overset{+}{_}$ AC $\overset{+}{\_}$ IP$^\omega_\neg$ $\overset{+}{\_} \Lambda$ *and A be an arbitrary formula in $\mathcal{L}$(E-HA$^\omega$). The following rule holds*

$$\text{H}^\omega \vdash A \;\Rightarrow\; \text{H}^\omega \vdash \underline{t} \; mrt \; A,$$

where $\underline{t}$ is a suitable tuple of terms of E-HA$^\omega$ *with $FV(\underline{t}) \subseteq FV(A)$ which can be extracted from a given proof of A.*

Proof: The treatment of the logical axioms is analogous to the one for the *mr*-interpretation in the proof of theorem 5.8. The same applies for the modus ponens. For the remaining rules the new second clause in the *mrt*-interpretation of the conclusion follows from the corresponding second clause(s) of the premise(s) using the same rule. Only in the case of the exportation rule one has to be a bit more careful because of the nested implications in the conclusion:
By induction hypothesis we have terms $\underline{s}$ such that

$$\underline{s}\ mrt\ (A \wedge B \rightarrow C).$$

Hence

$$\forall \underline{x}, \underline{y}(\underline{x}\ mrt\ A \wedge \underline{y}\ mrt\ B \rightarrow \underline{s}\underline{x}\underline{y}\ C) \wedge (A \wedge B \rightarrow C),$$

which is equivalent to

$$\forall \underline{x}(\underline{x}\ mrt\ A \rightarrow \forall \underline{y}(\underline{y}\ mrt\ B \rightarrow \underline{s}\underline{x}\underline{y}\ mrt\ C)) \wedge (A \rightarrow (B \rightarrow C)).$$

By lemma 5.6 we have $(\underline{x}\ mrt\ A) \rightarrow A$. Hence

$$\forall \underline{x}(\underline{x}\ mrt\ A \rightarrow \forall \underline{y}(\underline{y}\ mrt\ B \rightarrow \underline{s}\underline{x}\underline{y}\ mrt\ C) \wedge (B \rightarrow C)) \wedge (A \rightarrow (B \rightarrow C))$$

and hence

$$\forall \underline{x}(\underline{x}\ mrt\ A \rightarrow \underline{s}\underline{x}\ mrt\ (B \rightarrow C)) \wedge (A \rightarrow (B \rightarrow C)),$$

i.e.

$$\underline{s}\ mrt\ A \rightarrow (B \rightarrow C).$$

We leave it as an exercise to the reader to adopt the *mr*-interpretation of the non-logical axioms and rules from the proof of theorem 5.8 to the *mrt*-interpretation. For the interpretation of Λ and IP$^\omega_\neg$ we use proposition 5.7 (and – in the case of IP$^\omega_\neg$ – also lemma 5.6). □

Corollary 5.24. *Let* H$^\omega$:= E-HA$^\omega$ + AC + IP$^\omega_\neg$ + Λ. *Then the following rules hold:*

1)

$$\mathrm{H}^\omega \vdash A \vee B \Rightarrow \mathrm{H}^\omega \vdash A \text{ or } \mathrm{H}^\omega \vdash B,$$

for closed formulas $A \vee B$ (disjunction property DP*)*

2)

$$\mathrm{H}^\omega \vdash \exists x^\rho A(x) \Rightarrow \mathrm{H}^\omega \vdash A(t),$$

for a suitable term t^ρ of H$^\omega$ *with $FV(t) \subseteq FV(A) \setminus \{x^\rho\}$ (the special case of this property for closed formulas $\exists x^\rho A(x)$ is called existence property* EP*)*

3)

$$\mathrm{H}^\omega \vdash \forall x^\rho \exists y^\tau A(x,y) \Rightarrow \mathrm{H}^\omega \vdash \exists Y^{\tau\rho} \forall x^\rho A(x, Yx)$$

(closure of H$^\omega$ *under the rule of choice* ACR*).*

4)
$$\mathrm{H}^\omega \vdash (\neg A \to \exists x^\rho B(x)) \;\Rightarrow\; \mathrm{H}^\omega \vdash \exists x^\rho (\neg A \to B(x^\rho)),$$

where A is an arbitrary formula that doesn't contain x free (closure of H^ω *under the rule of independence of premise for negated formulas* $\mathrm{IPR}^\omega_\neg$*).*

2)–4) *also hold for tuples of variables $\underline{x}, \underline{y}$.*

Proof: 1) Suppose that $\mathrm{H}^\omega \vdash A \vee B$ for a closed formula $A \vee B$. By theorem 5.23 one finds closed terms $t^0, \underline{s}, \underline{r}$ such that

$$\mathrm{H}^\omega \vdash (t =_0 0 \to \underline{s}\ mrt\ A) \wedge (t \neq_0 0 \to \underline{r}\ mrt\ B).$$

In E-HA$^\omega$ the closed number term t^0 can be reduced (computed) to a numeral $\overline{n}$ and so

$$\mathrm{H}^\omega \vdash t =_0 \overline{n}.$$

The conclusion now follows from the fact that

$$\mathrm{H}^\omega \vdash \overline{n} = 0 \text{ or } \mathrm{H}^\omega \vdash \overline{n} \neq 0$$

and lemma 5.6.

2) By theorem 5.23 the assumptions yields terms $t^\rho, \underline{s}$ with $FV(t,\underline{s}) \subseteq FV(A) \setminus \{x\}$ such that

$$\mathrm{H}^\omega \vdash \underline{s}\ mrt\ A(t).$$

The claim now follows using lemma 5.6.

3) By 2) applied to the open formula $\exists y^\tau A(x^\rho, y^\tau)$ we get a term $t[x^\rho]^\tau$ such that

$$\mathrm{H}^\omega \vdash A(x, t[x]).$$

The conclusion follows by taking $Y := \lambda x^\rho . t[x^\rho]$.

4) Theorem 5.23 applied to

$$\mathrm{H}^\omega \vdash \neg A \to \exists x^\rho B(x)$$

yields terms $t^\rho, \underline{s}$ such that (using exercise 5.7)

$$\mathrm{H}^\omega \vdash \neg A \to \underline{s}\ mrt\ B(t)$$

and hence (by lemma 5.6)

$$\mathrm{H}^\omega \vdash \neg A \to B(t)$$

and so

$$\mathrm{H}^\omega \vdash \exists x(\neg A \to B(x)),$$

where x is not free in A. □

Remark 5.25. 1) Corollary 5.24 is proved analogously for

$$\mathrm{H}^\omega := \text{WE-HA}^\omega + \text{AC} + \mathrm{IP}^\omega_\neg + \Lambda$$

since the soundness proof for *mrt* also applies to the weakly extensional case.

2) For E-HA$^\omega$+AC+IP$^\omega_\neg$ this corollary can be obtained by ordinary modified realizability alone using theorem 5.12 and corollary 5.16.

Definition 5.26. The so-called Markov Principle in all finite types is the schema $\mathrm{M}^\omega := \bigcup_{\underline{\rho}\in\mathbf{T}} \{\mathrm{M}^{\underline{\rho}}\}$, where

$$\mathrm{M}^{\underline{\rho}}: \neg\neg\exists \underline{x}^{\underline{\rho}} A_0(\underline{x}) \rightarrow \exists \underline{x}^{\underline{\rho}} A_0(\underline{x}),$$

where A_0 is an arbitrary quantifier-free formula of WE-HA$^\omega$ and $\underline{x}$ is a tuple of variables of arbitrary types $\underline{\rho}$ ($A_0(\underline{x})$ may contain further free variables in addition to $\underline{x}$).

Even the special case M^0 of the Markov principle has no modified realizability interpretation by computable realizers (see exercise 2 below). Actually, one of the main early applications of modified realizability was to show the underivability of the Markov principle in Heyting arithmetic. In chapter 8 we will develop Gödel's functional interpretation which can be viewed as a much refined version of modified realizability and which does interpret M^ω in a simple way. Whereas modified realizability leaves $\exists$-free formulas unchanged, functional interpretation analyses $\exists$-free formulas further down to purely universal formulas.

5.2 Remarks on fragments of E-HA$^\omega$

The main theorems of this chapter, namely the soundness theorems for the *mr*- and *mrt*-interpretations (theorems 5.8, 5.23) as well as the main theorem on program extraction (theorem 5.13) can – with minor adaptations of their proofs – also be obtained for any of the fragments $\mathcal{T}_i^\omega := \widehat{\text{E-HA}}^\omega\!\upharpoonright$, E-G$_nA_i^\omega$ ($n \geq 2$) instead of E-HA$^\omega$. Then the terms extracted, of course, are closed terms of $\mathcal{T}_i^\omega$. For this it is sufficient to note that

- the *mr*,*mrt*-interpretation of the logical axioms and rules, as well as of the axioms of extensionality, axiom of choice and the independence of premise principles we considered can be carried out using only λ-terms (i.e. Π,Σ-combinators), constants $0,1$, definition by cases and the function $\overline{sg}$ which are available in $\mathcal{T}_i^\omega$,
- every quantifier-free formula $A_0(\underline{a})$ can be written as $t_{A_0}(\underline{a}) =_0 0$ (provably in $\mathcal{T}_i^\omega$) using only propositional logic and $x\cdot y =_0 0 \leftrightarrow x =_0 0 \vee y =_0 0$, $x+y =_0 0 \leftrightarrow x =_0 0 \wedge y =_0 0$, $\overline{sg}(x) =_0 0 \leftrightarrow x \neq 0$, and $|x-y| =_0 0 \leftrightarrow x =_0 y$ which all have *mr*,*mrt*-interpretations in $\mathcal{T}_i^\omega$ (except for the first one these equivalences are left unchanged by the interpretation),
- the axiom schema of quantifier-free induction can be realized using bounded search (again available in $\mathcal{T}_i^\omega$,) using the previous item,
- all other non-logical axioms can be written as purely universal axioms.

Using this one can show also the following:

Definition 5.27. 1) $\mathrm{IA}_{\neg}$ denotes the schema of induction for arbitrary negated formulas, i.e.

$$\mathrm{IA}_{\neg} : \neg A(0) \wedge \forall x^0(\neg A(x) \to \neg A(S(x))) \to \forall x^0 \neg A(x).$$

2) IA_{ef} denotes the schema of induction for arbitrary $\exists$-free formulas A_{ef}, i.e.

$$\mathrm{IA}_{ef} : A_{ef}(0) \wedge \forall x^0(A_{ef}(x) \to A_{ef}(S(x))) \to \forall x^0 A_{ef}(x),$$

where A_{ef} is $\exists$-free.

Proposition 5.28. $\widehat{\text{E-HA}}^{\omega}\!\upharpoonright + \mathrm{AC} + \mathrm{IP}^{\omega}_{\neg} + \mathrm{IA}_{\neg}$ *has a modified realizability interpretation in* $\widehat{\text{E-HA}}^{\omega}\!\upharpoonright + \mathrm{IA}_{ef}$ *(and hence a-fortiori in* $\widehat{\text{E-HA}}^{\omega}\!\upharpoonright + \mathrm{IA}_{\neg}$*) by closed terms of* $\widehat{\text{E-HA}}^{\omega}\!\upharpoonright$.

Proof: Exercise!

Remark 5.29. By the comment after proposition 3.21, the schema Σ^0_1-IA is already derivable in $\widehat{\text{E-HA}}^{\omega}\!\upharpoonright + \mathrm{AC}$ and hence also covered by proposition 5.28.

Corollary 5.30. *The provably recursive functions of*

$$\widehat{\text{E-HA}}^{\omega}\!\upharpoonright + \mathrm{AC} + \mathrm{IP}^{\omega}_{\neg} + \mathrm{IA}_{\neg}$$

are just the ordinary primitive recursive ones.

Definition 5.31. $\mathrm{IA}_{\Pi^0_\infty\text{-}\neg}$ denotes the schema of induction for formulas of the form

$$\forall x_1^0 \exists x_2^0 \forall x_3^0 \ldots \neg A(x_1, x_2, x_3, \ldots),$$

where A is an arbitrary formula.

Definition 5.32. A function is definable by Gödel primitive recursion of level n if it can be defined by $0, S, \Pi, \Sigma$ and R_ρ where $deg(\rho) \leq n$.

In the following proposition $\widehat{\text{E-HA}}^{\omega}\!\upharpoonright + R_1$ is the extension of $\widehat{\text{E-HA}}^{\omega}\!\upharpoonright$ by the constant R_1 and its defining axioms.

Proposition 5.33. $\widehat{\text{E-HA}}^{\omega}\!\upharpoonright + R_1 + \mathrm{AC} + \mathrm{IP}^{\omega}_{\neg} + \mathrm{IA}_{\Pi^0_\infty\text{-}\neg}$ *has a modified realizability interpretation in* $\widehat{\text{E-HA}}^{\omega}\!\upharpoonright + R_1 + \mathrm{IA}_{ef}$ *(and hence a-fortiori in* $\widehat{\text{E-HA}}^{\omega}\!\upharpoonright + R_1 + \mathrm{IA}_{\neg}$*) by closed terms of* $\widehat{\text{E-HA}}^{\omega}\!\upharpoonright + R_1$.

Proof: Exercise! □

Corollary 5.34. *The provably recursive functions of*

$$\widehat{\text{E-HA}}^{\omega}\!\upharpoonright + R_1 + \mathrm{AC} + \mathrm{IP}^{\omega}_{\neg} + \mathrm{IA}_{\Pi^0_\infty\text{-}\neg}$$

are definable by Gödel primitive recursion of level 1*, i.e. by closed terms in* T_1.

More information on the use of modified realizability for fragments of E-HA$^\omega$ can be found in [207] and [212]. For an adaptation of the *mr, mrt*-interpretations to systems of feasible arithmetic see [71].

5.3 Exercises, historical comments and suggested further reading

Exercises:

1) Prove proposition 5.7.
2) Prove lemma 5.18.
3) Show, using modified realizability, that M^0 (even with parameters of type 0 only) is not derivable in E-HA$^\omega$+AC+IP$^\omega_{ef}$.
4) (P. Freyd) Show that E-HA$^\omega$+AC proves the axiom schema of dependent choice DAC:= $\cup_{\rho\in\mathbf{T}}\{\mathrm{DAC}^\rho\}$, where

$$\mathrm{DAC}^\rho:\ \ \forall x^\rho \exists y^\rho A(x,y) \to \forall x^\rho \exists f^{\rho 0}\big(f(0)=x \wedge \forall z^0 A(f(z), f(S(z)))\big).$$

5) Prove proposition 5.28.
6) Prove proposition 5.33.

Historical comments and suggested further reading: Modified realizability was introduced by Kreisel in [244, 246]. One of its main purposes was to show the independence of the Markov principle from HA. Troelstra [366] (chapter III, section 4) provides a very concise and detailed treatment of modified realizability interpretation. For a general survey on various forms of realizability interpretations see Troelstra [369]. Applications of modified realizability in the context of bounded arithmetic are given in Cook-Urquhart [71]. For applications of modified realizability to the extraction of a program from a specific proof see Berger [18] and Berger-Schwichtenberg [25] where it shown that the so-called 'normalization-by-evaluation' algorithm (due to Berger-Schwichtenberg [24]) can be extracted from the standard Tait-Troelstra proof of strong normalization for the typed λ-calculus by an appropriate form of modified realizability. Recently, an automated extraction of this algorithm has been carried out in Berger et al. [20]. For applications of other forms of realizability for the extraction of programs from proofs see Hayashi [150]. In chapter 7 we will discuss applications of a monotone variant of modified realizability (Kohlenbach [212]) to semi-constructive systems, i.e. system based on E-HA$^\omega$ but with various highly non-constructive classical comprehension principles added. Recently, a new bounded modified realizability was introduced in Ferreira-Nunes [103] and has interesting connections to the monotone variant.

Chapter 6
Majorizability and the fan rule

6.1 A syntactic treatment of majorization and the fan rule

In this chapter we combine modified realizability with the structural property of majorizability which we already used in chapter 3 to prove results on the growth of the definable functions of G_nA^ω. We first show – following Howard [163] – that majorizability applies as well to the closed terms of WE-HA$^\omega$. We will indicate the far-reaching use one can make of majorizability combined with proof interpretations by showing the closure of E-HA$^\omega$ + AC + IP$^\omega_\neg$ under the so-called fan rule. Many more applications of this majorization technique combined with proof interpretations will be given throughout the rest of this book. We first recall the concept of majorizability from definition 3.34:

x^* maj_ρ x (read as 'x^* majorizes x') between functionals of type ρ is defined by induction on ρ:

$$\begin{cases} x^* \; maj_0 \; x :\equiv x^* \geq_0 x, \\ x^* \; maj_{\tau\rho} \; x :\equiv \forall y^*, y(y^* \; maj_\rho \; y \to x^* y^* \; maj_\tau \; xy). \end{cases}$$

Definition 6.1. Define $\varphi^{1(1)}$ by recursion (using only R_0) such that (provably in WE-HA$^\omega$)

$$\varphi(x^1, 0) =_0 x0$$
$$\varphi(x, z+1) =_0 \max_0(\varphi(x,z), x(z+1)),$$

where $\max_0$ is the usual (primitive recursively – using only R_0 – definable) maximum between natural numbers. We write $x^M := \lambda z^0.\varphi(x,z)$ (note that $x^M(z) = \max_{i \leq z}(x(i))$).

This definition easily extends to finite types by λ-abstraction: For x of type $\rho 0$ with $\rho = 0\rho_k \ldots \rho_1$ we define $x^M := \lambda z, \underline{v}.\varphi(\lambda z.xz\underline{v}, z)$, where $\underline{v} = v_1^{\rho_1}, \ldots, v_k^{\rho_k}$.

One easily proves the following

Lemma 6.2. WE-HA$^\omega \vdash \forall x^{\rho 0}\big(x^M 0 =_\rho x0 \wedge x^M(z+1) =_\rho \max_\rho(x^M z, x(z+1))\big)$, *where* $\max_{\tau\rho}(x_1,x_2) := \lambda y^\rho.\max_\tau(x_1 y, x_2 y)$ *for complex types.*

Remark 6.3. Using recursion of type ρ one can define x^M directly by iteration of $\max_\rho$. However our a bit more complicated approach shows that actually R_0 is sufficient.

Lemma 6.4. WE-HA$^\omega \vdash \forall x^{\rho 0}, \tilde{x}^{\rho 0}\big(\forall n^0(\tilde{x}n\ maj_\rho\ xn) \rightarrow \tilde{x}^M\ maj_{\rho 0} x\big)$.

Proof: Let $\rho = 0\rho_k \ldots \rho_1$ and $\underline{v} = v_1^{\rho_1}, \ldots, v_k^{\rho_k}$. One easily shows by (quantifier-free) induction on n that

$$\forall n^0\big(\forall m \le n(\tilde{x}^M n\underline{v} \ge_0 \tilde{x}m\underline{v})\big).$$

Together with the assumption that $\forall n(\tilde{x}n\ maj_\rho\ xn)$ this yields

$$\forall n, m, \underline{v}^*, \underline{v}\big(n \ge_0 m \wedge \underline{v}^*\ maj\ \underline{v} \rightarrow \tilde{x}^M n\underline{v}^* \ge_0 xm\underline{v}\big)$$

and hence $\tilde{x}^M\ maj_\rho\ x$ by lemma 3.35(iii) and remark 3.36.2). □

Corollary 6.5. WE-HA$^\omega \vdash \forall x^1(x^M\ maj_1\ x)$.

Proposition 6.6 (W.A. Howard [163]). *For each closed term* t^ρ *of* WE-HA$^\omega$ *one can construct a closed term* $t^{*\rho}$ *of* WE-HA$^\omega$ *such that*

$$\text{WE-HA}^\omega \vdash t^*\ maj_\rho\ t.$$

Proof: Induction on the structure of t:
Constants c: $0^0\ maj_0\ 0^0$, $S\ maj_1\ S$. Using lemma 3.35(iii),(i) we also have

$$\Pi_{\rho,\tau}\ maj\ \Pi_{\rho,\tau} \text{ and } \Sigma_{\delta,\rho,\tau}\ maj\ \Sigma_{\delta,\rho,\tau}.$$

$\underline{R}_{\underline{\rho}}$: Let $\underline{y}^*\ maj\ \underline{y}$, i.e. $\bigwedge_{i=1}^{k}(y_i^*\ maj\ y_i)$, and $\underline{z}^*\ maj\ \underline{z}$. By induction on x^0 one shows

$$\forall x^0\big(\underline{R}_{\underline{\rho}} x\underline{y}^*\underline{z}^*\ maj\ \underline{R}_{\underline{\rho}} x\underline{y}\underline{z}\big),$$

which, again by lemma 3.35(iii), implies

$$\forall x^0\big(\underline{R}_{\underline{\rho}} x\ maj\ \underline{R}_{\underline{\rho}} x\big).$$

Hence by lemma 6.4

$$R_i^* := (\underline{R}_i)^M_{\underline{\rho}}\ maj\ (\underline{R}_i)_{\underline{\rho}} \text{ for } i = 1, \ldots, k,$$

i.e. $\underline{R}^*\ maj\ \underline{R}$.
So for every constant c of WE-HA$^\omega$ we have a closed term t^* such that

$$\text{WE-HA}^\omega \vdash t^*\ maj\ c.$$

The proposition now follows from that fact that $t^*\ maj_{\tau\rho}\ t \wedge s^*\ maj_\rho\ s$ implies $t^*s^*\ maj_\tau\ ts$. □

Remark 6.7. Proposition 6.6 also holds with 's-maj' from definition 3.48 instead of 'maj' and one can use the same term t^* as constructed in the proof of proposition 6.6.

Theorem 6.8. H^ω:=E-HA$^\omega \underline{+}$ AC $\underline{+}$ IP$^\omega_\neg$. *Let s be a closed term, $A(x,y,z)$ a formula containing only x,y,z as free variables and $\deg(\tau) \le 2$. Then the following rule holds:*

$$\begin{cases} \mathrm{H}^\omega \vdash \forall x^1 \forall y \le_\rho sx \exists z^\tau A(x,y,z) \Rightarrow \\ \mathrm{H}^\omega \vdash \forall x^1 \forall y \le_\rho sx \exists z \le_\tau txA(x,y,z), \end{cases}$$

where t is a suitable closed term which can be extracted from a given proof of the assumption.
The theorem also holds for tuples of variables $\underline{x} = x_1^{\delta_1}, \dots, x_k^{\delta_k}, \underline{y} = y_1^{\rho_1}, \dots, y_m^{\rho_m}, \underline{s} = s_1, \dots, s_m, \underline{z} = z_1^{\tau_1}, \dots, z_n^{\tau_n}$ with $deg(\delta_i) \le 1$, $y_j \le_{\rho_j} s_j\underline{x}$ and $deg(\tau_l) \le 2$ for $i = 1, \dots, k$; $j = 1, \dots, m$; $l = 1, \dots, n$. Then we have a tuple of closed terms $\underline{t}$ instead of t.

Remark 6.9. 1) Note that in the previous theorem the bound tx on '$\exists z$' does not depend on y.
2) In the following, we will in order to keep the notational complexity down only treat here and in other results the case $k = m = n = 1$. The proof immediately extends to the case with tuples. Actually, the result for the special case also implies directly the one with tuples on the expense of using the coding of tuples of variables as mentioned in chapter 3.

Corollary 6.10 (Fan Rule, Kreisel, Troelstra [368]). *Let A be a formula of $\mathcal{L}$(E-HA$^\omega$) containing only free variables of type levels ≤ 1. Then for H^ω as above the following rule holds*

$$\begin{cases} \mathrm{H}^\omega \vdash \forall y^1 \exists n^0 A(y,n) \Rightarrow \\ \mathrm{H}^\omega \vdash \forall x^1 \exists m^0 \forall y \le_1 x \exists n \le_0 mA(y,n). \end{cases}$$

Proof of theorem 6.8: Suppose that

$$\mathrm{H}^\omega \vdash \forall x^1 \forall y \le_\rho sx \exists z^\tau A(x,y,z).$$

Then by corollary 5.24,2)–4) one can extract a closed term t such that

$$\mathrm{H}^\omega \vdash \forall x^1 \forall y \le_\rho sxA(x,y,txy).$$

By proposition 6.6 there are closed terms s^*, t^* such that

$$\text{E-HA}^\omega \vdash s^*\ maj\ s \wedge t^*\ maj\ t.$$

By corollary 6.5 we have in E-HA$^{\omega}$:

$$\forall x^1(s^*x^M \; maj_\rho \; sx)$$

and, therefore, by lemma 3.35.2

$$\forall x^1 \forall y \leq_\rho sx(s^*x^M \; maj_\rho \; y).$$

Hence

$$\forall x^1 \forall y \leq_\rho sx\big(t^*(x^M, s^*x^M) \; maj_\tau \; txy\big).$$

For simplicity lets now consider only the case $\tau = 2$:

$$\forall x^1 \forall y \leq_\rho sx \forall z^1 \big((t^*(x^M, s^*x^M))z^M \geq_0 txyz\big).$$

Hence

$$\forall x^1 \forall y \leq_\rho sx(\tilde{t}x \geq_2 txy),$$

where $\tilde{t} := \lambda x, z.[(t^*(x^M, s^*x^M))(z^M)]$. Thus $\tilde{t}$ satisfies the claim of the theorem. □

The proof of theorem 6.8 shows that the conditions on x (to be of type at most 1) and y (to be bounded by sx) where only used to construct majorants x^M and s^*x^M. The condition $deg(\tau) \leq 2$ was only needed to get from a majorant for txy a bound on txy. Hence we can also state the following variant of theorem 6.8:

Proposition 6.11. *Let* H$^{\omega}$ *be as in theorem 6.8 and* $A(x,y)$ *a formula containing only* x^ρ, y^τ *free where* ρ, τ *are arbitrary types. Then the following rule holds:*

$$\begin{cases} \mathrm{H}^\omega \vdash \forall x^\rho \exists y^\tau A(x,y) \Rightarrow \\ \mathrm{H}^\omega \vdash \forall x, x^*(x^* \; maj \; x \to \exists y(tx^* \; maj_\tau \; y \wedge A(x,y)), \end{cases}$$

where t is a suitable closed term which can be extracted from a given proof of the assumption.
The result also holds for tuples $\underline{x}, \underline{y}$ *of variables.*

Application of theorem 6.8: Consider the special representation of the unit interval $[0,1]$ given at the end of chapter 4. Let $\Phi^{1(1)(0)}_{(\cdot)}, \Phi^{1(1)}$ be closed terms of E-HA$^{\omega}$ which (provably in E-HA$^{\omega}$) represent functions $[0,1] \to \mathbb{R}$, i.e. which satisfy

$$\forall x, y \leq_1 N \forall n^0 \left(\tilde{x} =_\mathbb{R} \tilde{y} \to \Phi_n(\tilde{x}) =_\mathbb{R} \Phi_n(\tilde{y}) \wedge \Phi(\tilde{x}) =_\mathbb{R} \Phi(\tilde{y})\right)$$

and so, by proposition 4.23, are (uniformly) continuous. Then (switching back to the level of the mathematical meaning of the representation) theorem 6.8 implies the following rule:

$$\begin{cases} \text{E-HA}^\omega\text{+AC+IP}^\omega_\neg \vdash \forall k^0 \forall x \in [0,1] \exists n^0 \forall m \geq n \big(|\Phi_m(x) - \Phi(x)| <_{\mathbb{R}} 2^{-k} \big) \Rightarrow \\ \text{E-HA}^\omega\text{+AC+IP}^\omega_\neg \vdash \forall k^0 \exists n^0 \forall x \in [0,1] \forall m \geq n \big(|\Phi_m(x) - \Phi(x)| <_{\mathbb{R}} 2^{-k} \big), \end{cases}$$

i.e. provable pointwise convergence of Φ_n towards Φ on $[0,1]$ implies provable uniform convergence on $[0,1]$. It is well-known that already for E-HA$^\omega$-definable functions $\Phi_n, \Phi : [0,1] \to \mathbb{R}$ uniform convergence does not need to hold (not even ineffectively) if pointwise convergence is established using classical logic. Our application shows that such counterexamples do not exist if the pointwise convergence is proved essentially by constructive means. In chapter 7 we will show that this remains true even in the presence of various highly ineffective principles like full comprehension for negated formulas and others.
In chapter 10 we will prove versions of theorem 6.8 for systems based in full classical logic. Then, however, A has to be a purely existential formula which the formula in our application

$$(*) \ \forall m \geq n \big(|\Phi_m(x) - \Phi(x)| <_{\mathbb{R}} 2^{-k} \big)$$

is not.
Nevertheless, in the important special case where the pointwise convergence is monotone, e.g. decreasing, i.e. $\Phi_n \searrow \Phi$, $(*)$ can be written equivalently as the purely existential formula

$$|\Phi_n(x) - \Phi(x)| <_{\mathbb{R}} 2^{-k}$$

(recall that $<_{\mathbb{R}} \in \Sigma^0_1$). In this case, we will be able to extract moduli of uniform convergence even from ineffective proofs of pointwise convergence, which gives an effective rule-form of what is known as Dini's Theorem.

For further applications of modified realizability combined with majorization see [212] as well as chapter 7 which generalizes the results of this chapter to semi-constructive systems.

Remarks on fragments of E-HA$^\omega$:

Inspection of the proofs of this chapter easily shows that the results also hold for the fragments $\mathcal{T}_i^\omega := \widehat{\text{E-HA}}^\omega\!\restriction$, E-G$_nA_i^\omega$ $(n \geq 2)$ instead of E-HA$^\omega$. Here we use the remark at the end of chapter 5 and the fact that the majorization technique applies to these fragments as well. For E-G$_n$A$_i^\omega$ $(n = 1,2,3)$ we carried this out explicitly in chapter 3. For $\widehat{\text{(W)E-HA}}^\omega\!\restriction$ we just have to observe that quantifier-free induction suffices to prove that $R_0^M \ maj \ R_0$ and that R_0^M can be defined in $\widehat{\text{WE-HA}}^\omega\!\restriction$ using R_0. Fore more details see [201] and [207].

6.2 Exercises, historical comments and suggested further reading

Exercises:

1) Show that theorem 6.8 and corollary 6.10 both become false if E-HA$^{\omega}$ is replaced by E-PA$^{\omega}$ (or just E-HA$^{\omega}+\Sigma^0_1$-LEM) already for Π^0_1-formulas $A(y^1,n^0)$ (with only y,n free).
2) Show that theorem 6.8 becomes false (already for quantifier-free A_0) if 'x' is permitted to be of type 2 rather than 1 and that corollary 6.10 becomes false is A_0 is allowed to contain a parameter of type 2.
3) Study one of the usual counterexamples of sequences of functions $f_n \in C[0,1]$ which converge pointwise but not uniformly toward the constant-0 function and find out which part of the proof uses classical logic.

Historical comments and suggested further reading: Closure of E-HA$^{\omega}$ under the fan rule was first established in Troelstra [368] using a complicated technique of fan-computability. In Beeson [15] and Troelstra-van Dalen [371] different proofs (based on a forcing technique) are given. The proof presented here and the more general theorem 6.8 are taken from Kohlenbach [201].
Further applications of the combination of modified realizability with majorization can be found in Kohlenbach [201, 212] and (in a more general context) in Gerhardy-Kohlenbach [119]. Some of these applications will be treated in the next chapter.

Chapter 7
Semi-intuitionistic systems and monotone modified realizability

7.1 The soundness and bound extraction theorems

In the following we show that the results from chapter 6 even extend to the situation where large classes of non-effective sentences are added to the theories in question without having an impact on the extractability and complexity of uniform bounds. This rests on the following observation:

In the proof of theorem 6.8 the term t extracted by modified realizability (with truth) was only used to construct a majorizing term t^*. Only the latter term was then exploited for the construction of the uniform bound. This suggests a variant of the formulation in the soundness theorem where we extract instead of closed terms $\underline{t}$ such that

$$(*)\ \forall \underline{a}\big(\underline{t}\,\underline{a}\ mr\ A(\underline{a})\big)$$

closed terms $\underline{t}^*$ such that

$$(**)\ \exists \underline{z}\big(\underline{t}^*\ maj\ \underline{z} \wedge \forall \underline{a}(\underline{z}\,\underline{a}\ mr\ A(\underline{a}))\big).$$

We then say that '$\underline{t}^*$ satisfies the **monotone** modified realizability interpretation of $A(\underline{a})$' (here $\underline{a}$ are all the free variables of $A(\underline{a})$). Similarly for *mrt*.
The soundness theorems for *mr* and *mrt* proved in chapter 5 also hold with $(**)$ instead of $(*)$ which follows just by combining them with Howard's proposition 6.6. However, it is not actually necessary to first construct $\underline{t}$ by ordinary modified realizability and then – in a second step – to carry out the majorizing construction. In fact, inspection of the soundness proofs shows that one can **directly** construct $\underline{t}^*$ and prove property $(**)$ by induction on the given proof. This simplifies the construction and, moreover, allows one to extend the soundness theorem (in the form $(**)$) to further axioms as long as these axioms allow a monotone modified realizability interpretation (resp. monotone modified realizability-with-truth interpretation) by closed terms of E-HA$^{\omega}$. We will now show that this holds true for many highly non-effective axioms:

Let Θ be a set of sentences of the form $\exists \underline{v} \leq_{\underline{\sigma}} \underline{r} B_{ef}(\underline{v})$, where B_{ef} is $\exists$-free, and Ξ be a set of sentences of the form $\exists \underline{v} \leq_{\underline{\sigma}} \underline{r} \neg B(\underline{v})$. In both cases $\underline{r} = r_1^{\sigma_1}, \ldots, r_n^{\sigma_n}$ is a tuple of closed terms of E-HA$^\omega$ of arbitrary types and $\underline{v} \leq_{\underline{\sigma}} \underline{r} :\equiv \bigwedge_{i=1}^{n} (v_i \leq_{\sigma_i} r_i)$.

Theorem 7.1.

1) *Soundness of monotone modified realizability:*
Let H$^\omega$:= E-HA$^\omega$+ AC + IP$^\omega_{ef}$ + Θ *and* $A(\underline{a})$ *be an arbitrary formula in* $\mathcal{L}$(E-HA$^\omega$) *containing only* $\underline{a}$ *free. Then the following rule holds*

$$\mathrm{H}^\omega \vdash A(\underline{a}) \;\Rightarrow\; \text{E-HA}^\omega + \Theta \vdash \exists \underline{x}\, (\underline{t}^* \; maj \; \underline{x} \wedge \forall \underline{a}(\underline{x}\,\underline{a} \; mr \; A(\underline{a}))),$$

where $\underline{t}^*$ *is a suitable tuple of closed terms of* E-HA$^\omega$ *which can be extracted from a given proof of A.*

2) *Soundness of monotone modified realizability with truth:*
Let H$^\omega$:= E-HA$^\omega$ $\underline{+}$ AC $\underline{+}$ IP$^\omega_{\neg}$ $\underline{+}$ Ξ *and* $A(\underline{a})$ *be an arbitrary formula in* $\mathcal{L}$(E-HA$^\omega$) *containing only* $\underline{a}$ *free. Then the following rule holds*

$$\mathrm{H}^\omega \vdash A(\underline{a}) \;\Rightarrow\; \mathrm{H}^\omega \vdash \exists \underline{x}\, (\underline{t}^* \; maj \; \underline{x} \wedge \forall \underline{a}(\underline{x}\,\underline{a} \; mrt \; A(\underline{a}))),$$

where $\underline{t}^*$ *is a suitable tuple of closed terms of* E-HA$^\omega$ *which can be extracted from a given proof of A.*

Proof: 1) To simplify matters we use the deduction theorem to reduce the task of verifying the soundness of the monotone modified realizability interpretation to just the verification of the modus ponens rule. Any proof of the premise only uses finitely many Θ-sentences. For notational simplicity assume that only one such sentence Θ is used. Then we get

$$\mathrm{H}^\omega \setminus \{\Theta\} \vdash \Theta \rightarrow A(\underline{a}).$$

By the soundness theorem 5.8 for modified realizability we can extract closed terms $\underline{t}$ of E-HA$^\omega$ such that (using subsequent λ-abstraction)

$$\text{E-HA}^\omega \vdash \forall \underline{a}\, (\underline{t}\,\underline{a} \; mr \; (\Theta \rightarrow A(\underline{a}))).$$

By proposition 6.6 we can construct majorizing terms $\underline{t}^*$ for $\underline{t}$. Hence

$$\text{E-HA}^\omega \vdash \exists \underline{x}(\underline{t}^* \; \text{maj} \; \underline{x} \wedge \forall \underline{a}\, (\underline{x}\,\underline{a} \; mr \; (\Theta \rightarrow A(\underline{a})))),$$

i.e. $\underline{t}^*$ satisfies the monotone modified realizability interpretation of $\Theta \rightarrow A(\underline{a})$. Using lemma 3.35.2 we see that the monotone modified realizability interpretation of Θ is (provably in E-HA$^\omega + \Theta$) satisfied by any tuple of closed terms $\underline{r}^*$ such that

$$\text{E-HA}^\omega \vdash \bigwedge_{i=1}^{n} (r_i^* \; maj \; r_i).$$

One now easily verifies in E-HA$^\omega + \Theta$ that $\lambda \underline{a}.\underline{t}^*\underline{a}\,\underline{r}^*$ satisfies the monotone modified realizability interpretation of $A(\underline{a})$.

2) is proved analogously using the soundness theorem 5.23 for *mrt* and proposition 5.7 to show that Ξ has (provably in $\mathrm{E\text{-}HA}^\omega + \Xi$) a monotone *mrt*-interpretation by $\underline{r}^*$ constructed as above. □

Remark 7.2. 1) As mentioned already, an easier extraction of the terms $\underline{t}^*$ is obtained by treating directly all the rules by the monotone interpretations.
2) Theorem 7.1 holds analogously for 's-maj' instead of 'maj'.

Proposition 7.3. H^ω:=$\mathrm{E\text{-}HA}^\omega + \mathrm{AC} + \mathrm{IP}^\omega_\neg$. *Let s be a closed term,* $A(x^1, y^\rho, z^\tau)$ *a formula containing only* x, y, z *as free variables and* $\deg(\tau) \leq 2$. *Then the following rule holds:*

$$\begin{cases} \mathrm{H}^\omega + \Xi \vdash \forall x^1 \forall y \leq_\rho sx \exists z^\tau A(x,y,z) \ \Rightarrow \\ \mathrm{H}^\omega + \Xi \vdash \forall x^1 \forall y \leq_\rho sx \exists z \leq_\tau txA(x,y,z), \end{cases}$$

where t is a suitable closed term of $\mathrm{E\text{-}HA}^\omega$ *which can be extracted from a given proof of the assumption.*
As in theorem 6.8 the result also holds for tuples of variables.

Proof: The proposition follows similarly to theorem 6.8 using theorem 7.1.2). □

Remark 7.4. One easily obtains the variant of proposition 7.3 corresponding to proposition 6.11 (exercise).

As in the proof of corollary 6.10 one derives from proposition 7.3:

Corollary 7.5. *Let* H^ω *and* Ξ *be as in proposition 7.3. Then* $\mathrm{H}^\omega + \Xi$ *is closed under the fan rule.*

Definition 7.6. The schema of comprehension in all types for arbitrary negated formulas is given by

$$\mathrm{CA}^\omega_\neg : \ \exists \Phi^{0(\underline{\sigma})} \forall \underline{x}^{\underline{\sigma}} \big(\Phi(\underline{x}) =_0 0 \leftrightarrow \neg A(\underline{x}) \big),$$

where A is an arbitrary formula not containing Φ free and $\underline{\sigma}$ an arbitrary tuple of types.

Corollary 7.7. *Under the same assumptions as in proposition 7.3 we have the following rule:*

$$\begin{cases} \mathrm{H}^\omega + \mathrm{CA}^\omega_\neg \vdash \forall x^1 \forall y \leq_\rho sx \exists z^\tau A(x,y,z) \ \Rightarrow \\ \mathrm{H}^\omega + \mathrm{CA}^\omega_\neg \vdash \forall x^1 \forall y \leq_\rho sx \exists z \leq_\tau txA(x,y,z), \end{cases}$$

where t is a suitable closed term of $\mathrm{E\text{-}HA}^\omega$ *which can be extracted from a given proof of the assumption.*
As in theorem 6.8 the result also holds for tuples of variables.

Proof: The corollary follows from 7.3 observing that

1) we only have to consider closed instances of $\mathrm{CA}^\omega_\neg$ as parameters can be included together with $\underline{x}$ in the comprehension,

2) every closed instance of $\mathrm{CA}^{\omega}_{\neg}$ can be written as a sentence of the form Ξ as Φ can be 'normed' to a $0/1$-functional and E-HA$^{\omega}$ proves that

$$\forall \underline{x}^{\underline{\sigma}}\big(\Phi(\underline{x}) =_0 0 \leftrightarrow \neg A(\underline{x})\big) \leftrightarrow \neg\neg\forall \underline{x}^{\underline{\sigma}}\big(\Phi(\underline{x}) =_0 0 \leftrightarrow \neg A(\underline{x})\big)$$

using the stability of $=_0$.

□

Corollary 7.8. $\mathrm{H}^{\omega}+\mathrm{CA}^{\omega}_{\neg}$ *is closed under the fan rule.*

Some consequences of $\mathrm{CA}^{\omega}_{\neg}$: Relative to (weak fragments of) E-HA$^{\omega}$, $\mathrm{CA}^{\omega}_{\neg}$ implies

1) the schema of comprehension for arbitrary Harrop formulas

$$\mathrm{CA}^{\omega}_{\mathrm{Harrop}}:\ \exists \Phi \forall \underline{x}^{\underline{\sigma}}\big(\Phi(\underline{x}) =_0 0 \leftrightarrow A(\underline{x})\big),$$

where A is Harrop (which includes the case $\mathrm{CA}^{\omega}_{ef}$ where A is $\exists$-free),
2) as a special case of $\mathrm{CA}^{\omega}_{ef} \subseteq \mathrm{CA}^{\omega}_{\mathrm{Harrop}}$ we get

$$\exists \Phi^2 \forall f^1\big(\Phi(f) =_0 0 \leftrightarrow \forall x^0(f(x) =_0 0)\big),$$

3) the binary ('weak') König's lemma WKL and its uniform version UWKL, where WKL is defined as

$$\mathrm{WKL}:\ \forall f^1\Big(T(f) \wedge \forall x^0 \exists n^0(lth\, n = x \wedge fn = 0) \rightarrow \exists b \leq_1 \lambda k.1 \forall x^0\big(f(\overline{b}x) = 0\big)\Big),$$

where $T(f) :\equiv \forall n,m\big(f(n*m) = 0 \rightarrow fn = 0\big) \wedge \forall n,x\big(f(n*<x>) = 0 \rightarrow x \leq 1\big)$.

$T(f)$ asserts that f represents a 0,1-tree.
UWKL is defined as

UWKL:
$$\exists \Phi \leq_{1(1)} 1 \forall f^1\Big(T(f) \wedge \forall x^0 \exists n^0(lth\ n = x \wedge fn = 0) \rightarrow \forall x^0\big(f(\overline{\Phi(f)}x) = 0\big)\Big).$$

These principles are discussed in detail in chapters 9 and 10 below. We only mention here a result due to H. Ishihara ([171, 175]) which states that relative to e.g. E-HA$^{\omega}$, WKL implies the following contrapositive form – called fan principle – which in contrast to the ineffective WKL is accepted in intuitionistic analysis.

$$\mathrm{FAN_{KL}}:\ T(f) \wedge \forall b \leq_1 1 \exists x^0\big(f(\overline{b}x) \neq_0 0\big) \rightarrow \exists x^0 \forall b \leq_1 1 \exists \tilde{x} \leq x\big(f(\overline{b}\tilde{x}) \neq_0 0\big).$$

This and more general fan principles will be discussed in chapter 12.
4) the law-of-excluded-middle schema for negated formulas $\neg A \vee \neg\neg A$ (and consequently for Harrop and $\exists$-free formulas) as well as the independence-of-premise principle for these formulas.

As a consequence of the fact that UWKL is equivalent to a sentence of the form Ξ we obtain from corollary 7.5:

Corollary 7.9. *Let* H^ω *be as in proposition 7.3. Then* H^ω+UWKL *is closed under the fan rule.*

Let Σ^0_1-DNE denote the principle of double-negation-elimination for Σ^0_1-formulas containing only number parameters:

$$\Sigma^0_1\text{-DNE}: \ \neg\neg\exists x^0 P(\underline{a}^0, x^0) \to \exists x^0 P(\underline{a}^0, x^0),$$

where P is a primitive recursive predicate and all parameters are shown. It is clear that Σ^0_1-DNE is the special case M^0 of the Markov principle M^ω when restricted to the language of HA (identifying HA with its canonical embedding into E-HA$^\omega$).

Similarly we define the schemas of Π^0_1-LEM (resp. Σ^0_1-LEM) of the law-of-excluded-middle restricted to Π^0_1 (resp. Σ^0_1)-formulas containing only number parameters:

$$\Pi^0_n\text{-LEM}: \ A \vee \neg A, \ A \in \Pi^0_n \text{ containing only number parameters}$$

(analogously for Σ^0_1-LEM).
In the following, we implicitly refer to the obvious embedding of HA into E-HA$^\omega$.

Corollary 7.10. E-HA$^\omega$+AC+CA$^\omega_\neg \nvdash \Sigma^0_1$-DNE, Σ^0_1-LEM.

Proof: Let $T(x,y,z)$ denote the primitive recursive Kleene T-predicate.
If Σ^0_1-DNE would be derivable we would get

$$\text{E-HA}^\omega\text{+AC+CA}^\omega_\neg \vdash \forall x(\neg\neg\exists y T(x,x,y) \to \exists z T(x,x,z))$$

and so (using $\mathrm{IP}^\omega_\neg$ which is derivable from CA$^\omega_\neg$)

$$\text{E-HA}^\omega\text{+AC+CA}^\omega_\neg \vdash \forall x \exists z(\neg\neg\exists y T(x,x,y) \to T(x,x,z)).$$

By corollary 7.7 we can extract a closed term t^1 of E-HA$^\omega$ such that

$$\text{E-HA}^\omega\text{+AC+CA}^\omega_\neg \vdash \forall x \exists z \le tx(\neg\neg\exists y T(x,x,y) \to T(x,x,z)).$$

t denotes a computable function. Since T is decidable we can define a computable function f such that

$$\forall x(f(x) = 0 \leftrightarrow \exists z \le tx\, T(x,x,z))$$

which contradicts the undecidability of the (special) halting problem. The underivability of Σ^0_1-LEM follows a-fortiori. □

Corollary 7.11. HA+Π^0_1-LEM $\nvdash \Sigma^0_1$-DNE, Σ^0_1-LEM.

Proof: The corollary follows from the previous one by observing that HA+Π^0_1-LEM is (modulo the aforementioned embedding) a subsystem of E-HA$^\omega$+CA$^\omega_{ef}$ and hence of E-HA$^\omega$+CA$^\omega_\neg$. □

We now give another application of corollary 7.7: we show that the so-called weak Markov principle WMP is not derivable in E-HA$^\omega$+AC+CA$^\omega_\neg$.

Definition 7.12. 1) A real number $a \in \mathbb{R}$ is pseudo-positive if

$$\forall x \in \mathbb{R}(\neg\neg(0 < x) \vee \neg\neg(x < a)).$$

2) $a \in \mathbb{R}$ is positive if $a > 0$.

Remark 7.13. 1) Without loss of generality we may restrict $x \in \mathbb{R}$ in the definition of pseudo-positivity to $x \in [0,1]$.
2) '$x > y$' for $x, y \in \mathbb{R}$ is to be read as a positive existence statement '$\exists n \in \mathbb{N}(x \geq y + 2^{-n})$' which has – constructively – to be distinguished from the negative statement '$\neg(x \leq y)$'. Using the representation of real numbers from chapter 4 it actually becomes the following Σ^0_1-formula

$$\exists n^0 \big(\widehat{f_x}(n+1) -_{\mathbb{Q}} \widehat{f_y}(n+1) \geq_{\mathbb{Q}} \langle 2^{-n} \rangle\big),$$

where f_x (f_y) is a representative of x (y).

Definition 7.14. Weak Markov's principle is the statement

WMP: *Every pseudo-positive real number is positive.*

WMP has first been considered by Mandelkern in [271, 272] (in the former paper under the name 'almost separating principle' (ASP) and in the latter as 'weak limited principle of existence' (WLPE)). Under the currently common name of weak Markov's principle it has been studied by Ishihara ([172, 173]). WMP follows easily from the usual Markov principle as well as from an appropriate continuity principle and also from the extended Church's thesis ECT_0 (see [174]). So WMP holds both in Russian constructive mathematics as well as in intuitionistic mathematics (in the sense of [47]). For many years it was open whether WMP can be derived in systems used to formalize constructive mathematics in the sense of Bishop such as e.g. E-HA$^\omega$+AC. In [222] it was shown that it is even underivable in E-HA$^\omega$+AC+CA$^\omega_\neg$. We will present here the amazingly simple proof of this as an application of corollary 7.7.

Theorem 7.15. E-HA$^\omega$+AC+CA$^\omega_\neg \nvdash$ WMP.

Proof: Let H^ω:=E-HA$^\omega$+AC+CA$^\omega_\neg$ and assume that $\mathrm{H}^\omega \vdash$ WMP. Restricting w.l.o.g. a, x to $[0,1]$ and making use of our representation of $[0,1]$ from definition 4.24 and lemma 4.25.1) and 4.25.2) this gives us

$$\mathrm{H}^\omega \vdash \forall a^1 \big(\forall x^1 (\neg\neg(0 <_{\mathbb{R}} \tilde{x}) \vee \neg\neg(\tilde{x} <_{\mathbb{R}} \tilde{a})) \to \exists k^0 (\tilde{a} >_{\mathbb{R}} 2^{-k}\big).$$

Using lemma 4.25.3) and 4.25.4) this is equivalent to

$$\mathrm{H}^\omega \vdash \forall a \leq_1 N\big(\forall x^1 (\neg\neg(0 <_{\mathbb{R}} \tilde{x}) \vee \neg\neg(\tilde{x} <_{\mathbb{R}} \tilde{a})) \to \exists k^0 (\tilde{a} >_{\mathbb{R}} 2^{-k})\big),$$

where N is the function from lemma 4.25.4). This in turn is equivalent to

$$\mathrm{H}^{\omega} \vdash \forall a \leq_1 N\big(\forall x^1 \exists n \leq_0 1[(n =_0 0 \to \neg\neg(0 <_{\mathbb{R}} \tilde{x})) \wedge (n \neq 0 \to \neg\neg(\tilde{x} <_{\mathbb{R}} \tilde{a}))] \\ \to \exists k^0(\tilde{a} >_{\mathbb{R}} 2^{-k})\big),$$

which implies

$$\mathrm{H}^{\omega} \vdash \forall a \leq_1 N \forall Z \leq_2 1 \\ \big(\forall x^1[(Z(x) =_0 0 \to \neg\neg(0 <_{\mathbb{R}} \tilde{x})) \wedge (Z(x) \neq 0 \to \neg\neg(\tilde{x} <_{\mathbb{R}} \tilde{a}))] \to \exists k^0(\tilde{a} >_{\mathbb{R}} 2^{-k})\big).$$

The premise '$\forall x^1[(Z(x) = 0 \to \neg\neg(0 <_{\mathbb{R}} \tilde{x})) \wedge (Z(x) \neq 0 \to \neg\neg(\tilde{x} <_{\mathbb{R}} \tilde{a}))]$' is (relative to E-HA$^{\omega}$) equivalent to its double negation and hence is equivalent to a negated formula $\neg B$. Thus (using $\mathrm{IP}^{\omega}_{\neg}$ which follows from $\mathrm{CA}^{\omega}_{\neg}$)

$$\mathrm{H}^{\omega} \vdash \forall a \leq_1 N \forall Z \leq_2 1 \exists k^0 \\ \big(\forall x^1[(Z(x) = 0 \to \neg\neg(0 <_{\mathbb{R}} \tilde{x})) \wedge (Z(x) \neq 0 \to \neg\neg(\tilde{x} <_{\mathbb{R}} \tilde{a}))] \to \tilde{a} >_{\mathbb{R}} 2^{-k}\big).$$

Now corollary 7.7 yields a closed number term t^0 which can be reduced to a numeral $\overline{m}$ (see e.g. [366]) such that

$$\mathrm{H}^{\omega} \vdash \forall a \leq_1 N \forall Z \leq_2 1 \\ \big(\forall x^1[(Z(x) = 0 \to \neg\neg(0 <_{\mathbb{R}} \tilde{x})) \wedge (Z(x) \neq 0 \to \neg\neg(\tilde{x} <_{\mathbb{R}} \tilde{a}))] \to \tilde{a} >_{\mathbb{R}} 2^{-\overline{m}}\big).$$

Applying $\mathrm{AC}^{1,0}$ this implies

$$\mathrm{H}^{\omega} \vdash \forall a \leq_1 N \\ \big(\forall x^1 \exists n \leq_0 1[(n = 0 \to \neg\neg(0 <_{\mathbb{R}} \tilde{x})) \wedge (n \neq 0 \to \neg\neg(\tilde{x} <_{\mathbb{R}} \tilde{a}))] \to \tilde{a} >_{\mathbb{R}} 2^{-\overline{m}}\big).$$

The premise is equivalent to (our formalization of) $\tilde{a}$ being pseudo-positive. Hence we obtain

$$\mathrm{H}^{\omega} \vdash \forall a \in [0,1]\big(a \text{ pseudo-positive } \to a > 2^{-\overline{m}}\big)$$

which obviously is false. Thus we conclude that $\mathrm{H}^{\omega} \nvdash$ WMP. □

We now come back to the main topic of this section, namely the extraction of uniform bounds from proofs in semi-constructive systems, and summarize our results in the most useful form:

Let Ω be a set of sentences of the form

$$\forall \underline{u}^{\underline{\delta}}\big(C(\underline{u}) \to \exists \underline{v} \leq_{\underline{\sigma}} \underline{r}\,\underline{u} \neg B(\underline{u}, \underline{v})\big),$$

where B, C are arbitrary formulas. For a given such set Ω let $\tilde{\Omega}$ be the set of sentences

$$\exists \underline{V} \leq \underline{r} \forall \underline{u}\big(C(\underline{u}) \to \neg B(\underline{u}, \underline{V}\,\underline{u})\big).$$

Theorem 7.16 (Main theorem on uniform bound extraction by monotone mr). *Let* H^ω, A, s *and* τ *be as in proposition 7.3. Then we have the following rule:*

$$\begin{cases} \mathrm{H}^\omega + \Omega \vdash \forall x^1 \forall y \leq_\rho sx \exists z^\tau A(x,y,z) \Rightarrow \\ \mathrm{H}^\omega + \tilde{\Omega} \vdash \forall x^1 \forall y \leq_\rho sx \exists z \leq_\tau txA(x,y,z), \end{cases}$$

where t *is a suitable closed term of* E-HA$^\omega$ *which can be extracted from a given proof of the assumption.*

For H:=E-HA$^\omega$+AC+IP$^\omega_\neg$ *one can replace* $\tilde{\Omega}$ *in the conclusion by* Ω.

As in theorem 6.8 the result also holds for tuples of variables.

Proof: The theorem follows from proposition 7.3 by observing that

1) E-HA$^\omega + \Omega$ is contained in E-HA$^\omega + \tilde{\Omega}$ and
2) every sentence in $\tilde{\Omega}$ is – relative to E-HA$^\omega$ – equivalent to a sentence

$$\exists \underline{V} \leq \underline{r} \neg\neg \forall \underline{u} \big(C(\underline{u}) \to \neg B(\underline{u}, \underline{V}\,\underline{u})\big)$$

of the form Ξ.

The second claim in the theorem follows using the fact that (over E-HA$^\omega$+AC+IP$^\omega_{ef}$ and hence over E-HA$^\omega$+AC+IP$^\omega_\neg$) the formula $C(\underline{u})$ is equivalent to a formula of the form $\exists \underline{a}\, C_{ef}(\underline{u}, \underline{a})$ with $C_{ef}(\underline{u}, \underline{a}) :\equiv \underline{a}\ mr\ C(\underline{u})$ being $\exists$-free (see theorem 5.12). We now use the sentences

$$(*)\ \exists \underline{V} \leq \lambda a.r \forall \underline{u} \left(C_{ef}(\underline{u}, \underline{a}) \to \neg B(\underline{u}, \underline{V}\,\underline{a}\,\underline{u})\right)$$

instead of the sentences $\tilde{\Omega}$ which, again, are (over E-HA$^\omega$) equivalent to sentences of the form Ξ, namely

$$\exists \underline{V} \leq \lambda \underline{a}.\underline{r} \forall \underline{u} \neg\neg \big(C_{ef}(\underline{u}, \underline{a}) \to \neg B(\underline{u}, \underline{V}\,\underline{a}\,\underline{u})\big).$$

It remains to show that $(*)$ is equivalent to

$$(**)\ \forall \underline{u}^{\underline{\delta}} \big(C(\underline{u}) \to \exists \underline{v} \leq_{\underline{\sigma}} \underline{r}\,\underline{u} \neg B(\underline{u}, \underline{v})\big)$$

over E-HA$^\omega$+AC+IP$^\omega_\neg$: Over this theory, clearly $(**)$ is equivalent to

$$\forall \underline{u}, \underline{a} \exists \underline{v} \left(C_{ef}(\underline{u}, \underline{a}) \to \underline{v} \leq \underline{r}\,\underline{u} \wedge \neg B(\underline{u}, \underline{v})\right).$$

So, obviously, $(*)$ implies $(**)$. For the converse direction take $\underline{v}'$ with $v'_i :=$ $\min_{\sigma_i}(v_i, r_i\underline{u})$ to obtain (using the extensionality axiom)

$$\forall \underline{u}, \underline{a} \exists \underline{v} \leq \underline{r}\,\underline{u} \left(C_{ef}(\underline{u}, \underline{a}) \to \neg B(\underline{u}, \underline{v})\right).$$

Finally, apply AC. □

Remark 7.17. The case H:=E-HA$^\omega$+AC+IP$^\omega_\neg$ in the above theorem can actually be established using only *mr* instead of *mrt* since over this theory we have that $\exists \underline{x}\,(\underline{x}\ mr\ A) \leftrightarrow A$ (by theorem 5.12).

With a similar proof one can show that in theorem 7.16 one can allow Ω to depend on x, y : Let Ω be a formula of the form

$$\forall \underline{u}^{\underline{\delta}}\big(C(\underline{u},x,y) \to \exists \underline{v} \leq_{\underline{\sigma}} \underline{r}\,\underline{u} \neg B(\underline{u},\underline{v},x,y)\big)$$

containing only x, y free. Let $\tilde{\Omega}$ be the formula

$$\exists \underline{V} \leq \underline{r} \forall \underline{u}\big(C(\underline{u},x,y) \to \neg B(\underline{u},\underline{V}\,\underline{u},x,y)\big).$$

Theorem 7.18. *Let* H$^\omega$*,* A, s *and* τ *be as before. Then we have the following rule:*

$$\begin{cases} \mathrm{H}^\omega \vdash \forall x^1 \forall y \leq_\rho sx\big(\Omega(x,y) \to \exists z^\tau A(x,y,z)\big) \Rightarrow \\ \mathrm{H}^\omega \vdash \forall x^1 \forall y \leq_\rho sx\big(\tilde{\Omega}(x,y) \to \exists z \leq_\tau txA(x,y,z)\big), \end{cases}$$

where t is a suitable closed term of E-HA$^\omega$ *which can be extracted from a given proof of the assumption.*
As in theorem 6.8 the result also holds for tuples of variables.

7.2 Fragments, exercises, historical comments and suggested further reading

Proposition 7.3, corollaries 7.7,7.8 and theorem 7.16 are also valid with E-HA$^\omega$ replaced by $\mathcal{T}_i^\omega := \widehat{\text{E-HA}}^\omega \!\upharpoonright\!\!\backslash$, E-G$_nA_i^\omega$ ($n \geq 2$) instead of E-HA$^\omega$. This follows by an inspection of the proofs and the fact that monotone modified realizability applies to these systems as we sketched in the remark at the end of the previous chapter.

Exercises:

1) Prove the claims in remark 7.2.
2) Give the details of the proof of proposition 7.3.
3) Prove the claim in remark 7.4.
4) Argue using corollary 7.7 that not even $\neg A \vee \neg\neg A$ (for arbitrary A) suffices to construct a sequence (f_n) in $C[0,1]$ that pointwise but not uniformly converges to 0. Which instance of LEM is used in the usual classical counterexamples?

Historical comments and suggestions for further reading: The results presented in this chapter are mostly taken from Kohlenbach [212] which contains further related material. The corollaries 7.10, 7.11 were prompted by recent work of S. Hayashi on his so-called limit computable mathematics (see Hayashi-Nakata [151]). A thorough study of a hierarchy of restricted forms of the law-of-excluded middle schema LEM and related principles is given in Akama et al. [3]. In [361], Toftdal calibrates (over HA) various ineffective principles in mathematics in terms of restricted forms of LEM. Theorem 7.15 is taken from Kohlenbach [222].

Chapter 8
Gödel's functional ('Dialectica') interpretation

8.1 Introduction

The Gödel functional interpretation, introduced in [133], assigns to each formula $A(\underline{a})$ of WE-HA$^\omega$ a formula $A^D \equiv \exists \underline{x} \forall \underline{y} A_D(\underline{x}, \underline{y}, \underline{a})$, where A_D is quantifier-free (and hence decidable) and $\underline{x}, \underline{y}$ are tuples of variables of finite type.

In contrast to the no-counterexample interpretation, which we briefly discussed in chapter 2, the functional interpretation does not require A to be in prenex normal form and, therefore, is applicable in an intuitionistic context like WE-HA$^\omega$ where not every formula is provably equivalent to a prenex one.

Like modified realizability, but in contrast to the no-counterexample interpretation, functional interpretation has a nice behavior with respect to the logical deduction rules (modus ponens).

In contrast to the modified realizability interpretation, functional interpretation satisfies the Markov principle M$^\omega$ which is crucial in all applications to mathematical analysis discussed in chapters 16 and 18.

Consider a proof in WE-HA$^\omega$ of a theorem of the form

$$(+)\ \forall y A_0(y) \to \forall x B_0(x)$$

where A_0, B_0 are quantifier-free.
Since $(+)$ is $\exists$-free it is unchanged by modified realizability with the realizing tuple being empty. However, there is an obvious computational challenge provided by $(+)$ which becomes apparent if we consider the following strong version of $(+)$

$$(++)\ \forall x \exists y \big(A_0(y) \to B_0(x)\big).$$

The challenge is to extract a program t such that

$$(+++)\ \forall x\big(A_0(tx) \to B_0(x)\big).$$

Modified realizability doesn't accept this challenge as the passage from $(+)$ to $(++)$ requires the Markov principle (plus the decidability of quantifier-free formulas which holds in all our systems). However, as a consequence of the fact that functional interpretation does satisfy Markov's principle it interprets $(+)$ as $(++)$ and extracts from a given proof of $(+)$ a term t satisfying $(+++)$.

In chapter 10 we will introduce a translation of the classical variant WE-PA$^\omega$ of WE-HA$^\omega$ (i.e. WE-HA$^\omega$ plus the law-of-excluded-middle schema $A \vee \neg A$) into WE-HA$^\omega$, the so-called negative translation $A \mapsto A'$ due to [130]. We will see that the composition of $'$ and D, $A \mapsto (A')^D$ provides a very subtle constructive interpretation of A which faithfully reflects the proof-theoretic and computational strength of A (in contrast to the no-counterexample interpretation of (a prenex normal form of) A which in general is a much weaker interpretation and can be established as a particular corollary of the functional interpretation). The price to be paid for this is the necessity to use functionals of arbitrary finite types already for $A \in \mathcal{L}(\text{PA})$. Moreover, the functional interpretation is more involved than the modified realizability interpretation since (closely related to the fact that it satisfies the Markov principle) it analyses logically complex formulas down to the level of purely universal formulas whereas modified realizability stops at the $\exists$-free level. It is this feature which makes the composition of negative translation and functional interpretation a powerful tool of extractive proof theory for classical non-constructive proofs. In contrast to this the combination of negative translation and modified realizability interpretation $\underline{x}\ mr\ A'$ would be useless since A' always is an $\exists$-free formula and, therefore, $(\underline{x}\ mr\ A') \equiv A'$ where $\underline{x}$ is the empty tuple (one can, however, use modified realizability in connection with the negative translation if one applies the so-called Friedman-Dragalin A-translation as an intermediate step, see chapter 14).

There are two facts **any** interpretation which

1) satisfies the Markov principle (whose definition we recall here)

$$\text{M}^\omega:\ \neg\neg\exists \underline{x} A_0(\underline{x}) \to \exists \underline{x} A_0(\underline{x}),\ A_0 \text{ quantifier-free},$$

and

2) extracts computational witnesses from proofs

has to face:

1) full extensionality will not be permissible,
2) only a weakened form of IP^ω_{ef} will be satisfied.

Let us discuss the extensionality issue first.

By the extensionality axiom combined with M$^\omega$ we can prove in E-HA$^\omega$+M$^\omega$ in particular

$$(1)\ \forall \varphi^2, f^1, g^1 \exists x^0 (f(x) =_0 g(x) \to \varphi(f) =_0 \varphi(g))$$

and

$$(2)\ \forall \Phi^3, \varphi^2, \psi^2 \exists f^1 (\varphi(f) =_0 \psi(f) \to \Phi(\varphi) =_0 \Phi(\psi)).$$

As we discussed already it was shown in [163] that no term of E-HA$^\omega$ provides a witness for '$\exists x$' in (1) (uniformly in φ, f, g).
For (2) there is not even a functional definable in ZF set theory which realizes '$\exists f$' as a functional in Φ, φ, ψ as was shown again in [163].

So as a consequence of the fact that functional interpretation satisfies M$^\omega$ we have to restrict E-HA$^\omega$ to its weakly extensional version WE-HA$^\omega$. For applications to numerical analysis, this restriction will later turn out to be inessential due to an 'elimination of extensionality'-technique developed by Gandy and Luckhardt which we will briefly treat in chapter 10.

Let us now discuss the need to further restrict the independence of premise principle (the argument given below and further results in this direction are due to [179]): Suppose we would have a computational proof interpretation for
$\mathcal{T} :=$WE-HA$^\omega$+IP$^\omega_{ef}$+M$^\omega$. By M$^\omega$, $\mathcal{T}$ proves (here again T denotes the Kleene-T predicate)

$$\forall x(\neg\neg\exists y T(x,x,y) \to \exists z T(x,x,z))$$

and hence by IP$^\omega_{ef}$ (using that intuitionistically $\neg\neg\exists \leftrightarrow \neg\forall\neg$)

$$\forall x \exists z(\neg\neg\exists y T(x,x,y) \to T(x,x,z)).$$

But any witness function $f(x) := z$ for this statement would allow to solve the special halting problem and, therefore, cannot be computable.
This argument shows that already independence-of-premise for premises $\neg\neg\exists y A_0(y)$ is too strong in the presence of M$^\omega$ to allow a computational interpretation. We will see below, however, that functional interpretation does satisfy the independence-of-premise principle for purely universal premises.

Motivation of the functional interpretation:

The definition of A^D (like $\underline{x}\ mr\ A$) proceeds by induction on the logical structure of A (i.e. the length and the types of $\underline{x}, \underline{y}$ only depend on the logical structure of A).
The most interesting and difficult case again is the implication whose treatment we are going to motivate now:
Suppose we have already defined the functional interpretations $A^D \equiv \exists \underline{x} \forall \underline{y} A_D(\underline{x}, \underline{y})$ and $B^D \equiv \exists \underline{u} \forall \underline{v} B_D(\underline{u}, \underline{v})$. We are trying to define $(A \to B)^D$:
First consider

$$(A^D \to B^D) \equiv (\exists \underline{x} \forall \underline{y} A_D(\underline{x}, \underline{y}) \to \exists \underline{u} \forall \underline{v} B_D(\underline{u}, \underline{v})).$$

Our strategy to obtain from this a formula of the form $\exists \underline{a} \forall \underline{b}(A \to B)_D$ (with $(A \to B)_D$ quantifier-free) is to transform $(A^D \to B^D)$ into prenex normal form and then to apply the axiom of choice AC.

It is an easy exercise to verify that there are four different prenex normal forms of $A^D \to B^D$. We try to choose the most constructive (or rather: the least non-constructive) one:

For the first step we have two possibilities:

$$\exists \underline{x} \forall \underline{y} A_D(\underline{x}, \underline{y}) \to \exists \underline{u} \forall \underline{v} B_D(\underline{u}, \underline{v})$$

$$\mapsto \begin{cases} (1)\ \forall \underline{x}(\forall \underline{y} A_D(\underline{x}, \underline{y}) \to \exists \underline{u} \forall \underline{v} B_D(\underline{u}, \underline{v})) \\ (2)\ \exists \underline{u}(\exists \underline{x} \forall \underline{y} A_D(\underline{x}, \underline{y}) \to \forall \underline{v} B_D(\underline{u}, \underline{v})). \end{cases}$$

Here the choice is obvious: the passage to (1) is intuitionistically valid, whereas the passage to (2) not even holds in $\mathrm{IL}^\omega + \mathrm{IP}^\omega_{ef} + \mathrm{M}^\omega$.
From (1) there are two ways to proceed further:

$$(1) \mapsto \begin{cases} (1.1)\ \forall \underline{x} \exists \underline{u}(\forall \underline{y} A_D(\underline{x}, \underline{y}) \to \forall \underline{v} B_D(\underline{u}, \underline{v})) \\ (1.2)\ \forall \underline{x} \exists \underline{y}(A_D(\underline{x}, \underline{y}) \to \exists \underline{u} \forall \underline{v} B_D(\underline{u}, \underline{v})). \end{cases}$$

This time the choice is more difficult since both implications $(1) \to (1.1)$ and $(1) \to (1.2)$ are not provable in IL^ω. So we have to compromise our goal to use only strictly constructive transformation steps. However the first implication only requires a weak form of IP^ω_{ef} (for purely universal formulas A) in addition to IL^ω which has some constructive justification by the results of chapter 5. So lets choose (1.1).
From there we have two possibilities to finish our prenexation:

$$(1.1) \mapsto \begin{cases} (1.1.1)\ \forall \underline{x} \exists \underline{u} \forall \underline{v} \exists \underline{y}(A_D(\underline{x}, \underline{y}) \to B_D(\underline{u}, \underline{v})) \\ (1.1.2)\ \forall \underline{x} \exists \underline{u} \exists \underline{y} \forall \underline{v}(A_D(\underline{x}, \underline{y}) \to B_D(\underline{u}, \underline{v})). \end{cases}$$

Again the choice is not obvious: both implications are not provable in IL^ω.
Lets consider $(1.1) \to (1.1.1)$ first: The first step to

$$\forall \underline{x} \exists \underline{u} \forall \underline{v}(\forall \underline{y} A_D(\underline{x}, \underline{y}) \to B_D(\underline{u}, \underline{v}))$$

is perfectly valid from an intuitionistic point of view. However from there we – intuitionistically – only get (using the decidability of the quantifier-free formulas A_D, B_D, see the exercise below)

$$\forall \underline{x} \exists \underline{u} \forall \underline{v} \neg\neg \exists \underline{y}(A_D(\underline{x}, \underline{y}) \to B_D(\underline{u}, \underline{v}))$$

and so we need the Markov principle M^ω to obtain $(1.1.1)$.
For $(1.1) \to (1.1.2)$ the first step to

$$\forall \underline{x} \exists \underline{u} \exists \underline{y}(A_D(\underline{x}, \underline{y}) \to \forall \underline{v} B_D(\underline{u}, \underline{v}))$$

is not intuitionistically valid but again only the passage to the weaker

$$\forall \underline{x} \exists \underline{u} \neg\neg \exists \underline{y} (A_D(\underline{x},\underline{y}) \rightarrow \forall \underline{v} B_D(\underline{u},\underline{v})).$$

However this time not even M^ω suffices to get rid of $\neg\neg$ since '$A_D(\underline{x},\underline{y}) \rightarrow \forall \underline{v} B_D(\underline{u},\underline{v})$' is not quantifier-free.
So the implication '(1.1) $\rightarrow$ (1.1.1)' is less non-constructive than '(1.1) $\rightarrow$ (1.1.2)'. Hence we now 'officially' choose (1.1.1) as our prenex normal form of $A^D \rightarrow B^D$. Applying AC to (1.1.1) we finally obtain

$$(A \rightarrow B)^D :\equiv \exists \underline{U}, \underline{Y} \forall \underline{x}, \underline{v} \underbrace{\big(A_D(\underline{x}, \underline{Y}\,\underline{x}\,\underline{v}) \rightarrow B_D(\underline{U}\,\underline{x}, \underline{v})\big)}_{(A \rightarrow B)_D :\equiv}.$$

Despite of the fact that we had to make various compromises to end up with $(A \rightarrow B)^D$, this interpretation works while any of the remaining three prenex normal forms of $A^D \rightarrow B^D$ would result in a definition of $(A \rightarrow B)^D$ which even for $B :\equiv A$ in general would fail to have a constructive (computable) realization (exercise).

8.2 The soundness and program extraction theorems

We now give the complete definition of Gödel's functional interpretation:

Definition 8.1 (Gödel [133]). To every formula A of $\mathcal{L}(\text{WE-HA}^\omega)$ we assign a translation $A^D \equiv \exists \underline{x} \forall \underline{y} A_D(\underline{x},\underline{y})$ in the same language. The free variables of A^D are that of A. The types and length of $\underline{x},\underline{y}$ only depend on the logical structure of A. A_D is a quantifier-free formula (even without $\vee$).

(i) $A^D :\equiv A_D :\equiv A$ for prime formulas A.

Let $A^D \equiv \exists \underline{x} \forall \underline{y} A_D(\underline{x},\underline{y})$ and $B^D \equiv \exists \underline{u} \forall \underline{v} B_D(\underline{u},\underline{v})$. Then

(ii) $(A \wedge B)^D :\equiv \exists \underline{x},\underline{u} \forall \underline{y},\underline{v} [A \wedge B]_D$
$\qquad :\equiv \exists \underline{x},\underline{u} \forall \underline{y},\underline{v} [A_D(\underline{x},\underline{y}) \wedge B_D(\underline{u},\underline{v})]$,

(iii) $(A \vee B)^D :\equiv \exists z^0, \underline{x},\underline{u} \forall \underline{y},\underline{v} [A \vee B]_D$
$\qquad :\equiv \exists z^0, \underline{x},\underline{u} \forall \underline{y},\underline{v} [(z = 0 \rightarrow A_D(\underline{x},\underline{y})) \wedge (z \neq 0 \rightarrow B_D(\underline{u},\underline{v}))]$,

(iv) $(A \rightarrow B)^D :\equiv \exists \underline{U},\underline{Y} \forall \underline{x},\underline{v} (A \rightarrow B)_D :\equiv \exists \underline{U},\underline{Y} \forall \underline{x},\underline{v} \big(A_D(\underline{x}, \underline{Y}\,\underline{x}\,\underline{v}) \rightarrow B_D(\underline{U}\,\underline{x},\underline{v})\big)$,

(v) $(\exists z^\rho A(z))^D :\equiv \exists z, \underline{x} \forall \underline{y} (\exists z A(z))_D :\equiv \exists z, \underline{x} \forall \underline{y} A_D(\underline{x},\underline{y},z)$,

(vi) $(\forall z^\rho A(z))^D :\equiv \exists \underline{X} \forall z, \underline{y} (\forall z A(z))_D :\equiv \exists \underline{X} \forall z, \underline{y} A_D(\underline{X} z, \underline{y}, z)$,

Remark 8.2. $(A^D)^D \equiv A^D$.

Remark 8.3. As a consequence of the treatment of implication we obtain

(i) $(\neg A)^D \equiv \exists \underline{Y} \forall \underline{x} \neg A_D(\underline{x}, \underline{Y}\,\underline{x})$,

(ii) $(\neg\neg A)^D \equiv \exists \underline{X} \forall \underline{Y} \neg\neg A_D(\underline{X}\,\underline{Y}, \underline{Y}(\underline{X}\,\underline{Y})) \leftrightarrow \exists \underline{X} \forall \underline{Y} A_D(\underline{X}\,\underline{Y}, \underline{Y}(\underline{X}\,\underline{Y}))$, where the equivalence is provable in WE-HA^ω.

Definition 8.4. The independence-of-premise schema $\mathrm{IP}^{\omega}_{\forall}$ for universal premises is the union (for all types) of

$$\mathrm{IP}^{\rho}_{\forall} : (\forall \underline{x} A_0(\underline{x}) \to \exists y^{\rho} B(y)) \to \exists y^{\rho} (\forall \underline{x} A_0(\underline{x}) \to B(y)),$$

where y is not free in $\forall \underline{x} A_0(\underline{x})$.

Remark 8.5. There are even closed instances of $\mathrm{IP}^{0}_{\forall}$ in the language $\mathcal{L}(\mathrm{HA})$ that are not provable in HA. See Troelstra [366] (3.1.11).

Theorem 8.6 (soundness of functional interpretation, Gödel [133], Yasugi [380], Troelstra [366]).
Let $\mathcal{P}$ be an arbitrary set of purely universal sentences $\forall \underline{x}^{\underline{\sigma}} B_0(\underline{x})$ (B_0 quantifier-free) of $\mathcal{L}(\mathrm{WE\text{-}HA}^{\omega})$ and $A(\underline{a})$ be a formula of $\mathcal{L}(\mathrm{WE\text{-}HA}^{\omega})$ containing only $\underline{a}$ free.
Then the following rule holds:

$$\begin{cases} \mathrm{WE\text{-}HA}^{\omega} + \mathrm{AC} + \mathrm{IP}^{\omega}_{\forall} + \mathrm{M}^{\omega} + \mathcal{P} \vdash A(\underline{a}) \ \textit{implies that} \\ \mathrm{WE\text{-}HA}^{\omega} + \mathcal{P} \vdash \forall \underline{y} A_D(\underline{t}\,\underline{a}, \underline{y}, \underline{a}), \end{cases}$$

where $\underline{t}$ is a suitable tuple of closed terms of WE-HA$^{\omega}$ *which can be extracted from a given proof of the assumption.*

Proof: As in the proof of the soundness theorem for modified realizability we proceed by induction on the length of the derivation.

I) Logical axioms and rules:

1) $\underline{A \to A \wedge A}$:

$$(A \to A \wedge A)^D \equiv$$
$$\big(\exists \underline{x} \forall \underline{y} A_D(\underline{x}, \underline{y}, \underline{a}) \to \exists \underline{x}', \underline{x}'' \forall \underline{y}', \underline{y}'' \big(A_D(\underline{x}', \underline{y}', \underline{a}) \wedge A_D(\underline{x}'', \underline{y}'', \underline{a})\big)\big)^D \equiv$$
$$\exists \underline{Y}, \underline{X}', \underline{X}'' \forall \underline{x}, \underline{y}', \underline{y}'' \big(A_D(\underline{x}, \underline{Y}\,\underline{x}\,\underline{y}'\underline{y}'', \underline{a}) \to A_D(\underline{X}'\underline{x}, \underline{y}', \underline{a}) \wedge A_D(\underline{X}''\underline{x}, \underline{y}'', \underline{a})\big).$$

Hence

$$\underline{t}_{\underline{X}'} := \underline{t}_{\underline{X}''} := \lambda \underline{a}, \underline{x}.\underline{x}$$
$$\underline{t}_{\underline{Y}}\underline{a}\,\underline{x}\,\underline{y}'\underline{y}'' := \begin{cases} \underline{y}', \text{ if } t_{A_D}\underline{x}\,\underline{y}'\underline{a} \neq 0 \\ \underline{y}'', \text{ if } t_{A_D}\underline{x}\,\underline{y}'\underline{a} = 0, \end{cases}$$

(proposition 3.19) satisfy the functional interpretation of $A \to A \wedge A$ (here t_{A_D} is a closed term of WE-HA$^{\omega}$ such that WE-HA$^{\omega} \vdash t_{A_D}\underline{x}\,\underline{y}\,\underline{a} =_0 0 \leftrightarrow A_D(\underline{x}, \underline{y}, \underline{a})$, proposition 3.17).

2) $\underline{A \vee A \to A}$:

$$(A \vee A \to A)^D \equiv \exists \underline{Y}, \underline{Y}', \underline{X}'' \forall z^0, \underline{x}, \underline{x}', \underline{y}''$$
$$\big((z =_0 0 \to A_D(\underline{x}, \underline{Y}z\underline{x}\underline{x}'\underline{y}'', \underline{a})\big) \wedge \big(z \neq 0 \to A_D(\underline{x}', \underline{Y}'z\underline{x}\underline{x}'\underline{y}'', \underline{a})\big)$$
$$\to A_D(\underline{X}''z\underline{x}\underline{x}', \underline{y}'', \underline{a})\big).$$

Define

$$\underline{t}_{\underline{X}''} := \lambda \underline{a}, z, \underline{x}, \underline{x}'. \begin{cases} \underline{x}, \text{ if } z = 0 \\ \underline{x}', \text{ if } z \neq 0, \end{cases}$$
$$\underline{t}_{\underline{Y}} := \underline{t}_{\underline{Y}'} := \lambda \underline{a}, z, \underline{x}, \underline{x}', \underline{y}''.\underline{y}''.$$

These terms realize '$\exists \underline{Y}, \underline{Y}', \underline{X}''$', and hence satisfy the functional interpretation of the axiom.

3) $\underline{A \to A \vee B}$:

$$(A \to A \vee B)^D \equiv \big(\exists \underline{x} \forall \underline{y} A_D(\underline{x}, \underline{y}, \underline{a}) \to$$
$$\exists z^0, \underline{x}', \underline{u} \forall \underline{y}', \underline{v}\big((z = 0 \to A_D(\underline{x}', \underline{y}', \underline{a})) \wedge (z \neq 0 \to B_D(\underline{u}, \underline{v}, \underline{a}'))\big)\big)^D$$
$$\equiv \exists Z, \underline{X}', \underline{U}, \underline{Y} \forall \underline{x}, \underline{y}', \underline{v}\big(A_D(\underline{x}, \underline{Y}\,\underline{x}\,\underline{y}'\underline{v}, \underline{a}) \to$$
$$(Z\underline{x} = 0 \to A_D(\underline{X}'\underline{x}, \underline{y}', \underline{a})) \wedge (Z\underline{x} \neq 0 \to B_D(\underline{U}\,\underline{x}, \underline{v}, \underline{a}'))\big).$$

Hence $\underline{t}_{\underline{Y}} := \lambda \underline{\tilde{a}}, \underline{x}, \underline{y}', \underline{v}.\underline{y}'$, $t_Z := \lambda \underline{\tilde{a}}, \underline{x}.0^0$, $\underline{t}_{\underline{X}'} := \lambda \underline{\tilde{a}}, \underline{x}.\underline{x}$, $\underline{t}_{\underline{U}} := \lambda \underline{\tilde{a}}, \underline{x}.\underline{\mathcal{O}}$, where $\underline{\tilde{a}} = \{\underline{a}\} \cup \{\underline{a}'\}$, satisfy the functional interpretation of $A \to A \vee B$.

4) $\underline{A \wedge B \to A}$:

$$(A \wedge B \to A)^D \equiv \big(\exists \underline{x}, \underline{u} \forall \underline{y}, \underline{v}\big(A_D(\underline{x}, \underline{y}, \underline{a})\big) \wedge B_D(\underline{u}, \underline{v}, \underline{a}')\big) \to \exists \underline{x}' \forall \underline{y}' A_D(\underline{x}', \underline{y}', \underline{a})\big)^D$$
$$\equiv \exists \underline{X}', \underline{Y}, \underline{V} \forall \underline{x}, \underline{u}, \underline{y}'\, \big(A_D(\underline{x}, \underline{Y}\,\underline{x}\underline{u}\underline{y}', \underline{a}) \wedge B_D(\underline{u}, \underline{V}\,\underline{x}\underline{u}\underline{y}', \underline{a}') \to A_D(\underline{X}'\underline{x}\underline{u}, \underline{y}', \underline{a})\big).$$

Hence $\underline{t}_{\underline{X}'} := \lambda \underline{\tilde{a}}, \underline{x}, \underline{u}.\underline{x}$, $\underline{t}_{\underline{Y}} := \lambda \underline{\tilde{a}}, \underline{x}, \underline{u}, \underline{y}'.\underline{y}'$, $\underline{t}_{\underline{V}} := \lambda \underline{\tilde{a}}, \underline{x}, \underline{u}, \underline{y}'.\underline{\mathcal{O}}$, where $\underline{\tilde{a}} = \{\underline{a}\} \cup \{\underline{a}'\}$, satisfy the functional interpretation of $A \wedge B \to A$.

5) $\underline{A \vee B \to B \vee A}$:

$$(A \vee B \to B \vee A)^D \equiv \exists \underline{Z}', \underline{X}', \underline{U}', \underline{Y}, \underline{V} \forall z^0, \underline{x}, \underline{u}, \underline{y}', \underline{v}'$$
$$\big\{\big(z = 0 \to A_D(\underline{x}, \underline{Y}\,z\underline{x}\underline{u}\underline{y}'\underline{v}', \underline{a})\big) \wedge \big(z \neq 0 \to B_D(\underline{u}, \underline{V}z\underline{x}\,\underline{u}\underline{y}'\underline{v}', \underline{a}')\big)$$
$$\to \big(Z'z\underline{x}\,\underline{u} = 0 \to B_D(U'z\underline{x}\underline{u}, \underline{v}', \underline{a}')\big) \wedge \big(Z'z\underline{x}\underline{u} \neq 0 \to A_D(X'z\underline{x}\underline{u}, \underline{y}', \underline{a})\big)\big\}.$$

Define $\underline{t}_{\underline{U}'} := \lambda \underline{\tilde{a}}, z, \underline{x}, \underline{u}.\underline{u}$, $\underline{t}_{\underline{X}'} := \lambda \underline{\tilde{a}}, z, \underline{x}, \underline{u}.\underline{x}$, $\underline{t}_{\underline{Y}} := \lambda \underline{\tilde{a}}, z, \underline{x}, \underline{u}, \underline{y}', \underline{v}'.\underline{y}'$, $\underline{t}_{\underline{V}} := \lambda \underline{\tilde{a}}, z, \underline{x}, \underline{u}, \underline{y}', \underline{v}'.\underline{v}'$ and $t_{\underline{Z}'} := \lambda \underline{\tilde{a}}, z, \underline{x}, \underline{u}.\overline{sg}(z^0)$, where

$$\overline{sg}(z^0) := \begin{cases} 0, \text{ if } z \neq 0 \\ 1, \text{ if } z = 0 \end{cases} \quad \text{and } \underline{\tilde{a}} = \{\underline{a}\} \cup \{\underline{a}'\}.$$

6) $\underline{A \wedge B \to B \wedge A}$:

$$(A \wedge B \to B \wedge A)^D \equiv \exists \underline{U}', \underline{X}', \underline{Y}, \underline{V} \forall \underline{x}, \underline{u}, \underline{v}', \underline{y}'$$
$$\big(A_D(\underline{x}, \underline{Y}\,\underline{x}\,\underline{u}\,\underline{v}'\underline{y}', \underline{a}) \wedge B_D(\underline{u}, \underline{V}\,\underline{x}\,\underline{u}\,\underline{v}'\underline{y}', \underline{a}')$$
$$\to B_D(\underline{U}'\underline{x}\,\underline{u}, \underline{v}', \underline{a}') \wedge A_D(\underline{X}'\underline{x}\,\underline{u}, \underline{y}', \underline{a})\big).$$

Define

$$\underline{t}_{\underline{U}'} := \lambda \underline{\tilde{a}}, \underline{x}, \underline{u}.\underline{u},\ \underline{t}_{\underline{X}'} := \lambda \underline{\tilde{a}}, \underline{x}, \underline{u}.\underline{x},\ \underline{t}_{\underline{Y}} := \lambda \underline{\tilde{a}}, \underline{x}, \underline{u}, \underline{v}', \underline{y}'.\underline{y}',\ \underline{t}_{\underline{V}} := \lambda \underline{\tilde{a}}, \underline{x}, \underline{u}, \underline{v}', \underline{y}'.\underline{v}'.$$

7) $\underline{\bot \to A}$: $\big(\bot \to \exists \underline{x} \forall \underline{y} A_D(\underline{x}, \underline{y}, \underline{a})\big)^D \equiv \exists \underline{x} \forall \underline{y} (\bot \to A_D(\underline{x}, \underline{y}, \underline{a}))$. Take $\underline{t}_{\underline{x}} := \lambda \underline{a}.\underline{\mathcal{O}}$.
8) $\underline{\forall z A \to A[t/z]}$: Let $\underline{a}$ be all the free variables in $A[t/z]$.

$$\big(\forall z A \to A[t/z]\big)^D \equiv \big(\exists \underline{X} \forall z, \underline{y} A_D(\underline{X}z, \underline{y}, z) \to \exists \underline{x}' \forall \underline{y}' A_D(\underline{x}', \underline{y}', t)\big)^D$$
$$\equiv \exists \underline{X}', Z, \underline{Y} \forall \underline{X}, \underline{y}' \big(A_D(\underline{X}(Z\underline{X}\underline{y}'), \underline{Y}\,\underline{X}\underline{y}', Z\underline{X}\underline{y}') \to A_D(\underline{X}'\underline{X}, \underline{y}', t)\big).$$

Now take

$$\underline{t}_{\underline{X}'} := \lambda \underline{a}, \underline{X}.\underline{X}t,\ \ t_Z := \lambda \underline{a}, \underline{X}, \underline{y}'.t,\ \ \underline{t}_{\underline{Y}} := \lambda \underline{a}, \underline{X}, \underline{y}'.\underline{y}'.$$

9) $\underline{A[t/z] \to \exists z A}$: Let $\underline{a}$ be all the free variables in $A[t/z]$.

$$\big(A[t/z] \to \exists z A\big)^D \equiv \big(\exists \underline{x} \forall \underline{y} A_D(\underline{x}, \underline{y}, t) \to \exists z, \underline{x}' \forall \underline{y}' A_D(\underline{x}', \underline{y}', z)\big)^D$$
$$\equiv \exists Z, \underline{X}', \underline{Y} \forall \underline{x}, \underline{y}' \big(A_D(\underline{x}, \underline{Y}\,\underline{x}\,\underline{y}', t) \to A_D(\underline{X}'\underline{x}, \underline{y}', Z\underline{x})\big).$$

Define

$$t_Z := \lambda \underline{a}, \underline{x}.t,\ \ \underline{t}_{\underline{X}'} := \lambda \underline{a}, \underline{x}.\underline{x},\ \ \underline{t}_{\underline{Y}} := \lambda \underline{a}, \underline{x}, \underline{y}'.\underline{y}'.$$

10) **The modus ponens rule:** Assume

$$(1)\ \forall \underline{y} A_D(\underline{t}_1\underline{a}, \underline{y}, \underline{a})$$

and

$$(2)\ \forall \underline{x}, \underline{v}\big(A_D(\underline{x}, \underline{t}_2\underline{\tilde{a}}\,\underline{x}\,\underline{v}, \underline{a}) \to B_D(\underline{t}_3\underline{\tilde{a}}\,\underline{x}, \underline{v}, \underline{a}')\big),$$

where $\underline{\tilde{a}} = \{\underline{a}\} \cup \{\underline{a}'\}$.
We have to construct $\underline{t}_4$ such that

$$\forall \underline{v} B_D(\underline{t}_4\underline{a}', \underline{v}, \underline{a}').$$

Apply (2) to $\underline{x} := \underline{t}_1\underline{a}$. Then

$$(3)\ \forall \underline{v}\big(A_D(\underline{t}_1\underline{a}, \underline{t}_2(\underline{\tilde{a}}, \underline{t}_1\underline{a}, \underline{v}), \underline{a}) \to B_D(\underline{t}_3(\underline{\tilde{a}}, \underline{t}_1\underline{a}), \underline{v}, \underline{a}')\big).$$

Apply (1) to $\underline{y} := \underline{t}_2(\underline{\tilde{a}}, \underline{t}_1\underline{a}, \underline{v})$, then

$$(4)\ \forall \underline{v} A_D(\underline{t}_1 \underline{a}, \underline{t}_2(\tilde{\underline{a}}, \underline{t}_1 \underline{a}, \underline{v}), \underline{a}).$$

Hence

$$(5)\ \forall \underline{v} B_D(\underline{t}_3(\tilde{\underline{a}}, \underline{t}_1 \underline{a}), \underline{v}, \underline{a}').$$

Let $\underline{t}[\underline{a}']$ be the result of replacing all variables a_i in $\underline{t}_3(\tilde{\underline{a}}, \underline{t}_1 \underline{a})$ which do not occur in $\underline{a}'$ by $\mathcal{O}$ of appropriate type. Then $\underline{t}_4 := \lambda \underline{a}'.\underline{t}[\underline{a}']$ does the job.

11) **The syllogism rule:** For notational simplicity we omit the free parameters this time.
Assume

$$(1)\ \forall \underline{x}, \underline{v}\big(A_D(\underline{x}, \underline{t}_1 \underline{x}\,\underline{v}) \to B_D(\underline{t}_2 \underline{x}, \underline{v})\big)$$

and

$$(2)\ \forall \underline{u}, \underline{w}\big(B_D(\underline{u}, \underline{t}_3 \underline{u}\,\underline{w}) \to C_D(\underline{t}_4 \underline{u}, \underline{w})\big).$$

We have to construct $\underline{t}_5, \underline{t}_6$ such that

$$\forall \underline{x}, \underline{w}\big(A_D(\underline{x}, \underline{t}_5 \underline{x}\,\underline{w}) \to C_D(\underline{t}_6 \underline{x}, \underline{w})\big).$$

Apply (1) to $\underline{v} = \underline{t}_3(\underline{t}_2 \underline{x}, \underline{w})$ and (2) to $\underline{u} = \underline{t}_2 \underline{x}$. Then

$$(3)\ \forall \underline{x}, \underline{w}\big(A_D(\underline{x}, \underline{t}_1(\underline{x}, \underline{t}_3(\underline{t}_2 \underline{x}, \underline{w}))) \to C_D(\underline{t}_4(\underline{t}_2 \underline{x}), \underline{w})\big).$$

Hence

$$\underline{t}_5 := \lambda \underline{x}, \underline{w}.\underline{t}_1(\underline{x}, \underline{t}_3(\underline{t}_2 \underline{x}, \underline{w})),\ \underline{t}_6 := \lambda \underline{x}.\underline{t}_4(\underline{t}_2 \underline{x})$$

do the job.

12) **The importation and exportation rules:** Note that

$$(A \land B \to C)^D \equiv \exists \underline{P}, \underline{Y}, \underline{V} \forall \underline{x}, \underline{u}, \underline{q}\,\big(A_D(\underline{x}, \underline{Y}\,\underline{x}\,\underline{u}\,\underline{q}) \land B_D(\underline{u}, \underline{V}\,\underline{x}\,\underline{u}\,\underline{q}) \to C_D(\underline{P}\,\underline{x}\,\underline{u}, \underline{q})\big)$$

and

$$(A \to (B \to C))^D \equiv$$
$$\exists \underline{P}, \underline{Y}, \underline{V} \forall \underline{x}, \underline{u}, \underline{q}\,\big(A_D(\underline{x}, \underline{Y}\,\underline{x}\,\underline{u}\,\underline{q}) \to (B_D(\underline{u}, \underline{V}\,\underline{x}\,\underline{u}\,\underline{q}) \to C_D(\underline{P}\,\underline{x}\,\underline{u}, \underline{q}))\big).$$

Hence any solution of $(A \land B \to C)^D$ also is a solution of $(A \to (B \to C))^D$ and vice versa. So we simply can copy the solutions from the premise of these rules to obtain a solution for the conclusion.

13) **The expansion rule:** Again we omit the parameters for notational simplicity. Let $C^D \equiv \exists \underline{p} \forall \underline{q} C_D(\underline{p}, \underline{q})$. By induction hypothesis we have closed terms $\underline{t}_1, \underline{t}_2$ such that

$$(1)\ \forall \underline{x}, \underline{v}\big(A_D(\underline{x}, \underline{t}_1 \underline{x}\,\underline{v}) \to B_D(\underline{t}_2 \underline{x}, \underline{v})\big).$$

$$(C\vee A\to C\vee B)^D\equiv\Big(\exists z_1^0,\underline{p},\underline{x}\forall\underline{q},\underline{y}[(z_1=0\to C_D)\wedge(z_1\neq 0\to A_D)]\to$$
$$\exists z_2^0,\underline{p}',\underline{u}\forall\underline{q}',\underline{v}[(z_2=0\to C_D)\wedge(z_2\neq 0\to B_D)]\Big)^D$$
$$\equiv\exists Z_2,\underline{P}',\underline{U},\underline{Q},\underline{Y}\forall z_1^0,\underline{p},\underline{x},\underline{q}',\underline{v}$$
$$\Big([(z_1=0\to C_D(\underline{p},\underline{Q}z_1\underline{p}\underline{x}\underline{q}'\underline{v}))\wedge(z_1\neq 0\to A_D(\underline{x},\underline{Y}z_1\underline{p}\underline{x}\underline{q}'\underline{v}))]\to$$
$$[(Z_2z_1\underline{p}\underline{x}=0\to C_D(\underline{P}'z_1\underline{p}\underline{x},\underline{q}'))\wedge(Z_2z_1\underline{p}\underline{x}\neq 0\to B_D(\underline{U}z_1\underline{p}\underline{x},\underline{v}))]\Big).$$

Now take

$$t_{Z_2}:=\lambda z_1,\underline{p},\underline{x}.z_1,\ \ \underline{t}_{\underline{P}'}:=\lambda z_1,\underline{p},\underline{x}.\underline{p},\ \ \underline{t}_{\underline{U}}:=\lambda z_1,\underline{p},\underline{x}.\underline{t}_2\underline{x},$$
$$\underline{t}_{\underline{Q}}:=\lambda z_1,\underline{p},\underline{x},\underline{q}',\underline{v}.\underline{q}',\ \ \underline{t}_{\underline{Y}}:=\lambda z_1,\underline{p},\underline{x},\underline{q}',\underline{v}.\underline{t}_1\underline{x}\underline{v}.$$

Using (1) it follows that these terms provide a solution for $(C\vee A\to C\vee B)^D$.

14) **Quantifier rules:** Consider $\frac{B\to A}{B\to\forall zA}$, where z is not free in B. Let $\underline{a}$ be all the free variables of $B\to\forall zA$. By induction hypothesis we have closed terms $\underline{t}_1,\underline{t}_2$ such that

$$\forall\underline{a},z,\underline{u},\underline{y}\left(B_D(\underline{u},\underline{t}_1\underline{a}z\underline{u}\underline{y})\to A_D(\underline{t}_2\underline{a}z\underline{u},\underline{y},z)\right).$$
$$(B\to\forall zA)^D\equiv\exists\underline{X},\underline{V}\forall\underline{u},z,\underline{y}\big(B_D(\underline{u},\underline{V}\,\underline{u}z\underline{y})\to A_D(\underline{X}\,\underline{u}z,\underline{y},z)\big).$$

Now take

$$\underline{t}_{\underline{V}}:=\lambda\underline{a},\underline{u},z,\underline{y}.\underline{t}_1\underline{a}\,z\underline{u}\underline{y},\ \ \underline{t}_{\underline{X}}:=\lambda\underline{a},\underline{u},z.\underline{t}_2\underline{a}z\underline{u}.$$

The rule $\frac{A\to B}{\exists zA\to B}$ is treated analogously.

II) Axioms for $=_0,S,\Pi,\Sigma,\underline{R}$**:** These purely universal axioms are identical with their own functional interpretation.

III) The quantifier-free extensionality rule QF-ER: Both the premise and the conclusion are (modulo pulling out the universal quantifiers hidden in $=_\rho,=_\tau$ to the front) purely universal and so are identical to their functional interpretation. Note that we may assume that A_0 in QF-ER does not contain any $\vee$, since $A_0(\underline{x})$ can be written as $t_{A_0}\underline{x}=_0 0$ in WE-HA$^\omega$ using only instances of QF-ER which do not involve '$\vee$' to get $x=_0 y\to t[x]=_0 t[y]$ and to get functional completeness (lemma 3.15).

IV) The schema of induction: It is easier to use the equivalent induction rule: Let $B(y^0)^D\equiv\exists\underline{u}\forall\underline{v}B_D(\underline{u},\underline{v},y,\underline{a})$ and assume that we have already proved

$$\begin{cases}\forall\underline{v}B_D(\underline{t}_1\underline{a},\underline{v},0,\underline{a})\text{ and}\\ \forall\underline{u},\underline{w}\big(B_D(\underline{u},\underline{t}_2y\underline{a}\underline{u}\underline{w},y,\underline{a})\to B_D(\underline{t}_3y\underline{a}\underline{u},\underline{w},y+1,\underline{a})\big).\end{cases}$$

Define $\underline{t}$ by simultaneous primitive recursion in higher types such that

$$\begin{cases} \underline{t}(\underline{a},0) = \underline{t}_1\underline{a} \\ \underline{t}(\underline{a},y+1) = \underline{t}_3(y,\underline{a},\underline{t}(\underline{a},y)). \end{cases}$$

Then

$$\begin{cases} \forall \underline{v} B_D(\underline{t}(\underline{a},0),\underline{v},0,\underline{a}) \text{ and} \\ \forall \underline{w}\big(B_D(\underline{t}(\underline{a},y),\underline{t}_2(y,\underline{a},\underline{t}(\underline{a},y),\underline{w}),y,\underline{a}) \to B_D(\underline{t}(\underline{a},y+1),\underline{w},y+1,\underline{a})\big) \end{cases}$$

and therefore

$$\begin{cases} \forall \underline{v} B_D(\underline{t}(\underline{a},0),\underline{v},0,\underline{a}) \text{ and} \\ \forall \underline{v} B_D(\underline{t}(\underline{a},y),\underline{v},y,\underline{a}) \to \forall \underline{v} B_D(\underline{t}(\underline{a},y+1),\underline{v},y+1,\underline{a}). \end{cases}$$

Hence by the induction rule we obtain

$$\forall \underline{v} B_D(\underline{t}(\underline{a},y),\underline{v},y,\underline{a}).$$

V) The functional interpretations of AC, M^ω and $IP^\omega_\forall$ result in instances of $(A \to A)^D$ since for these principles the functional interpretations of the premise and the conclusion are identical (in the case of M^ω one has to use the stability of quantifier-free formulas). Hence we only need simple projection terms for their solution which can be verified already in WE-HA$^\omega$ without the use of AC, M^ω and $IP^\omega_\forall$ (here we may assume that the quantifier-free formulas A_0 in M^ω and $IP^\omega_\forall$ do not contain $\vee$ as $A_0(\underline{a})$ can be written as $t_{A_0}(\underline{a}) =_0 0$).

VI) In the universal axioms $\mathscr{P}$ we can assume again that B_0 does not contain $\vee$ so that they are identical to their functional interpretations. □

Warning: The soundness theorem does not hold for E-HA$^\omega$(see [163]). This is indeed a consequence of the fact that the soundness of the functional interpretation of a system implies that this system is closed under the Markov rule which E-HA$^\omega$ is not (see [179]).

Remark 8.7. Remarks 5.10 and 5.11 (with $\mathscr{P}$ instead of Δ_{ef}) apply to the soundness theorem for functional interpretation as well. As in the proof of theorem 5.8 we could also have extracted terms with free variables and only at the end (as in remark 5.9) perform a λ-abstraction rather than doing this at each step in the soundness proof. Actually this would be more efficient in an implementation of the extraction algorithms (see [156]).

Remark 8.8. Gödel actually established the conclusion of the soundness theorem in (an intensional variant of) a quantifier-free fragment qf-(WE-HA$^\omega$) of WE-HA$^\omega$ which results if quantifiers are omitted from the language, the axiom schema of induction is replaced by a quantifier-free rule of induction

$$\text{QF-IR}: \quad \frac{A_0(0)\,,\, A_0(x^0) \to A_0(x+1)}{A_0(x)}$$

and a substitution rule

$$\text{Sub}: \quad \frac{A}{A[t^\rho/x^\rho]}$$

(replacing $\forall$-elimination) is added. The universal axioms $\mathscr{P}$ are added as open formulas.
To verify $A_D(\underline{t}\,\underline{a},\underline{y},\underline{a})$ in qf-(WE-HA$^\omega$) requires a somewhat more complicated treatment of induction (see e.g. [366]).

Corollary 8.9. *Let A, B be formulas in $\mathscr{L}$(WE-HA$^\omega$). Then the following rule holds*

$$\text{WE-HA}^\omega + \text{AC} + \text{IP}^\omega_\forall + \text{M}^\omega \vdash A \leftrightarrow B \;\Rightarrow\; \text{WE-HA}^\omega \vdash A^D \leftrightarrow B^D.$$

Proof: By the soundness theorem 8.6 the assumption implies that

$$\text{WE-HA}^\omega \vdash (A \to B)^D \wedge (B \to A)^D$$

and hence the claim since

$$\text{WE-HA}^\omega \vdash (A \to B)^D \to (A^D \to B^D).$$

□

Definition 8.10 ([366]). The subset Γ_2 of formulas $\in \mathscr{L}$(WE-HA$^\omega$) is defined inductively by

1) Prime formulas are in Γ_2.
2) $A, B \in \Gamma_2 \Rightarrow A \wedge B, A \vee B, \forall x A, \exists x A \in \Gamma_2$.
3) If A is purely universal and $B \in \Gamma_2$, then $(\exists \underline{x} A \to B) \in \Gamma_2$.

Lemma 8.11. *For $A \in \Gamma_2$ one has* WE-HA$^\omega \vdash A^D \to A$.

Proof: Easy induction on the logical structure of A. □

Corollary 8.12. WE-HA$^\omega$+ AC + IP$^\omega_\forall$+ M$^\omega$ *is conservative over* WE-HA$^\omega$ *with respect to formulas $A \in \Gamma_2$.*

Proof: The corollary follows from theorem 8.6 and lemma 8.11. □

Proposition 8.13 (Characterization theorem for D, Yasugi [380], Troelstra [366]).

For all formulas A of WE-HA$^\omega$ *one has*

$$\text{WE-HA}^\omega + \text{AC} + \text{IP}^\omega_\forall + \text{M}^\omega \vdash A \leftrightarrow A^D.$$

Proof: Easy induction on the logical structure of A using exercise 5 below. □

Corollary 8.14. *Let $\mathscr{P}$ be an arbitrary set of purely universal sentences $\forall \underline{a}^{\underline{\sigma}} B_0(\underline{a})$ (B_0 quantifier-free) of* WE-HA$^\omega$. WE-HA$^\omega$+ AC + IP$^\omega_\forall$+ M$^\omega$ + $\mathscr{P}$ *has the disjunction property* DP, *the existence property* EP *and is closed under the rules of choice* ACR *and of independence-of-premise for purely universal formulas* IPR$^\omega_\forall$.

Proof: The corollary follows similarly to corollary 5.24 but with theorem 8.6 and proposition 8.13 instead of modified realizability with truth. □

Theorem 8.15 (Main theorem on program extraction by D-interpretation). *Let $\mathscr{P}$ be an arbitrary set of purely universal sentences $\forall \underline{a}^{\underline{\sigma}} P_0(\underline{a})$ (P_0 quantifier-free) of $\mathscr{L}$(WE-HA$^\omega$) and $A_0(x^\rho, u^\delta)$ be quantifier-free formula containing only x, u free and $B(x^\rho, y^\tau)$ an arbitrary formula containing only x, y free and ρ, δ, τ are arbitrary types. Then the following rule holds:*

$$\begin{cases} \text{WE-HA}^\omega + \text{AC} + \text{IP}^\omega_\forall + \text{M}^\omega + \mathscr{P} \vdash \forall x^\rho \left(\forall u^\delta A_0(x,u) \to \exists y^\tau B(x,y)\right) \\ \textit{then one can extract a closed term } t \textit{ of } \text{WE-HA}^\omega \textit{ s.t.} \\ \text{WE-HA}^\omega + \text{AC} + \text{IP}^\omega_\forall + \text{M}^\omega + \mathscr{P} \vdash \forall x^\rho \left(\forall u^\delta A_0(x,u) \to B(x,tx)\right). \end{cases}$$

In particular, if $\mathscr{S}^\omega \models \mathscr{P}$, then the conclusion holds in $\mathscr{S}^\omega$.
The result also holds for tuples of variables $\underline{x}, \underline{u}, \underline{y}$ where then t is a tuple of closed terms.

Proof: Let

$$\text{H}^\omega := \text{WE-HA}^\omega + \text{AC} + \text{IP}^\omega_\forall + \text{M}^\omega + \mathscr{P}$$

and assume that

$$\text{H}^\omega \vdash \forall x^\rho \left(\forall u^\delta A_0(x,u) \to \exists y^\tau B(x,y)\right).$$

The by IP$^\omega_\forall$ we get

$$\text{H}^\omega \vdash \forall x^\rho \exists y^\tau C(x,y),$$

where

$$C(x,y) :\equiv (\forall u^\delta A_0(x,u) \to B(x,y)).$$

Let $C^D(x,y) \equiv \exists \underline{a} \forall \underline{b} C_D(\underline{a}, \underline{b}, x, y)$. Then

$$\left(\forall x \exists y C(x,y)\right)^D \equiv \exists Y, \underline{A} \forall x, \underline{b} C_D(\underline{A}x, \underline{b}, x, Yx).$$

By theorem 8.6 there are closed terms $t, \underline{s}$ such that

$$\text{WE-HA}^\omega + \mathscr{P} \vdash \forall x, \underline{b} C_D(\underline{s}x, \underline{b}, x, tx).$$

Hence

$$\text{WE-HA}^\omega + \mathscr{P} \vdash \forall x \underbrace{\exists \underline{a} \forall \underline{b} C_D(\underline{a}, \underline{b}, x, tx)}_{\equiv C^D(x,tx)}.$$

By proposition 8.13 we have

$$\mathrm{H}^{\omega} \vdash C^{D}(x,tx) \leftrightarrow C(x,tx)$$

and hence

$$\mathrm{H}^{\omega} \vdash \forall x C(x,tx).$$

$\square$

8.3 Fragments, exercises, historical comments and suggested further reading

Fragments:

Using the remark made at the end of chapter 5 one easily verifies that the soundness theorem for the functional interpretation also holds if WE-HA$^{\omega}$ is replaced by $\mathcal{T}_i^{\omega} := \widehat{\text{WE-HA}}^{\omega}\!\upharpoonright\!\backslash, \mathrm{G}_n\mathrm{A}_i^{\omega}$ $(n \geq 2)$. Moreover, corollary 8.12 and proposition 8.13 and – most importantly – theorem 8.15 hold for these fragments as well (see [207] for more details). In [71] functional interpretation is applied to systems of bounded arithmetic corresponding to poly-time computability.

Exercises:

1) Let M$'$ be the schema

$$\neg\forall \underline{x} A_0(\underline{x}) \rightarrow \exists \underline{x} \neg A_0(\underline{x}) \quad (A_0 \text{ quantifier-free}).$$

Show that over WE-HA$^{\omega}$ the two schemas M$'$ and M$^{\omega}$ are equivalent.

2) Solve the functional interpretation of

$$\neg\neg(\exists x A_0(x) \vee \neg\exists x A_0(x))$$

by a closed term of WE-HA$^{\omega}$ (where the quantifier-free formula A_0 can be assumed not to contain $\vee$).

3) Solve the functional interpretation of

$$\neg\neg A \wedge \neg\neg B \rightarrow \neg\neg(A \wedge B).$$

4) (Troelstra [366]) In addition to the prenex normal form we used in the definition of $(A \rightarrow B)^D$, there are three more prenex normal forms of $(A^D \rightarrow B^D)$ which give rise to corresponding functional interpretations $(A \rightarrow B)^i$ $(i = 1,2,3)$ of $A \rightarrow B$. For all three of them already $(A \rightarrow A)^i$ fails to have a computable solution for suitable A.
Compute these interpretations $(A \rightarrow B)^i$ and give counterexamples to the computable solvability of $(A \rightarrow A)^i$ for each of them.

5) Let P, Q be arbitrary formulas and x a variable not free in Q. Prove:

a. $\mathrm{IL}+(\neg\neg P \to P) \vdash (\forall x P \to Q) \to \neg\neg\exists x(P \to Q)$
(Hint: First use $Q \vee \neg Q$ and then argue using $(A \to B) \to (\neg\neg A \to \neg\neg B)$ that actually the intuitionistic $\neg\neg(Q \vee \neg Q)$ is sufficient).
b. $\mathrm{IL}+(\neg\neg Q \to Q) \vdash \neg\neg\exists x(P \to Q) \to (\forall x P \to Q)$.

6) Prove the stronger form of the soundness theorem stated in remark 8.8.
7) Prove lemma 8.11.
8) Prove proposition 8.13.

Historical comments and suggested further reading: Gödel's functional interpretation as presented in this chapter was first published in Gödel [133] (an English translation with extended introductory notes by A.S. Troelstra can be found in Gödel [135]). However, Gödel arrived at this interpretation already around 1938 as is clear from his lectures [131] and, in particular, [132] which were published only posthumously in [136]. The same applies to Gödel's more detailed 1972 version [134] of [133] which first appeared in [135]. The system of primitive recursive functionals of finite type was anticipated in Hilbert [161] as we mentioned already at the end of chapter 3. Thorough treatments of functional interpretation can be found in Troelstra [366] (chapter 3, section 5), Schütte [324] and Luckhardt [266]. The latter covers in detail C. Spector's extension of functional interpretation to analysis by means of bar recursion as will be discussed in chapter 11. A more recent presentation of functional interpretation is given in Avigad-Feferman [7] which is a very readable and comprehensive treatment of the whole subject. A study of functional interpretation based on a natural deduction calculus is given in Jørgensen [179]. For general background information on Gödel's interpretation see also Kohlenbach [230]. A nice parametric version of the soundness proof of D that also works for various variants of functional interpretation as well as '*mr*' and a family of realizability interpretations 'in between' *mr* and D (due to Stein [346]) is given in Oliva [292]. Hernest [155] develops a so-called 'Light functional interpretation' (making use of U. Berger's concept of 'uniform quantifiers', Berger [19]) which allows one to drop certain contractions connected to computationally empty quantifiers. A nice application of this approach is given in Hernest [157]. A variant of functional interpretation which interprets $A \to A \wedge A$ in a simpler way so that the decidability of prime formulas is not needed was developed in Diller-Nahm [88]. For a recent discussion of this version see Diller [87]. In Jørgensen [179, 180] the possibility of a 'with truth'-variant (analogous to '*mrt*' discussed in chapter 5) of the Diller-Nahm variant is investigated and it is shown that such an interpretation would not be sound for the exportation rule (this shows that the claim made in Diller [87] that such an interpretation would be sound is not correct). However, there exists a so-called q-variant of '*mr*' that is closely related to '*mrt*' and which – for Kleene realizability – was first introduced in Kleene [193]. An analogous q-version of the Diller-Nahm interpretation is possible (see Stein [345]). Jørgensen [180] shows that this variant can be used to give proof (alternative to the use of '*mrt*') for the existence property of a version HA^{ω} of $\mathrm{WE\text{-}HA}^{\omega}$ that does not have an extensional treatment of equalities of higher type. A thorough investigation of functional interpretation from the perspective of categorical proof theory is given in Hyland [170]. Bishop [33] contains an interesting discussion about the functional interpretation of '$\rightarrow$' from a constructive point of view. Applications of functional interpretation to systems of bounded arithmetic are given in Cook-Urquhart [71]. A functional interpretation for constructive set theories has been developed in Burr [61]. A type-free form of functional interpretation applicable to Feferman's systems EM of explicit mathematics is developed in Beeson [13]. In Oliva [294], a decomposition of functional interpretation into a functional interpretation of linear logic and Girard's ([123]) embedding of intuitionistic logic into linear logic is discussed. A monotone version of functional interpretation (suitable for the extraction of uniform bounds) was introduced in Kohlenbach [206] and will be discussed in chapter 9 below. Recently, a bounded variant which is related in certain ways to the monotone version was developed in Ferreira-Oliva [104]. For applications of (the combination of negative translation and) functional interpretation to systems based on classical logic see chapter 10 and the references given there. A detailed analysis of functional interpretation in terms of proof complexity is given in Hernest-Kohlenbach [159]. In Schwichtenberg [329] functional interpretation is used to extract an algorithm which is close to Euclid's algorithm from a seemingly trivial proof. Semantical interpretations of Gödel's functional interpretation in terms of so-called 'Dialectica categories' have been developed in de Paiva [78] and further studied in Hyland [170] and Biering [31].

Chapter 9
Semi-intuitionistic systems and monotone functional interpretation

9.1 The soundness and bound extraction theorems

Most of the applications of functional interpretation to concrete proofs in numerical functional analysis which were obtained in recent years (see chapters 16 and 18 for a survey of some of them) use a combination of functional interpretation with Howard's majorizability construction similarly to the corresponding combination in the case of modified realizability. Analogously to the monotone modified realizability interpretation one can define a monotone functional interpretation which directly extract terms which majorize some functionals realizing the usual functional interpretation. More precisely one can prove a soundness theorem with the statement in the soundness theorem for functional interpretation replaced by

$$(+)\ \exists \underline{x}\big(\underline{t}^*\ maj\ \underline{x} \wedge \forall \underline{a}, \underline{y} A_D(\underline{x}(\underline{a}), \underline{y}, \underline{a})\big)$$

for suitable closed terms $\underline{t}^*$.

We then say that $\underline{t}^*$ satisfies the **monotone functional interpretation** of A. The soundness proof for the monotone functional interpretation proceeds by establishing $(+)$ by induction on the proof. It is similar to the usual soundness proof combined with some easy majorization arguments. The construction of terms $\underline{t}^*$ is even simpler than the construction of $\underline{t}$ in the usual functional interpretation. E.g. consider the case of the axiom $A \to A \wedge A$, which is – as we saw in chapter 8 – by far the most complicated axiom for the construction of terms satisfying the usual functional interpretation (a detailed complexity analysis is carried out in [159]), the monotone functional interpretation is simply satisfied by the terms

$$\underline{t}^*_{\underline{X}'} := \underline{t}^*_{\underline{X}''} := \lambda \underline{a}, \underline{x}.\underline{x},\ t^*_{Y_i} := \lambda \underline{a}, \underline{x}, \underline{y}', \underline{y}''.max(y'_i, y''_i),$$

which avoids the complicated construction of t_{A_D} needed to satisfy the usual functional interpretation of $A \to A \wedge A$.

The most important feature of monotone functional interpretation is that a large class of sentences Δ (including such important non-effective principles as the binary König's lemma which will be discussed further below in this chapter) have a trivial monotone functional interpretation and, therefore, can – similarly to the universal sentences $\mathscr{P}$ considered in chapter 8 – just added as axioms without having any impact neither on the extraction procedure nor on the extracted terms themselves.

Let Δ be a set of sentences of the form $\forall \underline{a}^{\underline{\delta}} \exists \underline{b} \leq_{\underline{\sigma}} \underline{r}\,\underline{a} \forall \underline{c}^{\underline{\gamma}} B_0(\underline{a},\underline{b},\underline{c})$, where B_0 is quantifier-free (which can as before be assumed not to contain $\vee$ since it can be written as $t_{B_0} =_0 0$) and does not contain any further free variables than those shown and $\underline{r}$ is a tuple of closed terms (of suitable types) of WE-HA$^\omega$. The types $\underline{\delta},\underline{\sigma},\underline{\gamma}$ are arbitrary. Here '$\underline{b} \leq_{\underline{\sigma}} \underline{r}\,\underline{a}$' stands for

$$\bigwedge_{i=1}^{k} (b_i \leq_{\sigma_i} r_i \underline{a}).$$

For a given set Δ we define $\tilde{\Delta}$ as the corresponding set of the Skolem normal forms of the sentences in Δ

$$\left\{ \tilde{\varphi} :\equiv \exists \underline{B} \leq \underline{r} \forall \underline{a}, \underline{c}\, B_0(\underline{a}, \underline{B}\,\underline{a}, \underline{c}) \ : \ \varphi :\equiv \forall \underline{a}^{\underline{\delta}} \exists \underline{b} \leq_{\underline{\sigma}} \underline{r}\,\underline{a} \forall \underline{c}^{\underline{\gamma}} B_0(\underline{a},\underline{b},\underline{c}) \in \Delta \right\}.$$

Note that for $\varphi \in \Delta$

$$\text{WE-HA}^\omega + \text{b-AC} \vdash \varphi \to \tilde{\varphi},$$

where b-AC $:\equiv \bigcup_{\delta,\rho \in \mathbf{T}} \left\{ \text{b-AC}^{\delta,\rho} \right\}$ with

$$\text{b-AC}^{\delta,\rho} :\equiv \forall Z^{\rho\delta} \left(\forall x^\delta \exists y \leq_\rho Zx\, A(x,y,Z) \to \exists Y \leq_{\rho\delta} Z \forall x A(x, Yx, Z) \right)$$

which in turn follows from AC (exercise).

We will denote the monotone functional interpretation by MD.

Theorem 9.1 (Soundness Theorem for MD).
Let Δ be as above and $A(\underline{a})$ be a formula of $\mathscr{L}$(WE-HA$^\omega$) containing only $\underline{a}$ free. Then the following rule holds:

$$\begin{cases} \text{WE-HA}^\omega + \text{AC} + \text{IP}^\omega_\forall + \text{M}^\omega + \Delta \vdash A(\underline{a}), \ \ \textit{then} \\ \text{WE-HA}^\omega + \tilde{\Delta} \vdash \exists \underline{x}(\underline{t}^*\ maj\ \underline{x} \wedge \forall \underline{a}, \underline{y} A_D(\underline{x}\,\underline{a}, \underline{y}, \underline{a})), \end{cases}$$

where $\underline{t}^*$ is a suitable tuple of closed terms of WE-HA$^\omega$ *which can be extracted from a given proof of the assumption.*

Proof: Induction on the length of the proof. For the axioms (except Δ), the result follows from the fact that the usual functional interpretation is satisfiable by suitable closed terms $\underline{t}$ and Howard's construction of majorizing functionals $\underline{t}^*$ of $\underline{t}$ (proposition 6.6). As mentioned above, the construction of $\underline{t}^*$ when done directly is sometimes even simpler than that of $\underline{t}$. This, in particular, applies to the axioms $A \to A \wedge A$ and $A \vee A \to A$ (for notational simplicity we drop the tuple notation and assume that $\underline{a}$ comprises all the free variables occurring in any formula involved with the understanding that – as in the soundness proof for functional interpretation – variables which do not occur in the end formula can be replaced by constants O^ρ of appropriate type):

1) $[A \to A \wedge A]^D \equiv \exists Y, X', X'' \forall x, y', y''$
$$\big(A_D(x, Yxy'y'', \underline{a}) \to A_D(X'x, y', \underline{a}) \wedge A_D(X''x, y'', \underline{a})\big).$$
Define
$$t_Y^* := \lambda \underline{a}, x, y', y''. \max(y', y''),\ t_{X'}^* := t_{X''}^* := \lambda \underline{a}, x.x.$$
'$\exists Y, X', X''$' is realized by
$$t_Y \underline{a} x y' y'' := \begin{cases} y', \text{ if } \neg A_D(x, y', \underline{a}) \\ y'', \text{ if } A_D(x, y', \underline{a}) \end{cases} \quad \text{and } t_{X'} := t_{X''} := \lambda \underline{a}, x.x.$$
Since $t_Y^*\ maj\ t_Y$, $t_{X'}^*\ maj\ t_{X'}$ and $t_{X''}^*\ maj\ t_{X''}$, the terms $t_Y^*, t_{X'}^*$ and $t_{X''}^*$ fulfill our claim.
2) $[A \vee A \to A]^D \equiv \exists Y, Y', X'' \forall z^0, x, x', y'' \{ \big(z =_0 0 \to A_D(x, Yzxx'y'', \underline{a})\big) \wedge$
$$\big(z \neq 0 \to A_D(x', Y'zxx'y'', \underline{a})\big) \to A_D(X''zxx', y'', \underline{a}) \}.$$
$$\text{Define} \begin{cases} t_{X''}^* := \lambda \underline{a}, z, x, x'. \max(x, x'), \\ t_Y^* := t_{Y'}^* := \lambda \underline{a}, z, x, x', y''.y''. \end{cases}$$
The terms $t_{X''}^*, t_Y^*$ and $t_{Y'}^*$ majorize the functionals
$$t_{X''} \underline{a} z^0 x x' := \begin{cases} x, \text{ if } z = 0 \\ x', \text{ if } z \neq 0, \end{cases}$$
$t_Y := t_Y^*$ and $t_{Y'} := t_{Y'}^*$, which realize '$\exists Y, Y', X''$', and hence satisfy the monotone functional interpretation of the axiom.
3) The interpretation of $A \vee B \to B \vee A$ is also simplified if only a majorizing functional has to be constructed:
$$(A \vee B \to B \vee A)^D \equiv \exists Z', X', U', Y, V \forall z^0, x, u, y', v'$$
$$\{\big(z = 0 \to A_D(x, Yzxuy'v', \underline{a})\big) \wedge \big(z \neq 0 \to B_D(u, Vzxuy'v', \underline{a})\big)$$
$$\to \big(Z'zxu = 0 \to B_D(U'zxu, v', \underline{a})\big) \wedge \big(Z'zxu \neq 0 \to A_D(X'zxu, y', \underline{a})\big)\}.$$
$t_{U'}^* := \lambda \underline{a}, z, x, u.u$, $t_{X'}^* := \lambda \underline{a}, z, x, u.x$, $t_Y^* := \lambda \underline{a}, z, x, u, y', v'.y'$, $t_V^* := \lambda \underline{a}, z, x, u, y', v'.v'$ are defined as in the usual functional interpretation, but $t_{Z'}^*$ is now simply $t_{Z'}^* := \lambda \underline{a}, z, x, u.1^0$ whereas the usual interpretation requires for the realization of "$\exists Z'$" the functional $t_{Z'} := \lambda \underline{a}, z, x, u.\overline{sg}(z^0)$, where $\overline{sg}(z^0) :=$

$$\begin{cases} 0, \text{ if } z \neq 0 \\ 1, \text{ if } z = 0. \end{cases}$$

It is clear that $t^*_{Z'}$ majorizes $t_{Z'}$.

4) The monotone functional interpretation of an axiom $\forall \underline{a}^{\underline{\delta}} \exists \underline{b} \leq_{\underline{\sigma}} \underline{r}\,\underline{a} \forall \underline{c}^{\underline{\gamma}} B_0(\underline{a},\underline{b},\underline{c})$ in Δ is satisfied (provably in WE-HA$^\omega + \exists \underline{B} \leq \underline{r} \forall \underline{a},\underline{c}\, B_0(\underline{a},\underline{B}\,\underline{a},\underline{c})$) by any tuple $\underline{r}^*$ of closed terms which (provably in WE-HA$^\omega$) majorize $\underline{r}$. Again, the construction of $\underline{r}^*$ is guaranteed by Howard's technique.
5) Modus ponens and syllogism: Let t_1^*, t_2^*, t_3^* be such that
(1) $\exists x_1 \big(t_1^* \; maj\; x_1 \wedge \forall y, \underline{a} A_D(x_1\underline{a}, y, \underline{a})\big)$ and
(2) $\exists x_2, x_3 \Big(t_2^* \; maj\; x_2 \wedge t_3^* \; maj\; x_3 \wedge \forall x, v, \underline{a}\big(A_D(x, x_2\underline{a}xv, \underline{a}) \to B_D(x_3\underline{a}x, v, \underline{a})\big)\Big)$.
Then $t_4^* := \lambda \underline{a}.t_3^*\underline{a}(t_1^*\underline{a}) \; maj\; \lambda \underline{a}.x_3\underline{a}(x_1\underline{a})$ and $\lambda \underline{a}.x_3\underline{a}(x_1\underline{a})$ realizes B^D. As in the proof of the soundness theorem for functional interpretation, we replace those free variables a_i from $\underline{a}$ that do not occur in B by $\mathcal{O}$ of appropriate type.
The rule $\frac{A \to B\;,\; B \to C}{A \to C}$ is treated similarly.
6) The monotone interpretation of the remaining logical rules uses the λ-terms of the usual functional interpretation since they preserve majorizability. The treatment of the quantifier-free extensionality rule as as trivial as for the usual functional interpretation.
7) Induction rule:

$$\frac{B(0)\;,\; \forall y^0 \big(B(y) \to B(y+1)\big)}{\forall y B(y)}.$$

Let $(B(y))^D \equiv \exists u \forall v B_D(u, v, y, \underline{a})$ and t_1^*, t_2^*, t_3^* be such that

$$\exists x_1 \big(t_1^* \; maj\; x_1 \wedge \forall v, \underline{a} B_D(x_1\underline{a}, v, 0, \underline{a})\big)$$

and

$$\exists x_2, x_3 \big(t_2^* \; maj\; x_2 \wedge t_3^* \; maj\; x_3 \wedge \forall u, w, y, \underline{a}\big(B_D(u, x_2 y\underline{a}uw, y, \underline{a}) \\ \to B_D(x_3 y\underline{a}u, w, y+1, \underline{a})\big)\big).$$

Define $t^* := t^M$, where t is defined by recursion

$$\begin{cases} t\underline{a}0 = t_1^*\underline{a} \\ t\underline{a}(y+1) = t_3^* y\underline{a}(t\underline{a}y). \end{cases}$$

One easily verifies (in WE-HA$^\omega$) that

$t^* \; maj\; x$, where x is defined by $\begin{cases} x\underline{a}0 = x_1\underline{a} \\ x\underline{a}(y+1) = x_3 y\underline{a}(x\underline{a}y) \end{cases}$

As in the soundness proof for functional interpretation it follows that

$$\forall y^0, v B_D(x\underline{a}y, v, y, \underline{a}).$$

□

Remark 9.2. Theorem 9.1 holds analogously for '*s-maj*' instead of '*maj*'.

Theorem 9.3 (Main theorem on uniform bound extraction by MD).
Let Δ as above and $A(x^1, y^\rho, z^\tau)$ an arbitrary formula containing only x, y, z free. Let $deg(\tau) \leq 2$ and $s^{\rho(1)}$ a closed term of WE-HA$^\omega$. *Then the following rule holds:*

$$\begin{cases} \text{WE-HA}^\omega + \text{AC} + \text{IP}^\omega_\forall + \text{M}^\omega + \Delta \vdash \forall x^1 \forall y \leq_\rho sx \exists z^\tau A(x,y,z) \\ \textit{then one can extract a closed term } t \textit{ of } \text{WE-HA}^\omega \textit{ s.t.} \\ \text{WE-HA}^\omega + \text{AC} + \text{IP}^\omega_\forall + \text{M}^\omega + \Delta \vdash \forall x^1 \forall y \leq_\rho sx \exists z \leq_\tau txA(x,y,z). \end{cases}$$

In particular, if $\mathcal{S}^\omega \models \Delta$, then the conclusion holds in $\mathcal{S}^\omega$. As in theorem 6.8 the result also holds for tuples of variables.
An analogous results holds for $\widehat{\text{WE-HA}}^\omega{\restriction}$ *instead of* WE-HA$^\omega$.

Proof: Without loss of generality we can assume that $\tau = 2$. As an abbreviation we define $\mathcal{T}$:=WE-HA$^\omega$+AC+IP$^\omega_\forall$+M$^\omega$. By the assumption and IP$^\omega_\forall$ we obtain

$$\mathcal{T} + \Delta \vdash \forall x, y \exists z (y \leq sx \to A(x,y,z)).$$

Monotone functional interpretation extracts a closed term t^* in WE-HA$^\omega$ such that

$$\mathcal{T} + \tilde{\Delta} \vdash \exists Z \big(t^* \; maj \; Z \wedge \forall x \forall y (y \leq sx \to A(x,y,Zxy))^D\big).$$

By proposition 8.13 we have $\mathcal{T} \vdash G^D \leftrightarrow G$ for all formulas G. Hence (using similar reasoning as in the proof of theorem 6.8)

$$\mathcal{T} + \tilde{\Delta} \vdash \exists Z \forall x \forall y \leq sx \big(\underbrace{\lambda w^1 . t^* x^M (s^* x^M) w^M}_{tx:=} \geq_2 Zxy \wedge A(x,y,Zxy)\big),$$

and thus

$$\mathcal{T} + \tilde{\Delta} \vdash \forall x \forall y \leq sx \exists z \leq_2 txA(x,y,z).$$

Since – using AC –

$$\mathcal{T} + \Delta \vdash \tilde{\Delta}$$

this implies the conclusion of the theorem. □

Corollary to the proof of theorem 9.3: Using lemma 8.11 it follows from the proof above that if $A \in \Gamma_2$ as defined in 8.10, then the conclusion of theorem 9.3 is provable in WE-HA$^\omega + \tilde{\Delta}$.

As in the proof of corollary 6.10 one concludes

Corollary 9.4. WE-HA$^\omega$+ AC + IP$^\omega_\forall$+ M$^\omega$ + Δ *is closed under the fan rule.*

Remark 9.5. One easily obtains the variant of theorem 9.3 corresponding to proposition 6.11 (exercise).

9.2 Applications of monotone functional interpretation

Application I:
Analogously to the application of theorem 6.8 given at the end of chapter 6 we obtain from theorem 9.3 the following rule for $\mathrm{H}^\omega := \mathrm{WE\text{-}HA}^\omega + \mathrm{AC} + \mathrm{IP}^\omega_{\forall} + \mathrm{M}^\omega + \Delta$:

$$\begin{cases} \mathrm{H}^\omega \vdash \forall k^0 \forall x \in [0,1] \exists n^0 \forall m \geq n (|\Phi_m(x) - \Phi(x)| <_{\mathbb{R}} 2^{-k}) \Rightarrow \\ \mathrm{H}^\omega \vdash \forall k^0 \exists n^0 \forall x \in [0,1] \forall m \geq n (|\Phi_m(x) - \Phi(x)| <_{\mathbb{R}} 2^{-k}), \end{cases}$$

where $\Phi^{1(1)(0)}_{(\cdot)}, \Phi^{1(1)}$ again are closed terms of WE-HA$^\omega$ which (provably in WE-HA$^\omega$) represent functions $[0,1] \to \mathbb{R}$. Thus even in the presence of M$^\omega$ we have that provable pointwise convergence of Φ_n towards Φ on $[0,1]$ implies provable uniform convergence on $[0,1]$.

Application II:

Definition 9.6 (Hereditarily extensional equality, Troelstra [366]). Between functionals x_1^ρ, x_2^ρ of type ρ we define the following relation by induction on ρ

$$\begin{cases} x_1 \approx_0 x_2 :\equiv (x_1 =_0 x_2), \\ x_1 \approx_{\tau\rho} x_2 :\equiv \forall y_1^\rho, y_2^\rho \big(y_1 \approx_\rho y_2 \to x_1 y_1 \approx_\tau x_2 y_2\big). \end{cases}$$

Lemma 9.7. 1) WE-HA$^\omega \vdash x_1 =_\rho \tilde{x}_1 \wedge x_2 =_\rho \tilde{x}_2 \wedge x_1 \approx_\rho x_2 \to \tilde{x}_1 \approx_\rho \tilde{x}_2$.
2) *Let* $\rho = \tau\rho_k \ldots \rho_1$. *Then*

$$\mathrm{WE\text{-}HA}^\omega \vdash x \approx_\rho \tilde{x} \leftrightarrow \forall y_1, \tilde{y}_1, \ldots, y_k, \tilde{y}_k \Big(\bigwedge_{i=1}^{k} (y_i \approx_{\rho_i} \tilde{y}_i) \to x y_1 \ldots x_k \approx_\tau \tilde{x} \tilde{y}_1 \ldots \tilde{y}_k \Big).$$

Proof: 1) Induction on ρ. 2) induction on k. □

Proposition 9.8 (Troelstra [366]). *Let* t^ρ *be a closed term of* WE-HA$^\omega$. *Then*

$$\mathrm{WE\text{-}HA}^\omega \vdash t \approx_\rho t.$$

Proof: Induction on the structure of t.
(i) Constants: Using the previous lemma one easily verifies that $0^0 \approx_0 0^0$, $S \approx_1 S$, $\Pi_{\rho,\tau} \approx \Pi_{\rho,\tau}$, $\Sigma_{\delta,\rho,\tau} \approx \Sigma_{\delta,\rho,\tau}$.

$\underline{R}_{\underline{\rho}}$: We show by induction on x^0 that $\underline{R}_{\underline{\rho}} x \approx \underline{R}_{\underline{\rho}} x$, i.e. $\bigwedge_{i=1}^{k} ((R_i)_{\underline{\rho}} x \approx (R_i)_{\underline{\rho}} x)$:

Suppose that $\underline{y}_1 \approx \underline{y}_2, \underline{z}_1 \approx \underline{z}_2$:
$\underline{R}_{\underline{\rho}} 0 \underline{y}_1 \underline{z}_1 = \underline{y}_1 \approx \underline{y}_2 = \underline{R}_{\underline{\rho}} 0 \underline{y}_2 \underline{z}_2 \Rightarrow \underline{R}_{\underline{\rho}} 0 \underline{y}_1 \underline{z}_2 \approx \underline{R}_{\underline{\rho}} 0 \underline{y}_2 \underline{z}_2$.

$\underline{R}_\rho(x+1)\underline{y}_1\underline{z}_1 = \underline{z}_1(\underline{R}_\rho x\underline{y}_1\underline{z}_1)x \overset{I.H.}{\approx} \underline{z}_2(\underline{R}_\rho x\underline{y}_2\underline{z}_2)x = \underline{R}_\rho(x+1)\underline{y}_2\underline{z}_2$
$\Rightarrow \underline{R}_\rho(x+1)\underline{y}_1\underline{z}_1 \approx \underline{R}_\rho(x+1)\underline{y}_2\underline{z}_2$.
Since $x_1 =_0 x_2 \leftrightarrow x_1 \approx_0 x_2$, we have by the $=_0$-axioms and the previous lemma

$$\forall x_1, x_2\big(x_1 \approx_0 x_2 \to \underline{R}_\rho x_1 \approx \underline{R}_\rho x_2\big),$$

i.e. $\underline{R}_\rho \approx \underline{R}_\rho$.
(ii) $t \approx_{\tau\rho} t \wedge s \approx_\rho s \to ts \approx_\tau ts$. □

Corollary 9.9. *Let $t^{1(1)}$ be a closed term of* WE-HA$^\omega$. *Then*

$$\text{WE-HA}^\omega \vdash \forall x^1, y^1(x =_1 y \to tx =_1 ty).$$

This also holds for tuples of variables $\underline{x}, \underline{y}$ of types of degree ≤ 1.

Proof: This follows from proposition 9.8 since WE-HA$^\omega \vdash x =_1 y \leftrightarrow x \approx_1 y$. □

Proposition 9.10. *Let $t^{1(1)}$ be closed. Then $t^{1(1)}$ is uniformly continuous on each set $\{x : x \leq_1 y\}$ with a modulus of uniform continuity which is definable in* WE-HA$^\omega$ *(uniformly in y), i.e. there is a closed term $\tilde{t}^{0(1)(0)}$ of* WE-HA$^\omega$*:*

$$\text{WE-HA}^\omega \vdash \forall k^0 \forall x, \tilde{x} \leq_1 y\Big(\bigwedge_{i=0}^{\tilde{t}ky}(xi =_0 \tilde{x}i) \to \bigwedge_{j=0}^{k}(txj =_0 t\tilde{x}j)\Big).$$

The proposition also holds for tuples $\underline{x}, \underline{y}$ with $x_i \leq_1 y_i$ for all components.

Proof: By the corollary above we have

$$\text{WE-HA}^\omega \vdash \forall x, \tilde{x}\big(\forall i(xi =_0 \tilde{x}i) \to \forall k \forall j \leq k(txj =_0 t\tilde{x}j)\big).$$

Hence

$$\text{WE-HA}^\omega + \text{M}^\omega \vdash \forall k \forall x, \tilde{x} \exists i\Big(xi =_0 \tilde{x}i \to \bigwedge_{j=0}^{k}(txj =_0 t\tilde{x}j)\Big)$$

and so a-fortiori

$$\text{WE-HA}^\omega + \text{M}^\omega \vdash \forall y^1 \forall k \forall x, \tilde{x} \leq_1 y \exists i\Big(xi =_0 \tilde{x}i \to \bigwedge_{j=0}^{k}(txj =_0 t\tilde{x}j)\Big).$$

By theorem 9.3 (and the corollary to its proof) one can extract a closed term $\tilde{t}$ of WE-HA$^\omega$ such that

$$\text{WE-HA}^\omega \vdash \forall y^1 \forall k \forall x, \tilde{x} \leq_1 y \exists i \leq \tilde{t}ky\Big(xi =_0 \tilde{x}i \to \bigwedge_{j=0}^{k}(txj =_0 t\tilde{x}j)\Big)$$

which finishes the proof. □

Application of monotone functional interpretation III:

We now use theorem 9.3 to prove that WE-HA$^\omega$ does not even for Π_1^0-axioms satisfy the deduction theorem:

Theorem 9.11. *There exists a Π_1^0-sentence A and a quantifier-free formula B such that*

$$\text{WE-HA}^\omega + A \vdash B, \text{ but } \text{WE-HA}^\omega \nvdash A \to B.$$

Proof: Let Con_{PA} denote the standard consistency predicate for Peano arithmetic PA, i.e.

$$Con_{\mathrm{PA}} \equiv \forall x \neg Prov_{\mathrm{PA}}(x, \ulcorner \overline{0} = \overline{1} \urcorner),$$

where $Prov_{\mathrm{PA}}$ is the usual primitive recursive proof predicate for PA. In the language $\mathcal{L}(\text{WE-HA}^\omega)$ of WE-HA$^\omega$ we can write Con_{PA} in the form

$$A :\equiv \forall x^0 (t_{\mathrm{PA}} x =_0 0),$$

where t_{PA} is a suitable closed term of WE-HA$^\omega$. By the definition of $=_1$ we, trivially, have

$$\text{WE-HA}^\omega + A \vdash t_{\mathrm{PA}} =_1 0^1,$$

where $0^1 := \lambda x^0.0^0$. An application of QF-ER to this yields

$$\text{WE-HA}^\omega + A \vdash x^2(t_{\mathrm{PA}}) =_0 x(0^1),$$

where x^2 is a free variable of type 2. Suppose now that

$$(+)\ \text{WE-HA}^\omega \vdash A \to x^2(t_{\mathrm{PA}}) =_0 x(0^1)$$

and so, in particular,

$$\text{WE-HA}^\omega \vdash A \to \forall x \leq_2 1^2 (x(t_{\mathrm{PA}}) =_0 x(0^1)).$$

Then, using (the special case M^0 of) M$^\omega$,

$$\text{WE-HA}^\omega + \text{M}^\omega \vdash \forall x \leq_2 1^2 \exists y^0 (t_{\mathrm{PA}} y =_0 0 \to x(t_{\mathrm{PA}}) =_0 x(0^1)),$$

where $1^2 := \lambda x^1.S0$.
By theorem 9.3 there exists a closed term s^0 of WE-HA$^\omega$ such that

$$\text{WE-HA}^\omega \vdash \forall y \leq_0 s(t_{\mathrm{PA}} y =_0 0) \to \forall x \leq_2 1(x(t_{\mathrm{PA}}) =_0 x(0^1)).$$

In WE-HA$^\omega$, every fixed closed term s^0 of type 0 can be reduced to a numeral (see e.g. [366]), i.e. there exists a number $n \in \mathbb{N}$ such that

$$\text{WE-HA}^\omega \vdash s =_0 \overline{n}.$$

By the Σ_1^0-completeness of WE-HA$^\omega$ we have

$$\text{WE-HA}^\omega \vdash \forall y \leq_0 \overline{n}(t_{\text{PA}} y =_0 0).$$

Hence

$$\text{WE-HA}^\omega \vdash \forall x \leq_2 1^2(x(t_{\text{PA}}) =_0 x(0^1)).$$

It is an easy exercise to conclude from this that

$$\text{WE-HA}^\omega \vdash t_{\text{PA}} =_1 0, \text{ i.e.}$$

$$\text{WE-HA}^\omega \vdash Con_{\text{PA}}.$$

This, however, contradicts Gödel's second incompleteness theorem, since WE-HA^ω is conservative over Heyting arithmetic HA (as follows by formalizing the model HEO of all hereditarily effective operations in HA, see [366]). Thus $(+)$ above is false and we conclude that the theorem holds taking $B :\equiv \big(x^2(t_{\text{PA}}) =_0 x(0)\big)$ and A as above. □

Corollary 9.12. *The deduction theorem for* WE-HA^ω *fails already for closed* Π^0_1*-axioms.*

Remark 9.13. Combined with the negative translation discussed in chapter 10 it follows that also WE-PA^ω does not satisfy the deduction theorem for Π^0_1-axioms.

9.3 Examples of axioms Δ: Weak König's lemma WKL

We first recall the definition of the weak König's lemma (in our language with function variables) from chapter 7:

$$\text{WKL: } \forall f^1 \Big(T(f) \wedge \forall x^0 \exists n^0 (lth\, n = x \wedge fn = 0) \to \exists b \leq_1 \lambda k.1 \forall x^0 \big(f(\overline{b}x) = 0\big)\Big),$$

where $T(f) :\equiv \forall n, m\big(f(n * m) = 0 \to fn = 0\big) \wedge \forall n, x\big(f(n * \langle x \rangle) = 0 \to x \leq 1\big)$.

$T(f)$ asserts that f represents a 0,1–tree. This definition was first given in [367].

WKL has (in the context of the language of 2nd order arithmetic with set variables) received quite some attention during the last 20 years. In the program of so-called reverse mathematics (see e.g. [107, 335, 336, 57, 58, 168, 331] and, in particular, [338] for a comprehensive account) it has been shown that WKL – relative to the second order fragment RCA_0 of $\widehat{\text{WE-PA}}^\omega{\upharpoonright}+\text{QF-AC}^{0,0}$ – allows one to derive many (ineffective) theorems in mathematics (in particular in analysis) and that conversely many of those theorems imply WKL. Moreover, H. Friedman proved that $\text{WKL}_0 := \text{RCA}_0\text{+WKL}$ is Π^0_2-conservative over primitive recursive arithmetic PRA. Later a proof-theoretic argument for this result based on cut elimination was given in [334]. In [203], this was extended to $\widehat{\text{WE-PA}}^\omega{\upharpoonright}+\text{QF-AC}^{0,0}+\text{WKL}$ using monotone functional interpretation (see [7] for a very readable account of that argument). In chapter 12 we will prove a further generalization of this result. From the applied perspective of proof mining these facts are of relevance because they allow one to extract primitive recursive algorithms and bounds from proofs in those parts of classical analysis which can be carried out by WKL. However, as we will see in chapter 10, for this purpose it is not necessary to eliminate WKL but only to observe that WKL can be written in the form of an axiom Δ. We will now show that this is also possible in the intuitionistic context of $\widehat{\text{WE-HA}}^{\omega}\!\restriction$. Hence we can apply theorem 9.3 to WKL. This is of interest as we on the one hand now can use the highly ineffective principle WKL whereas on the other hand maintain the 'intuitionistic' feature of our system that bounds are extractable for **arbitrary** formulas A (and not just purely existential formulas A to which we have to restrict things in the presence of full classical logic, see chapter 10). We will show that this even holds in the presence of the 'full' König's lemma KL where the condition on the tree being binary is replaced by 'finitely branching', i.e.

$$\text{KL: } \forall f^1 \Big(\tilde{T}(f) \wedge \forall x^0 \exists n^0 (lth\, n = x \wedge fn = 0) \to \exists b^1 \forall x^0 \big(f(\overline{b}x) = 0\big)\Big),$$

where $\tilde{T}(f) :\equiv \forall n,m\big(f(n*m)=0 \to fn=0\big) \wedge \forall n \exists m \forall x\big(f(n*<x>)=0 \to x \le m\big)$.

Classically, more precisely relative to RCA_0, KL is known to be equivalent to the schema of arithmetical comprehension and hence adds enormously to the provable recursive functions when added to e.g. $\widehat{\text{WE-PA}}^{\omega}\!\restriction$ (since now induction for all first order formulas becomes derivable). The reason for this is that to get a bound on '$\exists m$' in $\tilde{T}(f)$ one needs $\text{AC}^{0,0}$ for Π^0_1-formulas which – classically yields Π^0_1-comprehension and so – by iteration Π^0_∞-comprehension. Intuitionistically, however, even full AC is rather weak and allowed in the metatheorems of this chapter. Using AC we will in the end derive KL from WKL. Firstly, however, we have to consider WKL:

Since we will later discuss in more detail strengthened versions of Friedman's conservation result, we will now explicitly work in the fragment $\widehat{\text{WE-HA}}^{\omega}\!\restriction$. Except for the derivation of KL from WKL and AC, the proofs below can be carried out already in $\text{G}_3\text{A}^{\omega}_i$ but not in $\text{G}_2\text{A}^{\omega}_i$ since already the very formulation of WKL uses the (exponential) coding of sequences $(f(0),\ldots,f(n-1))$ into numbers $\overline{f}n$. In chapter 12 we will, therefore, introduce a 'non-standard' principle which allows already relatively to $\text{G}_2\text{A}^{\omega}$ to carry out many of the usual WKL applications with even simpler proofs and still guarantees the extractability of polynomial bounds.

In order to show that WKL can be written in the form of a sentence $\in \Delta$ we first observe that WKL is (already relative to $\widehat{\text{WE-HA}}^{\omega}\!\restriction$) equivalent to

$$(+)\ \forall f,g\Big(T(f) \wedge \forall x\big(lth(gx) = x \wedge f(gx) = 0\big) \to \exists b \le_1 \lambda k.1 \forall x^0 \big(f(\overline{b}x) = 0\big)\Big).$$

This follows from the fact that by $T(f)$ we have

$$\forall x^0 \exists n^0 \big(lth\, n = x \wedge fn = 0\big) \overset{\text{T(f)}}{\to} \forall x \exists n \le \overline{1^{00}}x \big(lth\, n = x \wedge fn = 0\big)$$

and so the proof of this equivalence does not need QF-AC^{0,0} but only (primitive recursive) bounded search.
(+) is a sentence having the logical form

$$(*)\ \forall x^1 \big(\forall n^0 A_0(n,x) \to \exists y \leq_1 sx \forall z^0 B_0(x,y,z)\big),$$

where A_0 and B_0 are quantifier-free formulas and, therefore, still not of the form of sentences permitted in Δ.
However, we will show now that $(+)$ (and hence WKL), in fact, is equivalent (provably in $\widehat{\text{WE-HA}}^{\omega}\!\upharpoonright$) to a sentence WKL$'$ of the form $\forall \underline{x}^1 \exists y \leq_1 \lambda k.1 \forall z^0 A_0^K(\underline{x},y,z)$ which is allowed in Δ. For this we need the following constructions:

$$1)\ \widehat{f}n := \begin{cases} fn \ \text{ if } fn \neq 0 \vee \Big(\forall k,l(k*l=n \to fk=0) \wedge \forall i < lth\,n\big((n)_i \leq 1\big)\Big), \\ 1^0 \ \text{ otherwise.} \end{cases}$$

$$2)\ f_g n := \begin{cases} fn \ \text{ if } f\big(g(lth\,n)\big)=0 \wedge lth\big(g(lth\,n)\big)=lth\,n, \\ 0^0 \ \text{ otherwise.} \end{cases}$$

Remark 9.14. $\widehat{f}$ (f_g) are definable in $\widehat{\text{WE-HA}}^{\omega}\!\upharpoonright$ uniformly as functionals in f (f and g).

As the next lemma shows, the construction $\widehat{f}$ guarantees that $\widehat{f}$ does represent a binary tree, i.e. $T(\widehat{f})$, while it doesn't change f if f already satisfies $T(f)$:

Lemma 9.15. 1) $\widehat{\text{WE-HA}}^{\omega}\!\upharpoonright \vdash \forall f\big(T(\widehat{f})\big)$,
2) $\widehat{\text{WE-HA}}^{\omega}\!\upharpoonright \vdash \forall f\big(T(f) \to f =_1 \widehat{f}\big)$.

Proof:

1) The fact that $\widehat{f}$ represents a tree follows from the following chain of implications:

$$\begin{aligned}
&\widehat{f}(n*m)=0 \to \widehat{f}(n*m)=f(n*m)=0 \to \\
&\forall k,l\big(k*l=n*m \to fk=0\big) \wedge \forall i < lth(n*m)\big((n*m)_i \leq 1\big) \to \\
&\forall k,l\big(k*l=n \to fk=0\big) \wedge \forall i < lth\,n\big((n)_i \leq 1\big) \to \widehat{f}n = fn = 0.
\end{aligned}$$

That $\widehat{f}$ represents a **binary** tree follows from

$$\begin{aligned}
&\widehat{f}(n*<x>)=0 \to \widehat{f}(n*<x>)=f(n*<x>)=0 \to \\
&\forall i < lth(n*<x>)\big((n*<x>)_i \leq 1\big) \to x \leq 1.
\end{aligned}$$

2) We assume that Tf. Then

$$fn=0 \to \forall k,l\big(k*l=n \to fk=0\big) \wedge \forall i < lth\,n\big((n)_i \leq 1\big).$$

Hence $\widehat{f}n = fn$ for all $n \in \mathbb{N}$.

□

We now show that

- f_g always satisfies the condition $\forall x \exists n (lth\, n = x \wedge f_g n = 0)$.
- If already f satisfies this condition and g is such that it selects a realizer for '$\exists n$', then the construction f_g does not change f.

Lemma 9.16. 1) $\widehat{\text{WE-HA}}^{\omega}\!\upharpoonright \vdash \forall f, g \forall x \exists n (lth\, n = x \wedge f_g n = 0)$,
2) $\widehat{\text{WE-HA}}^{\omega}\!\upharpoonright \vdash \forall f, g \Big(\forall x (lth(gx) = x \wedge f(gx) = 0) \to f_g =_1 f\Big)$.

Proof:

1) We leave the proof of this claim, which we will not need further below, as an exercise.
2) $\forall x (lth(gx) = x \wedge f(gx) = 0) \to \forall n \Big(lth(g(lth\, n)) = lth n \wedge f(g(lth\, n)) = 0\Big)$
$\to \forall n (f_g n = fn)$.

□

We are now in the position to define WKL' :

Definition 9.17. $\text{WKL}' :\equiv \forall f^1, g^1 \exists b \leq_1 \lambda k.1 \forall x^0 \Big(\widehat{(\widehat{f})_g}(\overline{b}x) =_0 0\Big)$.

The next proposition shows that an 'ε-weakening' (see chapter 10 for a discussion of this notion) of WKL' is provable already in $\widehat{\text{WE-HA}}^{\omega}\!\upharpoonright$ while WKL' itself is equivalent to WKL:

Proposition 9.18. 1) $\widehat{\text{WE-HA}}^{\omega}\!\upharpoonright \vdash \forall f, g, x \exists b \leq_1 \lambda k.1 \forall y \leq x \Big(\widehat{(\widehat{f})_g}(\overline{b}y) =_0 0\Big)$.
2) $\widehat{\text{WE-HA}}^{\omega}\!\upharpoonright \vdash \text{WKL} \leftrightarrow \text{WKL}'$.

Proof: 1) We show by induction on x that

$$(*)\ \forall x \exists n \Big(lth\, n = x \wedge \forall i < x ((n)_i \leq 1) \wedge \widehat{(\widehat{f})_g}(n) = 0\Big).$$

Note that the quantifier '$\exists n$' can be bounded by $\overline{1^1}x$. Hence we only need QF-IA which is available in $\widehat{\text{WE-HA}}^{\omega}\!\upharpoonright$. It is clear that $(*)$ implies 1): Let n (for given x) satisfy $(*)$ and define $b := \lambda i.(n)_i$. Then $\widehat{(\widehat{f})_g}(\overline{b}x) = 0$ which implies

$$\forall y \leq x (\widehat{(\widehat{f})_g}(\overline{b}y) = 0)$$

by lemma 9.15.1.
$\underline{x = 0}$: $lth\, n = x \leftrightarrow n = \langle\rangle = 0$.
Case (i): $\widehat{f}(g0) = 0 \wedge lth(g0) = 0$: Then $g0 = 0$ and so $\widehat{f}(0) = 0$ which implies $(\widehat{f})_g(0) = 0$ and, furthermore, $\widehat{(\widehat{f})_g}(0) = 0$.

Case (ii): $\widehat{f}(g0) \neq 0 \vee lth(g0) \neq 0$: Then $\left(\widehat{f}\right)_g(0) = 0$ which implies $\widehat{\left(\widehat{f}\right)_g}(0) = 0$.
$\underline{x \to x+1}$: The induction hypothesis yields the existence of an n_0 such that $lth\, n_0 = x$, $\widehat{\left(\widehat{f}\right)_g}(n_0) = 0$ and $\forall i < lth\, n_0\big((n_0)_i \leq 1\big)$. Now put $n_1 := n_0 * < 0 >$.
Case (i): $\widehat{f}\big(g(lth\, n_1)\big) = 0 \wedge lth\big(g(lth\, n_1)\big) = lth\, n_1$:
Then $n := g(lth\, n_1)$ fulfills the claim: first note that

$$lth\, n = lth\big(g(lth\, n_1)\big) = lth\, n_1 = x+1.$$

Moreover,

$$\begin{aligned}\widehat{f}n = 0 &\stackrel{l.9.15.1}{\to} \forall k,l\big(n = k*l \to \widehat{f}k = 0\big) \wedge \forall i < lth\big((n)_i \leq 1\big)\\ &\to \forall k,l\big(n = k*l \to \left(\widehat{f}\right)_g(k) = 0\big) \wedge \forall i < lth\big((n)_i \leq 1\big)\\ &\to \widehat{\left(\widehat{f}\right)_g}(n) = 0 \wedge \forall i < lth\, n\big((n)_i \leq 1\big).\end{aligned}$$

Case (ii): $\widehat{f}\big(g(lth\, n_1)\big) \neq 0 \vee lth\big(g(lth\, n_1)\big) \neq lth\, n_1$: Then $n := n_1$ fulfills the claim: By the case and the f_g-definition we get

$$(+)\ \left(\widehat{f}\right)_g(n_1) = 0.$$

From $\widehat{\left(\widehat{f}\right)_g}(n_0) = 0$ and the $\widehat{f}$-definition we, moreover, have

$$\forall k,l\big(n_0 = k*l \to \left(\widehat{f}\right)_g(k) = 0\big).$$

Together with (+) and $n_1 = n_0 * < 0 >$ this finally implies

$$\widehat{\left(\widehat{f}\right)_g}(n_1) = 0,$$

which concludes the proof of $(*)$ and hence of 1).

2) '$\to$': By lemma 9.15.1, $T\Big(\widehat{\left(\widehat{f}\right)_g}\Big)$ holds for all f,g. Using 1), WKL$'$ now immediately follows from WKL.
'$\leftarrow$': Assume that $T(f)$ and $\forall x \exists n\big(lth\ n = x \wedge fn = 0\big)$, which implies (using that by $T(f)$ the number n encodes a binary sequence and relying on the monotonicity properties of our sequence coding)

$$(++)\ \forall x \exists n \leq \overline{1^1}x\big(lth\ n = x \wedge fn = 0\big).$$

Define

$$gx := \begin{cases} min\ n \leq \overline{1^1}x[lth\ n = x \wedge fn = 0] & \text{if such an } n \text{ exists,}\\ 0^0 \text{ otherwise.}\end{cases}$$

g is primitive recursive in f and (++) implies $\forall x\big(lth(gx) = x \wedge f(gx) = 0\big)$. Applying lemma 9.16.2 we conclude that $f_g =_1 f$. Since $f =_1 \widehat{f}$ by lemma 9.15.2), this proves that $\widehat{\left(\widehat{f}\right)_g} =_1 f$. WKL$'$ now yields $\exists b \leq_1 \lambda k.1 \forall x^0 \big(f(\overline{b}x) = 0\big)$. □

Before we can treat KL we need one intermediate version of König's lemma:

Definition 9.19.
$\text{WKL}^* :\equiv \forall f^1, h^1 \Big(T^*(h,f) \wedge \forall x \exists n \big(lth\, n = x \wedge fn = 0\big) \to \exists b \leq_1 h \forall x^0 f(\overline{b}x) = 0\Big),$
where

$$T^*(f,h) :\equiv \begin{cases} \forall n,m \big(f(n*m) = 0 \to fn = 0\big) \\ \wedge \forall n,x \big(f(n* < x >) = 0 \to x \leq h\big(lth(n)\big)\big). \end{cases}$$

To see that WKL* is admissible as well one can either (using Troelstra (1973), 1.9.24) show that – relative to $\widehat{\text{WE-HA}}^{\omega}\upharpoonright$ – it is equivalent to WKL or – alternatively – check that similarly to WKL we can write WKL* directly as an axiom of the form Δ (exercise).

Lemma 9.20. $\widehat{\text{WE-HA}}^{\omega}\upharpoonright + \text{AC} \vdash \text{WKL}^* \to \text{KL}$.

Proof: By AC applied to

$$\forall n \exists m \forall x \big(f(n* < x >) = 0 \to x \leq m\big)$$

provides a function g such that

$$\forall n,x \big(f(n* < x >) = 0 \to x \leq g(n)\big).$$

Primitive recursively in g (using only R_0) one can define a function h such that

$$\forall n,x \big(f(n* < x >) = 0 \to x \leq h\big(lth(n)\big)\big).$$

Now apply WKL*. □

Corollary 9.21. *Theorem 9.3 holds true if 'Δ' is replaced by 'Δ+KL'.*

Proof: The corollary follows from the fact that $\widehat{\text{WE-HA}}^{\omega}\upharpoonright$+AC proves the equivalence of KL and WKL$'$ (as shown above), where WKL$'$ has the form of a sentence admissible in Δ. □

Corollary 9.22.

$$\text{WE-HA}^{\omega} + \text{AC} + \text{IP}^{\omega}_{\forall} + \text{M}^{\omega} + \text{KL} \nvdash \Pi^0_1\text{-LEM}.$$

Proof: Suppose

$$\text{WE-HA}^{\omega} + \text{AC} + \text{IP}^{\omega}_{\forall} + \text{M}^{\omega} + \text{KL} \vdash \Pi^0_1\text{-LEM}.$$

Then by M^0 (and hence by M^{ω}) we could even derive Σ^0_1-LEM. This would yield

$$\text{WE-HA}^{\omega} + \text{AC} + \text{IP}^{\omega}_{\forall} + \text{M}^{\omega} + \text{KL} \vdash \forall x^0 \exists y^0 \big(\exists z^0 T(x,x,z) \to T(x,x,y)\big).$$

Corollary 9.21 would now give us a closed term t of WE-HA$^\omega$ such that

$$\text{WE-HA}^\omega + \text{AC} + \text{IP}^\omega_\forall + \text{M}^\omega + \text{KL} \vdash \forall x^0 \exists y \leq_0 t(x)\big(\exists z^0 T(x,x,z) \to T(x,x,y)\big)$$

contradicting the undecidability of the (special) halting problem. □

Corollary 9.23.
Let $A_0(x^1, y^0)$ a quantifier-free formula of $\mathcal{L}$(WE-HA$^\omega$) containing only x, y free. Then the following rule holds:

$$\begin{cases} \text{WE-HA}^\omega + \text{AC} + \text{IP}^\omega_\forall + \text{M}^\omega + \text{KL} \vdash \forall x^1 \exists y^0 A_0(x,y) \\ \textit{then one can extract a closed term } t^2 \textit{ of } \text{WE-HA}^\omega \textit{ s.t.} \\ \text{WE-HA}^\omega + \text{b-AC+WKL} \vdash \forall x^1 A_0(x,t(x)). \end{cases}$$

So WE-HA$^\omega$+ AC + IP$^\omega_\forall$+ M$^\omega$+KL *has the same provably recursive functionals of type ≤ 2 as* WE-HA$^\omega$ *and, in particular, the same provably recursive functions as* HA.
Analogously one shows a corresponding result for $\widehat{\text{WE-HA}}^\omega\!\upharpoonright$ *instead of* WE-HA$^\omega$. *In particular, the provably recursive functions of* $\widehat{\text{WE-HA}}^\omega\!\upharpoonright$+ AC + IP$^\omega_\forall$+ M$^\omega$ + KL *are exactly the ordinary primitive recursive functions.*

Proof: By lemma 9.20, the comment made before that lemma and proposition 9.18.2 the assumption yields

$$\text{WE-HA}^\omega + \text{AC} + \text{IP}^\omega_\forall + \text{M}^\omega + \text{WKL}' \vdash \forall x^1 \exists y^0 A_0(x,y).$$

Hence by theorem 9.1 we get a closed term $\tilde{t}$ such that

$$\text{WE-HA}^\omega + \widetilde{\text{WKL}'} \vdash \forall x^1 \exists y \leq_0 \tilde{t}(x) A_0(x,y).$$

Let t_{A_0} be a closed term with

$$\text{WE-HA}^\omega \vdash \forall x,y\big(t_{A_0}(x,y) =_0 0 \leftrightarrow A_0(x,y)\big).$$

Now define $t(x) := \min y \leq \tilde{t}(x)[t_{A_0}(x,y) =_0 0]$. Then

$$\text{WE-HA}^\omega + \widetilde{\text{WKL}'} \vdash \forall x^1 A_0(x,t(x))$$

and hence (using proposition 9.18.2)

$$\text{WE-HA}^\omega + \text{b-AC+WKL} \vdash \forall x^1 A_0(x,t(x)).$$

□

Remark 9.24. Together with lemma 10.32 and corollary 10.34 (both to be proved in chapter 10) it follows from the proof above that if

$$\text{WE-HA}^\omega + \text{AC} + \text{IP}^\omega_\forall + \text{M}^\omega \vdash \text{ KL} \to \forall x^1 \exists y^0 A_0(x,y),$$

then even

$$\text{WE-HA}^\omega \vdash \forall x^1 A_0(x,t(x)).$$

Remark 9.25. Theorem 9.3 and its corollaries above crucially depend on the fact that WE-HA$^\omega$ only contains the quantifier-free extensionality rule. With full extensionality, already E-HA$^\omega$+AC+M$^\omega$+WKL has a much bigger class of provably recursive functions than WE-HA$^\omega$. This is due to the fact that AC+WKL allow one to derive the uniform weak König's lemma UWKL (mentioned already in chapter 7) which, in the presence of full extensionality and M$^\omega$ suffices to get the comprehension functional

$$\exists \varphi^2 \forall f^1 \left(\varphi(f) =_0 0 \leftrightarrow \exists x^0 (f(x) =_0 0)\right) \text{ (see chapter 10 for details).}$$

If on the other hand

- AC is restricted to AC^{0,1},
- M$^\omega$ is restricted to M^1,
- IP$^\omega_\forall$ is restricted to the case where both the variables $\underline{x}$ in the premise as well as the variable y which is moved over that premise are of types ≤ 1,
- the type ρ in theorem 9.3 is ≤ 1,

then an elimination-of-extensionality procedure to be discussed in chapter 10 allows one to reduce the context with E-HA$^\omega$ to that with WE-HA$^\omega$.
In particular

$$\text{E-HA}^\omega + \text{AC}^{0,1} + \text{M}^1 + \text{KL} \nvdash \Pi^0_1\text{-LEM}$$

and a-fortiori

$$\text{E-HA}^\omega + \text{AC}^{0,1} + \text{M}^1 + \text{LLPO} \nvdash \Pi^0_1\text{-LEM}$$

where LLPO is the so-called 'lesser limited principle of omniscience'

$$\text{LLPO} :\equiv \forall f^1, g^1 \left(\neg(\exists n(f(n)=0) \wedge \exists n(g(n)=0)) \to \forall n(f(n) \neq 0) \vee \forall (g(n) \neq 0)\right)$$

(see e.g. [47]). For this last result we use the well-known fact that KL and even WKL implies intuitionistically LLPO. Alternatively, one can easily verify directly that LLPO has a trivial monotone functional interpretation, see [218].

9.4 WKL as a universal sentence Δ

We now show that over WE-HA$^\omega$+WKL every sentence A of the form

$$\forall x^1 \exists y \leq_1 sx \forall z^{0/1} A_0(x,y,z) \text{ } (A_0 \text{ quantifier-free, } s \text{ closed})$$

is equivalent to a suitable Π^0_1-sentence

$$\forall n^0 B_0(n)$$

which can be constructed from A.

Lemma 9.26. *Let $A_0(\underline{x}, y^1)$ be a quantifier-free formula whose free variables are included in $\underline{x} = x_1, \ldots, x_n$ and y^1, where the types of the x_i are of degree ≤ 1 ($i = 1, \ldots, n$). Then the following holds*

$$\text{WE-HA}^\omega \vdash \Box y^1 A_0(\underline{x}, y) \leftrightarrow \Box k^0 A_0(\underline{x}, \lambda m.(k)_m),$$

where $\Box \in \{\forall, \exists\}$.

Proof: We may assume that $n = 1$ and that x has type 1 (since tuples of variables of types of degree ≤ 1 can be encoded into a single variable of type 1). By proposition 3.17 we can construct a closed term t_{A_0} of WE-HA$^\omega$ such that

$$\text{WE-HA}^\omega \vdash \forall x^1, y^1 \left(t_{A_0} xy =_0 0 \leftrightarrow A_0(x,y)\right).$$

By proposition 9.10 applied to $\max_1(x,y)$ we obtain a closed term $\widehat{t}_{A_0}$ such that

$$\text{WE-HA}^\omega \vdash \forall x^1, y^1 \forall z \leq_1 y \Big(\bigwedge_{i=0}^{\widehat{t}_{A_0}(\max(x,y))} (zi =_0 yi) \rightarrow t_{A_0} xz =_0 t_{A_0} xy \Big).$$

Hence

$$\text{WE-HA}^\omega \vdash t_{A_0}(x, \lambda m.(\overline{y}(\widehat{t}_{A_0}(\max(x,y)) + 1))_m) =_0 t_{A_0} xy$$

and so

$$\text{WE-HA}^\omega \vdash A_0(x, \lambda m.(\overline{y}(\widehat{t}_{A_0}(\max(x,y)) + 1))_m) \leftrightarrow A_0(x,y).$$

$\Box$

Lemma 9.27. *For each sentence of the form*

$$\forall \underline{x}^{\underline{\rho}} \exists y \leq_1 s\underline{x} A_0(\underline{x}, y)$$

(where $\underline{\rho}$ is a tuple of types of degree ≤ 1) one can construct a closed term χ of WE-HA$^\omega$ *such that*

$$\text{WE-HA}^\omega \vdash \forall \underline{x} [(\exists y \leq s\underline{x} A_0(\underline{x}, y) \leftrightarrow A_0(\underline{x}, \chi\underline{x})) \wedge \chi\underline{x} \leq_1 s\underline{x}].$$

Proof: Again we may assume that $\underline{x} = x$, where x is of type 1. By the proof of lemma 9.26 we have (provably in WE-HA$^\omega$) that

$$A_0(x,y) \rightarrow A_0(x, \lambda m.(\overline{y}(\widehat{t}(\max_1(x, sx))))_m)$$

for a suitable closed term $\widehat{t}$ of WE-HA$^\omega$.
Define $tx := \widehat{t}(\max_1(x, sx))$.
If $y \leq_1 sx$, then (using basic monotonicity properties of our sequence coding from

chapter 3) we get

$$\bar{y}(tx) \leq_0 (\overline{sx})(tx) =: \Phi x.$$

Hence

$$\exists y \leq_1 sxA_0(x,y) \to \exists k \leq_0 \Phi x \big[A_0(x, \lambda m.(k)_m) \land \forall i < lth(k)\big((k)_i \leq sxi\big)\big].$$

Now define

$$\chi_0 x := \begin{cases} \min k \leq \Phi x[A_0(x,\lambda m.(k)_m) \land \forall i < lth(k)\big((k)_i \leq sxi\big)], \text{ if existent} \\ 0^0, \text{ otherwise} \end{cases}$$

and in turn

$$\chi x := \lambda m.(\chi_0 x)_m.$$

□

Proposition 9.28. *Let* $\forall x^1 \exists y \leq_1 sx \forall z^1 A_0(x,y,z)$ *be a sentence of* $\mathscr{L}(\text{WE-HA}^\omega)$ *(with s closed). Then*

$$\text{WE-HA}^\omega + \text{WKL} \vdash \forall x^1, k^0 \exists y \leq_1 sx \bigwedge_{i=0}^{k} A_0(x,y,\lambda m.(i)_m) \leftrightarrow \forall x^1 \exists y \leq_1 sx \forall z^1 A_0.$$

Proof: '⇐' is trivial.
'⇒': By lemma 9.26 it suffices to consider

$$\forall x \exists y \leq sx \forall k^0 A_0(x,y,k).$$

$\forall x,k \exists y \leq sx \bigwedge_{i=0}^{k} A_0(x,y,i)$ implies

$$(1)\ \forall x,k \exists n \underbrace{\Big(lth\, n = k \land \forall j < k\big((n)_j \leq sxj\big) \land \exists y \leq sx \bigwedge_{i=0}^{k} A_0(x, n * \lambda m.y(m+k), i) \Big)}_{\overline{A}_0(x,k,n) :\equiv}.$$

By lemma 9.27 (and lemma 3.17 plus the fact that the primitive recursive functionals in the sense of Gödel, i.e. in the sense of WE-HA$^\omega$, are closed under bounded quantification) $\overline{A}_0$ is quantifier–free definable in WE-HA$^\omega$. Therefore – by lemma 3.17 – we can define (uniformly in x) in WE-HA$^\omega$ a function f_x such that

$$f_x n := \begin{cases} 0 \text{ if } \overline{A}_0(x, lth\, n, n), \\ 1 \text{ otherwise.} \end{cases}$$

For all x we have $T^*(f_x, sx)$. Furthermore, (1) implies

$$\forall x, k \exists n (lth\, n = k \land f_x n = 0).$$

Therefore, WKL* (which relative to WE-HA$^\omega$ follows from WKL using [366] (1.9.24)) applied to f_x, sx yields

$$(2)\ \forall x \exists y_0 \leq sx \forall k \Big(\exists y \leq sx \bigwedge_{i=0}^{k} A_0(x, \overline{y}_0 k * \lambda m.y(m+k), i) \Big).$$

It remains to show that $\forall k A_0(x, y_0, k)$: Assume there exists a k_0 such that

$$\neg A_0(x, y_0, k_0).$$

Since A_0 is quantifier–free, there exists (again be lemma 3.17) a closed term t of WE-HA$^\omega$ such that

$$\text{WE-HA}^\omega \vdash \forall x, y, k \big(txyk =_0 0 \leftrightarrow \neg A_0(x, y, k) \big).$$

By proposition 9.10 there exists a modulus of uniform continuity $\tilde{t}$ for t when restricted to $y \leq sx$ (given by a closed term of WE-HA$^\omega$):

$$(3)\ \text{WE-HA}^\omega \vdash \forall x \forall y, \tilde{y} \leq sx \forall k \big(tx(\overline{y}(\tilde{t}xk) * \lambda m.\tilde{y}(m + \tilde{t}xk))k =_0 txyk \big).$$

Define $n_0 := \tilde{t}xk_0$. Since $txy_0k_0 = 0$, (3) implies

$$\forall i \geq n_0 \forall y \leq sx \big(\neg A_0(x, \overline{y}_0 i * \lambda m.y(m+i), k_0) \big).$$

Define $i := max(n_0, k_0)$. (2) yields

$$\exists y \leq sx A_0(x, \overline{y}_0 i * \lambda m.y(m+i), k_0),$$

which is a contradiction. Hence

$$\neg \exists k \neg A_0(x, y_0, k)$$

and so (by intuitionistic logic)

$$\forall k \neg\neg A_0(x, y_0, k)$$

and finally (using 3.18)

$$\forall k A_0(x, y_0, k).$$

□

Corollary 9.29. *For each sentence of the form*

$$\forall x^1 \exists y \leq_1 sx \forall z^{0/1} A_0(x, y, z)$$

in $\mathscr{L}$(WE-HA$^\omega$) *one can construct a corresponding* Π^0_1*-sentence*

$$\forall n^0 B_0(n)$$

in $\mathscr{L}$(WE-HA$^\omega$) *such that*

$$\text{WE-HA}^\omega + \text{WKL} \vdash \forall x^1 \exists y \leq_1 sx \forall z^{0/1} A_0(x,y,z) \leftrightarrow \forall n^0 B_0(n).$$

Proof: By proposition 9.28 it suffices to construct a sentence $\forall n^0 B_0(n)$ such that

$$\text{WE-HA}^\omega \vdash \forall x^1, k^0 \exists y \leq_1 sx \bigwedge_{i=0}^{k} A_0(x,y,\lambda m.(i)_m) \leftrightarrow \forall n^0 B_0(n).$$

By lemma 3.17 and the fact that the primitive recursive functionals in the sense of Gödel (i.e. in the sense of WE-HA^ω) are closed under bounded quantification it follows that $\bigwedge_{i=0}^{k} A_0(x,y,\lambda m.(i)_m)$ can be written as a (prime and hence) quantifier-free formula $A_0'(x^1,y^1,k^0)$. The claim now follows by lemma 9.27 and lemma 9.26. □

9.5 Fragments, exercises, historical comments and suggested further reading

Remarks on fragments:

As we discussed in the remark at the end of chapter 6, the majorization technique is available for the fragments $\mathcal{T}_i^\omega := \widehat{\text{WE-HA}}^\omega\!\restriction$, $\text{G}_n\text{A}_i^\omega$ $(n \geq 2)$ of WE-HA^ω as well. Together with the fact that functional interpretation can be applied to these systems (see the remark at the end of chapter 8) one easily verifies that theorems 9.1 and 9.3 hold if WE-HA^ω is replaced by one of these fragments. corollary 9.21 holds for $\widehat{\text{WE-HA}}^\omega\!\restriction$ instead of WE-HA^ω. If KL in corollary 9.21 is replaced by WKL, it also holds for $\text{G}_n\text{A}_i^\omega$ for $n \geq 3$. Lemmas 9.26 and 9.27, proposition 9.28 and corollary 9.29 also hold for $\widehat{\text{WE-HA}}^\omega\!\restriction$ and $\text{G}_n\text{A}_i^\omega$ for $n \geq 3$.

Exercises:

1) a. Prove that WE-HA^ω+b-AC $\vdash \varphi \to \tilde{\varphi}$ for φ of the form Δ and the corresponding Skolem normal form $\tilde{\varphi} \in \tilde{\Delta}$.
   b. Prove that WE-HA^ω+AC $\vdash$ b-AC.
2) Prove remark 9.5.
3) Prove lemma 9.16.1).

Historical comments and suggested further reading: Monotone functional interpretation was introduced in Kohlenbach [206] but the systematic use of combinations of functional interpretation with majorizability is due already to Kohlenbach [200, 201, 203] which contain much more information about it. An early use of a combination of functional interpretation with a variant of Howard's majorizability in the context of a theory for inductive definitions ($\text{ID}_1^c(O)$) has been given by Zucker in [383]. Theorem 9.3 is from Kohlenbach [212]. Application II basically is contained in Kohlenbach [201] (the fact that closed terms $t^{1(1)}$ of WE-HA^ω have –

restricted to the Cantor space – a modulus of uniform continuity functional given by a term t^* in WE-HA$^\omega$ is due to Kreisel and Schwichtenberg [325]). Using an implementation of monotone functional interpretation in the MINLOG system, Hernest gave an automated synthesis of moduli of uniform continuity based on application II in [158]. Application 'III' is taken from Kohlenbach [217]. Most of the material from sections 9.3 and 9.4 is taken from Kohlenbach [203]. In Kohlenbach-Oliva [236] it is shown that often monotone functional interpretation just creates the enrichment of data used prominently in Bishop's constructive analysis (Bishop [32], Bishop-Bridges [34]) and it is argued that the monotone functional interpretation can be viewed as a proper 'numerical implication' (rather than Gödel's original interpretation of the implication as suggested in Bishop [33]). A new so-called 'bounded functional interpretation' was recently introduced by Ferreira and Oliva in [104]. It has many similarities to monotone functional interpretation (and can be used to prove some results similar to those first obtained by monotone functional interpretation), but also has important differences in the treatment of bounded quantifiers. This makes it possible to use bounded functional interpretation to prove interesting results for systems of feasible analysis which could not be obtained by monotone functional interpretation so far (see Ferreira-Oliva [105]). The issue to what extent some basic analysis can be formalized in such systems of feasible analysis is addressed in Fernandes-Ferreira [102]. For a unified treatment of the soundness theorem for modified realizability, functional interpretation (and variants thereof) as well as their bounded and monotone versions see Oliva [292].

Chapter 10
Systems based on classical logic and functional interpretation

10.1 The negative translation

There are several interpretations – so-called 'negative' or 'double-negation' translations – of classical logic as well as many theories based on classical logic into their intuitionistic variant. All these translations $A \mapsto A'$ have in common that A' is (or is intuitionistically equivalent to) a negative formula.
The first such translation is due to Gödel [130] (although G. Gentzen independently discovered a similar translation). There is some preceding work by Kolmogorov [237] and Glivenko [124]. Two further variants of Gödel's translation are due to Kuroda [259] and it is one of these which we will adopt here:

Definition 10.1. Let A be a formula in a theory based on $\mathscr{L}(\mathrm{IL}^{\omega}_{\_=})$. A' is defined as $A' :\equiv \neg\neg A^*$, where A^* is defined by induction on the logical structure of A:

(i) $A^* :\equiv A$, if A is a prime formula,
(ii) $(A \Box B)^* :\equiv (A^* \Box B^*)$, where $\Box \in \{\wedge, \vee, \rightarrow\}$,
(iii) $(\exists x^\rho A)^* :\equiv \exists x^\rho A^*$,
(iv) $(\forall x^\rho A)^* :\equiv \forall x^\rho \neg\neg A^*$.

Remark 10.2. The Kuroda negative translation A' from 10.1 of A is intuitionistically equivalent to a negative formula B_{neg}, i.e.

$$\mathrm{IL}^{\omega}_{\_=} \vdash A' \leftrightarrow B_{neg}$$

for a suitable negative formula B_{neg}.

Proof: Exercise! $\Box$

Proposition 10.3. (i) $\mathrm{PL}^{\omega}_{\_=} \vdash A \Rightarrow \mathrm{IL}^{\omega}_{\_=} \vdash A'$,
(ii) (W)E-PA$^\omega \vdash A \Rightarrow$ (W)E-HA$^\omega \vdash A'$.

Proof: (i) Induction on the length of the derivation: (I) We first treat the axioms:
(a) For instances F of all intuitionistic axioms except $\forall x A \rightarrow A[t/x]$ the translation

F^* again is an instance of the same axiom (since $*$ commutes with $\wedge, \vee, \rightarrow$ and $\exists$) and so $F' \equiv \neg\neg F^*$ follows intuitionistically.
(b) For $\forall x A \rightarrow A[t/x]$, the negative translation yields $\neg\neg(\forall x \neg\neg A^* \rightarrow A^*[t/x])$ which intuitionistically is equivalent to $\forall x \neg\neg A^* \rightarrow \neg\neg A^*[t/x]$ which is an instance of the same axiom.
(c) $(A \vee \neg A)' \equiv \neg\neg(A^* \vee \neg A^*)$ which is provable in $\mathrm{IL}^{\omega}_{=}$ (it is an easy exercise that $\neg\neg(A \vee \neg A)$ holds intuitionistically for arbitrary formulas A).
(II) We now treat the rules. Modus ponens $\frac{A, A \rightarrow B}{B}$: By induction hypothesis we have

$$\neg\neg A^* \text{ and } \neg\neg(A^* \rightarrow B^*).$$

By intuitionistic logic we get $\neg\neg A^* \rightarrow \neg\neg B^*$ and hence by modus ponens $\neg\neg B^*$, i.e. B'.
The syllogism, exportation, importation rules and the $\exists$-rule are treated similarly using again the intuitionistic laws $\neg\neg(A \rightarrow B) \leftrightarrow (A \rightarrow \neg\neg B) \leftrightarrow (\neg\neg A \rightarrow \neg\neg B)$.
For the expansion rule one additionally needs that intuitionistically
$A \vee \neg\neg B \rightarrow \neg\neg(A \vee B)$.
Now consider the $\forall$-rule $\frac{A \rightarrow B}{A \rightarrow \forall x B}$: By induction hypothesis we have $\neg\neg(A^* \rightarrow B^*)$ and therefore (by intuitionistic logic) $A^* \rightarrow \neg\neg B^*$. By the $\forall$-rule we obtain $A^* \rightarrow \forall x \neg\neg B^*$, i.e. $(A \rightarrow \forall x B)^*$ and, therefore, a-fortiori $(A \rightarrow \forall x B)'$.
(ii) We only have to extend the proof of (i) by the treatment of the non-logical axioms and rules: The negative translation of the purely universal $=_0, S, \Pi, \Sigma, R$-axioms trivially follows from the axioms themselves (note that $\mathrm{WE\text{-}HA}^{\omega} \vdash \neg\neg A_0 \leftrightarrow A_0$ and so the translation of purely universal axioms is in fact equivalent – relative to $\mathrm{WE\text{-}HA}^{\omega}$ – to the axioms since intuitionistically $\neg\neg\forall x \neg\neg A(x) \leftrightarrow \forall x \neg\neg A(x)$).
Similarly the negative translation of (E) (resp. of the premise and the conclusion of QF-ER) can be seen to be equivalent to (E) in $\mathrm{WE\text{-}HA}^{\omega}$.
The induction rule: By the induction hypothesis we have

$$\neg\neg(A(0))^* \text{ and } \neg\neg\big((A(x))^* \rightarrow (A(x+1))^*\big).$$

Hence also $\neg\neg(A(x))^* \rightarrow \neg\neg(A(x+1))^*$. Thus by the induction rule we obtain $\neg\neg(A(x))^*$, i.e. $(A(x))'$. □

Definition 10.4. The schema QF-AC of quantifier-free choice in all finite types is the restriction of AC to quantifier-free formulas $A_0 \equiv A$. For convenience we formulate this schema for tuples:

$$\text{QF-AC}: \ \forall \underline{x} \exists \underline{y} A_0(\underline{x}, \underline{y}) \rightarrow \exists \underline{Y} \forall \underline{x} A_0(\underline{x}, \underline{Y}\,\underline{x}),$$

where A_0 is quantifier-free and $\underline{x}, \underline{y}$ are tuples of variables of arbitrary types.
For single variables x^{ρ} and y^{τ} of types ρ and τ respectively, we denote the corresponding special case of QF-AC by $\text{QF-AC}^{\rho,\tau}$.

Remark 10.5. Since one can show in $\mathrm{WE\text{-}HA}^{\omega}$ that finite tuples $\underline{x}$ of variables (of different types) can be coded together into a single variable x whose type depends on the types of $\underline{x}$ (see [366] for details on that) the version with single variables in fact implies the one with tuples. However, the direct use of tuples is much simpler. Moreover, in the context of the extensions of the systems to the new ground type X treated in chapter 17 such a contraction of tuples is no longer possible.

Proposition 10.6. *Let $\mathcal{P}$ be an arbitrary set of purely universal sentences $\forall \underline{z}^{\underline{\sigma}} B_0(\underline{z})$ (B_0 quantifier-free) in $\mathcal{L}(\text{WE-PA}^\omega)$. Then*

$$\text{WE-PA}^\omega + \text{QF-AC} + \mathcal{P} \vdash A \Rightarrow \text{WE-HA}^\omega + \text{QF-AC} + \mathcal{P} + \text{M}^\omega \vdash A'.$$

Proof: Since WE-HA$^\omega$ proves that $P \leftrightarrow P'$ for all $P \in \mathcal{P}$ we only have to extend the proof of proposition 10.3 by showing that

$$\text{WE-HA}^\omega + \text{QF-AC} + \text{M}^\omega \vdash (\text{QF-AC})'.$$

We have in WE-HA$^\omega$

$$\Big(\forall \underline{x}\exists \underline{y} A_0(\underline{x},\underline{y}) \to \exists \underline{Y}\forall \underline{x} A_0(\underline{x},\underline{Y}\,\underline{x})\Big)' \leftrightarrow$$
$$\neg\neg\Big(\forall \underline{x}\neg\neg\exists \underline{y} A_0(\underline{x},\underline{y}) \to \exists \underline{Y}\forall \underline{x}\neg\neg A_0(\underline{x},\underline{Y}\,\underline{x})\Big),$$

which clearly is implied by

$$(*)\ \forall \underline{x}\neg\neg\exists \underline{y} A_0(\underline{x},\underline{y}) \to \exists \underline{Y}\forall \underline{x} A_0(\underline{x},\underline{Y}\,\underline{x}),$$

which in turn follows from M$^\omega$ and QF-AC. □

10.2 Combination of negative translation and functional interpretation

We now combine the negative translation with functional interpretation. In the following we denote the combination of negative translation and functional interpretation 'ND-interpretation'.

Theorem 10.7 (soundness of ND). *Let $\mathcal{P}$ be an arbitrary set of purely universal sentences $\forall \underline{z}^{\underline{\sigma}} B_0(\underline{z})$ (B_0 quantifier-free) of $\mathcal{L}(\text{WE-PA}^\omega)$ and $A(\underline{a})$ be an arbitrary formula of $\mathcal{L}(\text{WE-PA}^\omega)$ containing only $\underline{a}$ free. Then the following rule holds*

$$\begin{cases} \text{WE-PA}^\omega + \text{QF-AC} + \mathcal{P} \vdash A(\underline{a}) \\ \Rightarrow \text{ ND } \textit{extracts closed terms } \underline{t} \textit{ of } \text{WE-HA}^\omega \textit{ such that} \\ \text{WE-HA}^\omega + \mathcal{P} \vdash \forall \underline{y}(A')_D(\underline{t}\,\underline{a},\underline{y},\underline{a}). \end{cases}$$

Proof: The theorem follows from proposition 10.6 together with theorem 8.6. □

Theorem 10.8 (Main theorem on program extraction by ND). *Let $\mathcal{P}$ be an arbitrary set of purely universal sentences $\forall \underline{z}^{\underline{\sigma}} B_0(\underline{z})$ (B_0 quantifier-free) of $\mathcal{L}(\text{WE-PA}^\omega)$*

and $A_0(x^\rho, y^\tau)$ *be a (quantifier-free) formula of* $\mathcal{L}(\text{WE-PA}^\omega)$ *which only contains* x^ρ, y^τ *as free variables. Then the following rule holds:*

$$\begin{cases} \text{WE-PA}^\omega + \text{QF-AC} + \mathcal{P} \vdash \forall x^\rho \exists y^\tau A_0(x,y) \\ \Rightarrow \text{ND} \textit{ extracts a closed term } t \textit{ of } \text{WE-HA}^\omega \textit{ such that} \\ \text{WE-HA}^\omega + \mathcal{P} \vdash \forall x A_0(x,tx). \end{cases}$$

In particular, if $\mathcal{S}^\omega \models \mathcal{P}$*, then the conclusion holds in* $\mathcal{S}^\omega$.
The result also applies to tuples $\underline{x}, \underline{y}$ *where we then have a tuple* $\underline{t}$ *of closed terms.*

Proof: Again we use proposition 10.6 as in the proof above.

$$\begin{array}{l} \text{WE-PA}^\omega + \text{QF-AC} + \mathcal{P} \vdash \forall x \exists y A_0(x,y) \overset{prop.10.6}{\Rightarrow} \\ \text{WE-HA}^\omega + \text{QF-AC} + \text{M}^\omega + \mathcal{P} \vdash \forall x \neg\neg\exists y A_0(x,y) \overset{\text{M}^\omega}{\Rightarrow} \\ \text{WE-HA}^\omega + \text{QF-AC} + \text{M}^\omega + \mathcal{P} \vdash \forall x \exists y A_0(x,y) \overset{thm.8.6}{\Rightarrow} \\ \text{WE-HA}^\omega + \mathcal{P} \vdash \forall x A_0(x,tx) \text{ for a suitable closed term } t \end{array}$$

(note that A_0 can be treated as a prime formula in WE-HA^ω). □

As an application of theorem 10.8 we obtain the no-counterexample interpretation of PA by terms of WE-HA^ω:

Proposition 10.9. *Let* $A \in \mathcal{L}(\text{PA})$ *be a prenex sentence. Then the following rule holds:*

$$\begin{cases} \text{PA} \vdash A \\ \Rightarrow \textit{ one can extract closed terms } \underline{\Phi} \textit{ of } \text{WE-HA}^\omega \textit{ such that} \\ \text{WE-HA}^\omega \vdash \underline{\Phi} \textit{ n.c.i. } A. \end{cases}$$

Proof: Apply theorem 10.8 to the Herbrand normal form A^H of A. □

Remark 10.10. As mentioned already in chapter 3 in connection with HA, PA is strictly speaking not a subsystem of WE-PA^ω since we have included symbols for all primitive recursive functions as primitive notions in PA whereas they are defined notions in WE-PA^ω. However PA is a subsystem of a corresponding definitorial extension of WE-PA^ω to which theorem 10.8 applies as well.

Lemma 10.11. *Let* A_{ef} *be an* $\exists$*-free formula of* $\mathcal{L}(\text{WE-PA}^\omega)$*. Then*

$$(1)\ \text{WE-PA}^\omega + \text{QF-AC} \vdash A_{ef} \leftrightarrow A_{ef}^D$$

and

$$(2)\ \text{WE-HA}^\omega \vdash A_{ef}^D \leftrightarrow \exists \underline{x} \forall \underline{y} A^*(\underline{x}\,\underline{y}, \underline{y})$$

for a suitable quantifier-free A^* *(without* $\vee$*).*

Proof: Both claims are proved simultaneously by induction on the complexity of A_{ef} using corollary 8.9:
1) A prime: Then $A \equiv A^D \equiv A_D$, $\underline{x}, \underline{y}$ are the empty tuples and we take $A^* :\equiv A_D$.
2) $\forall z A(z)$: WE-HA$^\omega$ proves

$$(\forall z A(z))^D \overset{\text{def.}}{\leftrightarrow}$$
$$(\forall z A^D(z))^D \overset{(2)-\text{I.H.},8.9}{\leftrightarrow} \left(\forall z \exists \underline{x} \forall \underline{y} A^*(\underline{x}\,\underline{y}, \underline{y}, z)\right)^D \overset{\text{def.}}{\leftrightarrow} \exists \underline{X} \forall z, \underline{y} A^*(\underline{X} z \underline{y}, \underline{y}, z)$$

and so we can put $(\forall z A(z))^* :\equiv A^*(z)$.
WE-PA$^\omega$+QF-AC proves

$$(\forall z A(z))^D \overset{\text{above}}{\leftrightarrow} \exists \underline{X} \forall z, \underline{y} A^*(\underline{X} z \underline{y}, \underline{y}, z) \overset{\text{QF-AC}}{\leftrightarrow}$$
$$\forall z, \underline{y} \exists \underline{x} A^*(\underline{x}, \underline{y}, z) \overset{\text{QF-AC}}{\leftrightarrow} \forall z \exists \underline{x} \forall \underline{y} A^*(\underline{x}\,\underline{y}, \underline{y}, z) \overset{(1),(2)-\text{I.H.}}{\leftrightarrow} \forall z A(z).$$

3) $A \wedge B$: WE-HA$^\omega$ proves

$$(A \wedge B)^D \overset{\text{def.}}{\leftrightarrow} (A^D \wedge B^D)^D \overset{(2)-\text{I.H.},8.9}{\leftrightarrow} \left(\exists \underline{x} \forall \underline{y} A^*(\underline{x}\,\underline{y}, \underline{y}) \wedge \exists \underline{u} \forall \underline{v} B^*(\underline{u}\,\underline{v}, \underline{v})\right)^D \overset{\text{def.}}{\leftrightarrow}$$
$$\exists \underline{x}, \underline{u} \forall \underline{y}, \underline{v} (A^*(\underline{x}\,\underline{y}, \underline{y}) \wedge B^*(\underline{u}\,\underline{v}, \underline{v})) \overset{!}{\leftrightarrow} \exists \underline{X}, \underline{U} \forall \underline{y}, \underline{v} (A^*(\underline{X}\,\underline{y}\,\underline{v}, \underline{y}) \wedge B^*(\underline{U}\,\underline{y}\,\underline{v}, \underline{v}))$$

and so we can put

$$(A \wedge B)^*(\underline{\alpha}, \underline{\beta}, \underline{y}, \underline{v}) :\equiv A^*(\underline{\alpha}, \underline{y}) \wedge B^*(\underline{\beta}, \underline{v}).$$

Ad !: '$\rightarrow$': $\underline{X} := \lambda \underline{y} \lambda \underline{v}.\underline{x}\,\underline{y}$, $\underline{U} := \lambda \underline{y} \lambda \underline{v}.\underline{u}\,\underline{v}$. '$\leftarrow$': $\underline{x} := \lambda \underline{y}.\underline{X}\,\underline{y}\,\underline{O}$, $\underline{u} := \lambda \underline{v}.\underline{U}\,\underline{O}\,\underline{v}$.
WE-PA$^\omega$+QF-AC proves

$$(A \wedge B)^D \overset{\text{above}}{\leftrightarrow} \exists \underline{x}, \underline{u} \forall \underline{y}, \underline{v} (A^*(\underline{x}\,\underline{y}, \underline{y}) \wedge B^*(\underline{u}\,\underline{v}, \underline{v})) \leftrightarrow$$
$$\exists \underline{x} \forall \underline{y} A^*(\underline{x}\,\underline{y}, \underline{y}) \wedge \exists \underline{u} \forall \underline{v} B^*(\underline{u}\,\underline{v}, \underline{v}) \overset{(1),(2)-\text{I.H.}}{\leftrightarrow} A \wedge B.$$

4) $A \rightarrow B$: WE-HA$^\omega$ proves

$$(A \rightarrow B)^D \overset{\text{def.}}{\leftrightarrow} (A^D \rightarrow B^D)^D \overset{(2)-\text{I.H.},8.9}{\leftrightarrow} \left(\exists \underline{x} \forall \underline{y} A^*(\underline{x}\,\underline{y}, \underline{y}) \rightarrow \exists \underline{u} \forall \underline{v} B^*(\underline{u}\,\underline{v}, \underline{v})\right)^D \overset{\text{def.}}{\leftrightarrow}$$
$$\exists \underline{U}, \underline{Y} \forall \underline{x}, \underline{v} (A^*(\underline{x}(\underline{Y}\,\underline{x}\,\underline{v}), \underline{Y}\,\underline{x}\,\underline{v}) \rightarrow B^*(\underline{U}\,\underline{x}\,\underline{v}, \underline{v})).$$

So we can put

$$(A \rightarrow B)^*(\underline{\alpha}, \underline{\beta}, \underline{x}, \underline{v}) :\equiv (A^*(\underline{x}\,\underline{\beta}, \underline{\beta}) \rightarrow B^*(\underline{\alpha}, \underline{v})).$$

WE-PA$^\omega$+QF-AC proves

$$(A \to B)^D \overset{\text{above}}{\leftrightarrow} \exists \underline{U}, \underline{Y} \forall \underline{x}, \underline{v} (A^*(\underline{x}(\underline{Y}\,\underline{x}\,\underline{v}), \underline{Y}\,\underline{x}\,\underline{v}) \to B^*(\underline{U}\,\underline{x}\,\underline{v}, \underline{v})) \overset{\text{QF-AC}}{\leftrightarrow}$$
$$\forall \underline{x}, \underline{v} \exists \underline{u}, \underline{y} (A^*(\underline{x}\,\underline{y}, \underline{y}) \to B^*(\underline{u}, \underline{v})) \overset{\text{class.logic}}{\leftrightarrow} (\exists \underline{x} \forall \underline{y} A^*(\underline{x}\,\underline{y}, \underline{y}) \to \forall \underline{v} \exists \underline{u} B^*(\underline{u}, \underline{v})) \overset{\text{QF-AC}}{\leftrightarrow}$$
$$(\exists \underline{x} \forall \underline{y} A^*(\underline{x}\,\underline{y}, \underline{y}) \to \exists \underline{u} \forall \underline{v} B^*(\underline{u}\,\underline{v}, \underline{v})) \overset{(1),(2)-\text{I.H.}}{\leftrightarrow} (A \to B).$$

□

Remark 10.12. 1) In lemma 10.11, WE-HA$^\omega$ can be replaced by any fragment containing the combinators and the minimal amount of arithmetic needed to carry out the functional interpretation of the logical axioms and rules. WE-PA$^\omega$ can be replaced by the classical variant of such a fragment. In fact, instead of full classical logic only M$^\omega$+IP$^\omega_\forall$ is needed. We can even drop IP$^\omega_\forall$ since every formula in the negative fragment is intuitionistically equivalent to one in the fragment based on $\wedge, \neg, \forall$ only and for the latter already M$^\omega$ suffices.
2) For A_{ef} not containing $\wedge$ even the stronger statement $(A_{ef})_D(\underline{x}, \underline{y}) \equiv A^*(\underline{x}\,\underline{y}, \underline{y})$ holds. However, as observed by L. Leuştean, this is no longer true in the presence of $\wedge$ and so the corresponding claim in [366] (p. 241) (as well as the original proof in [244]) is not correct as it stands.

As a consequence of lemma 10.11 we obtain that the combination of negative translation and functional interpretation $(A')^D$ is much closer related to A than the no-counterexample interpretation (or the Skolem normal form) is (for prenex arithmetical A), since the equivalence of A and $(A')^D$ can be proved using only quantifier-free choice (although in higher types) whereas the no-counterexample interpretation of A only implies A in the presence of (number-theoretic) choice for arithmetical formulas:

Proposition 10.13 (Characterization theorem for ND, Kreisel [244]). *Let A be an arbitrary formula of $\mathcal{L}$(WE-PA$^\omega$). Then*

$$\text{WE-PA}^\omega + \text{QF-AC} \vdash A \leftrightarrow (A')^D.$$

Proof: Let B_{neg} be as in remark 10.2 a negative formula such that

$$\text{WE-HA}^\omega \vdash A' \leftrightarrow B_{neg}.$$

Corollary 8.9 then yields

$$\text{WE-HA}^\omega \vdash (A')^D \leftrightarrow B^D_{neg}.$$

Since B_{neg} is $\exists$-free the claim now follows from lemma 10.11. □

We now compare the three classical $\exists\forall$-normal forms considered in this book: Let A be a formula of $\mathcal{L}$(WE-PA$^\omega$) in prenex normal form. Let A^S be the Skolem normal form of A, $A^{n.c.i} :\equiv \exists \underline{\Phi}(\underline{\Phi}\ n.c.i.\ A)$ be its no-counterexample interpretation and $(A')^D$ be its ND-interpretation (so far we defined the Skolem normal form and the no-counterexample interpretation only for prenex arithmetical formulas but the definitions immediately extend to higher types; note that $A^S \equiv A^D$ since A is in prenex normal form).

Theorem 10.14.
$$\text{WE-HA}^\omega \vdash A^S \to (A')^D \to A^{n.c.i}.$$

Proof: Proof of the first implication: let B_{neg} as in 10.2. Since

$$\text{WE-PA}^\omega \vdash A^S \to B_{neg}$$

we can conclude from lemma 10.11 that

$$\text{WE-PA}^\omega + \text{QF-AC} \vdash A^S \to \forall \underline{y} \exists \underline{x} B^*(\underline{x}, \underline{y}),$$

where B^* is quantifier-free and

$$\text{WE-HA}^\omega \vdash B^D_{neg} \leftrightarrow \exists \underline{X} \forall \underline{y} B^*(\underline{X}\,\underline{y}, \underline{y}).$$

Let $A^S \equiv \exists \underline{a} \forall \underline{b} A^S_{qf}(\underline{a}, \underline{b})$ with A^S_{qf} quantifier-free. Then

$$\text{WE-PA}^\omega + \text{QF-AC} \vdash \forall \underline{a}, \underline{y} \exists \underline{b}, \underline{x} (A^S_{qf}(\underline{a}, \underline{b}) \to B^*(\underline{x}, \underline{y})).$$

Theorem 10.8 now yields closed terms $\underline{t}$ such that

$$\text{WE-HA}^\omega \vdash \forall \underline{a}, \underline{y} (\forall \underline{b} A^S_{qf}(\underline{a}, \underline{b}) \to B^*(\underline{t}\underline{a}\underline{y}, \underline{y})).$$

Hence
$$\text{WE-HA}^\omega \vdash A^S \to B^D_{neg}.$$

Arguing as in the proof of proposition 10.13 one shows that

$$\text{WE-HA}^\omega \vdash (A')^D \leftrightarrow B^D_{neg}.$$

Hence the first implication follows.
Proof of the second implication: By proposition 10.13 we have

$$\text{WE-PA}^\omega + \text{QF-AC} \vdash (A')^D \to A.$$

Since
$$\text{WE-PA}^\omega \vdash A \to A^H,$$

where $A^H \equiv \forall \underline{c} \exists \underline{d} A^H_{qf}(\underline{c}, \underline{d})$ (with A^H_{qf} being quantifier-free) is the Herbrand normal form of A, we obtain

$$\text{WE-PA}^\omega + \text{QF-AC} \vdash (A')^D \to \forall \underline{c} \exists \underline{d} A^H_{qf}(\underline{c}, \underline{d}).$$

The rest of the proof now proceeds as above (using that $(A')^D$ is in $\exists\forall$-form). □

Remark 10.15. In theorem 10.14 WE-HA$^\omega$ can actually be replaced by a fragment sufficient for the functional interpretation of the logical axioms and rules only (which only requires a very minor amount of arithmetic). This is the case because proposition 10.13 holds for the classical variant of such a fragment instead of WE-PA$^\omega$.
The proof of theorem 10.14 actually shows that there are closed terms in that fragment of WE-HA$^\omega$ that transform realizers of A^S into realizers of $(A')^D$ and closed terms that transform realizers of $(A')^D$ into realizers of $A^{n.c.i}$. These terms built up out of typed λ-terms with the few basic operations like $\min, \max, |\cdot - \cdot|, \overline{sg}$ needed to define characteristic terms for quantifier-free formulas plus $0, 1$.

Whereas A^S raises the maximal type degree of quantified variables in A at most by 1 and $A^{n.c.i}$ at most by 2 (so for arithmetical formulas A, i.e. formulas which only contain quantifiers over type-0 variables, A^S only contains quantifiers of degree ≤ 1 and $A^{n.c.i.}$ only quantifiers of degree ≤ 2) the type increase in forming $(A')^D$ is unbounded (even for the class of arithmetical formulas). The benefit, however, is that $(A')^D$ is the only $\exists\forall$-normal form which satisfies proposition 10.13:

Proposition 10.16. WE-PA$^\omega$+QF-AC *does neither derive* $A \to A^S$ *nor* $A^{n.c.i.} \to A$ *in general.*

Proof: It is clear that for general A, the schemas $A \to A^S$ and $A^H \to A$ are equivalent to full AC over WE-PA$^\omega$ which is well-known to be unprovable in WE-PA$^\omega$+QF-AC. The proposition now follows from the fact that WE-PA$^\omega$+QF-AC $\vdash A^H \to A^{n.c.i.}$. □

Corollary 10.17. *The converse directions of the implications in theorem 10.14 are not even provable in* WE-PA$^\omega$+QF-AC.

Proof: The claim follows from proposition 10.13 and proposition 10.16. □

Looking back to the treatment of the special case

$$A :\equiv \forall x \exists y \forall z A_0(x,y,z) \text{ and } B :\equiv \forall u \exists v B_0(u,v)$$

of the modus ponens problem in chapter 2, we see that the solution we finally aimed at in this case is precisely what the ND-interpretation does in this case: A' and $(A \to B)'$ are (intuitionistically equivalent to)

$$\forall x \neg\neg \exists y \forall z A_0(x,y,z) \text{ and } \forall u \neg\neg \exists v B_0(u,v)$$

and the functional interpretation of proofs of these formulas provides precisely functionals $\Phi_0, \ldots, \Phi_3$ whose existence we stipulated in the discussion of the modus ponens problem in chapter 2, i.e. functionals $\Phi_0, \Phi_1, \Phi_2, \Phi_3$ such that

$$\forall x, g A_0(x, \Phi_0(x,g), g(\Phi_0(x,g)))$$

and

$$\Phi_1(u,Y), \Phi_2(u,Y), \Phi_3(u,Y) \text{ realizing } x, g, v \text{ in } (*),$$

where

$$(*)\ \forall u, Y \exists x, g, v\big(A_0(x, Y(x,g), g(Y(x,g))) \to B_0(u,v)\big).$$

As discussed in chapter 2, the Skolem normal form (which coincides with ND for B) is to strong to be useful for A since in general a classical proof of A will not provide an effective Skolem function. The no-counterexample interpretation (which – essentially – coincides with ND for A) is too weak an interpretation of (any prenexation of) $A \to B$ to allow for a simple solution of the modus ponens problem.

Here we have a situation where the application of negative translation actually strengthens a statement from a computational point of view: suppose we had an intuitionistic proof (say in HA or WE-HA$^\omega$) of

$$(+)\ \forall x \exists y \forall z A_0(x,y,z) \to \forall u \exists v B_0(u,v).$$

Then functional interpretation would have extracted from this WE-HA$^\omega$-functionals $\varphi_1, \varphi_2, \varphi_3$ such that

$$\forall f, u(A_0(\varphi_1(u,f), f(\varphi_1(u,f)), \varphi_2(u,f)) \to B_0(u, \varphi_3(u,f))$$

which allows for a direct application of the modus ponens if we are provided with a function f satisfying the Skolem normal form of A (which as we discussed above will in general not be computable). Similarly, modified realizability applied to $(+)$ would provide a functional ψ (given by a closed term of WE-HA$^\omega$) such that

$$\forall f, u(\forall x, z A_0(x, f(x), z) \to B_0(u, \psi(u,f))$$

which again requires a Skolem function for A to discharge the premise. In contrast to this, applying negative translation first, yielding

$$(+)'\ \forall x \neg\neg \exists y \forall z A_0(x,y,z) \to \forall u \neg\neg \exists v B_0(u,v),$$

works for our benefit even in cases where $(+)$ is already provable intuitionistically: the fact that the conclusion gets slightly weakened to $\forall u \neg\neg \exists v B_0(u,v)$ does not cause a problem for functional interpretation (which immediately removes these double negations by trivially interpreting the Markov principle) though it would for modified realizability which would produce the empty realizer. However, the severe weakening of the premise to $\forall x \neg\neg \exists y \forall z A_0(x,y,z)$ **strengthens** the implication and we now only need a functional satisfying the functional interpretation of $\forall x \neg\neg \exists y \forall z A_0(x,y,z)$, i.e. Φ_0 above since in this case the functional interpretation coincides with the no-counterexample interpretation of A, to get a realizing functional for the conclusion. In further comparing functional interpretation with modified realizability let us note that even applied to $(+)$ (instead of $(+)'$) it is functional interpretation which gives the much stronger result: whereas the modified realizability of $(+)$ really requires a 'full' Skolem function f for A to drop the premise, the functional interpretation of $(+)$ (which coincides with the no-counterexample interpretation of $(A \to B)^{pr} :\equiv \forall u \exists x \forall y \exists z, v(A_0(x,y,z) \to B_0(u,v)))$ only requires to be able to compute a sequence $(f_u)_{u \in \mathbb{N}}$ of functions satisfying

$$\forall u A_0(\varphi_1(u, f_u), f_u(\varphi_1(u, f_u)), \varphi_2(u, f_u)).$$

As mentioned already in chapter 2 such an f_u can in fact be constructed by bar recursion (to be discussed in chapter 11) from a functional Φ_0 satisfying the no-counterexample interpretation of A and φ_1, φ_2. This is achieved by solving (in the parameter u) the following equations for x, f, g

$$x = \varphi_1(u, f), \; \Phi_0(x, g) = f(\varphi_1(u, f)), \; g(\Phi_0(x, g)) = \varphi_2(u, f).$$

So in the end even applying functional interpretation to $(+)$ instead of $(+)'$ would allow us to construct a solution using just the information Φ_0 on (the proof of) A. However, the use of bar recursion will in general increase the complexity of the so obtained solution significantly. The same is true for yet another alternative to be discussed in chapter 14 below.

Remark 10.18. Further interesting observations concerning the weakness of a constructive interpretation of implications of the form $\Pi^0_3 \to \Pi^0_3$ and $\Pi^0_3 \to \Pi^0_2$ can be found in [252] and [251].

In chapter 18 the special instance of the modus ponens just discussed will play an important role in the form of a lemma

$$\forall k \in \mathbb{N} \exists n \in \mathbb{N} \forall m \in \mathbb{N} \; (|a_{n+m} - a_n| \le 2^{-k}) \to \forall k \in \mathbb{N} \exists n \in \mathbb{N}(a_n \le 2^{-k}),$$

where $(a_n)_{n \in \mathbb{N}}$ is a nonincreasing sequence of non-negative real numbers (namely the sequence of distances $d(x_n, f(x_n))$ of certain iteration sequences (x_n) of nonexpansive functions f, see chapter 17). By discharging the valid assumption

$$\forall k \in \mathbb{N} \exists n \in \mathbb{N} \forall m \in \mathbb{N} \; (|a_{n+m} - a_n| \le 2^{-k})$$

it is proved that (a_n) converges towards 0. Clearly, any Cauchy rate of (a_n) is a rate of convergence for $a_n \to 0$. Using ND, however, we see that we can extract from the proof a rate of convergence for $a_n \to 0$ using only a functional Φ satisfying the no-counterexample interpretation of $\forall k \in \mathbb{N} \exists n \in \mathbb{N} \forall m \in \mathbb{N} \; (|a_{n+m} - a_n| \le 2^{-k})$. If we use a combination of negative translation and **monotone** functional interpretation (to be discussed next) we actually only need a majorant Φ^* for such a Φ and we may (as proved in proposition 2.26) take e.g.

$$\Phi^*(g, k) := \tilde{g}^{(2^k)}(0),$$

where $g^{(n+1)}(0) := g(g^{(n)}(0))$ and $g^{(0)}(0) := 0$ and $\tilde{g}(n) := n + g(n)$. Strictly speaking, Φ^* has a further function argument h coming from the universal quantifier hidden in '$\le_{\mathbb{R}} 2^{-k}$', however the majorant we give does not depend on h. The details are left as an exercise. So although the convergence is proved using the non-computational (by E. Specker's result discussed in chapter 2) principle that bounded monotone sequences are Cauchy one can use negative translation combined with (monotone) functional interpretation to extract a simple subrecursive rate of convergence towards 0 which, taking advantage of the fact that Φ^* does not depend on

(a_n), will have strong uniformity features (see chapter 18 for the actual extraction).

We next continue the discussion of the infinite pigeonhole principle (IPP) from chapter 2: as we saw already, the no-counterexample interpretation of (IPP) is too weak to be useful due to the weakness of its Herbrand normal form. Let us instead consider the ND-interpretation of (IPP): The negative translation of (IPP) is (modulo the use of $\neg\neg\forall\neg\neg \leftrightarrow \forall\neg\neg$ and $\neg\neg s =_0 t \leftrightarrow s =_0 t$)

$$(\text{IPP})^N :\equiv \forall n \in \mathbb{N} \forall f : \mathbb{N} \to C_n \neg\neg\exists i \leq n \forall k \in \mathbb{N} \neg\neg\exists m \geq k \left(f(m) = i \right).$$

Functional interpretation now yields the following ND-interpretation of (IPP) which we had already anticipated in chapter 2:

$$\begin{aligned}(\text{IPP})^{ND} \equiv \forall n \in \mathbb{N} \forall f : \mathbb{N} \to C_n \forall K : C_n \times \mathbb{N}^{\mathbb{N}} \to \mathbb{N} \exists i \leq n \exists g : \mathbb{N} \to \mathbb{N} \\ \left(g(K(i,g)) \geq K(i,g) \wedge f(g(K(i,g))) = i \right).\end{aligned}$$

Strictly speaking, the real combination of negative translation and functional interpretation $(\text{IPP}')^D$ of (IPP) is the $\exists\forall$-form resulting from a final application of QF-AC to $(\text{IPP})^{ND}$ which we omit here for the sake of better readability. Moreover, officially (to fit into our finite types over $\mathbb{N}$), '$\forall f : \mathbb{N} \to C_n\, A(f)$' is represented as '$\forall f^1\, A(f_n)$' with $f_n(k) := \min\{f(k), n\}$ and '$\forall K : C_n \times \mathbb{N}^{\mathbb{N}} \to \mathbb{N}$' as '$\forall K : \mathbb{N} \times \mathbb{N}^{\mathbb{N}} \to \mathbb{N}$', i.e. as '$\forall K^{0(1)(0)}$'.
The functional interpretation of (IPP) requires functionals $I(n, f, K)$ and $G(n, f, K)$ realizing '$\exists i$' and '$\exists g$' in $(\text{IPP})^{ND}$. We now give the solution for $n = 1$:
Define

$$\begin{aligned} g_0 &:=_1 \lambda a^0.\max(a, K(1, \lambda b^0.\max(a,b))), \\ c &:=_0 K(0, g_0),\ g_1 :=_1 \lambda b^0.\max(c, b). \end{aligned}$$

Then

$$\begin{aligned} g_0(K(0,g_0)) &= \max(K(0,g_0), K(1, \lambda b.\max(K(0,g_0), b))) \\ &= \max(c, K(1,g_1)) = g_1(K(1,g_1)). \end{aligned}$$

Moreover, for $i \in \{0,1\}$:

$$g_i(K(i,g_i)) \geq K(i,g_i).$$

Now let

$$I(f,K) := i := f(g_0(K(0,g_0)))\ (= f(g_1(K(1,g_1)))).$$

Then

$$i = f(g_i(K(i,g_i))) \wedge g_i(K(i,g_i)) \geq K(i,g_i).$$

Hence $I(f,K)$ and $G(f,K) := g_{I(f,K)}$ satisfy the ND-interpretation of the special case of (IPP) where $n = 1$. The general case can be treated by an appropriate iteration of this procedure using R_1. The reader will appreciate the complexity of the solution by writing down the solution for $n = 2$ (exercise). In order to get a manageable description of the solution it is very convenient to use (as observed by Oliva in

[293]) a finite form of Spector's principle of bar recursion to be treated in the chapter 11. Bar recursion of lowest type suffices to give an ND-interpretation of countable choice for Π^0_1-formulas (which actually implies countable choice for arithmetical formulas). The finite version of this choice principle gives the collection principle for Π^0_1-formulas needed to prove (IPP). Hence it is not surprising that using a finite version of bar recursion (which in principle could be re-written as a messy R_1-recursion) helps to spell out the ND-interpretation of the full (IPP)-principle which we will provide in chapter 11.

We now make uses of monotone functional interpretation combined with negative translation. Let NMD denote the combination of negative translation and monotone functional interpretation.

As in the previous chapter, Δ denotes a set of sentences of the form

$$\forall \underline{a}^{\underline{\delta}} \exists \underline{b} \leq_{\underline{\sigma}} \underline{r}\underline{a} \forall \underline{c}^{\underline{\gamma}} B_0(\underline{a}, \underline{b}, \underline{c}),$$

where B_0 is quantifier-free and does not contain any further free variables than those shown and $\underline{r}$ is a tuple of closed terms (of suitable types) of WE-HA$^\omega$. The types $\underline{\delta}, \underline{\sigma}, \underline{\gamma}$ are arbitrary.
$\tilde{\Delta}$ is defined as the corresponding set of sentences

$$\exists \underline{B} \leq \underline{r} \forall \underline{a}, \underline{c} B_0(\underline{a}, \underline{B}\,\underline{a}, \underline{c}).$$

One easily observes that for $B \in \Delta$ we have WE-HA$^\omega \vdash B \rightarrow B'$. Hence 10.6 extends to axioms Δ:

Proposition 10.19.

$$\text{WE-PA}^\omega + \text{QF-AC} + \Delta \vdash A \;\Rightarrow\; \text{WE-HA}^\omega + \text{QF-AC} + \Delta + \text{M}^\omega \vdash A'.$$

The above proposition combined with the soundness theorem 9.1 for MD immediately yields the following

Theorem 10.20 (Soundness Theorem for NMD).

$$\begin{cases} \text{WE-PA}^\omega + \text{QF-AC} + \Delta \vdash A(\underline{a}), \textit{ then} \\ \text{WE-HA}^\omega + \tilde{\Delta} \vdash \exists \underline{x}(\underline{t}^* \; maj \; \underline{x} \wedge \forall \underline{a}, \underline{y} (A')_D(\underline{x}\,\underline{a}, \underline{y}, \underline{a})), \end{cases}$$

where $\underline{t}^*$ is a suitable tuple of closed terms of WE-HA$^\omega$ *which can be extracted by* NMD *from a given proof of the assumption.*

Combined with the majorization technique used already in the previous chapter we obtain

Theorem 10.21 (Main theorem on uniform bound extraction by NMD).
Let $A_0(x^1, y^\rho, z^\tau)$ be a (quantifier-free) formula of $\mathcal{L}$(WE-PA$^\omega$) containing only x, y, z as free variables, $deg(\tau) \leq 2$ and s be a closed term. Then

$$\begin{cases} \text{WE-PA}^\omega + \text{QF-AC} + \Delta \vdash \forall x^1 \forall y \leq_\rho sx \exists z^\tau A_0(x,y,z) \\ \Rightarrow \text{NMD } \textit{extracts a closed term } t \textit{ of } \text{WE-HA}^\omega \textit{ such that} \\ \text{WE-HA}^\omega + \tilde{\Delta} \vdash \forall x^1 \forall y \leq_\rho sx \exists z \leq_\tau tx A_0(x,y,z). \end{cases}$$

In particular, if $\mathcal{S}^\omega \models \Delta$, then the conclusion holds in $\mathcal{S}^\omega$. As in theorem 6.8 the result also applies to tuples of variables.

Proof:

$$\text{WE-PA}^\omega + \text{QF-AC} + \Delta \vdash \forall x^1 \forall y \leq_\rho sx \exists z^\tau A_0(x,y,z)$$

implies by proposition 10.19

$$\text{WE-HA}^\omega + \text{QF-AC} + \Delta + \text{M}^\omega \vdash \forall x^1 \forall y \leq_\rho sx \neg\neg \exists z^\tau A_0(x,y,z)$$

and so using M^ω

$$\text{WE-HA}^\omega + \text{QF-AC} + \Delta + \text{M}^\omega \vdash \forall x^1 \forall y \leq_\rho sx \exists z^\tau A_0(x,y,z).$$

Theorem 9.3 together with the corollary to its proof now yields a closed term t such that

$$\text{WE-HA}^\omega + \tilde{\Delta} \vdash \forall x^1 \forall y \leq_\rho sx \exists z \leq_\tau tx A_0(x,y,z)).$$

□

The vast potential for applications of this theorem to problems in analysis will be discussed in detail in chapter 15.

Remark 10.22. One easily obtains the variant of theorem 10.21 corresponding to proposition 6.11 (also combined with the format in theorem 10.21): exercise!

Remark 10.23. As mentioned already, theorem 10.21 also holds for tuples of variables as long as they satisfy the conditions stated. For a tuple $\exists \underline{z}$ where some of the z_i's have types of degree ≤ 2 while others have not one can still extract bounds for the former.

Remark 10.24. In theorem 10.21 one may also have more general axioms Δ^* of the form

$$\forall \underline{a}_1^{\underline{\delta}_1} \exists \underline{b}_1 \leq_{\underline{\sigma}_1} \underline{r}_1 \underline{a}_1 \forall \underline{a}_2^{\underline{\delta}_2} \exists \underline{b}_2 \leq_{\underline{\sigma}_2} \underline{r}_2 \underline{a}_1 \underline{a}_2 \ldots \forall \underline{a}_n^{\underline{\delta}_n} \exists \underline{b}_n \leq_{\underline{\sigma}_n} \underline{r}_n \underline{a}_1 \ldots \underline{a}_n \forall \underline{c}^{\underline{\gamma}} B_0$$

with $\tilde{\Delta}^*$ be

$$\exists \underline{B}_1 \leq \underline{r}_1, \ldots, \underline{B}_n \leq \underline{r}_n \forall \underline{a}_1, \ldots, \underline{a}_n, \underline{c} B_0$$

in the conclusion (exercise).

Corollary 10.25. *Let Δ as before and $A_0(x^1, z^0)$ be a (quantifier-free) formula of* WE-PA$^\omega$ *containing only x, z free. Then the following rule holds:*

$$\begin{cases} \text{WE-PA}^\omega + \text{QF-AC} + \Delta \vdash \forall x^1 \exists z^0 A_0(x,z) \\ \Rightarrow \textit{using } \text{NMD} \textit{ one extracts a closed term } t \textit{ of } \text{WE-HA}^\omega \textit{ such that} \\ \text{WE-HA}^\omega + \tilde{\Delta} \vdash \forall x^1 A_0(x,tx). \end{cases}$$

Proof: The corollary follows from theorem 10.21, the fact that one can construct a characteristic term t_{A_0} for A_0 (lemma 3.17) and (primitive recursive) bounded search. □

Theorem 10.26. *Let Δ be a sentence of the form*

$$\forall a^\delta \exists b \leq_\sigma ra \forall c^\gamma B_0(a,b,c)$$

where $deg(\gamma) \leq 2$ and define the so-called ε-weakening of its Skolem normal form $\tilde{\Delta}$ by

$$\tilde{\Delta}_\varepsilon :\equiv \forall c^\gamma \exists B \leq_{\sigma\delta} r \forall a \forall \tilde{c} \leq_\gamma c\, B_0(a,Ba,\tilde{c}).$$

Then for A_0 and τ as in theorem 10.21 the following rule holds:

$$\begin{cases} \text{WE-PA}^\omega + \text{QF-AC} \vdash \Delta \rightarrow \forall x^1 \forall y \leq_\rho sx \exists z^\tau A_0(x,y,z) \\ \Rightarrow \text{NMD} \textit{ extracts a closed term } t \textit{ of } \text{WE-HA}^\omega \textit{ such that} \\ \text{WE-HA}^\omega \vdash \tilde{\Delta}_\varepsilon \rightarrow \forall x^1 \forall y \leq_\rho sx \exists z \leq_\tau tx A_0(x,y,z). \end{cases}$$

Proof:

$$\text{WE-PA}^\omega + \text{QF-AC} \vdash \Delta \rightarrow \forall x^1 \forall y \leq_\rho sx \exists z^\tau A_0(x,y,z)$$

implies

$$\text{WE-PA}^\omega + \text{QF-AC} \vdash \tilde{\Delta} \rightarrow \forall x^1 \forall y \leq_\rho sx \exists z^\tau A_0(x,y,z)$$

and hence

$$\text{WE-PA}^\omega + \text{QF-AC} \vdash \forall x^1 \forall y \leq sx \forall B \leq r \exists a,c,z \big(B_0(a,Ba,c) \rightarrow A_0(x,y,z)\big).$$

Theorem 10.21 (with $\Delta = \emptyset$) together with remark 10.23 now yields closed terms t,t' such that

$$\text{WE-HA}^\omega \vdash \forall x^1 \forall y \leq sx \forall B \leq r \exists a \exists c \leq t'x \exists z \leq tx \big(B_0(a,Ba,c) \rightarrow A_0(x,y,z)\big)$$

and hence

$$\text{WE-HA}^\omega \vdash \forall x^1 \big(\exists B \leq r \forall a \forall c \leq t'x B_0(a,Ba,c) \rightarrow \forall y \leq sx \exists z \leq tx A_0(x,y,z)\big).$$

Thus

$$\text{WE-HA}^\omega \vdash \forall x^1 \big(\forall c \exists B \leq r \forall a \forall \tilde{c} \leq c B_0(a,Ba,\tilde{c}) \rightarrow \forall y \leq sx \exists z \leq tx A_0(x,y,z)\big)$$

and, therefore,

$$\text{WE-HA}^{\omega} \vdash \tilde{\Delta}_{\varepsilon} \to \forall x^1 \forall y \leq_{\rho} sx \exists z \leq_{\tau} txA_0(x,y,z).$$

□

Remark 10.27. 1) For arbitrary γ we still can reduce Δ to

$$\forall c^{\gamma} \exists B \leq_{\sigma\delta} r \forall a \forall \tilde{c}(c \; maj_{\gamma} \tilde{c} \to B_0(a, Ba, \tilde{c}))$$

since the condition $deg(\gamma) \leq 2$ was used only to construct a bound for $\tilde{c}$ from a majorant of $\tilde{c}$. The reduction of Δ to $\tilde{\Delta}_{\varepsilon}$ is also possible for arbitrary τ except that in this case we will not get a bound on z (but only a majorant).

2) Instead of a single sentence Δ we may also have a finite conjunction of such sentences and the sentences themselves may have tuples of variables as in theorem 10.21 above, i.e.

$$\forall \underline{a}^{\underline{\delta}} \exists \underline{b} \leq_{\underline{\sigma}} \underline{r}\underline{a} \forall \underline{c}^{\underline{\gamma}} B_0(\underline{a}, \underline{b}, \underline{c})$$

as long as all $deg(\gamma_i) \leq 2$. If some of the γ_i are not of degree ≤ 2 one can still bound the c_i's of the other types which are of degree ≤ 2.

3) Δ needs to be an implicative premise rather than an axiom in the sense of WE-PA$^{\omega}$+QF-AC+Δ since WE-PA$^{\omega}$ does (due to the fact that it contains the quantifier-free extensionality rule QF-ER) not satisfy the deduction theorem.

4) Analogously to theorem 7.18 one can even allow the sentence Δ to depend on the parameters x and y. For many more results in this direction see [203].

Remark 10.28. Theorem 10.21 resp. theorem 10.26 (taken together with the remark above) implies the seemingly stronger version where we allow $\tilde{\Delta}$ to be used in the premise rather than only Δ: $\tilde{\Delta}$ itself has the form of just another sentence Δ with the initial universal quantifier being empty but with increased types and c replaced by the pair a, c and $\tilde{\tilde{\Delta}} \equiv \tilde{\Delta}$.

In view of this remark one may ask why we formulated theorems 10.21 and 10.26 with Δ rather than $\tilde{\Delta}$. The reason is that in concrete applications to mathematics, the ineffective principles used typically come in the form Δ with all types of degree ≤ 1. By an elimination procedure for the full extensionality axioms to be discussed below we can in such situations even allow E-PA$^{\omega}$ instead of WE-HA$^{\omega}$ to be used (which also satisfies the deduction theorem so that we then actually can treat Δ as an ordinary axiom). For the $\tilde{\Delta}$ form the elimination of extensionality would not be available in such cases since B usually is already of type 2. In fact, we will see that in general $\tilde{\Delta}$ (e.g. UWKL, see below) is much stronger in the presence of extensionality than Δ (e.g. WKL).

The usefulness of theorem 10.26 becomes clear by the following

Proposition 10.29. *Let Δ and $\tilde{\Delta}_{\varepsilon}$ be as in theorem 10.26 and*

$$\Delta_{\varepsilon} :\equiv \forall a, c \exists b \leq ra \forall \tilde{c} \leq c \, B_0(a, b, \tilde{c}).$$

Then

$$\text{WE-HA}^{\omega} \vdash \Delta_{\varepsilon} \Leftrightarrow \text{WE-HA}^{\omega} \vdash \tilde{\Delta}_{\varepsilon}.$$

Proof: The proposition is immediate from that fact that $\tilde{\Delta}_\varepsilon$ follows from Δ_ε by the rule of choice ACR and the fact that WE-HA$^\omega$ is closed under ACR (see corollary 5.24 and remark 5.25). □

Proposition 10.30. *Let Δ be as in theorem 10.26 with δ, σ, γ of degree ≤ 1 and Δ_ε as in proposition 10.29. Then*

$$\text{WE-HA}^\omega \vdash \Delta_\varepsilon \leftrightarrow \tilde{\Delta}_\varepsilon.$$

Proof: The proposition follows by applying lemma 9.27 twice. □

The relevance of theorem 10.26 in connection with propositions 10.29 and 10.30 comes from the fact that for many ineffective theorems of the form Δ in analysis Δ_ε has a simple constructive proof so that one can eliminate Δ altogether from the proof of the conclusion. An example for this is e.g. the weak König's lemma WKL which can, as we saw in the previous chapter, be written in the form WKL$'$ of a sentence Δ such that $(\text{WKL}')_\varepsilon$ is provable even in $\widehat{\text{WE-HA}}^\omega{\restriction}$ (see proposition 9.18). Hence we get the following

Corollary 10.31. *For A_0 and τ as in theorem 10.21 the following rule holds:*

$$\begin{cases} \text{WE-PA}^\omega + \text{QF-AC} \vdash \text{WKL} \rightarrow \forall x^1 \forall y \leq_\rho sx \exists z^\tau A_0(x,y,z) \\ \Rightarrow \text{NMD } \textit{extracts a closed term } t \textit{ of } \text{WE-HA}^\omega \textit{ such that} \\ \text{WE-HA}^\omega \vdash \forall x^1 \forall y \leq_\rho sx \exists z \leq_\tau txA_0(x,y,z). \end{cases}$$

As in theorem 6.8 the result also holds for tuples of variables.

10.3 Application: Uniform weak König's lemma UWKL

Let us recall the definition of the binary ('weak') König's lemma WKL from chapter 9:

$$\text{WKL: } \forall f^1 \Big(T(f) \wedge \forall x^0 \exists n^0 (lth\ n = x \wedge fn = 0) \rightarrow \exists b \leq_1 \lambda k.1 \forall x^0 \big(f(\overline{b}x) = 0 \big) \Big).$$

Defining

$$T^\infty(f) :\equiv T(f) \wedge \forall x^0 \exists n^0 (lth\ n = x \wedge fn = 0)$$

(i.e. $T^\infty(f)$ expresses that f represents an infinite binary tree), WKL writes as follows:

$$\forall f^1 \Big(T^\infty(f) \rightarrow \exists b \leq_1 \lambda k.1 \forall x^0 \big(f(\overline{b}x) = 0 \big) \Big).$$

From chapter 7 we recall the definition of the **uniform weak König's lemma UWKL**:

$$\mathrm{UWKL} :\equiv \exists \Phi \leq_{1(1)} 1 \forall f^1 \Big(T^\infty(f) \to \forall x^0 \big(f((\overline{\Phi f})x) = 0 \big) \Big).$$

In chapter 9 we saw that WKL can be written equivalently as WKL$'$ which has the form of an axiom Δ. With WKL$'$, also $\widetilde{\mathrm{WKL}'}$ has the form of an axiom Δ.

Lemma 10.32.

$$\widehat{\mathrm{WE\text{-}HA}}^\omega{\restriction} \vdash \mathrm{UWKL} \leftrightarrow \widetilde{\mathrm{WKL}'}.$$

Proof: $\widetilde{\mathrm{WKL}'}$ is the statement

$$\exists B \leq 1 \forall f^1, g^1, x^0 \Big(\widehat{(\hat{f})}_g ((\overline{Bfg})x) =_0 0 \Big).$$

'$\Rightarrow$': Let Φ be a functional satisfying UWKL. Define $B(f,g) := \Phi\Big(\widehat{(\hat{f})_g}\Big)$. By lemma 9.15.1 and proposition 9.18.1 $T^\infty\Big(\widehat{(\hat{f})_g}\Big)$. Hence by UWKL

$$\forall x^0 \Big(\widehat{(\hat{f})}_g ((\overline{Bfg})x) =_0 0 \Big).$$

'$\Leftarrow$': Primitive recursively in f we define

$$\tilde{f}(x) := \begin{cases} \min n \leq \overline{1^1} x [lth(n) = x \wedge f(n) = 0] \text{ if existent,} \\ 0^0 \text{ otherwise.} \end{cases}$$

Then by lemmas 9.15.2 and 9.16.2 (and monotonicity properties of our sequence coding)

$$\forall f^1 \big(T^\infty(f) \to f =_1 \widehat{(\hat{f})_{\tilde{f}}} \big).$$

Let B satisfy $\widetilde{\mathrm{WKL}}'$, then $\Phi(f) := B(f, \tilde{f})$ satisfies UWKL. $\Box$

Corollary 10.33. *In theorem 10.21 we can replace 'Δ' and '$\tilde{\Delta}$' by 'Δ+UWKL' and '$\tilde{\Delta}$+UWKL'. In particular with A_0, s, τ as in theorem 10.21:*

$$\begin{cases} \mathrm{WE\text{-}PA}^\omega + \mathrm{QF\text{-}AC} + \mathrm{UWKL} \vdash \forall x^1 \forall y \leq_\rho sx \exists z^\tau A_0(x,y,z) \\ \Rightarrow \text{ NMD } \textit{extracts a closed term } t \textit{ of } \mathrm{WE\text{-}HA}^\omega \textit{ such that} \\ \mathrm{WE\text{-}HA}^\omega + \mathrm{UWKL} \vdash \forall x^1 \forall y \leq_\rho sx \exists z \leq_\tau tx A_0(x,y,z). \end{cases}$$

As in theorem 6.8 the result also holds for tuples of variables.
Analogously for $\widehat{\mathrm{WE\text{-}PA}}^\omega{\restriction}$, $\widehat{\mathrm{WE\text{-}HA}}^\omega{\restriction}$ *instead of* WE-PA$^\omega$, WE-HA$^\omega$.

Proof: The corollary follows from the fact that (by lemma 10.32) UWKL is equivalent to a sentence of the form Δ (with $\underline{a}$ being the empty tuple) namely $\widetilde{\mathrm{WKL}}'$ together with theorem 10.21 and the fact that $\widetilde{\widetilde{\mathrm{WKL}}}' \equiv \widetilde{\mathrm{WKL}}'$. $\Box$

Analogously to corollary 10.31 we obtain (using again lemma 10.32) the following: under the slight restriction that UWKL must not be used in the proof of the premise of an instance of the extensionality rule QF-ER and therefore can be moved to the right of '$\vdash$' as an implicative assumption, one can actually eliminate UWKL from the proof of the conclusion of corollary 10.33, thereby establishing a strong conservation result for UWKL (see [221]):

Corollary 10.34. *With A_0, s, τ as in theorem 10.21:*

$$\begin{cases} \text{WE-PA}^\omega + \text{QF-AC} \vdash \text{UWKL} \to \forall x^1 \forall y \leq_\rho sx \exists z^\tau A_0(x,y,z) \\ \Rightarrow \text{NMD } \textit{extracts a closed term } t \textit{ of } \text{WE-HA}^\omega \textit{ such that} \\ \text{WE-HA}^\omega \vdash \forall x^1 \forall y \leq_\rho sx \exists z \leq_\tau tx A_0(x,y,z). \end{cases}$$

As in theorem 6.8 the result also holds for tuples of variables.
Analogously for $\widehat{\text{WE-PA}}^\omega{\restriction}$, $\widehat{\text{WE-HA}}^\omega{\restriction}$ *instead of* WE-PA^ω, WE-HA^ω.

Remark 10.35. For general axioms Δ allowing arbitrary types, the truth of theorems 10.21 and 10.26 and corollaries 10.33 and 10.34 crucially depends on the fact that in WE-PA^ω we permit only a restricted use of extensionality. Further below, we will show that indeed corollary 10.33 fails for E-PA^ω as UWKL implies in the presence of full extensionality a strong comprehension axiom. If, however, the degrees of the types $\underline{\sigma}$ in all axioms Δ are restricted to ≤ 1 (which still covers WKL but no longer UWKL), QF-AC is restricted to $\text{QF-AC}^{1,0}+\text{QF-AC}^{0,1}$ and the degree of the type ρ of y is ≤ 1, then an elimination of extensionality procedure sketched below allows one to derive a version of theorem 10.21 for E-PA^ω (see theorem 10.47).

10.4 Elimination of extensionality

In this section we discuss a syntactic method for the elimination of the extensionality axiom from proofs of theorems having appropriate type restrictions on their variables. The technique goes back essentially to R. Gandy but was carried out for the systems at hand in Luckhardt [266]. We give a simplification of the treatment from [266]: the basic idea is to restrict all quantifiers to the hereditarily extensional functionals x^ρ where 'hereditarily extensional' is understood as $x \approx_\rho x$ as defined in 9.6. However, for reasons similar to the use of strong majorizability in the definition of the model $\mathcal{M}^\omega$ we use a strong variant of $\approx_\rho$ which ensures that $x \approx y$ implies $x \approx x$. We denote this variant by $=^e_\rho$:

Definition 10.36. Between functionals of type ρ we define a relation $=^e_\rho$ by induction on ρ :

$$\begin{cases} x =^e_0 y :\equiv x =_0 y, \\ x =^e_{\tau\rho} y :\equiv \forall u^\rho, v^\rho (u =^e_\rho v \to xu =^e_\tau xv \wedge xu =^e_\tau yv). \end{cases}$$

The next seven lemmas are all provable in WE-HA$^\omega$.

Lemma 10.37. $x =^e_\rho y \to x =^e_\rho x$.

Proof: For both $\rho = 0$ and $\rho \neq 0$ the claim is immediate. □

Lemma 10.38. $x_1 =^e_\rho x_2 \wedge x_2 =^e_\rho x_3 \to x_1 =^e_\rho x_3$.

Proof: Induction on ρ. The case $\rho = 0$ is trivial.
Let $x_1 =^e_{\tau\rho} x_2, x_2 =^e_{\tau\rho} x_3$ and assume that $u =^e_\rho v$. Then $x_1 u =^e_\tau x_1 v$ and $x_2 u =^e_\tau x_3 v$. By lemma 10.37 we also have $u =^e_\rho u$ and so $x_1 u =^e_\tau x_2 u$. Hence by I.H. $x_1 u =^e_\tau x_3 v$. In total this yields $x_1 =^e_{\tau\rho} x_3$. □

Lemma 10.39. $x =^e_\rho y \to y =^e_\rho x$.

Proof: Induction on ρ. The case $\rho = 0$ is trivial.
Let $x =^e_{\tau\rho} y$ and assume that $u =^e_\rho v$. By I.H. we have $v =^e_\rho u$ and so $xv =^e_\tau yu$. Again by I.H. this gives (1) $yu =^e_\tau xv$. Lemma 10.37 implies $u =^e_\rho u$ and so $xu =^e_\tau yu$. By I.H. this yields $yu =^e_\tau xu$. Since also $xu =^e_\tau yv$, lemma 10.38 yields (2) $yu =^e_\tau yv$. (1), (2) give $y =^e_{\tau\rho} x$. □

Lemma 10.40. 1) $x_1 =_\rho \tilde{x}_1 \wedge x_2 =_\rho \tilde{x}_2 \wedge x_1 =^e_\rho x_2 \to \tilde{x}_1 =^e_\rho \tilde{x}_2$.
2) *Let* $\rho = \tau\rho_k \ldots \rho_1$.

$$x =^e_\rho \tilde{x} \leftrightarrow \forall y_1, \tilde{y}_1, \ldots, y_k, \tilde{y}_k \Big(\bigwedge_{i=1}^{k} (y_i =^e_{\rho_i} \tilde{y}_i) \to x\underline{y} =^e_\tau x\underline{\tilde{y}} \wedge x\underline{y} =^e_\tau \tilde{x}\underline{\tilde{y}} \Big).$$

Proof: 1) Induction on ρ.
2) Induction on k. The case $k = 1$ is immediate from the definition.
$k \mapsto k+1$:

$$x =^e_\rho \tilde{x} \overset{\text{I.H.}}{\Leftrightarrow}$$
$$\forall y_1, \tilde{y}_1, \ldots, y_k, \tilde{y}_k$$
$$\Big(\bigwedge_{i=1}^{k} (y_i =^e_{\rho_i} \tilde{y}_i) \to x y_1 \ldots y_k =^e_{\tau\rho_{k+1}} x\tilde{y}_1 \ldots \tilde{y}_k \wedge x y_1 \ldots y_k =^e_{\tau\rho_{k+1}} \tilde{x}\tilde{y}_1 \ldots \tilde{y}_k \Big)$$
$$\Leftrightarrow$$
$$\forall y_1, \tilde{y}_1, \ldots, y_k, \tilde{y}_k \Big(\bigwedge_{i=1}^{k} (y_i =^e_{\rho_i} \tilde{y}_i) \to \forall y_{k+1}, \tilde{y}_{k+1} (y_{k+1} =^e_{\rho_{k+1}} \tilde{y}_{k+1} \to$$
$$x y_1 \ldots y_{k+1} =^e_\tau x y_1 \ldots y_k \tilde{y}_{k+1} \wedge x\underline{y} =^e_\tau x\underline{\tilde{y}} \wedge x\underline{y} =^e_\tau \tilde{x}\underline{\tilde{y}}) \Big)$$
$$\overset{\text{lemma10.37}}{\Leftrightarrow} \forall y_1, \tilde{y}_1, \ldots, y_{k+1}, \tilde{y}_{k+1} \Big(\bigwedge_{i=1}^{k+1} (y_i =^e_{\rho_i} \tilde{y}_i) \to x\underline{y} =^e_\tau x\underline{\tilde{y}} \wedge x\underline{y} =^e_\tau \tilde{x}\underline{\tilde{y}} \Big).$$

□

Lemma 10.41. *Let* ρ *be of degree* ≤ 1. *Then* $x =^e_\rho x$.

Proof: Immediate from lemma 10.40.2. □

Notation: In the following, for $\underline{u} = u_1^{\rho_1}, \ldots, u_k^{\rho_k}$ and $\underline{v} = v_1^{\rho_1}, \ldots, v_k^{\rho_k}$, we abbreviate $\bigwedge_{i=1}^{k} (u_i =_{\rho_i}^e v_i)$ by $\underline{u} =^e \underline{v}$.

Lemma 10.42. *Let $t[\underline{x}]$ be term of* WE-HA$^\omega$ *of arbitrary type with at most $\underline{x}$ as free variables. Then*

$$\underline{x} =^e \underline{x} \to t[\underline{x}] =^e t[\underline{x}].$$

Proof: Induction on $t[\underline{x}]$: The case of the variables $\underline{x}$ is clear by assumption. For the constants c^ρ one shows $c =_\rho^e c$ using lemma 10.40 similarly to the proof of $c \approx_\rho c$ in proposition 9.8.
The induction step follows from

$$s[\underline{x}] =_{\tau\rho}^e s[\underline{x}] \wedge t[\underline{x}] =_\rho^e t[\underline{x}] \to (s[\underline{x}])(t[\underline{x}]) =_\tau^e (s[\underline{x}])(t[\underline{x}]).$$

□

Lemma 10.43. $x =_\rho^e y \wedge \forall \underline{v}(\underline{v} =^e \underline{v} \to y\underline{v} =_0 z\underline{v}) \to x =_\rho^e z.$

Proof: Assume that $x =_\rho^e y$ and $\forall \underline{v}(\underline{v} =^e \underline{v} \to y\underline{v} =_0 z\underline{v})$. Assume furthermore that $\underline{u} =^e \underline{v}$. Then by lemma 10.40.2 $x\underline{u} =_0 y\underline{v}$ and (1) $x\underline{u} =_0 x\underline{v}$. By lemmas 10.39 and 10.37 we have $\underline{v} =^e \underline{v}$ and so $y\underline{v} =_0 z\underline{v}$ by the assumption. Hence (2) $x\underline{u} =_0 z\underline{v}$. (1), (2) imply $x =_\rho^e z$ using again lemma 10.40.2. □

Definition 10.44. For every formula A of $\mathscr{L}$(E-PA$^\omega$) we define a translation A_e by relativizing all quantifiers to hereditarily extensional functionals in the sense of $=^e$:

(i) $A_e :\equiv A$, if A is a prime formula,
(ii) $(A \Box B)_e :\equiv (A_e \Box B_e)$, where $\Box \in \{\wedge, \vee, \to\}$,
(iii) $\left(\exists x^\rho A\right)_e :\equiv \exists x^\rho \left(x =_\rho^e x \wedge A_e\right)$,
(iv) $\left(\forall x^\rho A\right)_e :\equiv \forall x^\rho \left(x =_\rho^e x \to A_e\right)$.

The following result is (for a slightly more complicated definition of A_e) due to [266]:

Proposition 10.45. *The following rule holds:*

$$\begin{cases} \text{E-PA}^\omega + \text{QF-AC}^{0,1} + \text{QF-AC}^{1,0} \vdash A(\underline{a}) \Rightarrow \\ \text{WE-PA}^\omega + \text{QF-AC}^{0,1} + \text{QF-AC}^{1,0} \vdash \underline{a} =^e \underline{a} \to A_e(\underline{a}), \end{cases}$$

where $\underline{a}$ are all the free variables of A.

Proof: Induction on the derivation: (i) for the propositional axioms A the claim is trivial since A_e is an instance of the same axiom (as the relativization commutes with all propositional connectives). Hence $\underline{a} =^e \underline{a} \to A_e$ follows logically.
(ii) For the quantifier axioms the result follows from lemma 10.42.
(iii) For the logical rules, the correctness of the e-translation follows by simple logical manipulations.

(iv) The $=_0$-axioms and the definition axioms for the constants are all purely universal and so get at most weakened under the relativization.
The induction axiom gets translated into another instance of the induction axiom using that trivially $x =_0^e x$.
(v) The extensionality axiom (for $\rho = 0\rho_k \dots \rho_1$)

$$(E^\rho) : \forall z^\rho, x_1, \dots, x_k, y_1, \dots, y_k (\bigwedge_{i=1}^{k} (x_i =_{\rho_i} y_i) \to z\underline{x} =_0 z\underline{y})$$

translates into

$$\begin{aligned}&\forall z^\rho, x_1, \dots, x_k, y_1, \dots, y_k \\ &\quad (z =^e z \wedge \bigwedge_{i=1}^{k} (x_i =^e x_i) \wedge \bigwedge_{i=1}^{k} (y_i =^e y_i) \wedge \bigwedge_{i=1}^{k} \forall \underline{v}_i (\underline{v}_i =^e \underline{v}_i \to x_i \underline{v}_i =_0 y_i \underline{v}_i) \\ &\qquad\qquad\qquad\qquad\qquad\qquad\qquad\qquad \to z\underline{x} =_0 z\underline{y}).\end{aligned}$$

Now assume $z =^e z$, $\bigwedge_{i=1}^{k} (x_i =^e x_i)$, $\bigwedge_{i=1}^{k} (y_i =^e y_i)$, and $\bigwedge_{i=1}^{k} \forall \underline{v}_i (\underline{v}_i =^e \underline{v}_i \to x_i \underline{v}_i =_0 y_i \underline{v}_i)$.
Then by lemma 10.43 $\bigwedge_{i=1}^{k} (x_i =^e y_i)$ and so using lemma 10.40.2 $z\underline{x} =_0 z\underline{y}$. Hence we have shown that

$$\text{WE-PA}^\omega \vdash (E^\rho)_e.$$

(vi) $(\text{QF-AC}^{0,1})_e$ follows from $\text{QF-AC}^{0,1}$ using lemma 10.41. For $\text{QF-AC}^{1,0}$ one uses that for type-2 functionals $Y =_2^e Y$ just means that Y is extensional, where the existence of an extensional choice functional follows by taking the least value: if $\forall x^1 \exists y^0 A_0(x,y,\underline{a})$, then by $\text{QF-AC}^{1,0}$ $\exists Y^2 \forall x^1 A_0(x, Yx, \underline{a})$. Now define (by primitive recursive bounded search in $x,\underline{a}$) $Y^*(x) := \min y \leq Y(x)[A_0(x,y,\underline{a})]$. It is easy to show that (under the assumption $\underline{a} =^e \underline{a}$) the functional Y^* is extensional, i.e. $Y^* =_2^e Y^*$. Since $A_0 \equiv (A_0)_e$ it follows that $(\text{QF-AC}^{1,0})_e$. $\Box$

Remark 10.46. 1) The proposition above also holds if the rule QF-ER is removed from WE-PA^ω provided that the equality axiom $x =_0 y \to t[x] =_0 t[y]$ is added instead.
2) The proposition also holds for the intuitionistic systems E-HA^ω and WE-HA^ω instead of E-PA^ω and WE-PA^ω respectively.

Using proposition 10.45 we now prove the following theorem, where ENMD denotes the combination of elimination of extensionality, negative translation and subsequent monotone functional interpretation:

Theorem 10.47 (Main theorem on uniform bound extraction by ENMD).
Let Δ be a set of sentences of the form as in theorem 10.21 but with the condition that the degrees of the types $\underline{\sigma}$ are all ≤ 1 and those of $\underline{\gamma}$ all ≤ 2. Let $A_0(x^1, y^1, z^\tau)$ be a (quantifier-free) formula of $\mathcal{L}(\text{E-PA}^\omega)$ containing only x,y,z as free variables, $deg(\tau) \leq 2$ and s be a closed term. Then

$$\begin{cases} \text{E-PA}^\omega + \text{QF-AC}^{1,0} + \text{QF-AC}^{0,1} \ + \Delta + \text{WKL} \vdash \forall x^1 \forall y \leq_1 sx \exists z^\tau A_0(x,y,z) \\ \Rightarrow \text{ENMD } \textit{extracts a closed term } t \textit{ of } \text{WE-HA}^\omega \textit{ such that} \\ \text{WE-HA}^\omega + \tilde{\Delta}_\varepsilon \vdash \forall x^1 \forall y \leq_1 sx \exists z \leq_\tau tx A_0(x,y,z). \end{cases}$$

In particular, if $\mathcal{S}^\omega \models \Delta$, then the conclusion holds in $\mathcal{S}^\omega$. As in theorem 6.8 the result also applies to tuples of variables satisfying the type restrictions.

Proof: We may assume that Δ is finite and so can form the conjunction of its elements which we also denote by Δ. By the deduction theorem we have

$$\text{E-PA}^\omega + \text{QF-AC}^{1,0} + \text{QF-AC}^{0,1} \vdash \Delta \wedge \text{WKL} \to \forall x^1 \forall y \leq_1 sx \exists z^\tau A_0(x,y,z)$$

and so by proposition 10.45

$$\text{WE-PA}^\omega + \text{QF-AC}^{1,0} + \text{QF-AC}^{0,1} \vdash \left(\Delta \wedge \text{WKL} \to \forall x^1 \forall y \leq_1 sx \exists z^\tau A_0(x,y,z)\right)_e.$$

By the restrictions on the types of the positively occurring $\forall$-quantifiers and the negatively occurring $\exists$-quantifiers and lemma 10.41 we have

$$\text{WE-PA}^\omega \vdash (\ldots)_e \to (\ldots).$$

Hence

$$\text{WE-PA}^\omega + \text{QF-AC}^{1,0} + \text{QF-AC}^{0,1} \vdash \Delta \wedge \text{WKL} \to \forall x^1 \forall y \leq_1 sx \exists z^\tau A_0(x,y,z).$$

Now by theorem 10.26 and the proof of corollary 10.31 the claim in the theorem follows. □

Remark 10.48. By proposition 10.30 we can replace in theorem 10.47 $\tilde{\Delta}_\varepsilon$ by Δ_ε if all the quantified variables in Δ have types of degree ≤ 1.

In one of the applications to approximation theory which will be given in chapter 16 below, we will analyze a proof of a $\forall\exists$-theorem which is based on a lemma having the logical form

$$(\%)\ \forall x^1 (\forall w^0 B_0(x,w) \to \exists y \leq_1 sx \forall z^0 C_0(x,y,z)),$$

where B_0 and C_0 are quantifier-free. The logical form looks similar to that of lemmas Δ as in the previous theorem, for which we have shown that their proofs do not contribute to the construction of extractable bounds so that they can be taken simply as axioms. Nevertheless, due to the additional universal premise '$\forall w^0 B_0(x,w)$' this is no longer true for lemmas of the form $(\%)$: Let T be the Kleene-T-predicate and consider the following sentence which holds by classical logical (and hence is provable in WE-PA$^\omega$):

$$\forall m^0 \left(\forall n^0 \neg T(\overline{e},m,n) \to 0 = 1\right) \to \forall k^0 \exists l^0 T(\overline{e},k,l),$$

where e is the code of a total recursive function which growths faster than any WE-PA$^\omega$-definable function. Note that by treating '$\exists y \leq sx$' and '$\forall z^0$' as dummy quantifiers (and via the obvious embedding of the type 0 into the type 1) the premise

$$\forall m^0 \big(\forall n^0 \neg T(\overline{e},m,n) \to 0 = 1\big)$$

has the logical form (%). Suppose we could extract from the proof a closed term t of WE-PA$^\omega$ such that

$$\forall m^0 \big(\forall n^0 \neg T(\overline{e},m,n) \to 0 = 1\big) \to \forall k^0 \exists l \leq_0 tk\, T(\overline{e},k,l)$$

is true, then by the truth of the premise (since e encodes a total recursive function) the conclusion

$$\forall k \exists l \leq_0 tk\, T(\overline{e},k,l)$$

would be true in contradiction to the construction of e.
Given this situation, the next theorem gives the best possible answer to the question to what extent proofs of lemmas (%) do contribute to the extractable bound for the conclusion:

Theorem 10.49. *Let Δ be as in proposition 10.26 with δ,σ,γ types of degree ≤ 1 and $B_0(x,w), C_0(x,y,z)$ be quantifier-free formulas of $\mathcal{L}$(E-PA$^\omega$) containing only x,w and x,y,z free respectively.*
Assume that

$$(i)\ \text{E-PA}^\omega + \text{QF-AC}^{1,0} + \text{QF-AC}^{0,1} + \Delta \vdash$$
$$\forall x^1 \big(\forall w^0 B_0(x,w) \to \forall z^0 \exists y \leq_1 sx \bigwedge_{j=0}^{z} C_0(x,y,j)\big)$$

and

$$(ii)\ \text{E-PA}^\omega + \text{QF-AC}^{1,0} + \text{QF-AC}^{0,1} + \Delta \vdash$$
$$\forall x \big(\forall w^0 B_0(x,w) \to \exists y \leq_1 sx \forall z^0 C_0(x,y,z)\big) \to \forall u^1; v \leq_1 tu \exists k^0 D_0(u,v,k)\big).$$

Then:
From (i) one can extract a closed term χ of WE-HA$^\omega$ *such that*

$$(i)^*\ \text{WE-HA}^\omega + \Delta_\varepsilon \vdash \forall x,z \big(\bigwedge_{i=0}^{\chi xz} B_0(x,i) \to \exists y \leq sx \bigwedge_{j=0}^{z} C_0(x,y,j)\big).$$

From (ii) – using χ – one can extract a closed term Ψ of WE-HA$^\omega$ *such that*

$$(ii)^*\ \text{WE-HA}^\omega + \Delta_\varepsilon \vdash \forall u \forall v \leq tu \bigvee_{k=0}^{\Psi u} D_0(u,v,k).$$

Proof: From (i) one concludes that

$$\text{E-PA}^{\omega}+\text{QF-AC}^{1,0}+\text{QF-AC}^{0,1}+\Delta \vdash \forall x^1, z^0 \exists w^0 \Big(B_0(x,w) \to \exists y \leq_1 sx \bigwedge_{j=0}^{z} C_0 \Big).$$

By lemma 9.27 there exists (effectively) a quantifier-free formula $F_0 \in \mathcal{L}(\text{WE-PA}^{\omega})$ such that

$$\text{WE-HA}^{\omega} \vdash F_0(x,w,z) \leftrightarrow \Big(B_0(x,w) \to \exists y \leq sx \bigwedge_{j=0}^{z} C_0 \Big).$$

Theorem 10.47 and proposition 10.30 now yield a closed term χ of WE-HA$^{\omega}$ such that

$$\text{WE-HA}^{\omega} + \Delta_{\varepsilon} \vdash \forall x, z \Big(\bigwedge_{i=0}^{\chi xz} B_0(x,i) \to \exists y \leq sx \bigwedge_{j=0}^{z} C_0 \Big)$$

which implies $(i)^*$.

(ii) implies (using the deduction theorem for E-PA$^{\omega}$ and elimination of extensionality)

$$\begin{aligned}&\text{WE-PA}^{\omega}+\text{QF-AC} \vdash \\ &\quad \Delta \to \forall x \exists y \leq sx \forall z \exists w \Big(\bigwedge_{i=0}^{w} B_0(x,i) \to \bigwedge_{j=0}^{z} C_0(x,y,j) \Big) \to \forall u; v \leq tu \exists k D_0.\end{aligned}$$

Hence

$$\begin{aligned}&\text{WE-PA}^{\omega}+\text{QF-AC} \vdash \\ &\quad \Delta \to \exists Y \leq s, W^{001} \forall x, z \Big(\bigwedge_{i=0}^{Wxz} B_0(x,i) \to \bigwedge_{j=0}^{z} C_0(x,Yx,j) \Big) \to \forall u; v \leq tu \exists k D_0.\end{aligned}$$

Therefore

$$\begin{aligned}&\text{WE-PA}^{\omega}+\text{QF-AC} \vdash \\ &\quad \Delta \to \forall Y \leq s; W, u; v \leq tu \exists x, z, k \Big(\Big(\bigwedge_{i=0}^{Wxz} B_0(x,i) \to \bigwedge_{j=0}^{z} C_0(x,Yx,j) \Big) \to D_0(u,v,k) \Big).\end{aligned}$$

Substituting χ for W yields

$$\begin{aligned}&\text{WE-PA}^{\omega}+\text{QF-AC} \vdash \\ &\quad \Delta \to \forall Y \leq s; u; v \leq tu \exists x, z, k \Big(\Big(\bigwedge_{i=0}^{\chi xz} B_0(x,i) \to \bigwedge_{j=0}^{z} C_0(x,Yx,j) \Big) \to D_0(u,v,k) \Big).\end{aligned}$$

By theorem 10.26 and proposition 10.30 we extract closed terms Φ, Ψ of WE-HA$^{\omega}$ from the given proof such that

$$(*)\ \text{WE-HA}^\omega + \Delta_\varepsilon \vdash \forall Y \le s; u; v \le tu\Big(\forall x \forall z \le \Phi u\Big(\bigwedge_{i=0}^{\chi xz} B_0(x,i) \to \bigwedge_{j=0}^{z} C_0(x,Yx,j)\Big) \to \bigvee_{k=0}^{\Psi u} D_0(u,v,k)\Big).$$

Hence also

$$\text{WE-HA}^\omega + \Delta_\varepsilon \vdash \forall u\Big(\exists Y \le s \forall x; z \le \Phi u\Big(\bigwedge_{i=0}^{\chi xz} B_0(x,i) \to \bigwedge_{j=0}^{z} C_0(x,Yx,j)\Big) \to \forall v \le tu \bigvee_{k=0}^{\Psi u} D_0(u,v,k)\Big).$$

Using lemma 9.27 this yields

$$(**)\ \text{WE-HA}^\omega + \Delta_\varepsilon \vdash \forall u\Big(\forall x \exists y \le sx \forall z \le \Phi u\Big(\bigwedge_{i=0}^{\chi xz} B_0(x,i) \to \bigwedge_{j=0}^{z} C_0(x,y,j)\Big) \to \forall v \le tu \bigvee_{k=0}^{\Psi u} D_0(u,v,k)\Big).$$

It remains to show that

$$\text{WE-HA}^\omega \vdash \forall x,z\Big(\bigwedge_{i=0}^{\chi xz} B_0(x,i) \to \exists y \le sx \bigwedge_{j=0}^{z} C_0(x,y,j)\Big) \to \forall u,x \exists y \le_1 sx \forall z \le_0 \Phi u\Big(\bigwedge_{i=0}^{\chi xz} B_0(x,i) \to \bigwedge_{j=0}^{z} C_0(x,y,j)\Big).$$

(Together with $(**)$ and $(i)^*$ this implies $(ii)^*$).
Assume $(+)\ \forall x,z\Big(\bigwedge_{i=0}^{\chi xz} B_0(x,i) \to \exists y \le sx \bigwedge_{j=0}^{z} C_0(x,y,j)\Big)$.
We have to show that

$$\forall u,x \exists y \le_1 sx \forall z \le_0 \Phi u\Big(\bigwedge_{i=0}^{\chi xz} B_0(x,i) \to \bigwedge_{j=0}^{z} C_0(x,y,j)\Big).$$

Since this is trivial if $\forall z \le \Phi u \neg \bigwedge_{i=0}^{\chi xz} B_0(x,i)$ (take $y := 0^1$), we may assume that

$$\exists z \le \Phi u \bigwedge_{i=0}^{\chi xz} B_0(x,i).$$

Define primitive recursively in x,u (in the sense of WE-HA^ω) $z_{u,x}$ such that $z_{u,x} = max\Big\{z \le_0 \Phi u \mid \bigwedge_{i=0}^{\chi xz} B_0(x,i)\Big\}$. Then $\bigwedge_{i=0}^{\chi x z_{u,x}} B_0(x,i)$. By (+) there exists an $y \le sx$ with $(++)\ \bigwedge_{j=0}^{z_{u,x}} C_0(x,y,j)$. We show

$$(+++)\ \forall z \leq_0 \Phi u\Big(\bigwedge_{i=0}^{\chi xz} B_0(x,i) \to \bigwedge_{j=0}^{z} C_0(x,y,j)\Big):$$

Case 1: $z \leq z_{u,x}$. Then by (++) $\bigwedge_{j=0}^{z} C_0(x,y,j)$.

Case 2: $\Phi u \geq z > z_{u,x}$. By the maximality of $z_{u,x}$ it follows that $\neg \bigwedge_{i=0}^{\chi xz} B_0(x,i)$ and hence $\bigwedge_{i=0}^{\chi xz} B_0(x,i) \to \bigwedge_{j=0}^{z} C_0(x,y,j)$. $\square$

Remark 10.50. 1) The above theorem is useful in analyzing proofs which can be split into the two parts

$$(i)\ \forall \alpha \exists \beta \leq r\alpha \forall n A_0 \to \forall x\big(\forall w B_0 \to \exists y \leq sx \forall z C_0\big) \text{ and}$$

$$(ii)\ \forall x\big(\forall w B_0 \to \exists y \leq sx \forall z C_0\big) \to \forall u \forall v \leq tu \exists k D_0:$$

One analyses separately the proof of (i), which, in particular, is a proof of

$$(i)^*\ \forall \alpha \exists \beta \leq r\alpha \forall n A_0 \to \forall x\big(\forall w B_0 \to \forall z \exists y \leq sx \bigwedge_{j=0}^{z} C_0\big),$$

and the proof of (ii) and combines the results to a bound for '$\exists k$'.
In section 2 of chapter 16 we will give an application of this strategy in the context of the logical analysis of the standard proof of the uniqueness of best Chebycheff approximation.

2) Theorem 10.49 gives an alternative proof of the fact that E-PA$^\omega$+QF-AC^{1,0}+QF-AC^{0,1}+WKL is conservative over WE-HA$^\omega$ w.r.t. $\forall u^1 \forall v \leq_1 tu \exists k^0 A_0$-sentences (analogously for the systems with restricted induction), since $\widehat{\text{WE-HA}}^\omega\!\upharpoonright$ proves

$$\text{WKL} \leftrightarrow \forall f,g\Big(Tf \wedge \forall x(lth(gx)=x \wedge f(gx)=0) \to \exists b \leq_1 \lambda k.1 \forall x\big(f(\overline{b}x)=0\big)\Big)$$

as well as

$$\forall f,g\Big(Tf \wedge \forall x\big(lth(gx)=x \wedge f(gx)=0\big) \to \forall x \exists b \leq \lambda k.1 \bigwedge_{j=0}^{x} \big(f(\overline{b}j)=0\big)\Big).$$

10.5 Fragments of (W)E-PA$^\omega$

As we discussed already at the end of the previous chapter, monotone functional interpretation applies also to the fragments $\mathcal{T}_i^\omega := \widehat{\text{WE-HA}}^\omega\!\upharpoonright$, G$_nA_i^\omega$ ($n \geq 2$) of WE-HA$^\omega$. Since also the negative translation is applicable (as is the extensionality elimination) we can obtain the main results of this chapter also for $\mathcal{T}$ instead of (W)E-PA$^\omega$ where $\mathcal{T}$ is the variant of $\mathcal{T}_i$ with classical logic. In particular, we obtain

Theorem 10.51. *Theorems 10.21 and 10.26 also hold for* $\mathrm{G}_n\mathrm{A}^\omega$, $\mathrm{G}_n\mathrm{A}^\omega_i$ $(n \geq 2)$ *resp.* $\widehat{\text{WE-PA}}^\omega\!\upharpoonright$, $\widehat{\text{WE-HA}}^\omega\!\upharpoonright$ *instead of* WE-PA$^\omega$, WE-HA$^\omega$, *where then t is a closed term of* $\mathrm{G}_n\mathrm{A}^\omega_i$ *resp.* $\widehat{\text{WE-HA}}^\omega\!\upharpoonright$.

Adapting the proof of theorem 10.47 to E-G$_2$A$^\omega$ and using corollary 3.42 and proposition 3.43 we get

Theorem 10.52 (Polynomial bound extraction by ENMD).
Let Δ be a set of sentences in $\mathcal{L}$(E-G$_2$A$^\omega$) as in theorem 10.47 and $A_0(x^0, y^1, z^0)$ be a (quantifier-free) formula of $\mathcal{L}$(E-G$_2$A$^\omega$) containing only x, y, z as free variables and s be a closed term. Then

$$\begin{cases} \text{E-G}_2\text{A}^\omega + \text{QF-AC}^{1,0} + \text{QF-AC}^{0,1} + \Delta \vdash \forall x^0 \forall y \leq_1 sx \exists z^0 A_0(x,y,z) \\ \Rightarrow \text{ ENMD } \textit{extracts a polynomial } p \textit{ such that} \\ \text{G}_2\text{A}^\omega_i + \tilde{\Delta}_\varepsilon \vdash \forall x^0 \forall y \leq_1 sx \exists z \leq_0 p(x) A_0(x,y,z). \end{cases}$$

Proof: See Kohlenbach [207] for full details. □

We now derive some further results for (W)$\widehat{\text{E-PA}}^\omega$:

Proposition 10.53 (Feferman [98]).

$$\begin{cases} \widehat{\text{WE-PA}}^\omega\!\upharpoonright + \text{QF-AC} \vdash A(\underline{a}) \\ \Rightarrow \exists \textit{ closed terms } \underline{t} \textit{ of } \widehat{\text{WE-HA}}^\omega\!\upharpoonright \textit{ such that} \\ \widehat{\text{WE-HA}}^\omega\!\upharpoonright \vdash \forall \underline{y} (A')_D(\underline{t}\,\underline{a}, \underline{y}, \underline{a}) \end{cases}$$

(Here $\underline{a}$ are all of the free variables of $A(\underline{a})$.

Proof: The proof of theorem 10.7 easily relativizes to $\widehat{\text{WE-PA}}^\omega\!\upharpoonright$. □

As in theorem 10.8 we get the following immediate consequence of proposition 10.53:

Corollary 10.54 (Feferman [98]). *Let $A_0(x, y)$ be a (quantifier-free) formula of $\mathcal{L}(\widehat{\text{WE-PA}}^\omega\!\upharpoonright)$ which only contains x, y as free variables. Then the following rule holds:*

$$\begin{cases} \widehat{\text{WE-PA}}^\omega\!\upharpoonright + \text{QF-AC} \vdash \forall x^\rho \exists y^\tau A_0(x,y) \\ \Rightarrow \textit{ one can extract a closed term } t \textit{ of } \widehat{\text{WE-HA}}^\omega\!\upharpoonright \textit{ such that} \\ \widehat{\text{WE-HA}}^\omega\!\upharpoonright \vdash \forall x A_0(x, tx). \end{cases}$$

The result also holds for tuples of variables $\underline{x}^{\underline{\rho}}, \underline{y}^{\underline{\tau}}$ where then t is a tuple of closed terms.

One can show that the function(al)s of types of degree ≤ 2 definable by closed terms in $\widehat{\text{WE-HA}}^{\omega}\!\upharpoonright$ are just the usual primitive recursive ones in the sense of Kleene (see [98]). From the proof of this fact combined with the previous proposition plus elimination of extensionality one gets

Proposition 10.55 (Feferman [98]). *Let $R(x,y)$ be a primitive recursive relation (in the sense of* PRA*). Then the following rule holds:*

$$\begin{cases} \widehat{\text{E-PA}}^{\omega}\!\upharpoonright + \text{QF-AC}^{1,0} + \text{QF-AC}^{0,1} \vdash \forall x^0 \exists y^0 R(x,y) \\ \Rightarrow \exists \textit{ primitive recursive function } p \textit{ such that} \\ \text{PRA} \vdash R(x, px). \end{cases}$$

Let PA_1 be the restriction of PA to induction for Σ^0_1-formulas only.

Corollary 10.56. *(Parsons,Mints,Takeuti,...)* PA_1 *is Π^0_2-conservative over* PRA.

Analogously to proposition 10.9 we get

Proposition 10.57. *Let A be a prenex sentence of $\mathcal{L}(\text{PA}_1)$. Then the following rule holds:*

$$\begin{cases} \text{PA}_1 \vdash A \\ \Rightarrow \textit{ one can extract primitive recursive functionals } \underline{\Phi} \textit{ in the sense of Kleene} \\ \widehat{\text{WE-HA}}^{\omega}\!\upharpoonright \vdash \underline{\Phi} \textit{ n.c.i. } A. \end{cases}$$

Conservative fragments of analysis

As we mentioned already in the previous chapter, weak König's lemma WKL has received quite some attention in the last 20 years as despite of its ability to derive large portions of classical mathematics it is proof-theoretically weak by the aforementioned theorem of H. Friedman stating that RCA_0+ WKL is Π^0_2-conservative over PRA, where RCA_0 basically is the second order fragment of $\widehat{\text{WE-PA}}^{\omega}\!\upharpoonright$+ QF-$\text{AC}^{0,0}$

Adapting theorem 10.47 to the restricted context of $\widehat{\text{E-PA}}^{\omega}\!\upharpoonright$ yields the following generalization of Friedman's result:

Theorem 10.58. *Let $A_0(x,y,z)$ be a (quantifier-free) formula of $\mathcal{L}(\widehat{\text{E-PA}}^{\omega}\!\upharpoonright)$ containing only x^1, y^1, z^0 as free variables and let s be a closed term. Then the following rule holds:*

$$\begin{cases} \widehat{\text{E-PA}}^{\omega}\!\restriction + \text{QF-AC}^{1,0} + \text{QF-AC}^{0,1} + \text{WKL} \vdash \forall x^1 \forall y \leq_1 sx \exists z^0 A_0(x,y,z) \\ \Rightarrow \textit{one can extract a primitive recursive (in Kleene's sense) } \Phi \textit{ such that} \\ \widehat{\text{WE-HA}}^{\omega}\!\restriction \vdash \forall x^1 \forall y \leq_1 sx \exists z \leq_0 \Phi x A_0(x,y,z). \end{cases}$$

An analogous result holds for E-PA$^{\omega}$, WE-HA$^{\omega}$ *instead of* $\widehat{\text{E-PA}}^{\omega}\!\restriction$, $\widehat{\text{WE-HA}}^{\omega}\!\restriction$. *Then* Φ *will be a primitive recursive functional in the sense of Gödel's T (i.e. given by a closed term of* WE-HA$^{\omega}$*).*

In chapter 12 we will obtain this theorem again as a corollary to the even stronger theorem 12.8.

Corollary 10.59. $\widehat{\text{E-PA}}^{\omega}\!\restriction + \text{QF-AC}^{1,0} + \text{QF-AC}^{0,1} + \text{WKL}$ *is* Π^0_2*-conservative over* PRA.

10.6 The computational strength of full extensionality

As we have shown above, over the weakly extensional systems WE-PA$^{\omega}$ and $\widehat{\text{WE-PA}}^{\omega}\!\restriction$ the uniform weak König's lemma UWKL is just as weak (proof theoretically and w.r.t. the provable recursive function(al)s of type ≤ 2) as WKL. In this section we establish that this radically changes if we switch to the fully extensional context of E-PA$^{\omega}$ or $\widehat{\text{E-PA}}^{\omega}\!\restriction$. Already over the corresponding intuitionistic systems E-HA$^{\omega}$ and $\widehat{\text{E-HA}}^{\omega}\!\restriction$ the uniform weak König's lemma UWKL is much stronger than WKL w.r.t. proof-theoretic strength. In the presence of M^0 for numbers – and a-fortiori for the classical systems – UWKL is also stronger w.r.t. the class of provably recursive functions. In order to establish this we first show that in the presence of full extensionality UWKL is equivalent to a strong uniform version of Π^0_1-comprehension:

Proposition 10.60.
$\widehat{\text{E-HA}}^{\omega}\!\restriction \vdash \text{UWKL} \leftrightarrow \exists \varphi^2 \forall f^1 \big(\varphi f =_0 0 \leftrightarrow \forall x^0 (fx =_0 0)\big)$.

Proof: 1) '$\rightarrow$': We use an argument from higher type recursion theory due to T.J. Grilliot [143] which is also known as 'Grilliot's trick' and can be formalized in our weak context already. The argument shows how to construct the comprehension functional out of a functional that is effectively discontinuous. In order to apply it we first show that any Φ satisfying UWKL is – provably in $\widehat{\text{E-HA}}^{\omega}\!\restriction$ – (effectively) discontinuous, i.e.

$$\widehat{\text{E-HA}}^{\omega}\!\restriction \vdash \begin{cases} \forall \Phi^{1(1)} \big(\forall f^1 \big(T^{\infty}(f) \rightarrow \forall x^0 (f((\overline{\Phi f})x) =_0 0)\big) \rightarrow \\ \qquad \exists g^{1(0)}_{(\cdot)}, g^1 \big(T^{\infty}(g) \wedge \forall i T^{\infty}(g_i) \wedge \forall i \forall j \geq i (g_j(i) =_0 g(i)) \\ \qquad\qquad \wedge \forall i, j \big(\Phi(g_i, 0) = \Phi(g_j, 0) \neq \Phi(g, 0)\big)\big)\big). \end{cases}$$

Furthermore, $g_{(\cdot)}$ and g can be computed uniformly in Φ by closed terms of $\widehat{\text{E-HA}}^{\omega}\!\upharpoonright$.
Let g be primitive recursively defined such that

$$g(k) = \begin{cases} 0, \text{ if } \forall m < lth(k)\big((k)_m = 0\big) \vee \forall m < lth(k)\big((k)_m = 1\big) \\ 1, \text{otherwise.} \end{cases}$$

g represents a tree with two infinite paths, corresponding to an infinite sequence of 0's and an infinite sequence of 1's. So obviously

$$\widehat{\text{E-HA}}^{\omega}\!\upharpoonright \vdash T^{\infty}(g).$$

Suppose now that we have (by UWKL) a functional $\Phi^{1(1)}$ be such that

$$\forall f^1 \big(T^{\infty}(f) \to \forall x(f((\overline{\Phi f})x) =_0 0)\big).$$

Case 1: $\Phi(g,0) = 0$. Let $\lambda i,k.g_i(k)$ be primitive recursively defined such that

$$g_i(k) = \begin{cases} 0, \text{ if } [lth(k) \leq i \wedge \forall m < lth(k)\big((k)_m = 0\big)] \vee [\forall m < lth(k)\big((k)_m = 1\big)] \\ 1, \text{otherwise.} \end{cases}$$

g_i represents the same tree as g except that the left branch has been truncated at level i. So again we easily verify within $\widehat{\text{E-HA}}^{\omega}\!\upharpoonright$ that $\forall i T^{\infty}(g_i)$. It is immediate from the definitions of g_i and g that

$$\forall k \forall l \geq lth(k)\big(g_l(k) = g(k)\big).$$

It is easy to check that our sequence coding from chapter 3 has the property that $lth(k) \leq k$. Thus

$$\forall k \forall l \geq k\big(g_l(k) = g(k)\big).$$

Clearly, $\lambda x.1$ is the only infinite path of the binary tree represented by g_i. Hence

$$\forall i\big(\Phi(g_i,0) = 1\big).$$

Case 2: $\Phi(g,0) = 1$. The proof is analogous to case 1 with

$$g_i(k) := \begin{cases} 0, \text{ if } [lth(k) \leq i \wedge \forall m < lth(k)\big((k)_m = 1\big)] \vee [\forall m < lth(k)\big((k)_m = 0\big)] \\ 1, \text{otherwise.} \end{cases}$$

This finishes the proof of the effective discontinuity of Φ. Using the aforementioned Grilliot's trick we now show the provably in $\widehat{\text{E-HA}}^{\omega}\!\upharpoonright$ we can define primitive recursively in Φ a functional φ^2 such that

$$(+)\forall f^1(\varphi f =_0 0 \leftrightarrow \forall x(fx =_0 0)).$$

By primitive recursion define a closed term $t^{1(1)}$ of $\widehat{\text{E-HA}}^{\omega}\!\upharpoonright$ such that (provably in $\widehat{\text{E-HA}}^{\omega}\!\upharpoonright$) we have

$$thi = \begin{cases} g_j(i), \text{ for the least } j < i \text{ such that } h(j) > 0, \text{ if such a } j \text{ exists} \\ g(i), \text{ otherwise.} \end{cases}$$

Together with $\forall i \forall j \geq i\big(g_j(i) = g(i)\big)$ this implies

$$\exists j(h(j) > 0) \rightarrow th =_1 g_j \text{ for the least such } j$$

and

$$\forall j(h(j) = 0) \rightarrow th =_1 g.$$

Applying the extensionality axiom to the type-2-functional Φ we get

$$(*)\ \forall j(h(j) = 0) \leftrightarrow \Phi(th,0) =_0 \Phi(g,0).$$

Hence $\varphi := \lambda h^1.|\Phi(th,0) - \Phi(g,0)|$ fulfills our claim.
We conclude the proof by combining the two constructions of φ corresponding to the two cases above into a single functional which defines φ primitive recursively in Φ: Let χ be a closed term such that

$$\widehat{\text{E-HA}}^{\omega}\!\upharpoonright \vdash \forall x^0\big((x =_0 0 \rightarrow \chi x =_{1(1)} t) \wedge (x \neq 0 \rightarrow \chi x =_{1(1)} \tilde{t})\big),$$

where t is defined as above with g_i from case 1 whereas $\tilde{t}$ is defined analogously but with g_i as in case 2. Then define $\varphi := \lambda h^1.|\Phi((\chi(\Phi(g,0))(h),0) - \Phi(g,0)|$.
2) '$\leftarrow$': Primitive recursively in φ one can easily compute a functional Φ which selects an infinite branch of an infinite binary tree, say the leftmost one. $\square$

Remark 10.61. 1) Grilliot's argument plays an important role in the context of the Kleene/Kreisel countable functionals (see [287], whose formulation of it we adopted here) and will be used again in chapter 12. Other proof-theoretic applications of this argument can be found in [225].
2) It is the direction '$\rightarrow$' of $(*)$ which needs (E). The direction '$\leftarrow$' can be shown using only QF-ER, since free variables are allowed to occur in premises A_0 of QF-ER.
3) Further refinements of proposition 10.60 can be found in [320].

Corollary to the proof of proposition 10.60: One can construct closed terms t_1, t_2 of $\widehat{\text{E-HA}}^{\omega}\!\upharpoonright$ such that

$$\widehat{\text{E-HA}}^{\omega}\!\upharpoonright \vdash \begin{cases} \forall \Phi^{1(1)}\big(\forall f^1\big(T^{\infty}(f) \rightarrow \forall x^0\big(f((\overline{\Phi f})x) = 0\big)\big) \rightarrow \\ \qquad \forall f^1\big((t_1\Phi)f =_0 0 \leftrightarrow \forall x(fx = 0)\big)\big) \end{cases}$$

and

$$\widehat{\mathrm{WE\text{-}HA}}^{\omega}\!\upharpoonright\vdash\begin{cases}\forall\varphi^2\big(\forall f^1(\varphi f=0\leftrightarrow\forall x(fx=0))\rightarrow\\ \qquad\forall f^1\big(T^{\infty}(f)\rightarrow\forall x^0\big(f((\overline{t_2\varphi f})x)=0\big)\big)\big).\end{cases}$$

Corollary 10.62.
$\widehat{\mathrm{E\text{-}HA}}^{\omega}\!\upharpoonright+\mathrm{M}^0\vdash\mathrm{UWKL}\leftrightarrow\exists\tilde{\varphi}^2\forall f^1\big(\tilde{\varphi}f=_0 0\leftrightarrow\exists x^0(fx=_0 0)\big)$.

While over WE-PA$^{\omega}$ (resp.$\widehat{\mathrm{WE\text{-}PA}}^{\omega}\!\upharpoonright$) the addition of UWKL does not add to the proof-theoretic strength (corollary 10.34), which is that of PA (resp. PRA) by well-known results e.g. from [98], this changes in the presence of full extensionality:

Proposition 10.63. 1) $\widehat{\mathrm{E\text{-}HA}}^{\omega}\!\upharpoonright+$ UWKL *and* $\widehat{\mathrm{E\text{-}PA}}^{\omega}\!\upharpoonright+$ UWKL *have the same proof-theoretic strength as* PA.
2) E-HA$^{\omega}$+UWKL *and* E-PA$^{\omega}$+UWKL *prove the consistency of* PA *and have the same proof-theoretic strength as* $(\Pi^0_1\text{-CA})_{<\varepsilon_0}$ *(as defined in [98]).*

Proof: 1) By the corollary above,
$\widehat{\mathrm{E\text{-}PA}}^{\omega}\!\upharpoonright$+UWKL$=\widehat{\mathrm{E\text{-}PA}}^{\omega}\!\upharpoonright+\exists\tilde{\varphi}\forall f^1\big(\tilde{\varphi}f=_0 0\leftrightarrow\exists x^0(fx=_0 0)\big)$. The latter systems allows one (by iterated use of $\tilde{\varphi}$) to prove the schema of arithmetical comprehension which in turn yields (with QF-IA) the schema for induction for all arithmetical formulas. Hence PA can be viewed as a subsystem of $\widehat{\mathrm{E\text{-}PA}}^{\omega}\!\upharpoonright$+UWKL and so his negative translation as a subsystem of $\widehat{\mathrm{E\text{-}HA}}^{\omega}\!\upharpoonright$+UWKL.
That PA is also an upper bound for the proof-theoretic strength of the systems in question follows from classical results due to Feferman (see [98]).
2) follows analogously to 1) using the classical results from [98] on the strength of

$$\exists\tilde{\varphi}^2\forall f^1\big(\tilde{\varphi}f=_0 0\leftrightarrow\exists x^0(fx=_0 0)\big)$$

over E-PA$^{\omega}$ and subsequent negative translation for the intuitionistic case. □

Although already $\widehat{\mathrm{E\text{-}HA}}^{\omega}\!\upharpoonright$+UWKL has the same proof-theoretic strength as PA its provably recursive functions are still the same ones as that of PRA. This follows from proposition 7.3 (relativized to $\widehat{\mathrm{E\text{-}HA}}^{\omega}\!\upharpoonright$ as indicated in the remark at the end of that chapter) and the fact that UWKL is just an axiom of the form Ξ. However, as soon as M$^{\omega}$ or just M^0 is added this changes drastically:

Proposition 10.64. *The provably recursive functions of* $\widehat{\mathrm{E\text{-}HA}}^{\omega}\!\upharpoonright$+UWKL+M^0 *and* $\widehat{\mathrm{E\text{-}PA}}^{\omega}\!\upharpoonright$+UWKL *are precisely the ones of* PA, *i.e. the* $\alpha(<\varepsilon_0)$*-recursive functions.*

Proof: By the negative embedding of PA into $\widehat{\mathrm{E\text{-}HA}}^{\omega}\!\upharpoonright$+UWKL from the previous proof, the negative translation of every Π^0_2-sentence A provable in PA can be proved in $\widehat{\mathrm{E\text{-}HA}}^{\omega}\!\upharpoonright$+UWKL. Hence A itself can be proved in $\widehat{\mathrm{E\text{-}HA}}^{\omega}\!\upharpoonright$+UWKL+M^0. Again the upper bound follows from [98]. □

10.7 Exercises, historical comments and suggested further reading

Exercises:

1) Prove the claim in remark 10.2.
2) (Krivine, [348, 347]) Consider PA in the language fragment with $\vee, \neg, \forall$ only (which in the classical context of PA is not a real restriction). The following negative translation $A \mapsto A^K$ is due to Krivine:

$$P^- :\equiv \neg P, \text{ for prime formulas } P,$$

$$(A \vee B)^- :\equiv A^- \wedge B^-,$$

$$(\neg A)^- :\equiv \neg A^-,$$

$$(\forall x A)^- :\equiv \exists x A^-.$$

Now define $A^K :\equiv \neg A^-$.
Show that $\mathrm{IL} \vdash A' \leftrightarrow A^K$ and, consequently,

$$\mathrm{PA} \vdash A \;\Rightarrow\; \mathrm{HA} \vdash A^K.$$

3) The following variant (due to Shoenfield [332]) of the combination of functional interpretation with negative translation gives a direct interpretation of PA into the WE-PA$^\omega$. Again we only consider the fragment with $\vee, \neg, \forall$: to each formula A of $\mathcal{L}$(PA) we assign (omitting the obvious embedding of PA into WE-PA$^\omega$) a formula $A^{Sh} \equiv \forall \underline{u} \exists \underline{x} A_{Sh}(\underline{u}, \underline{x})$ on $\mathcal{L}$(WE-HA$^\omega$) with A_{Sh} quantifier-free as follows (with $B^{sh} \equiv \forall \underline{v} \exists \underline{y} B_{Sh}(\underline{v}, \underline{y})$)

$$P^{Sh} :\equiv P_{Sh} :\equiv P \text{ for prime formulas } P,$$

$$(\neg A)^{Sh} :\equiv \forall \underline{f} \exists \underline{u} \neg A_{Sh}(\underline{u}, \underline{f}\underline{u}),$$

$$(A \vee B)^{Sh} :\equiv \forall \underline{u}, \underline{v} \exists \underline{x}, \underline{y} (A_{Sh}(\underline{x}, \underline{u}) \vee B_{Sh}(\underline{v}, \underline{y})),$$

$$(\forall z A)^{Sh} :\equiv \forall z, \underline{u} \exists \underline{x} A_{Sh}(z, \underline{u}, \underline{x}).$$

Show that from a proof of a sentence A in PA one can extract closed terms $\underline{t}$ of WE-HA$^\omega$ such that WE-HA$^\omega \vdash \forall \underline{u} A_{Sh}(\underline{u}, \underline{t}\underline{u})$.
4) ([347]) Let $A^K \equiv \neg A^-$ be Krivine's negative translation from above. Show that for every formula A of $\mathcal{L}$(PA) the following holds

$$(1)\; (A^-)_D(\underline{u}, \underline{x}) \Leftrightarrow \neg A_{Sh}(\underline{u}, \underline{x}), \text{ where } (A^-)^D \equiv \exists \underline{u} \forall \underline{x} (A^-)_D(\underline{u}, \underline{x})$$

$$(2)\; (A^K)_D(\underline{f}, \underline{u}) \Leftrightarrow A_{Sh}(\underline{u}, \underline{f}\underline{u}), \text{ where } (A^K)^D \equiv \exists \underline{f} \forall \underline{u} (A^K)_D(\underline{f}, \underline{u}),$$

where $\Leftrightarrow$ stands for provability in WE-HA$^{\omega}$.
Roughly speaking (2) says: Shoenfield = Gödel $\circ$ Krivine. Use this to conclude the second exercise from the first and the soundness of Gödel's functional interpretation.
5) Solve the ND-interpretation of the special case of (IPP) where $n = 2$.
6) Prove the claim in remark 10.24.

Historical comments and suggested further reading:
As mentioned in the 2nd exercise, the combination of negative translation and functional interpretation has been studied as a single interpretation in Shoenfield [332]. An extension of functional interpretation to full classical analysis was given by C. Spector using so-called bar recursive functionals (Spector [343], Howard [162], Luckhardt [266]). This will be treated in the next chapter. For an interesting alternative to functional interpretation in this context see Berardi et al. [17] and Berger-Oliva [22]. In Friedrich [109], Spector's approach is extended to still stronger systems using bar recursive functionals of infinite types.

A different extension of functional interpretation to classical analysis was achieved by Girard in [121] using his well-known system F of polymorphic functionals.

The combination of negative translation and monotone functional interpretation is due to Kohlenbach [206]. The material on WKL and UWKL is taken mainly from Kohlenbach [203] and [221]. Some further conservation results concerning WKL can be found in Simpson [338] and Simpson et al. [339]. For conservation results of suitable forms of WKL over systems for feasible analysis see Ferreira-Oliva [105]. Theorem 10.49 (without Δ) is from Kohlenbach [203] while theorem 10.52 is from Kohlenbach [207].

Many important applications of functional interpretation to classical systems using certain ineffective functionals (representing strong forms of comprehension) are carried out in Feferman [98].

In Parsons [299] fragments of Peano arithmetic are investigated using functional interpretation (see also Avigad-Feferman [7]).

For a nice comparison of functional interpretation and modified realizability see Jørgensen [179] and Oliva [292].

For further applications of functional interpretation combined with majorization to classical systems and the development of monotone functional interpretation see Kohlenbach [203, 206, 207, 208, 213, 210]. For applications to specific proofs in the context of approximation theory see Kohlenbach [204, 205, 206] and Kohlenbach-Oliva [235] as well as chapter 16 below. For applications in fixed point theory see Briseid [50, 51, 52, 53], Gerhardy [116], Kohlenbach [219, 220, 224, 227], Kohlenbach-Lambov [231], Kohlenbach-Leuştean [232, 233, 234], Leuştean [263] and chapter 18 below. For a recent application of functional interpretation in ergodic theory see Avigad et al. [8]. In Hertz [160] the ND-interpretation is used to unwind two proofs of the Hilbert basis theorem. One proof (based on Simpson [337]) even yields results which are the optimal (w.r.t. to the complexity class they belong to). In Mints [281], Shoenfield's variant of functional interpretation is used to unwind a non-effective proof for cut elimination.

Chapter 11
Functional interpretation of full classical analysis

11.1 Functional interpretation of full comprehension

In this section we carry out the proof of C. Spector's ([343]) fundamental result that the functional interpretation of the negative translation of the schema of full comprehension over numbers

$$\mathrm{CA}^0 : \exists f^1 \forall x^0 \big(f(x) =_0 0 \leftrightarrow A(x) \big),$$

where $A(x)$ is an arbitrary formula of $\mathcal{L}(\text{WE-PA})^\omega$ (not containing f free), can be solved by the so-called bar recursive functionals.
Together with (the proof of) theorem 8.6 this gives (via negative translation) a functional interpretation of WE-PA$^\omega$+QF-AC+CA^0 and (with a preceding elimination of extensionality) of E-PA$^\omega$+QF-AC^{1,0}+QF-AC^{0,1}+CA^0, i.e. a system which contains (via the identification of subsets of $\mathbb{N}$ with their characteristic functions) full 2nd order arithmetic (in the sense of [338]) and is known to formalize most parts of classical analysis.

In the presence of classical logic and QF-AC^{0,0} the schema of comprehension CA^0 is equivalent to the following schema

$$\mathrm{AC}^{0,0} : \forall x^0 \exists y^0 A(x,y) \rightarrow \exists f^1 \forall x^0 A(x, f(x)),$$

where, again, $A(x,y)$ is an arbitrary formula (not containing f free).

Proposition 11.1. *Over* WE-PA$^\omega$+QF-AC^{0,0} *the schemes* CA^0 *and* AC^{0,0} *are equivalent.*

Proof: '⇒': By CA^0 (using pairing) applied to $A(x,y)$ we get a function g such that

$$\forall x^0, y^0 (g(x,y) =_0 0 \leftrightarrow A(x,y)).$$

Hence $\forall x \exists y A(x,y)$ yields $\forall x \exists y (g(x,y) = 0)$ and so by QF-AC^{0,0}

$$\exists f^1 \forall x^0 \big(g(x, f(x)) = 0 \big),$$

i.e.

$$\exists f^1 \forall x^0 A(x, f(x)).$$

'$\Leftarrow$': CA^0 follows from $AC^{0,0}$ applied to

$$\forall x^0 \exists n^0 (n =_0 0 \leftrightarrow A(x)),$$

which is a consequence of the law-of-excluded-middle and $0 \neq 1$. □

Spector even proved that the negative translation of the stronger axiom schema of countable choice $AC^0 := \bigcup_{\rho \in \mathbf{T}} \{AC^{0,\rho}\}$, where

$$AC^{0,\rho} : \forall x^0 \exists y^\rho A(x,y) \to \exists f^{\rho(0)} \forall x^0 A(x, f(x)),$$

has a functional interpretation in the bar recursive functionals.
In order to approach this, let us first consider the Kuroda negative translation (from chapter 10) $(AC^0)'$ of AC^0 which – over WE-HA$^\omega$ – is equivalent to

$$\forall x^0 \neg\neg \exists y^\rho A^*(x,y) \to \neg\neg \exists f^{\rho(0)} \forall x^0 \neg\neg A^*(x, f(x)),$$

where A^* is defined as in definition 10.1.
This is derivable in WE-HA$^\omega$+AC^0+DNS, where

$$\text{DNS}: \; \forall x^0 \neg\neg A(x) \to \neg\neg \forall x^0 A(x)$$

is the so-called double-negation-shift schema which is intuitionistically underivable. The problem of solving the functional interpretation of $(AC^0)'$, therefore, reduces to finding a solution for the functional interpretation DNS^D of DNS. We first have to compute DNS^D: Let $\exists \underline{a} \forall \underline{b} A_D(x, \underline{a}, \underline{b})$ be the functional interpretation of $A(x)$. Then

$$\begin{aligned}
&\text{DNS}^D \equiv \\
&\Big(\forall x^0 \neg\neg \exists \underline{a} \forall \underline{b} A_D(x,\underline{a},\underline{b}) \to \neg\neg \forall y^0 \exists \underline{u} \forall \underline{v} A_D(y,\underline{u},\underline{v})\Big)^D \equiv \\
&\Big(\forall x^0 \exists \underline{A} \forall \underline{B} \neg\neg A_D(x, \underline{A}\,\underline{B}, \underline{B}(\underline{A}\,\underline{B})) \to \neg\neg \exists \underline{u} \forall y^0, \underline{v} A_D(y, \underline{u}y, \underline{v})\Big)^D \equiv \\
&\left(\begin{array}{l} \exists \underline{A} \forall x^0, \underline{B} \neg\neg A_D(x, \underline{A}x\underline{B}, \underline{B}(\underline{A}x\underline{B})) \to \\ \qquad\qquad \exists \underline{U} \forall Y, \underline{V} \neg\neg A_D(Y(\underline{U}Y\underline{V}), \underline{U}Y\underline{V}(Y(\underline{U}Y\underline{V})), \underline{V}(\underline{U}Y\underline{V})) \end{array}\right)^D \equiv \\
&\left(\begin{array}{l} \forall \underline{A} \exists \underline{U} \forall Y, \underline{V} \exists x, \underline{B} \big(\neg\neg A_D(x, \underline{A}x\underline{B}, \underline{B}(\underline{A}x\underline{B})) \to \\ \qquad\qquad \neg\neg A_D(Y(\underline{U}Y\underline{V}), \underline{U}Y\underline{V}(Y(\underline{U}Y\underline{V})), \underline{V}(\underline{U}Y\underline{V}))\big) \end{array}\right)^D.
\end{aligned}$$

So by the final step in the functional interpretation of an implication, noticing that $\underline{U}$ only occurs with the arguments $Y, \underline{V}$, we have to construct terms $t_x, t_{\underline{U}}$ and $t_{\underline{B}}$ that only contain $\underline{A}, Y, \underline{V}$ free and satisfy

$$(*)\; \forall \underline{A}, Y, \underline{V} \big(\neg\neg A_D(t_x, \underline{A}(t_x, t_{\underline{B}}), t_{\underline{B}}(\underline{A}(t_x, t_{\underline{B}}))) \to \neg\neg A_D(Y(t_{\underline{U}}), t_{\underline{U}}(Y(t_{\underline{U}})), \underline{V}(t_{\underline{U}}))\big).$$

We achieve this by constructing such terms satisfying the following system of equations (for all $\underline{A}, Y, \underline{V}$):

$$(+)\begin{cases} t_x = Y(\underline{t_U}) \\ \underline{A}(t_x, \underline{t_B}) = \underline{t_U}(Y(\underline{t_U})) \\ \underline{t_B}(\underline{A}(t_x, \underline{t_B})) = \underline{V}(\underline{t_U}). \end{cases}$$

This even provides a solution to $(*)$ which is **independent** from the prime formulas occurring in A (and the free variables of A).

We now extend WE-HA$^\omega$ by new constants $\underline{B}^{\underline{\rho},\underline{\tau}}$ for simultaneous bar recursion together with the following defining axioms (writing n' for $n+1$):

$$(\mathrm{BR}_{\underline{\rho},\underline{\tau}}) :$$
$$\begin{cases} y(\overline{\underline{x}, n^0}) <_0 n \to B_i^{\underline{\rho},\underline{\tau}} y\underline{z}\underline{u}n\underline{x} =_{\tau_i} z_i n(\overline{\underline{x}, n}) \\ y(\overline{\underline{x}, n}) \geq_0 n \to B_i^{\underline{\rho},\underline{\tau}} y\underline{z}\underline{u}n\underline{x} =_{\tau_i} u_i(\lambda \underline{D}^{\underline{\rho}}.\underline{B}^{\underline{\rho},\underline{\tau}} y\underline{z}\underline{u}n'(\overline{\underline{x}, n} * \underline{D}))n(\overline{\underline{x}, n}) \end{cases}$$

for $i = 1, \ldots, k$, where

$$(\overline{\underline{x}, n})_j(k^0) =_{\rho_j} \begin{cases} x_j(k), \text{ if } k < n \\ 0^{\rho_j}, \text{ otherwise} \end{cases}$$

and

$$(\overline{\underline{x}, n} * \underline{D})_j(k^0) =_{\rho_j} \begin{cases} x_j(k), \text{ if } k < n \\ D_j, \text{ if } k = n \\ 0^{\rho_j}, \text{ otherwise.} \end{cases}$$

The union of all these axioms for all $\underline{\rho}, \underline{\tau} \in \mathbf{T}$ is denoted by (BR). Using the fact (discussed already in chapter 3) that we could contract tuples of variables of any types into single variables we could actually also assume without loss of generality that we have only single variables and terms in $(+)$ and $(*)$ as we could have contracted the tuples $\underline{a}, \underline{b}, \underline{u}, \underline{v}$ stemming from the functional interpretation of A into single variables a, b, u, v.

So it would suffice to solve

$$(+)\begin{cases} t_x =_0 Y(t_U) \\ A(t_x, t_B) =_\rho t_U(Y(t_U)) \\ t_B(A(t_x, t_B)) =_\tau V(t_U) \end{cases}$$

using the bar recursor $B_{\rho,\rho 0}$, where $B_{\rho,\tau}$ is characterized by

$$(\mathrm{BR}_{\rho,\tau}) : \begin{cases} y(\overline{x,n^0}) <_0 n \rightarrow B_{\rho,\tau}yzunx =_\tau zn(\overline{x,n}) \\ y(\overline{x,n}) \geq_0 n \rightarrow B_{\rho,\tau}yzunx =_\tau u(\lambda D^\rho . B_{\rho,\tau}yzun'(\overline{x,n} * D))n(\overline{x,n}), \end{cases}$$

where

$$(\overline{x,n})(k^0) =_\rho \begin{cases} x(k), \text{ if } k < n \\ 0^\rho, \text{ otherwise} \end{cases}$$

and

$$(\overline{x,n} * D)(k^0) =_\rho \begin{cases} x(k), \text{ if } k < n \\ D, \text{ if } k = n \\ 0^\rho, \text{ otherwise.} \end{cases}$$

We will, however, not follow this approach since we will consider extensions of our systems to new types in chapter 17 where the contraction of tuples would only be possible on the expense of adding certain product types. So officially we use simultaneous bar recursion as a primitive concept and solve directly the version of $(+)$ with tuples. Nevertheless, from now on we will – for notational simplicity – omit the tuple-notation in $(+)$ and the schema of simultaneous bar recursion, keeping in mind, however, that we have instead of single terms s,t actually tuples of terms $\underline{t},\underline{s}$ and that an equation $t = s$ stands for the conjunction of the equations

$$t_1 = s_1, \ldots, t_n = s_n$$

for tuples $t_1^{\rho_1}, \ldots, t_n^{\rho_n}$ and $s_1^{\rho_1}, \ldots, s_n^{\rho_n}$ of terms of length n.

Following closely [343] we actually will use only the following special form of bar recursion:

$$\Phi_\rho yunxm =_\rho \begin{cases} xm, \text{ if } m <_0 n \\ 0^\rho, \text{ if } m \geq_0 n \wedge y(\overline{x,n}) < n \\ \Phi_\rho yun'(\overline{x,n} * D_0)m, \text{ otherwise,} \end{cases}$$

where

$$D_0 =_\rho un(\lambda D^\rho . \Phi_\rho yun'(\overline{x,n} * D)).$$

Φ_ρ is primitive recursively definable from $B_{\rho,\rho(0)}$ by

$$\Phi_\rho yunx =_{\rho 0} B_{\rho,\rho 0} yz\tilde{u}nx,$$

where

$$znxm :=_\rho \begin{cases} xm, \text{ if } m < n \\ 0^\rho, \text{ otherwise} \end{cases}$$

and

$$\tilde{u}vnxm :=_\rho \begin{cases} xm, \text{ if } m < n \\ v(unv)m, \text{ otherwise.} \end{cases}$$

The following lemmas are first proved informally. We will then show how they can be derived in WE-HA$^\omega$+ (BR) and even qf-(WE-HA$^\omega$)+(BR), where qf-(WE-HA)$^\omega$ is a quantifier-free fragment of WE-HA$^\omega$ in the sense of remark 8.8.

Lemma 11.2. *Define*

$$x_0 := \Phi_\rho 0^0 0^{\rho 0} := \Phi_\rho yu 0^0 0^{\rho 0}.$$

Then

$$x_0 =_{\rho 0} \Phi_\rho z(\overline{x_0, z}) \textit{ for all } z^0.$$

Proof: We prove the claim by induction on z :
For case $z = 0$ it holds by definition since $\overline{x_0, 0^0} =_{\rho 0} 0^{\rho 0}$.
$z \mapsto z'$: By the induction hypothesis we have

$$x_0 =_{\rho 0} \Phi_\rho z(\overline{x_0, z}).$$

Case 1: $y(\overline{x_0, z}) < z$. Then

$$x_0 =_{\rho 0} \Phi_\rho z(\overline{x_0, z}) =_{\rho 0} \overline{x_0, z}$$

and, therefore,

$$\overline{x_0, z} =_{\rho 0} \overline{x_0, z'}.$$

Hence

$$y(\overline{x_0, z'}) =_0 y(\overline{x_0, z}) < z < z'$$

and so

$$\Phi_\rho z'(\overline{x_0, z'}) =_{\rho 0} \overline{x_0, z'} =_{\rho 0} \overline{x_0, z} =_{\rho 0} x_0.$$

Case 2: $y(\overline{x_0, z}) \geq z$. Then

$$x_0 =_{\rho 0} \Phi_\rho z(\overline{x_0, z}) =_{\rho 0} \Phi_\rho z'(\overline{x_0, z} * D_0).$$

In particular, $x_0 z =_\rho D_0$ and, therefore,

$$x_0 =_{\rho 0} \Phi_\rho z'(\overline{x_0, z'})$$

since $\overline{x_0, z'} =_{\rho 0} \overline{x_0, z} * D_0$. □

Lemma 11.3. *For x_0 as in lemma 11.2 and $n :=_0 yx_0$ we have that $y(\overline{x_0, n}) \geq n$.*

Proof: Assume that $y(\overline{x_0, z}) < n$. Then, by lemma 11.2,

$$x_0 =_{\rho 0} \Phi_\rho n(\overline{x_0, n}) =_{\rho 0} \overline{x_0, n}$$

and so

$$n > y(\overline{x_0,n}) = yx_0 = n$$

which is a contradiction. □

Lemma 11.4. *For* x_0, n *as in lemmas 11.2 and 11.3 we have*

$$x_0 n =_\rho un(\lambda D^\rho.\Phi_\rho n'(\overline{x_0,n} * D)).$$

Proof: Using lemmas 11.2 and 11.3 we get

$$x_0 n \overset{l.11.2}{=}_\rho \Phi_\rho n(\overline{x_0,n})n \overset{l.11.3}{=}_\rho \Phi_\rho n'(\overline{x_0,n} * D_0)n =_\rho D_0 =_\rho un(\lambda D^\rho.\Phi_\rho n'(\overline{x_0,n} * D)).$$

□

It is straightforward to see that the proof of lemma 11.2 can be formalized in E-HA$^\omega$+ (BR). The formalization of the proofs of lemmas 11.3 and 11.4 is then trivial. However, to see that the proof of lemma 11.2 (and consequently also of lemmas 11.3 and 11.4) can be formalized WE-HA$^\omega$+ (BR) is already not that easy and for qf-(WE-HA$^\omega$)+(BR) it is not trivial at all. Actually, there does not seem to exist a detailed proof in the literature: in [343] (p.16) it only is remarked that the proofs 'can be converted to formal proofs in Σ_4 without excessive effort'. In [266] (p.80), where a thorough treatment of Spector's result is given, it even is claimed that 'it seems that so far nobody has observed that a precise formalization of all details in Spector's proof ... using only $=_0$ requires an additional constructive ω-rule.' In [109] it is stated (without proof) in footnote 3 that 'Die in [266] verwendete zusätzliche ω-Regel ist vermeidbar, wenn man den in [266], S. 81–83, gegebenen informalen Beweis der entscheidenden (BR)-Eigenschaften formalisiert.'

Lemma 11.5. *Lemmas 11.2–11.4 can be carried out in* WE-HA$^\omega$+ (BR) *and even in* qf-(WE-HA$^\omega$)+ (BR).

Proof: As mentioned already, the only problematic case is the proof of lemma 11.2 which uses both higher induction as well as the full extensionality axiom to establish in the induction step that

$$\overline{x_0,z} =_{\rho 0} \overline{x_0,z'} \to y(\overline{x_0,z}) =_0 y(\overline{x_0,z'})$$

and

$$x_0 z =_\rho D_0 \to \Phi_\rho z'(\overline{x_0,z'}) =_{\rho 0} \Phi_\rho z'(\overline{x_0,z} * D_0).$$

The use of the extensionality axiom can be reduced to that of the quantifier-free rule of extensionality by proving by induction on z a claim different from the one in the proof of lemma 11.2, namely

$$(\%)\ \forall \varphi^{0\rho 0}(\varphi x_0 =_0 \varphi(\Phi_\rho z(\overline{x_0,z}))).$$

$(\%)$ implies the original claim by using $\varphi_{\underline{v}} x :=_0 x\underline{v}$, where for $\rho = 0\rho_k,\ldots,\rho_1$ we have $\underline{v} = v_1^{\rho_1},\ldots v_k^{\rho_k}$.
In this way the proof of lemma 11.2 can be carried out in WE-HA$^\omega$+(BR). As mentioned in remark 8.8, WE-HA$^\omega$ has a functional interpretation in qf-(WE-HA)$^\omega$.

The proof of this fact immediately extends to WE-HA$^\omega$+(BR) as (BR) are purely universal (resp. open) axioms which are interpreted by themselves. Since the statement of lemma 11.2 can be written as an open formula, functional interpretation applied to the proof of lemma 11.2 in WE-HA$^\omega$+(BR) yields a proof of this lemma already in qf-(WE-HA)$^\omega$+(BR). □

Theorem 11.6 (Spector [343]). *Let $A(\underline{a})$ be an arbitrary formula of $\mathcal{L}$(WE-PA$^\omega$) containing only the free variables $\underline{a}$. Then the following rule holds:*

$$\begin{cases} \text{WE-PA}^\omega + \text{QF-AC} + \text{AC}^0 \vdash A(\underline{a}) \ \textit{implies that} \\ \text{WE-HA}^\omega + (\text{BR}) \vdash \forall \underline{y} (A')_D(\underline{t}\,\underline{a}, \underline{y}, \underline{a}), \end{cases}$$

where $\underline{t}$ is a suitable tuple of closed terms of WE-HA$^\omega$+ (BR) *which can be extracted from a given proof of the assumption and A' denotes the negative translation of A.*
The verification can even be carried out in qf-(WE-HA$^\omega$)+(BR).

Proof: As discussed at the beginning of this chapter, WE-PA$^\omega$+QF-AC+AC^0 has a negative interpretation in WE-HA$^\omega$+QF-AC+AC^0+M$^\omega$+DNS and hence in WE-HA$^\omega$+AC+M$^\omega$+DNS. It, therefore, suffices to extend the proof of theorem 8.6 by showing that the functional interpretation of DNS can be solved by closed terms of qf-(WE-HA$^\omega$)+ (BR) (provably in that very theory). As mentioned already we will establish this by solving the systems of equations (+). The following terms (containing only A, Y, V free) provide a solution:

$$(1)\ t_x :=_0 Yx_0,\ t_U :=_{\rho 0} x_0,\ t_B :=_{\tau\rho} \lambda D^\rho.V(E_{Yx_0}D),$$

where

$$(2)\ x_0 :=_{\rho 0} \Phi_\rho Y u 0^0 0^{\rho 0}$$

and

$$(3)\ E_n :=_{\rho 0 \rho} \lambda D.\Phi_\rho Y u n'(\overline{x_0, n} * D)$$

with

$$(4)\ unv :=_\rho An(\lambda D.V(v(D))).$$

By lemma 11.4 we have

$$(5)\ x_0(Yx_0) = u(Yx_0)(E_{Yx_0}),$$

and lemma 11.2 yields (for $z := Yx_0$)

$$(6)\ x_0 =_{\rho 0} \Phi_\rho Y u z'(\overline{x_0, z} * (u(Yx_0)(E_{Yx_0}))) = E_{Yx_0}(u(Yx_0)(E_{Yx_0})).$$

Moreover, by (1), we get

$$(7)\ \underline{t_x} = Yx_0 = \underline{Y(t_U)}.$$

(1), (4) and (5) yield

$$(8)\ \underline{A(t_x,t_B)} \overset{(4),(1)}{=} u(t_x)(E_{Yx_0}) \overset{(1)}{=} u(Yx_0)(E_{Yx_0}) \overset{(5)}{=} x_0(Yx_0) \overset{(1)}{=} \underline{t_U(Y(t_U))}.$$

Finally, using (1), (6) and (8), we also solve the third equation of $(+)$:

$$(9)\ \underline{t_B(A(t_x,t_B))} \overset{(1)}{=} V(E_{Yx_0}(A(t_x,t_B))) \overset{(8)}{=} V(E_{Yx_0}(u(t_x)(E_{Yx_0}))) \overset{(6)}{=} Vx_0 \overset{(1)}{=} \underline{V(t_U)}.$$

By lemma 11.5 it is clear that the whole argument can be carried out in qf-(WE-HA$^\omega$)+(BR). □

Corollary 11.7. *Let $A(\underline{a})$ be an arbitrary sentence of $\mathscr{L}(\text{WE-PA}^\omega)$. Then the following rule holds:*

$$\begin{cases} \text{E-PA}^\omega + \text{QF-AC}^{1,0} + \text{AC}^0 \vdash A \ \ \textit{implies that} \\ \text{WE-HA}^\omega + (\text{BR}) \vdash \forall \underline{y}((A_e)')_D(\underline{t},\underline{y}), \end{cases}$$

where $\underline{t}$ is a suitable tuple of closed terms of WE-HA$^\omega$+ (BR) *which can be extracted from a given proof of the assumption. Here A_e is the elimination-of-extensionality translation of A from definition 10.44 and $(A_e)'$ the negative translation of A_e*
The verification can even be carried out in qf-(WE-HA$^\omega$)+(BR).

Proof: The corollary follows from theorem 11.6 combined with the elimination of extensionality procedure from chapter 10 (proposition 10.45) observing that $(\text{AC}^0)_e$ is again an instance of AC^0. □

11.2 Functional interpretation of dependent choice

We now prove (following [162] and – in particular – [266]) that (BR) even suffices for the functional interpretation of the negative translation of dependent choice DC. where DC:= $\{\text{DC}^{\underline{\rho}} : \underline{\rho} \in \mathbf{T}\}$ is defined as follows

$$\text{DC}^{\underline{\rho}} : \forall x^0, \underline{y}^{\underline{\rho}} \exists \underline{z}^{\underline{\rho}} A(x,\underline{y},\underline{z}) \to \exists \underline{f}^{\underline{\rho}(0)} \forall x^0 A(x,\underline{f}(x),\underline{f}(S(x))).$$

As before, we will most of the time suppress the tuple notation.

Remark 11.8. The formulation of DC above (first considered in [167] under the name (A.1), see also [266] where our formulation of DC is called 'dependent ω-choice' ωAC) combines the usual formulation of dependent choice DAC:= $\{\text{DAC}^\rho :$ $\rho \in \mathbf{T}\}$, where

$$\text{DAC}^\rho : \forall x^\rho \exists y^\rho A(x,y) \to \forall x^\rho \exists f^{\rho(0)} [f(0) =_\rho x \wedge \forall z^0 A(f(z), f(S(z)))]$$

and countable choice

$$\text{AC}^0 : \forall x^0 \exists y^\rho A(x,y) \to \exists f^{\rho(0)} \forall x^0 A(x, f(x))$$

which are both provable from DC relative to WE-PA$^\omega$.

Proof:

1) DC $\Rightarrow$ AC^0: Assume

$$\forall x^0 \exists z^\rho A(x,z).$$

Then by DC (treating '$\forall y^\rho$' as a dummy quantifier) we obtain a functional $f^{\rho(0)}$ such that

$$\forall x^0 A(x, f(S(x))).$$

Now define $f'(x) := f(S(x))$.

2) DC $\Rightarrow$ DAC: Assume

$$(*)\ \forall a^\rho \exists b^\rho A(a,b)$$

and define (for any y^ρ)

$$A_y(x^0,a,b) :\equiv \begin{cases} x =_0 0 \to A(y,b) \\ x \neq_0 0 \to A(a,b). \end{cases}$$

Then

$$\forall x^0 \forall a^\rho \exists b^\rho A_y(x,a,b)$$

(for $x = 0$ take b for y as a in $(*)$, and for $x \neq 0$ the claim is immediate from $(*)$). Hence we can apply DC to obtain (for any y) a functional $f^{\rho(0)}$ such that

$$\forall x^0 A_y(x, f(x), f(S(x))).$$

So, in particular,

$$A_y(0, f(0), f(S(0))), \text{ i.e. } A(y, f(1))$$

and

$$\forall x > 0 A_y(x, f(x), f(S(x))), \text{ i.e. } \forall x > 0 A(f(x), f(S(x))).$$

Now define $f'(0) := y$ and $f'(S(x)) := f(S(x))$.
Then

$$f'(0) = y,\ A(f'(0), f'(1)) \text{ and } \forall x > 0 A(f'(x), f'(S(x))),$$

i.e.

$$f'(0) = y \wedge \forall x^0 A(f'(x), f'(S(x))).$$

□

Theorem 11.9 (Howard [162], Luckhardt [266]). *Let $A(\underline{a})$ be an arbitrary formula of $\mathcal{L}$(WE-PA$^\omega$) containing only the free variables $\underline{a}$. Then the following rule holds:*

$$\begin{cases} \text{WE-PA}^\omega + \text{QF-AC} + \text{DC} \vdash A(\underline{a}) \ \textit{implies that} \\ \text{WE-HA}^\omega + (\text{BR}) \vdash \forall \underline{y}\, (A')_D(\underline{t}\,\underline{a}, \underline{y}, \underline{a}), \end{cases}$$

where $\underline{t}$ is a suitable tuple of closed terms of WE-HA$^\omega$+ (BR) *which can be extracted from a given proof of the assumption and A' denotes the negative translation*

of A.
The verification can even be carried out in qf-(WE-HA$^\omega$)+(BR).

Proof: As in the proof of theorem 11.6 it suffices to consider DC. By proposition 10.13, every formula $A(\underline{z})$ is (relative to WE-PA$^\omega$+ QF-AC) equivalent to a formula of the form $\exists \underline{a} \forall \underline{b} A_0(\underline{z}, \underline{a}, \underline{b})$, where A_0 is quantifier-free. It, therefore, suffices to consider

$$\text{DC-}\exists\forall: \ \forall x^0, \underline{y}^{\underline{\rho}} \exists \underline{z}^{\underline{\rho}}, \underline{a} \forall \underline{b} A_0(x, \underline{y}, \underline{z}, \underline{a}, \underline{b}) \to \exists \underline{f}^{\underline{\rho}0} \forall x^0 \exists \underline{a} \forall \underline{b} A_0(x, \underline{f}(x), \underline{f}(x'), \underline{a}, \underline{b}).$$

By introducing 'dummy' variables '$\underline{\tilde{a}}$' corresponding to '$\underline{a}$' this follows from DC-$\forall$ defined as

$$\forall x^0, y^\rho, \underline{\tilde{a}} \exists z^\rho, \underline{a} \forall \underline{b} A_0(x, y, \underline{\tilde{a}}, z, \underline{a}, \underline{b}) \to \exists f, \underline{g} \forall x^0, \underline{b} A_0(x, f(x), \underline{g}(x), f(x'), \underline{g}(x'), \underline{b}).$$

Hence we can restrict ourselves in fact to DC-$\forall$ whose negative translation is (using some renaming of variables to fit with the proof of theorem 11.6) equivalent to (suppressing once again the tuple notation)

$$(*): \ \forall x^0, c^\rho \neg\neg \exists a^\rho \forall b^\tau A_0(x, c, a, b) \to \neg\neg \exists u^{\rho 0} \forall y^0, v^\tau A_0(y, uy, uy', v).$$

So it suffices to solve the functional interpretation of $(*)$ which can be achieved in almost the same way as for DNS:

$$(*)^D \equiv$$
$$\left(\begin{array}{l} \forall \underline{A} \exists \underline{U} \forall Y, \underline{V} \exists x, \underline{B} (\neg\neg A_D(x, c, \underline{A}xc\underline{B}, \underline{B}(\underline{A}xc\underline{B})) \to \\ \qquad \neg\neg A_D(Y(\underline{UYV}), \underline{UYV}(Y(\underline{UYV})), \underline{UYV}(Y(\underline{UYV})'), \underline{V}(\underline{UYV}))) \end{array} \right)^D.$$

So (arguing as in the proof of theorem 11.6) we have to construct terms t_x, t_c, t_U and t_B that only contain A, Y, V free and satisfy

$$(+) \begin{cases} t_x =_0 Y(t_U) \\ t_c =_\rho t_U(Y(t_U)) \\ A(t_x, t_c, t_B) =_\rho t_U(Y(t_U)') \\ t_B(A(t_x, t_c, t_B)) =_\tau V(t_U). \end{cases}$$

Define

$$(1) \ Y'x :=_0 (Yx)', \ t_x :=_0 Yx_0, \ t_U :=_{\rho 0} x_0, \ t_B :=_{\tau\rho} \lambda D^\rho.V(E_{Y'x_0}D),$$

where

$$(2) \ x_0 :=_{\rho 0} \Phi_\rho Y' u 0^0 0^{\rho 0}$$

and

$$(3)\ t_c :=_\rho t_U(Y(t_U)),\ E_n :=_{\rho 0 \rho} \lambda D^\rho.\Phi_\rho Y' un'(\overline{x_0, n} * D)$$

with

$$(4)\ unv :=_\rho A(n \dot{-} 1, v0^\rho(n \dot{-} 1), \lambda D.V(v(D))).$$

By lemma 11.4 we have

$$(5)\ x_0(Y'x_0) = u(Y'x_0)(E_{Y'x_0}).$$

Define $z := Yx_0$. Then $z' = Y'x_0$.
Lemma 11.2 yields

$$(6)\ x_0 =_{\rho 0} \Phi_\rho Y' uz''(\overline{x_0, z} * (u(Y'x_0)(E_{Y'x_0})) = E_{Y'x_0}(u(Y'x_0)(E_{Y'x_0})).$$

The first two equations of $(+)$ follow immediately from (1) and (3).
(1), (4), (5) and the Φ-definition yield

$$\begin{aligned}
&\underline{A(t_x, t_c, t_B)} \overset{(1)}{=} A(z, x_0 z, t_B) \overset{\Phi\text{-def.}}{=} \\
(7)\ &A(z, \Phi_\rho Y' uz''(\overline{x_0, z'} * 0^\rho)(z), \lambda D^\rho.V(\Phi_\rho Y' uz''(\overline{x_0, z'} * D))) \overset{(4)}{=} \\
&uz'(\lambda D.\Phi_\rho Y' uz''(\overline{x_0, z'} * D)) \overset{(5)}{=} x_0(Y'x_0) \overset{(1)}{=} \underline{t_U(Y(t_U)')}.
\end{aligned}$$

$$(8)\ \underline{t_B(A(t_x, t_c, t_B))} \overset{(1)}{=} V(E_{z'}(A(t_x, t_c, t_B))) \overset{(8)}{=} V(E_{z'}(u(z')(E_{z'}))) \overset{(6)}{=} Vx_0 \overset{(1)}{=} \underline{V(t_U)}.$$

□

As shown in [266], the elimination-of-extensionality result from proposition 10.45 can also be applied to DC since

$$\text{WE-PA}^\omega + \text{DC} \vdash (\text{DC})_e,$$

where this time some classical logic has to be used. We leave the proof as an exercise and conclude:

Corollary 11.10. *Corollary 11.7 also holds with* AC^0 *replaced by* DC.

11.3 Functional interpretation of arithmetical comprehension

An important special case of comprehension over numbers is the schema of so-called arithmetical comprehension

$$\text{CA}^0_{ar} : \exists f^1 \forall x^0 (f(x) =_0 0 \leftrightarrow A_{ar}(x)),$$

where $A_{ar}(x)$ is an arbitrary **arithmetical** formula of $\mathcal{L}(\text{WE-PA}^\omega)$ (not containing f free but otherwise with arbitrary parameters of arbitrary types). Here a formula is called arithmetical if it contains only quantifiers for variables of type 0. This is equivalent to saying that $A_{ar} \in \Pi^0_\infty$ (where the arithmetical hierarchy Π^0_n is understood to permit parameters of higher types).

As was observed already by H. Weyl in [378] and further elaborated by P. Lorenzen [265] as well as in the context of reverse mathematics [338] and arithmetical comprehension suffices to carry out a substantial amount of ordinary mathematics (in particular analysis of separable spaces and algebra for countable structures). See also Feferman [100] for further modern developments along this theme.

It is easy to see that over WE-PA$^\omega$ (and weak fragments thereof) CA^0_{ar} is equivalent to the seemingly weaker single axiom

$$\Pi^0_1\text{-CA}: \ \forall f^{0(0)(0)} \exists g^1 \forall x^0 (g(x) =_0 0 \leftrightarrow \forall y^0 (f(x,y) =_0 0)).$$

Clearly, Π^0_1-CA is a special case of CA^0_{ar}. The converse follows by making iterated use of Π^0_1-CA (to climb-up the arithmetical hierarchy), λ-abstraction and pairing as well as the fact that we have characteristic functions for quantifier-free formulas in WE-PA$^\omega$ (proposition 3.17):

Proposition 11.11. *Each instance of the schema of arithmetical comprehension* CA^0_{ar} *is derivable in* WE-PA$^\omega + \Pi^0_1$-CA.

Proof: Exercise! □

Just as before in the case of full comprehension one shows that Π^0_1-CA (and hence CA^0_{ar}) is implied over WE-PA$^\omega$ by

$$\Pi^0_1\text{-AC}: \ \forall f^{0(0)(0)(0)} \big(\forall x^0 \exists y^0 \forall z^0 f(x,y,z) =_0 0 \rightarrow \exists g^1 \forall x^0, z^0 f(x,g(x),z) =_0 0\big).$$

Conversely, Π^0_1-AC follows from Π^0_1-CA and QF-AC^{0,0}.
Using this fact we conclude

Proposition 11.12. *Each instance of the schema of number choice* $\mathrm{AC}^{0,0}_{ar}$ *for arithmetical formulas*

$$\mathrm{AC}^{0,0}_{ar}: \ \forall x^0 \exists y^0 A_{ar}(x,y) \rightarrow \exists f^1 \forall x^0 A_{ar}(x, f(x)), \ A_{ar} \ \textit{arithmetical},$$

is derivable in WE-PA$^\omega + \Pi^0_1$-AC.

Proof: Exercise! □

In the following, let $(\mathrm{BR}_{0,1})$ be the restriction of (BR) to the bar recursor constant $B_{0,1}$.

Theorem 11.13. *Let* $A(\underline{a})$ *be an arbitrary formula of* $\mathcal{L}$(WE-PA$^\omega$) *containing only the free variables* $\underline{a}$. *Then the following rule holds:*

$$\begin{cases} \text{WE-PA}^\omega + \text{QF-AC} + \mathrm{AC}^{0,0}_{ar} \vdash A(\underline{a}) \ \textit{implies that} \\ \text{WE-HA}^\omega + (\mathrm{BR}_{0,1}) \vdash \forall \underline{y} (A')_D(\underline{t}\,\underline{a}, \underline{y}, \underline{a}), \end{cases}$$

where $\underline{t}$ *is a suitable tuple of closed terms of* WE-HA$^{\omega}$+ (BR$_{0,1}$) *which can be extracted from a given proof of the assumption and* A' *denotes the negative translation of* A.
The verification can be be carried out in qf-(WE-HA$^{\omega}$) +(BR$_{0,1}$).

Proof: By proposition 11.12 it is sufficient to consider Π^0_1-AC. Since '$\exists y^0 \forall z^0 f(x,y,z) =_0 0$' is its own functional interpretation, we only have (arguing as in the proof of theorem 11.6) to interpret instances of DNS where a and b (namely x, y) are both of type 0. Inspection of the proof of theorem 11.6 reveals that the functional interpretation of this special case can be solved already by (BR$_{0,1}$). $\square$

Next, we prove the counterpart of theorem 11.13 for the fragments $\widehat{\text{WE-PA}}^{\omega}{\restriction}$ and $\widehat{\text{WE-HA}}^{\omega}{\restriction}$ of WE-PA$^{\omega}$ and WE-HA$^{\omega}$:

Theorem 11.14. *Let* $A(\underline{a})$ *be an arbitrary formula of* $\mathscr{L}(\widehat{\text{WE-PA}}^{\omega}{\restriction})$ *containing only the free variables* $\underline{a}$. *Then the following rule holds:*

$$\begin{cases} \widehat{\text{WE-PA}}^{\omega}{\restriction} + \text{QF-AC} + \text{AC}^{0,0}_{ar} \vdash A(\underline{a}) \ \textit{implies that} \\ \widehat{\text{WE-HA}}^{\omega}{\restriction} + (\text{BR}_{0,1}) \vdash \forall \underline{y}\, (A')_D(\underline{t}\,\underline{a}, \underline{y}, \underline{a}), \end{cases}$$

where $\underline{t}$ *is a suitable tuple of closed terms of* $\widehat{\text{WE-HA}}^{\omega}{\restriction}$+ (BR$_{0,1}$) *which can be extracted from a given proof of the assumption and* A' *denotes the negative translation of* A.
The verification can be be carried out in qf-($\widehat{\text{WE-HA}}^{\omega}{\restriction}$) + (BR$_{0,1}$).

Proof: It is clear from the proof of theorem 11.13 that the solution of the functional interpretation of (the negative translation of) AC$^{0,0}_{ar}$ only needs closed terms in $\widehat{\text{WE-HA}}^{\omega}{\restriction}$+ (BR$_{0,1}$). So together with proposition 10.53 it follows that the functional interpretation of $\widehat{\text{WE-PA}}^{\omega}{\restriction}$ + QF-AC + AC$^{0,0}_{ar}$ can be carried out with closed terms in $\widehat{\text{WE-HA}}^{\omega}{\restriction}$+ (BR$_{0,1}$). What remains to be checked is that the verification of the solution in the case of AC$^{0,0}_{ar}$ can indeed be carried out already in $\widehat{\text{WE-HA}}^{\omega}{\restriction}$ + (BR$_{0,1}$) rather than WE-HA$^{\omega}$+(BR$_{0,1}$). Even for the case $\rho = 0$ at hand, the induction formula used in the formalization of the proof of lemma 11.2 as indicated in the proof of lemma 11.5 is too complicated (due to the quantifier '$\forall \varphi^1$') to have a functional interpretation in $\widehat{\text{WE-HA}}^{\omega}{\restriction}$.
However, for $\rho = 0$ we can use the original induction formula

$$(*)\ \forall z^0 (x_0 =_1 \Phi_1 z(\overline{x_0, z}))$$

from the proof of lemma 11.2 since in this case all the uses of extensionality can be derived from QF-ER. This follows from the fact that

$$x_0 z =_0 0^0 \rightarrow \overline{x_0, z} =_1 \overline{x_0, z'},$$

where

$$x_0 z =_0 0^0$$

is a quantifier-free formula.
The proof of $(*)$ is an instance of Π^0_1-IA which has a functional interpretation in qf-$(\widehat{\text{WE-HA}^\omega}\restriction)$ using R_0 as can easily be verified. □

Proposition 11.15. *Let $A \in \mathcal{L}(\text{PA})$ be a prenex sentence. Then the following rule holds:*

$$\begin{cases} \text{PA} \vdash A \\ \Rightarrow \text{ one can extract closed terms } \underline{\Phi} \text{ of } \widehat{\text{WE-HA}^\omega}\restriction + (\text{BR}_{0,1}) \text{ such that} \\ \widehat{\text{WE-HA}^\omega}\restriction + (\text{BR}_{0,1}) \vdash \underline{\Phi} \text{ n.c.i. } A. \end{cases}$$

Proof: PA can be embedded into $\widehat{\text{WE-PA}^\omega}\restriction + \text{QF-AC} + \text{AC}^{0,0}_{ar}$ just as in remark 10.10 since the schema of induction for arithmetical formulas is derivable from quantifier-free induction and CA^0_{ar}. The proposition now follows by applying theorem 11.14 to the Herbrand normal form A^H of A. □

Proposition 10.9 and proposition 11.15 provide two alternative ways of extracting functionals satisfying the no-counterexample interpretation of PA. In fact, both ways are optimal w.r.t. the class of type-2 functionals used: let T denote the set of all closed terms of WE-PA^ω, i.e. those terms which define in $\mathcal{S}^\omega$ the class of primitive recursive functionals in the sense of Gödel [133]. Let T_0 denote the closed terms of $\widehat{\text{WE-PA}^\omega}\restriction$, i.e. those terms that define in $\mathcal{S}^\omega$ the class of primitive recursive functionals in the sense of Kleene [195]. In [215] it is shown that T and $T_0 + B_{0,1}$ (i.e. the closed terms of $\widehat{\text{WE-PA}^\omega}\restriction + (\text{BR}_{0,1})$) define the same class of functionals of type 2 in $\mathcal{S}^\omega$, where in the case of closed terms t^2 in $\widehat{\text{WE-PA}^\omega}\restriction + (\text{BR}_{0,1})$ we take as our interpretation $[t]_{\mathcal{M}^\omega} \in S_2$ since $\mathcal{M}^\omega$ is a model of bar recursion whereas $\mathcal{S}^\omega$ is not (see the next section). Together with the previous theorem this implies that the provable recursive function(al)s of type 2 of $\widehat{\text{WE-PA}^\omega}\restriction + \text{QF-AC}+\text{AC}^{0,0}_{ar}$ are primitive recursive in the sense of Gödel. Using the fact that $\widehat{\text{WE-PA}^\omega}\restriction + \text{QF-AC}+\text{AC}^{0,0}_{ar}$ proves induction for arbitrary arithmetical formulas one – conversely – can show that all primitive recursive functionals of type 2 (in the sense of Gödel) are provably recursive in $\widehat{\text{WE-PA}^\omega}\restriction + \text{QF-AC}+\text{AC}^{0,0}_{ar}$. For more details we refer to [215].

In Safarik [319], a detailed analysis of the ND-interpretation of the Bolzano-Weierstraß property of sequences in $[0,1]^d$ in terms of bar recursion of lowest type is given. The Bolzano-Weierstraß property can be proved using an extension Π^0_2-WKL of WKL where the tree predicate is Π^0_2 rather than quantifier-free as in WKL. Using Σ^0_1-CA (or equivalently Π^0_1-CA) this can be reduced to Π^0_1-WKL which in turn reduces to WKL (the latter fact is well-known, see e.g. [338]). By Howard [164], WKL has a functional interpretation using a weak binary form of bar recursion of type 0 which is trivially majorizable. As we saw above, the functional interpretation of Π^0_1-

CA only requires $B_{0,1}$ relative to T_0 (i.e. relative to the closed terms of $\widehat{\text{WE-PA}}^{\omega}\!\upharpoonright$). Based on this, Safarik constructs an explicit solution of the monotone functional interpretation (more precisely the NMD-interpretation) of the Bolzano-Weierstraß property for $[0,1]^d$ by terms in $T_0 + B_{0,1}$ containing only an unnested use of $B_{0,1}$. Together with results in Schwichtenberg [327] on the closure of Gödel's T under the rule of bar recursion of type 0 and results from [215], Safarik shows using this that the (monotone) functional interpretation of proofs of Π^0_2-theorems using fixed single instances of the Bolzano-Weierstraß principle relative to $\widehat{\text{WE-PA}}^{\omega}\!\upharpoonright$ yields a T_1-definable bound (and hence witness by bounded search). In chapter 13 below we will show (following Kohlenbach [208, 210]) that over $\text{G}_\infty\text{A}^\omega$ one even obtains T_0-definable bounds.

11.4 Functional interpretation of (IPP) by finite bar recursion

Following Oliva [293], we now show how a finite version $B^{0(0)(0)}_{fin}$ of $B_{0,1}$ can be used to give a short description of the solution for the ND-interpretation of (IPP) discussed already in chapter 10: recall the formulation of $(\text{IPP})^{ND}$:

$$\forall n \in \mathbb{N} \forall f : \mathbb{N} \to C_n \forall K : C_n \times \mathbb{N}^{\mathbb{N}} \to \mathbb{N} \exists i \leq n \exists g : \mathbb{N} \to \mathbb{N}$$
$$\big(g(K(i,g)) \geq K(i,g) \wedge f(g(K(i,g))) = i\big),$$

where, again, '$\forall f : \mathbb{N} \to C_n\ A(f)$' is represented as '$\forall f^1\ A(f_n)$' with $f_n(k) := \min\{f(k), n\}$ and '$\forall K : C_n \times \mathbb{N}^{\mathbb{N}} \to \mathbb{N}$' as '$\forall K : \mathbb{N} \times \mathbb{N}^{\mathbb{N}} \to \mathbb{N}$', i.e. as '$\forall K^{0(1)(0)}$'. As shown in chapter 10 we have to find $x_0, \ldots, x_n \in \mathbb{N}$ and $g_0, \ldots, g_n \in \mathbb{N}^{\mathbb{N}}$ such that

$$(+)\ x_i = K(i, g_i) \wedge g_i(x_i) = \max\{x_0, \ldots, x_n\} \ \ (0 \leq i \leq n).$$

Then we can put

$$I(n,f,K) := i := f(g_0(K(0,g_0))) = \ldots = f(g_n(K(n,g_n))) \wedge G(n,f,K) := g := g_i.$$

Solution of $(+)$: let $B_{fin} : (\mathbb{N} \times \mathbb{N}^{\mathbb{N}} \to \mathbb{N}) \times \mathbb{N} \times \mathbb{N} \to \mathbb{N}$
(i.e. of type $0(0)(0)(0(1)(0))$) satisfy

$$(\text{BR}_{fin}) :\ B_{fin}(K,n,s) =_0 \begin{cases} \langle\rangle, \text{ if } lth(s) \geq n+1 \\ \langle c_s \rangle * B_{fin}(K,n,s * \langle c_s \rangle), \text{ otherwise,} \end{cases}$$

where

$$c_s := K(lth(s), h_s) \wedge h_s := \lambda x.M(s * \langle x \rangle * B_{fin}(K,n,s * \langle x \rangle)),$$

with

$$M(k) := \max\{(k)_0, \ldots, (k)_{lth(k)-1}\}.$$

In the following we omit the argument K.
Now define $\langle x_0,\dots,x_n\rangle := B_{fin}(n,\langle\rangle)$ (i.e. $x_i := (B_{fin}(n,\langle\rangle))_i$ for $i \le n$).
By induction on i one shows that
Claim:

$$\forall i \le n+1\,(\langle x_i,\dots,x_n\rangle = B_{fin}(n,\langle x_0,\dots,x_{i-1}\rangle))$$

(here for $i = n+1$ resp. $i = 0$, $\langle x_i,\dots,x_n\rangle$ resp. $\langle x_0,\dots,x_{i-1}\rangle$ denote $\langle\rangle$).
Proof of claim: $i = 0$: Immediate from the definition of $\langle x_0,\dots,x_n\rangle$.
$i \mapsto i+1 \le n+1$: By I.H.

$$\langle x_i,\dots,x_n\rangle = \langle c_{\langle x_0,\dots,x_{i-1}\rangle}\rangle * B_{fin}(n,\langle x_0,\dots,x_{i-1}\rangle * \langle c_{\langle x_0,\dots,x_{i-1}\rangle}\rangle).$$

Hence $x_i = c_{\langle x_0,\dots,x_{i-1}\rangle}$ and $B_{fin}(n,\langle x_0,\dots,x_i\rangle) = \langle x_{i+1},\dots x_n\rangle$.
End of proof of claim.
By the claim we, in particular, have that $x_i = c_{\langle x_0,\dots,x_{i-1}\rangle}$ for $0 \le i \le n$.
Now define $g_i := h_{\langle x_0,\dots,x_{i-1}\rangle}$ for $0 \le i \le n$. Then

$$x_i = c_{\langle x_0,\dots,x_{i-1}\rangle} = K(i,h_{\langle x_0,\dots,x_{i-1}\rangle}) = K(i,g_i)$$

and

$$\begin{aligned} &g_i(x_i) = h_{\langle x_0,\dots,x_{i-1}\rangle}(x_i) = \\ &M(\langle x_0,\dots,x_{i-1}\rangle * \langle x_i\rangle * B_{fin}(n,\langle x_0,\dots,x_{i-1}\rangle * \langle x_i\rangle)) \\ &\overset{\text{Claim}}{=} M(\langle x_0,\dots,x_n\rangle) = \max\{x_0,\dots,x_n\} \end{aligned}$$

for all $0 \le i \le n$.

11.5 Models of bar recursion

WE-HA$^\omega$+(BR) proves that

$$(1)\ \forall y^{0(\rho 0)}, x^{\rho 0} \exists n^0 (y(\overline{x,n}) < n).$$

In fact, by a closed term of WE-HA$^\omega$+(BR) one can define a μ-operator satisfying the following axioms

$$(2)\ y(\overline{x,\mu xy}) < \mu xy \wedge (n < \mu xy \to y(\overline{x,n}) \ge n)$$

for all y,x,n of appropriate types (exercise).
As mentioned already in an exercise to chapter 3, there is no such functional in $\mathscr{S}^\omega$ for the simple reason that (1) does not hold in $\mathscr{S}^\omega$. Hence $\mathscr{S}^\omega$ a-fortiori is not a model of bar recursion.

As shown by Scarpellini ([321]), the type structure $\mathscr{C}^\omega$ is a model of E-PA$^\omega$+(BR).

For the details we refer to [321] and only indicate here why (1) holds in $\mathcal{C}^\omega$: Let $x \in C_{\rho 0}, \widehat{x} \in C_\rho$. Furthermore, let (k_n) be a sequence in C_1 such that $k_n \to_0 k$ for some $k \in \mathbb{N}$. Then $\forall n \geq n_0(k_n = k)$ for some n_0. Hence

$$(\overline{x,n})(k_n) = (\overline{x,n})(k) = x(k) = x(k_n)$$

for all $n \geq \max(k+1, n_0)$. This implies that

$$\overline{x,n} \to_\rho \widehat{x}.$$

Thus

$$y(\overline{x,n}) \to_0 y\widehat{x}, \text{ i.e. } \exists N \forall n \geq N(y(\overline{x,n}) = y\widehat{x}).$$

Now (1) is satisfied with $n := \max(N, y\widehat{x}+1)$.

For the type structure ECF^ω the following was proved first in [366]:

Proposition 11.16. ECF^ω *is a model of* E-PA^ω+(BR). *Moreover, as interpretations of the closed terms of* E-PA^ω+(BR) *we can choose computable associates* $\alpha \in \mathbb{N}^\mathbb{N}$.

For many years all models of bar recursion (see the historical comments at the end of this chapter) made use of continuity to establish (1). However, in 1985 M. Bezem proved ([27]) that his model $\mathcal{M}^\omega$ of all strongly majorizable functionals is a model of bar recursion (for bar recursion of type 0 this essentially is due to W.A. Howard already) despite of the fact that it contains discontinuous functionals such as

$$\varphi^2(f^1) := \begin{cases} 1, \text{ if } \exists n(f(n) = 0) \\ 0, \text{ otherwise.} \end{cases}$$

Since (an extension of) this model will play a crucial role in chapter 17 we give here a detailed proof of this result:

Theorem 11.17 (Bezem [27]). $\mathcal{M}^\omega$ *is a model of* E-PA^ω+(BR).

Proof: By proposition 3.69 it suffices to show that $\mathcal{M}^\omega \models$ (BR). The proof will be based on the following form of dependent choice which in the intuitionistic literature is discussed under the name of 'bar induction' (see [266] and [366] for a thorough discussion of various forms of bar induction):

$$(\mathrm{BI}) : \begin{cases} \forall x \in M_{\rho 0} \exists n_0 \in \mathbb{N} \forall n \geq n_0\, Q(\overline{x,n}; n) \wedge \\ \forall x \in M_{\rho 0}, n \in \mathbb{N} \big(\forall D \in M_\rho\, Q(\overline{x,n} * D; n') \to Q(\overline{x,n}; n)\big) \\ \to \forall x \in M_{\rho 0}, n \in \mathbb{N}\; Q(\overline{x,n}; n). \end{cases}$$

To show that (BI) holds in $\mathcal{M}^\omega$ we argue by contraposition: Suppose that

$$\exists x_0 \in M_{\rho 0}, n_0 \neg Q(\overline{x_0, n_0}; n_0)$$

and

$$\forall x \in M_{\rho 0}, n \in \mathbb{N}\big(\forall D \in M_\rho\, Q(\overline{x,n} * D; n') \to Q(\overline{x,n}; n)\big),$$

i.e.

$$\forall x \in M_{\rho 0}, n \in \mathbb{N}\big(\neg Q(\overline{x,n}; n) \to \exists D \in M_\rho \neg Q(\overline{x,n} * D; n')\big).$$

Then (using dependent choice on the meta-level) there exists a sequence $\tilde{x}$ (extending $\bar{x}_0 n_0$) in $M_\rho^{M_0}$ with

$$\forall n \geq n_0 \neg Q(\overline{\tilde{x}, n}; n).$$

By corollary 3.68 $\tilde{x} \in M_{\rho 0}$ which contradicts the first premise of (BI). Hence $\mathscr{M}^\omega$ is a model of (BI).
We now define the following candidate for our majorant of $B_{\rho,\tau}$:

$$B^*_{\rho,\tau} := \lambda y, z, u, n, x.(B_{\rho,\tau} y^m z^m u_z)^M nx,$$

where (using for z the notation z^m instead of z^M since the latter is already defined differently)

$$y^m(x) := y(x^M),\; z^m nx := znx^M,\; u_z := \lambda v, n, x. \max(znx^M, uvnx^M)$$

using the construction $x \mapsto x^M$ from definition 3.65 for sequence types $\sigma 0$. As an intermediate functional we consider

$$B^p_{\rho,\tau} := \lambda y, z, u, n, x. B_{\rho,\tau} y^m z^m u_z nx.$$

Let $y^*, y \in M_{0(\rho 0)}, z^*, z \in M_{\tau(\rho 0)0}, u^*, u \in M_{\tau(\rho 0)0(\tau\rho)}$ be such that y^* *s-maj* y, z^* *s-maj* z and u^* *s-maj* u.
Define

$$Q(x;n) :\equiv$$
$$\forall \tilde{x} \in M_{\rho 0}\big(\forall k < n(xk\, s\text{-}maj_\rho\, \tilde{x}k) \to B^p y^* z^* u^* nx\; s\text{-}maj_\tau\; B^p y^* z^* u^* n\tilde{x}, B^p yzun\tilde{x}, Byzun\tilde{x}\big).$$

We will show – using (BI) – that $\forall x \in M_{\rho 0}, n\, Q(\overline{x,n}; n)$. Since for all $x \in M_{\rho 0}$ there exists an $x^* \in M_{\rho 0}$ with x^* *s-maj* x (lemma 3.63.1) and, therefore, $\forall k(x^*k \text{ } s\text{-}maj \text{ } xk)$ this implies that

$$\forall x \in M_{\rho 0}(B^p y^* z^* u^* nx, B^p yzunx, Byzunx \in M_\tau)$$

and

$$\forall x^*, x \in M_{\rho 0}, n(x^* \; s\text{-}maj\; x \to B^p y^* z^* u^* nx^* \; s\text{-}maj\; B^p y^* z^* u^* nx, B^p yzunx, Byzunx).$$

This yields

$$\forall n(B^p y^* z^* u^* n \; s\text{-}maj\; B^p yzun, Byzun).$$

Lemma 3.66 implies that

$$B^* y^* z^* u^* = (B^p y^* z^* u^*)^M \; s\text{-}maj\; (B^p yzu)^M = B^* yzu, Byzu.$$

Using again lemma 3.63 we conclude that

$$B^* \ s\text{-}maj \ B \in M.$$

It, therefore, remains to show that $\forall x \in M_{\rho 0}, n \, Q(\overline{x,n};n)$:
(i) $\forall x \in M_{\rho 0} \exists n_0 \forall n \geq n_0 \, Q(\overline{x,n};n)$:
Let $x \in M_{\rho 0}$. Then $\exists x^* \in M_{\rho 0}$ such that x^* $s\text{-}maj$ x. Take $n_0 := y^*(x^*)^M + 1$. Then for all $n \geq n_0$

$$(y^*)^m(\overline{x,n}) = y^*(\overline{x,n}^M) < n$$

since $(x^*)^M$ $s\text{-}maj$ $(\overline{x,n})^M$ by lemma 3.66 and y^* $s\text{-}maj$ y^* by lemma 3.63.1).
Now let $n \geq n_0$ and $\tilde{x} \in M_{\rho 0}$ with $\forall k < n (xk \ s\text{-}maj \ \tilde{x}k)$. Then (using again lemma 3.66 as well as lemma 3.63.1)

$$n > (y^*)^m(\overline{x,n}) = y^*(\overline{x,n}^M) \geq \underbrace{y^*(\overline{\tilde{x},n}^M)}_{=(y^*)^m(\overline{\tilde{x},n})}, \underbrace{y(\overline{\tilde{x},n}^M)}_{=y^m(\overline{\tilde{x},n})}, y(\overline{\tilde{x},n})$$

and so

$$B^p y^* z^* u^* n(\overline{x,n}) = B(y^*)^m (z^*)^m u^*_{z^*} n(\overline{x,n}) = z^* n(\overline{x,n}^M)$$
$$\begin{cases} s\text{-}maj \ z^* n(\overline{\tilde{x},n}^M) = B^p y^* z^* u^* n(\overline{\tilde{x},n}) \\ s\text{-}maj \ zn(\overline{\tilde{x},n}^M) = B^p yzun(\overline{\tilde{x},n}) \\ s\text{-}maj \ zn(\overline{\tilde{x},n}) = Byzun(\overline{\tilde{x},n}). \end{cases}$$

(ii) $\forall x \in M_{\rho 0}, n \big(\forall D \in M_\rho \, Q(\overline{x,n} * D; n') \to Q(\overline{x,n};n)\big)$:
Let $n \in \mathbb{N}$ and $x, \tilde{x} \in M_{\rho 0}$ with $\forall k < n (xk \ s\text{-}maj \ \tilde{x}k)$. Also let $D^*, D \in M_\rho$ with D^* $s\text{-}maj$ D. Using that (with lemma 3.63.1)

$$\forall k \big((\overline{x,n} * D^*)(k) \ s\text{-}maj \ (\overline{x,n} * D)(k), (\overline{\tilde{x},n} * D)(k)\big)$$

it follows from $Q(\overline{x,n} * D^*; n')$ that

$$B^p y^* z^* u^* n'(\overline{x,n} * D^*) \begin{cases} s\text{-}maj \ B^p y^* z^* u^* n'(\overline{x,n} * D) \\ s\text{-}maj \ B^p y^* z^* u^* n'(\overline{\tilde{x},n} * D) \\ s\text{-}maj \ B^p yzun'(\overline{\tilde{x},n} * D) \\ s\text{-}maj \ Byzun'(\overline{\tilde{x},n} * D). \end{cases}$$

Hence

$$\lambda D. B^p y^* z^* u^* n'(\overline{x,n} * D) \begin{cases} s\text{-}maj \ \lambda D. B^p y^* z^* u^* n'(\overline{\tilde{x},n} * D) \\ s\text{-}maj \ \lambda D. B^p yzun'(\overline{\tilde{x},n} * D) \\ s\text{-}maj \ \lambda D. Byzun'(\overline{\tilde{x},n} * D). \end{cases}$$

By lemma 3.63.1) it, in particular, follows that all these functionals are in $\mathcal{M}^\omega$.

Case 1: $(y^*)^m(\overline{x,n}) = y^*(\overline{x,n}^M) < n$. Then

$$n > (y^*)^m(\overline{\tilde{x},n}), y^M(\overline{\tilde{x},n}), y(\overline{\tilde{x},n}).$$

Hence

$$B^p y^* z^* u^* nx = z^* n(\overline{x,n}^M) \begin{cases} s\text{-}maj\ z^* n(\overline{\tilde{x},n}^M) = B^p y^* z^* u^* n\tilde{x} \\ s\text{-}maj\ zn(\overline{\tilde{x},n}^M) = B^p yzun\tilde{x} \\ s\text{-}maj\ zn(\overline{\tilde{x},n}) = Byzun\tilde{x}. \end{cases}$$

Case 2: $(y^*)^m(\overline{x,n}) = y^*(\overline{x,n}^M) \geq n$. Then (using lemma 3.63.1)

$$\begin{aligned} &B^p y^* z^* u^* n(\overline{x,n}) = u^*_{z^*}(\lambda D.B^p y^* z^* u^* n'(\overline{x,n} * D))n(\overline{x,n}) = \\ &\max(z^* n(\overline{x,n}^M), u^*(\lambda D.B^p y^* z^* u^* n'(\overline{x,n} * D))n(\overline{x,n}^M) \\ &\begin{cases} s\text{-}maj\ \max(z^* n(\overline{\tilde{x},n}^M), u^*(\lambda D.B^p y^* z^* u^* n'(\overline{\tilde{x},n} * D))n(\overline{\tilde{x},n}^M)), z^* n(\overline{\tilde{x},n}^M) \\ s\text{-}maj\ \max(zn(\overline{\tilde{x},n}^M), u(\lambda D.B^p yzun'(\overline{\tilde{x},n} * D))n(\overline{\tilde{x},n}^M)), zn(\overline{\tilde{x},n}^M) \\ s\text{-}maj\ zn(\overline{\tilde{x},n}), u(\lambda D.Byzun'(\overline{\tilde{x},n} * D))n(\overline{\tilde{x},n}). \end{cases} \end{aligned}$$

Hence

$$B^p y^* z^* u^* n(\overline{x,n}) \begin{cases} s\text{-}maj\ B^p y^* z^* u^* n(\overline{\tilde{x},n}) \\ s\text{-}maj\ B^p yzun(\overline{\tilde{x},n}) \\ s\text{-}maj\ Byzun(\overline{\tilde{x},n}) \end{cases}$$

and so $Q(\overline{x,n};n)$.
By (BI), (i) and (ii) imply that

$$\forall x \in M_{\rho 0}, n\, Q(\overline{x,n}, n)$$

which finishes the proof. □

Corollary to the proof: For any closed term t^ρ of E-PA$^\omega$+(BR) one can construct a closed term t^* of E-PA$^\omega$+(BR) such that

$$[t^*]_{\mathcal{M}^\omega}\ s\text{-}maj_\rho\ [t]_{\mathcal{M}^\omega}.$$

In view of remark 3.62 we can also write this as

$$\mathcal{M}^\omega \models t^*\ s\text{-}maj_\rho\ t,$$

where now $s\text{-}maj_\rho$ refers to the syntactic relation from definition 3.48.

Corollary 11.18. *Let $A(\underline{a})$ be an arbitrary formula of $\mathcal{L}$(WE-PA$^\omega$) containing only the free variables $\underline{a}$. Then the following rule holds:*

$$\begin{cases} \text{WE-PA}^\omega + \text{QF-AC} + \text{DC} \vdash A(\underline{a}) \text{ implies that} \\ \mathcal{M}^\omega \models \forall \underline{y} (A')_D(\underline{t}\,\underline{a}, \underline{y}, \underline{a}), \end{cases}$$

where $\underline{t}$ is a suitable tuple of closed terms of WE-HA$^\omega$+ (BR) *which can be extracted from a given proof of the assumption and A' denotes the negative translation of A.*

Proof: By theorem 11.9 and theorem 11.17. □

Proposition 11.19. *Proposition 3.71 extends for $\mathcal{C}^\omega$,* ECF$^\omega$ *and $\mathcal{M}^\omega$ (but not for $\mathcal{S}^\omega$) to closed terms of type degree ≤ 2 of* E-PA$^\omega$+(BR).

Proof: One only has to extend the proof of proposition 3.71 by showing that

$$[B_{\rho,\tau}]_{\mathcal{C}^\omega} \approx [B_{\rho,\tau}]_{\text{ECF}^\omega} \approx [B_{\rho,\tau}]_{\mathcal{M}^\omega}$$

for the corresponding logical relations $\approx$. This is done using (BI) in a way similar (but much simpler) to the use of (BI) in the majorization proof of $B_{\rho,\tau}$. Exercise. □

Corollary 11.20. *Let t^2 be a closed term of* E-PA$^\omega$+(BR) *of type 2. Then $[t]_{\mathcal{M}^\omega} \in S_2$ is a computable functional, i.e. there exists a code e of an oracle Turing machine such that*

$$\forall f \in \mathbb{N}^{\mathbb{N}} (\{e\}(f) \simeq [t]_{\mathcal{M}^\omega}(f)).$$

Proof: Immediate from propositions 11.16, 11.19 and the comment after definition 3.58. □

11.6 Exercises, historical comments and suggested further reading

Exercises:

1) ([266]) Consider the following variants of (BR):

$$(\text{BR}^-_{\rho,\tau}):\ \begin{cases} y(\overline{x,n^0}) <_0 n \rightarrow \text{B}^-_{\rho,\tau} yzunx =_\tau z \\ y(\overline{x,n}) \geq_0 n \rightarrow \text{B}^-_{\rho,\tau} yzunx =_\tau u(\lambda D^\rho . \text{B}^-_{\rho,\tau} yzun'(\overline{x,n} * D)), \end{cases}$$

and

$$(\text{BR}^+_{\rho,\tau}):\ \begin{cases} y(\overline{x,n^0}) <_0 n \rightarrow \text{B}^+_{\rho,\tau} yzunx =_\tau znx \\ y(\overline{x,n}) \geq_0 n \rightarrow \text{B}^+_{\rho,\tau} yzunx =_\tau u(\lambda D^\rho . \text{B}^+_{\rho,\tau} yzun'(\overline{x,n} * D))nx, \end{cases}$$

Show that over WE-HA$^\omega$ the schemata (BR$^-$), (BR) and (BR$^+$) are equivalent (though the type τ might change).

2) Prove proposition 11.11.
3) Prove proposition 11.12.
4) Compare Spector's solution of the functional interpretation of DNS with the functional interpretation of the intuitionistically valid schema

$$\neg\neg A \wedge \neg\neg B \rightarrow \neg\neg(A \wedge B)$$

which only requires typed λ-terms (see the exercises to chapter 8).

5) Show (using induction on k) that

$$(*)\ \text{WE-HA}^\omega \vdash \forall k^0(\forall x \leq_0 k\neg\neg A(x,k) \rightarrow \neg\neg\forall x \leq_0 k A(x,k)).$$

Compare the functional interpretation of $(*)$ (by closed terms of WE-HA$^\omega$) with Spector's interpretation of DNS.

6) Show that the functional μ satisfying

$$x(\overline{y,\mu xy}) < \mu xy \wedge (n < \mu xy \rightarrow x(\overline{y,n}) \geq n)$$

is definable in WE-HA$^\omega$+(BR).

7) ([266]) Show that WE-PA$^\omega$+DC $\vdash$ (DC)$_e$.
8) ([164]) Solve the functional interpretation of the negative translation of WKL in $\widehat{\text{WE-HA}}^\omega\!\upharpoonright$+ (BR$_{0,1}$) (since WKL is provable in $\widehat{\text{WE-PA}}^\omega\!\upharpoonright$+ QF-AC+AC$^{0,0}_{ar}$ theorem 11.14 applies). Formulate the special 'binary' version of $B_{0,1}$ sufficient for this.
9) Complete the proof of proposition 11.19.
10) Construct majorants I^*, G^* for the solution I, G to the functional interpretation of (IPP) as given in section 11.4.

Historical comments and suggested further reading: Bar recursion was first considered in Spector's fundamental paper [343]. Spector's work was further improved and extended in Howard [162] and Luckhardt [266]. A nice motivation for Spector's solution of the functional interpretation of DNS by bar recursion is due to Oliva [293] on which also the material in section 11.4 is based. An intensional functional interpretation of analysis by bar recursion was given in Diller-Vogel [89]. A powerful extension of bar recursion to infinite types was developed in Friedrich [109] where – via a game quantifier interpretation as intermediate step – even a functional interpretation of comprehension over functions (of type 1) rather than numbers was given. In Kohlenbach [226] and Gerhardy-Kohlenbach [120] an extension of bar recursion to new types is used to interpret formal systems with abstract metric, hyperbolic and normed spaces added as 'Urelemente' (see chapter 17 below). More information on the proof theory of bar recursion can be found in Kreisel [248]. A detailed ordinal analysis of bar recursion of type 0 is given in Howard [164, 165]. Other ordinal information on (BR) is provided in Vogel [374]. In Schwichtenberg

[327] it is shown that the primitive recursive functionals in the sense of Gödel are closed under the rule of bar recursion of type 0 and 1. Bar recursion in the context of fragments of PA$^{\omega}$ is used in Kohlenbach [215, 216] and Oliva [291].
The first model of bar recursion was given by Scarpellini in [321] (in two variants) by his type structure of sequentially continuous functionals. Term models for bar recursion were first defined in Luckhardt [266] (constructively) and Tait [352] (classically). Inspired by [266], Scarpellini gave a more constructive refinement of his first model in [322]. In Troelstra [366] it is shown that the extensional Kleene/Kreisel continuous functionals ECF form a model of bar recursion. In the important paper Bezem [27], the model of strongly majorizable functionals is introduced and shown to be a model of bar recursion. Bezem's proof proceeds by first showing that the usual bar recursors are primitive recursively equivalent to a variant of bar recursion and then majorizing the latter. In Kohlenbach [200] majorization is shown directly for Spector's bar recursors by first establishing a so-called pointwise majorizability and then using a general procedure for converting pointwise majorants to strong majorants. The proof given in this chapter resulted from that strategy where the latter procedure is directly implemented into the definition of B^*.
Strong normalization of the bar recursive functionals is proved in Vogel [373] and (without using infinite terms) in Bezem [28]. Simultaneous bar recursion is considered in Luckhardt [266] and Diller-Vogel [89]. In Bezem [29] the equivalence of a number of known variants of bar recursion is established. Kohlenbach [200] (chapter III) introduces various new forms of bar recursion which are not equivalent to (BR) and based on different bar conditions. Yet another new version of bar recursion (called 'modified bar recursion') is used in Berger-Oliva [22] to solve the Friedman A-translation of the modified realizability interpretation of the negative translation of dependent choice (see also chapter 14). That version is motivated by constructions in [17]. The relationship between modified bar recursion, Spector's bar recursion and the variant introduced by Kohlenbach in [200] (chapter III) is clarified in Berger-Oliva [23]. Yet another approach to the axiom of dependent choice is Krivine's realizability developed in [253] which has some relations to the approach in Berger-Oliva [22] (see Oliva-Streicher [295] for a discussion of this point).
Other approaches to systems based on restricted forms of comprehension based on ineffective uses of functional interpretation resp. on cut-elimination can be found in Feferman [98] and Takeuti [356] respectively. An alternative functional interpretation of classical analysis based on a polymorphic λ-calculus was given in Girard [121].

Chapter 12
A non-standard principle of uniform boundedness

12.1 The Σ^0_1-boundedness principle

In intuitionistic mathematics certain axioms are used which are classically inconsistent but are valid under an appropriate constructive interpretation. Examples are axioms stating that all functions $f : \mathbb{N}^{\mathbb{N}} \to \mathbb{N}^{\mathbb{N}}$ resp. $f : 2^{\mathbb{N}} \to \mathbb{N}$ are continuous resp. uniformly continuous. Such axioms sometimes allow one to simplify proofs even for classically valid theorems. One of the principles behind such axioms is the so-called fan principle which has been considered in various different forms in intuitionistic mathematics. The simplest form is just the contraposition of the binary König's lemma which can be written in the following form (where we denote for simplicity the constant-1 function $\lambda x^0.1^0$ by 1):

$$\mathrm{FAN_{KL}} : \; T(f) \wedge \forall b \leq_1 1 \exists x^0 \big(f(\overline{b}x) \neq_0 0\big) \to \exists x^0 \forall b \leq_1 1 \exists \tilde{x} \leq x \big(f(\overline{b}\tilde{x}) \neq_0 0\big),$$

where $T(f)$ expresses that f is the characteristic function of a binary tree and is defined as in chapter 7.

In this form the fan principle, of course, is classically valid as WKL is. However, in intuitionistic mathematics one also considers the following much more general form

$$\mathrm{FAN} : \forall f \leq_1 1 \exists x^0 A(f,x) \to \exists x^0 \forall f \leq_1 1 \exists \tilde{x} \leq_0 x A(f,\tilde{x})$$

where the property

$$f(\overline{b}x) \neq_0 0$$

is replaced by an arbitrary formula A.

In the presence of a minimal amount of arithmetic, FAN is inconsistent with classical logic. FAN applied to the logically (classically) valid sentence

$$\forall f \leq_1 1 \exists x^0 \big(\exists y^0 (f(y) =_0 0) \to f(x) =_0 0\big)$$

yields

$$\exists x^0 \forall f \leq_1 1 \big(\exists y (f(y) =_0 0) \leftrightarrow \exists y \leq x (f(y) =_0 0)\big)$$

which obviously is false.
In contrast to $\mathrm{FAN_{KL}}$, the property $A(f,x) :\equiv \big(\exists y(f(y) =_0 0) \to f(x) =_0 0\big)$ to which FAN is applied in this counterexample is not quantifier-free (and not decidable). This leads us to consider the restriction of FAN to quantifier-free formulas

$$\text{QF-FAN} : \forall f \leq_1 1 \exists x^0 A_0(f,x) \to \exists x^0 \forall f \leq_1 1 \exists \tilde{x} \leq_0 x A_0(f,\tilde{x}),$$

where A_0 is a quantifier-free formula (of say E-PA^ω or $\widehat{\text{E-PA}}^\omega\!\upharpoonright$).
Is QF-FAN classically consistent? The answer to this question depends on whether higher type parameters are allowed to occur in A_0 or not. If A_0 contains at most parameters g of type ≤ 1, then one classically verify QF-FAN as follows: Let $A_0(g^1, f^1, x)$ be a quantifier-free formula containing only g, f, x free. Consider the function

$$F(f,g) := \begin{cases} \min x[A_0(g, \min_1(f, 1^1), x)], \text{ if existent} \\ \text{undefined, otherwise,} \end{cases}$$

where $\min_1(f,1)(x) := \min(f(x),1)$.
If $\forall f \leq_1 1 \exists x A_0(g,f,x)$ then $F(\cdot,g)$ is a total functional of type-2 which is computable in g since A_0 is decidable. By standard recursion theoretic arguments $F(\cdot,g)$, therefore, is uniformly continuous on the Cantor space $2^{\mathbb{N}}$ of all $0/1$-functions f. Hence F is bounded on that space which implies the conclusion of QF-FAN.
This argument breaks down for general A_0 with parameters of arbitrary types and in fact QF-FAN can be classically refuted as follows: Let $F : \mathbb{N}^{\mathbb{N}} \to \mathbb{N}$ be an arbitrary function. By extensionality and classical logic (in fact only the Markov principle for numbers is needed here) we get

$$\forall f, g \leq_1 1 \exists x^0 \big(f(x) =_0 g(x) \to F(f) =_0 F(g)\big).$$

QF-FAN yields (using that we can encode '$\forall f, g \leq_1 1$' as '$\forall h \leq_1 1$' by putting $f(x) := h(2x), g(x) := h(2x+1)$)

$$\exists x \forall f, g \leq_1 1 \big(\forall y \leq x (f(y) =_0 g(y)) \to F(f) =_0 F(g)\big),$$

i.e. the uniform continuity of F on the Cantor space which, of course, can classically be refuted by taking e.g.

$$F(f) := \begin{cases} 0, \text{ if } \exists x(f(x) =_0 0) \\ 1, \text{ otherwise.} \end{cases}$$

So even QF-FAN is classically false. However, whereas FAN was already in conflict with pure classical logic (and a trivial amount of arithmetic), the counterexample for QF-FAN required the definition of a comprehension functional F which is not available in the classical systems like E-PA^ω+QF-AC or even E-PA^ω+QF-AC+AC^0 which we have considered so far. This suggests the possibility to add QF-FAN consistently to these systems thereby exploiting some of the uses of non-

classical axioms even in a classical setting. Below we will show that the addition of QF-FAN to e.g. E-PA$^\omega$+QF-AC or $\widehat{\text{E-PA}}^\omega\!\restriction$+QF-AC not only is consistent but has no impact on the provably recursive functionals of type-2.
In fact, we will use an extension of QF-FAN where the bound x in the conclusion is obtained as a function in a number parameter of A. Intuitionistically, one usually assumes the full axiom of choice for numbers

$$\text{AC}^{0,0}: \forall x^0 \exists y^0 A(x,y) \to \exists f \forall x A(x, f(x))$$

which allows one to derive this form from our formulation. However, AC^{0,0}, which is intuitionistically weak (as we saw in chapters 5 and 8) would classically imply the full comprehension schema

$$\exists f^1 \forall x^0 \big(f(x) =_0 0 \leftrightarrow A(x)\big)$$

with dramatic effects on the computational content of the proof (see chapter 11). On the other hand, a generalization of FAN usually considered in intuitionistic mathematics (see e.g. [366] (1.9.24))

$$\text{FAN}_c: \forall f \leq_1 1 \exists x^0 A(f,x) \to \exists y \forall f \leq_1 1 \exists x \forall g \leq_1 1 \big(\overline{f}y =_0 \overline{g}y \to A(g,x)\big)$$

which has a uniform continuity principle built-in would (restricted to quantifier-free A_0) classically be redundant for QF-FAN since – as we saw above – QF-FAN allows already (in the presence of classical logic) to derive the uniform continuity of all type-2 functionals on the Cantor space (which together with QF-AC^{1,0} would imply this extension). Finally, we observe that we can allow A_0 in QF-FAN to be a Σ_1^0-formula as we can code further existential quantifiers and x together. Moreover, instead of the special bound 1 one may have any bounding function. So the right formulation of QF-FAN is the following schema which we call **uniform Σ_1^0-boundedness** in order to avoid confusion with other more common forms of FAN as discussed above.

Definition 12.1. The schema of **uniform Σ_1^0-boundedness** is defined as

$$\Sigma_1^0\text{-UB}: \begin{cases} \forall y^{1(0)} \big(\forall k^0 \forall x \leq_1 yk \exists z^0\, A(x,y,k,z) \\ \qquad\qquad \to \exists \chi^1 \forall k^0 \forall x \leq_1 yk \exists z \leq_0 \chi k\, A(x,y,k,z)\big), \end{cases}$$

where $A \equiv \exists \underline{l} A_0(\underline{l})$ and $\underline{l}$ is a tuple of variables of type 0 and A_0 is a quantifier–free formula (which may contain parameters of arbitrary types).

Proposition 12.2. $\widehat{\text{E-PA}}^\omega\!\restriction + \Sigma_1^0\text{-UB} \vdash \text{WKL}$.

Proof: Let f be such that $T(f)$. We prove WKL by contraposition. Assume

$$\forall b \leq_1 1 \exists x^0 \big(f(\overline{b}x) \neq_0 0\big).$$

Then by Σ_1^0-UB

$$\exists x^0 \forall b \leq_1 1 \exists \tilde{x} \leq x \big(f(\overline{b}\tilde{x}) \neq_0 0 \big)$$

and therefore by $T(f)$

$$\exists x \forall n^0 (lth(n) =_0 x \to f(n) \neq_0 0).$$

□

Proposition 12.3.

$$\widehat{\text{E-PA}}^{\omega}\!\upharpoonright + \Sigma^0_1\text{–UB} \vdash$$
$$\forall \Phi^{1(1)} \forall y^1 \exists \chi^1 \forall k^0 \forall z_1, z_2 \leq_1 y \Big(\bigwedge_{i \leq_0 \chi k} (z_1 i =_0 z_2 i) \to \bigwedge_{j \leq_0 k} (\Phi z_1 j =_0 \Phi z_2 j) \Big),$$

i.e. Φ is uniformly continuous (w.r.t. the topology induced by the metric on the Baire space) on $\{z : z \leq_1 y\}$ and has a modulus of uniform continuity χ.

Proof: $\forall z_1, z_2 \leq_1 y (z_1 =_1 z_2 \to \Phi z_1 =_1 \Phi z_2)$ implies

$$\forall z_1, z_2 \leq_1 y \forall k^0 \exists n^0 \Big(\bigwedge_{i \leq_0 n} (z_1 i =_0 z_2 i) \to \bigwedge_{j \leq_0 k} (\Phi z_1 j =_0 \Phi z_2 j) \Big).$$

Using Σ^0_1-UB (and the aforementioned encoding of '$\forall z_1, z_2 \leq y$' into a single quantifier) we obtain

$$\exists \chi^1 \forall k^0 \forall z_1, z_2 \leq_1 y \Big(\bigwedge_{i \leq_0 \chi k} (z_1 i =_0 z_2 i) \to \bigwedge_{j \leq_0 k} (\Phi z_1 j =_0 \Phi z_2 j) \Big).$$

□

Corollary 12.4.

$$\text{E-PA}^{\omega} + \text{QF-AC} + \text{WKL} \nvdash \Sigma^0_1\text{-UB}.$$

Proof: The full set-theoretic type structure $\mathcal{S}^{\omega}$ is a model of E-PA$^{\omega}$+QF-AC+WKL but – because of proposition 12.3 – not of Σ^0_1-UB. □

Definition 12.5. 1) F denotes the sentence

$$F :\equiv \forall \Phi^{2(0)}, y^{1(0)} \exists y_0 \leq_{1(0)} y \forall k^0 \forall z \leq_1 yk \big(\Phi kz \leq_0 \Phi k(y_0 k) \big).$$

2) F^- is the following weakening of F :

$$F^- :\equiv \forall \Phi^{2(0)}, y^{1(0)} \exists y_0 \leq_{1(0)} y \forall k^0, z^1, n^0 \Big(\bigwedge_{i <_0 n} (zi \leq_0 yki) \to \Phi k(\overline{z,n}) \leq_0 \Phi k(y_0 k) \Big),$$

where, for z^1, $(\overline{z,n})(k^0) :=_0 zk$, if $k <_0 n$ and $:= 0^0$, otherwise.

Proposition 12.6. 1) $\widehat{\text{WE-PA}}^{\omega}\!\upharpoonright$+QF-AC$^{1,0} \vdash F \to \Sigma^0_1$-UB.

2) $\widehat{\text{WE-PA}}^{\omega}\restriction + \text{QF-AC}^{1,0} \vdash F^- \to \Sigma_1^0\text{-UB}^-$, *where*

$$\Sigma_1^0\text{-UB}^- : \begin{cases} \forall y^{1(0)} \big(\forall k^0 \forall x \le_1 yk \exists z^0\, A(x,y,k,z) \to \exists \chi^1 \forall k^0, x^1, n^0 \\ \qquad \big(\bigwedge_{i<n} (xi \le_0 yki) \to \exists z \le_0 \chi k\, A((\overline{x,n}), y, k, z)\big)\big). \end{cases}$$

Here $A \equiv \exists \underline{l} A_0(\underline{l})$ and $\underline{l}$ is a tuple of variables of type 0 and A_0 is a quantifier-free formula (which may contain parameters of arbitrary types).

Proof: 1) The assumption

$$\forall k^0 \forall x \le_1 yk \exists z^0\, A(x,y,k,z)$$

implies that

$$\forall k^0 \forall x^1 \exists z^0, v^0, \underline{l} \big(xv \le_0 ykv \to A_0(\underline{l}, x, y, k, z)\big).$$

Now applying $\text{QF-AC}^{1,0}$ and the fact that k,x as well as $z,v,\underline{l}$ can be coded together, one obtains the existence of a functional $\Phi^{2(0)}$ such that

$$\forall k^0 \forall x \le_1 yk\, A(x,y,k,\Phi kx).$$

By applying F to Φ and y we get a $y_0^{1(0)}$ such that

$$\forall k^0 \forall x \le_1 yk(\Phi kx \le_0 \Phi k(y_0 k)).$$

So the proposition holds with $\chi k := \Phi k(y_0 k)$.
The proof of 2) is analogous. □

Proposition 12.7. $\widehat{\text{E-PA}}^{\omega}\restriction + \text{QF-AC}^{1,0} + F^- \vdash F$.

Proof: We reason in $\widehat{\text{E-PA}}^{\omega}\restriction + \text{QF-AC}^{1,0} + F^-$. Clearly, F follows from F^- provided we can establish

$$(1)\ \forall \Phi^2 \forall f^1 \exists n^0 \big(\Phi(f) =_0 \Phi(\overline{f,n})\big).$$

Suppose, therefore, that on the contrary there would exist Φ^2 and f such that

$$(2)\ \forall n^0 \big(\Phi(f) \ne_0 \Phi(\overline{f,n})\big).$$

Then taking $f_i := \overline{f, i+1}$ we get

$$(3)\ \forall i^0 \forall j \ge i \big(f_j(i) =_0 f(i)\big)$$

and

$$(4)\ \forall i^0 \big(\Phi(f) \ne_0 \Phi(f_i)\big).$$

Now one can use again Grilliot's trick (which we already used in the proof of proposition 10.60 in chapter 10) to derive from (3) and (4)

$$(5)\ \exists \varphi^2 \forall g^1 \big(\varphi(g) =_0 0 \leftrightarrow \exists x (gx =_0 0)\big)$$

Proof of (5)*:* We can construct a closed term $t^{1(1)(1)}$ of $\widehat{\text{E-PA}}^{\omega}\!\upharpoonright$ such that (provably in $\widehat{\text{E-PA}}^{\omega}\!\upharpoonright$) we have

$$tfgi =_0 \begin{cases} f_j(i), \text{ for the least } j < i \text{ such that } g(j) > 0, \text{ if such a } j \text{ exists} \\ f(i), \text{ otherwise.} \end{cases}$$

Together with (3) this yields

$$\exists j (g(j) > 0) \to tfg =_1 f_j \text{ for the least such } j$$

and

$$\forall j (g(j) =_0 0) \to tfg =_1 f.$$

Hence using the extensionality axiom of $\widehat{\text{E-PA}}^{\omega}\!\upharpoonright$ and (4) we get

$$\forall j (g(j) =_0 0) \leftrightarrow \Phi(tfg) =_0 \Phi(f).$$

So $\varphi := \lambda g^1 . \overline{sg}(|\Phi(t(f, \overline{sg} \circ g)) - \Phi(f)|)$ where $\overline{sg}(x) := 0$ for $x \neq 0$ and $\overline{sg}(x) := 1$ otherwise, does the job.
End of proof of (5).
The proof now is concluded by observing that (5) contradicts F^- (relative to $\widehat{\text{E-PA}}^{\omega}\!\upharpoonright$+QF-AC^{1,0}), since F^- implies that every Φ^2 is bounded on the set of all functions $\overline{g,n}$ with $g <_1 1, n \in \mathbb{N}$, whereas QF-AC^{1,0} together with (5) yields the existence of a functional μ such that

$$(6)\ \forall g^1 \big(\exists x^0 (gx =_0 0) \to g(\mu(g)) =_0 0\big).$$

It is obvious that μ is unbounded on this very set. □

Theorem 12.8. *Let* $A_0(x^1, y^1, z^0, v^\tau) \in \mathcal{L}(\text{E-PA}^\omega)$ *be a quantifier-free formula containing only* x, y, z, v *as free variables (where* τ *is an arbitrary type) and* s *a closed term of* E-PA$^\omega$*. Then the following rule holds:*

$$\begin{cases} \text{E-PA}^\omega + \text{QF-AC}^{1,0} + \text{QF-AC}^{0,1} + \Sigma^0_1\text{-UB} \vdash \forall x^1 \forall y \leq_1 sx \exists z^0, v^\tau A_0(x,y,z,v) \\ \Rightarrow \textit{ one can extract a closed term } \Psi \textit{ in } \text{E-PA}^\omega \textit{ s.t.} \\ \text{WE-HA}^\omega \vdash \forall x^1 \forall y \leq_1 sx \exists z \leq_0 \Psi x \exists v^\tau A_0(x,y,z,v). \end{cases}$$

The result also holds in the presence of additional axioms Δ *as in theorem 10.47. Then the conclusion is provable in* WE-HA$^\omega + \tilde{\Delta}_\varepsilon$*.*
The result also holds for tuples of variables $\underline{x}^{\underline{\delta}}, \underline{y}^{\underline{\rho}}$ *(with* $deg(\underline{\delta}, \underline{\rho}) \leq 1$*),* $\underline{z}^0, \underline{v}^{\underline{\tau}}$ *and terms* $\underline{s}$*.*
An analogous result holds for $\widehat{\text{E-PA}}^{\omega}\!\upharpoonright, \widehat{\text{WE-HA}}^{\omega}\!\upharpoonright$ *instead of* E-PA$^\omega$ *and* WE-HA$^\omega$*.*

Proof: We can assume that Δ is finite. The finite conjunction of the axioms in Δ (resp. in $\tilde{\Delta}_\varepsilon$) we also denote by Δ (resp. by $\tilde{\Delta}_\varepsilon$). In the following we abbreviate 'QF-AC^{1,0}+QF-AC^{0,1}' by 'QF-AC'. The assumption and propositions 12.7, 12.6 imply

$$\text{E-PA}^\omega + \text{QF-AC} \vdash \Delta \wedge F^- \to \forall x^1 \forall y \leq_1 sx \exists z^0, v^\tau\, A_0(x,y,z,v).$$

By the elimination of extensionality (proposition 10.45) we get

$$\text{WE-PA}^\omega + \text{QF-AC} \vdash \Delta \wedge F^- \to \forall x^1 \forall y \leq_1 sx \exists z^0, v^\tau\, A_0(x,y,z,v)$$

and therefore

$$\begin{aligned} &\text{WE-PA}^\omega + \text{QF-AC} \vdash \Delta \to \big(\exists Y \leq \lambda \Phi^{2(0)}, y^{1(0)}.y \forall \Phi^{2(0)}, \tilde{y}^{1(0)}, k^0, \tilde{z}^1, n^0 \\ &\quad \big(\bigwedge_{i<n} (\tilde{z}i \leq \tilde{y}ki) \to \Phi k(\overline{\tilde{z},n}) \leq_0 \Phi k(Y\Phi\tilde{y}k)\big) \to \forall x^1 \forall y \leq_1 sx \exists z^0, v^\tau A_0(x,y,z,v)\big), \end{aligned}$$

and thus

$$\text{WE-PA}^\omega + \text{QF-AC} \vdash \Delta \to \forall Y \leq \lambda \Phi, y.y \forall x^1 \forall y \leq_1 sx \exists \Phi, \tilde{y}, k, \tilde{z}, n, z, v(\ldots).$$

By theorem 10.26 and remarks 10.23, 10.27.2 we can extract closed terms Ψ_1, Ψ_2 of WE-PA$^\omega$ such that

$$\text{WE-HA}^\omega + \tilde{\Delta}_\varepsilon \vdash \forall Y \leq \lambda \Phi, y.y \forall x^1 \forall y \leq_1 sx \exists \Phi, \tilde{y}, k, \tilde{z} \exists n \leq_0 \Psi_1 x \exists z \leq_0 \Psi_2 x \exists v(\ldots).$$

Moving quantifiers inwards yields

$$\begin{aligned} &\text{WE-HA}^\omega + \tilde{\Delta}_\varepsilon \vdash \forall x \big(\exists Y \leq \lambda \Phi^{2(0)}, y^{1(0)}.y \forall \Phi, \tilde{y}^{1(0)}, k^0, \tilde{z}^1 \forall n \leq_0 \Psi_1 x \\ &\quad \big(\bigwedge_{i<n} (\tilde{z}i \leq \tilde{y}ki) \to \Phi k(\overline{\tilde{z},n}) \leq \Phi k(Y\Phi\tilde{y}k)\big) \to \forall y \leq_1 sx \exists z \leq_0 \Psi_2 x \exists v A_0(x,y,z,v)\big). \end{aligned}$$

To conclude the proof we have to show that

$$\begin{aligned} \text{WE-HA}^\omega \vdash \forall n_0 \exists Y \leq \lambda \Phi^{2(0)}, y^{1(0)}.y \forall \Phi, \tilde{y}^{1(0)}, k^0, \tilde{z}^1 \forall n \leq_0 n_0 \\ \big(\bigwedge_{i<n} (\tilde{z}i \leq \tilde{y}ki) \to \Phi k(\overline{\tilde{z},n}) \leq \Phi k(Y\Phi\tilde{y}k)\big): \end{aligned}$$

Taking advantage of the fact that our definition of $\overline{f}x$ implies that

$$\bigwedge_{i<n} (\tilde{z}i \leq_0 \tilde{y}ki) \to \overline{\tilde{z}}n \leq_0 (\overline{\tilde{y}k})n_0 \text{ for } n \leq_0 n_0$$

we now define

$$\tilde{Y} := \lambda \Phi, \tilde{y}, k, n_0. \max_{j \leq_0 (\overline{\tilde{y}k})n_0} \Phi k\big(\overline{\min_1(\lambda i.(j)_i, \tilde{y}k), n_0}\big).$$

One easily shows (using the fact that $\Phi_{\langle\cdot\rangle} \in \text{WE-HA}^\omega$) that $\tilde{Y}$ is definable as well.

$$\widehat{Y} := \lambda \Phi, \tilde{y}, k, n_0. \min_{j \le_0 (\overline{\tilde{y}k}) n_0} \left[\Phi k\big(\overline{\min_1(\lambda i.(j)_i, \tilde{y}k), n_0}\big) =_0 \tilde{Y}\Phi\tilde{y}kn_0 \right].$$

For every n_0 we now put

$$Y := \lambda \Phi, \tilde{y}, k.\big(\overline{\min_1\,(\lambda i.(\widehat{Y}\Phi\tilde{y}kn_0)_i, \tilde{y}k), n_0}\big).$$

The proof for $\widehat{\text{E-PA}}^{\omega}\!\restriction, \widehat{\text{WE-HA}}^{\omega}\!\restriction$ is analogous. □

Remark 12.9. Note that in the theorem above, the conclusion is valid in $\mathcal{S}^{\omega}$ although the proof of the premise may have used Σ^0_1-UB which is not valid in $\mathcal{S}^{\omega}$. The theorem also gives a (relative) consistency proof for

$$\text{E-PA}^{\omega}{+}\text{QF-AC}^{1,0}{+}\text{QF-AC}^{0,1} + \Sigma^0_1\text{-UB}.$$

Remark 12.10. For

$$(\text{WE-PA}^{\omega} + \text{QF-AC}^{1,0} + \text{QF-AC}^{0,1}) \oplus \Sigma^0_1\text{-UB}^-$$

instead of

$$\text{E-PA}^{\omega} + \text{QF-AC}^{1,0} + \text{QF-AC}^{0,1} + \Sigma^0_1\text{-UB}$$

the proof of this theorem is much simpler and does not need elimination of extensionality since we do not need proposition 12.7. Here '$(\ldots) \oplus \Sigma^0_1\text{-UB}^-$' means that $\Sigma^0_1\text{-UB}^-$ must not be used in the proof of the premise of an application of the quantifier-free rule of extensionality QF-ER. WE-PA^{ω} satisfies the deduction theorem w.r.t $\oplus$ but not w.r.t $+$. In this context we may even use QF-AC for all types as the restriction to the types $1,0$ resp. $0,1$ was caused only by the necessity to apply the elimination of extensionality procedure.

As a corollary to theorem 12.8 and proposition 12.2 we obtain another proof of theorem 10.58.

Let us now switch to an intuitionistic context. In such a context we can treat much more general forms of uniform boundedness:

- Instead of considering only Σ^0_1-formulas we can allow arbitrary formulas.
- Since the extensionality axiom is treated trivially by monotone modified realizability and we, therefore, do not need to apply any elimination of extensionality procedure, we can, furthermore, consider $\forall x \le_\rho y$ for all types ρ rather than just for $\rho = 1$.

Definition 12.11. 1) The generalization of the axiom F to arbitrary types ρ is defined as follows

$$F^{\rho} :\equiv \forall \Phi^{0\rho 0}, y^{\rho 0} \exists y_0 \le_{\rho 0} y \forall k^0 \forall z \le_\rho yk\big(\Phi kz \le_0 \Phi k(y_0 k)\big).$$

The union of these axioms for all types ρ we denote by F^{ω}.

2) The full uniform boundedness principles in all types is defined in the following way

$$\mathrm{UB}^\rho : \forall y^{\rho 0}\big(\forall k^0 \forall x \leq_\rho yk \exists z^0 A(x,y,k,z) \to \exists \chi^1 \forall k^0 \forall x \leq_\rho yk \exists z \leq_0 \chi k A(x,y,k,z)\big).$$

The union of all these axioms for arbitrary formulas A we denote by UB^ω.

Remark 12.12. 1) For $\rho = 1$, F^ρ coincides with F.
2) For $\rho = 1$ and $A \in \Sigma^0_1$, UB^ρ coincides with Σ^0_1-UB.

Lemma 12.13. $\mathrm{E\text{-}HA}^\omega + \mathrm{AC} + F^\rho \vdash \mathrm{UB}^\rho$.

Proof: Exercise! Hint: Use that over $\mathrm{E\text{-}HA}^\omega$ we can write $\forall x \leq_\rho yk \exists z A(x,y,k,z)$ equivalently as $\forall x^\rho \exists z A(\min_\rho(x,yk),y,k,z)$. □

Theorem 12.14. *Let $A(x^1,y^\rho,z^\tau)$ be a formula of $\mathcal{L}(\mathrm{E\text{-}HA}^\omega)$ (containing only x,y,z free) which is in Γ_1 (see definition 5.19) such that all positively (resp. negatively) occurring $\forall$-quantifiers (resp. $\exists$-quantifiers) are of types of degree ≤ 1 while all other quantifiers are of types of degree ≤ 2. Assume that $deg(\rho), deg(\tau) \leq 2$ and let s be a closed term. Then the following rule holds:*

$$\begin{cases} \mathrm{E\text{-}HA}^\omega + \mathrm{AC} + \mathrm{IP}^\omega_{ef} + \mathrm{UB}^\omega \vdash \forall x^1 \forall y \leq_\rho sx \exists z^\tau A(x,y,z) \\ \Rightarrow \text{ one can extract a closed term } t \text{ of } \mathrm{E\text{-}HA}^\omega \text{ such that} \\ \mathcal{S}^\omega \models \forall x^1 \forall y \leq_\rho sx \exists z \leq_\tau tx\, A(x,y,z). \end{cases}$$

The theorem also holds in the presence of further $\mathcal{S}^\omega$-valid axioms of the form Θ as in theorem 7.1.1) provided that all positively (resp. negatively) occurring $\exists$-quantifiers (resp. $\forall$-quantifiers) are of types of degree ≤ 1 while all other quantifiers are of types of degree ≤ 2.

Proof: We may assume that $\tau = 2$. By lemma 12.13 the premise of the rule implies that

$$\mathrm{E\text{-}HA}^\omega + \mathrm{AC} + \mathrm{IP}^\omega_{ef} + F^\omega \vdash \forall x^1 \forall y \leq_\rho sx \exists z^\tau A(x,y,z)$$

and so a-fortiori

$$\mathrm{E\text{-}HA}^\omega + \mathrm{AC} + \mathrm{IP}^\omega_{ef} + \tilde{F}^\omega \vdash \forall x^1 \forall y \leq_\rho sx \exists z^\tau A(x,y,z),$$

where

$$\tilde{F}^\rho :\equiv \exists Y_0 \leq \lambda \Phi, y.y \forall \Phi, y, k^0 \forall z \leq_\rho yk\big(\Phi kz \leq_0 \Phi k(Y_0(\Phi,y,k))\big).$$

$\tilde{F}^\rho$ is of the form of the axioms Θ considered in theorem 7.1.1). Hence we can apply that theorem to extract closed terms $\underline{t}^*$ such that, provably in $\mathrm{E\text{-}HA}^\omega + \tilde{F}^\omega$, these terms satisfy the monotone modified realizability interpretation of the conclusion, i.e.

$$\mathrm{E\text{-}HA}^\omega + \tilde{F}^\omega \vdash \exists \underline{u}\big(\underline{t}^* \; maj\; \underline{u} \wedge \underline{u}\; mr\; \forall x^1 \forall y \leq_\rho sx \exists z^\tau A(x,y,z)\big).$$

Since $A \in \Gamma_1$, lemma 5.20 yields from this

$$\text{E-HA}^{\omega} + \tilde{F}^{\omega} \vdash \exists u_1 \left(t_1^* \; maj \; u_1 \wedge \forall x^1 \forall y \leq_\rho sx \, A(x,y,u_1 xy)\right).$$

Arguing as in the proof of theorem 6.8 it follows that

$$\text{E-HA}^{\omega} + \tilde{F}^{\omega} \vdash \forall x^1 \forall y \leq_\rho \exists z \leq_\tau tx \, A(x,y,z),$$

where

$$tx := \lambda z^1 . t_1^*(x^M)(s^* x^M) z^M$$

for some majorizing term s^* of s. The model of strongly majorizable functionals $\mathcal{M}^{\omega}$ (see definition 3.61) is a model of $\tilde{F}^{\omega}$ (exercise). Hence

$$\mathcal{M}^{\omega} \models \forall x^1 \forall y \leq_\rho sx \exists z \leq_\tau tx \, A(x,y,z).$$

Using the restrictions on the types of the quantifiers in A and remark 3.72 it follows that (note that for $y \in S_\rho$, $y \leq sx$ implies that $y \in M_\rho$ since $s^* x^M$ is a majorant and $deg(\rho) \leq 2$ so that $M_\rho \subseteq S_\rho$)

$$\mathcal{S}^{\omega} \models \forall x^1 \forall y \leq_\rho sx \exists z \leq_\tau tx \, A(x,y,z).$$

The extension by axioms Θ follows immediately from the proof above and the fact that the type restrictions imply that Θ also holds in $\mathcal{M}^{\omega}$. □

Remark 12.15. Other results on even more general forms of UB have been obtained by F. Ferreira and P. Oliva using their novel bounded functional interpretation ([104]).

12.2 Applications of Σ_1^0-boundedness

In this section we show how Σ_1^0-UB can be used to give very short proofs of important theorems in analysis (whose proofs based on, say, WKL would be much more involved). Moreover, whereas proofs based on WKL rely on complicated representations of e.g. $f \in C[0,1]$ we can treat such functions now directly as type-2 objects and can even avoid to state any continuity assumptions in most applications. This is due to the fact that Σ_1^0-UB implies that every function $f : [0,1] \to \mathbb{R}$ is uniformly continuous as we show in the first application (illustrating again the 'non-standard' character of this principle).

In the following d is an arbitrary but fixed natural number ≥ 1.

Application 1:

Proposition 12.16. E-PA$^{\omega}$ + Σ_1^0-UB *proves:*
Every function $F : [0,1]^d \to \mathbb{R}$ *is uniformly continuous and possesses a modulus of uniform continuity.*

Proof: Referring to the representation of $[0,1]^d$ and $\mathbb{R}$ from chapter 4, the assertion above has the following form when expressed in $\mathcal{L}(\text{E-PA}^\omega)$:

'If $\Phi^{1(1)}$ represents a function $[0,1]^d \to \mathbb{R}$, i.e.

$$\forall x_1^1, x_2^1 (\bigwedge_{i=1}^{d} (0 \le_{\mathbb{R}} \nu_i^d(x_1), \nu_i^d(x_2) \le_{\mathbb{R}} 1 \wedge \nu_i^d(x_1) =_{\mathbb{R}} \nu_i^d(x_2)) \to \Phi x_1 =_{\mathbb{R}} \Phi x_2),$$

then Φ is uniformly continuous on $[0,1]^d$ and possesses a modulus of uniform continuity.'

Here ν^d, ν_i^d are the coding functions for d-tuples of number-theoretic functions from definition 3.30. Making use of the representation of $[0,1]$ from lemma 4.25, we can restrict ourselves to representatives x^1 of elements of $[0,1]^d$ which satisfy $\nu_i^d(x) \le_1 N$ for $i = 1,\dots,d$ (where $N := \lambda n.j(2^{n+3}, 2^{n+2}-1)$).

$$\forall x_1, x_2 \le_1 \nu^d(N,\dots,N)(\bigwedge_{i=1}^{d} (\widetilde{\nu_i^d(x_1)} =_{\mathbb{R}} \widetilde{\nu_i^d(x_2)}) \to \Phi\tilde{x}_1 =_{\mathbb{R}} \Phi\tilde{x}_2)$$

is equivalent to

$$\forall x_1, x_2 \le_1 \nu^d(N,\dots,N) \forall k^0 \exists n^0 \underbrace{\left(\|\tilde{x}_1 -_{\mathbb{R}^d} \tilde{x}_2\|_{\max} \le_{\mathbb{R}} 2^{-n} \to |\Phi\tilde{x}_1 -_{\mathbb{R}} \Phi\tilde{x}_2|_{\mathbb{R}} <_{\mathbb{R}} 2^{-k}\right)}_{\equiv: A \in \Sigma^0_1},$$

where $\|\cdot\|_{\max}$ denotes the maximum metric on $\mathbb{R}^d$ and $\tilde{x}$ is an abbreviation for $\nu^d(\widetilde{\nu_1^d(x_1)},\dots,\widetilde{\nu_d^d(x_d)})$. From chapter 4 we recall that $\le_{\mathbb{R}} \in \Pi^0_1$ and $<_{\mathbb{R}} \in \Sigma^0_1$. Since x_1, x_2 can be coded together, Σ^0_1-UB yields (using the monotonicity of A with respect to n)

$$\exists \chi^1 \forall x_1, x_2 \le_1 \nu^d(N,\dots,N) \forall k^0 \left(\|\tilde{x}_1 -_{\mathbb{R}^d} \tilde{x}_2\|_{\max} \le 2^{-\chi k} \to |\Phi\tilde{x}_1 -_{\mathbb{R}} \Phi\tilde{x}_2|_{\mathbb{R}} < 2^{-k}\right).$$

□

Remark 12.17. 1) This result generalizes also to variable rectangles $[a_1,b_1] \times \ldots \times [a_d,b_d]$ instead of $[0,1]^d$ (where $a_i < b_i$ for $i = 1,\dots,d$).
2) Instead of $\|\cdot\|_{\max}$ we can also use e.g. the Euclidean metric on $\mathbb{R}^d$ thereby obtaining a modulus of continuity w.r.t. this metric. However, since both norms on $\mathbb{R}^d$ are constructively equivalent, a modulus of uniform continuity w.r.t. one norm can be easily transformed into a modulus for the other norm.

Remark 12.18. 1) Proposition 12.16 also holds with WE-HA$^\omega$+M^0 instead of E-PA$^\omega$, where M^0 is the restriction of the Markov principle M$^\omega$ to $\underline{x} = x^0$.
2) If F is assumed to be pointwise continuous, then Σ^0_1-UB$^-$ (in the sense of remark 12.10) suffices for the proof of proposition 12.16.

Corollary 12.19. E-PA$^\omega + \Sigma^0_1$-UB *proves: Every* $\Phi^{1(1)}$ *which represents a function* $F : \mathbb{R}^d \to \mathbb{R}$ *is pointwise continuous on* $\mathbb{R}^d$ *and possesses a modulus of pointwise continuity operation.*

Proof: As a corollary to the proof of proposition 12.16 above we obtain a function $\chi^{1(0)}$ such that $\chi(m)$ is a modulus of uniform continuity for Φ on $[-m,m]^d$ by applying Σ^0_1-UB to

$$\forall m^0 \forall x_1, x_2 \leq_1 \nu^d(N(m),\ldots,N(m)) \forall k^0 \exists n^0 \underbrace{\big(\|x_1 -_{\mathbb{R}^d} x_2\|_{\max} \leq 2^{-n} \to |\Phi x_1 -_{\mathbb{R}} \Phi x_2|_{\mathbb{R}} < 2^{-k}\big)}_{\in \Sigma^0_1},$$

where $N(m) := \lambda n.j(m2^{n+3}, 2^{n+2}-1)$ is the boundedness function from our representation of $[-m,m]$.

Define $\xi^{0(1)} \in$ E-PA$^\omega$ by $\xi(x^1) := \max_0(\lceil(\widehat{\nu^d_1(x)})(0)\rceil + 2, \ldots, \lceil(\widehat{\nu^d_d(x)})(0)\rceil + 2)$. The natural number $\xi(x^1)$ is an upper bound for $\|x^1\|_{\max} + 1$. Finally define $\omega(x^1) := \lambda k^0.\chi(\xi(x),k)$. Since $\|x-y\|_{\max} \leq 2^{-\omega(x,k)}$ implies that $\|x\|_{\max}, \|y\|_{\max} \leq \xi(x)$, $\omega(x)$ not only is a modulus of uniform continuity on $\{y \in \mathbb{R}^d : \|y\|_{\max} \leq \xi(x)\}$ but also a modulus of pointwise continuity of $\Phi : \mathbb{R}^d \to \mathbb{R}$ in (the point represented by) x. □

Remark 12.20. Note that the above modulus of pointwise continuity $\omega(x^1, k^0)$ depends on the representation x^1 of a given point in $\mathbb{R}^d$, i.e. it is not extensional w.r.t. $=_{\mathbb{R}^d}$. So ω is only an operation and not a function of x as an element of $\mathbb{R}^d$ (but a function of $x \in \mathbb{N}^{\mathbb{N}}$ as an representative of such an element).

Application 2: Sequential form of the Heine-Borel covering property of $[0,1]^d$ and other compact spaces

Consider the open ball with center $x_0 \in \mathbb{R}^d$ and radius $\varepsilon > 0$ (w.r.t. the Euclidean norm):

$$B_\varepsilon(x_0) := \{y \in \mathbb{R}^d : \|x_0 - y\|_E < \varepsilon\}.$$

The next proposition even holds in the intuitionistic context of E-HA$^\omega$ (in fact also WE-HA$^\omega$).

Proposition 12.21. E-HA$^\omega + \Sigma^0_1$-UB *proves that every sequence of open balls in* $\mathbb{R}^d$ *which covers* $[0,1]^d$ *contains a finite subcover.*

Proof: To show

$$(1) \begin{cases} \forall f : \mathbb{N} \to \mathbb{R}_+ \setminus \{0\} \forall g : \mathbb{N} \to [0,1]^d \big(\forall x \in [0,1]^d \exists k \in \mathbb{N}(x \in B_{fk}(gk)) \\ \qquad \to \exists k_0 \forall x \in [0,1]^d \exists k \leq k_0 (x \in B_{fk}(gk))\big). \end{cases}$$

Reasoning as in application 1 above we see that (1) has – when formalized in $\mathcal{L}$(E-HA$^\omega$) – the following logical form

$$(2)\begin{cases}\forall f^{1(0)},g^{1(0)}(\forall l^0(f(l)>_{\mathbb{R}}0)\land\forall x\leq_1 v^d(N,\ldots,N)\exists k^0(\|\tilde{x}-_{\mathbb{R}^d}g(k)\|_E<_{\mathbb{R}}fk)\\ \qquad\to\exists k_0^0\forall x\leq_1 v^d(N,\ldots,N)\exists k\leq_0 k_0(\|\tilde{x}-_{\mathbb{R}^d}g(k)\|_E<_{\mathbb{R}}f(k))),\end{cases}$$

which immediately follow using Σ^0_1-UB and the fact that $<_{\mathbb{R}}\in\Sigma^0_1$. □

Similarly one shows this result for $[a_1,b_1]\times\ldots\times[a_d,b_d]$ and also for other compact spaces as e.g. $K_{c,\lambda}:=\{f\in C[0,1]:\|f\|_\infty\leq c\land f$ has Lipschitz constant $\lambda\}$.

Remark 12.22. A slightly more complicated argument shows that Σ^0_1-UB$^-$ (in the sense of remark 12.10) suffices for the proof of proposition 12.21. Also E-HA$^\omega$ can be replaced by WE-HA$^\omega$.

Application 3: Attainment of the maximum value for $f\in C([0,1]^d,\mathbb{R})$

Proposition 12.23. E-PA$^\omega+\Sigma^0_1$-UB *proves:*
Every function $F:[0,1]^d\to\mathbb{R}$ *attains it maximum value on* $[0,1]^d$.

Proof: By proposition 12.16, we may use that F is (uniformly) continuous. We proceed by contradiction: suppose that

$$(1)\ \exists F:[0,1]^d\to\mathbb{R}\big(F\,\text{continuous}\land\forall x\in[0,1]^d\exists r\in[0,1]^d\cap\mathbb{Q}^d(F(x)<F(r))\big).$$

Bearing in mind that F again is represented as some extensional (w.r.t. $=_{[0,1]^d},=_{\mathbb{R}}$) functional $\Phi^{1(1)}$, the proposition $\forall x\in[0,1]^d\exists r\in[0,1]^d\cap\mathbb{Q}^d(F(x)<F(r))$ has the following logical form

$$(2)\ \forall x\leq_1 v^d(N,\ldots,N)\underbrace{\exists n^0(\Phi\tilde{x}<_{\mathbb{R}}\Phi(\lambda k^0.q(n)))}_{\in\Sigma^0_1},$$

where q is a primitive recursive enumeration of (the codes of) $[0,1]^d\cap\mathbb{Q}^d$. (2) implies

$$(3)\ \forall x\leq_1 v^d(N,\ldots,N)\exists n^0,l^0(\Phi\tilde{x}<_{\mathbb{R}}\Phi(\lambda k^0.q(n))-2^{-l}).$$

Σ^0_1-UB applied to (3) yields

$$\exists n_0\forall x\leq_1 v^d(M,\ldots,M)\forall l^0\exists n\leq_0 n_0(\Phi\tilde{x}<_{\mathbb{R}}\Phi(\lambda k^0.q(n))-2^{-n_0}).$$

In E-PA$^\omega$ one can show that there exists an $n_1\leq n_0$ be such that

$$(4)\ \Phi(\lambda k^0.q(n_1))=_{\mathbb{R}}\max_{\mathbb{R}}(\Phi(\lambda k^0.q(0)),\ldots,\Phi(\lambda k^0.q(n_0))).$$

Since there exists an x^1 such that $x\leq_1 v^d(N,\ldots,N)$ and $\tilde{x}=_{[0,1]^d}\lambda k^0.q(n_1)$ (namely, e.g. $\tilde{x}:=_1 v^d(v^d_1(\widetilde{\lambda k^0.q(n_1)}),\ldots,v^d_d(\widetilde{\lambda k^0.q(n_1)})))$ we obtain (using the extensionality of Φ) a contradiction. Hence (moving back to the informal mathematical formulation)

$$(5)\forall F:[0,1]^d \to \mathbb{R}\big(F \text{ pointwise cont.} \to \exists x \in [0,1]^d \forall r \in [0,1]^d \cap \mathbb{Q}^d (F(x) \geq F(r))\big),$$

which implies

$$(6)\ \forall F:[0,1]^d \to \mathbb{R}(F \text{ pointwise cont.} \to \exists x \in [0,1]^d \forall y \in [0,1]^d (F(x) \geq F(y))).$$

□

Remark 12.24. The proof of proposition 12.23 also works with WE-PA$^\omega$, and if F is assumed to be pointwise continuous, then Σ_1^0-UB$^-$ suffices (in the sense of remark 12.10).

Application 4: Dini's theorem

Maybe the most striking application of Σ_1^0-UB is the following trivial derivation of Dini's theorem:

Proposition 12.25. E-HA$^\omega$ + Σ_1^0-UB *proves: Every sequence* (F_n) *of functions:* $F_n : [0,1]^d \to \mathbb{R}$ *which non-decreases pointwise to a function* $F : [0,1]^d \to \mathbb{R}$ *converges uniformly on* $[0,1]^d$ *to* F*, and there exists a modulus of uniform convergence.*

Proof: By the assumption we have

$$\forall k^0 \forall x \in [0,1]^d \exists n^0 (F(x) - F_n(x) <_{\mathbb{R}} 2^{-k}).$$

As in the proof of proposition 12.23 Σ_1^0-UB can be used to show that

$$\exists \chi^1 \forall k^0 \forall x \in [0,1]^d \exists n \leq_0 \chi(k)(F(x) - F_n(x) <_{\mathbb{R}} 2^{-k}).$$

Since $(F_n)_{n\in\mathbb{N}}$ is nondecreasing by assumption this implies

$$(*)\ \exists \chi^1 \forall k^0 \forall x \in [0,1]^d \forall n \geq_0 \chi(k)(F(x) - F_n(x) <_{\mathbb{R}} 2^{-k}),$$

which concludes the proof. □

Remark 12.26. The proof of proposition 12.25 also works with WE-HA$^\omega$. If F and F_n are assumed to be pointwise continuous, then Σ_1^0-UB$^-$ suffices (in the sense of remark 12.10).

Applications 1–4 generalize to other compact spaces K instead of $[0,1]^d$ as long as these space have a representation in the sense of chapter 4 where the simplified form of this representation discussed at the end of that chapter is sufficient.

Application 5: Existence of the inverse function of a strictly monotone function

Proposition 12.27. E-PA$^\omega$ + Σ_1^0-UB *proves:*
Every strictly increasing function $F : [0,1] \to \mathbb{R}$ *possesses a strictly increasing inverse function* $F^{-1} : [F(0), F(1)] \to [0,1]$ *which is uniformly continuous on* $[F(0), F(1)]$ *and has a modulus of uniform continuity.*

Proof: Let $\Phi^{1(1)}$ be the representation of a function strictly increasing function $F:[0,1]\to\mathbb{R}$. By application 1, F is uniformly continuous so that the intermediate value theorem applies.
The strict monotonicity of F implies

$$(1)\ \forall x,y\in[0,1]\forall k^0\exists n^0(x\geq y+2^{-k}\to F(x)>F(y)+2^{-n}).$$

Modulo our representation of $[0,1]$, F, $\geq_{\mathbb{R}}$ and $>_{\mathbb{R}}$, (1) has the logical form

$$\forall x,y\leq_1 N\forall k^0\exists n^0\big(\underbrace{\tilde{x}\geq_{\mathbb{R}}\tilde{y}+_{\mathbb{R}}2^{-k}\to\Phi\tilde{x}>_{\mathbb{R}}\Phi\tilde{y}+2^{-n}}_{\equiv:A\in\Sigma^0_1}\big).$$

By Σ^0_1-UB we obtain (using the monotonicity of A w.r.t. n) a modulus of uniform strict monotonicity, i.e.

$$(2)\ \exists\chi^1\forall x,y\leq_1 N\forall k^0\big(\tilde{x}\geq_{\mathbb{R}}\tilde{y}+_{\mathbb{R}}2^{-k}\to\Phi\tilde{x}>_{\mathbb{R}}\Phi\tilde{y}+2^{-\chi k}\big).$$

Observing that χ is a modulus of uniform continuity for the inverse function F^{-1} the rest of the proof is now straightforward and left to the reader. □

Remark 12.28. Since the proof of 12.27 actually only uses the constructive ε-version of the intermediate value theorem, the result can be proved in E-HA$^\omega+\Sigma^0_1$-UB if F is assumed to be pointwise continuous. If F, moreover, is assumed to have moduli of uniform continuity and of uniform strict monotonicity, Σ^0_1-UB is not needed (see also [328]).

Remark 12.29. The proof of proposition 12.27 also works with WE-PA$^\omega$, and if F is assumed to be pointwise continuous, then Σ^0_1-UB$^-$ suffices (in the sense of remark 12.10).

Application 6: No injection of $\mathbb{N}^{\mathbb{N}}$ into $\mathbb{N}$

That there is no injection $\Phi^2:\mathbb{N}^{\mathbb{N}}\to\mathbb{N}$ can be formalized as follows

$$(\text{no-injection}):\forall\Phi^2\,\exists f^1,g^1\,\big(\exists n^0(f(n)\neq_0 g(n))\wedge\Phi(f)=_0\Phi(g)\big).$$

Proposition 12.30. WE-PA$^\omega+\Sigma^0_1$-UB $\vdash$ (no-injection).

Proof: Assume that the negation of '(no-injection)' would hold, i.e.

$$\exists\Phi^2\forall f^1,g^1(\exists n^0(f(n)\neq_0 g(n))\to\Phi(f)\neq_0\Phi(g))$$

and hence a-fortiori for some Φ^2

$$\forall f,g\leq_1 1\,(\exists n^0(f(n)\neq_0 g(n)\to\Phi(f)\neq_0\Phi(g)).$$

By Σ^0_1-UB (applied to $\forall f\leq_1 1\exists n\,(n=\Phi(f))$)

$$(K)\ \exists K^0 \forall f \leq_1 1\,(K \geq \Phi(f)).$$

No consider all the (pairwise different) binary functions f_n defined as $f_n(k) := 1$ for $k \leq n$ and $f_n(k) = 0$ for $k > n$. Then $\Phi(f_i) \neq \Phi(f_j)$ for (in particular) all $0 \leq i < j \leq K+1$ which – by the (finite) pigeonhole principle – contradicts (K). □

From proposition 12.30 and theorem 12.8 we immediately can conclude that E-PA$^\omega$+QF-AC^{1,0}+QF-AC^{0,1} plus the principle (no-injection) proves the same Π^0_2-sentences as WE-HA$^\omega$.

Without appeal to Σ^0_1-UB one apparently needs a (non-trivial) use of countable choice AC^0 applied to

$$\forall \Phi^2, k^0 \exists f^1 (\exists g^1 (k =_0 \Phi(g) \to k =_0 \Phi(f))$$

to prove '(no-injection)' which can be interpreted via bar recursion as we saw in chapter 11 (the functional interpretation of '(no-injection)' has been carried out by P. Oliva in [293]).

Remark 12.31. The proof of proposition 12.30 actually establishes that already $2^{\mathbb{N}}$ has no injection into $\mathbb{N}$.

12.3 Remarks on the fragments E-G$_n$A$^\omega$

The proof of theorem 12.8 can easily be adapted to the systems E-G$_n$A$^\omega$ instead of E-PA$^\omega$ for $n \geq 3$. Let us now consider the case of E-G$_2$A$^\omega$: one observes that the formulation of Σ^0_1-UB as well as the proofs of propositions 12.6, 12.7 can be carried out with E-G$_2$A$^\omega$. This is the case because we don't use the (exponential) sequence coding $\overline{f}n := \langle f(0), \ldots, f(n-1)\rangle$ which codes an initial segment of f into a number. Instead we use the function $\overline{f,n}$ which can be defined (uniformly in f,n) just using definition of cases which is available in E-G$_2$A$^\omega$. Inspecting the proof of theorem 12.8 we realize that the sequence coding $\overline{f}n$ is used only in the final part of the proof to eliminate the 'remains' of F^- after the bound Ψ_2 has been extracted already. As a result, we do get a polynomial bound when working in E-G$_2$A$^\omega$. It is only the verification of that bound which uses the exponential sequence coding and, therefore, has to take place in G$_3$A$^\omega_i$. In particular, we can extend the proof of theorem 10.52 to obtain the following theorem:

Theorem 12.32. *Let* $A_0(x^0, y^1, z^0) \in \mathcal{L}$(E-G$_2A^\omega$) *be a quantifier-free formula containing only* x, y, z *as free variables and* s *a closed term of* E-G$_2$A$^\omega$. *Then the following rule holds:*

$$\begin{cases} \text{E-}G_2A^\omega + \text{QF-AC}^{1,0} + \text{QF-AC}^{0,1} + \Sigma^0_1\text{-UB} \vdash \forall x^0 \forall y \leq_1 sx \exists z^0\, A_0(x,y,z) \\ \Rightarrow \textit{one can extract a polynomial } p \textit{ s.t.} \\ G_3A^\omega_i \vdash \forall x^0 \forall y \leq_1 sx \exists z \leq_0 p(x)\, A_0(x,y,z). \end{cases}$$

The result also holds in the presence of additional axioms Δ *as in theorem 10.47. Then the conclusion is provable in* $G_3A^\omega_i + \tilde{\Delta}_\varepsilon$.

It is this feature of Σ^0_1-UB which makes this principle so superior in proof mining to the weak König's lemma WKL: not only are proofs of usual WKL-applications much easier when Σ^0_1-UB is used instead but also, Σ^0_1-UB often allows one to carry out such proofs even relative to E-G_2A^ω where WKL is not available as the very formulation of WKL is based on exponential sequence codings. Indeed, the applications 1–5 above all can be carried out relative to E-G_2A^ω if the representation of real numbers is based on Cauchy sequences of rationals with rate of convergence $1/(n+1)$ instead of 2^{-n} (at a few places the proofs have to be slightly modified, see [213] for details).

We conclude this chapter with a reformulation of WKL which can be formulated already in the language of E-G_2A^ω :

We first generalize WKL to a sequential version WKL_{seq} which states that for every sequence of infinite 0,1-trees there exists a sequence of infinite branches:

Definition 12.33.

$$\text{WKL}_{seq} :\equiv \begin{cases} \forall f^{1(0)} \big(\forall k^0 (T(fk) \wedge \forall x^0 \exists n^0 (lth\ n =_0 x \wedge fkn =_0 0)) \\ \qquad \rightarrow \exists b \leq_{1(0)} \lambda k^0, i^0.1 \forall k^0, x^0 (fk((\overline{bk})x) =_0 0)\big). \end{cases}$$

We now introduce a different formulation WKL^2_{seq} of WKL_{seq} which avoids the coding of finite sequences (of variable length) as numbers and can be used in G_2A^ω :

Definition 12.34.

$$\text{WKL}^2_{seq} :\equiv \begin{cases} \forall \Phi^{0010} \big(\forall k^0, x^0 \exists b \leq_1 \lambda n^0.1^0 \bigwedge_{i=0}^{x} (\Phi k(\overline{b,i})i =_0 0) \\ \qquad \rightarrow \exists b \leq_{1(0)} \lambda k^0, n^0.1 \forall k^0, x^0 (\Phi k(\overline{bk,x})x =_0 0)\big). \end{cases}$$

Over $G_3A^\omega_i$ (where the sequence coding functional $\Phi_{\langle\cdot\rangle}$ is available) the new formulation WKL^2_{seq} is equivalent to WKL_{seq} :

Proposition 12.35. $G_3A^\omega_i \vdash \text{WKL}^2_{seq} \leftrightarrow \text{WKL}_{seq}$.

Proof: '$\rightarrow$': Consider the functional

$$\Phi k^0 b^1 x^0 := fk(\overline{b}x)$$

and assume that

$$\forall k^0 T(fk) \text{ and } \quad (+)\ \forall k,x\exists n(lth\ n = x \wedge fkn = 0).$$

Then, putting $b := \lambda i.(n)_i$ for 'n' from (+), we get

$$\forall k,x\exists b \leq \lambda n.1 \bigwedge_{i=0}^{x} (\Phi k(\overline{b,i})i =_0 0).$$

We now can apply WKL^2_{seq} and obtain

$$\exists b \leq \lambda k,n.1 \forall k,x(\Phi k(\overline{bk,x})x =_0 0),$$

i.e.

$$\exists b \leq \lambda k,n.1 \forall k,x\big(fk((\overline{bk})x) =_0 0\big).$$

'$\leftarrow$': Define

$$fkn := \begin{cases} \Phi k(\lambda i.(n)_i)(lth\ n), \text{ if } \forall j \leq lth\ n\big((\Phi k(\overline{\lambda i.(n)_i, j})j =_0 0\big) \wedge (n)_j \leq 1\big) \\ 1^0, \text{ otherwise.} \end{cases}$$

The assumption

$$\forall k,x\exists b \leq_1 \lambda n^0.1^0 \bigwedge_{i=0}^{x} \big(\Phi k(\overline{b,i})i =_0 0\big)$$

implies that

$$\forall k,x\exists n(lth\ n = x \wedge fkn = 0).$$

Furthermore, since $T(fk)$ for all k (by f-definition), we can use WKL_{seq} which gives

$$\exists b \leq_{1(0)} \lambda k,n.1 \forall k^0,x^0\big(fk((\overline{bk})x) =_0 0\big),$$

i.e.

$$\exists b \leq \lambda k,n.1 \forall k,x(\Phi k(\overline{bk,x})x =_0 0).$$

□

Theorem 12.36. $\text{G}_2\text{A}^\omega + \text{QF-AC}^{0,1} \vdash \Sigma^0_1\text{-UB}^- \to \text{WKL}^2_{seq}$.

Proof: We prove the contraposite form of WKL^2_{seq} and assume

$$\forall b \leq_{1(0)} \lambda k^0,i^0.1 \exists k^0,x^0\big(\Phi k(\overline{bk,x})x \neq_0 0\big).$$

Since the type $1(0)$ can be coded in type 1, we can apply $\Sigma^0_1\text{-UB}^-$ and obtain

$$(*)\ \exists x_0 \forall b \leq_{1(0)} \lambda k,i.1 \exists k,x \leq_0 x_0 \big(\Phi k(\underbrace{\overline{(\overline{bk,x_0}),x}}_{=_1 \overline{bk,x}})x \neq_0 0\big).$$

Now suppose that

$$\forall k^0, x^0 \exists b^1 \big(\bigwedge_{i=0}^{x} (bi \leq_0 1 \land \Phi k(\overline{b,i})i =_0 0) \big).$$

Applying QF-AC^{0,1} gives

$$\forall x^0 \exists b^{1(0)} \forall k^0 \big(\bigwedge_{i=0}^{x} (bki \leq_0 1 \land \Phi k(\overline{bk,i})i =_0 0) \big).$$

Observe that $\overline{bk,i} =_1 \overline{(\overline{bk,x}),i}$ for $i \leq x$ and $\overline{bk,x} \leq_1 \lambda i.1$ if $\bigwedge_{i=0}^{x} (bki \leq_0 1)$.
Hence

$$\forall x^0 \exists b \leq_{1(0)} \lambda k, i.1 \forall k \bigwedge_{i=0}^{x} \big(\Phi k(\overline{bk,i})i = 0 \big),$$

which contradicts $(*)$. □

Together with theorem 12.32 this theorem implies the following

Corollary 12.37.

$$\begin{cases} \text{E-G}_2\text{A}^\omega + \text{QF-AC}^{1,0} + \text{QF-AC}^{0,1} + \text{WKL}^2_{seq} \vdash \forall x^0 \forall y \leq_1 sx \exists z^0 A_0(x,y,z) \\ \Rightarrow \textit{one can extract a polynomial } p \textit{ s.t.} \\ \text{G}_3\text{A}^\omega_i \vdash \forall x^0 \forall y \leq_1 sx \exists z \leq_0 p(x)\, A_0(x,y,z), \end{cases}$$

where $s \in \mathrm{G_2R}^\omega$ and A_0 is a quantifier-free formula of $\mathrm{G_2A}^\omega$ *which contains only x,y,z as free variables.*

12.4 Exercises, historical comments and suggested further reading

Exercises:

1) Show that E-PA$^\omega$+AC^{1,0} + Σ^0_1-UB$^-$ is inconsistent.
2) Show that $\mathcal{M}^\omega \not\models \Sigma^0_1\text{-UB}^-$, but $\mathcal{C}^\omega \models \Sigma^0_1\text{-UB}$.
3) Let Σ^0_2-UB be defined as Σ^0_1-UB but with A being a Σ^0_2-formula (with parameters of arbitrary types). Show that E-PA$^\omega$ + Σ^0_2-UB is inconsistent.
4) Prove lemma 12.13.
5) Show that $\mathcal{M}^\omega \models \tilde{F}^\rho$.
6) Complete the proof of proposition 12.27.
7) Verify the statements in remarks 12.18, 12.22, 12.24, 12.26 and 12.29.

Historical comments and suggested further reading:

A special case of F and F^- was first considered in Kohlenbach [206] where it was shown that this weaker form of F^- can be eliminated by monotone functional interpretation. In the form discussed in this chapter, F and F^- as well as Σ^0_1-UB and a corresponding weakening Σ^0_1-UB$^-$ were studied in Kohlenbach [207]. In that paper a proof-theoretic elimination procedure of Σ^0_1-UB$^-$ was established and a combination of monotone functional interpretation with a model theoretic argument was used to show that Σ^0_1-UB does not create new provably recursive functionals of type-2 even relative to weak systems as G_2A^ω. In Kohlenbach [223], it was observed that in the presence of full extensionality, F^- already implies F. In Kohlenbach [213], is was shown that Σ^0_1-UB allows one (even relative to systems as weak as G_2A^ω) to give short proofs of important analytic theorems, such as Dini's theorem, the existence of an inverse for strongly monotone functions $f \in C[0,1]$, the attainment of the maximum of such functions, the (sequential) Heine-Borel compactness of $[0,1]^d$ and others. Since the use of Σ^0_1-UB makes it superfluous to state the (uniform) continuity assumptions explicitly, these proofs – as we saw above – are particularly simple and don't require any encoding of functions $f \in C[0,1]$ as type-1 objects which would be necessary to obtain the same results by WKL (see Simpson [338] and Kohlenbach [223]). The results in section 12.3 are taken from Kohlenbach [207]. In Kohlenbach [210], a combination of (function parameter-free) Π^0_1-comprehension and Σ^0_1-UB was used to derive fixed instances of the Bolzano-Weierstraß principle and the Ascoli-lemma relative to G_2A^ω which allowed a precise calibration of the contribution of the use of such instances to the provably recursive functionals (see also chapter 13).

For applications of F in an (semi-)intuitionistic context see [212]. More general forms of F and Σ^0_1-UB are discussed in Kohlenbach [212] and Ferreira-Oliva [104]. The latter paper contains a systematic approach to conservation results for generalized fan principles including Σ^0_1-UB over classical systems based on a new bounded functional interpretation. In Kohlenbach [228], Σ^0_1-UB is extended to the new types for abstract metric and hyperbolic spaces which we will discuss in chapter 17 below. In [6], Avigad defines a nonstandard extension of a system of elementary recursive arithmetic in higher types (roughly corresponding to G_3A^ω) which has certain features that are similar to the Σ^0_1-UB consequences shown in this chapter. E.g. this system also proves that all functions $f : [0,1] \to \mathbb{R}$ are uniformly continuous.

Chapter 13
Elimination of monotone Skolem functions

13.1 Skolem functions of type degree 1 in fragments of finite type arithmetic

Let us begin with a

Notational convention: throughout this chapter, when dealing with the systems $\mathrm{G}_n\mathrm{A}^\omega$,$\mathrm{G}_n\mathrm{A}^\omega_i$ from chapter 3, we will not include all the arbitrary universal axioms '11)' but only universal sentence that w.r.t. to the canonical embedding of $\mathrm{G}_n\mathrm{A}^\omega$ into $\widehat{\text{WE-PA}}^\omega\!\upharpoonright\!\backslash$ (i.e. with the standard primitive recursive definitions of the constants and the relation $\leq$ from $\mathcal{L}(\mathrm{G}_n\mathrm{A}^\omega)$ that are not included as primitive notions in $\mathcal{L}(\widehat{\text{WE-PA}}^\omega\!\upharpoonright\!\backslash)$) are provable in $\widehat{\text{WE-PA}}^\omega\!\upharpoonright\!\backslash$. Most notably this yields the schema of quantifier-free induction QF-IA. In this way these systems will (modulo that aforementioned canonical embedding) be subsystems of $\widehat{\text{WE-PA}}^\omega$, $\widehat{\text{WE-HA}}^\omega$. Since we allow in the main results of this chapter arbitrary further axioms of the form Δ (as introduced in theorem 10.21) this, anyway, covers any additional universal axioms we might want to use as well. The reason why we included arbitrary universal axioms when defining the systems $\mathrm{G}_n\mathrm{A}^\omega$ in chapter 3 was that for the cases where $n = 1,2$ it would be quite tedious to verify whether certain basic arithmetical facts are provable. However in this chapter we mainly deal with the case $n \geq 3$ or even $n = \infty$ where these things are much easier to verify.

Let A be a sentence in prenex normal form (in the language of some first order theory $\mathcal{T}$) and A^H its Herbrand normal form, where the index functions $f_1,\ldots,f_n$ used to built the Herbrand normal form are new (i.e. they do neither occur in A nor in any axiom of $\mathcal{T}$). Let, furthermore, $\mathcal{T}[\underline{f}]$ be the theory which results from $\mathcal{T}$ by adding the function symbols $\underline{f}$ to the language **without** extending any non-logical axiom schema of $\mathcal{T}$ to the new language. As mentioned already in chapter 2, $A \rightarrow A^H$ already holds in first order logic, while $A^H \rightarrow A$ is not logically valid (in first order logic). However, for first order theories $\mathcal{T}$ one easily verifies (interpreting the index functions by appropriate choice functions) that

$$(+)\ \mathscr{T}[\underline{f}] \models A^H \Rightarrow \mathscr{T} \models A$$

and so, by the completeness theorem for first order logic,

$$(++)\ \mathscr{T}[\underline{f}] \vdash A^H \Rightarrow \mathscr{T} \vdash A.$$

Moreover, from a constructive (syntactic) proof of Herbrand's theorem one obtains a procedure that transforms a $\mathscr{T}[\underline{f}]$-proof of A^H into a $\mathscr{T}$-proof of A. If no $\underline{f}$-equality axioms are used in the $\mathscr{T}[\underline{f}]$-proof of A^H, this procedure follows the strategy illustrated at an example in chapter 2 (see the discussion before Herbrand's theorem 2.18 in chapter 2). If $=$-axioms of the form

$$\bigwedge_{j=1}^{i}(s_j = t_j) \rightarrow f_i(s_1,\ldots,s_i) = f_i(t_1,\ldots,t_i)$$

are used in the given proof for some $1 \leq i \leq n$, then things are already more involved (see e.g. [332] for a textbook treatment).

We will show next, that $(++)$ is no longer true in general for systems with function quantifiers since then new function symbols f may automatically become subject to axioms of the form $\forall g A$ via $\forall g$-elimination resulting in $A[f/g]$. Similarly, for systems with free function variables only (but no function quantifiers) but the substitution rule

$$(\text{Sub}): \quad \frac{A}{A[\varphi/f]},$$

where f is an n-ary function variable and φ an n-ary function term, instead. In fact, this failure already shows up as soon as we have axiom schemata such as QF-IA around in which function variables are allowed to occur. In particular, this applies to our systems $\mathrm{G}_n\mathrm{A}^\omega$ and $\widehat{\mathrm{(W)E\text{-}PA}}^\omega\!\restriction$.

Let Π^0_∞-IA$^-$ denote the schema of induction for all arithmetical formulas (i.e. formulas only containing quantifiers over variable of type 0) of $\mathscr{L}(\mathrm{G}_\infty\mathrm{A}^\omega)$ which only have parameters of type 0. Then (via a canonical embedding) PA is a subsystem of $\mathrm{G}_\infty\mathrm{A}^\omega + \Pi^0_\infty$-IA$^-$. Since we have function variables available in $\mathscr{L}(\mathrm{G}_\infty\mathrm{A}^\omega)$ we formulate the Herbrand normal form A^H of a sentence

$$\exists x_1^0 \forall y_1^0 \ldots \exists x_n^0 \forall y_n^0 A_0(x_1,y_1,\ldots,x_n,y_n)$$

as in the case of the no-counterexample interpretation (see chapter 2) with universally quantifier function variables, i.e.

$$A^H :\equiv \forall f_1,\ldots,f_n \exists y_1,\ldots,y_n A_0(x_1,f_1(x_1),\ldots,x_n,f_n(x_1,\ldots,x_n)).$$

Proposition 13.1. *Let $A \in \Pi^0_\infty$ be a theorem of* PA. *Then one can construct a sentence $\tilde{A} \in \Pi^0_\infty$ in the language $\mathscr{L}$*(PA) *of* PA *such that (modulo the embedding of* PA *into $\mathrm{G}_\infty\mathrm{A}^\omega + \Pi^0_\infty$-IA$^-$)*

$$\mathrm{G}_\infty\mathrm{A}^\omega \vdash \tilde{A}^H \ \ \textit{and}\ \ \mathrm{G}_\infty\mathrm{A}^\omega \vdash A \leftrightarrow \tilde{A}.$$

In fact, the proof of $A \leftrightarrow \tilde{A}$ *only uses classical logic and* $0 \neq 1$.

Proof: Let A be a sentence in $\mathscr{L}(\mathrm{PA})$ such that $\mathrm{PA} \vdash A$. Now let $\widehat{F}_1,\ldots,\widehat{F}_k$ denote the arithmetical instances (even without function parameters) of the induction schema IA in $\mathscr{L}(\mathrm{G}_\infty\mathrm{A}^\omega)$ resulting from the instances of the induction schema in $\mathscr{L}(\mathrm{PA})$ used in the proof under the canonical embedding of PA into $\mathrm{G}_\infty\mathrm{A}^\omega + \Pi^0_\infty\text{-IA}^-$. Consider their respective universal closures $\tilde{F}_1,\ldots,\tilde{F}_k$. Then the following holds

$$\mathrm{G}_\infty\mathrm{A}^\omega \vdash \bigwedge_{i=1}^{k} \tilde{F}_i \rightarrow A.$$

The formulas $\widehat{F}_i$ have the form

$$\widehat{F}_i \equiv F_i(0) \wedge \forall x\,(F_i(x) \rightarrow F_i(x+1)) \rightarrow \forall x\,F_i(x).$$

Let B be any prenex normal form of $\big(\bigwedge_{i=1}^{k} (y_i =_0 0 \leftrightarrow F_i(x_i)) \rightarrow A\big)$.

Then

$$\tilde{A} :\equiv \exists \underline{a}, x_1,\ldots,x_k \forall y_1,\ldots,y_k\, B(x_1,\ldots,x_k,y_1,\ldots,y_k,\underline{a})$$

is a prenex normal form of

$$\forall \underline{a}, x_1,\ldots,x_k \exists y_1,\ldots,y_k \bigwedge_{i=1}^{k} (y_i = 0 \leftrightarrow F_i(x_i)) \rightarrow A,$$

where $\underline{a}$ comprises all the (number) parameters of the induction formulas F_i for $i = 1,\ldots,k$.
Because of

$$\mathrm{G}_\infty\mathrm{A}^\omega \vdash \forall \underline{a}, x_1,\ldots,x_k \exists y_1,\ldots,y_k \bigwedge_{i=1}^{k} (y_i = 0 \leftrightarrow F_i(x_i)),$$

we obtain

$$\mathrm{G}_\infty\mathrm{A}^\omega \vdash A \leftrightarrow \tilde{A}.$$

Now define

$$C :\equiv \forall f_1,\ldots,f_k \exists \underline{a}, x_1,\ldots,x_k\, B(x_1,\ldots,x_k, f_1\underline{a}x_1\ldots x_k,\ldots,f_k\underline{a}x_1\ldots x_k,\underline{a}).$$

C is a partial Herbrand normal form of $\tilde{A}$ and so implies the full Herbrand normal form $\tilde{A}^H$ even logically.
It, therefore, remains to show that $\mathrm{G}_\infty\mathrm{A}^\omega \vdash C$:
Assume $\forall \underline{a}, x_1,\ldots,x_k \bigwedge_{i=1}^{k} \big(f_i\underline{a}x_1\ldots x_k = 0 \leftrightarrow F_i(x_i)\big)$. Quantifier-free induction applied to

$$A_0(x_i) :\equiv \big(f_i(\underline{a},0,\ldots,0,x_i,0,\ldots,0)=0\big)$$

(instead of using IA applied to F_i) yields $\tilde{F}_i$.
Hence it follows that

$$\mathrm{G}_\infty\mathrm{A}^\omega \vdash \forall \underline{f}\big(\forall \underline{a},x_1,\ldots,x_k \bigwedge_{i=1}^{k}\big(f_i\underline{a}x_1\ldots x_k=0 \leftrightarrow F_i(x_i)\big) \to A\big),$$

i.e. $\mathrm{G}_\infty\mathrm{A}^\omega \vdash C$.

□

Corollary 13.2. *For each $n\in\mathbb{N}$ one can construct a sentence $A\in\Pi^0_\infty$ such that*

$$\mathrm{G}_\infty\mathrm{A}^\omega \vdash A^H, \textit{ but } \mathrm{G}_\infty\mathrm{A}^\omega+\Sigma^0_n\text{-IA} \subseteq \widehat{\text{WE-PA}}^\omega{\restriction}+\Sigma^0_n\text{-IA} \nvdash A.$$

Proof: It is well-known (see e.g. [298](theorem 5),[299] and [355]) that for $n\geq 1$ there are functions that are provably recursive in $\widehat{\text{WE-PA}}^\omega{\restriction}+\Sigma^0_{n+1}$-IA that grow faster than any function provably recursive in $\widehat{\text{WE-PA}}^\omega{\restriction}+\Sigma^0_n$-IA. So there is a Π^0_2-sentence $A\equiv\forall x\exists y T(\overline{e},x,y)$ (for some $e\in\mathbb{N}$) that is provable in $\widehat{\text{WE-PA}}^\omega{\restriction}+\Sigma^0_{n+1}$-IA but not in $\widehat{\text{WE-PA}}^\omega{\restriction}+\Sigma^0_n$-IA. Now take the corresponding sentence $\tilde{A}$ according to proposition 13.1. It follows that $\mathrm{G}_\infty\mathrm{A}^\omega\vdash\tilde{A}^H$, but $\widehat{\text{WE-PA}}^\omega{\restriction}+\Sigma^0_n$-IA $\nvdash\tilde{A}$. □

Remark 13.3. 1) The proof above can be used to show that corollary 13.2 remains valid even if we add to $\widehat{\text{WE-PA}}^\omega{\restriction}+\Sigma^0_n$-IA all $\mathcal{S}^\omega$-true universal sentences as axioms since the latter do not contribute to the provably recursive functions as follows by the ND-interpretation.

2) Applying the elimination of the primitive recursive function symbols (included in our formulation of PA) in terms of formulas expressing their graphs and $0,S,+,\cdot$ only (on which the standard proof of the fact that our version of PA is a conservative extension over that with $0,S,+,\cdot$ only is based), one can improve corollary 13.2 to get A in this restricted language with even $\mathrm{G}_2\mathrm{A}^\omega$ proving A^H while still having $\widehat{\text{WE-PA}}^\omega{\restriction}+\Sigma^0_n$-IA $\nvdash A$.

The reason for the provability of $\tilde{A}^H$ in proposition 13.1 is that the schema of quantifier-free induction is applicable to the index functions used in defining $\tilde{A}^H$. In fact, the proof did not use any of the higher type functionals. In particular, proposition 13.1 is also valid for a fragment of $\mathrm{G}_\infty\mathrm{A}^\omega$ which no higher type variables and quantifiers and only free function variables (in QF-IA) plus a substitution rule (Sub) that would allow to replace a function variable by any function term (see [208] for more details).

13.2 Elimination of Skolem functions for monotone formulas

Corollary 13.2 shows that for theories like $G_\infty A^\omega$ and (by remark 13.3) G_nA^ω (with $n \geq 2$) with or without further universal axioms added the Herbrand normal form A^H of an arithmetical formula A in general is much weaker than A with respect to provability in G_nA^ω.

In this section we will show the rather nontrivial fact that this phenomenon does not occur if A satisfies a certain monotonicity condition. This will enable us to replace certain implicative assumptions on the existence of Skolem functions in provable theorems by weaker purely arithmetical assumptions which do not involve Skolem functions (see corollary 13.12 below).

In the subsequent sections this is used to calibrate the contribution of single uses of the monotone convergence principle and the Bolzano-Weierstraß principle to the growth of bounds extractable from provable $\forall\exists$-statements.

Definition 13.4. Let $A \in \mathcal{L}(G_nA^\omega)$ be a formula having the form

$$A \equiv \forall u^1 \forall v \leq_\tau tu \exists y_1^0 \forall x_1^0 \ldots \exists y_k^0 \forall x_k^0 \exists w^\gamma A_0(u,v,y_1,x_1,\ldots,y_k,x_k,w),$$

where A_0 is quantifier–free and contains only $u,v,\underline{y},\underline{x},w$ free. Furthermore let t be $\in G_nR^\omega$ and τ,γ are arbitrary finite types.

1) A is called monotone if

$$Mon(A) :\equiv \begin{cases} \forall u^1 \forall v \leq_\tau tu \forall x_1,\tilde{x}_1,\ldots,x_k,\tilde{x}_k,y_1,\tilde{y}_1,\ldots y_k,\tilde{y}_k \\ \Big(\bigwedge\limits_{i=1}^{k} (\tilde{x}_i \leq_0 x_i \wedge \tilde{y}_i \geq_0 y_i) \wedge \exists w^\gamma A_0(u,v,y_1,x_1,\ldots,y_k,x_k,w) \\ \qquad \rightarrow \exists w^\gamma A_0(u,v,\tilde{y}_1,\tilde{x}_1,\ldots,\tilde{y}_k,\tilde{x}_k,w)\Big). \end{cases}$$

2) The Herbrand normal form A^H of A is defined to be

$$A^H :\equiv \forall u^1 \forall v \leq_\tau tu \forall h_1^{\rho_1},\ldots,h_k^{\rho_k} \exists y_1^0,\ldots,y_k^0,w^\gamma$$
$$\underbrace{A_0(u,v,y_1,h_1y_1,\ldots,y_k,h_ky_1\ldots y_k,w)}_{A_0^H :\equiv}, \text{ where } \rho_i = 0\underbrace{(0)\ldots(0)}_{i}.$$

Remark 13.5. Note that our above notion of Herbrand normal form of A is nothing else but the usual Herbrand normal form of the following prenex normal form

$$\forall u^1 \forall v^\tau \exists y_1^0 \forall x_1^0 \ldots \exists y_k^0 \forall x_k^0 \exists w^\gamma \exists \underline{z}(v\underline{z} \leq_0 tu\underline{z} \rightarrow A_0(u,v,y_1,x_1,\ldots,y_k,x_k,w)),$$

of A.

Definition 13.6. An n-ary number theoretic function $f^{0(0)\ldots(0)}$ is called monotone if

$$\forall x_1,\ldots,x_n,y_1,\ldots,y_n\big(\bigwedge_{i=1}^{n}(x_i \geq_0 y_i) \to f\underline{x} \geq_0 f\underline{y}\big).$$

Theorem 13.7. *Let $n \geq 1$ and $\Psi_1,\ldots,\Psi_k \in \mathrm{G}_n\mathrm{R}^\omega$. Then*

$$\mathrm{G}_n\mathrm{A}^\omega + Mon(A) \vdash \forall u^1 \forall v \leq_\tau tu \forall h_1,\ldots,h_k\Big(\bigwedge_{i=1}^{k}(h_i \; monotone) \\ \to \exists y_1 \leq_0 \Psi_1 u\underline{h} \ldots \exists y_k \leq_0 \Psi_k u\underline{h} \exists w^\gamma A_0^H\Big) \to A.$$

Remark 13.8. In theorem 13.7 one may also have tuples '$\exists \underline{w}$' instead of '$\exists w^\gamma$' in A.

Proof of theorem 13.7: Let us assume the premise of the implication to be proved. In fact, by taking $\Psi u\underline{h} := \max_0(\Psi_1 u\underline{h},\ldots,\Psi_k u\underline{h})$, we may assume that

$$(0)\forall u^1 \forall v \leq_\tau tu \forall h_1,\ldots,h_k\big(\bigwedge_{i=1}^{k}(h_i \text{ monotone}) \to \exists y_1,\ldots,y_k \leq_0 \Psi u\underline{h} \exists w^\gamma A_0^H\big).$$

Note that the function variables $u,h_1,\ldots,h_k$ all have types of degree ≤ 1. Hence – by corollary 3.42 – one can construct a term $\Psi^*[u,\underline{h}]^0$ such that

1) $\Psi^*[u,\underline{h}]$ is built up from $u,\underline{h},0^0,A_0,\ldots,A_n$ only (by application).
2) $\lambda u,\underline{h}.\Psi^*[u,\underline{h}] \; maj \; \Psi$.

1) in particular implies:
1*) Every occurrence of an $h_j \in \{h_1,\ldots,h_k\}$ in $\Psi^*[u,\underline{h}]$ has the form $h_j(r_{n_1},\ldots,r_{n_j})$, i.e. h_j occurs only with a full stock of arguments but not as a function argument in the form $s(h_j r_{n_1}\ldots r_{n_l})$ for some $l < j$.
By 2), $\forall u^1(u^M \; maj \; u)$ (where $u^M x := \max_{i\leq x} ui$) and ($h_i$ monotone $\to h_i \; maj \; h_i$) we have
2*) $\mathrm{G}_n\mathrm{A}^\omega \vdash \forall u \forall h_1,\ldots,h_k\big(\bigwedge_{i=1}^{k}(h_i \text{ monotone}) \to \Psi^*[u^M,\underline{h}] \geq_0 \Psi u\underline{h}\big)$.
Note that the replacement of h_i by $h_i^M := \lambda x_1,\ldots,x_i.\max_{\substack{\tilde{x}_1\leq x_1\\ \vdots \\ \tilde{x}_i\leq x_i}} h(\tilde{x}_1,\ldots,\tilde{x}_i)$, which would make the monotonicity assumption on h_i superfluous, would destroy property 1*) on which the proof below is based. This is the reason why we have to assume h_i to be monotone. In order to overcome this assumption we will use essentially the monotonicity of A.
Let $r_1,\ldots,r_l$ be all the subterms of $\Psi^*[u^M,\underline{h}]$ that occur as an argument of some of the functions $h_1,\ldots,h_k$ in $\Psi^*[u^M,\underline{h}]$ plus the term $\Psi^*[u^M,\underline{h}]$ itself.
Let $\widehat{r}_j[a_1,\ldots,a_{q_j}]$ be the term which results from r_j by replacing every occurrence of a maximal $h_1,\ldots,h_k$-subterm (i.e. a maximal subterm which has the form $h_i(s_1,\ldots,s_i)$ for an $i = 1,\ldots,k$) by a new variable and let $a_1,\ldots,a_{q_j}$ denote these variables. We now define

$$\tilde{r}_j a_1 \dots a_{q_j} := \max\Big(\max_{\substack{\tilde{a}_1 \le a_1 \\ \vdots \\ \tilde{a}_{q_j} \le a_{q_j}}} \widehat{r}_j[\tilde{a}_1,\dots,\tilde{a}_{q_j}], a_1,\dots,a_{q_j}\Big),$$

with $\tilde{r}_j := \widehat{r}_j$ in the case where $\widehat{r}_j$ does not contain any variable $a_1,\dots,a_{q_j}$ (which is the case when r_j does not contain any $h_1,\dots,h_k$-term). Note that $\tilde{r}_j$ can be defined in $\mathrm{G}_n\mathrm{R}^\omega$ from $\widehat{r}_j$ by successive use of Φ_1.
By the construction of $\tilde{r}_j$ we get

$$\mathrm{G}_n\mathrm{A}^\omega \vdash \big(\tilde{r}_j \; maj \; \lambda \underline{a}.\widehat{r}_j[a_1,\dots,a_{q_j}]\big) \wedge \forall \underline{a}(\tilde{r}_j\underline{a} \ge_0 a_1,\dots,a_{q_j}).$$

Since $\Psi^*[u^M,\underline{h}]$ is built up from $\widehat{r}_j,\underline{h}$ only (by substitution) and (h_i monotone $\to$ $h_i \; maj \; h_i$), $u^M \; maj \; u$, this implies

$$\mathrm{G}_n\mathrm{A}^\omega \vdash \forall u, h_1,\dots,h_k\big(\bigwedge_{i=1}^{k}(h_i \text{ monotone}) \to \overline{\Psi}[u^M,\underline{h}] \ge_0 \Psi^*[u^M,\underline{h}] \ge_0 \Psi u\underline{h}\big),$$

where $\overline{\Psi}[u^M,\underline{h}]$ is built up as $\Psi^*[u^M,\underline{h}]$ but with $\tilde{r}_j(a_1,\dots,a_{q_j})$ instead of $\widehat{r}_j[a_1,\dots,a_{q_j}]$.
Summarizing the situation achieved so far we have obtained a term $\overline{\Psi}[u^M,\underline{h}]$ such that

(α) $\forall u^1 \forall v \le_\tau tu \forall \underline{h}\big(\underline{h} \text{ monotone} \to \exists y_1,\dots,y_k \le_0 \overline{\Psi}[u^M,\underline{h}] \exists w^\gamma A_0^H\big)$.
(β) $h_1,\dots,h_k$ occur in $\overline{\Psi}[u^M,\underline{h}]$ only as in 1^*), i.e. with all places for arguments filled and not as function arguments themselves.
(γ) For $\overline{\Psi}[u^M,\underline{h}]$ and all subterms s which occur as an argument of a function $h_1,\dots,h_k$ in $\overline{\Psi}[u^M,\underline{h}]$ we have $\mathrm{G}_n\mathrm{A}^\omega \vdash \widehat{s}[a_1,\dots,a_q] \ge_0 a_1,\dots,a_q$, where $\widehat{s}$ results by replacing every occurrence of a maximal $h_1,\dots,h_k$-subterm in s by a new variable a_l.

(α) follows from (0), i.e.

$$\mathrm{G}_n\mathrm{A}^\omega \vdash (0) \to (\alpha),$$

and (β), (γ) do not depend on any assumption.

For the rest of the proof we only use (α)-(γ) and $Mon(A)$:

From now on let $r_1,\dots,r_l$ denote all subterms of $\overline{\Psi}[u^M,\underline{h}]$ which occur as an argument of a function $\in \{h_1,\dots,h_k\}$ in $\overline{\Psi}[u^M,\underline{h}]$ plus $\overline{\Psi}[u^M,\underline{h}]$ itself. $M := \{r_1,\dots,r_l\}$ (This set formation is meant w.r.t. syntactic identity $\equiv$ of terms and not $=_0$, i.e. '$s \in M$' means '$s \equiv r_1 \vee \dots \vee s \equiv r_l$').
We now show that we can reduce '$\exists y_1,\dots,y_k \le \overline{\Psi}[u^M,\underline{h}]$' in ($\alpha$) to a disjunction with fixed length, namely to the disjunction over M. Moreover, it suffices to assume that the functions $h_1,\dots,h_k$ are monotone on M :

$$(1)\begin{cases}\forall u^1\forall v\le_\tau tu\forall \underline{h}\big(\underline{h}\text{ monotone on }M\rightarrow\exists s_1,\ldots,s_k\in M\exists w^\gamma\\ \qquad\qquad A_0(u,v,s_1,h_1s_1,\ldots,s_k,h_ks_1\ldots s_k,w)\big).\end{cases}$$

Proof of (1): Let $h_1,\ldots,h_k$ be monotone on M. We order the terms r_i w.r.t. $\le_0$. The order of the resulting ordered tuple depends of course on $u,h_1,\ldots,h_k$. For notational simplicity we assume that $r_1\le_0\ldots\le_0 r_l$. We now define (again depending on $u,\underline{h}$) functions $\tilde{h}_1,\ldots,\tilde{h}_k$ by

$$\tilde{h}_iy_1^0\ldots y_i^0:=h_i(r_{j_{y_1}},\ldots,r_{j_{y_i}}),\text{ where}$$

$$j_{y_q}:=\begin{cases}1,\text{ if }y_q\le_0 r_1\\ j+1,\text{ if }r_j<_0 y_q\le_0 r_{j+1}\\ l,\text{ if }r_l<_0 y_q.\end{cases}$$

Since l (and therefore the number of cases in this definition of $\tilde{h}_i$) is a fixed number depending only on the term structure of $\overline{\Psi}[u^M,\underline{h}]$ but not on $u,\underline{h}$, the functions $\tilde{h}_i$ can be defined uniformly in $u,\underline{h}$ within $\mathrm{G}_n\mathrm{A}^\omega$ using proposition 3.29. Note that on arguments from M, the values of $\tilde{h}_i$ and h_i coincide.
By the definition of $\tilde{h}_i$ and the assumption that $h_1,\ldots,h_k$ are monotone on M we conclude

$$(a)\ \tilde{h}_1,\ldots,\tilde{h}_k\text{ are monotone everywhere.}$$

By (β) we know that $h_1,\ldots,h_k$ occur in $\overline{\Psi}[u^M,\underline{h}]$ only in the form $h_i(s_1,\ldots,s_i)$ for certain terms $s_1,\ldots,s_i\in M$. Hence we can define the $\underline{h}$-**depth** of a term $s\in M$ as the maximal number of nested occurrences of $h_1,\ldots,h_k$ in s and show by induction on this rank (on the meta-level):

$$(b)\begin{cases}\bigwedge_{i=1}^{l}(r_i=_0\tilde{r}_i),\text{ where }\tilde{r}_i\text{ results if in }r_i\in M\text{ the functions }h_1,\ldots,h_k\\ \text{are replaced by }\tilde{h}_1,\ldots,\tilde{h}_k\text{ everywhere.}\\ \text{In particular }\overline{\Psi}[u^M,\underline{\tilde{h}}]=_0\overline{\Psi}[u^M,\underline{h}].\end{cases}$$

By (a) we can apply (α) to $\tilde{h}_1,\ldots,\tilde{h}_k$ which – using (b) – yields the following: for all $u^1,v\le tu$ and all $\underline{h}$ which are monotone on M) we have that

$$(c)\ \exists y_1,\ldots,y_k\le_0\overline{\Psi}[u^M,\underline{h}]\exists w^\gamma A_0(u,v,y_1,\tilde{h}_1y_1,\ldots,y_k,\tilde{h}_ky_1\ldots y_k,w).$$

Let $y_1,\ldots,y_k\le_0\overline{\Psi}[u^M,\underline{h}]$ be such that (c) is fulfilled. Because of $\tilde{h}_iy_1\ldots y_i=h_i(r_{j_{y_1}},\ldots,r_{j_{y_i}})$ this implies

$$(d)\ \exists w^\gamma A_0(u,v,y_1,h_1r_{j_{y_1}},\ldots,y_k,h_kr_{j_{y_1}}\ldots r_{j_{y_k}},w).$$

With $y_q \le r_{j_{y_q}}$ for $q = 1,\ldots.k$ (since $y_q \le \overline{\Psi}[u^M,\underline{h}] \le r_l$ –because of $\overline{\Psi}[u^M,\underline{h}] \in M$ and the y_q-assumption– the case '$y_q > r_l$' does not occur) and $Mon(A)$ we conclude

$$\exists w^\gamma A_0(u,v,r_{j_{y_1}},h_1 r_{j_{y_1}},\ldots,r_{j_{y_k}},h_k r_{j_{y_1}}\ldots r_{j_{y_k}},w)$$

and therefore

$$(e)\ \exists s_1,\ldots,s_k \in M \exists w^\gamma A_0(u,v,s_1,h_1 s_1,\ldots,s_k,h_k s_1\ldots s_k,w).$$

This concludes the proof of (1) which can easily be carried out in $\mathrm{G}_n\mathrm{A}^\omega$ (assuming $Mon(A),(\alpha)$ and using (β)), i.e.

$$\mathrm{G}_n\mathrm{A}^\omega \vdash Mon(A)\wedge(\alpha)\to(1).$$

We now define $N := \bigcup\limits_{i=1}^{k} N_i$, where $N_i := \{h_i(s_1,\ldots,s_i) : s_1,\ldots,s_i \in M\}$ (Again this set is meant w.r.t. syntactic identity $\equiv$ between terms). With the terms in N we associate new number variables according to their $\underline{h}$-depth as follows: Let p the maximal $\underline{h}$-depth of all terms $\in N$.

1. Let $t \in N$ be a term with $\underline{h}$-depth$(t) = p$. Then $t \mapsto y_i^1$, if $t \in N_i$.
2. Let $t \in N$ be a term with $\underline{h}$-depth$(t) = p-1$. Then $t \mapsto y_i^2$, if $t \in N_i$.

⋮

p. Let $t \in N$ be a term with $\underline{h}$-depth$(t) = 1$. Then $t \mapsto y_i^p$, if $t \in N_i$.

This association of variables to the terms in N has the following properties:

(i) Terms $s_1,s_2 \in N$ with different $\underline{h}$-depth have different variables associated with.
(ii) If $s_1,s_2 \in N$ have the same $\underline{h}$-depth, then the variables associated with s_1 and s_2 are equal iff $s_1,s_2 \in N_i$ for an $i = 1,\ldots,k$, i.e. s_1 and s_2 start with a common function variable h_i.

For $r \in M \cup N$ we define $\widehat{r}$ as the term which results if every maximal $\underline{h}$-subterm occurring in r is replaced by its associated variable. Thus $\widehat{r}$ does not contain $h_1,\ldots,h_k$. For $r \in N$, $\widehat{r}$ is just the variable associated with r. $\widehat{M} := \{\widehat{r} : r \in M\}$.

We now show that (1) implies a certain index function-free (i.e. $h_1,\ldots,h_k$-free) disjunction (see (4) below):
For q with $2 \le q \le p$ let $\widehat{r}_1^q,\ldots,\widehat{r}_{n_q}^q$ be all terms $\in \widehat{M}$ whose smallest upper index i of a variable y_j^i occurring in them equals q (i.e. there occurs a variable y_j^q in the term and for all variables y_m^i occurring in the term, $i \ge q$ holds). Since for $r \in M$ the $\underline{h}$-depth of $h_1(r) \in N$ is strictly greater than those of subterms of r, there are no terms $\widehat{r} \in \widehat{M}$ containing a variable y_j^1. $\widehat{r}_1^{p+1},\ldots,\widehat{r}_{n_{p+1}}^{p+1}$ denote those terms $\in \widehat{M}$ which do not contain any variable y_j^i at all.
To begin with we show that (1) implies that for all u and for all $v \le tu$ the following holds

$$(2)\begin{cases}\forall y_1^1,\ldots,y_k^1;\ldots;y_1^p,\ldots,y_k^p \\ \qquad\Big((+)\to \bigvee_{s_1,\ldots,s_k\in M} \exists w^\gamma A_0(u,v,\widehat{s}_1,\widehat{h_1 s_1},\ldots,\widehat{s}_k,\widehat{h_k s_1\ldots s_k},w)\Big),\end{cases}$$

where

$$(+):\equiv\begin{cases}\bigwedge_{\substack{q=1,\ldots,p-1\\ l=1,\ldots,p-q}} \big(y_1^q,\ldots,y_k^q > \widehat{r}_1^{q+l},\ldots,\widehat{r}_{n_{q+l}}^{q+l},y_1^{q+l},\ldots,y_k^{q+l}\big)\wedge \\ \bigwedge_{q=1,\ldots,p} \big(y_1^q,\ldots,y_k^q > \widehat{r}_1^{p+1},\ldots,\widehat{r}_{n_{p+1}}^{p+1}\big).\end{cases}$$

Here $a_1,\ldots,a_k > b_1,\ldots,b_l$ means $\bigwedge_{\substack{1\le i\le k\\ 1\le j\le l}} (a_i > b_j)$.

Assume that there are values $y_1^1,\ldots,y_k^1;\ldots;y_1^p,\ldots,y_k^p$ such that $(+)$ holds and

$$\bigwedge_{\widehat{s}_1,\ldots,\widehat{s}_k\in\widehat{M}} \neg\exists w^\gamma A_0(u,v,\widehat{s}_1,\widehat{h_1 s_1},\ldots,\widehat{s}_k,\widehat{h_k s_1\ldots s_k},w).$$

We construct (working in $\mathrm{G}_n\mathrm{A}^\omega$) functions $h_1,\ldots,h_k$ which are monotone on M and satisfy

$$\forall s_1,\ldots,s_k\in M\neg\exists w A_0(u,v,s_1,h_1 s_1,\ldots,s_k,h_k s_1\ldots s_k,w)$$

yielding a contradiction to (1): Define for $i=1,\ldots,k$

$$h_i(x_1,\ldots,x_i) :=$$
$$\begin{cases} y_i^{\min_{1\le l\le i}(q_l)-1}, \text{ if } \bigvee_{\widehat{r}_{j_1}^{q_1},\ldots,\widehat{r}_{j_i}^{q_i}\in\widehat{M}} \big((x_1,\ldots,x_i) =_0 (\widehat{r}_{j_1}^{q_1},\ldots,\widehat{r}_{j_i}^{q_i})\big) \\ 0^0, \text{ otherwise.}\end{cases}$$

Note that for $\widehat{r}_{j_i}^{q_i}\in\widehat{M}$ we have $q_i\ge 2$ since e.g. $h_1 r_{j_i} (\in N)$ has an $\underline{h}$-depth which is strictly greater than those of subterms in r_{j_i}. Hence $\min_{1\le l\le i}(q_l)-1\ge 1$.

To see that this definition is well-defined and does what we want we have to verify that $(+)$ implies the following:

(i) The h_i are well-defined i-ary functions : $\mathbb{N}^i\to\mathbb{N}$ and the definition above can be carried out in $\mathrm{G}_n\mathrm{A}^\omega$.
(ii) $\widehat{r}=_0 r$ for all $r\in M\cup N$ (for these $h_1,\ldots,h_k$).
(iii) $h_1,\ldots,h_k$ are monotone on $\widehat{M}$ (and hence –by (ii)– on M).

Ad (i): Consider $(\widehat{r}_{j_1}^{q_1},\ldots,\widehat{r}_{j_i}^{q_i})$ and $(\widehat{r}_{\tilde{j}_1}^{\tilde{q}_1},\ldots,\widehat{r}_{\tilde{j}_i}^{\tilde{q}_i})$. We show that $y_i^{\min_{1\le l\le i}(q_l)-1}\neq y_i^{\min_{1\le l\le i}(\tilde{q}_l)-1}$ implies $(\widehat{r}_{j_1}^{q_1},\ldots,\widehat{r}_{j_i}^{q_i})\neq(\widehat{r}_{\tilde{j}_1}^{\tilde{q}_1},\ldots,\widehat{r}_{\tilde{j}_i}^{\tilde{q}_i})$:

We may assume $\min_{1\le l\le i}(q_l) < \min_{1\le l\le i}(\tilde{q}_l)$. Let l_0 be such that $q_{l_0} = \min_{1\le l\le i}(q_l) \wedge 1 \le l_0 \le i$. $\widehat{r}^{q_{l_0}}_{j_{l_0}}$ contains a variable $y_d^{q_{l_0}}$ for some $d = 1,\ldots,k$. By the property (γ) of $\overline{\Psi}[u^M,\underline{h}]$ this implies

$$\widehat{r}^{q_{l_0}}_{j_{l_0}} \ge y_d^{q_{l_0}} \overset{(+),q_{l_0}<\tilde{q}_{l_0}}{>} \widehat{r}^{\tilde{q}_{l_0}}_{\tilde{j}_{l_0}} \text{ and thus } (\widehat{r}^{q_1}_{j_1},\ldots,\widehat{r}^{q_i}_{j_i}) \neq (\widehat{r}^{\tilde{q}_1}_{\tilde{j}_1},\ldots,\widehat{r}^{\tilde{q}_i}_{\tilde{j}_i}).$$

Hence h_i can be defined in $\mathrm{G}_n\mathrm{A}^\omega$ by a definition by cases which are pairwise exclusive.

Ad (ii): (ii) follows from the definition of $h_1,\ldots,h_k$ by induction on the $\underline{h}$-depth of r.

Ad (iii): Assume $\bigwedge_{l=1}^{i}(\widehat{r}^{q_l}_{j_l} \le_0 \widehat{r}^{\tilde{q}_l}_{\tilde{j}_l})$. Let l_0 $(1\le l_0\le i)$ be such that $q_{l_0} = \min_{1\le l\le i}(q_l)$. By contraposition of the implication established in the proof of (i) one has: $\min_{1\le l\le i}(q_l) \ge \min_{1\le l\le i}(\tilde{q}_l)$.

Case 1: $\min_{1\le l\le i}(q_l) = \min_{1\le l\le i}(\tilde{q}_l)$. Then (by h_i-definition)

$$h_i(\widehat{r}^{q_1}_{j_1},\ldots,\widehat{r}^{q_i}_{j_i}) = y_i^{\min(q_l)-1} = y_i^{\min(\tilde{q}_l)-1} = h_i(\widehat{r}^{\tilde{q}_1}_{\tilde{j}_1},\ldots,\widehat{r}^{\tilde{q}_i}_{\tilde{j}_i}).$$

Case 2: $q_{l_0} = \min_{1\le l\le i}(q_l) > \min_{1\le l\le i}(\tilde{q}_l) = \tilde{q}_{\tilde{l}_0}$ (where $1\le l_0,\tilde{l}_0\le i$). Then

$$h_i(\widehat{r}^{q_1}_{j_1},\ldots,\widehat{r}^{q_i}_{j_i}) = y_i^{q_{l_0}-1} \overset{(+)}{<} y_i^{\tilde{q}_{\tilde{l}_0}-1} = h_i(\widehat{r}^{\tilde{q}_1}_{\tilde{j}_1},\ldots,\widehat{r}^{\tilde{q}_i}_{\tilde{j}_i}).$$

Hence $h_1,\ldots,h_k$ are monotone on $\widehat{M}$ and therefore (by (ii)) on M, which concludes the proof of (2) from (1) in $\mathrm{G}_n\mathrm{A}^\omega$ (using $(\beta),(\gamma)$). Over $\mathrm{G}_n\mathrm{A}^\omega$, (1) in turn follows from $Mon(A)\wedge(\alpha)$ (using (β)), and (0), i.e.

$$F :\equiv \forall u^1 \forall v \le_\tau tu \forall \underline{h}\big(\underline{h} \text{ monotone } \to \exists y_1,\ldots,y_k \le_0 \Psi u\underline{h}\exists w^\gamma A_0^H\big).$$

Put together, we so far have established the following $\underline{h}$-free finite 'Herbrand' disjunction

$$(3)\begin{cases} \mathrm{G}_n\mathrm{A}^\omega + Mon(A) \vdash \\ F \to \Big[v \le tu \wedge (+) \to \bigvee_{s_1,\ldots,s_k\in M} \exists w^\gamma A_0(u,v,\widehat{s}_1,\widehat{h_1 s_1},\ldots,\widehat{s}_k,\widehat{h_k s_1\ldots s_k},w)\Big]. \end{cases}$$

Since $\widehat{s}_1 \equiv \widehat{s'}_1 \wedge \ldots \wedge \widehat{s}_i \equiv \widehat{s'}_i$ implies $\widehat{h_i s_1\ldots s_i} \equiv \widehat{h_i s'_1\ldots s'_i}$ for $s_1,s'_1,\ldots,s'_i,s'_i \in M$ (where again '$\equiv$' denotes syntactic identity between terms) we, actually, can contract the above disjunction to

$$(4)\begin{cases} \mathrm{G}_n\mathrm{A}^\omega + Mon(A) \vdash \\ F \to \Big[v \le tu \wedge (+) \to \bigvee\limits_{\widehat{s}_1,\dots,\widehat{s}_k \in \widehat{M}} \exists w^\gamma A_0(u,v,\widehat{s}_1,\widehat{h_1 s_1},\dots,\widehat{s}_k,\widehat{h_k s_1 \dots s_k},w)\Big]. \end{cases}$$

The remainder of this proof in devoted to establish that (4) indeed implies

$$(5)\ \mathrm{G}_n\mathrm{A}^\omega + Mon(A) \vdash F \to A,$$

what we have to show.

We achieve this by an appropriate order of quantifier introductions, suitable quantifier shifts (made possible my $Mon(A)$) and contraction. We start with the variables with smallest upper index, i.e. $y_1^1,\dots,y_k^1$. Among these variables we first take that of maximal lower index, i.e. y_k^1: We split the conjunctive premise

$$(+) \equiv \begin{cases} \bigwedge\limits_{\substack{q=1,\dots,p-1\\ l=1,\dots,p-q}} (y_1^q,\dots,y_k^q > \widehat{r}_1^{q+l},\dots,\widehat{r}_{n_{q+l}}^{q+l},y_1^{q+l},\dots,y_k^{q+l}) \wedge \\ \bigwedge\limits_{q=1,\dots,p} (y_1^q,\dots,y_k^q > \widehat{r}_1^{p+1},\dots,\widehat{r}_{n_{p+1}}^{p+1}) \end{cases}$$

of our implication as well as its disjunctive conclusion

$$A^d :\equiv \bigvee_{\widehat{s}_1,\dots,\widehat{s}_k \in \widehat{M}} \exists w^\gamma A_0(u,v,\widehat{s}_1,\widehat{h_1 s_1},\dots,\widehat{s}_k,\widehat{h_k s_1 \dots s_k},w)$$

into the part in which y_k^1 occurs and into its y_k^1-free part:

$$(6)\begin{cases} F \to \Big[v \le tu \wedge \bigwedge\limits_{l=1,\dots,p-1} (y_k^1 > \widehat{r}_1^{1+l},\dots,\widehat{r}_{n_{1+l}}^{1+l},\widehat{r}_1^{p+1},\dots,\widehat{r}_{n_{p+1}}^{p+1},y_1^{1+l},\dots,y_k^{1+l}) \\ \quad \wedge \underbrace{\bigwedge{}'(\dots)}_{y_k^1\text{-free part of }(+)} \to \bigvee\limits_j \exists w^\gamma A_0(u,v,\widehat{s}_1^j,\widehat{h_1 s_1^j},\dots,\widehat{s}_k^j,y_k^1,w) \vee \underbrace{\bigvee\limits_{j'}(\dots)}_{y_k^1\text{-free part of }A^d} \Big]. \end{cases}$$

y_k^1 does not occur at any place other than indicated. Hence $\forall$-introduction applied to y_k^1 yields:

$$(7)\ F \to \forall y_k^1 \big[v \le tu \wedge \bigwedge_l (y_k^1 > \dots) \wedge \bigwedge{}'(\dots) \to \bigvee_j \exists w^\gamma A_0(\dots,y_k^1,w) \vee \bigvee_{j'}(\dots)\big],$$

where y_k^1 does not occur at any place other than indicated.
We now instantiate '$\forall y_k^1$' in (7) by

$$\tilde{y}_k^1 := \max_{1\le l\le p-1} \max\big(y_k^1,\widehat{r}_1^{1+l},\dots \widehat{r}_{n_{1+l}}^{1+l},\widehat{r}_1^{p+1},\dots,\widehat{r}_{n_{p+1}}^{p+1},y_1^{1+l},\dots,y_k^{1+l}\big)+1.$$

This results in

$$F \to \left[v \le tu \wedge \bigwedge{}'(\dots) \to \bigvee_j \exists w^\gamma A_0(\dots, \tilde{y}_k^1, w) \vee \bigvee_{j'}(\dots)\right].$$

$Mon(A)$ and $\bigvee_j \exists w^\gamma A_0(\dots, \tilde{y}_k^1, w)$ imply $\bigvee_j \exists w^\gamma A_0(\dots, y_k^1, w)$, since $\tilde{y}_k^1 \ge y_k^1$. Now $\forall$-introduction applied to y_k^1 and shifting $\forall y_k^1$ in front of $\bigvee_j$, which is possible since y_k^1 occurs only in this disjunction, yields

$$(8)\ F \to \left[v \le tu \wedge \bigwedge{}'(\dots) \to \forall y_k^1 \bigvee_j \exists w^\gamma A_0(\dots, y_k^1, w) \vee \bigvee_{j'}(\dots)\right].$$

From $Mon(A)$ it follows that

$$\forall y_k^1 \bigvee_j \exists w^\gamma A_0(\dots, y_k^1, w)$$

in fact yields

$$\bigvee_j \forall y_k^1 \exists w^\gamma A_0(\dots, y_k^1, w).$$

For this we proceed by contraposition: suppose that $\bigwedge_j \exists y_k^1 \forall w^\gamma \neg A_0(\dots, y_k^1, w)$. Then $\exists y \bigwedge_j \exists y_k^1 \le_0 y \forall w^\gamma \neg A_0(\dots, y_k^1, w)$. By $Mon(A)$ this implies $\exists y \bigwedge_j \forall w^\gamma \neg A_0(\dots, y, w)$.

So in total, (8) implies (since y_k^1 does not occur in $\widehat{s_k^j}$)

$$(9)\ \begin{cases} F \to \left[v \le tu \wedge \bigwedge{}'(\dots) \to \right. \\ \quad \left. \bigvee_j \exists x \forall y \exists w A_0(u, v, \widehat{s_1^j}, \widehat{h_1 s_1^j}, \dots, \widehat{h_{k-1} s_1^j \dots s_{k-1}^j}, x, y, w) \vee \bigvee_{j'}(\dots)\right]. \end{cases}$$

Next we apply the same procedure to the variable y_{k-1}^1 and then to y_{k-2}^1 and so on until all the variables $y_1^1, \dots, y_k^1$ with upper index 1 are bounded. We then continue with y_k^2, y_{k-1}^2 and so on. This corresponds to the sequence of quantifications used in Herbrand's theorem for first order logic to show that there is a direct proof from an index function-free Herbrand disjunction of a first order formula A to A itself. Taking always variables of minimal upper index guarantees that a $\forall$-introduction only gets applied to a variable that at this stage only occurs in places where it is universal quantified in the original formula A. By quantifying among these variables first the one with maximal lower index one ensures that a universal quantifier is introduced only once the quantifiers to the right-hand side of it in A have already been introduced. In addition to these two reasons for the special order of quantifications there is in our situation another (essentially used) property which is fulfilled only if variables which have minimal upper index are quantified first: If y_j^i has minimal index i (under all variables which still have to be quantified), then y_j^i occurs in the still remaining part of the implicative assumption (+) only in the form '$y^i_j > (\ldots y^i_j$-free...)'. Hence we are in the same situation as we were in the beginning w.r.t. y^1_k and, therefore, are able to eliminate this part of (+) which is connected with y^i_j altogether using $Mon(A)$ (just as we did for y^1_k above). Once all the universal quantifier for the y^i_j and all the existential quantifiers for the Herbrand terms $\widehat{s^j}_i$ are introduced, we have obtained

$$(10)\ F \to \big[v \le tu \to \bigvee \exists x^0_1 \forall y^0_1 \ldots \exists x^0_k \forall y^0_k \exists w^\gamma A_0(u,v,x_1,y_1,\ldots,x_k,y_k,w),$$

where $\bigvee$ is a disjunction of identical copies of

$$\exists x^0_1 \forall y^0_1 \ldots \exists x^0_k \forall y^0_k \exists w^\gamma A_0(u,v,x_1,y_1,\ldots,x_k,y_k,w).$$

Hence contraction of $\bigvee$ yields

$$(11)\ F \to \big[v \le tu \to \exists x^0_1 \forall y^0_1 \ldots \exists x^0_k \forall y^0_k \exists w^\gamma A_0(u,v,x_1,y_1,\ldots,x_k,y_k,w)$$

and so – using $\forall$-introduction applied to u,v – we finally obtain

$$(12)\ F \to A.$$

□

Corollary to the proof of theorem 13.7: The proof of theorem 13.7 does not depend at all on the structure of the quantifier-free part A_0 of A or on the special type of parameters u^1 and $v \le_\tau tu$ in A_0. In fact as long as the bounds Ψ_i in addition to the index functions $\underline{h}$ only depend on parameters $\underline{\alpha}$ whose types have degrees ≤ 1, the proof goes through. Moreover, $Mon(A)$ can be taken as an implicative assumption. So in total one even obtains

$$\mathrm{G}_n\mathrm{A}^\omega \vdash \forall X, \underline{\alpha}\Big(Mon(A(X)) \wedge \forall \underline{\alpha} \forall h_1,\ldots,h_k\Big(\bigwedge_{i=1}^{k}(h_i \text{ monotone}) \\ \to \exists y_1 \le_0 \Psi_1 \underline{\alpha}\,\underline{h} \ldots \exists y_k \le_0 \Psi_k \underline{\alpha}\,\underline{h} \exists w^\gamma A(X)^H_0\Big) \to A(X)\Big),$$

where

$$A(X) :\equiv \exists y^0_1 \forall x^0_1 \ldots \exists y^0_k \forall x^0_k \exists w^\gamma (X(y_1,x_1,\ldots,y_k,x_k,w) =_0 0).$$

Here X is a functional variable of type $0(\gamma)(0)\ldots(0)$ and $\underline{\alpha}$ is a tuple of variables of type degree ≤ 1. Of course, in applications such bounding terms Ψ_i will be available only for special X, e.g. for X being the characteristic term of the quantifier-free part A_0 of A in theorem 13.7. Then from a proof $\mathrm{G}_n\mathrm{A}^\omega$+QF-AC+$\Delta$ of A^H one can extract by NMD uniform bounds depending (in addition to $\underline{h}$) only on the type-1 parameter u but not on the bounded parameter $v \le_\tau tu$, see (the proof of) theorem 13.10 below. This is the reason why we focussed on that format in theorem 13.7.

Remark 13.9. By the corollary to the proof of theorem 13.7 above this theorem immediately applies to sentences A formulated in extensions $\mathcal{T}$ of the system $\mathrm{G}_n\mathrm{A}^\omega$

as long as the bounding terms Ψ_i belong to some class S that shares certain crucial features of the terms in $G_\infty R^\omega$, namely:

1) Every term $\Psi^\rho \in S$ with $\deg(\rho) \le 2$ has a majorant $\Psi^*[\underline{h}^1]$ such that

   (i) $\mathscr{T} \vdash \lambda \underline{h}.\Psi^*[\underline{h}]\ maj\ \Psi$,
   (ii) $\Psi^*[\underline{h}]$ is built up only from $\underline{h}$ and terms $\in S$ of type level ≤ 1 (by substitution).

2) S is (provably in $\mathscr{T}$) closed under the successor, definition by cases, λ-abstraction and contains the variable maximum functional Φ_1.

Condition 1) puts of an upper bound on the allowed complexity of S. E.g. 1) is not satisfied if S contains the iteration functional Φ_{it} defined by $\Phi_{it}0yf =_0 y$, $\Phi_{it}x'yf =_0 f(\Phi_{it}xyf)$ definable by R_0 and hence available already in $\widehat{\text{WE-PA}}^\omega\!\upharpoonright$. In fact, in section 13.4 we will show that theorem 13.7 becomes false already if G_nR^ω is replaced by the primitive recursive functionals in the sense of Kleene, i.e. the closed terms of $\widehat{\text{WE-PA}}^\omega\!\upharpoonright$. The crucial structural difference between Φ_{it} and functionals $\Psi \in G_\infty R^\omega$ is the following one: each term $\Psi^{001} \in G_\infty R^\omega$ can be majorized by a term $\Psi^*[x^0,h^1]$ which uses h only at a **fixed** number of arguments, i.e. there exists a fixed number n (which depends only on the structure of Ψ^* but not on x,h) such that for all h,x the value of $\Psi^*[x,h]$ only depends on (at most) n-many h-values. Let us illustrate this by an example: Φ defined by $\Phi hx = \max(h0,\dots,hx)$ depends on $x+1$–many h–values but is majorized by Φ^* defined by $\Phi^*hx := hx$ which for every x depends only on one h-value, namely on hx. If a term Ψ has a majorant which satisfies 1) we say that **Ψ is majorizable with finite support**. One easily convinces oneself that Φ_{it} is not majorizable with finite support.
2) is a lower bound on the complexity of $\mathscr{T},S$, which also is essential. E.g. take $\mathscr{T} := \mathscr{L}^2$ and $S := \{0^0\}$, where $\mathscr{L}^2$ is first order logic with $=_0, \le_0$ extended by quantification over functions and two constants $0^0, 1^0$. Consider now

$$G :\equiv \exists x^0 \forall y^0 \exists z^0, f^1 (F_0(f,z) \to A_0(x,y)),$$

where $F_0(f,z) :\equiv (fz = 0 \wedge 0 \neq 1)$ and $A_0(x,y) :\equiv (y \neq 0 \wedge x = x \to \bot)$. Then

$$\mathscr{L}^2 \vdash \forall g^1 \exists x,z \le_0 0 \exists f (F_0(f,z) \to A_0(x,gx)) \wedge Mon(G), \text{ but } \mathscr{L}^2 \nvdash G,$$

i.e. theorem 13.7 fails for $\mathscr{L}^2,S$. If however $\mathscr{L}^2$ is extended by λ-abstraction, then G becomes derivable since we can form $f := \lambda x^0.1^0$ now.

Theorem 13.10. *Let $n \ge 1$ and A be as in theorem 13.7 and Δ be as in theorem 10.21, i.e. a set of sentences $\forall x^\delta \exists y \le_\rho sx \forall z^\eta G_0(x,y,z)$ where s is a closed term of G_nA^ω and G_0 a quantifier-free formula, and let A' denote the (Kuroda) negative translation of A. Then the following rule holds:*

$$\begin{cases} \mathrm{G}_n\mathrm{A}^\omega + \text{QF-AC} + \Delta \vdash A^H \wedge Mon(A) \Rightarrow \\ \mathrm{G}_n\mathrm{A}^\omega + \tilde{\Delta} \vdash A \;\; \textit{and by monotone functional interpretation} \\ \textit{one can extract a tuple } \underline{\Psi} \in \mathrm{G}_n\mathrm{R}^\omega \textit{ such that} \\ \mathrm{G}_n\mathrm{A}^\omega_i + \tilde{\Delta} \vdash \underline{\Psi} \textit{ satisfies the monotone functional interpretation of } A', \end{cases}$$

where $\tilde{\Delta} := \{\exists Y \leq_{\rho\delta} s\forall x^\delta, z^\eta G_0(x, Yx, z) : \forall x^\delta \exists y \leq_\rho sx\forall z^\eta G_0(x,y,z) \in \Delta\}$.

Proof: By theorem 10.51, NMD applied to a proof of A^H in $\mathrm{G}_n\mathrm{A}^\omega + \Delta +$ QF-AC extracts uniform bounds $\Psi_1, \ldots, \Psi_k \in \mathrm{G}_n\mathrm{R}^\omega$ on $\exists y_1, \ldots, y_k$ (not depending on v) such that

$$(1)\ \mathrm{G}_n\mathrm{A}^\omega_i + \tilde{\Delta} \vdash \forall u \forall v \leq tu \forall \underline{h} \exists y_1 \leq_0 \Psi_1 u\underline{h} \ldots \exists y_k \leq_0 \Psi_k u\underline{h} \exists w A_0^H,$$

where

$$\tilde{\Delta} := \{\exists Y \leq_{\rho\delta} s\forall x^\delta, z^\eta G_0(x, Yx, z) : \forall x^\delta \exists y \leq_\rho sx\forall z^\eta G_0(x,y,z) \in \Delta\}.$$

Since $Mon(A)$ is implied by the monotone functional interpretation of its negative translation, the soundness theorem 10.20 for NMD applied to the assumption

$$\mathrm{G}_n\mathrm{A}^\omega + \Delta + \text{QF-AC} \vdash Mon(A)$$

yields

$$(2)\ \mathrm{G}_n\mathrm{A}^\omega_i + \tilde{\Delta} \vdash Mon(A).$$

By (1) and (2) we are now in the position to apply theorem 13.7 and obtain

$$\mathrm{G}_n\mathrm{A}^\omega + \tilde{\Delta} \vdash A.$$

The second part of the theorem follows again by NMD (theorem 10.51) since $\tilde{\Delta}$ is just another set of axioms of the form Δ (just with higher types and the missing initial universal quantifier treated as 'dummy quantifier').

□

Remark 13.11. 1) As a corollary to the proof above (using theorem 10.26 and theorem 10.51) it follows that if the axioms Δ (which we may assume to be finitely many) are as in theorem 10.26 and one has

$$\mathrm{G}_n\mathrm{A}^\omega + \text{QF-AC} \vdash \Delta \rightarrow A^H \wedge Mon(A)$$

then the conclusions of theorem 13.10 hold with $\tilde{\Delta}_\varepsilon$ (defined as in theorem 10.26) instead of $\tilde{\Delta}$. If Δ is even as in proposition 10.30, one can reduce Δ in fact to Δ_ε. If only for some of the axioms Δ this is the case then we can replace those by the corresponding ε-weakening (resp. the ε-weakening of their $\tilde{\Delta}$-form).

2) If the types τ, γ in the sentence A are of degree ≤ 1, QF-AC is restricted to QF-$\text{AC}^{1,0}$ and QF-$\text{AC}^{0,1}$ and Δ is has the type restriction as in theorem 10.47, then

theorem 13.10 even holds for E-G_nA^ω instead of G_nA^ω. Since E-G_nA^ω satisfies the deduction theorem, the previous point applies provided that Δ satisfies the respective type restrictions.

For our applications in the next sections we need the following corollary of theorem 13.10:

Corollary 13.12. *Let* $n \geq 1$ *and* $\forall x^0 \exists y^0 \forall z^0 A_0(u^1, v^\tau, x, y, z) \in \mathcal{L}(G_nA^\omega)$ *be a formula which contains only* u, v *as free variables and satisfies provably in* $G_nA^\omega + \Delta +$ QF-AC *the following monotonicity property:*

$$(*)\, \forall u, v, x, \tilde{x}, y, \tilde{y}(\tilde{x} \leq_0 x \wedge \tilde{y} \geq_0 y \wedge \forall z^0 A_0(u, v, x, y, z) \to \forall z^0 A_0(u, v, \tilde{x}, \tilde{y}, z)),$$

(i.e. $Mon(\exists x \forall y \exists z \neg A_0)$*). Furthermore, let* $B_0(u, v, w^\gamma) \in \mathcal{L}(G_nA^\omega)$ *be a (quantifier-free) formula which contains only* u^1, v^τ, w^γ *as free variables where* $deg(\gamma) \leq 2$*. Then the following rule holds:*

$$\begin{cases} \textit{If } G_nA^\omega + \Delta + \text{QF-AC} \vdash \\ \qquad \forall u^1 \forall v \leq_\tau tu\big(\exists f^1 \forall x, z A_0(u, v, x, fx, z) \to \exists w^\gamma B_0(u, v, w)\big) \wedge (*) \\ \textit{then} \\ G_nA^\omega + \tilde{\Delta} \vdash \forall u^1 \forall v \leq_\tau tu\big(\forall x \exists y \forall z A_0(u, v, x, y, z) \to \exists w^\gamma B_0(u, v, w)\big) \\ \textit{and one can extract a term } \chi \in G_nR^\omega \textit{ such that} \\ G_nA_i^\omega + \tilde{\Delta} \vdash \forall u^1 \forall v \leq_\tau tu \forall \Psi^* \big(\Psi^* \ maj\ \Psi^* \wedge \forall x^0, g^1 \exists y \leq_0 \Psi^* xg A_0(u, v, x, y, gy) \\ \qquad\qquad \to \exists w \leq_\gamma \chi u \Psi^*\, B_0(u, v, w)\big). \end{cases}$$

Proof: We may assume that $\gamma = 2$. The monotonicity property $(*)$ already logically implies $Mon(G)$, where

$$G :\equiv \forall u^1 \forall v \leq_\tau tu \exists x^0 \forall y^0 \exists z^0, w^2 \big(A_0(u, v, x, y, z) \to B_0(u, v, w)\big).$$

Moreover,

$$\forall u^1 \forall v \leq_\tau tu\big(\exists f^1 \forall x, z A_0(u, v, x, fx, z) \to \exists w^\gamma B_0(u, v, w)\big)$$

is logically equivalent to G^H. Hence the assumption of the rule to be proved yields that

$$G_nA^\omega + \Delta + \text{QF-AC} \vdash G^H + Mon(G).$$

From this we conclude by theorem 13.10 that

$$G_nA^\omega + \tilde{\Delta} \vdash G, \text{ i.e.}$$

$$G_nA^\omega + \tilde{\Delta} \vdash \forall u^1 \forall v \leq_\tau tu\big(\forall x \exists y \forall z A_0(u, v, x, y, z) \to \exists w^\gamma B_0(u, v, w)\big).$$

By classical logic and QF-$AC^{0,0}$ this implies

$$\mathrm{G}_n\mathrm{A}^\omega + \tilde{\Delta} + \text{QF-AC}^{0,0} \vdash$$
$$\forall u^1 \forall v \leq_\tau tu \left(\forall x^0, g^1 \exists y A_0(u,v,x,y,gy) \to \exists w B_0(u,v,w)\right)$$

and hence a fortiori

$$\mathrm{G}_n\mathrm{A}^\omega + \tilde{\Delta} + \text{QF-AC}^{0,0} \vdash$$
$$\forall u^1 \forall v \leq_\tau tu \forall \Psi \left(\forall x^0, g^1 \exists y \leq_0 \Psi xg A_0(u,v,x,y,gy) \to \exists w B_0(u,v,w)\right).$$

The proof is concluded using (the proof of) theorem 10.21 (adapted to $\mathrm{G}_n\mathrm{A}^\omega$) and remarks 10.22 and 10.28 (note that $\forall x^0, g^1 \exists y \leq_0 \Psi xg A_0(u,v,x,y,gy)$ can be written as a purely universal formula and hence the whole implication '$(\ldots)$' as an $\exists$-formula). □

Remark 13.13. The comments made in remark 13.11 apply analogously to corollary 13.12.

The monotonicity property $(*)$ is crucial for corollary 13.12 to be true: it is well-known that using Σ^0_1-IA or – equivalently – Π^0_1-IA (with number parameters only) every primitive recursive function (in the sense of Kleene) can be proved to be total relative to $\mathrm{G}_3\mathrm{A}^\omega$. Now let $n \geq 3$ and consider a Π^0_2-sentence $B :\equiv \forall u^0 \exists w^0 B_0(u,w)$ expressing the totality of A_{n+1} which grows faster than any function definable in $\mathrm{G}_n\mathrm{R}^\omega$. Let $A(x) :\equiv \forall y^0 A_0(x,y,a^0)$ be the (function parameter-free) induction formula sufficient to show B, where A_0 is a quantifier-free formula in the language of $\mathrm{G}_n\mathrm{A}^\omega$ and a encodes the tuple of number parameters of the induction formula. Now consider the logically valid sentence

$$\forall x, a \exists y \forall z \left(A_0(x,z,a) \vee \neg A_0(x,y,a)\right)$$

which – coding x, a together – becomes

$$(+)\ \forall x \exists y \forall z \left(A_0(j_1x,z,j_2x) \vee \neg A_0(j_1x,y,j_2x)\right),$$

where

$$C_0(x,y,z) :\equiv A_0(j_1x,z,j_2x) \vee \neg A_0(j_1x,y,j_2a)$$

is quantifier-free. Now let f be a Skolem function for $(+)$. Then with

$$g(x) := \begin{cases} 0, \text{ if } A_0(j_1x, fx, j_2x) \\ 1, \text{ otherwise,} \end{cases}$$

$\tilde{g}(x) := g(j(x,a))$ is the characteristic function of $A(x)$. Hence plugging $\tilde{g}$ into the second order axiom of quantifier-free induction from $\mathrm{G}_n\mathrm{A}^\omega$ one obtains the instance of Π^0_1-IA needed to proof B. So in total

$$\mathrm{G}_n\mathrm{A}^\omega \vdash \exists f^1 \forall x, z\, C_0(x, fx, z) \to \forall u^0 \exists w^0 B_0(u,w).$$

If corollary 13.12 would hold in the absence of the monotonicity condition $(*)$ this would (using that, trivially, $(+)$ is provable in $\mathrm{G}_n\mathrm{A}^\omega$) imply that

$$\mathrm{G}_n\mathrm{A}^\omega \vdash \forall u^0 \exists w^0 B_0(u,w)$$

and hence (applying NMD to $\mathrm{G}_n\mathrm{A}^\omega$) a $\mathrm{G}_n\mathrm{R}^\omega$-definable bound on A_{n+1} which is a contradiction. In section 13.4 below we will show that the correct contribution of fixed instances of Π^0_1-CA **can** be calibrated using corollary 13.12 if – instead of $(+)$ – one uses a monotone version of $(+)$ which no longer is provable in $\mathrm{G}_n\mathrm{A}^\omega$ but only in $\widehat{\text{WE-PA}}^\omega\!\upharpoonright + \text{QF-AC}^{0,0}$ (see proposition 13.19 below).

From the fact that corollary 13.12 fails in the absence of the monotonicity property $(*)$ it also follows that theorem 13.10 becomes false if the property $Mon(A)$ is dropped. In turn this yields that theorem 13.7 fails without the axiom $Mon(A)$.

Corollary 13.12 makes it possible in many cases to reduce the use of an **analytical** premise

$$\exists f^1 \forall x, z A_0(x, fx, z)$$

in a proof to an **arithmetical** premise

$$\forall x^0 \exists y^0 \forall z^0 A_0(x,y,z)$$

provided that the latter is monotone. The main benefit of this for the growth of bounds extractable from given proofs rests on the following fact: direct monotone functional interpretation MD applied to a $\mathrm{G}_n\mathrm{A}^\omega + \Delta +$ QF-AC-proof of a sentence

$$\forall u^1 \forall v \leq_\tau tu \big(\exists f^1 \forall x, z A_0(u,v,x,fx,z) \to \exists w^0 B_0(u,v,w)\big)$$

extracts a functional $\Psi \in \mathrm{G}_n\mathrm{R}^\omega$ which only in a Skolem function f (satisfying $\forall x,z A_0(u,v,x,fx,z)$) as oracle produces a bound on '$\exists w^0$'. However, in the applications below (PCM,BW), f will be noncomputable in general as the principles considered all imply Π^0_1-CA. Combined with negative translation one only needs a solution of the functional interpretation of

$$\neg\neg\exists f \forall x, z A_0(u,v,x,fx,z)$$

as input, but even this needs bar recursion $B_{0,1}$ as we saw in chapter 11. In contrast to this the bound χ in corollary 13.12 only depends on a functional which satisfies the monotone functional interpretation of the negative translation of $\forall x \exists y \forall z A_0(x,y,z)$, i.e. a functional solving the no-counterexample interpretation of $\forall x \exists y \forall z A_0(x,y,z)$. In our applications below such a functional is easily be constructed as a closed term of $\widehat{\text{WE-HA}}^\omega\!\upharpoonright$.

In the remaining sections of this chapter we apply the results presented in this section in order to determine the impact on the rate of growth of uniform bounds for provably $\forall u^1 \forall v \leq_\tau tu \exists w^\gamma A_0$–sentences which may result from the use of sequences (which however may depend on the parameters of the proposition to be proved) of instances of:

1) The principle of convergence for bounded monotone sequences of real numbers (PCM).
2) Π_1^0-CA and Π_1^0-AC.
3) The Bolzano-Weierstraß property (BW) for bounded sequences in $\mathbb{R}^d$.

Further principle which can be treated in this way are the Ascoli lemma and the existence of lim sup and lim inf for bounded sequences in $\mathbb{R}$. However, for these we refer to the literature ([210]).

13.3 The principle of convergence for bounded monotone sequences of real numbers (PCM)

Let $a^{1(0)}$ be (a representative in the sense of chapter 4) of a nonincreasing sequence of nonnegative real numbers, i.e.

$$(0)\ \forall n^0(0 \le_{\mathbb{R}} a(n+1) \le_{\mathbb{R}} an).$$

The convergence of this sequence can be expressed as

$$(1)\ \exists b^1 \forall k^0 \exists n^0 \forall m \ge n(|a(m) -_{\mathbb{R}} b|_{\mathbb{R}} <_{\mathbb{R}} 2^{-k}).$$

Clearly, for b satisfying (1) one has (using (0))

$$(2)\ \forall n^0(a_n \ge_{\mathbb{R}} b).$$

Over, say G_3A^{ω}+QF-AC^{0,0}, (1) is equivalent to

$$(3)\ \exists h^1 \forall k^0, m^0(m \ge hk \to |a(m) -_{\mathbb{R}} a(hk)|_{\mathbb{R}} <_{\mathbb{R}} 2^{-k}).$$

$(1) \Rightarrow (3)$: Since $<_{\mathbb{R}} \in \Sigma_1^0$ we can apply Σ_1^0-AC^{0,0} (and hence QF-AC^{0,0} by coding of pairs) to

$$(4)\ \forall k^0 \exists n^0(a(n) -_{\mathbb{R}} b <_{\mathbb{R}} 2^{-k})$$

and obtain a function h^1 such that

$$(5)\ \forall k^0(a(hk) -_{\mathbb{R}} b <_{\mathbb{R}} 2^{-k}).$$

From $(0), (2)$ it follows that h satisfies (3).
$(3) \Rightarrow (1)$: If h satisfies (3), then $(a(hk))_k$ is a Cauchy sequence with rate 2^{-k} and so (using lemma 4.3) has a limit b^1 which, clearly, also is the limit of $(a(k))_k$. Hence (1) follows.

We define the principle of monotone convergence PCM in the form (3) (where it will be convenient to replace $<_{\mathbb{R}}$ by $\leq_{\mathbb{R}}$). In order to simplify the logical form of PCM we use the construction $\tilde{a}(n) := \max_{\mathbb{R}}(0, \min_{i \leq n}(a(i))$ which ensures that $\tilde{a}$ is nonincreasing and bounded from below by 0. If a already fulfills these properties nothing is changed by the passage from a to $\tilde{a}$).

From general results in reverse mathematics (see e.g. [338]) it follows that the full 2nd order closure

$$\text{PCM} :\equiv \forall a^{1(0)}\, \text{PCM}(a)$$

of

$$\text{PCM}(a^{1(0)}) :\equiv \exists h^1 \forall k^0, m^0 \big(m \geq_0 hk \to |\tilde{a}(m) -_{\mathbb{R}} \tilde{a}(hk)|_{\mathbb{R}} \leq_{\mathbb{R}} 2^{-k}\big).$$

is (over a weak base system) equivalent to arithmetical comprehension. A more refined treatment (given in [216]) shows the existence of closed terms ξ_1, ξ_2 in $\mathrm{G}_3\mathrm{R}^{\omega}$ such that, provably in $\mathrm{G}_3\mathrm{A}^{\omega}$+QF-AC^{0,0},

$$\forall f^{1(0)} \big(\, \text{PCM}(\xi_1(f)) \to \Pi^0_1\text{-CA}(f)\big)$$

and

$$\forall a^{1(0)} \big(\Pi^0_1\text{-CA}(\xi_2(a)) \to\, \text{PCM}(a)\big),$$

where

$$\Pi^0_1\text{-CA}(f^{1(0)}) :\equiv \exists g^1 \forall x^0 \big(gx =_0 0 \leftrightarrow \forall y^0 (fxy =_0 0)\big)$$

(the actual result proved in [216] is still more refined than this).
So fixed instances PCM(t) of PCM only yield fixed instances of Π^0_1-CA (or, equivalently, Σ^0_1-CA) but not of higher arithmetical comprehension which would require an iterated use of PCM. Since this form of comprehension plugged into the schema of quantifier-free induction gives any instance of Π^0_1- and Σ^0_1-induction one has to expect arbitrary primitive recursive (in the sense of Kleene) complexity for Π^0_2-sentences proved over $\mathrm{G}_\infty\mathrm{A}^{\omega}$ from fixed instances of PCM. We show below that this lower bound also is an upper bound.
PCM$(a^{1(0)})$ is the Skolem normal form of the arithmetical principle

$$\text{PCM}_{ar}(a^{1(0)}) :\equiv \forall k^0 \exists n^0 \forall m^0 \big(m \geq_0 n \to |\tilde{a}(n) -_{\mathbb{R}} \tilde{a}(m)|_{\mathbb{R}} \leq_{\mathbb{R}} 2^{-k}\big),$$

which expresses the Cauchy property of the sequence $(\tilde{a}(n))_n$.

Remark 13.14. The restriction to the lower bound 0 is (convenient but) not essential: If $\forall n^0 (c \leq_{\mathbb{R}} a(n+1) \leq_{\mathbb{R}} an)$ we may define $a'(n) := a(n) -_{\mathbb{R}} c$. PCM applied to a' implies PCM for a. Everything holds analogously for nondecreasing sequences which are bounded from above.

We now show that the contribution of single instances PCM(a) of PCM to the growth of uniform bounds is (at most) given by a primitive recursive (in the sense of Kleene) functional, i.e. a closed term of $\widehat{\text{WE-HA}}^{\omega}{\restriction}$:

Proposition 13.15. *Let $B_0(u^1, v^{\tau}, w^{\gamma}) \in \mathcal{L}(\mathrm{G}_\infty\mathrm{A}^{\omega})$ be a quantifier–free formula which contains only $u^1, v^{\tau}, w^{\gamma}$ free, where $\gamma \leq 2$. Furthermore let $\xi, t \in \mathrm{G}_\infty\mathrm{R}^{\omega}$ and*

Δ be as in theorem 13.10. Then the following rule holds

$$\begin{cases} \mathrm{G}_\infty\mathrm{A}^\omega + \Delta + \text{QF-AC} \vdash \forall u^1 \forall v \leq_\tau tu\big(\text{PCM}\,(\xi uv) \to \exists w^\gamma B_0(u,v,w)\big) \Rightarrow \\ \mathrm{G}_\infty\mathrm{A}^\omega + \tilde{\Delta} \vdash \forall u^1 \forall v \leq_\tau tu\big(\text{PCM}_{ar}\,(\xi uv) \to \exists w^\gamma B_0(u,v,w)\big) \\ \textit{and one can extract a closed term } \Phi \textit{ of } \widehat{\text{WE-HA}}^\omega\!\restriction \textit{ such that} \\ \widehat{\text{WE-HA}}^\omega\!\restriction + \tilde{\Delta} \vdash \forall u^1 \forall v \leq_\tau tu \exists w \leq_\gamma \Psi(u) B_0(u,v,w). \end{cases}$$

Proof: One easily observes that

$$\mathrm{G}_\infty\mathrm{A}^\omega \vdash \forall a^{1(0)} \forall k, \tilde{k}, n, \tilde{n}\big(\tilde{k} \leq_0 k \wedge \tilde{n} \geq_0 n \wedge \forall m \geq_0 n(\tilde{a}(n) -_{\mathbb{R}} \tilde{a}(m) \leq_{\mathbb{R}} 2^{-k}) \\ \to \forall m \geq_0 \tilde{n}(\tilde{a}(\tilde{n}) -_{\mathbb{R}} \tilde{a}(m) \leq_{\mathbb{R}} 2^{-\tilde{k}})\big).$$

Clearly, this sentence as well as the premise of the rule to be proved can already be established with $\mathrm{G}_n\mathrm{A}^\omega$ instead of $\mathrm{G}_\infty\mathrm{A}^\omega$ for sufficiently large $n \in \mathbb{N}$. Hence we can apply corollary 13.12 and conclude that the assumption of the proposition implies

$$\mathrm{G}_\infty\mathrm{A}^\omega + \tilde{\Delta} \vdash \forall u^1 \forall v \leq_\tau tu\big(\text{PCM}_{ar}\,(\xi uv) \to \exists w^\gamma B_0(u,v,w)\big).$$

$\forall a^{1(0)}\ \text{PCM}_{ar}(a)$ is provable using Σ^0_1-IA and hence in $\widehat{\text{WE-PA}}^\omega\!\restriction$+QF-AC^{0,0} (using proposition 3.21). Thus we obtain in total

$$\widehat{\text{WE-PA}}^\omega\!\restriction + \tilde{\Delta} + \text{QF-AC}^{0,0} \vdash \forall u^1 \forall v \leq_\tau tu \exists w B_0(u,v,w).$$

By the main theorem on uniform bound extraction by NMD (theorem 10.21) adapted to $\widehat{\text{WE-PA}}^\omega\!\restriction$ this yields the extractability of a closed term Ψ of $\widehat{\text{WE-HA}}^\omega\!\restriction$ such that (note that $\tilde{\Delta}$ again is of the general form Δ with $\tilde{\tilde{\Delta}} = \tilde{\Delta}$)

$$\widehat{\text{WE-HA}}^\omega\!\restriction + \tilde{\Delta} \vdash \forall u^1 \forall v \leq_\tau tu \exists w \leq_\gamma \Psi(u) B_0(u,v,w).$$

□

Remark 13.16. 1) Instead of proving $\forall a^{1(0)}\text{PCM}_{ar}(a)$ by Σ^0_1-IA and then using NMD as we did in the proof above one can alternatively also directly produce a functional in $\widehat{\text{WE-HA}}^\omega\!\restriction$ solving the NMD-interpretation of $\forall a^{1(0)}\text{PCM}_{ar}(a)$ as we – essentially – did already in chapter 10 and then use the second conclusion of corollary 13.12.

2) The variations of the above result corresponding to remark 13.11 hold as well.

Proposition 13.15 also holds for fixed sequences $\widetilde{(a_l)}$ of nonincreasing sequences $\tilde{a}(\cdot)$ of real numbers in $\mathbb{R}_+$ instead of fixed nonincreasing sequences. Consider

$$\text{PCM}^*(a^{1(0)(0)}_{(\cdot)}) :\equiv \exists h^{1(0)} \forall l^0, k^0 \forall m \geq_0 hkl\big(\widetilde{(a_l)}(hkl) -_{\mathbb{R}} \widetilde{(a_l)}(m) \leq_{\mathbb{R}} 2^{-k}\big),$$

$\text{PCM}^*(a)$ is implied by

$$\mathrm{PCM}^+(a) :\equiv \exists h^1 \forall k^0 \forall m \geq_0 hk \forall l \leq_0 k\big(\widetilde{(a_l)}(hk) -_{\mathbb{R}} \widetilde{(a_l)}(m) \leq_{\mathbb{R}} 2^{-k}\big)$$

which is the Skolem normal form of the monotone formula

$$\mathrm{PCM}^+_{ar}(a) :\equiv \forall k^0 \exists n \forall m \geq_0 n \forall l \leq_0 k\big(\widetilde{(a_l)}(n) -_{\mathbb{R}} \widetilde{(a_l)}(m) \leq_{\mathbb{R}} 2^{-k}\big).$$

Proposition 13.17. *Proposition 13.15 also holds with* $\mathrm{PCM}^+(\xi uv)$ *and* $\mathrm{PCM}^+_{ar}(\xi uv)$ *instead of* $\mathrm{PCM}(\xi uv)$ *and* $\mathrm{PCM}_{ar}(\xi uv)$.

Proof: Exercise! □

13.4 Π_1^0-CA and Π_1^0-AC

As mentioned in the previous section the use of fixed instances of Π_1^0-CA in a proof can be reduced to that of instances of PCM. In this section we give a more direct treatment of Π_1^0-CA.

Definition 13.18. Define

$$A_0^C(f^{1(0)}, x^0, y^0, z^0) :\equiv \forall \tilde{x} \leq_0 x \exists \tilde{y} \leq_0 y \forall \tilde{z} \leq_0 z\big(f\tilde{x}\tilde{y} \neq_0 0 \vee f\tilde{x}\tilde{z} =_0 0\big).$$

A_0^C can be expressed as a quantifier-free formula in $\mathcal{L}(\mathrm{G_1A}^\omega)$.

Proposition 13.19. *For* $n \geq 3$ *one has:*

1) $\mathrm{G}_n\mathrm{A}_i^\omega$ *proves*

$$\forall f, x, \tilde{x}, y, \tilde{y}\big(\tilde{x} \leq_0 x \wedge \tilde{y} \geq_0 y \wedge \forall z^0 A_0^C(f,x,y,z) \rightarrow \forall z^0 A_0^C(f,\tilde{x},\tilde{y},z)\big),$$

i.e. $\forall x \exists y \forall z A_0^C(f,x,y,z)$ *satisfies the monotonicity condition* $(*)$ *in corollary 13.12.*

2) *The Skolem normal form of* $\forall x \exists y \forall z A_0^C(f,x,y,z)$ *implies* $\Pi_1^0\text{-CA}(f)$:

$$G_nA_i^\omega \vdash \forall f^{1(0)}\big(\exists g^1 \forall x^0, z^0 A_0^C(f,x,gx,z) \rightarrow \Pi_1^0\text{-CA}(f)\big).$$

3) *The functional defined by* $\Phi x^0 h^1 := \max_{i \leq x+1} h^{(i)}(0)$ *satisfies* NMD *(i.e. the monotone functional interpretation of the negative translation) of* $\forall x \exists y \forall z A_0^C(f,x,y,z)$:

$$\widehat{\text{WE-HA}}^\omega \upharpoonright \vdash \Phi\ maj\ \Phi \wedge \forall f^{1(0)}, x^0, h^1 \exists y \leq_0 \Phi x h A_0^C(f,x,y,hy).$$

4) $\widehat{\text{WE-PA}}^\omega \upharpoonright + \text{QF-AC}^{0,0} \vdash \forall f^{1(0)}, x^0 \exists y^0 \forall z^0 A_0^C(f,x,y,z)$.

Proof: 1) follows immediately from the definition of A_0^C.
2) Let g be such that $\forall x, z \forall \tilde{x} \leq x \exists \tilde{y} \leq gx \forall \tilde{z} \leq z\big(f\tilde{x}\tilde{y} \neq 0 \vee f\tilde{x}\tilde{z} = 0\big)$. Now take $\tilde{x} := x$ and $\tilde{z} := z$. It follows that

$$\forall x, z \exists \tilde{y} \le gx\big(fx\tilde{y} \neq 0 \lor fxz = 0\big)$$

and so in turn

$$\forall x\big(\forall \tilde{y} \le gx(fx\tilde{y} = 0) \leftrightarrow \forall z(fxz = 0)\big).$$

Thus

$$hx := \begin{cases} 0, \text{ if } \forall \tilde{y} \le gx(fx\tilde{y} = 0) \\ 1, \text{ otherwise} \end{cases}$$

satisfies Π^0_1-CA(f).
3) We proceed by contradiction. Let f, x, h be such that

$$(*)\ \forall y \le \Phi xh \exists \tilde{x} \le x \forall \tilde{y} \le y \exists \tilde{z} \le hy\big(f\tilde{x}\tilde{y} = 0 \land f\tilde{x}\tilde{z} \neq 0\big).$$

Case 1: $\exists i < x+1\big(h(h^{(i)}0) \le h^{(i)}0\big)$:
$(*)$ applied to $y := h^{(i)}0 \le \Phi xh$ yields an $\tilde{x} \le x$ such that

$$(**)\ \forall \tilde{y} \le h^{(i)}0 \exists \tilde{z} \le h(h^{(i)}0)\big(f\tilde{x}\tilde{y} = 0 \land f\tilde{x}\tilde{z} \neq 0\big)$$

and thus for $\tilde{y} := 0$ one has a $\tilde{z} \le h(h^{(i)}0)$ such that $f\tilde{x}\tilde{z} \neq 0$. But on the other hand –again by $(**)$– one has $f\tilde{x}\tilde{z} = 0$ (since $\tilde{z} \le h(h^{(i)}0) \le h^{(i)}0$) which is a contradiction.
Case 2: $\forall i < x+1\big(h(h^{(i)}0) > h^{(i)}0\big)$:
By the (finite) pigeonhole principle, $(*)$ implies that there exists $i < j \le x+1$ and $\tilde{x} \le x$ such that

$$(1)\ \forall \tilde{y} \le h^{(i)}0 \exists \tilde{z} \le h(h^{(i)}0)\big(f\tilde{x}\tilde{y} = 0 \land f\tilde{x}\tilde{z} \neq 0\big)$$

and

$$(2)\ \forall \tilde{y} \le h^{(j)}0 \exists \tilde{z} \le h(h^{(j)}0)\big(f\tilde{x}\tilde{y} = 0 \land f\tilde{x}\tilde{z} \neq 0\big).$$

Hence $\exists \tilde{z} \le h(h^{(i)}0)\big(f\tilde{x}\tilde{z} \neq 0\big)$ by (1) (take $\tilde{y} := 0$) and $\forall \tilde{y} \le h^{(j)}0\big(f\tilde{x}\tilde{y} = 0\big)$ by (2) which is a contradiction since by the case (and $i < j \le x+1$) $h(h^{(i)}0) = h^{(i+1)}0 \le h^{(j)}0$.
Put together we have proved that $\forall f, x, h \exists y \le_0 \Phi xh A^C_0(f, x, y, hy)$. We leave it as an exercise to formalize the above proof in $\widehat{\text{WE-HA}}^{\omega}\!\upharpoonright$.
It remains to show that Φ *maj* Φ: Assume that $\tilde{h}$ *maj*$_1$ h. By quantifier-free induction on x one shows that $\forall x\big(\tilde{h}^{(x)}0 \ge h^{(x)}0\big)$. Using this it follows (using again quantifier-free induction on $\tilde{x}$) that

$$\forall \tilde{x}, x(\tilde{x} \ge x \to \Phi\tilde{x}\tilde{h} \ge \Phi xh).$$

and hence the claim.
4) follows from 3).

□

Corollary 13.12 combined with proposition 13.19 yields

Proposition 13.20. *Let $B_0(u^1, v^\tau, w^\gamma) \in \mathcal{L}(\mathrm{G}_\infty\mathrm{A}^\omega)$ be a quantifier-free formula which contains only u^1, v^τ, w^γ free, where $\gamma \le 2$. Furthermore let $\xi, t \in \mathrm{G}_\infty\mathrm{R}^\omega$ and Δ be as in theorem 13.10. Then the following rule holds*

$$\begin{cases} \mathrm{G}_\infty\mathrm{A}^\omega + \Delta + \text{QF-AC} \vdash \forall u^1 \forall v \le_\tau tu\big(\Pi_1^0\text{-CA}(\xi uv) \to \exists w^\gamma B_0(u,v,w)\big) \\ \text{then one can extract a closed term } \Phi \text{ of } \widehat{\text{WE-HA}}^\omega{\restriction} \text{ such that} \\ \widehat{\text{WE-HA}}^\omega{\restriction} + \tilde{\Delta} \vdash \forall u^1 \forall v \le_\tau tu \exists w \le_\gamma \Psi(u)\, B_0(u,v,w). \end{cases}$$

Remark 13.21. The variations of the above result corresponding to remark 13.11 hold as well.

As in the case of $\mathrm{PCM}^*(a)$ one might consider instead of fixed instances of Π_1^0-CA fixed sequences of such instances, i.e. fixed instances of

$$\Pi_1^0\text{-CA}^*(f) :\equiv \forall l^0 \exists g^1 \forall x^0 \big(g(x) =_0 0 \leftrightarrow \forall y^0 (flxy =_0 0)\big).$$

However, this is trivial in this case since (provably in $\mathrm{G}_3\mathrm{A}^\omega$)

$$\Pi_1^0\text{-CA}(\varphi(f)) \to \Pi_1^0\text{-CA}^*(f),$$

where $\varphi(f) := f(j_1x, j_2x, y)$.

We now sketch a proof for the fact that proposition 13.20 (and hence also theorem 13.10 as well as theorem 13.7) fails (even for $\Delta = \emptyset$) if we add to $\mathrm{G}_\infty\mathrm{A}^\omega$ either the schema Σ_1^0-IA of Σ_1^0-induction (with function parameters), which can also be written as a single second order axiom

$$\Sigma_1^0\text{-IA}: \begin{cases} \forall f(\exists y(f(0,y)=0) \wedge \forall x(\exists y(f(x,y)=0) \to \exists y(f(x+1,y)=0)) \\ \qquad \to \forall x \exists y(f(x,y)=0)) \end{cases}$$

as we did in chapter 2, or the (Kleene)-primitive recursor constant R_0 for primitive recursion of type 0 with its defining axioms. As the proof of proposition 3.21 in chapter 3 shows, Σ_1^0-IA follows in the presence of $\text{QF-AC}^{0,0}$ and R_0. Hence we only have to produce a counterexample for the case where Σ_1^0-IA is added: let $B \equiv \forall u^0 \exists w^0 B_0(u,w)$ be a Π_2^0-sentence in the language of $\mathrm{G}_\infty\mathrm{A}^\omega$ expressing the totality of the Ackermann function. B can be proved using Π_2^0-induction or – equivalently – Σ_2^0-induction with number parameters only. However, any such instance of Σ_2^0-induction can be reduced to Σ_1^0-IA by absorbing the inner universal quantifier in the induction formula by Π_1^0-CA(ξ) for suitable closed ξ (using the above reduction of Π_1^0-CA* to Π_1^0-CA). So if proposition 13.20 would hold for $\mathrm{G}_\infty\mathrm{A}^\omega$ being replaced by $\mathrm{G}_\infty\mathrm{A}^\omega + \Sigma_1^0$-IA, we would get a $\widehat{\text{WE-HA}}^\omega{\restriction}$-definable (and hence primitive recursive in the sense of Kleene) bound on the Ackermann function which contradicts the well-known fact that the Ackermann function grows faster than any primitive recursive function.

Remark 13.22. The counterexample just given crucially uses that function variables are allowed to occur in Σ^0_1-IA. Any instance of Σ^0_1-IA containing only number variables in fact follows from Π^0_1-CA(ξ) for suitable closed ξ together with the quantifier-free induction schema from $G_\infty A^\omega$ and hence is permissible to be used in connection with proposition 13.20. In fact also the fixed function(al) parameters u, v are allowed to occur in Σ^0_1-IA instances (see below where we even include such instances of Δ^0_2-induction).

We now consider sequences of Π^0_1-instances of AC_{ar}:

$$\Pi^0_1\text{-AC}(f^{1(0)(0)(0)}) :\equiv \forall l^0 \big(\forall x^0 \exists y^0 \forall z^0 (flxyz =_0 0) \to \exists g^1 \forall x^0, z^0 (flx(gx)z =_0 0)\big).$$

Π^0_1-AC(f) can be reduced to Π^0_1–CA(g) uniformly by

Proposition 13.23.

$$G_2A^\omega + \text{QF-AC}^{0,0} \vdash \forall f^{1(0)(0)(0)} \big(\Pi^0_1\text{-CA}(f') \to \Pi^0_1\text{-AC}(f)\big),$$

where $f' := \lambda v^0, z^0. f(\nu^3_1(v), \nu^3_2(v), \nu^3_3(v), z)$.

Proof: By Π^0_1-CA(f') there exists a function h^1 such that

$$\forall v^0 (hv = 0 \leftrightarrow \forall z (f'vz = 0)).$$

$\tilde{h}lxy := h(\nu^3(l,x,y))$. Then

$$\forall l, x, y(\tilde{h}lxy = 0 \leftrightarrow \forall z(flxyz = 0)).$$

$\text{QF-AC}^{0,0}$ applied to $\forall x \exists y(\tilde{h}lxy = 0)$ yields $\exists g \forall x, z(flx(gx)z = 0)$.

□

As a consequence of proposition 13.23 we obtain

Proposition 13.24. *Proposition 13.20 also holds with* Π^0_1-AC(ξuv) *instead of* Π^0_1-CA(ξuv).

Definition 13.25. 1) The principle of Δ^0_2-IA(f, g) of Δ^0_2-induction for sequences of Δ^0_2-formulas given by f, g is defined as follows

$$\Delta^0_2\text{-IA}(f,g) :\equiv \begin{cases} \forall l^0 \Big(\forall x^0 (\exists u^0 \forall v^0 (flxuv =_0 0) \leftrightarrow \forall \tilde{u}^0 \exists \tilde{v}^0 (glx\tilde{u}\tilde{v} =_0 0)\big) \to \\ \big[\exists u \forall v (fl0uv = 0) \wedge \forall x (\exists u \forall v (flxuv = 0) \to \exists u \forall v (flx'uv = 0)) \\ \qquad \to \forall x \exists u \forall v (flxuv = 0)\big]\Big). \end{cases}$$

2) The principle Π^0_1-CP of bounded collection for a sequence of Π^0_1-formulas given by f is defined as follows

$$\Pi^0_1\text{-CP}(f) :\equiv$$
$$\forall l^0, x^0 \big(\forall \tilde{x} < x \exists y^0 \forall z^0 (flx\tilde{x}yz =_0 0) \to \exists y_0 \forall \tilde{x} < x \exists y <_0 y_0 \forall z (flx\tilde{x}yz = 0)\big).$$

Δ_2^0-IA(f,g) follows from appropriate instances Π_1^0-CA$(\xi_1(f))$ and Π_1^0-CA$(\xi_2(f))$ of Π_1^0-CA (provably in G_3A^ω, exercise) which can be encoded into a single instance Π_1^0-CA$(\xi_3(f))$. Similarly, Π_1^0-CP(f) can be reduced to Π_1^0-AC$(\xi_4(f))$ for appropriate $\xi_4 \in G_3R^\omega$. Hence in the previous results we can also allow instances Δ_2^0-IA$(\tilde{\xi}_1 uv, \tilde{\xi}_2 uv)$ and Π_1^0-CP$(\widehat{\xi} uv)$ in addition to Π_1^0-CA(ξuv) and Π_1^0-AC$(\xi' uv)$ to be used in a proof. So as long as it is prevented to have the comprehension resp. choice functions from Π_1^0-CA(ξuv) and Π_1^0-AC$(\xi' uv)$ to occur in instances of Δ_2^0-IA or Σ_1^0-IA (except for QF-IA where we allow arbitrary function parameters) we may use these principles as well. For precise formulations and additional information we refer to [208]. In [211] the whole hierarchies Π_n^0-CA and Π_n^0-AC and the corresponding arithmetical principles Δ_{n+1}^0-IA and Π_n^0-CP are calibrated w.r.t. their contribution to the complexity of extractable bounds.

13.5 The Bolzano-Weierstraß property for bounded sequences in $\mathbb{R}^d$

We now consider the Bolzano-Weierstraß principle for sequences in $[-1,1]^d \subset \mathbb{R}^d$. The restriction to the special bound 1 is convenient but not essential: If $(x_n) \subset \mathbb{R}^d$ is bounded by $C > 0$, we define $x'_n := \frac{1}{C} \cdot x_n$ and apply the Bolzano-Weierstraß principle to this sequence. For simplicity we formulate the Bolzano- Weierstraß principle w.r.t. the maximum norm $\|\cdot\|_{\max}$. This of course implies the principle for the Euclidean norm $\|\cdot\|_E$ since $\|\cdot\|_E \leq \sqrt{d} \cdot \|\cdot\|_{\max}$.
We start with the investigation of the following formulation of the Bolzano-Weierstraß principle:

$$\mathrm{BW} \;:\; \forall (x_n) \subset [-1,1]^d \exists x \in [-1,1]^d \forall k^0, m^0 \exists n >_0 m \big(\|x - x_n\|_{\max} \leq 2^{-k}\big),$$

i.e. (x_n) possesses a limit point x.
Using the representation of $[-1,1]$ from chapter 4, the principle BW has the form $\forall x_1^{1(0)}, \ldots, x_d^{1(0)}\ \mathrm{BW}(x_1, \ldots, x_d)$, where

$$\mathrm{BW}(\underline{x}) :\equiv \exists a_1, \ldots, a_d \leq_1 M \forall k^0, m^0 \exists n >_0 m \bigwedge_{i=1}^{d} \big(|\tilde{a}_i -_{\mathbb{R}} \widetilde{x_i n}| \leq_{\mathbb{R}} 2^{-k}\big).$$

Here M and $y^1 \mapsto \tilde{y}$ are the constructions from the representation of $[-1,1]$ in chapter 4. We now prove (using the sequence coding from chapter 3 the following results can be formulated even for variable dimension d though – for notational simplicity – we only formulate things for fixed d).

Lemma 13.26.

$$(*)\ G_3A^\omega + \text{QF-AC}^{1,0} \vdash F^- \to \forall x_1^{1(0)}, \ldots, x_d^{1(0)} \big(\Pi_1^0\text{-CA}(\chi \underline{x}) \to \mathrm{BW}(\underline{x})\big),$$

for a suitable $\chi \in \mathrm{G}_3\mathrm{R}^\omega$.

Proof: $\mathrm{BW}(\underline{x})$ is equivalent to

$$(1)\ \exists a_1,\ldots,a_d \leq_1 M \forall k^0 \exists n >_0 k \bigwedge_{i=1}^{d} \left(|\tilde{a}_i -_{\mathbb{R}} \widetilde{x_i n}| \leq_{\mathbb{R}} 2^{-k}\right)$$

which in turn is equivalent to

$$(2)\ \exists a_1,\ldots,a_d \leq_1 M \forall k^0 \exists n >_0 k \bigwedge_{i=1}^{d} \left(|\tilde{a}_i k -_{\mathbb{Q}} (\widetilde{x_i n})(k)| \leq_{\mathbb{Q}} 3 \cdot 2^{-k}\right).$$

Assume $\neg(2)$, i.e.

$$(3)\ \forall a_1,\ldots,a_d \leq_1 M \exists k^0 \forall n >_0 k \bigvee_{i=1}^{d} \left(|\tilde{a}_i k -_{\mathbb{Q}} (\widetilde{x_i n})(k)| >_{\mathbb{Q}} 3 \cdot 2^{-k}\right).$$

Let $\chi \in \mathrm{G}_2\mathrm{R}^\omega$ be such that

$$\begin{aligned} &\mathrm{G}_2\mathrm{A}^\omega \vdash \forall x_1^{1(0)},\ldots,x_d^{1(0)} \forall l^0, n^0 \big(\chi \underline{x} l n =_0 0 \leftrightarrow \\ &\quad \big[n >_0 \nu_{d+1}^{d+1}(l) \to \bigvee_{i=1}^{d} |\nu_i^{d+1}(l) -_{\mathbb{Q}} (\widetilde{x_i n})(\nu_{d+1}^{d+1}(l))| >_{\mathbb{Q}} 3 \cdot 2^{-\nu_{d+1}^{d+1}(l)+1}\big]\big). \end{aligned}$$

Π_1^0–$\mathrm{CA}(\chi\underline{x})$ yields the existence of a function h such that

$$(4)\ \forall l_1^0,\ldots,l_d^0,k^0 \left(h l_1 \ldots l_d k =_0 0 \leftrightarrow \forall n >_0 k \bigvee_{i=1}^{d} \left(|l_i -_{\mathbb{Q}} (\widetilde{x_i n})(k)| >_{\mathbb{Q}} 3 \cdot 2^{-k}\right)\right).$$

Using h, (3) has the form

$$(5)\ \forall a_1,\ldots,a_d \leq_1 M \exists k^0 \left(h(\tilde{a}_1 k,\ldots,\tilde{a}_d k, k) =_0 0\right).$$

By Σ_1^0-UB^- (which follows from $\text{QF-AC}^{1,0}$ and F^- by adapting the proof of proposition 12.6.2) to $\mathrm{G}_3\mathrm{A}^\omega$) we obtain

$$(6)\ \exists k_0 \forall a_1,\ldots,a_d \leq_1 M \forall m^0 \exists k \leq_0 k_0 \forall n >_0 k \bigvee_{i=1}^{d} \left(|(\widetilde{\overline{a_i}, m})(k) -_{\mathbb{Q}} (\widetilde{x_i n})(k)| >_{\mathbb{Q}} 3 \cdot 2^{-k}\right)$$

and therefore

$$(7)\ \exists k_0 \forall a_1,\ldots,a_d \leq_1 M \forall m^0 \forall n >_0 k_0 \bigvee_{i=1}^{d} \left(|(\widetilde{\overline{a_i}, m}) -_{\mathbb{R}} \widetilde{x_i n}| >_{\mathbb{R}} 2^{-k_0}\right).$$

Since $|\widetilde{a_i, m+3} -_{\mathbb{R}} \tilde{a}_i| \leq_{\mathbb{R}} 2^{-m}$ (see the definition of $y \mapsto \tilde{y}$ from chapter 4, after remark 4.26) it follows that

$$(8)\ \exists k_0 \forall a_1,\ldots,a_d \leq_1 M \forall n >_0 k_0 \bigvee_{i=1}^{d} \left(|\tilde{a}_i -_{\mathbb{R}} \widetilde{x_i n}| >_{\mathbb{R}} 2^{-k_0-1}\right), \text{ i.e.}$$

$$(9)\ \exists k_0 \forall (a_1,\ldots,a_d) \in [-1,1]^d \forall n >_0 k_0 \left(\|\underline{a} - \underline{x}n\|_{\max} > 2^{-k_0-1}\right).$$

By applying this to $\underline{a} := \underline{x}(k_0+1)$ yields the contradiction

$$\|\underline{x}(k_0+1) - \underline{x}(k_0+1)\|_{\max} > 2^{-k_0-1},$$

which concludes the proof. □

Proposition 13.27. *Let $B_0(u^1, v^\tau, w^\gamma) \in \mathcal{L}(\mathrm{G}_\infty\mathrm{A}^\omega)$ be a quantifier-free formula which contains only u^1, v^τ, w^γ free, where $\gamma \leq 2$. Furthermore let $\underline{\xi}, t \in \mathrm{G}_\infty\mathrm{R}^\omega$ and Δ be as in theorem 13.10. Then the following rule holds*

$$\begin{cases} \mathrm{G}_\infty\mathrm{A}^\omega + \Delta + \text{QF-AC} \vdash \forall u^1 \forall v \leq_\tau tu\big(\mathrm{BW}(\underline{\xi}uv) \to \exists w^\gamma B_0(u,v,w)\big) \\ \text{then one can extract a closed term } \Phi \text{ of } \widehat{\text{WE-HA}}^\omega\!\upharpoonright \text{ such that} \\ \widehat{\text{WE-HA}}^\omega\!\upharpoonright + \tilde{\Delta} \vdash \forall u^1 \forall v \leq_\tau tu \exists w \leq_\gamma \Psi(u) B_0(u,v,w). \end{cases}$$

Proof: The proof follows from proposition 13.20 combined with lemma 13.26 and remark 13.11.1) using that F^- is of the form of an axiom Δ as in theorem 10.26 with (as shown in the proof of theorem 12.8)

$$\widehat{\text{WE-HA}}^\omega\!\upharpoonright \vdash (\widetilde{F^-})_\varepsilon.$$

□

Remark 13.28. The variations of the above result corresponding to remark 13.11 hold as well.

In a somewhat more involved way one can show the above proposition also with $\mathrm{BW}(\underline{\xi}uv)$ replaced by $\mathrm{BW}^{seq}(\underline{\xi}uv)$ where $\mathrm{BW}^{seq}(\underline{x})$ expresses that the sequence in $[-1,1]^d$ represented by $\underline{x}$ has a convergent subsequence, i.e.

$$\mathrm{BW}^{seq}(\underline{x}) :\equiv \exists a_1,\ldots,a_d \leq_1 M \exists f^1 \left(\forall k^0 (f(k) < f(k+1) \wedge \left(|\tilde{a}_i -_{\mathbb{R}} \widetilde{x_i(fk)}| \leq_{\mathbb{R}} 2^{-k}\right)\right).$$

Note that the straightforward proof of BW^{seq} from BW involves an application of the recursor R_0 which is not allowed in our context as discussed in the previous section. However, this can be avoided by a use of a fixed sequence of instances of Σ^0_1-induction which does not involve a from BW so that we can apply the comments made at the end of the previous section. For details we refer to [210] which also provides analogous results for the Ascoli lemma stating that a sequence of functions in

$C[0,1]$ which has a common modulus of uniform continuity (i.e. which is equicontinuous) and has a common bound in the uniform norm has a limit point (and a convergent subsequence). Only fixed instances of the existence of the limit superior for sequences in $[0,1]$ cause (relative to $G_\infty A^\omega$) growth rates of bounds beyond the primitive recursive ones. In [210] an upper bound on the complexity is given: one can always extract bounds definable in T_1, i.e. by primitive recursion in the sense of Gödel with the recursors R_ρ restricted to the type degree 1. In [216] this is shown to be best possible.

13.6 Exercises, historical comments and suggested further reading

Exercises:

1) Prove proposition 13.17.
2) Complete the proof of proposition 13.19.3).
3) Show that proposition 13.23 also holds with 'Π^0_1-CA(ξuv)' be replaced by

$$\text{`}\Pi^0_1\text{-CA}(\xi_1 uv) \wedge \Pi^0_1\text{-CP}(\xi_2 uv) \wedge \Delta^0_2\text{-IA}(\xi_3 uv)\text{'},$$

where $\xi_1, \xi_2, \xi_3 \in G_\infty R^\omega$.

Historical comments and suggested further reading: Most of the results in sections 13.1, 13.2 and 13.4 are due to Kohlenbach [208] which contains many more results. The treatment of PCM in section 13.3 and of BW in section 13.5 is taken from Kohlenbach [210]. That paper also contains a similar study of the version of BW asserting the existence of a convergent subsequence as well as the Ascoli lemma. Moreover, it is shown in [210] that over $G_\infty A^\omega$ the use of fixed instances of the principle asserting the existence of the limit superior of bounded sequences of real numbers in proofs of sentences as considered in this chapter can be reduced to Σ^0_2-induction. Hence by Parsons [299] the extractability of uniform bounds in T_1 from such proofs in guaranteed. That this result is optimal is shown in Kohlenbach [216]. In Kohlenbach [211], vast extensions of the results presented in section 13.4 are obtained. In particular, the whole hierarchy of principles Δ^0_{n+1}-CA,Π^0_n-AC (for all $n \in \mathbb{N}$) as well as generalized forms of uniform bounded principles are treated and a number of consequences for fragments of first order arithmetic are derived. In Kohlenbach [216] it is shown that even over G_2A^ω, PCM (then formulated with rate $1/(k+1)$ instead of 2^{-k}) is equivalent to Σ^0_1-IA. Concrete applications of this chapters technique of arithmetizing uses of analytical principles based on sequential compactness in proofs can be found in Kohlenbach [227] (see also theorem 18.58 below) and Avigad et al. [8].

Chapter 14
The Friedman A-translation

14.1 The A-translation

In [108] H. Friedman introduced a strikingly simple device to establish closure under the Markov rule in the form

$$T_i \vdash \neg\neg\exists x P(\underline{a}, x) \Rightarrow T_i \vdash \exists x P(\underline{a}, x)$$

(P is a prime formula) which works for many intuitionistic theories T_i. In theories in which every quantifier-free formula $A_0(\underline{x})$ can be written as a prime formula $t_{A_0}(\underline{x}) = 0$ this implies the usual form of the Markov rule. As we have seen one can use functional interpretation to obtain closure under the Markov rule of those theories to which functional interpretation applies. The important feature of Friedman's so-called A-translation however is that it is much easier to apply and also works for some systems like intuitionistic Zermelo-Fraenkel set theory ZFI for which no functional interpretation has been developed yet.
Combined with the negative translation, Friedman's A-translation, therefore, can be used to show Π^0_2-conservation of many classical theories T over their intuitionistic counterpart T_i.
As a corollary of this one gets that T has the same provably recursive functions as T_i.
By combining negative translation and A-translation with modified realizability one obtains an alternative method (to the use of negative translation and functional interpretation) for unwinding proofs of Π^0_2-sentences in e.g. PA. Note that the direct combination of negative translation and modified realizability without the intermediate step of the A-translation would be useless since the modified realizability interpretation is trivial for negative formulas which result under negative translation.

However one should mention also some serious limitations of the approach based on the A-translation:

1) the A-translation is not sound for QF-ER and so doesn't apply to our systems WE-HA^ω and $\widehat{\text{WE-HA}}^{\omega}\upharpoonright$ while functional interpretation does.
2) A-translation only shows closure under the Markov **rule** but doesn't establish conservation results for the Markov **principle** with respect to general classes of formulas (which functional interpretation does). In particular it is not sound for the negative translation of QF-AC (which follows from QF-AC only in the presence of the Markov principle) and therefore cannot be used to show that e.g. the provably recursive functions of $\text{WE-PA}^\omega +$ QF-AC are just the ones definable by closed terms of WE-HA^ω even if we omit the extensionality rule QF-ER.
   However, [75] introduces a refinement of the A-translation which is able to treat Markov's principle and quantifier-free choice for numbers.
3) The combination of negative translation, A-translation and modified realizability is not known to be faithful for subsystems PA_n of PA with restricted induction (respectively for corresponding finite type extensions of PA_n) whereas negative translation combined with functional interpretation is (see [299] for the latter). Indeed, we will show below that the approach via the A-translations in some important cases seems to require realizers which are no longer primitive recursive (in the sense of Kleene) whereas negative translation and functional interpretation extract primitive recursive realizers.
   Again, the refinement of the A-translation due to [75] improves this situation (see [5]).

Remark 14.1. The A-translation was independently also discovered by A. Dragalin in [91].

In this chapter we will establish the A-translation only for HA since this suffices to illustrate the general pattern. For extensions to other systems like ZFI the reader should consult Friedman's original paper [108].

Definition 14.2 (Friedman [108], Dragalin [91]). Let $A \in \mathcal{L}(\text{HA})$ be a formula of HA. For every formula $F \in \mathcal{L}(\text{HA})$ (such that A doesn't contain free variables which are bound in F) we define the A-translation F^A of F as follows: F^A results if all prime formulas P in F are replaced by $P \vee A$.

Proposition 14.3 (Friedman [108]). $\text{HA} \vdash F \Rightarrow \text{HA} \vdash F^A$.

Proof: Easy induction on the length of the derivation. □

Corollary 14.4 (Friedman [108]). $\text{HA} \vdash \forall x \neg\neg\exists y A_0(x,y) \Rightarrow \text{HA} \vdash \forall x \exists y A_0(x,y)$.

Proof: In HA, $A_0(x,y)$ can be written as a prime formula $t_{A_0}(x,y) = 0$. Hence (since $\neg G$ is an abbreviation for $G \to 0 = 1$)

$$\text{HA} \vdash \forall x \neg\neg\exists y A_0(x,y)$$

implies

$$\text{HA} \vdash \big(\exists y (t_{A_0}(x,y) = 0) \to 0 = 1\big) \to 0 = 1.$$

By the A-translation for $A :\equiv \exists y(t_{A_0}(x,y)=0)$ we get

$$\mathrm{HA} \vdash \big((\exists y(t_{A_0}(x,y)=0) \to 0=1\big) \to 0=1\big)^{\exists y(t_{A_0}(x,y)=0)},$$

i.e.

$$\mathrm{HA} \vdash \big(\exists y(t_{A_0}(x,y)=0 \vee \exists y(t_{A_0}(x,y)=0)) \to \exists y(t_{A_0}(x,y)=0)\big) \to \exists y(t_{A_0}(x,y)=0),$$

and hence

$$\mathrm{HA} \vdash \big((\exists y t_{A_0}(x,y)=0 \vee \exists y(t_{A_0}(x,y)=0)) \to \exists y(t_{A_0}(x,y)=0)\big) \to \exists y(t_{A_0}(x,y)=0).$$

Since $G \vee G \to G$ holds by intuitionistic logic, we get

$$\mathrm{HA} \vdash \exists y(t_{A_0}(x,y)=0)$$

and hence

$$\mathrm{HA} \vdash \forall x \exists y A_0(x,y).$$

□

We now present an example: in the applications to be treated in chapter 18 we will make heavy use of the ND-interpretation of proofs which involve a modus ponens instance $\frac{A\ ,\ A\to B}{B}$ with $A \in \Pi^0_3$ and $B \in \Pi^0_2$, where, in particular, the case

$$A :\equiv \forall k \exists n \forall m(|r_{n+m} -_{\mathbb{R}} r_n| \leq_{\mathbb{R}} 2^{-k})$$

with (r_n) being a monotone sequence of real or rational numbers in $[0,N]$ for some $N \in \mathbb{N}$ is of interest (see the discussion in chapter 2). As shown already in chapter 2 (see the discussion in chapter 10), the ND-interpretation of this A can be realized by a primitive recursive functional (in the sense of Kleene) which does not seem to be the case for the modified realizability interpretation of the B-translation (where we treat B as an open Σ^0_1-formula) of the negative translation A' of A which, however, can be solved with a realizer defined by recursion of type-1:

For simplicity we restrict ourselves to the case of nonincreasing sequences (r_n) of rational numbers in $[0,1]$. The negative translation of A is (equivalent over $\mathrm{HA}[(r_n)]$, i.e. $\mathrm{HA}[(r_n)]$ extended by a function constant representing a nonincreasing sequence of rational numbers in $[0,1]$, to)

$$(*)\ \forall k \neg\neg \exists n \forall m(|r_{n+m} -_{\mathbb{Q}} r_n| \leq_{\mathbb{Q}} 2^{-k}).$$

Let $B :\equiv \exists v^0 B_0(v)$, where B_0 is quantifier-free (we again may assume that B_0 even is atomic). The B-translation $(*)^B$ of $(*)$ is given by

$$\forall k \Big\{ \Big(\exists n \forall m(|r_{n+m} -_{\mathbb{Q}} r_n| \leq_{\mathbb{Q}} 2^{-k} \vee \exists v_0 B_0(v_0)) \to \exists v_1 B_0(v_1)\Big) \to \exists v B_0(v) \Big\}.$$

Modified realizability applied to this asks for a functional Φ of type degree 3 that (given k^0, φ^2) produces $v := \Phi(k, \varphi)$ with

$$\forall n^0, g^1 \left(\forall m^0(|r_{n+m} -_{\mathbb{Q}} r_n| \leq_{\mathbb{Q}} 2^{-k} \vee B_0(g(m))) \rightarrow B_0(\varphi n g)\right) \rightarrow B_0(v)$$

(here we disregard for simplicity the '$\exists z^0$'-part in the *mr*-interpretation of '$\vee$' since $|r_{n+m} -_{\mathbb{Q}} r_n| \leq_{\mathbb{Q}} 2^{-k}$ and $B_0(v_0)$ are primitive recursively decidable).
One easily verifies that we can take

$$\Phi k \varphi :=_0 \Psi(2^k)0^0,$$

where

$$\Psi 0 :=_1 0^1,\ \Psi(x+1) :=_1 \lambda y.\varphi y(\Psi x)$$

is defined by the recursor R_1 for type-1 recursion (which is powerful enough to define the Ackermann function, see chapter 3).

Of course, that a program involves R_1 does not per se imply that it has to be inefficient. In fact, results in Berger et al. [26] indicate that applied to special Π^0_2-consequences B and small inputs such programs might well be efficient (see also Raffalli [307] for a particularly interesting program extracted – by a related technique close to Krivine's realizability [253] – from a proof of Dickson's lemma).

Remark 14.5. The combination of negative translation and A-translation has the following variant: Let the R-translation be the result of replacing in a formula every occurrence of $\perp$ (i.e. in HA '$0 = 1$') by R (so, in contrast to the A-translation, prime formulas P different from $\perp$ are not affected). The resulting interpretation is not sound for intuitionistic logic and hence not for HA since it does not translate the 'ex falso quodlibet'-axiom $\perp \rightarrow F$ correctly, but it is sound for MA which results from HA by dropping 'ex falso quodlibet' from the logical axioms, i.e. by formulating HA with so-called minimal logic instead of intuitionistic logic. Since the original Gödel negative translation (but not the Kuroda variant) actually translates PA not only into HA but even into MA (see [371]) we can use the combination of Gödel's negative translation and R-translation as well (see e.g. [21]). This prima facie is a simplification as we insert R at fewer places than the A-translation would do. However, there is a warning in place here: in the example above we made use of the fact that over HA (and even a small fragment thereof) we can prove the stability of prime formulas and hence change the 'official' Gödel negative translation (written here in an equivalent form over minimal logic)

$$(1)\ \forall k \neg\neg\exists n \forall m \neg\neg(|r_{n+m} -_{\mathbb{Q}} r_n| \leq_{\mathbb{Q}} 2^{-k})$$

of

$$(2)\ \forall k \exists n \forall m(|r_{n+m} -_{\mathbb{Q}} r_n| \leq_{\mathbb{Q}} 2^{-k})$$

first into

$$(3)\ \forall k \neg\neg\exists n \forall m(|r_{n+m} -_{\mathbb{Q}} r_n| \leq_{\mathbb{Q}} 2^{-k})$$

and then apply the A-translation to the latter (which results in only 3 occurrences of B instead of 5, see above).
However, this 'pre-processing' is not possible if we use the R-translation since $\neg\neg(s =_0 t) \to s =_0 t$ cannot be proved in MA and in fact does not have a valid R-translation. So we need to apply the simpler R-translation to the more complicated formula (1) resulting in 4 occurrences of B and hence in the end in a more complicated (though over a weak fragment of HA using the stability of $=_0$) equivalent translation compared to the A-translation of (3).

14.2 Historical comments and suggested further reading

The A-translation was introduced in Friedman [108] and independently by Dragalin in [91]. For more information on the A-translation see Leivant [261]. Applications of (refined) combinations of negative translation, A-translation and realizability can be found e.g. in [25, 21, 26, 283]. For the interesting refinement of the A-translation mentioned above see Coquand-Hofmann [75] and – for applications of that refinement – Avigad [5]. In Berger-Oliva [22], a modified realizability interpretation of the A-translation of the negative translation of the axiom of dependent choice is given by a novel form of bar recursion called modified bar recursion. Modified bar recursion allows one to define Spector's bar recursion but the converse is not true. In fact, modified bar recursion is not S1-S9 computable (in the sense of Kleene) over the continuous functionals whereas Spector's bar recursion is. Bezem's model of strongly majorizable functionals discussed in chapters 3 and 11 is also a model of modified bar recursion but the construction of the majorant is ineffective (and uses discontinuous functionals). Since the interpretation given in Berger-Oliva [22] uses a continuity axiom it is not clear whether the result of the interpretation does hold in Bezem's model. For all this see Berger-Oliva [22, 23].

Chapter 15
Applications to analysis: general metatheorems I

15.1 A general metatheorem for Polish spaces

In this chapter we show how some of the main results from chapters 8–12 can be combined with the representation of Polish spaces from chapter 4 to establish general metatheorems on the extractability of effective uniform bounds from proofs in analysis. 'Uniform' here means the independence of the bounds from parameters in compact metric spaces. In chapter 16 we will apply these results to concrete proofs in best approximation theory and extract effective rates of so-called strong unicity for both best Chebycheff as well as L_1-approximations of continuous functions $f \in C[0,1]$ by polynomials of degree $\leq n$ (and more general so-called Haar spaces in the case of Chebycheff approximation). In chapter 17, the metatheorems will be much generalized to guarantee even strongly uniform bounds which are independent from parameters in abstract bounded metric spaces. These more general theorems will then be applied to proofs in metric fixed point theory in chapter 18.

In many problems in numerical (functional) analysis the task is to construct solutions $x \in K$ for equations

$$A(x) :\equiv (F(x) = 0),$$

where K is a compact metric space and $F : K \to \mathbb{R}$ a continuous function.

Typically, this involves two steps:

1) one constructs approximate solutions $x_n \in K$ satisfying

$$A_n(x_n) :\equiv \left(|F(x_n)| < 2^{-n}\right),$$

2) using the compactness of K and the continuity of F one then concludes that either $(x_n)_{n\in\mathbb{N}}$ itself or some subsequence converges to a solution of $A(x)$.

If F has exactly one root $\widehat{x} \in K$ the – in general non-computational – step of selecting a convergent subsequence is not necessary but still an effective rate of the convergence $x_n \to \widehat{x}$ is missing. Further below we will see how proof-theoretic analysis of a given proof for the uniqueness of $\widehat{x}$ can be applied to extract such an effective rate of convergence under quite general circumstances and give some applications to approximation theory.

If the solution $\widehat{x}$ is not necessarily unique, one often cannot effectively obtain a solution but weaker tasks like obtaining so-called effective rates of asymptotic regularity (see below) might be solvable. In chapter 18 we will illustrate this in the area of fixed point theory.
In the following, when we write e.g. $A :\equiv \forall n^0 \forall x \in X, y \in K \exists m^0 A_1(n,x,y,m)$ for a Polish space X and a compact metric space K in standard representation by $\mathbb{N}^{\mathbb{N}}$ resp. $\{f \in \mathbb{N}^{\mathbb{N}} : f \leq_1 M\}$ for some fixed M and a purely existential property A_1 so that A formalized has the form

$$\forall n^0, x^1 \forall y \leq_1 M \exists m^0 A_1(n,x,y,m),$$

we assume that $A_1(n,x,y,m)$ is extensional in x,y with respect to $=_X$ and $=_K$ and so really expresses a property of elements of X,K. Although the proof of this extensionality is not needed for our extraction, we will assume this extensionality to be provable in the theory at hand as this will be the case in all our applications.

The applications to approximation theory (i.e. the extraction of rates of convergence towards a unique solution) fall under the schema of the following

Theorem 15.1 (General metatheorem on proof mining: the compact case). *Let X be a Polish space, K a compact metric space and $A_1(n^0,x^1,y^1,m^0)$ a purely existential formula of $\mathcal{L}$(E-PA$^\omega$) (where the types of the existential quantifiers are of degree ≤ 1) having only n,x,y,m as free variables. We assume that X,K are explicitly representable in* E-PA$^\omega$ *in the sense of chapter 4. Let A_1 be provably extensional in $x \in X, y \in K$, i.e. assume that* E-PA$^\omega$+QF-AC^{1,0}+QF-AC^{0,1}+WKL *proves that*

$$\forall n^0, m^0, x_1^1, x_2^1, y_1^1, y_2^1 (x_1 =_X x_2 \wedge y_1 =_K y_2 \wedge A_1(n,x_1,y_1,m) \rightarrow A_1(n,x_2,y_2,m)).$$

If a sentence

$$(*)\ \forall n \in \mathbb{N} \forall x \in X \forall y \in K \exists m \in \mathbb{N} A_1(n,x,y,m)$$

is proved in E-PA$^\omega$+QF-AC^{1,0}+QF-AC^{0,1}+WKL, *then one can extract a uniform bound $\Phi(n,x)$ for $\exists m$, i.e.*

$$\text{WE-HA}^\omega \vdash \forall n \in \mathbb{N} \forall x \in X \forall y \in K \exists m \leq \Phi(n,x) A_1(n,x,y,m),$$

where Φ is a closed term of E-PA$^\omega$. *More precisely, the bound $\Phi(n,x)$ depends on x via a representation $f_x \in \mathbb{N}^{\mathbb{N}}$ of $x \in X$ in the sense of chapter 4.*
Furthermore, the extensionality of A_1 w.r.t. $x \in X, y \in K$ is provable in WE-HA$^\omega$.
Supplement: We may add arbitrary axioms Γ of the form

$$(**)\ \forall x \in X' \exists y \in K' \forall w \in W (F'(x,y,w) =_{\mathbb{R}} 0)$$

to E-PA$^\omega$, *where X',W are a Polish spaces, K' a compact metric space and $F' : X' \times K' \times W \rightarrow \mathbb{R}$ a continuous function (all assumed to be explicitly definable in*

E-PA$^\omega$). *Then the verification takes place in* WE-HA$^\omega + \Gamma_\varepsilon$, *where*

$$\Gamma_\varepsilon :\equiv \forall x \in X' \forall k,n \in \mathbb{N} \exists y \in K' \big(\bigwedge_{i=0}^{n} |F'(x,y,w_i)| < 2^{-k} \big)$$

is the 'ε-weakening' of Γ (here (w_n) is the enumeration of the countable dense subset on which the representation of W is based).
An analogous result holds for

$$\widehat{\text{E-PA}}^{\omega}\!\upharpoonright + \text{QF-AC}^{1,0} + \text{QF-AC}^{0,1} + \text{WKL} + \Gamma \text{ and } \widehat{\text{WE-HA}}^{\omega}\!\upharpoonright + \Gamma_\varepsilon$$

instead of

$$\text{E-PA}^{\omega} + \text{QF-AC}^{1,0} + \text{QF-AC}^{0,1} + \text{WKL} + \Gamma \text{ and } \text{WE-HA}^{\omega} + \Gamma_\varepsilon.$$

Then Φ is a closed term of $\widehat{\text{E-PA}}^{\omega}\!\upharpoonright$.

A crucial feature of the bound $\Phi(n,x)$ is that it does not depend on $y \in K$!

In many cases, $\exists m A_1$ is monotone in m (see sections 15.2–15.4 below). Then $\Phi(n,x)$ even provides a uniform realizer for $\exists m$.

Remark 15.2. 1) By the dependence of the bound Φ on the choice of a representation ('name') $f_x \in \mathbb{N}^\mathbb{N}$ of $x \in X$, Φ is not an extensional function on X but an intensional operation which, however, is extensional on the space $\mathbb{N}^\mathbb{N}$ of names of X-elements. More precisely, one should, therefore, have formulated (and sometimes we do so when this is important) the conclusion of theorem 15.1 as

$$\text{WE-HA}^\omega \vdash \forall n \in \mathbb{N} \forall x \in X \forall f_x^1 (f_x \text{ represents } x$$
$$\to \forall y \in K \exists m \leq \Phi(n, f_x^1) A_1(n,x,y,m)),$$

where we write 'x' in '$A_1(n,x,y,m)$' to indicate that the formula is extensional w.r.t. $=_X$ but we write 'f_x' in '$\Phi(n,f_x)$' to indicate that the value of $\Phi(n,f_x)$ does depend on the given representative f_x of x. For simplicity, however, we often simply write $\Phi(n,x)$ and just remark that the bound depends on the chosen representative of x.

For the case of $C[0,1]$ this intensionality means that Φ will depend on $f \in C[0,1]$ endowed with a modulus of uniform continuity $\omega_f \in \mathbb{N}^\mathbb{N}$ of f, i.e.

$$\forall k \in \mathbb{N} \forall x,y \in [0,1] \big(|x -_\mathbb{R} y|_\mathbb{R} <_\mathbb{R} 2^{-\omega_f(k)} \to |f(x) -_\mathbb{R} f(y)|_\mathbb{R} <_\mathbb{R} 2^{-k} \big).$$

Note that by the effective Weierstraß theorem (see e.g. [303]) this input is equivalent to the 'official' representation of $f \in C[0,1]$ from chapter 4 by a 2^{-n} Cauchy sequence (w.r.t. to the uniform norm) of polynomials with rational coefficients converging to f.

2) The extraction algorithm in theorem 15.1 is based on the combination of negative translation and monotone functional interpretation discussed in chapters 9 and 10 and used in the proof of theorems 10.47, 10.52. This algorithm, of course, only serves as a general guideline and is only applied step-by-step if nothing better is available. In practice, we will make use of obvious mathematical simplifications and optimizations all the time.
3) Instead of a single compact metric space K one can also have certain uniformly (in parameters of type 0, 1) representable families of such spaces as e.g. $[-m,m]^{(d)}$, where m ranges over $\mathbb{N}$ (here one uses the representation of $[-m,m]$ given after remark 4.26 in chapter 4). Then the extracted bound will additionally depend on these parameters, i.e. on m in the example just mentioned.
4) The supplement to theorem 15.1 allows one in concrete applications to treat lemmas Γ having the form $(**)$ in a given proof simply as axioms which means that their proofs (which might be the most tedious part of the overall proof) do not need to be analyzed at all. It is this feature (again a consequence of the modularity of proof interpretations such as functional interpretation) which is most crucial for the applicability of the theorem to rather involved (and not fully formalized) proofs in mathematics. In all the applications presented in this book we make heavy use of this fact.
5) Whereas the extra axioms Γ (like WKL which just happens to be a special case of such an axiom Γ by proposition 9.18.2) from chapter 9) in general are ineffective, Γ_ε usually is constructively provable, e.g. for WKL_ε this follows from proposition 9.18.1). So, as a by-product, the theorem also yields a constructivization of the original proof. However, there is an important remark in order here: the construction of the bound Φ follows the original non-constructive proof and is first verified non-constructively. The reduction to a constructive verifying system is achieved by subsequent proof theoretic procedures (which again are based on monotone functional interpretation) and is completely superfluous for constructing Φ. So to say it again, Φ is **not** extracted by first transforming the given proof into a constructive one (e.g. by eliminating WKL) and then analyzing the latter. Such a procedure would be much too complicated in practice.

Proof of theorem 15.1: The proof of theorem 15.1 proceeds (for E-PA$^\omega$+QF-AC^{1,0}+QF-AC^{0,1}+WKL) by applying the representations from chapter 4 to the Polish space X and the compact metric space K which reduces the statement to theorem 10.47. The provability of the extensionality of A_1 in WE-HA$^\omega$ also follows immediately from theorem 10.47.
As a consequence of the general logical form of these representations we also see that the additional axioms Γ have the logical form of the axioms Δ permitted in theorem 10.47. In fact, using the continuity of $F'(x,y,w)$ in w (proposition 4.18), $(**)$ can be stated equivalently as

$$\forall x \in X' \exists y \in K' \forall n \in \mathbb{N}(F'(x,y,w_n) =_{\mathbb{R}} 0)$$

and – in turn – as

$$\forall x \in X' \exists y \in K' \forall n,k \in \mathbb{N}(|F'(x,y,w_n)| \leq_{\mathbb{R}} 2^{-k}).$$

Using the representation from chapter 4 this sentence has the following logical form

$$(+)\ \forall x^1 \exists y \leq_1 N \forall n^0, k^0, l^0 A_0(x,y,w_n,k,l),$$

where $\forall l^0 A_0(x,y,w_n,k,l)$ is the Π_1^0-formula expressing that $(|F'(x,y,w_n)| \leq_{\mathbb{R}} 2^{-k})$. Here N is some closed function term of WE-HA$^\omega$ resulting from the representation of K.

$(+)$ has the logical form of an axiom Δ in theorem 10.47.

Moreover, since all variables in $(+)$ have types of degree ≤ 1, proposition 10.30 applies. Hence the use of $(+)$ can be reduced to that of its ε-weakening

$$(+)_\varepsilon\ \forall x^1, m^0 \exists y \leq_1 N \forall n,k,l \leq m A_0(x,y,w_n,k,l)$$

which is implied by

$$\forall x^1, m^0 \exists y \leq_1 N \forall n,k \leq m \forall l A_0(x,y,w_n,k,l),$$

i.e. by

$$\forall x \in X' \forall k,n \in \mathbb{N} \exists y \in K' \Big(\bigwedge_{i=0}^{n} |F'(x,y,w_i)| \leq 2^{-k} \Big).$$

□

We conclude this section by showing that both components of the constructive notion of compactness of K, namely total boundedness and completeness, are necessary for theorem 15.1 to hold in general. The following examples hold e.g. for $\mathcal{T}^\omega :=$E-PA$^\omega$+QF-AC^{0,0} :

Necessity of total boundedness: Let B be the closed unit ball in $C[0,1]$ (the latter represented as in chapter 4). B is bounded and constructively representable in (weak fragments of) $\mathcal{T}^\omega$ but not totally bounded. By the Weierstraß approximation theorem

$$\mathcal{T}^\omega \vdash \forall f \in B \exists n \in \mathbb{N}\big(n \text{ code of } p \in \mathbb{Q}[X] \text{ s.t. } \|p-f\|_\infty < \frac{1}{2}\big)$$

but there is **no uniform bound** on $\exists n$: take the sequence of functions $f_k := \sin(kx)$. Here by a 'n code of $p \in \mathbb{Q}[X]$' we mean that n encodes the finite tuple of (codes of the) rational coefficients of p.

Necessity of completeness: The set $[0,2]_{\mathbb{Q}}$ of all rational numbers $0 \leq q \leq 2$ is totally bounded and constructively representable and

$$\mathcal{T}^\omega \vdash \forall q \in [0,2]_{\mathbb{Q}}\ \exists n \in \mathbb{N}(|q -_{\mathbb{R}} \sqrt{2}|_{\mathbb{R}} >_{\mathbb{R}} 2^{-n}).$$

However, there clearly is **no uniform** bound on $\exists n \in \mathbb{N}$.

Remark 15.3. One might think to represent $[0,2]_{\mathbb{Q}}$ with the discrete metric w.r.t. which it **is** a Polish space. However, with this metric it no longer is totally bounded.

Necessity of the intensionality of the bound Φ: Consider the following obvious fact

$$\mathcal{T}^\omega \vdash \forall x \in \mathbb{R} \exists n^0 ((n)_{\mathbb{R}} >_{\mathbb{R}} x).$$

Suppose there would exist an $=_{\mathbb{R}}$-extensional computable functional $\Phi : \mathbb{N}^{\mathbb{N}} \to \mathbb{N}$ bounding (and hence realizing) '$\exists n$'. Then Φ would represent a computable and hence continuous (see chapter 4) function $\mathbb{R} \to \mathbb{N}$ so that Φ had to be a constant function which cannot be true.

Necessity of $A_\exists$ being an '$\exists$-formula':

Let (f_n) be the usual sequence of spike-functions in $C[0,1]$, s.t. (f_n) converges pointwise but not uniformly towards 0. Then

$$\mathcal{T}^\omega \vdash \forall x \in [0,1] \forall k \in \mathbb{N} \exists n \in \mathbb{N} \forall m \in \mathbb{N} (|f_{n+m}(x)| \leq 2^{-k}),$$

but there is no uniform bound (not even an ineffective one) on '$\exists n$'.

15.2 Applications to uniqueness proofs

Let X be a Polish space, K a compact metric space and $F : X \times K \to \mathbb{R}$ a continuous function. Suppose that all these objects are explicitly representable in e.g. $\mathcal{T}^\omega :=$E-PA$^\omega$+QF-AC^{1,0}+QF-AC^{0,1}+WKL and assume that we can prove in $\mathcal{T}^\omega$ that for every $x \in X$, $F(x,\cdot)$ has at most one root in K, i.e.

$$(1)\ \forall x \in X \forall y_1, y_2 \in K \Big(\bigwedge_{i=1}^{2} F(x, y_i) = 0 \to y_1 = y_2 \Big).$$

Let d_K denote the metric on K. (1) can be rewritten as

$$(1)^*\ \forall x \in X \forall k \in \mathbb{N} \forall y_1, y_2 \in K \exists l \in \mathbb{N} \Big(\bigwedge_{i=1}^{2} (|F(x,y_i)| \leq_{\mathbb{R}} 2^{-l}) \to d_K(y_1,y_2) <_{\mathbb{R}} 2^{-k} \Big),$$

where $\big(\bigwedge_{i=1}^{2} (|F(x,y_i)| \leq_{\mathbb{R}} 2^{-l}) \to d_K(y_1,y_2) <_{\mathbb{R}} 2^{-k}\big) \in \Sigma^0_1$, since $\leq_{\mathbb{R}} \in \Pi^0_1$ and $<_{\mathbb{R}} \in \Sigma^0_1$ given the representation of reals from chapter 4 as Cauchy sequences of rational numbers with fixed rate of convergence.
Thus by the general logical metatheorem 15.1 one can extract from such a proof an explicit bound $\Phi(x,k)$ (given by a closed term of the underlying arithmetical system E-PA$^\omega$), which will depend on the chosen representation f^1_x of x such that

$$(2)\ \forall x \in X \forall k \in \mathbb{N} \forall y_1, y_2 \in K \Big(\bigwedge_{i=1}^{2} (|F(x,y_i)| < 2^{-\Phi(x,k)}) \to d_K(y_1,y_2) < 2^{-k} \Big).$$

Moreover, by theorem 15.1, (2) can be proved without using WKL and even in the intuitionistic and weakly extensional variant WE-HA$^\omega$ of E-PA$^\omega$ (and hence in constructive analysis in the sense of Bishop).

Put together, we have shown the following

Theorem 15.4. *Let X be a Polish space, K a compact metric space (both* E-PA$^\omega$ *definable) and $F : X \times K \to \mathbb{R}$ a function given by a closed term of* E-PA$^\omega$ *(in the sense of chapter 4 and therefore automatically continuous).*

If a sentence

$$\forall x \in X \forall y_1, y_2 \in K\Big(\bigwedge_{i=1}^{2}(F(x,y_i) =_{\mathbb{R}} 0) \to y_1 =_K y_2\Big)$$

is proved in E-PA$^\omega$+QF-AC^{1,0}+QF-AC^{0,1}+WKL*, then one can extract a closed term Φ of* E-PA$^\omega$ *(depending on a representative f_x^1 of x in the sense of chapter 4) such that*

$$\text{WE-HA}^\omega \vdash \forall x \in X \forall k \in \mathbb{N} \forall y_1, y_2 \in K\Big(\bigwedge_{i=1}^{2}(|F(x,y_i)| < 2^{-\Phi(x,k)}) \to d_K(y_1,y_2) < 2^{-k}\Big).$$

As before we may add arbitrary axioms Γ of the form

$$\forall x \in X' \exists y \in K' \forall w \in W(F'(x,y,w) =_{\mathbb{R}} 0)$$

to $\mathcal{T}^\omega$, where X', W are a Polish spaces, K' a compact metric space and $F' : X' \times K' \times W \to \mathbb{R}$ a continuous function (all assumed to be explicitly definable in E-PA$^\omega$*). Then the verification of the bound Φ can be carried out in* WE-HA$^\omega + \Gamma_\varepsilon$*, where Γ_ε is defined as in theorem 15.1.*
An analogous result holds for $\widehat{\text{E-PA}}^\omega\!\upharpoonright$ *and* $\widehat{\text{WE-HA}}^\omega\!\upharpoonright$ *instead of* E-PA$^\omega$ *and* WE-HA$^\omega$.

As in theorem 15.1, Φ does **not** depend on $y_1, y_2 \in K$ (but will depend on a given representative f_x^1 of x). Because of this fact, $\Phi(x,k)$ – which we call a **modulus of uniqueness** – can be used to compute the unique root (if existent) from any algorithm $\chi(x,k)$ computing approximate so-called $\varepsilon(= 2^{-k})$-roots of $F(x,\cdot)$:

$$(3)\ \forall x \in X \forall k \in \mathbb{N}\big(\chi(x,k) \in K \wedge |F(x,\chi(x,k))| < 2^{-k}\big).$$

One easily verifies that (2) and (3) imply that $(\chi(x,\Phi(x,k)))_{k\in\mathbb{N}}$ is a Cauchy sequence in K which converges with rate of convergence 2^{-k} to the unique root $y \in K$ of $F(x,\cdot)$. So $\Psi(x) := y = \lim\limits_{k\to\infty} \chi(x,\Phi(x,k))$ can be computed with arbitrarily prescribed precision (which can also be proved in WE-HA$^\omega(+\Gamma_\varepsilon)$, see theorem 15.5 below) and the computational complexity of Ψ can be estimated in terms of the complexities of Φ and χ. From the uniqueness of the solution, the extensionality of $\Psi(x)$ in x follows, i.e. Ψ represents a function $X \to K$.

In numerical analysis (in particular in best approximation theory) such moduli of uniqueness have been considered in various special situations under the names of 'strong unicity' or 'rate of strong uniqueness'. We will study this in detail in the next chapter.
For another interesting general use of the concept of modulus of uniqueness see Lambov [260].

Theorem 15.5. *Under the same assumptions as in theorem 15.4, if a sentence*

$$\forall x \in X \forall y_1, y_2 \in K\Big(\bigwedge_{i=1}^{2}(F(x,y_i) =_{\mathbb{R}} 0) \to y_1 =_K y_2\Big)$$

is proved in E-PA$^\omega$+QF-AC^{1,0}+QF-AC^{0,1}+WKL, *then one can extract a closed term* Ψ *of* E-PA$^\omega$ *such that*

$$\text{WE-HA}^\omega \vdash \begin{cases} \forall x \in X, k \in \mathbb{N} \exists y \in K(|F(x,y)| \le 2^{-k}) \to \\ \Psi \in C(X,K) \wedge \forall x \in X(F(x,\Psi(x)) =_{\mathbb{R}} 0_{\mathbb{R}}). \end{cases}$$

Here '$\Psi \in C(X,K)$*' expresses that the term* $\Psi^{1(1)}$ *represents a continuous function* $X \to K$.
As before we may add arbitrary axioms Γ *of the form*

$$\forall x \in X' \exists y \in K' \forall w \in W(F'(x,y,w) =_{\mathbb{R}} 0)$$

to $\mathcal{T}^\omega$, *where* X', W *are a Polish spaces,* K' *a compact metric space and* $F' : X' \times K' \times W \to \mathbb{R}$ *a continuous function (all assumed to be explicitly definable in* E-PA$^\omega$*). Then the verification of the bound* Φ *can be carried out in* WE-HA$^\omega + \Gamma_\varepsilon$, *where* Γ_ε *is defined as in theorem 15.1.*
An analogous result holds for $\widehat{\text{E-PA}}^\omega\!\upharpoonright$ *and* $\widehat{\text{WE-HA}}^\omega\!\upharpoonright$ *instead of* E-PA$^\omega$ *and* WE-HA$^\omega$.

Proof: By the assumption that X, K and F are E-PA$^\omega$-definable there are closed terms M^1 and Φ_F of E-PA$^\omega$ so that the premise

$$\forall x \in X, k \in \mathbb{N} \exists y \in K(|F(x,y)| \le 2^{-k})$$

has the form

$$\forall x^1, k^0 \exists y \le_1 M(|\Phi_F(x,y)|_{\mathbb{R}} \le_{\mathbb{R}} 2^{-k})$$

and, therefore (using lemma 4.2),

$$\forall x^1, k^0 \exists y \le_1 M(|\widehat{\Phi_F(x,y)}|_{\mathbb{R}}(k) \le_{\mathbb{Q}} 2^{-k+1}).$$

Since

$$A_0(x^1, y^1, k^0) :\equiv (|\widehat{\Phi_F(x,y)}|_{\mathbb{R}}(k) \le_{\mathbb{Q}} 2^{-k+1})$$

is quantifier-free we can use lemma 9.27 to construct a closed term $\chi^{1(0)(1)}$ of WE-PA$^\omega$ such that

$$\forall x^1, k^0(\chi(x,k) \leq_1 M \wedge |\widehat{\Phi_F(x,\chi(x,k))}|_{\mathbb{R}}(k) \leq_{\mathbb{Q}} 2^{-k+1}).$$

Hence for $\tilde{\chi}(x,k) := \chi(x,k+2)$ we obtain

$$\forall x^1, k^0(\tilde{\chi}(x,k) \leq_1 M \wedge |\Phi_F(x,\tilde{\chi}(x,k))|_{\mathbb{R}} <_{\mathbb{R}} 2^{-k}),$$

i.e., switching back from the representations to the level of the objects being represented,

$$(1)\ \forall x \in X, k \in \mathbb{N}(\tilde{\chi}(x,k) \in K \wedge |F(x,\tilde{\chi}(x,k))| < 2^{-k})$$

(strictly speaking we have to write here $\tilde{\chi}(f_x,k)$ for some representative f_x^1 of x but we avoid this to improve the readability). By theorem 15.4, we can extract a closed term Φ of WE-HA$^\omega$ such that (provably in WE-HA$^\omega$)

$$(2)\ \forall x \in X \forall k \in \mathbb{N} \forall y_1, y_2 \in K \Big(\bigwedge_{i=1}^{2} (|F(x,y_i)| < 2^{-\Phi(x,k)}) \to d_K(y_1,y_2) < 2^{-k}\Big)$$

(again, Φ strictly speaking depends on a representative f_x^1 of x).
Hence $(\tilde{\chi}(x,\Phi(x,k)))_{k\in\mathbb{N}}$ is (a representation of) a Cauchy sequence in K with Cauchy rate 2^{-k}.
Now define

$$\Psi'(x,k) := (\widehat{\Psi_1(x,k)})(k+5), \text{ where } \Psi_1(x,k) := \tilde{\chi}(x,\Phi(x,k+5)).$$

Then, by lemma 4.5, $\Psi'(x)$ represents the limit of the Cauchy sequence

$$(\tilde{\chi}(x,\Phi(x,k)))_{k\in\mathbb{N}}$$

in K which – using the construction from section 4.3 of chapter 4 – can be converted into a representative $\Psi(x)$ such that $\Psi(x) =_K \Psi'(x)$ and $\Psi(x) \leq_1 M$. That $\Psi(x)$ satisfies the theorem is now proved using the uniform continuity of $F(x,\cdot) : K \to \mathbb{R}$ (proposition 4.23): choose $k \in \mathbb{N}$ arbitrarily and let $l \in \mathbb{N}$ be such that

$$(3)\ \forall y_1, y_2 \in K \big(d_K(y_1,y_2) < 2^{-l+1} \to |F(x,y_1) - F(x,y_2)| < 2^{-k}\big).$$

By (1) we have that

$$(4)\ |F(x,\tilde{\chi}(x,\max(k,\Phi(x,l))))| < 2^{-k}.$$

Furthermore, using (2),

$$(5)\ d_K(\tilde{\chi}(x,\max(k,\Phi(x,l))),\tilde{\chi}(x,\Phi(x,l))) < 2^{-l}.$$

Moreover,

$$(6)\ d_K(\Psi(x),\tilde{\chi}(x,\Phi(x,l))) < 2^{-l}.$$

By (5), (6) we get

$$(7)\ d_K(\Psi(x),\tilde{\chi}(x,\max(k,\Phi(x,l)))) < 2^{-l+1}.$$

(3), (4) and (7) imply that

$$(8)\ |F(x,\Psi(x))| < 2^{-k+1}.$$

Since k was arbitrary, we have shown that

$$(9)\ F(x,\Psi(x)) =_{\mathbb{R}} 0.$$

Now let x_1^1, x_2^1 be such that $x_1 =_X x_2$. From (9) it follows that

$$\bigwedge_{i=1}^{2} F(x_i,\Psi(x_i)) =_{\mathbb{R}} 0.$$

Since, by assumption, F is extensional w.r.t. $=_X$ (provably in WE-HA$^\omega$) this yields

$$F(x_1,\Psi(x_1)) =_{\mathbb{R}} 0 =_{\mathbb{R}} F(x_1,\Psi(x_2)).$$

So by (2)

$$\Psi(x_1) =_K \Psi(x_2),$$

i.e. Ψ represents a function $X \to K$. Since Ψ is a closed term of WE-HA$^\omega$ it follows from proposition 4.18 that this function is continuous. □

Remark 15.6. In the proof above we actually only need the pointwise continuity of $F(x,\cdot) : K \to \mathbb{R}$ in the point $\Psi(x)$ so that we could have used also proposition 4.18 instead of proposition 4.23. However, for the proof of the former we referred to a result from the literature whereas proposition 4.23 follows easily from results proved in this book which makes the proof above (except for the pointwise continuity of $\Psi : X \to K$) self-contained.

Remark 15.7. In the results above we may even have functions $F : X \times Y \to \mathbb{R}$, where X, Y are general Polish spaces and can allow certain constructively definable families $(K_x)_{x\in X}$ of compact subspaces of Y which are effectively parametrized by $x \in X$ instead of a fixed K. See Kohlenbach [204] for details.

In the uniqueness proofs from best approximation theory which will be analyzed in the next chapter, WKL actually is used in the form of the principle

$$\forall f \in C[0,1] \exists x \in [0,1] \big(f(x) = \inf_{y\in[0,1]} f(y)\big) \tag{15.1}$$

which – despite of the fact that f by our representation of $f \in C[0,1]$ is given with a modulus of uniform continuity – is equivalent to WKL.

We now show that the conditions for the conversion of ineffective existence into effective existence in the results above, namely the uniqueness of the solution, the compactness of K and the fact that the conclusion '$F(f,x) =_{\mathbb{R}} 0$' is purely universal, are all necessary even for $X := \mathbb{N}$ and K being a bounded subset of $\mathbb{N}$:

1) **Necessity of uniqueness:** There is a quantifier-free formula $A_0(u,v,w)$ in $\mathscr{L}(\mathrm{PA})$ containing only u,v,w free such that

$$\mathrm{PA} \vdash \forall u \exists v \leq 1 \forall w A_0(u,v,w),$$

but there is no recursive $f : \mathbb{N} \to \mathbb{N}$ so that

$$\forall u, w A_0(u, f(u), w)$$

is true.

Proof: From classical recursion theory we know that there are recursive enumerable (r.e.) subsets sets

$$A = \{u : \exists v(\alpha(u,v) = 0)\} \text{ and } B = \{u : \exists v(\beta(u,v) = 0)\}$$

of $\mathbb{N}$ with α, β being primitive recursive such that

a) PA$\vdash$ $\forall u\Big(\forall w\big(\alpha(u,w) \neq 0\big) \vee \forall w\big(\beta(u,w) \neq 0\big)\Big)$, i.e. $A \cap B = \emptyset$ provably in PA, and
b) there exists no recursive function $f : \mathbb{N} \to \mathbb{N}$ such that

$$\forall u\Big([f(u) = 0 \to \forall w\big(\alpha(u,w) \neq 0\big)] \wedge [f(u) \neq 0 \to \forall w\big(\beta(u,w) \neq 0\big)]\Big)$$

(i.e. A, B are recursively inseparable r.e. sets).

E.g. we may take α and β as the characteristic functions of

$$T(u,u,v) \wedge U(v) = 0$$

and

$$T(u,u,v) \wedge U(v) = 1,$$

respectively, where T is the Kleene T-predicate and U is the primitive recursive function reading off the output of the computation (carried out by the Turing machine with code u applied to u) with code v, both known from the well-known Kleene normal form theorem.
a) implies that PA $\vdash \forall u \exists v \leq 1 \forall w A_0(u,v,w)$, where
$A_0(u,v,w) :\equiv [v = 0 \to \alpha(u,w) \neq 0] \wedge [v \neq 0 \to \beta(u,w) \neq 0]$. Now let $f : \mathbb{N} \to \mathbb{N}$ be such that $\forall u, w A_0(u, f(u), w)$, i.e.

$$\forall u[f(u) = 0 \to \forall w\big(\alpha(u,w) \neq 0\big)] \wedge [f(u) \neq 0 \to \forall w\big(\beta(u,w) \neq 0\big)].$$

Then, by b), f is not recursive. □

2) **Necessity of the boundedness of v:** There exists a quantifier-free formula $A_0(u,v,w)$ of $\mathscr{L}(\mathrm{PA})$ containing only u,v,w free such that

$$\mathrm{PA} \vdash \forall u \exists! v \forall w A_0(u,v,w),$$

but there is no recursive f so that

$$\forall u, w A_0(u, f(u), w)$$

is true.

Proof: Define

$$A_0(u,v,w) :\equiv T(u,u,v) \vee [\neg T(u,u,w) \wedge v = 0],$$

where again T denotes Kleene's T-predicate. Then, using that

$$T(u,u,v_1) \wedge T(u,u,v_2) \to v_1 = v_2,$$

one proves that

$$\mathrm{PA} \vdash \forall u \exists! v \forall w A_0(u,v,w).$$

Now let$f : \mathbb{N} \to \mathbb{N}$ be such that

$$\forall u, w A_0(u, f(u), w).$$

Because of

$$T(u,u,f(u)) \leftrightarrow \exists v T(u,u,v),$$

the recursive undecidability of the special halting problem now implies that f is not recursive. $\square$

3) **Necessity of the assumption that $\forall w A_0$ is Π^0_1** : There exists a quantifier-free formula $A_0(u,v,w,z)$ of $\mathscr{L}(\mathrm{PA})$ containing only u,v,w,z free such that

$$\mathrm{PA} \vdash \forall u \exists! v \leq 1 \forall w \exists z A_0(u,v,w,z),$$

but for no recursive $f : \mathbb{N} \to \mathbb{N}$

$$\forall u, w \exists z A_0(u, f(u), w, z)$$

is true.

Proof: We consider again the first example:

$$\mathrm{PA} \vdash \forall u \exists v \leq 1 \forall w A_0'(u,v,w),$$

where

$$A_0'(u,v,w) :\equiv [v = 0 \to \alpha(u,w) \neq 0] \wedge [v \neq 0 \to \beta(u,w) \neq 0].$$

Hence

$$\mathrm{PA} \vdash \forall u \exists ! v \leq 1 \Big(\forall w A_0'(u,v,w) \wedge (v \neq 0 \to \neg \forall w A_0'(u,0,w)) \Big)$$

and therefore

$$\mathrm{PA} \vdash \forall u \exists ! v \leq 1 \forall w \exists z A_0(u,v,w,z),$$

where

$$A_0(u,v,w,z) :\equiv A_0'(u,v,w) \wedge (v \neq 0 \to \neg A_0'(u,0,z)).$$

Let $f : \mathbb{N} \to \mathbb{N}$ be such that $\forall u, w \exists z A_0(u, f(u), w, z)$.
Then

$$\forall u \big([f(u) = 0 \to \forall w(\alpha(u,w) \neq 0)] \wedge [f(u) \neq 0 \to \forall w(\beta(u,w) \neq 0)] \big).$$

Hence (as in the first example) f is not recursive. □

Remark 15.8. The proof above shows that even for $A \in \Delta_2^0$ (instead of $A \in \Pi_1^0$) it may happen that

$$\mathrm{PA} \vdash \forall u \exists ! v \leq 1 A(u,v), \text{ but } (\neg \exists f \text{ rec.: } \forall u A(u, f(u)) \text{ is true}),$$

since

$$A(u,v) :\equiv \forall w A_0'(u,v,w) \wedge (v \neq 0 \to \neg \forall w A_0'(u,0,w)) \in \Delta_2^0.$$

15.3 Applications to monotone convergence theorems

For definiteness let us fix $\mathcal{T}^\omega :=$E-PA$^\omega$+QF-AC^{1,0}+QF-AC^{0,1}+WKL.

Let X be a Polish space and K be a compact metric space which both are $\mathcal{T}^\omega$-definable. Moreover, let $F : X \times K \times \mathbb{N} \to \mathbb{R}_+$ be a $\mathcal{T}^\omega$-definable function such that for any $x \in X$ and $y \in K$ the sequence $(F(x,y,n))_{n \in \mathbb{N}}$ is nonincreasing. Suppose that $(F(x,y,n))_{n \in \mathbb{N}}$ converges (provably in $\mathcal{T}^\omega$) to zero, i.e.

$$\mathcal{T}^\omega \vdash \forall x \in X \forall y \in K \forall k \in \mathbb{N} \exists n \in \mathbb{N} \forall m \geq n \, (F(x,y,m) <_{\mathbb{R}} 2^{-k}).$$

Thus a-fortiori

$$\mathcal{T}^\omega \vdash \forall x \in X \forall y \in K \forall k \in \mathbb{N} \exists n \in \mathbb{N} (F(x,y,n) <_{\mathbb{R}} 2^{-k})$$

which has the logical form required in theorem 15.1. Hence we can apply this theorem and extract a modulus $\delta(x,k)$ (depending on a representation f_x^1 of x) such that

$$\forall x \in X \forall y \in K \forall k \in \mathbb{N} \exists n \leq \delta(x,k)(F(x,y,n) <_{\mathbb{R}} 2^{-k}).$$

Since the sequence $(F(x,y,n))_{n \in \mathbb{N}}$ is assumed to be nonincreasing this yields

$$\forall x \in X \forall y \in K \forall k \in \mathbb{N} \forall n \geq \delta(x,k)(F(x,y,n) <_{\mathbb{R}} 2^{-k}),$$

i.e. theorem 15.1 extracts from proofs of the pointwise convergence of K-families of monotone sequences of reals rates of K-uniform convergence.
In chapter 18 this will be applied in the context of metric fixed point theory, where such monotone convergence results show up in the context of asymptotic regularity: For a given function $F : K \to K$ and a starting point $x \in K$, let x_n denote the n-th iteration of F on x, i.e. $x_n :\equiv F^n(x)$. If $(d(x_n, x_{n+1}))_{n\in\mathbb{N}} \overset{n\to\infty}{\to} 0$ for all $x \in K$, then F is called asymptotically regular.
In many cases (e.g. for nonexpansive functions F as considered in chapter 17 below) the sequence $(d(x_n, x_{n+1}))_{n\in\mathbb{N}}$ is nonincreasing so that, by the discussion above, theorem 15.1 can be used to extract uniform rates of convergence. In chapter 18 we will actually obtain various such uniform convergence results even for just bounded (convex) sets which need not be compact. This follows from more general metatheorems which we will prove in chapter 17.

The monotonicity in these convergence statements is used only to be able to write the convergence in the logical form required in theorem 15.1. This is crucial for applications in a context based on classical logic in which one applies monotone functional interpretation to the negative translation of formulas. Without monotonicity the negative translation of

$$\exists n \in \mathbb{N} \forall m \geq n \, (F(x,y,m) <_{\mathbb{R}} \varepsilon)$$

would yield

$$\neg\neg\exists n \in \mathbb{N} \forall m \geq n \, (F(x,y,m) <_{\mathbb{R}} \varepsilon)$$

from which monotone functional interpretation no longer extracts a modulus of convergence. In the semi-intuitionistic contexts of chapters 7 and 9, however, one can extract moduli of convergence even without any monotonicity assumptions.

15.4 Applications to proofs of contractivity

Let us take again $\mathcal{T}^\omega$:=E-PA$^\omega$+QF-AC^{1,0}+QF-AC^{0,1}+WKL.
Let (K,d) be a $\mathcal{T}^\omega$-definable compact metric space. A function $F : K \to K$ is defined to be *contractive* if

$$\forall x,y \in K (x \neq y \to d(F(x),F(y)) < d(x,y)).$$

Edelstein's fixed point theorem [92] says that F has a unique fixed point and that the Picard iteration $(F^n(x))_{n\in\mathbb{N}}$ converges to this fixed point for any $x \in K$. Let $F : K \to K$ be $\mathcal{T}^\omega$-definable and provably in $\mathcal{T}^\omega$ be contractive. Then by theorem 15.1 we can extract an (additive) modulus of uniform contractivity $\eta : \mathbb{N} \to \mathbb{N}$ satisfying

$$\forall x,y \in K \forall k \in \mathbb{N}(d(x,y) > 2^{-k} \to d(F(x),F(y)) + 2^{-\eta(k)} < d(x,y)).$$

The concept of contractivity can be written also in the trivially equivalent form

$$\forall x,y \in K(x \neq y \to \exists n \in \mathbb{N}(d(F(x),F(y)) < (1-2^{-n}) \cdot d(x,y))),$$

in which case the interpretation yields a (multiplicative) modulus of uniform contractivity $\tilde{\eta} : \mathbb{N} \to \mathbb{N}$ satisfying

$$\forall x,y \in K \forall k \in \mathbb{N}(d(x,y) > 2^{-k} \to d(F(x),F(y)) < (1-2^{-\tilde{\eta}(k)}) \cdot d(x,y)).$$

Remark 15.9. We may in fact consider the more general case of functions $F : X \times K \to K$, where X is a Polish space, in which case the modulus η will also depend on (a representative of) $x \in X$. Similarly in section 16.75 below.

Such a modulus $\alpha(k) :\equiv 1 - 2^{-\tilde{\eta}(k)}$ has in fact been considered in the literature by Rakotch [308] and – in the context of Bishop style constructive analysis – in [48]. Using the boundedness of K, we can easily produce an η out of a given α and vice-versa.
As shown in [236, 119], it is exactly such a modulus which is needed to obtain a rate of convergence in Edelstein's fixed point theorem [92, 308]. As in the case of moduli of uniqueness it is crucial here that η does not depend on x, y.
Numerous variants of the notion of 'contractive mapping' have been considered in the literature. The main purpose of these variants is to obtain generalizations of Edelstein's classical fixed point theorem to more general classes of functions. Under monotone functional interpretation, these notions again give rise to appropriate moduli and we expect that in many cases explicit rates of convergence can be provided in terms of the corresponding moduli of contractivity. For a survey of 25 notions of contractivity and generalizations of Edelstein's result see [315]. This line of work is further continued in [69, 275, 316], to list only a few references. The most general among the notions of contractivity considered in Rhoades [315] is the concept of generalized p-contractive mappings. A fixed point theorem for continuous generalized p-contractive mappings on compact metric spaces is proved in Kincses-Totik [182]. Using logical proof analysis, a fully effective version of a generalization of this result to complete bounded metric spaces is given in Briseid [51] (see theorem 17.122 in chapter 17 below). Yet another notion of contractivity, the so-called asymptotic contractions, is studied in Kirk [188] together with a fixed point theorem. Building on Gerhardy [116] (which in turn uses logical proof analysis) a fully effective version is this result (which originally was proved using ultrapowers) is presented in Briseid [52].

15.5 Remarks on fragments of $\mathscr{T}^\omega$

In Kohlenbach [209] a system **PBA** of **polynomially bounded analysis** was designed (where axioms Γ of the form $(**)$ as in theorem 15.1 are freely used whenever convenient) which guarantees that $\Phi(n,x)$ in theorem 15.1 will be a polynomial in n and f_x^M (in the sense of definition 3.44), where f_x is a representative of x. Basically, this system is

$$\text{E-G}_2\text{A}^\omega + \text{QF-AC}^{1,0} + \text{QF-AC}^{0,1} + \Delta + \Sigma^0_1\text{-UB},$$

where Δ consists of all true (in the sense of the full set-theoretic type structure $\mathcal{S}^\omega$) sentences of the form $\forall x^\delta \exists y \leq_\rho tx \forall z^\tau A_0(x,y,z)$ with A_0 quantifier-free, t closed and δ, ρ, τ of degree ≤ 1.

EBA:=PBA+exp is the system

$$\text{E-G}_3\text{A}^\omega + \text{QF-AC}^{1,0} + \text{QF-AC}^{0,1} + \Delta + \Sigma^0_1\text{-UB}.$$

It guarantees that $\Phi(n,x)$ is elementary recursive (again in the sense of definition 3.44) in n, f_x^M for any representative f_x of x.

As a matter of fact, these systems contain quite some parts of analysis including many non-computational principles related to Heine-Borel compactness (which logically corresponds to the use of weak König's lemma WKL, see [338]). In particular, this captures:

1) Basic properties of the operations $+,-,\cdot,(\cdot)^{-1},|\cdot|,\max,\min$ and the relations $=,\leq,<$ for rational numbers and real numbers (which are given by Cauchy sequences of rationals with fixed Cauchy rate $1/(n+1)$ of convergence).
2) Basic properties of maximum and sum for sequences of real numbers of variable length.
3) Basic properties of continuous functions $f:[a,b]^d \to \mathbb{R}$, $\sup\limits_{x\in[a,b]} f(x)$ and $\int_a^x f(x)dx$ for $f \in C[a,b]$ where $a<b$ and $x \in [a,b]$.
4) The Leibniz criterion, the quotient criterion, the comparison test for series of real numbers. The convergence of the geometric series together with its summation formula. The non-convergence of the harmonic series. (But not: The Cauchy property of bounded monotone sequences in $\mathbb{R}$ or the Bolzano-Weierstraß property for bounded sequences in $\mathbb{R}$, see chapter 13 and Kohlenbach [210]).
5) Characteristic properties of the trigonometric functions sin, cos, tan, arcsin, arccos, arctan and of the restrictions $\exp_k$ and $\ln_k$ of exp, ln to $[-k,k]$ for every **fixed** number k (in **EBA** we have the unrestricted functions exp, ln).
6) The fundamental theorem of calculus.
7) The equivalence (local and global) of sequential continuity and ε-δ-continuity for $f:\mathbb{R}\to\mathbb{R}$.
8) The mean value theorem of differentiation.
9) The mean value theorem for integrals.
10) The Cauchy-Peano existence theorem.
11) Brouwer's fixed point theorem for continuous functions $f:[a,b]^d \to [a,b]^d$.
12) The attainment of the maximum of $f \in C([a,b]^d,\mathbb{R})$ on $[a,b]^d$.

13) The uniform continuity (together with the existence of a modulus of uniform continuity) of pointwise continuous functions $f : [a,b]^d \to \mathbb{R}$.
14) The sequential form of the Heine-Borel covering property of $[a,b]^d \subset \mathbb{R}^d$.
15) Dini's theorem: Every sequence (G_n) of continuous functions $G_n : [a,b]^d \to \mathbb{R}$ which increases pointwise to a continuous function $G : [a,b]^d \to \mathbb{R}$ converges uniformly on $[a,b]^d$ to G and there exists a modulus of uniform convergence.
16) Every strictly increasing continuous function $G : [a,b] \to \mathbb{R}$ possesses a continuous strictly increasing inverse function $G^{-1} : [Ga,Gb] \to [a,b]$.

Principles based on **sequential compactness**, like the monotone convergence principle PCM or the Bolzano-Weierstraß principle BW, cannot be included into the realm of polynomially bounded analysis as even fixed instances PCM(t) of PCM amount to corresponding instances of Π^0_1-CA and hence (over $\mathrm{G}_n\mathrm{A}^\omega$) to all first order instances of Σ^0_1-IA which suffices to introduce all the primitive recursive functions. As shown in chapter 13, however, over $\mathrm{G}_n\mathrm{A}^\omega$ the contribution of fixed (sequences of) instances of these principle also is not stronger than this. Only the use of fixed instances of the existence of the limit superior for bounded sequences in $\mathbb{R}$, in fact, causes a growth beyond primitive recursive growth as such instances can be used to prove corresponding instances of Σ^0_2-induction. This and many more results, not covered in chapter 13, can be found in Kohlenbach [210, 216].

15.6 Historical comments and suggested further reading

Most of the results from sections 15.1 and 15.2 were first proved in Kohlenbach [204]. The material in sections 15.3 and 15.4 is based on Kohlenbach-Oliva [236] where further discussions can be found. The results mentioned in section 15.5 are proved in Kohlenbach [209, 210].

Chapter 16
Case study I: Uniqueness proofs in approximation theory

16.1 Uniqueness proofs in best approximation theory

An area in which complicated uniqueness proofs feature prominently is approximation theory and, in particular, the topic of best approximation. Here the setting is as follows: Let $(X,\|\cdot\|)$ be a real normed linear space and $E \subseteq X$ be a finite dimensional subspace. An element $y_b \in E$ is said to approximate $x \in X$ best if $\|x-y_b\| = \inf_{y\in E} \|x-y\| =:\text{dist}(x,E)$. We have the following easy (but ineffective) existence theorem:

Theorem 16.1. *Let $(X,\|\cdot\|)$ be a real normed space and $E \subseteq X$ a finite dimensional subspace of X. Then $\forall x \in X \exists y_b \in E \big(dist(x,E) = \|x-y_b\|\big)$.*

Proof: Assume $\|x-y\| \le \|x-0\| (= \|x\|)$, where 0 is the zero vector in E $(x \in X, y \in E)$. Then $\|y\| \le 2\|x\|$. Hence $\text{dist}(x,E) = \text{dist}(x,K_x)$, where $K_x := \{y \in E : \|y\| \le 2\|x\|\}$ is compact in E (since E is finite dimensional). The theorem now follows from the fact that the continuous function $F : E \to \mathbb{R}$, $F(y) := \|x-y\|$ attains its infimum on K_x. $\square$

In addition to the importance of concept of 'modulus of uniqueness' for the computation of the unique solution (discussed already in section 15.2) the special relevance in the context of best approximation theory is further witnessed by the following proposition which shows that any such modulus, in particular, yields a modulus of pointwise continuity for the corresponding projection operator:

Proposition 16.2. *Let $(X,\|\cdot\|)$ be a real normed linear space, $E \subseteq X$ a finite dimensional subspace. Assume that every $x \in X$ possesses a uniquely determined best approximation in E and that the operation Φ for $x \in X, q \in \mathbb{Q}^*_+$ satisfies $\Phi(x,q) \in \mathbb{Q}^*_+$ and*

$$\forall y, y_b \in E(\|x-y\| \le \text{dist}(x,E) + \Phi(x,q) \wedge \|x-y_b\| = dist(x,E) \to \|y-y_b\| \le q).$$

Then the following holds

1) $\frac{1}{2}\cdot\Phi$ *is a modulus of pointwise continuity for the projection* $\mathcal{P}: X\to E$ *which maps* $x\in X$ *to its best approximation* $y_b\in E$*, i.e.*

$$\forall x,x_0\in X, q\in\mathbb{Q}^*_+\big(\|x-x_0\|\le\frac{1}{2}\Phi(x_0,q)\to\|\mathcal{P}(x)-\mathcal{P}(x_0)\|\le q\big).$$

2) *If* Φ *is linear in* q*,i.e.* $\Phi(x,q)=q\cdot\gamma(x)$ *with* $\gamma\in\mathbb{Q}^*_+$*, then* $\gamma(x)$ *is a 'constant of strong unicity' ([286]), i.e.*

$$\forall x\in X, y\in E\big(\|x-y\|\ge\|x-y_b\|+\gamma(x)\cdot\|y-y_b\|\big),$$

where y_b *is the best approximation of* x *in* E*.*

3) *For* $\gamma(x)$ *as in '2)' we get that* $\lambda(x):=\frac{2}{\gamma(x)}$ *is a pointwise Lipschitz constant for* $\mathcal{P}$*, i.e.*

$$\forall x,x_0\in X\big(\|\mathcal{P}(x)-\mathcal{P}(x_0)\|\le\lambda(x_0)\cdot\|x-x_0\|\big).$$

Proof: 1) It easily follows that $\mathrm{dist}(\cdot,E)$ is a nonexpansive (i.e. Lipschitz continuous with Lipschitz constant 1) function in x, i.e.

$$(*)\ \|x-x_0\|\le\varepsilon\to|\mathrm{dist}(x,E)-\mathrm{dist}(x_0,E)|\le\varepsilon.$$

Let us assume that $\|x-x_0\|\le\frac{1}{2}\Phi(x_0,q)$. We then obtain

$$(1)\begin{cases}\|x_0-\mathcal{P}(x)\|\le\|x_0-x\|+\|x-\mathcal{P}(x)\|=\mathrm{dist}\,(x,E)+\|x_0-x\|\\ \overset{(*)}{\le}\ \mathrm{dist}\,(x_0,E)+\frac{1}{2}\Phi(x_0,q)+\|x-x_0\|\le\mathrm{dist}\,(x_0,E)+\Phi(x_0,q)\end{cases}$$

and

$$(2)\ \|x_0-\mathcal{P}(x_0)\|=\mathrm{dist}(x_0,E).$$

Together with the assumption on Φ it follows that $\|\mathcal{P}(x)-\mathcal{P}(x_0)\|\le q$.
2) The assumption on Φ implies that

$$\|y-y_b\|\ge q\to\|x-y\|\ge\underbrace{\mathrm{dist}(x,E)}_{=\|x-y_b\|}+\gamma(x)\cdot q\cdot\frac{n}{n+1}\ \text{for all } n\in\mathbb{N}, q\in\mathbb{R}^*_+.$$

$q:=\|y-y_b\|$ yields

$$\forall n\in\mathbb{N}\big(\|x-y\|\ge\|x-y_b\|+\gamma(x)\cdot\|y-y_b\|\cdot\frac{n}{n+1}\big).$$

Hence

$$\|x-y\|\ge\|x-y_b\|+\gamma(x)\cdot\ \|y-y_b\|.$$

3) follows easily from 1). □

Remark 16.3. 1) The fact that in the previous proposition it suffices to have the property of Φ being a modulus of uniqueness for y,y_b rather then for general y_1,y_2 is up to a factor $1/2$ equivalent to the official definition (except that for convenience we use $\leq$ instead of $<$) but sometimes allows one to save that factor yielding a slight numerical improvement.

2) Part 3) in the previous proposition is (for Chebycheff approximation) due to [66] (p. 82).

Let us now move to the case of best polynomial approximation of continuous functions $f \in C[0,1]$ by polynomials of degree $\leq n$ (P_n). We will consider approximations both w.r.t. to the uniform norm $\|f\|_\infty := \sup\limits_{x \in [0,1]} |f(x)|$ (Chebycheff approximation) as well as the norm $\|f\|_1 := \int\limits_0^1 |f(x)dx$ (L_1-approximation or approximation in the mean). Both cases of best approximation have been studied extensively since Chebycheff's classical work around 1850. The following discussion applies to $\|\cdot\| := \|\cdot\|_\infty$ as well as to $\|\cdot\| := \|\cdot\|_1$. In both cases the best approximation of f by an element of P_n is unique:

$$(4)\ \forall n \in \mathbb{N} \forall f \in C[0,1] \forall p_1, p_2 \in P_n \Big(\bigwedge_{i=1}^{2} (\|f - p_i\| = dist(f, P_n)) \to p_1 \equiv p_2 \Big),$$

where $dist(f,P_n) := \inf\limits_{p \in P_n} \|f - p\|$. Moreover, the usual classical proofs for this uniqueness rely for both norms on WKL and classical logic (see sections 16.2, 16.3 and 16.4 below).

As indicated already above, in (4) we can without loss of generality replace the non-compact subspace P_n of $C[0,1]$ with the compact convex subset $\tilde{K}_{f,n} := \{p \in P_n : \|p\| \leq 2\|f\|\}$ and $\text{dist}(f,P_n) = \text{dist}(f,\tilde{K}_{f,n})$ can easily be seen to be computable (uniformly in f as represented above, i.e. given together with a modulus of uniform continuity, and n).

Moreover, the existence of best Chebycheff resp. L_1-approximations now easily follows from the fact that the continuous function $\lambda p.\|f - p\|$ attains its infimum on $\tilde{K}_{f,n}$. Note, that this existence proof also relies on WKL (which is needed to show that continuous functions on compact subsets of $\mathbb{R}^{n+1}$ attain their infimum). Whereas the use of WKL cannot be eliminated from the existence proof just presented, it can be eliminated from the uniqueness proof and, moreover, the metatheorem 15.4 on the extractability of an effective modulus of uniqueness is applicable with $F(f,n,p) := \|f - p\| - dist(f,P_n)$. This means that the extractabilty of a primitive recursive (in the sense of Gödel) modulus of uniqueness Φ (given by a closed term of E-PA$^\omega$) satisfying

$$(5) \begin{cases} \forall n, k \in \mathbb{N} \forall f \in C[0,1] \forall p_1, p_2 \in \tilde{K}_{f,n} \\ \quad \Big(\bigwedge_{i=1}^{2} (\|f - p_i\| - dist(f,P_n) < 2^{-\Phi(f,n,k)}) \to \|p_1 - p_2\| < 2^{-k} \Big) \end{cases}$$

is guaranteed, where Φ depends on f via the representation of $C[0,1]$ from chapter 4, i.e. via the enrichment by a modulus of uniform continuity ω_f of f.

The same is true for the slightly altered compact subspace $K_{f,n} := \{p : \|p\| \le \frac{5}{2}\|f\|\}$ which we will use (instead of $\tilde{K}_{f,n}$) in sections 16.2 and 16.4 below (with $\|\cdot\|$ being $\|\cdot\|_\infty$ and $\|\cdot\|_1$ respectively) for the following reason: a simple trick (see the proofs of theorems 16.30 and 16.71 below) allows one to extend a modulus of uniqueness for $K_{f,n}$ to P_n so that (with a slightly altered Φ)

$$(6)\begin{cases} \forall n,k \in \mathbb{N}\forall f \in C[0,1]\forall p_1,p_2 \in P_n \\ \quad \Big(\bigwedge_{i=1}^{2}(\|f-p_i\| - dist(f,P_n) < 2^{-\Phi(f,n,k)}) \to \|p_1-p_2\| < 2^{-k}\Big). \end{cases}$$

As we discussed above, the modulus of uniqueness can be used to compute the best approximation (it can also be used to give a constructive existence proof for the best approximation in WE-HA$^\omega$ (see theorem 16.13 below and [204]).

Even in the L_1-case we have to represent $C[0,1]$ as the space $(C[0,1],\|\cdot\|_\infty)$ to apply the logical metatheorem mentioned above since $(C[0,1],\|\cdot\|_1)$ is not a Polish space. As we discussed already, this amounts to enriching the input f by a modulus of uniform continuity ω_f so that Φ again will depend on ω_f.

Note that if $C[0,1]$ is replaced by the (pre-)compact (w.r.t. $\|\cdot\|_\infty$) set $\mathcal{K}_{\omega,M}$ of all functions $f \in C[0,1]$ which have the common modulus of uniform continuity ω and the common bound $\|f\|_\infty \le M$, then the same logical metatheorem guarantees (using the representation of that subspace as given at the end of chapter 4) the extractability of a modulus of uniqueness Φ which only depends on $\mathcal{K}_{\omega,M}$ i.e. on ω, M (in addition to n,k). Moreover, even the M-dependency can be eliminated as the approximation problem for f can be reduced to that for $\tilde{f}(x) := f(x) - f(0)$ so that only a bound $N \ge \sup_{x\in[0,1]} |f(x) - f(0)|$ is required, which can easily be computed from ω (e.g take $N := \lceil \frac{1}{\omega(1)} \rceil$). Therefore, from the logical metatheorem and the fact that the uniqueness proofs both for the Chebycheff approximation as well as for the L_1-approximation can be formalized in E-PA$^\omega$+WKL we obtain already the extractability of primitive recursive (in the sense of Gödel) moduli of uniqueness Φ which only depend on ω_f, n and k: **a-priori information.**

$(P_n,\|\cdot\|_\infty)$ can be identified with $(\mathbb{R}^{n+1},\|\cdot\|_\infty)$ via coefficients, where

$$\|(c_0,\ldots,c_n)\|_\infty := \sup_{x\in[0,1]} |c_n x^n + \ldots c_1 x + c_0|.$$

Similarly, $(P_n,\|\cdot\|_1)$ can be identified with $(\mathbb{R}^{n+1},\|\cdot\|_1)$.
Since all norms on the finite dimensional space $\mathbb{R}^{n+1}$ are equivalent, a rate of convergence w.r.t. $\|\cdot\|_\infty$ resp. $\|\cdot\|_1$ yields a rate of convergence w.r.t. $\|\cdot\|_{\max}$ (and conversely) by a multiplication with a suitable constant which can be determined by the so-called Markov inequality (for a proof of this non-trivial inequality see e.g. [66]):

Proposition 16.4 ([274]). *Markov's inequality states that for any polynomial p of degree $\le n$, $\|p'\|_\infty \le 2n^2\|p\|_\infty$, where p' denotes the first derivative of p.*

Using this inequality one can show that for any polynomial $p \in P_n$

$$\|p\|_\infty \le 2(n+1)^2\|p\|_1.$$

Hence, any upper bound on $\|p_1 - p_2\|_1$ gives also an upper bound on $\|p_1 - p_2\|_\infty$ and we can use this to get a bound on the coefficients of $p_1 - p_2$. Namely, if $p_1(x) - p_2(x) = a_n x^n + \ldots + a_1 x + a_0$ and $\|p_1 - p_2\|_1 \le M$ then $|a_i| \le \frac{(2(n+1)^2)^{i+1}}{i!}M$ (see remark 16.59 in section 16.4). In terms of the uniform norm we get the following estimate from Markov's inequality

$$|a_i| \le \frac{(2n^2)^i}{i!}\|p\|_\infty.$$

So in both cases ($\|\cdot\|_\infty$ and $\|\cdot\|_1$), the modulus of uniqueness can be used to compute the coefficients of the best approximating polynomial p_b with arbitrary prescribed precision.

In the next three sections we will extract explicit moduli of uniqueness from

- ([200, 204]) the most common uniqueness proof for the best polynomial Chebycheff approximation given in 1919 by de La Vallée Poussin [304] (and in a fully detailed form in [284]);
- ([200, 205]) a slightly older argument for the uniqueness of best Chebycheff approximation due to Young [381] from 1907 for general Haar spaces (with full details in [313]) and – in a variant – due to Borel [39] in 1905 for the polynomial case;
- ([235]) a proof for the uniqueness of best polynomial L_1-approximation (of continuous functions) from 1965 due to Cheney [65], which avoids the use of measure theory made in the original uniqueness proof from 1921 due to Jackson [178].

We like to point the attention of the reader to the following main general outcomes of this extended case study:

1) Both proofs for the uniqueness of best Chebycheff approximation – the one from 1919 ([304]) as well as the one from 1907 ([381]) – yield moduli of uniqueness which (when given a lower bound $0 < l \le \text{dist}(f, P_n)$ on the distance) are linear in the error q and hence provide effective estimates for what is called 'constant of strong unicity' in Chebycheff approximation theory (see above). Moreover, the estimates depend on f only via a modulus of uniform continuity and a norm upper bound and so are uniform for classes $\mathcal{K}$ of equicontinuous functions with a common norm upper bound (which can be removed by the shift $f \mapsto \tilde{f}$ above) and a common (strictly positive) lower bound l on the distance. The ineffective existence of a constant of strong unicity was first proved in 1963 ([286]) and the existence of a common such constant for classes $\mathcal{K}$ only in 1976 ([153]).

2) Both uniqueness proofs mentioned in 1) above use the ineffective so-called alternation theorem. Whereas the proof from [304] uses a simple interpolation argument to conclude from the alternation theorem the uniqueness, the proof from [381, 313] needs a more complicated argument and, additionally, the intermediate value theorem. Nevertheless, that latter proof is much easier to analyze and gives the numerically better results. This can be explained on the logical level by the fact that using the existence of a best approximation the proof can be formulated in a way which uses the alternation theorem only in the form of a sentence Γ (of the kind included in theorems 15.1 and 15.4) which can be treated just as an axiom whereas the alternation theorem itself has a more complicated form, namely that of a sentence (%) as defined just before theorem 10.49 in chapter 10. As a result of this, the analysis of the second proof avoids to have to analyze the proof of the alternation theorem itself which, however, is crucial for the analysis of the uniqueness proof from [304] which proceeds according to theorem 10.49 (see section 16.3 for a detailed discussion of this point). Using already the uniqueness (and existence) of the best approximation, the alternation theorem and its variant of the form Γ used in the proof from [381, 313] are trivially equivalent. However, the proofs we discuss here are about to establish at all the uniqueness so that this observation does not help to logically improve the situation concerning the proof from [304].
3) In connection with 2) it is interesting to note that the proof of first result towards strong unicity in the literature, namely the proof of the local Lipschitz property of the Chebycheff projection operator due to [106] (which – as was observed much later in [37] – can also be used to obtain strong unicity), implicitly refers back to that older – more complicated – argument for the uniqueness from [381]. So one might speculate that the reason for the long time it took in the history of approximation theory to discover strong unicity has been that the more complicated (but comparatively easy to analyze) older uniqueness proof had been 'forgotten' after the discovery of the simpler (but very complicated to analyze quantitatively) new argument. Nevertheless, it is remarkable that by logical analysis even that latter argument can be seen to contain implicitly the strong unicity result.
4) The effective bounds extracted from the uniqueness proof from [381] numerically improve the previously known estimates due to D. Bridges [44, 45, 46] and K.-I. Ko [198].
5) The logical analysis of the uniqueness proof for best L_1-approximation due to [65] in 1965 yields a modulus of uniqueness of (essentially) the form $d_n \cdot \varepsilon \cdot \omega(c_n \cdot \varepsilon)$ where ω is a modulus of uniform continuity of the function $f \in C[0,1]$ to be approximated, n the degree of the polynomials and d_n, c_n constants depending only on n. The existence of such constants was established ineffectively first in 1978 by Kroó [254] (see also [256]) improving a result from 1975 due to Björnestal [35] which, moreover, shows that the (nonlinear) ε-dependency of the form above is optimal (see section 16.4 for a more precise and much more detailed discussion). So again, logical analysis of a given proof leads even to qualitative results which were obtained only much later. The explicit effective description of the modulus with all the constants involved obtained in [235] by

proof-theoretic analysis has been the first of this kind and was used subsequently for the first complexity upper bound for best L_1-approximations in Oliva [290].

In this chapter we only consider rather classical uniqueness proofs for problems where many additional qualitative as well as numerical features have been known meanwhile. This makes it possible to demonstrate the power of the logical analysis of such proofs by showing that (as discussed above) several of these subsequent results follow directly by a logical analysis of the classical (much earlier) uniqueness proofs and that even new quantitative estimates can be obtained as well. Of course, one could analyze in a similar way more recent uniqueness and strong uniqueness proofs for more advanced approximation problems e.g. for best uniform approximation by various forms of spline functions (see e.g. [288] and [382]) or one-sided uniform as well as L_1-approximation (see e.g. [301] for the latter). To carry out a logical analysis of such more advanced uniqueness proofs would be an interesting research project.

For the moduli and bounds extracted in the rest of this chapter we usually switch from error estimates given as '2^{-k}' or '2^{-n}' to the more flexible formulation with '$q \in \mathbb{Q}^*_+$' or '$\varepsilon \in \mathbb{Q}^*_+$' which still makes the effective nature of the bounds explicit. If the moduli involved as impute data such as the modulus of uniform continuity ω_f of f are taken as functions $\mathbb{R}^*_+ \to \mathbb{R}^*_+$ (rather than $\mathbb{Q}^*_+ \to \mathbb{Q}^*_+$) then our moduli of uniqueness etc. also become (in the error argument ε) functions $\mathbb{R}^*_+ \to \mathbb{R}^*_+$ and we sometimes use this formulation when comparing our effective bounds with results from the literature.

16.2 Best Chebycheff approximation I

As discussed above, the existence of a polynomial $p_b \in P_n$ such that

$$\|f - p_b\|_\infty = \mathrm{dist}_\infty(f, P_n) := \inf\{\|f - p\|_\infty : p \in P_n\}$$

follows from a simple (though ineffective) compactness argument. The uniqueness of p_b is much more difficult to prove and was first established rigorously in [183] (based on prior work by Chebycheff [64]). The most common uniqueness proof in the literature appeared first in de La Vallée Poussin's monograph [304]. A more explicit presentation of this proof is given in [284] on which the following outline is based. In this section we show that theorem 15.4 can be applied to this proof and carry out its proof-theoretic analysis and the extraction of a modulus of uniqueness with all numerical details. In the next section we will present a different – and slightly more complicated – uniqueness proof which will result in numerically better bounds and an easier extraction. Finally, we will explain these empirical findings by logical metatheorems.

The results in this section are mostly taken from Kohlenbach [204].

Let (A^*) denote the sentence

$$(A^*)\ \forall f \in C[0,1], a,b \in [0,1], a < b \exists x_0 \in [a,b](f(x_0) = \sup_{x\in[a,b]} f(x))$$

and take $f \in C[0,1], n \in \mathbb{N}$ as fixed but arbitrary in the following. $E_{n,f} := \text{dist}_\infty(f, P_n)$. We show that the uniqueness proof from de La Vallée Poussin/Natanson ([304, 284]) can be formalized in WE-PA$^\omega$ + (A).
(A^*) is equivalent (over WE-PA$^\omega$) to the sentence (15.1), i.e.

$$(A)\ \forall f \in C[0,1] \exists x_0 \in [0,1](f(x_0) = \sup_{x\in[0,1]} f(x))$$

(just apply (A) to $\tilde{f}(\lambda) := f((1-\lambda)a + \lambda b)$ to get (A^*)) which has the form

$$\forall x \in X \exists y \in K(F(x,y) =_{\mathbb{R}} 0),$$

where X is Polish and K compact so is a sentence Γ as permitted in theorems 15.1 and 15.4.
For brevity we will write for the rest of this section just '$\|\cdot\|$' instead of '$\|\cdot\|_\infty$'. For $E_{n,f} = 0$, the uniqueness of the best approximation is trivial since then $p_b = f$ for any best approximation $p_b \in P_n$. Hence, we may assume without loss of generality that $E_{n,f} > 0$. Let $p_b \in P_n$ be a best approximation of f, i.e. $\|f - p_b\| = E_{n,f}$. From (A) it follows that $|p_b(x) - f(x)| \in C[0,1]$ attains its maximum on $[0,1]$, i.e. $\exists x_0 \in [0,1](|p_b(x_0) - f(x_0)| = E_{n,f})$. A point having this property is called an extremal point (abbreviation: (e)-point) of $p_b - f$. An (e)-point x_0 is a (+)-point if $p_b(x_0) - f(x_0) = E_{n,f}$ and a (–)-point if $p_b(x_0) - f(x_0) = -E_{n,f}$.

Lemma 16.5 (A). *Let $f \in C[0,1]$ and $p_b \in P_n$ be a best approximation of f. Then there exist both (+)-points and (–)-points of $p_b - f$.*

Proof: Assume without loss of generality that no (–)-point exists, i.e.

$$\forall x \in [0,1]\big(p_b(x) - f(x) > -E_{n,f}\big).$$

Applying (A) we conclude that

$$\inf_{x\in[0,1]} \big(p_b(x) - f(x)\big) > -E_{n,f}.$$

Now define

$$h := \frac{1}{2}\big(\inf_{x\in[0,1]} \big(p_b(x) - f(x)\big) + E_{n,f}\big).$$

Then $h > 0$ and $\inf_{x\in[0,1]} \big(p_b(x) - f(x)\big) = -E_{n,f} + 2h$ and so $-E_{n,f} + 2h \le p_b(x) - f(x) \le E_{n,f}$. Hence $-E_{n,f} + h \le p_b(x) - h - f(x) \le E_{n,f} - h$ for all $x \in [0,1]$. This, however, implies that $p_b - h \in P_n$ is a better approximation of f than p_b, contradicting the assumption that p_b was a best approximating polynomial in P_n. □

Remark 16.6. It is clear that the proof of lemma 16.5 can be formalized in WE-PA$^\omega$ + (A) (and hence e.g. in WE-PA$^\omega$+WKL+QF-AC^{0,0}). The use of (A) is unavoidable since, conversely, lemma 16.5 implies (A) already for $n = 0$ (relative to WE-PA$^{\omega}$): Let m resp. M denote the infimum resp. supremum of f on $[0,1]$. Then $E_{0,f} = \text{dist}(f, P_0) = \frac{1}{2}(M - m)$ and $\frac{1}{2}(M + m)$ is the best approximation of f in P_0. Lemma 16.5 yields the existence of an $x_0 \in [0,1]$ such that $f(x_0) - \frac{1}{2}(M+m) = \frac{1}{2}(M - m)$, i.e. $f(x_0) = M$ (this argument is taken from [46]).

We now prove the fundamental alternation theorem

Theorem 16.7 (Chebycheff [64], Kirchberger [183]). *Let $f \in C[0,1]$ and $p_b \in P_n$ be a best approximation of f. Then there exists a strictly increasing sequence of $n+2$ points $x_1 < x_2 < \ldots < x_{n+2}$ in $[0,1]$ which are alternating (+)-points and (−)-points in, i.e.*

$$\bigwedge_{i=1}^{n+2} \left(p_b(x_i) - f(x_i) = (-1)^{i+j} E_{n,f} \right)$$

where $j = 0$ or $j = 1$. Any such system of points is called an alternant of $p_b - f$.

Proof (Natanson [284]): For $E_{n,f} = 0$ the statement is trivial and so we may assume that $E_{n,f} > 0$. Since $f \in C[0,1]$ is uniformly continuous on $[0,1]$ we can construct (for suitable $s \in \mathbb{N}$) points $\xi_0 = 0 < \xi_1 < \xi_2 < \ldots < \xi_s = 1$ such that

$$\sup_{x \in I_i} \left(p_b(x) - f(x) \right) - \inf_{x \in I_i} \left(p_b(x) - f(x) \right) < \frac{1}{2} E_{n,f},$$

where $I_i = [\xi_i, \xi_{i+1}]$ for $i = 0, \ldots, s-1$. I_i is called an (e)-interval if it contains an (e)-point of $p_b - f$. If I_i is an (e)-interval then $p_b(x) - f(x) \neq 0$ for all $x \in I_i$. An (e)-interval I_i is called (+)-interval if $\forall x \in I_i (p_b(x) - f(x) > 0)$ and (−)-interval if $\forall x \in I_i (p_b(x) - f(x) < 0)$. We number the (e)-intervals consecutively from the left to the right $I_{j_1}, \ldots, I_{j_N}$ and assume without loss of generality that I_{j_1} is a (+)-interval. As a result of this we end up with the following schema of alternating blocks of (+)- and (−)-intervals:

$$(*) \begin{cases} I_{j_1}, \ldots, I_{j_{k_1}} & (+)\text{-intervals} \\ I_{j_{k_1+1}}, \ldots, I_{j_{k_2}} & (-)\text{-intervals} \\ \vdots & \\ I_{j_{k_{m-1}+1}}, \ldots, I_{j_{km}} & (-1)^{-m-1}\text{-intervals } (k_m = N). \end{cases}$$

By lemma 16.5 we know that $m \geq 2$. We now show that in fact $m \geq n+2$.
Assume: $(**)$ $m < n+2$. $p_b - f$ has different signs in $I_{j_{k_1}}$ and $I_{j_{k_1+1}}$. Hence the right boundary of $I_{j_{k_1}}$ does not coincide with the left boundary of $I_{j_{k_1+1}}$. Therefore, there exists a point $z_1 \in [0,1]$ which is strictly greater than all points of $I_{j_{k_1}}$ and strictly less than all points of $I_{j_{k_1+1}}$ (for short we write: $I_{j_{k_1}} < z_1 < I_{j_{k_1+1}}$). In the same way,

one shows the existence of points $z_2, z_3, \ldots, z_{m-1} \in [0,1]$ such that $I_{j_{k_i}} < z_i < I_{j_{k_i+1}}$ for $i = 2, \ldots, m-1$. Now define

$$\rho(x) := (z_1 - x)(z_2 - x) \cdot \ldots \cdot (z_{m-1} - x).$$

From $(**)$ it follows that $\rho(x) \in P_n$. Since no interval $I_{j_1}, \ldots, I_{j_N}$ contains one of the zeroes $z_1, \ldots, z_{m-1}$ of $\rho(x)$, $\rho(x)$ has constant sign on each I_{j_i} $(1 \le i \le N)$ which, moreover, coincides the sign of $p_b(x) - f(x)$ on I_{j_i}. Now consider an interval I_i of the initial partition $I_0, \ldots, I_{s-1}$ of $[0,1]$ which is not an (e)-interval, i.e. $\forall x \in I_i (|p_b(x) - f(x)| < E_{n,f})$. (A) implies that $\sup_{x \in I_i} |p_b(x) - f(x)| < E_{n,f}$ and so we conclude that

$$E^* := \max\{\sup_{x \in I_i} |p_b(x) - f(x)| : I_i \text{ is not an (e)-interval}\} < E_{n,f}.$$

Let $R := \sup_{x \in [0,1]} |\rho(x)|$. For sufficiently small $\lambda > 0$ one has that $\lambda R < E_{n,f} - E^*$ and $\lambda R < \frac{1}{2} E_{n,f}$. Define $Q(x) := p_b(x) - \lambda \rho(x) \in P_n$.
We now show that $\forall x \in [0,1] \big(|Q(x) - f(x)| < E_{n,f}\big)$ (using (A) this implies $\|Q - f\| < E_{n,f}$ which contradicts the definition of $E_{n,f}$):
Case 1: I_i is not an (e)-interval:
$|Q(x) - f(x)| \le |p_b(x) - f(x)| + \lambda |\rho(x)| \le E^* + \lambda R < E_{n,f}$ for all $x \in I_i$.
Case 2: I_i is an (e)-interval: Let $x \in I_i$. $p_b(x) - f(x)$ and $\lambda \cdot \rho(x)$ have the same sign and $|p_b(x) - f(x)| > \lambda \cdot |\rho(x)|$, since $|p_b(x) - f(x)| \ge \frac{1}{2} E_{n,f}$ and $\lambda \cdot |\rho(x)| < \frac{1}{2} E_{n,f}$. It follows that

$$|Q(x) - f(x)| = |p_b(x) - f(x) - \lambda \rho(x)| = |p_b(x) - f(x)| - \lambda \cdot |\rho(x)|.$$

Hence $|Q(x) - f(x)| \le E_{n,f} - \lambda \cdot |\rho(x)| < E_{n,f}$, since $\rho(x) \neq 0$ on (e)-intervals. □

Corollary 16.8. *There exists at most one (and so – by the existence proof above – exactly one) polynomial of best approximation for $f \in C[0,1]$ in P_n.*

Proof: Suppose that $p_1, p_2 \in P_n$ are best approximations of f, i.e. $\|p_1 - f\| = \|p_2 - f\| = E_{n,f}$. It is straightforward to show that $p(x) := \frac{p_1(x) + p_2(x)}{2} \in P_n$ is a best approximation of f as well. By theorem 16.7 there exists an alternant $x_1 < x_2 < \ldots < x_{n+2}$ for $p(x) - f(x)$ in $[0,1]$. If x_k is a (+)-point of $p(x) - f(x)$, then $\frac{p_1(x_k) - f(x_k)}{2} + \frac{p_2(x_k) - f(x_k)}{2} = E_{n,f}$. Since $p_2(x_k) - f(x_k) \le E_{n,f}$ it follows that $\frac{p_1(x_k) - f(x_k)}{2} + \frac{E_{n,f}}{2} \ge E_{n,f}$. Hence $p_1(x_k) - f(x_k) \ge E_{n,f}$ and, therefore, $p_1(x_k) - f(x_k) = E_{n,f}$ (since p_1 is a best approximation). By a similar argument one shows that $p_2(x_k) - f(x_k) = E_{n,f}$. Thus $p_1(x_k) = p_2(x_k)$ for the (+)-points of the alternant $x_1, \ldots, x_{n+2}$. For the (–)-points among $x_1, \ldots, x_{n+2}$ this is established analogously. Thus it follows that $p_1, p_2 \in P_n$ coincide on $n+2$ distinct points, which implies that $p_1 \equiv p_2$. □

Remark 16.9. The argument used in the proof of corollary 16.8 to derive the uniqueness of the best approximation p_b from the existence of $n+2$ distinct points x_i such that $|p_b(x_i) - f(x_i)| = E_{n,f}$ first appeared in [362]. Note that it would have been sufficient to have an alternant of length $n+1$ in the alternation theorem to conclude the uniqueness of p_b.

We now indicate, how the uniqueness proof can be formalized in WE-PA$^\omega + (A)$. Relative to theorem 16.7, the corollary 16.8 is easily be seen to be provable in WE-PA$^\omega$, e.g. using any standard proof of the fact that a nonvanishing polynomial $p \in P_n$ can have at most n roots (see further below for a specific proof of this in our quantitative analysis). Therefore, it remains to show that the proof of theorem 16.7 can be formalized within WE-PA$^\omega + (A)$: The proof of theorem 16.7 uses essentially (A) and classical logic. Furthermore for the definition of E^* and the schema $(*)$ we need the following schema of bounded comprehension:

$$(+)\ \forall k^0, x_0^1, \ldots, x_k^1 \exists i_0^0, \ldots, i_{k-1}^0 \bigwedge_{j=0}^{k-1} \left(i_j = 0 \leftrightarrow A(x_j, x_{j+1})\right),$$

which can be expressed in $\mathcal{L}$(WE-PA$^\omega$) via primitive recursive coding of sequences of type-1 objects as follows

$$(++)\ lth_1(x) =_0 k+1 \to \exists i^0 \left(lth(i) = k \wedge \forall j^0 \left(j < k \to [(i)_j =_0 0 \leftrightarrow A((x)_j, (x)_{j+1})]\right)\right)$$

for all k^0, x^1, where $A \in \mathcal{L}$(WE-PA$^\omega$). Using such codings we can in general talk about e.g. sequences $x_1, \ldots, x_{n+2}$ in $[0,1]$ (such as alternants etc.) where n is a variable. (++) can easily be proved in WE-PA$^\omega$ using classical logic and induction on k. By lemma 16.5 and (++) it follows that for every $n \in \mathbb{N}, f \in C[0,1]$ the schema $(*)$ (in the proof of theorem 16.7) and E^* exist, which – by (A) – leads to a contradiction to the assumption $(**)$.

Remark 16.10. Since the points $\xi_0, \ldots, \xi_s$ constructed in the beginning of the proof of theorem 16.7 can be chosen to be rational, it actually suffices to use $(++)$ for objects x^0 of type 0 instead of 1.

Putting things together we have shown that

$$\begin{aligned}&\text{WE-PA}^\omega + (A) \vdash \forall f \in C[0,1], n \in \mathbb{N}, (c_0, \ldots, c_n), (\tilde{c}_0, \ldots, \tilde{c}_n) \in \mathbb{R}^{n+1} \\ &\qquad \left(\|f - (c_n x^n + \ldots + c_0)\| = E_{n,f} = \|f - (\tilde{c}_n x^n + \ldots + \tilde{c}_0)\| \to \bigwedge_{i=0}^{n} c_i = \tilde{c}_i\right).\end{aligned}$$

As discussed above, P_n can be replaced by $K_{f,n} := \left\{p \in P_n : \|p\| \le \frac{5}{2}\|f\|\right\}$, and a modulus of uniqueness for $p_1, p_2 \in K_{f,n}$ can – see below – be extended to a modulus on P_n. $p \in K_{f,n}$ implies that the coefficients $c_0, \ldots, c_n$ of p are bounded $|c_i| \le \chi(f,n)$, where χ is a primitive recursive function on f and n (see the comments after proposition 16.4). Applying theorem 15.4 (though with a compact metric space $A_{f,n}$ which is parametrized by n and any upper bound on $\|f\|$) to the result above yields a modulus of uniqueness Ψ on

$$A_{f,n} := \left\{ (c_0, \ldots, c_n) : \bigwedge_{i=0}^{n} |c_i| \le \chi(f,n) \right\} \subset \mathbb{R}^{n+1},$$

provable in WE-HA$^\omega$ + (A^ε), where

$$(A^\varepsilon): \ \forall f \in C[0,1] \forall k \in \mathbb{N} \exists x_0 \in [0,1] \big(|f(x_0) - \sup_{x \in [0,1]} f(x)| < 2^{-k} \big).$$

This modulus extends to $\mathbb{R}^{n+1}$ by taking $\Psi'(f,n,k) := \max\{k+3, \Psi(f,n,k)\}$: If at least one of p_1, p_2 (say p_1) is not in $K_{f,n}$ then $\|f - p_1\| > \frac{3}{2} E_{n,f}$. If now

$$\|f - p_i\| \le E_{n,f} + 2^{-\Psi'(f,n,k)} \le E_{n,f} + 2^{-k-3} \text{ for } i = 1,2,$$

then $E_{n,f} < 2^{-k-2}$. Hence

$$\|f - p_i\| \le 2^{-k-2} + 2^{-k-3} < 2^{-k-1} \text{ for } i = 1,2$$

and so $\|p_1 - p_2\| < 2^{-k}$.

Since this 'ε-version' of (A) is provable in WE-HA$^\omega$ we can conclude

Theorem 16.11. WE-HA$^\omega$ *proves that for all* $f \in C[0,1], n \in \mathbb{N}, k \in \mathbb{N}, (c_0, \ldots, c_n), (\tilde{c}_0, \ldots, \tilde{c}_n) \in \mathbb{R}^{n+1}$

$$\begin{cases} \|f - (c_n x^n + \ldots + c_1 x + c_0)\|, \|f - (\tilde{c}_n x^n + \ldots + \tilde{c}_1 x + \tilde{c}_0)\| \le E_{n,f} + 2^{-\Psi(f,n,k)} \to \\ \bigwedge_{i=0}^{n} |c_i - \tilde{c}_i| \le 2^{-k}, \end{cases}$$

where the functional Ψ^{0001} *is given by a closed term of* WE-HA$^\omega$ *and represents an operation on the standard representation of* $C[0,1]$ *and* n,k, *i.e. primitive recursive in* f,n,k *where* f *is given together with a modulus of uniform continuity on* $[0,1]$.

Remark 16.12. As mentioned already, using the special representation of the space of functions $K_{\omega,M}$ given at the end of chapter 4 it is clear a-priori that we can extract a modulus of uniqueness which depends on f **only** via a modulus of uniform continuity ω of f and a (rational) upper bound on $\|f\|$. Even the latter can be removed by replacing f by $\tilde{f}(x) := f(x) - f(0)$ (without having to change the modulus of uniqueness) and observing that an upper bound for $\tilde{f}$ can be computed from ω alone (see the corollary to the proof of theorem 16.34 below for details).

From (the proof of) theorem 15.5 it follows that we can (using the modulus of uniqueness Ψ above) construct an algorithm Φ for the computation of the coefficients of the best approximation with prescribed precision. Moreover, this algorithm can be verified in WE-HA$^\omega$ and so, in particular, yields a proof on the existence of the best Chebycheff approximation in WE-HA$^\omega$ (we have not verified in detail that Markov's inequality can be proved in this system, but by fully elementary and constructive arguments one can construct a bound χ on the coefficients, though of lesser quality than that resulting from the Markov inequality):

Theorem 16.13.
WE-HA$^\omega \vdash \forall f \in C[0,1], n \in \mathbb{N}\big(\Phi fn \in A_{f,n} \subset \mathbb{R}^{n+1} \wedge \|f - \widehat{\Phi fn}\| = E_{n,f}\big)$,
where $\widehat{(c_0,\ldots,c_n)}$ *denotes the polynomial* $c_n x^n + \ldots + c_0$ *and* Φ *is a closed term of* WE-HA$^\omega$.

Remark 16.14. By the uniqueness of the best approximation $\Phi(f,n)$ is extensional in f and thus $f \mapsto \Phi(f,n)$ represents a (by proposition 4.18 pointwise continuous) function $C[0,1] \to \mathbb{R}^{n+1}$. Moreover, by (the proof of) proposition 16.2.1, $\Psi(f,n,k)+1$ is a modulus of pointwise continuity for this function.

We now carry out the extraction of an explicit modulus Ψ of uniqueness from the above uniqueness proof. The result is not only primitive recursive in the sense of Gödel's (WE-)HA$^\omega$ (and even in the sense of Kleene's $(\widehat{\text{WE-)HA}}^\omega\restriction)$ but will be a very elementary operation in f,n and k and – if in addition a positive rational lower bound $0 < q \le E_{n,f}$ for $E_{n,f}$ is given – even linear in the error $\varepsilon = 2^{-k}$, i.e. a constant of strong unicity (in the sense of [286]).

The key part of the extraction of the modulus is the analysis of the proof of the alternation theorem 16.7 which yields a new quantitative version of the ε-weakening of this theorem: we construct an effective operation χ which is linear in $\varepsilon \in \mathbb{Q}^*_+$ such that

$$\forall f \in C[0,1], p \in P_n, q, \varepsilon \in \mathbb{Q}^*_+\big(\varepsilon, q < E_{n,f} \wedge \|f - p\| \le E_{n,f} + (\chi fnkq)\cdot\varepsilon \to \exists \text{ an } \varepsilon\text{-alternant of length } k \text{ for } p - f\big),$$

where $k = 2,\ldots,n+2$ (see corollary 16.27 below).
The moduli Ψ resp. χ are simple constructions in the data f (where f is always endowed with a modulus of uniform continuity on $[0,1]$) and n,ε resp. in f,n,ε and a positive rational lower bound q on $E_{n,f}$ (which is only needed to get a χ which is linear in ε, see theorem 16.26). In particular Ψ,χ do not depend on the best approximation or the alternation points (as is guaranteed by theorem 15.1 since they live in the compact spaces $K_{f,n}$ and $[0,1]$ respectively). Thus, these moduli are a-priori estimates.

Notation. 16.15 $K_{f,n} := \{p \in P_n : \|p\| \le \frac{5}{2}\|f\|\}$.
Let $f \in C[0,1], n \in \mathbb{N}, p \in P_n$ *and* $1 \le k \le n+2$. $(x_1,\ldots,x_k) \in [0,1]^k$ *is called an alternant of length k for* $p - f$ *if*

$$x_1 < x_2 < \ldots < x_k \wedge \bigwedge_{i=1}^{k}\big((-1)^{i+j}(p(x_i) - f(x_i)) = E_{n,f}\big) \text{ for } j = 0 \text{ or } j = 1.$$

As before, $\omega_f : \mathbb{Q}^*_+ \to \mathbb{Q}^*_+$ *is called modulus of uniform continuity of* $f \in C[0,1]$ *on* $[0,1]$ *if*

$$\forall q \in \mathbb{Q}^*_+, x, y \in [0,1]\big(|x-y| < \omega_f(q) \to |f(x) - f(y)| < q\big).$$

The above given uniqueness proof (from [304, 284]) can be split into the following parts:

1) $\forall p_1, p_2 \in K_{f,n}\big(\|f-p_1\| = E_{n,f} = \|f-p_2\| \to \|f - \frac{p_1+p_2}{2}\| = E_{n,f}\big)$,
2) $\forall p \in K_{n,f}\big(\|f-p\| = E_{n,f} \to \exists (x_1,\ldots,x_{n+1}) \in [0,1]^{n+1}$
$\big((x_1,\ldots,x_{n+1})$ is an alternant for $p-f$ of length $n+1)\big)$,
3) $\forall p_1, p_2 \in K_{f,n}, (x_1,\ldots,x_{n+1}) \in [0,1]^{n+1}\big(\|p_1-f\| = E_{n,f} = \|p_2-f\|$
$\wedge (x_1,\ldots,x_{n+1})$ is an alternant for $\frac{p_1+p_2}{2} - f \to \bigwedge_{i=1}^{n+1} p_1(x_i) = p_2(x_i)\big)$,
4) $\forall p_1, p_2 \in K_{f,n}, (x_1,\ldots,x_{n+1}) \in [0,1]^{n+1}\big(\bigwedge_{i=1}^{n}(x_{i+1} > x_i) \wedge \bigwedge_{i=1}^{n+1}(p_1(x_i) = p_2(x_i))$
$\to p_1 \equiv p_2\big)$,
5) $\forall p \in P_n\big(\|f-p\| = E_{n,f} \to p \in K_{f,n}\big)$.

Claim: 1)–4) $\to \forall p_1, p_2 \in K_{f,n}\big(\|f-p_1\| = E_{n,f} = \|f-p_2\| \to p_1 \equiv p_2\big)$
$\overset{5)}{\to} \forall p_1, p_2 \in P_n\big(\|f-p_1\| = E_{n,f} = \|f-p_2\| \to p_1 \equiv p_2\big)$.

Proof: 1) $\to \|f - \frac{p_1+p_2}{2}\| = E_{n,f} \overset{2)}{\to} \exists (x_1,\ldots,x_{n+1}) \in [0,1]^{n+1} : (x_1,\ldots,x_{n+1})$ is an alternant for $\frac{p_1+p_2}{2} - f \overset{3)}{\to} \exists (x_1,\ldots,x_{n+1}) \in [0,1]^{n+1}\big(\bigwedge_{i=1}^{n}(x_{i+1} > x_i) \wedge \bigwedge_{i=1}^{n+1} p_1(x_i) = p_2(x_i)\big) \overset{4)}{\to} p_1 \equiv p_2$. □

Remark 16.16. 1) The alternation theorem 16.7 actually yields the existence of an alternant of length $n+2$ in 2). However, in the proof of corollary 16.8 only $n+1$ points are necessary. This can be utilized in the extraction of Ψ and yields a numerical improvement.
2) The restriction to $K_{n,f}$ instead of P_n and the use of 5) are necessary only for the extraction of Ψ (in order to apply theorems 15.1 and 15.4) but not for the proof of the uniqueness.

The most difficult part of the proof of the uniqueness is the proof of the alternation theorem 2). Let us assume that $E_{n,f} > 0$, then

$$\bigwedge_{i=1}^{n}(x_{i+1} \geq x_i) \wedge \exists j \in \{0,1\} \bigwedge_{i=1}^{n+1}\big((-1)^{i+j}(p(x_i) - f(x_i)) = E_{n,f}\big)$$

implies that $(x_1,\ldots,x_{n+1})$ is an alternant of length $n+1$ for $p-f$, i.e.

$$\bigwedge_{i=1}^{n}(x_{i+1} > x_i) \wedge \exists j \in \{0,1\} \bigwedge_{i=1}^{n+1}\big((-1)^{i+j}(p(x_i) - f(x_i)) = E_{n,f}\big).$$

Obviously, the alternation theorem (as well as the uniqueness of the best approximation) holds trivially in the case $E_{n,f} = 0$. That is why we can without loss of generality replace '2)' by

$$2.1)\ \forall p \in K_{f,n}\Big(\|p - f\| = E_{n,f} \to \exists (x_1,\ldots,x_{n+1}) \in [0,1]^{n+1}, j \in \{0,1\}$$
$$\big(\bigwedge_{i=1}^{n}(x_{i+1} \geq x_i) \wedge \bigwedge_{i=1}^{n+1}((-1)^{i+j}(p(x_i) - f(x_i)) = E_{n,f})\big)\Big).$$

Modulo the standard representation from chapter 4 of the spaces involved and the coding of finite tuples of variables of types $0,1$, the formula 2.1) has the form

$$(*)\ \forall x^1\left(\forall k^0 A_0(k,x) \to \exists y \leq sx \forall z^0 B_0(x,y,z)\right),$$

where A_0, B_0 are quantifier-free. Because of the premise '$\forall k^0 A_0(k,x)$', $(*)$ is not an admissible sentence Γ in theorems 15.1 and 15.4 (in contrast to (A)), i.e. it is not enough to consider the proof of '$(*) \to$ uniqueness of best approximation'. We also have to analyze the proof of '$A \to (*)$' in WE-PA$^\omega$. In theorem 10.49 we have shown that it is sufficient to analyze the proof of '$(A) \to \varepsilon$-weakening of $(*)$', where the ε-weakening of $(*)$ is

$$\forall x^1, z^0\left(\forall k^0 A_0(k,x) \to \exists y \leq sx \bigwedge_{i=0}^{z} B_0(x,y,i)\right).$$

From the proof of this weakening one can extract a functional χ such that

$$\forall x, z\Big(\bigwedge_{k=0}^{\chi xz} A_0(k,x) \to \exists y \leq sx \bigwedge_{i=0}^{z} B_0(x,y,i)\Big).$$

Applying the functional obtained from the proof of '$(*) \to$ uniqueness' to χ then yields the modulus of uniqueness (see theorem 10.49). In our situation, this strategy means that we have to analyze the proof of the following ε-weakening of the alternation theorem (under the assumption (A)):

$$(+)\ \forall p \in K_{f,n}\Big(\|f - p\| = E_{n,f} \to \forall q, r \in \mathbb{Q}^*_+ \exists (x_1,\ldots,x_{n+1}) \in [0,1]^{n+1}, j \in \{0,1\}$$
$$\big(\bigwedge_{i=1}^{n+1}(|(-1)^{i+j}(p(x_i) - f(x_i)) - E_{n,f}| < q) \wedge \bigwedge_{i=1}^{n}(x_i < x_{i+1} + r)\big)\Big).$$

For $q < E_{n,f}$, $l < E_{n,f} - q$ and $r := \omega_{p-f}(2l)$ (where ω_{p-f} is a modulus of uniform continuity of $p - f$)

$$\bigwedge_{i=1}^{n+1}(|(-1)^{i+j}(p(x_i) - f(x_i)) - E_{n,f}| < q) \wedge \bigwedge_{i=1}^{n}(x_i < x_{i+1} + r)$$

implies

$$\bigwedge_{i=1}^{n+1}(|(-1)^{i+j}(p(x_i) - f(x_i)) - E_{n,f}| < q) \wedge \bigwedge_{i=1}^{n}(x_i < x_{i+1}).$$

Since $(+)$ is trivial for $E_{n,f} = 0$ it is equivalent to

$$\widehat{2}.1)\ \forall p \in K_{f,n}\Big(\|f-p\| = E_{n,f} \to \forall q \in \mathbb{Q}^*_+ \exists (x_1,\ldots,x_{n+1}) \in [0,1]^{n+1}, j \in \{0,1\} \\ \big(\bigwedge_{i=1}^{n+1} (|(-1)^{i+j}(p(x_i)-f(x_i)) - E_{n,f}| < q) \wedge \bigwedge_{i=1}^{n} (x_{i+1} > x_i)\big)\Big).$$

Moreover, the argument used for this shows that (for all $p \in K_{f,n}, (x_1,\ldots,x_{n+1}) \in [0,1]^{n+1}, j \in \{0,1\}$)

$$2.2)\ \bigwedge_{i=1}^{n}(x_{i+1} \geq x_i) \wedge \bigwedge_{i=1}^{n+1} (|(-1)^{i+j}(p(x_i)-f(x_i)) - E_{n,f}| < E_{n,f}) \to \bigwedge_{i=1}^{n}(x_{i+1} > x_i).$$

By simple manipulations (e.g. writing $\forall q \in \mathbb{Q}^*_+ \big(\|f-p\| \leq E_{n,f}+q\big)$ instead of $\|f-p\| = E_{n,f}$) and a suitable (partial) prenexation, 1), $\widehat{2}.1)$, 2.2), 3) and 4) are equivalent to

$$1^*)\begin{cases} \forall p_1,p_2 \in K_{f,n}, q \in \mathbb{Q}^*_+ \exists r \in \mathbb{Q}^*_+ \\ \qquad \big(\bigwedge_{i=1}^{2} (\|f-p_i\| \leq E_{n,f}+r) \to \|f - \frac{p_1+p_2}{2}\| < E_{n,f}+q\big). \end{cases}$$

$$2^*.1)\begin{cases} \forall p \in K_{n,f}, q \in \mathbb{Q}^*_+ \exists r \in \mathbb{Q}^*_+ \\ \Big(\|f-p\| \leq E_{n,f}+r \to \exists (x_1,\ldots,x_{n+1}) \in [0,1]^{n+1}, j \in \{0,1\} \\ \quad \big(\bigwedge_{i=1}^{n+1} (|(-1)^{i+j}(p(x_i)-f(x_i)) - E_{n,f}| < q) \wedge \bigwedge_{i=1}^{n} (x_{i+1} > x_i)\big)\Big). \end{cases}$$

$$2^*.2)\begin{cases} \forall p \in K_{f,n}, (x_1,\ldots,x_{n+1}) \in [0,1]^{n+1}, q \in \mathbb{Q}^*_+, j \in \{0,1\} \exists l \in \mathbb{Q}^*_+ \\ \Big(\bigwedge_{i=1}^{n} (x_{i+1} \geq x_i) \wedge \bigwedge_{i=1}^{n+1} (|(-1)^{i+j}(p(x_i)-f(x_i)) - E_{n,f}| + q \leq E_{n,f}) \\ \qquad\qquad \to \bigwedge_{i=1}^{n} (x_{i+1} - x_i > l)\Big). \end{cases}$$

$$3^*)\begin{cases} \forall p_1,p_2 \in K_{f,n}, (x_1,\ldots,x_{n+1}) \in [0,1]^{n+1}, q \in \mathbb{Q}^*_+, j \in \{0,1\} \exists r,l \in \mathbb{Q}^*_+ \\ \Big(\bigwedge_{i=1}^{2} (\|f-p_i\| \leq E_{n,f}+r) \wedge \bigwedge_{i=1}^{n+1} \Big(|(-1)^{i+j}(\frac{p_1(x_i)+p_2(x_i)}{2} - f(x_i)) - E_{n,f}| \leq l\Big) \\ \to \bigwedge_{i=1}^{n+1} (|p_1(x_i)-p_2(x_i)| < q)\Big). \end{cases}$$

$$4^*)\begin{cases} \forall p_1,p_2 \in K_{f,n}, (x_1,\ldots,x_{n+1}) \in [0,1]^{n+1}, r,q \in \mathbb{Q}^*_+ \exists l \in \mathbb{Q}^*_+ \\ \qquad \big(\bigwedge_{i=1}^{n} (x_{i+1}-x_i) \geq r) \wedge \bigwedge_{i=1}^{n+1} (|p_1(x_i)-p_2(x_i)| \leq l) \to \|p_1-p_2\| < q\big). \end{cases}$$

Modulo the Σ^0_1-collection principle (Σ^0_1-CP) (which can be proved in WE-PA$^\omega$ and – in the presence of QF-AC^{0,0} – also in $\widehat{\text{WE-PA}}^\omega\!\upharpoonright$ and even G$_n$A$^\omega$, see exercise 3.7.8) and the representation of $\mathbb{Q}, [0,1]^n$ etc. from chapter 4 the matrix in each of the sentences 1*)–4*) is equivalent to a $\exists \underline{x}^{0/1} A_0$-formula A. Moreover, A is monotone in the $\exists$-data '$\exists r \in \mathbb{Q}^*_+$' and '$\exists l \in \mathbb{Q}^*_+$'. Hence we can apply theorem

15.1 in order to extract functionals (switching tacitly from '2^{-k}' etc. to '$q \in \mathbb{Q}_+^*$') $\Phi_1, \Phi_2, \tilde{\Phi}_2, \Phi_3, \tilde{\Phi}_3, \Phi_4$ which realize '$\exists r$' and '$\exists l$' and depend only on f (more precisely: a representative of f in the sense of the representation of $C[0,1]$ from chapter 4) and n, q (resp. f, n, r, q in 4^*)). For notational simplicity, we omit the arguments f and n of the functionals in the following:

$$1^{**})\begin{cases}\forall p_1, p_2 \in K_{f,n}, q \in \mathbb{Q}_+^* \\ \big(\bigwedge\limits_{i=1}^{2}(\|f - p_i\| \le E_{n,f} + \Phi_1(q)) \to \|f - \frac{p_1+p_2}{2}\| < E_{n,f} + q\big).\end{cases}$$

$$2^{**}.1)\begin{cases}\forall p \in K_{f,n}, q \in \mathbb{Q}_+^* \\ \Big(\|f - p\| \le E_{n,f} + \Phi_2(q) \to \exists (x_1, \ldots, x_{n+1}) \in [0,1]^{n+1}, j \in \{0,1\} \\ \big(\bigwedge\limits_{i=1}^{n+1}(|(-1)^{i+j}(p(x_i) - f(x_i)) - E_{n,f}| < q) \wedge \bigwedge\limits_{i=1}^{n}(x_{i+1} > x_i)\big)\Big),\end{cases}$$

$$2^{**}.2)\begin{cases}\forall p \in K_{f,n}, (x_1, \ldots, x_{n+1}) \in [0,1]^{n+1}, j \in \{0,1\}, q \in \mathbb{Q}_+^* \Big(\bigwedge\limits_{i=1}^{n}(x_{i+1} \ge x_i) \wedge \\ \bigwedge\limits_{i=1}^{n+1}(|(-1)^{i+j}(p(x_i) - f(x_i)) - E_{n,f}| + q \le E_{n,f}) \to \bigwedge\limits_{i=1}^{n}(x_{i+1} - x_i > \tilde{\Phi}_2(q))\Big).\end{cases}$$

$$3^{**})\begin{cases}\forall p_1, p_2 \in K_{f,n}, (x_1, \ldots, x_{n+1}) \in [0,1]^{n+1}, j \in \{0,1\}, q \in \mathbb{Q}_+^* \\ \Big(\bigwedge\limits_{i=1}^{2}(\|f - p_i\| \le E_{n,f} + \Phi_3(q)) \\ \wedge \bigwedge\limits_{i=1}^{n+1}(|(-1)^{i+j}(\frac{p_1(x_i)+p_2(x_i)}{2} - f(x_i)) - E_{n,f}| \le \tilde{\Phi}_3(q)) \\ \to \bigwedge\limits_{i=1}^{n+1}(|p_1(x_i) - p_2(x_i)| < q)\Big).\end{cases}$$

$$4^{**})\begin{cases}\forall p_1, p_2 \in K_{f,n}, (x_1, \ldots, x_{n+1}) \in [0,1]^{n+1}, r, q, \in \mathbb{Q}_+^* \\ \big(\bigwedge\limits_{i=1}^{n}(x_{i+1} - x_i \ge r) \wedge \bigwedge\limits_{i=1}^{n+1}(|p_1(x_i) - p_2(x_i)| \le \Phi_4(r,q)) \to \|p_1 - p_2\| < q\big),\end{cases}$$

where $\Phi_i(q), \Phi_4(r,q) \in \mathbb{Q}_+^*$ for all $q, r \in \mathbb{Q}_+^*$; i=1,2,3.

For $f \in C[0,1], n \in \mathbb{N}, q, l \in \mathbb{Q}_+^*$ we now define

$$\Phi(l,q) := \\ \min\Big\{\Phi_1\Big[\Phi_2\Big(\min\Big(\tfrac{l}{4}, \tilde{\Phi}_3\big(\Phi_4(\tilde{\Phi}_2(\tfrac{3l}{4}), q)\big)\Big)\Big)\Big], \Phi_3\big(\Phi_4(\tilde{\Phi}_2(\tfrac{3l}{4}), q)\big)\Big\} \in \mathbb{Q}_+^*.$$

Proposition 16.17.

$$\forall p_1, p_2 \in K_{f,n}, l, q \in \mathbb{Q}_+^* \\ \big(l \le E_{n,f} \wedge \bigwedge\limits_{i=1}^{2}(\|f - p_i\| \le E_{n,f} + \Phi(l,q)) \to \|p_1 - p_2\| \le q\big).$$

Proof: By 1^{**}), the assumption yields

$$\|f - \frac{p_1+p_2}{2}\| \le E_{n,f} + \Phi_2(c), \text{ where } c := \min\Big(\frac{l}{4}, \tilde{\Phi}_3\big(\Phi_4(\tilde{\Phi}_2(\frac{3l}{4}),q)\big)\Big).$$

Hence 2^{**}) yields the existence of $(x_1,\ldots,x_{n+1}) \in [0,1]^{n+1}, j \in \{0,1\}$ with

$$(1) \bigwedge_{i=1}^{n+1} \Big(|(-1)^{i+j}(\frac{p_1(x_i)+p_2(x_i)}{2} - f(x_i)) - E_{n,f}| < c\Big)$$

and

$$(2) \bigwedge_{i=1}^{n+1} (x_{i+1} - x_i > \tilde{\Phi}_2(\frac{3l}{4})),$$

where (2) follows since

$$|(-1)^{i+j}(\frac{p_1(x_i)+p_2(x_i)}{2} - f(x_i)) - E_{n,f}| < \frac{l}{4} \le \frac{E_{n,f}}{4}$$

implies that

$$|(-1)^{i+j}(\frac{p_1(x_i)+p_2(x_i)}{2} - f(x_i)) - E_{n,f}| + \frac{3l}{4} \le E_{n,f}.$$

By 3^{**}) we obtain from (1) that

$$(3) \bigwedge_{i=1}^{n+1} \Big(|p_1(x_i) - p_2(x_i)| < \Phi_4\big(\tilde{\Phi}_2(\frac{3l}{4}),q\big)\Big)$$

since by Φ-definition

$$\|f - p_{1/2}\| \le E_{n,f} + \Phi_3\big(\Phi_4(\tilde{\Phi}_2(\frac{3l}{4}),q)\big).$$

Finally, 4^{**}) together with (2) and (3) implies that

$$\|p_1 - p_2\| < q.$$

□

Corollary to the proof of proposition 16.17: In 1^{**})–4^{**}) it is sufficient to have '$\le$' and '$\ge$' in the conclusion instead of '$<$' and '$>$'respectively (which was used only to get the logical form required in theorem 15.1). In the following we, therefore, only realize these statements up to this change.

Φ can easily be modified to yield a modulus Φ^* which no longer depends on l and thus avoids the assumption '$l \le E_{n,f}$': $\Phi^*(q) := \min\left(\frac{q}{4}, \Phi(\frac{q}{4}, q)\right)$:

Proposition 16.18.

$$\forall p_1, p_2 \in K_{f,n}, q \in \mathbb{Q}_+^* \left(\|f - p_1\|, \|f - p_2\| \le E_{n,f} + \Phi^*(q) \to \|p_1 - p_2\| \le q\right).$$

Proof: Case 1: $E_{n,f} \ge \frac{q}{4}$. Then the assertion follows immediately from proposition 16.17.
Case 2: $E_{n,f} < \frac{q}{4}$. Then $\|f - p_1\|, \|f - p_2\| \le E_{n,f} + \Phi^*(q) < \frac{q}{4} + \frac{q}{4} = \frac{q}{2} \to$

$$\|p_1 - p_2\| \le \|p_1 - f\| + \|f - p_2\| < \frac{q}{2} + \frac{q}{2} = q. \qquad \square$$

The extraction of $\Phi_1, \Phi_3, \tilde{\Phi}_3$ as well as the functionals themselves are very simple. Φ_4 can be constructed using the Lagrange interpolation formula and majorization. The extraction of $\tilde{\Phi}_2$ again uses majorization in an essential way. The main part of our extraction, however, is the construction of Φ_2. While the other steps in the uniqueness proof are constructive, the proof of the alternation theorem (from which Φ_2 is extracted) is essentially nonconstructive both by the use of classical logic as well as the ineffective theorem (A).

Definition 16.19. For $n \in \mathbb{N}, f \in C[0,1], p \in P_n$ and $\varepsilon \in \mathbb{Q}_+^*$ we define:

1) $x \in [0,1]$ is an ε-(e)-point ('ε-extremal-point') of $p - f$ if $E_{n,f} - \varepsilon \le |p(x) - f(x)| \le E_{n,f} + \varepsilon$.
2) $x \in [0,1]$ is an ε-(+)-point (ε-(−)-point) of $p - f$ if $E_{n,f} - \varepsilon \le p(x) - f(x) \le E_{n,f} + \varepsilon$ $(-E_{n,f} - \varepsilon \le p(x) - f(x) \le -E_{n,f} + \varepsilon)$.
3) $(x_1, \ldots, x_k) \in [0,1]^k$ $(1 \le k \le n+2)$ is an ε-alternant of $p - f$ (having length k) if $x_1 < \ldots < x_k$ and $\bigwedge_{i=1}^{k} |(-1)^{i+j}(p(x_i) - f(x_i)) - E_{n,f}| \le \varepsilon$ for $j = 0$ or $j = 1$.
4) $p \in P_n$ is an ε-best approximation of f if $\|p - f\| \le E_{n,f} + \varepsilon$.

Remark 16.20. 1) Although, formally, definition 16.19.3) is defined for all $\varepsilon \ge 0$, it is useful only in the case $0 \le \varepsilon < E_{n,f}$ since only then the sign of $p(x_i) - f(x_i)$ alternates for $i = 1, \ldots, k$.
2) For $\varepsilon = 0$, definition 16.19 coincides with the usual definitions of (e)-point, (+), (−)-point, alternant and best approximation.
3) A definition which is similar to 16.19 can be found in [43].

Extraction of $\Phi_1, \Phi_2, \tilde{\Phi}_2, \Phi_3, \tilde{\Phi}_3, \Phi_4$:

Extraction of Φ_1: The extraction is trivial and we can define $\Phi_1(q) := q$:
Assume that

$$\|f - p_{1/2}\| \le E_{n,f} + q.$$

Then

$$-E_{n,f} - q \le f(x) - p_i(x) \le E_{n,f} + q \text{ for } i = 1,2.$$

Thus

$$-2E_{n,f} - 2q \le 2f(x) - \big(p_1(x) + p_2(x)\big) \le 2E_{n,f} + 2q$$

and so

$$-E_{n,f} - q \le f(x) - \frac{p_1(x) + p_2(x)}{2} \le E_{n,f} + q$$

for all $x \in [0,1]$ which implies that

$$\|f - \frac{p_1 + p_2}{2}\| \le E_{n,f} + q.$$

□

Extraction of $\Phi_3, \tilde{\Phi}_3$: Again the extraction is easy: $\Phi_3(q) := \frac{q}{4}, \tilde{\Phi}_3(q) := \frac{q}{4}$. Assume $\|f - p_1\|, \|f - p_2\| \le E_{n,f} + \frac{q}{4}$. Let $(x_1, \ldots, x_{n+1}) \in [0,1]^{n+1}$ be a $\frac{q}{4}$-alternant of $\frac{p_1+p_2}{2} - f$ and let x_i be a $\frac{q}{4}$-(+)-point. Then

$$E_{n,f} - \frac{q}{4} \le \frac{p_1(x_i) + p_2(x_i)}{2} - f(x_i) = \frac{p_1(x_i) - f(x_i)}{2} + \frac{p_2(x_i) - f(x_i)}{2}$$

which (using that $p_2(x_i) - f(x_i) \le E_{n,f} + \frac{q}{4}$) implies

$$E_{n,f} - \frac{q}{4} \le \frac{p_1(x_i) - f(x_i)}{2} + \frac{E_{n,f} + \frac{q}{4}}{2}$$

and so

$$2E_{n,f} - \frac{q}{2} \le p_1(x_i) - f(x_i) + E_{n,f} + \frac{q}{4}.$$

Hence (using that $p_1(x_i) - f(x_i) \le E_{n,f} + \frac{q}{4}$)

$$(1)\ E_{n,f} - \frac{q}{2} - \frac{q}{4} \le p_1(x_i) - f(x_i) \le E_{n,f} + \frac{q}{4}.$$

In the same way one shows that

$$(2)\ E_{n,f} - \frac{q}{2} - \frac{q}{4} \le p_2(x_i) - f(x_i) \le E_{n,f} + \frac{q}{4}.$$

(1) and (2) imply $|p_1(x_i) - p_2(x_i)| \le q$. An analogous reasoning applies if x_i is a $\frac{q}{4}$-(−)-point. □

Extraction of Φ_4: By the Lagrange interpolation formula we have for all $p \in P_n$

$$(*)\ p(x) = \sum_{i=1}^{n+1} l_i(x) p(x_i), \text{ where } l_i(x) := \frac{\prod_{j=1, j\neq i}^{n+1} (x - x_j)}{\prod_{j=1, j\neq i}^{n+1} (x_i - x_j)} \quad (1 \le i \le n+1).$$

Since $x_{i+1} - x_i \geq r$ (by the assumption) it follows that

$$(**)\ |l_i(x)| \leq \frac{1}{\prod\limits_{j=1, j\neq i}^{n+1} r \cdot |i-j|} \leq \frac{1}{r^n (i-1)!(n-i+1)!} \text{ for all } x \in [0,1].$$

For $p_1 - p_2$ such that $\bigwedge\limits_{i=1}^{n+1} \big(|p_1(x_i) - p_2(x_i)| \leq q\big)$, $(*)$ implies

$$|p_1(x) - p_2(x)| \leq \sum_{i=1}^{n+1} |l_i(x)| \cdot |p_1(x_i) - p_2(x_i)| \leq q \cdot \sum_{i=1}^{n+1} |l_i(x)|$$
$$(***) \overset{(**)}{\leq} q \cdot \sum_{i=1}^{n+1} \frac{1}{r^n (i-1)!(n-i+1)!}.$$

One easily verifies that

$$(i-1)!(n-i+1)! \geq \lfloor \frac{n}{2} \rfloor ! \lceil \frac{n}{2} \rceil !\ (1 \leq i \leq n+1).$$

Hence

$$|p_1(x) - p_2(x)| \leq q \cdot \frac{n+1}{\lfloor \frac{n}{2} \rfloor ! \lceil \frac{n}{2} \rceil ! r^n}.$$

So

$$\Phi_4(r,q) := \frac{\lfloor \frac{n}{2} \rfloor ! \lceil \frac{n}{2} \rceil ! r^n}{n+1} \cdot q$$

does the job. Note that Φ_4 does not depend on f! This is due to the fact that $(***)$ holds for all $p_1, p_2 \in P_n$ and not only for $p_1, p_2 \in K_{f,n} \subset P_n$. Put together the reasoning above yields

$$\bigwedge_{i=1}^{n} (x_{i+1} - x_i \geq r) \wedge \bigwedge_{i=1}^{n+1} \big(|p_1(x_i) - p_2(x_i)| \leq \Phi_4(r,q)\big) \to \|p_1 - p_2\| \leq q$$

for all $p_1, p_2 \in P_n, (x_1, \ldots, x_{n+1}) \in [0,1]^{n+1}, r, q \in \mathbb{Q}^*_+$. □

Extraction of $\tilde{\Phi}_2$: Assume

$$\bigwedge_{i=1}^{k-1} (x_{i+1} \geq x_i) \wedge \bigwedge_{i=1}^{k} \big(|(-1)^{i+j}(p(x_i) - f(x_i)) - E_{n,f}| + q \leq E_{n,f}\big),$$

where $j = 0$ or $j = 1$. It is clear that

$$\bigwedge_{i=1}^{k-1} \big(|(p-f)(x_{i+1}) - (p-f)(x_i)| \geq 2q\big).$$

Let ω_{p-f} be a modulus of uniform continuity for $p - f$ on $[0,1]$. Then

$$\bigwedge_{i=1}^{k-1}(x_{i+1} - x_i \geq \omega_{p-f}(2q)).$$

Define $\widehat{\Phi}_2(q,p) := \omega_{p-f}(2q)$. In the following, we use majorization to obtain a $\tilde{\Phi}_2$ from $\widehat{\Phi}_2$ such that

$$\forall p \in K_{f,n}, q \in \mathbb{Q}^*_+ \big(0 < \tilde{\Phi}_2(q) \leq \widehat{\Phi}_2(q,p)\big).$$

$\widehat{\Phi}_2$ depends on p via ω_{p-f}. Hence we obtain $\tilde{\Phi}_2$ by replacing ω_{p-f} in $\widehat{\Phi}_2$ by a common modulus of uniform continuity for all $p - f$ with $p \in K_{f,n}$: Let $p(x) = c_n x^n + \ldots c_1 x + c_0 \in P_n, x, y \in [0,1]$. By the Markov inequality (proposition 16.4) $p \in K_{f,n}$ implies that $\|p'\| \leq 2n^2\big(\frac{5}{2}\|f\|\big) = 5n^2\|f\|$, where p' is the derivative of p. By the mean value theorem this yields that p is Lipschitz continuous on $[0,1]$ with the Lipschitz constant $5n^2\|f\|$. Thus

$$\omega_n(q) := \frac{q}{5n^2\|f\|}$$

is a uniform modulus of continuity for all $p \in K_{f,n}$ on $[0,1]$ if $n \geq 1$ (for $n = 0$ define $\omega_n(q) := 1$). Let ω_f be a modulus of uniform continuity for $f \in C[0,1]$. Then

$$\tilde{\omega}_{f,n}(q) := \min\big(\omega_n(\tfrac{q}{2}), \omega_f(\tfrac{q}{2})\big)$$

is a common modulus of uniform continuity for $\{p - f \mid p \in K_{f,n}\}$ on $[0,1]$. Hence $\tilde{\Phi}_2(q) := \tilde{\omega}_{f,n}(2q)$ fulfills 2^{**}.2). More precisely, this is true not for $\tilde{\Phi}_2$ but for its variant with $\|f\|$ being replaced by some rational upper bound $\chi(f) \in \mathbb{Q}^*_+$ for $\|f\|$, since $\tilde{\Phi}_2(q) \notin \mathbb{Q}$ in general. It is clear that such a bound $\chi(f)$ for $\|f\|$ can easily be computed (since $f \in C[0,1]$ is endowed with a modulus of uniform continuity ω_f). Note that $\chi(f)$ may depend on ω_f. In fact, **any** rational upper bound on $\|f\|$ can be used. □

Extraction of Φ_2: For the extraction of Φ_2 we need the following majorization:

Lemma 16.21. *Let $z_1, \ldots, z_n \in [0,1]$ $(n \geq 1)$ be real numbers with (for $n > 1$) $\bigwedge_{i=1}^{n-1}(z_{i+1} - z_i \geq 2\alpha)$ for a fixed $\alpha \in \mathbb{Q}^*_+, \alpha \leq 1$ and define*

$$K := \left\{x \in [0,1] : \bigwedge_{i=1}^{n}(|z_i - x| \geq \frac{\alpha}{2})\right\}.$$

Then for $p(x) := (z_1 - x)(z_2 - x) \cdot \ldots \cdot (z_n - x) \in P_n$ the following holds:

$$\inf_{x \in K}|p(x)| \geq \prod_{i=1}^{n}|2i - \frac{1}{2} - n| \cdot \alpha^n = \prod_{i=1}^{\lceil \frac{n}{2} \rceil}(2i - \frac{3}{2}) \cdot \prod_{i=1}^{\lfloor \frac{n}{2} \rfloor}(2i - \frac{1}{2}) \cdot \alpha^n$$

(Note that the right side does not depend on the z_i).

Proof: For $n = 1$ the lemma is trivial. Assume $n > 1$ and define

$$\tilde{z}_i := 2(i-1)\alpha\ (1 \leq i \leq n).$$

Now we put

$$\tilde{p}(x) := (\tilde{z}_1 - x)(\tilde{z}_2 - x) \cdot \ldots \cdot (\tilde{z}_n - x)$$

and define

$$\tilde{K} := \left\{ x \in [0,1] : \bigwedge_{i=1}^{n} (|\tilde{z}_i - x| \geq \frac{\alpha}{2}) \right\}.$$

i) Claim: $\inf_{x\in\tilde{K}} |\tilde{p}(x)| \leq \inf_{x\in K} |p(x)|$. The claim is proved by showing $\forall x \in K \exists \tilde{x} \in \tilde{K}\big(|\tilde{p}(\tilde{x})| \leq |p(x)|\big)$. This is easily verified by treating the cases $0 \leq x \leq z_1 - \frac{\alpha}{2}$, $z_n + \frac{\alpha}{2} \leq x \leq 1$ and $\exists i(1 \leq i \leq n-1 \wedge z_i + \frac{\alpha}{2} \leq x \leq z_{i+1} - \frac{\alpha}{2})$ separately.

ii) Claim: $\inf_{x\in\tilde{K}} |\tilde{p}(x)| = \prod_{i=1}^{n} |2i - \frac{1}{2} - n| \cdot \alpha^n = \prod_{i=1}^{\lceil \frac{n}{2} \rceil} (2i - \frac{3}{2}) \cdot \prod_{i=1}^{\lfloor \frac{n}{2} \rfloor} (2i - \frac{1}{2}) \cdot \alpha^n$.

Case 1: n even. We show that $\inf_{x\in\tilde{K}} |\tilde{p}(x)| = |\tilde{p}(\tilde{z}_{\frac{n}{2}} + \frac{\alpha}{2})| = |\tilde{p}(\tilde{z}_{\frac{n}{2}+1} - \frac{\alpha}{2})|$. Let $x_0 \in \tilde{K}$ and assume $x_0 \leq \tilde{z}_{\frac{n}{2}} - \frac{\alpha}{2}$. Then $|\tilde{p}(x_0')| < |\tilde{p}(x_0)|$ for $x_0' := x_0 + 2\alpha \in \tilde{K}$ since $\bigwedge_{i=1}^{n-1} (\tilde{z}_i - x_0| = |\tilde{z}_{i+1} - x_0'|)$ and $|\tilde{z}_n - x_0| > |\tilde{z}_1 - x_0'|$. Analogously for $x_0 \geq \tilde{z}_{\frac{n}{2}+1} + \frac{\alpha}{2}$ and $x_0' := x_0 - 2\alpha$. One easily verifies for $i = 0, \ldots, \frac{n}{2} - 1$:
$|(\tilde{z}_{\frac{n}{2}-i} - x)(\tilde{z}_{\frac{n}{2}+1+i} - x)|$ attains its minimum on $I := [\tilde{z}_{\frac{n}{2}} + \frac{\alpha}{2}, \tilde{z}_{\frac{n}{2}+1} - \frac{\alpha}{2}]$ at the point $x = \tilde{z}_{\frac{n}{2}} + \frac{\alpha}{2}$. Hence $|\tilde{p}(x)|$ attains its minimum on I at this point. Together with the reasoning above we can conclude that $|\tilde{p}(\tilde{z}_{\frac{n}{2}} + \frac{\alpha}{2})| = \inf_{x\in\tilde{K}} |\tilde{p}(x)|$.

Since

$$\tilde{z}_{\frac{n}{2}} + \frac{\alpha}{2} = (n - \frac{3}{2})\alpha \wedge |\tilde{z}_i - (n - \frac{3}{2})\alpha| = \alpha|2i - \frac{1}{2} - n|$$

it follows that

$$|\tilde{p}(\tilde{z}_{\frac{n}{2}} + \frac{\alpha}{2})| = \prod_{i=1}^{n} |2i - \frac{1}{2} - n| \cdot \alpha^n.$$

Furthermore, we have

$$\prod_{i=\frac{n}{2}+1}^{n} |2i - \frac{1}{2} - n| = \prod_{i=1}^{\frac{n}{2}} (2i - \frac{1}{2}) \text{ and } \prod_{i=1}^{\frac{n}{2}} |2i - \frac{1}{2} - n| = \prod_{i=1}^{\frac{n}{2}} (2i - \frac{3}{2}).$$

Case 2: n odd. We show $|\tilde{p}(\tilde{z}_{\lceil \frac{n}{2} \rceil} - \frac{\alpha}{2})| = |\tilde{p}(\tilde{z}_{\lceil \frac{n}{2} \rceil} + \frac{\alpha}{2})| = \inf_{x\in\tilde{K}} |\tilde{p}(x)|$ (This implies (as in case 1) the claim since $\tilde{z}_{\lceil \frac{n}{2} \rceil} - \frac{\alpha}{2} = (n - \frac{3}{2})\alpha$).

a) $0 \leq x < \tilde{z}_{\lceil \frac{n}{2} \rceil} - \frac{\alpha}{2} \rightarrow |\tilde{p}(x)| > |\tilde{p}(\tilde{z}_{\lceil \frac{n}{2} \rceil} - \frac{\alpha}{2})|$:
$\tilde{p}(x) = \tilde{q}(x) \cdot (\tilde{z}_n - x)$, where $\tilde{q}(x) = (\tilde{z}_1 - x) \cdot \ldots \cdot (\tilde{z}_{n-1} - x)$. $0 \leq x < \tilde{z}_{\lceil \frac{n}{2} \rceil} - \frac{\alpha}{2}$ implies (by case 1, since $n-1$ is even)

$|\tilde{q}(x)| \overset{!}{\geq} |\tilde{q}(\tilde{z}_{\lceil \frac{n}{2} \rceil} - \frac{\alpha}{2})| \wedge |\tilde{z}_n - x| > |\tilde{z}_n - (\tilde{z}_{\lceil \frac{n}{2} \rceil} - \frac{\alpha}{2})|$ and therefore $|\tilde{p}(x)| > |\tilde{p}(\tilde{z}_{\lceil \frac{n}{2} \rceil} - \frac{\alpha}{2})|$.
(! holds since $\lceil \frac{n}{2} \rceil = \frac{n-1}{2} + 1$ and $|\tilde{q}(\tilde{z}_{\frac{n-1}{2}+1} - \frac{\alpha}{2})| = \inf\limits_{x \in \tilde{K}} |\tilde{q}(x)|$).
b) $1 \geq x > \tilde{z}_{\lceil \frac{n}{2} \rceil} + \frac{\alpha}{2} \to |\tilde{p}(x)| > |\tilde{p}(\tilde{z}_{\lceil \frac{n}{2} \rceil} + \frac{\alpha}{2})|$ follows analogously (consider $\tilde{p}(x) = (\tilde{z}_1 - x) \cdot \hat{\tilde{q}}$). □

Remark 16.22. 1) The proof of lemma 16.21 shows that the estimate is optimal (as an estimate which depends on α and n only), since the polynomial $\tilde{p}(x)$ is admissible.
2) The (less good) estimate $\inf\limits_{x \in K} |p(x)| \geq \left(\frac{\alpha}{2}\right)^n$ is trivial.

The following lemma is easy:

Lemma 16.23. *Let $0 < \alpha \leq \frac{1}{n}$. Then for all $j, k \leq n$:*

$$k < j \to \prod_{i=1}^{\lceil \frac{j}{2} \rceil} (2i - \frac{3}{2}) \cdot \prod_{i=1}^{\lfloor \frac{j}{2} \rfloor} (2i - \frac{1}{2}) \cdot \alpha^j < \prod_{i=1}^{\lceil \frac{k}{2} \rceil} (2i - \frac{3}{2}) \cdot \prod_{i=1}^{\lfloor \frac{k}{2} \rfloor} (2i - \frac{1}{2}) \cdot \alpha^k \leq 1.$$

Lemma 16.24. *Assume that $E_{n,f} > 0$ and let $q \in \mathbb{Q}_+^*$ be such that $E_{n,f} \geq q$. Then $\tilde{\omega}_{f,n}(\frac{q}{2}) \leq \frac{1}{2(n+1)}$ (here and in the following $\tilde{\omega}_{f,n}$ denotes the common modulus of uniform continuity for all $p - f$ with $p \in K_{f,n}$ from the construction of $\tilde{\Phi}_2$, i.e. $\tilde{\omega}_{f,n}$, in particular, is a modulus of uniform continuity of $p_b - f$ with $p_b \in K_{f,n}$ being the best approximation of f).*

Proof: We divide $[0,1]$ into subintervals of length $\tilde{\omega}_{f,n}(\frac{q}{2})$ (the last one may have shorter length). Since $q \leq E_{n,f}/2$, the amplitude of $p_b - f$ on each of these subintervals is at most $\frac{E_{n,f}}{2}$ (where $p_b \in K_{f,n}$ is the best approximation of f in P_n). By the alternation theorem the number of these subintervals is at least $2(n+1)$. □

Lemma 16.25. *For $f \in C[0,1], n \in \mathbb{N}, p \in P_n, 0 \leq \varepsilon < E_{n,f}$ assume that $\|p - f\| \leq E_{n,f} + \varepsilon$. Then there exist both ε-(+)-points and ε-(−)-points of $p - f$.*

Proof: The proof is very similar to the proof of lemma 16.5. □

Theorem 16.26. *Let $0 \leq \varepsilon < \frac{E_{n,f}}{4}$, $0 < q \leq E_{n,f}$ ($\varepsilon, q \in \mathbb{Q}$) and $2 \leq k \leq n+2$. If $p_\varepsilon \in P_n$ is a $\prod\limits_{i=1}^{\lfloor \frac{k-2}{2} \rfloor} (2i - \frac{1}{2}) \cdot \prod\limits_{i=1}^{\lceil \frac{k-2}{2} \rceil} (2i - \frac{3}{2}) \cdot \min\left(\frac{1}{n}, \omega_{f,p_\varepsilon}(\frac{q}{2})\right)^{k-2} \cdot \varepsilon$-best approximation of $f \in C[0,1]$, where ω_{f,p_ε} is a modulus of uniform continuity of $p_\varepsilon - f$, then there exists an ε-alternant having length k for $p_\varepsilon - f$.*

Proof: We may assume that $\omega_{f,p_\varepsilon}(\frac{q}{2}) \leq \frac{1}{n}$ (For otherwise we define $\omega'_{f,p_\varepsilon}(q) := \min\left(\frac{1}{n}, \omega_{f,p_\varepsilon}(q)\right)$). Define $C_l := \prod\limits_{i=1}^{\lfloor \frac{l}{2} \rfloor} (2i - \frac{1}{2}) \cdot \prod\limits_{i=1}^{\lceil \frac{l}{2} \rceil} (2i - \frac{3}{2}) \cdot \left(\omega_{f,p_\varepsilon}(\frac{q}{2})\right)^l$ for $0 \leq l \leq n$. We divide $[0,1]$ into subintervals $I_0, \ldots, I_{s-1}$ of length $\omega_{f,p_\varepsilon}(\frac{q}{2})$ (I_s may have shorter length). The amplitude of $p_\varepsilon - f$ on $I_i = [\xi_i, \xi_{i+1}]$ is $\leq \frac{q}{2} \leq \frac{E_{n,f}}{2}$. I_i is called ε-(e)-interval if it contains an ε-(e)-point of $p_\varepsilon - f$. Since $\varepsilon < \frac{E_{n,f}}{2}$, $(p_\varepsilon - f)(x)$ is distinct from 0 on every ε-(e)-interval and therefore has constant sign. An ε-(e)-interval I_i is an ε-(+)-interval (ε-(–)-interval) if $p_\varepsilon(x) - f(x) > 0$ ($p_\varepsilon(x) - f(x) < 0$) for all $x \in I_i$ (Thus an ε-(+)-interval contains an ε-(+)-point but no ε-(–)-point). Enumerating all ε-(e)-intervals from left to right $(I_{j_1}, \ldots, I_{j_N})$ we obtain the following schema (assuming without loss of generality that I_{j_1} is an ε-(+)-interval):

$$(*)\begin{cases} I_{j_1}, \ldots, I_{j_{k_1}} & \varepsilon\text{-(+)-intervals} \\ I_{j_{k_1+1}}, \ldots, I_{j_{k_2}} & \varepsilon\text{-(–)-intervals} \\ \vdots & \\ I_{j_{k_{m-1}+1}}, \ldots, I_{j_{km}} & \varepsilon\text{-}(-1)^{m-1}\text{-intervals} \ \ (k_m = N). \end{cases}$$

The assumption on p_ε implies that $\|f - p_\varepsilon\| \leq E_{n,f} + C_{k-2} \cdot \varepsilon \overset{\text{lemma16.23}}{\leq} E_{n,f} + \varepsilon$. By lemma 16.25, the schema $(*)$ consists of at least two groups (with at least one ε-(+)-interval and one ε-(–)-interval), i.e. $m \geq 2$. We now show that in fact $m \geq k$: Assume that on the contrary $m < k$ $(**)$. $p_\varepsilon - f$ has different sign on $I_{j_{k_1}}$ and $I_{j_{k_1+1}}$. Hence the right endpoint of $I_{j_{k_1}}$ is distinct from the left endpoint of $I_{j_{k_1+1}}$. Thus there exists at least one interval $I_{i_1} = [\xi_{i_1}, \xi_{i_1+1}]$ of $I_0, \ldots, I_{s-1}$ which lies between $I_{j_{k_1}}$ and $I_{j_{k_1+1}}$ and hence is not an ε-(e)-interval. Define $z_1 := \frac{\xi_{i_1} + \xi_{i_1+1}}{2}$. Analogously, we define $z_2, \ldots, z_{m-1} \in [0,1]$ by $z_l := \frac{\xi_{i_l} + \xi_{i_l+1}}{2}$, where $I_{i_l} = [\xi_{i_l}, \xi_{i_l+1}]$ is an interval located in between $I_{j_{k_l}}$ and $I_{j_{k_l+1}}$ $(1 \leq l \leq m-1)$.

Define $\rho(x) := (z_1 - x)(z_2 - x) \cdot \ldots \cdot (z_{m-1} - x)$. By our assumption $(**)$ it follows that $\rho(x) \in P_n$ (since $m - 1 \leq k - 2 \leq n$). $z_1, \ldots, z_{m-1}$ are the only zeroes of $\rho(x)$. Since none of the ε-(e)-intervals contains one of these zeroes, $\rho(x)$ has constant sign on each I_{j_i}. This sign equals the sign of $p_\varepsilon(x) - f(x)$ on I_{j_i}.
$(***)$ $R := \sup_{x \in [0,1]} |\rho(x)| \leq 1$ (since $z_1, \ldots, z_{m-1} \in [0,1]$). By the definition of z_l we have $z_{l+1} - z_l \geq 2\omega_{f,p_\varepsilon}(\frac{q}{2})$. Furthermore, $|z_l - x| \geq \frac{\omega_{f,p_\varepsilon}(\frac{q}{2})}{2}$ $(1 \leq l \leq m-1)$ for every x which lies in an ε-(e)-interval. Hence lemma 16.21 implies:

$$(****)\begin{cases} \text{If x is element of an } \varepsilon\text{-(e)-interval, then} \\ |\rho(x)| \geq C_{m-1} \overset{l.16.23}{\geq} C_{k-2} \text{ and, therefore, } |\varepsilon\rho(x)| = \varepsilon|\rho(x)| \geq \varepsilon \cdot C_{k-2}. \end{cases}$$

Let I_i be an interval of $I_0, \ldots, I_{s-1}$ which is not an ε-(e)-interval. Then

$$E^i := \sup_{x \in I_i} |p_\varepsilon(x) - f(x)| < E_{n,f} - \varepsilon.$$

Let E^* be the maximum of these E^i (for all non-ε-(e)-intervals I_i). Then $E^* < E_{n,f} - \varepsilon$. Choose $\lambda > 0$ so small that $\lambda R < E_{n,f} - E^* - \varepsilon$ and $\lambda R \leq \frac{E_{n,f}}{4} - \varepsilon R$. The latter can be achieved since $\frac{E_{n,f}}{4} - \varepsilon R > 0$ holds because of $R \leq 1$ and $\varepsilon < \frac{E_{n,f}}{4}$. Define $Q(x) := p_\varepsilon(x) - (\lambda + \varepsilon)\rho(x) \in P_n$. We have to show that

$$\forall x \in [0,1](|Q(x) - f(x)| < E_{n,f})$$

and so – using (A) – that $\|Q - f\| < E_{n,f}$ which contradicts the definition of $E_{n,f}$:
Case 1: I_i is not an ε-(e)-interval:

$$\begin{aligned} x \in I_i \to |Q(x) - f(x)| &\leq |p_\varepsilon(x) - f(x)| + (\varepsilon + \lambda)|\rho(x)| \leq E^* + \varepsilon|\rho(x)| + \lambda|\rho(x)| \\ &\overset{(***)}{<} E^* + \varepsilon|\rho(x)| + E_{n,f} - E^* - \varepsilon \leq \varepsilon \cdot R - \varepsilon + E_{n,f} \overset{(***)}{\leq} E_{n,f}. \end{aligned}$$

Case 2: I_i is an ε-(e)-interval, $x \in I_i$. Then $p_\varepsilon(x) - f(x)$ and $(\varepsilon + \lambda)\rho(x)$ have the same sign and $|p_\varepsilon(x) - f(x)| > (\varepsilon + \lambda)|\rho(x)|$ since $|p_\varepsilon(x) - f(x)| \geq E_{n,f} - \varepsilon - \frac{E_{n,f}}{2} > \frac{E_{n,f}}{4}$ and $(\varepsilon + \lambda)|\rho(x)| \leq \varepsilon R + \lambda R \leq \frac{E_{n,f}}{4}$. Hence

$$\begin{aligned} |Q(x) - f(x)| &= |p_\varepsilon(x) - f(x) - (\varepsilon + \lambda)\rho(x)| = |p_\varepsilon(x) - f(x)| - (\varepsilon + \lambda)|\rho(x)| \\ &\leq E_{n,f} + \varepsilon \cdot C_{k-2} - \underbrace{\varepsilon|\rho(x)|}_{\overset{(****)}{\geq} \varepsilon \cdot C_{k-2}} - \underbrace{\lambda|\rho(x)|}_{>0} < E_{n,f}. \end{aligned}$$

□

Corollary 16.27 (ε-alternation theorem). *Suppose that* $0 < q \leq E_{n,f}$, $0 \leq \varepsilon < E_{n,f}$ $(q, \varepsilon \in \mathbb{Q})$ *and* $2 \leq k \leq n+2$. *Define*

$$\chi(f,n,q,k) := \begin{cases} 1, & \text{if } k = 2 \\ \frac{1}{4} \prod_{i=1}^{\lfloor \frac{k-2}{2} \rfloor} (2i - \frac{1}{2}) \cdot \prod_{i=1}^{\lceil \frac{k-2}{2} \rceil} (2i - \frac{3}{2}) \cdot \left(\tilde{\omega}_{f,n}(\frac{q}{2})\right)^{k-2}, & \text{if } k > 2. \end{cases}$$

If $p \in P_n$ *is a* $\left(\chi(f,n,q,k)\right) \cdot \varepsilon$*-best approximation of* $f \in C[0,1]$, *then there exists an* ε*-alternant of length* k *for* $p - f$ *in* $[0,1]$ *(note that* $\chi(f,n,q,k)$ *does not depend on* p*). If* $\varepsilon < \frac{E_{n,f}}{4}$ *then the factor* $\frac{1}{4}$ *can be omitted.*

Proof: If $k = 2$, then the corollary follows from lemma 16.25. Thus we may assume $k > 2$:
Case 1: $p \in K_{f,n}$. By construction, $\tilde{\omega}_{f,n}$ is a modulus of uniform continuity for $p - f$ on $[0,1]$. Since $\frac{\varepsilon}{4} < \frac{E_{n,f}}{4}$ the corollary follows from theorem 16.26 using that by lemma 16.24 we have $\tilde{\omega}_{f,n}(\frac{q}{2}) \leq \frac{1}{n}$.

Case 2: $p \notin K_{f,n}$. In this case we have $\|p\| > \frac{5}{2}\|f\|$ and so

$$(+)\ \|f-p\| > \frac{3}{2}\|f\| \geq \frac{3}{2}E_{n,f}.$$

Hence p cannot be a $\big(\chi(f,n,q,k)\big) \cdot \varepsilon$-best approximation of f since

$$\|p-f\| \leq E_{n,f} + \frac{1}{4}(\dots)\cdot\varepsilon \overset{16.23}{\leq} E_{n,f} + \frac{\varepsilon}{4}$$

would imply (by $(+)$) $\frac{\varepsilon}{4} > \frac{E_{n,f}}{2}$, contradicting $\varepsilon < E_{n,f}$. □

Remark 16.28. 1) Corollary 16.27 immediately implies that also $\tilde{\chi}(f,n,\varepsilon,k) := \big(\chi(f,n,\varepsilon,k)\big)\cdot\varepsilon$ is an alternation modulus, i.e. if $p \in P_n$ is a $\tilde{\chi}(f,n,\varepsilon,k)$-best approximation of f then there exists an ε-alternant of length k for $p-f$ in $[0,1]$. $\tilde{\chi}$ no longer depends on q but also is not linear in ε anymore.
2) For $\varepsilon = 0$, the proof of theorem 16.26 transforms into the classical proof of the usual alternation theorem 16.7.

By remark 16.28.1), $\tilde{\chi}$ fulfills the demands for Φ_2. Nevertheless, we use χ instead of $\tilde{\chi}$ for the construction of the modulus of uniqueness Φ although it depends in addition to f,n,ε also on an estimate $0 < q \leq E_{n,f}$. The reason for this is that χ is linear in ε and so the whole modulus of uniqueness will be linear in ε which is an important property in view of proposition 16.2. Furthermore, the construction of Φ^* from Φ in proposition 16.18 yields a modulus which no longer depends on q (but also is not linear). Thus we define

$$\Phi_2(r,q) :=$$
$$\tfrac{1}{4}\prod_{i=1}^{\lfloor\frac{n\dot{-}1}{2}\rfloor}(2i-\tfrac{1}{2})\cdot\prod_{i=1}^{\lceil\frac{n\dot{-}1}{2}\rceil}(2i-\tfrac{3}{2})\cdot\big(\tilde{\omega}_{f,n}\big(\tfrac{r}{2}\big)\big)^{n\dot{-}1}\cdot q = \tfrac{1}{4}\tilde{C}_{n\dot{-}1}\cdot\big(\tilde{\omega}_{f,n}\big(\tfrac{r}{2}\big)\big)^{n\dot{-}1}\cdot q$$

for $r,q \in \mathbb{Q}_+^*$, where

$$\tilde{C}_l := \prod_{i=1}^{\lfloor\frac{l}{2}\rfloor}(2i-\frac{1}{2})\cdot\prod_{i=1}^{\lceil\frac{l}{2}\rceil}(2i-\frac{3}{2})\cdot\Big(\tilde{\omega}_{f,n}(\frac{r}{2})\Big)^{l}.$$

By corollary 16.27, we have for all $f \in C[0,1], n \in \mathbb{N}, p \in P_n$ and $r,q \in \mathbb{Q}_+^*$ such that $q < E_{n,f}, r \leq E_{n,f}$:

$$\|f-p\| \leq E_{n,f} + \Phi_2(r,q) \rightarrow \exists q\text{–alternant of length } n+1 \text{ for } p-f.$$

Lemma 16.29. *Let $\tilde{\omega}_{f,n}$ be the modulus from the construction of $\tilde{\Phi}_2$. Then for $0 < q \leq E_{n,f}$:*

$$\tilde{k}(q) := \frac{\lfloor\frac{n}{2}\rfloor!\lceil\frac{n}{2}\rceil!}{n+1}\cdot\left(\tilde{\omega}_{f,n}\left(\frac{3q}{2}\right)\right)^n \leq 1.$$

Proof: The lemma follows easily from the fact that $\tilde{\omega}_{f,n}(\frac{3q}{2}) \le \frac{1}{n+1}$, which is proved similarly to lemma 16.24. □

Now we are ready to define our modulus of uniqueness (using proposition 16.17 and the fact that fact $\Phi_1(q) = q$):

$$\Phi(l,r,q) := \min\left\{\Phi_2\Big(r, \min\big(\frac{l}{4}, \tilde{\Phi}_3(\Phi_4(\tilde{\Phi}_2(\frac{3l}{4}),q))\big)\Big), \Phi_3\big(\Phi_4(\tilde{\Phi}_2(\frac{3l}{4}),q)\big)\right\}.$$

By (the corollary to the proof of) proposition 16.17, Φ is a modulus of uniqueness for $p_1, p_2 \in K_{f,n}$ if $0 < r, l \le E_{n,f}$. By lemmas 16.23 and 16.24, we have

$$\begin{aligned}&\Phi_2\Big(r, \min\big(\tfrac{l}{4}, \tilde{\Phi}_3(\Phi_4(\tilde{\Phi}_2(\tfrac{3l}{4}),q))\big)\Big) \le \min\big(\tfrac{l}{4}, \tilde{\Phi}_3(\Phi_4(\tilde{\Phi}_2(\tfrac{3l}{4}),q))\big)\\ &\le \tilde{\Phi}_3(\Phi_4(\tilde{\Phi}_2(\tfrac{3l}{4}),q)) = \Phi_3(\Phi_4(\tilde{\Phi}_2(\tfrac{3l}{4}),q))\end{aligned}$$

since $\tilde{\Phi}_3(q) = \frac{q}{4} = \Phi_3(q)$. Hence

$$\begin{aligned}\Phi(l,r,q) &= \Phi_2\Big(r, \min\big(\tfrac{l}{4}, \tilde{\Phi}_3(\Phi_4(\tilde{\Phi}_2(\tfrac{3l}{4}),q))\big)\Big)\\ &= \tfrac{1}{4}\cdot \tilde{C}_{n\dot{-}1}\cdot\big(\tilde{\omega}_{f,n}(\tfrac{r}{2})\big)^{n\dot{-}1}\cdot \min\big(\tfrac{l}{4}, \tfrac{1}{4}\tilde{k}(l)\cdot q\big).\end{aligned}$$

Furthermore, $\min\big(\frac{l}{4}, \frac{1}{4}\tilde{k}(l)\cdot q\big)$ can be replaced by $\min\big(\frac{E_{n,f}}{4}, \frac{1}{4}\tilde{k}(l)\cdot q\big)$ since the minimum with $\frac{l}{4}$ is only used in proposition 16.17 to derive

$$|(-1)^{i+j}(\frac{p_1(x_i)+p_2(x_i)}{2} - f(x_i)) - E_{n,f}| + \frac{3l}{4} \le E_{n,f}$$

from

$$|--\text{''}--| < \frac{l}{4} \le \frac{E_{n,f}}{4}.$$

But the former follows also from $|--\text{''}--| < \frac{E_{n,f}}{4}$ and $l \le E_{n,f}$.
Replacing $\frac{1}{4}\tilde{k}(l)$ by $\frac{1}{10}\tilde{k}(l)$ makes $\Phi(l,r,q)$ only smaller and, therefore, a fortiori yields a modulus of uniqueness. Finally we identify r and l. Thus Φ is redefined as follows

$$\tilde{\Phi}(l,q) := \frac{1}{4}\tilde{C}_{n\dot{-}1}\cdot\left(\tilde{\omega}_{f,n}(\frac{l}{2})\right)^{n\dot{-}1}\cdot \min\big(\frac{E_{n,f}}{4}, \frac{1}{10}\tilde{k}(l)\cdot q\big).$$

Now suppose that $\frac{1}{10}q < \frac{E_{n,f}}{4}$. Then (by lemma 16.29)

$$\frac{1}{10}\tilde{k}(l)\cdot q < \frac{E_{n,f}}{4} \text{ and so } \tilde{\Phi}(l,q) = \frac{1}{4}\tilde{C}_{n\dot{-}1}\cdot\left(\tilde{\omega}_{f,n}\left(\frac{l}{2}\right)\right)^{n\dot{-}1}\cdot\frac{1}{10}\tilde{k}(l)\cdot q.$$

In this case we can omit the factor $\frac{1}{4}$ by corollary 16.27.
In the following, we show that the resulting functional is in fact a modulus of uniqueness for arbitrary $q > 0$ and $p_1, p_2 \in P_n$:

Theorem 16.30. *Proof–theoretic analysis of the uniqueness proof from [304] and [284] yields the following quantitative version of uniqueness: For $f \in C[0,1], n \in \mathbb{N}, l \in \mathbb{Q}^*_+$ define*

$$\Psi(f,n,l) :=$$

$$\frac{1}{10(n+1)} \prod_{i=1}^{\lfloor \frac{n \dot- 1}{2} \rfloor} (2i - \tfrac{1}{2}) \cdot \prod_{i=1}^{\lceil \frac{n \dot- 1}{2} \rceil} (2i - \tfrac{3}{2}) \cdot \lfloor \tfrac{n}{2} \rfloor ! \lceil \tfrac{n}{2} \rceil ! \left(\tilde{\omega}_{f,n}\left(\tfrac{l}{2}\right)\right)^{n \dot- 1} \cdot \left(\tilde{\omega}_{f,n}\left(\tfrac{3l}{2}\right)\right)^{n},$$

where

$$\tilde{\omega}_{f,n}(q) := \min\left(\frac{q}{10n^2\|f\|_\infty}, \omega_f\left(\frac{q}{2}\right)\right) \text{ for } n \geq 1 \text{ and } := 1 \text{ otherwise.}$$

Then for all $f \in C[0,1], n \in \mathbb{N}$ and all $p_1, p_2 \in P_n$ the following holds:

$$\forall l, q \in \mathbb{Q}^*_+ \left(l \leq E_{n,f} \wedge \bigwedge_{i=1}^{2} (\|f - p_i\|_\infty \leq E_{n,f} + \left(\Psi(f,n,l)\right) \cdot q) \to \|p_1 - p_2\|_\infty \leq q\right).$$

In instead of $\|f\|_\infty$ one may use any strictly positive upper bound $M \geq \|f\|_\infty$ instead.

Proof: As before, we simply write $\|\cdot\|$ instead of $\|\cdot\|_\infty$.
Case 1: $\frac{1}{10}q < \frac{E_{n,f}}{4}$: If $p_1, p_2 \in K_{f,n}$, then the theorem follows from the reasoning above. Thus suppose w.l.o.g. that $p_1 \notin K_{f,n}$. Then $\|p_1\| \geq \frac{5}{2}\|f\|$ and therefore $\|p_1 - f\| \geq \frac{3}{2}\|f\| \geq \frac{3}{2}E_{n,f}$. Hence $\|p_1 - f\| \leq E_{n,f} + \left(\Psi(f,n,l)\right) \cdot q$ implies $\left(\Psi(f,n,l)\right) \cdot q \geq \frac{E_{n,f}}{2}$, which contradicts the fact that – using lemmas 16.23, 16.24 and 16.29 – $\left(\Psi(f,n,l)\right) \cdot q \leq \frac{1}{10}q < \frac{E_{n,f}}{4}$. Thus p_1 cannot be a $\left(\Psi(f,n,l)\right) \cdot q$-best approximation of f.
Case 2: $\frac{1}{10}q \geq \frac{E_{n,f}}{4}$, i.e. $E_{n,f} \leq \frac{2}{5}q$. Hence

$$\|f - p_1\|, \|f - p_2\| \leq E_{n,f} + \left(\Psi(f,n,l)\right) \cdot q \leq E_{n,f} + \frac{1}{10}q$$

implies that

$$\|p_1 - p_2\| \leq \|p_1 - f\| + \|f - p_2\| \leq 2E_{n,f} + \frac{1}{5}q \leq \frac{4}{5}q + \frac{1}{5}q = q.$$

For the case $n = 0$ note that we can take the trivial modulus of uniqueness $q/2$. □

Corollary 16.31. *Let $\Psi(f,n,l)$ be defined as in theorem 16.30 and $l \in \mathbb{Q}^*_+$ such that $0 < l \leq E_{n,f}$.*

1) *$\Psi(f,n,l)$ is a constant of strong unicity, i.e.*

$$\forall f \in C[0,1], n \in \mathbb{N}, p \in P_n\big(\|f-p\|_\infty \geq \|f-p_b\|_\infty + \big(\Psi(f,n,l)\big)\cdot\|p-p_b\|_\infty\big),$$

where p_b is the best approximation of f in P_n.

2) $\lambda_l(f,n) := \frac{2}{\Psi(f,n,l)}$ *is a pointwise Lipschitz constant for the Chebycheff projection $\mathscr{P}$:*

$$\forall f, f_0 \in C[0,1], n \in \mathbb{N}\big(\|\mathscr{P}(f,n)-\mathscr{P}(f_0,n)\|_\infty \leq \lambda_l(f_0,n)\cdot\|f-f_0\|_\infty\big).$$

3) $\tilde{\Psi}(f,n,q) := \min\big(\frac{q}{4}, \Psi(f,n,\frac{q}{4})\cdot q\big)$ *is a modulus of uniqueness:*

$$\forall p_1, p_2 \in P_n, q \in \mathbb{Q}^*_+\big(\bigwedge_{i=1}^{2}(\|f-p_i\|_\infty \leq E_{n,f}+\tilde{\Psi}(f,n,q)) \to \|p_1-p_2\|_\infty \leq q\big).$$

4) $\frac{1}{2}\tilde{\Psi}(f,n,q)$ *($\tilde{\Psi}$ as in 3)) is a modulus of pointwise continuity of $\mathscr{P}$.*

Note that $\tilde{\Psi}$ is a constructive operation in f (endowed with a modulus of uniform continuity) and n,q and does not depend on any estimate $0 < l \leq E_{n,f}$.

Proof: 1) and 2) follow immediately from theorem 16.30 and proposition 16.2.
3) follows from theorem 16.30 and the proof of proposition 16.18.
4) follows from 3) and proposition 16.2. □

Remark 16.32. Since theorem 16.30 (and corollary 16.31) is provable in WE-PA$^\omega$ + (A) it follows that by theorem 15.1 that it can already be proved in WE-HA$^\omega$ plus the ε-weakening A_ε of A and so (since the latter is provable in WE-HA$^\omega$) in WE-HA$^\omega$ alone.

Let A be a subset of $C[0,1]$ and assume that $\omega_{A,n}$ is a common modulus of uniform continuity for all $p-f$ where $f \in A$ and $p \in K_{f,n}$. Assume, furthermore, that $0 < l_A \leq \inf_{f\in A} E_{n,f}$. Then the modulus Ψ_A obtained from Ψ by replacing $\tilde{\omega}_{f,n}$ by $\omega_{A,n}$ and l by l_A is a common constant of strong unicity for all $f \in A$.
In particular, if $K \subset C[0,1]$ is totally bounded, i.e. pre-compact, w.r.t. $\|\cdot\|_\infty$, then there exists a common modulus of uniform continuity ω_K for all $f \in K$ and a constant $M_K \in \mathbb{Q}^*_+$ such that $\|f\|_\infty \leq M_K$ for all $f \in K$. Define

$$\omega_{K,n}(q) := \min\big(\omega_{n,M_K}(\frac{q}{2}), \omega_K(\frac{q}{2})\big), \text{ where } \omega_{n,M_K}(q) := \frac{q}{5n^2M_K}.$$

As in the case of $\tilde{\omega}_{f,n}$ it follows that $\omega_{K,n}$ is a common modulus of uniform continuity for all $p-f$ with $f \in K$ and $p \in K_{f,n}$. If, moreover, K is compact and $K \cap P_n = \emptyset$, then $\inf_{f\in K} E_{n,f} > 0$. Put together we have shown the following theorem:

Theorem 16.33. 1) *Let $K \subset C[0,1]\setminus P_n$ compact (w.r.t. $\|\cdot\|_\infty$) with a common modulus of uniform continuity ω_K for all $f \in K$ and $M_K \in \mathbb{Q}^*_+$ such that $\forall f \in K(\|f\|_\infty \leq M_K)$ and define $\omega_{K,n}$ as above. Let Ψ_K denote the result obtained from Ψ (defined as in theorem 16.30) by replacing $\tilde{\omega}_{f,n}$ by $\omega_{K,n}$ and l by $l_K \in \mathbb{Q}^*_+$ such that $0 < l_K \leq \inf_{f\in K} E_{n,f}$. Then $(\Psi_K(n,l_K)$ (resp. $\lambda_K := \frac{2}{\Psi_K(n,l_K)}$) is a common constant of strong unicity (resp. Lipschitz constant) for all $f \in K$.*

2) *Let $K \subset C[0,1]$ be pre-compact with ω_K and M_K as above and define $\tilde{\Psi}_K(n,q) := \min\left(\frac{q}{4}, \Psi_K(n,\frac{q}{4}) \cdot q\right)$ (resp. $\frac{1}{2}\tilde{\Psi}_K(n,q)$) is a common modulus of uniqueness (pointwise continuity for the Chebycheff projection) for all $f \in K$. This result also applies if $\inf_{f\in K} E_{n,f} = 0$ since we do not use any positive lower bound on $\inf_{f\in K} E_{n,f}$.*

As mentioned already at the end of the previous section, the (ineffective) existence of a constant of strong unicity was first proved in 1963 [286] and the existence of a common constant in the sense of theorem 16.33 even only in 1976 [153].
Let $\gamma_n^*(f)$ be the largest constant of strong unicity for given $f \in C[0,1], n \in \mathbb{N}$. In [152] it is proved that $\{1/\gamma_n^*(f) : f \in B\}$ is unbounded for all $n \geq 1$, where B is the closed unit ball in $C[0,1]$. So there is no common constant of strong unicity for all of B. In [302] it is shown that $\liminf_{n\to\infty} \gamma_n^*(f) = 0$ for certain classes of functions. In [152] it was conjectured that this would be the case for all functions $f \in C[0,1]$ except polynomials (note that our lower estimate on $\gamma_n^*(f)$ does indeed have this property which – of course – does not prove the conjecture). This conjecture was finally established (without providing any bounds) in 1999 by W. Gehlen ([112]). In [113], Gehlen even proved that the smallest Lipschitz constants $\lambda_n^*(f)$ for the Chebycheff projection operator satisfy $\limsup_{n\to\infty} \lambda_n^*(f) = \infty$, whenever f is not a polynomial.

General remarks on the extraction: Theorem 16.11 (based on theorem 15.4 and hence in turn on theorem 15.1) guaranteed the extractability of a primitive recursive (in the sense of WE-HA$^\omega$) modulus of uniqueness without having to invest any new mathematical ideas. In the above analysis we, nevertheless, used general mathematical knowledge at various places to achieve obvious numerical improvements. This applies e.g. to the proof of lemma 16.21 which led to a somewhat better estimate than the direct one from remark 16.22.2). Lemmas 16.24 and 16.29 can be avoided by just replacing $\tilde{\omega}_{f,n}(q/2)$ by $\min(1/n, \tilde{\omega}_{f,n}(q/2))$ and $\tilde{k}(q)$ by $\min(1, \tilde{k}(q))$ (which is, however, superfluous as these lemmas show). Also the original uniqueness proof didn't specify any particular argument for the interpolation property of the polynomials in P_n. For the quantitative analysis it is clear that it is reasonable to use the Lagrange interpolation formula since it explicitly provides the unique interpolating polynomial. For an explicit version of the compactness of $K_{f,n}$ (in terms of a common modulus of uniform continuity) one naturally uses the (as a uniform bound) optimal Markov inequality. Both this inequality as well as the Lagrange interpolation formula can be written as purely universal sentences and hence can be treated as axioms (without having to analyze their proofs at all). Since the fact that without any additional ideas an effective modulus of uniqueness can be extracted from the above discussed uniqueness proof is already established by theorem 16.11 to carry out an explicit extraction only makes sense if one aims at obtaining a modulus as good as possible from the given proof, i.e. one that does not seem to allow for a substantial improvement unless a genuinely new uniqueness argument is used. Such a new argument resulting in a significantly better modulus (and an easier extraction) is analyzed in the next section.

Theorem 16.11 not only guarantees the extractability of an effective modulus but also the constructive verification (in the sense of WE-HA$^\omega$) of the resulting modulus. In our reasoning above, however, we made free use of classical reasoning, whenever this was useful. Nevertheless, it is clear how to 'constructivize' the treatment either directly or by the use of theorem 15.1: e.g. already lemma 16.25 requires the principle (A), but if we strengthen the assumption that p is an ε-best approximation to the assumption that it is a δ-best approximation for some $\delta < \varepsilon$ (e.g. $\varepsilon/2$) then the proof can be carried out constructively. This, however, would make the proofs only more complicated without any improvement of the numerical content. Moreover, despite of the fact that the modulus Ψ in theorem 16.30 was obtained making free use of classical reasoning, one can (as we did in remark 16.32) appeal to theorem 15.1 to get the assurance that it also can be verified constructively whatever the benefits of this information might be.

16.3 Best Chebycheff approximation II

In this section (which is based largely on Kohlenbach [205]) we analyze a second proof for the uniqueness of the best Chebycheff approximation due to [381] and [313]. It turns out that the analysis is much simpler although the uniqueness proof is somewhat more complicated and again uses the difficult alternation theorem (and hence (A), i.e. WKL). However, this time the alternation theorem is used in the form of a sentence Γ (in the sense of theorem 15.1) so that – guaranteed by that general metatheorem – we can treat it as an axiom and don't need to analyze its proof. Furthermore, the numerical bounds on strong uniqueness are significantly better than the ones obtained in the previous section. For the polynomial case (we will also treat the case of general so-called Haar spaces below) we get:

Theorem 16.34. *Proof-theoretic analysis of the uniqueness proof from [381] (with full details in [313]) gives the following result: Let* $E_{n,f} := \inf\limits_{p\in P_n} \|f-p\|_\infty$ *and*

$$\tilde{\Phi}(\omega,n,\varepsilon) := \min\left(\varepsilon/4, \frac{\lfloor \frac{n}{2} \rfloor ! \lceil \frac{n}{2} \rceil !}{2(n+1)} \cdot (\omega_n(\varepsilon/2))^n \cdot \varepsilon\right),$$

with

$$\omega_n(\varepsilon) := \begin{cases} \min\left(\omega\left(\frac{\varepsilon}{2}\right), \frac{\varepsilon}{8n^2\lceil \frac{1}{\omega(1)} \rceil}\right), & \text{if } n \geq 1 \\ 1 & \text{if } n = 0. \end{cases}$$

Then $\tilde{\Phi}$ *is a common modulus of uniqueness for all* $f \in C[0,1]$ *which have the modulus of uniform continuity* ω, *i.e.*

$$\forall n \in \mathbb{N}; p_1, p_2 \in P_n; \varepsilon \in \mathbb{Q}_+^* \left(\bigwedge_{i=1}^{2} (\|f-p_i\|_\infty \leq E_{n,f} + \tilde{\Phi}(\omega,n,\varepsilon)) \to \|p_1-p_2\|_\infty \leq \varepsilon \right).$$

Moreover if $E_{n,f} > 0$ *and* $l \in \mathbb{Q}_+^*$ *such that* $l \leq E_{n,f}$ *and*

$$\Phi(\omega,n,l) := \frac{\lfloor \frac{n}{2} \rfloor ! \lceil \frac{n}{2} \rceil !}{2(n+1)} \cdot (\omega_n(2l))^n$$

then $\Phi(\omega,n,l)$ *is a constant of strong unicity for* f *(see the previous section). Instead of '*$\lceil 1/\omega(1) \rceil$*' we may have an arbitrary upper bound* $\mathbb{Q}_+^* \ni M \geq \|f\|_\infty$.

We will now extract this bound as well as a generalization to arbitrary so-called Haar subspaces of $C[0,1]$ instead of P_n. It will turn out that this improves numerically the only prior known bounds (due to [43, 45] and – implicitly – [198, 199]). As mentioned already in the previous section, ineffectively the existence of constants of strong unicity was first proved in [286] and the existence of uniform such constants in [153]. In the following, $(\phi_1,\ldots,\phi_n)$ always is a tuple of n linearly independent functions in $C[0,1]$.

Definition 16.35. $(\phi_1,\ldots,\phi_n)$ is called a Chebycheff system (of dimension n over $[0,1]$ if every non-trivial generalized polynomial $c_1\phi_1 + \ldots + c_n\phi_n$ has at most $n-1$ roots. In this case

$$H_{\underline{\phi}} := Lin_{\mathbb{R}}(\phi_1,\ldots,\phi_n) := \{c_1\phi_1 + \ldots + c_n\phi_n \mid c_1,\ldots,c_n \in \mathbb{R}\}$$

is called a Haar subspace of $C[0,1]$.

The following proposition is an easy exercise in linear algebra:

Proposition 16.36. *The following statements are equivalent:*

1) $(\phi_1,\ldots,\phi_n)$ *is an n-dimensional Chebycheff system over* $[0,1]$.
2) *For any n pairwise distinct points* $x_1,\ldots,x_n \in [0,1]$ *we have that*

$$Det(A(\underline{x}) := \begin{vmatrix} \phi_1(x_1) & \phi_2(x_1) & \ldots & \phi_n(x_1) \\ \phi_1(x_2) & \phi_2(x_2) & \ldots & \phi_n(x_2) \\ \cdots & \cdots & \cdots & \cdots \\ \phi_1(x_n) & \phi_2(x_n) & \ldots & \phi_n(x_n) \end{vmatrix} \neq 0.$$

3) *For any n pairwise distinct points* $x_1,\ldots,x_n \in [0,1]$ *and arbitrary points* $y_1,\ldots,y_n \in \mathbb{R}$ *there exists exactly one* $\psi \in H_{\underline{\phi}}$ *such that*

$$\psi(x_i) = y_i \quad for \;\; i = 1,\ldots,n.$$

Example 16.37. (see e.g. [70]) The following tuples are Chebycheff systems (in particular over $[0,1]$):

1) $(1,x,x^2,\ldots,x^n)$ is a Chebycheff system of dimension $n+1$ over any interval,
2) $(1,e^x,e^{2x},\ldots,e^{nx})$ is a Chebycheff system of dimension $n+1$ over any interval,

3) $(1, \sin(x), \ldots, \sin(nx), \cos(x), \ldots, \cos(nx))$ is a Chebycheff system of dimension $2n+1$ over any interval $[a, a+2\pi)$.

The uniqueness proof we analyze in this section is a slight simplification of a classical proof which goes back to [381] and relies on the alternation theorem which in turn relies on (and also implies) WKL (see the previous section).

We now sketch this uniqueness proof: let $f \in C[0,1]$, $(\phi_1, \ldots, \phi_n)$ be a Chebycheff system over [0,1], $H := Lin_{\mathbb{R}}(\phi_1, \ldots, \phi_n)$, $E_{H,f} :=$dist(f, H). Assume that $\psi_1, \psi_2 \in H$ are best approximations of f in H, i.e. $\|f - \psi_1\|_\infty = E_{H,f} = \|f - \psi_2\|_\infty$. The alternation theorem extends to arbitrary Chebycheff systems (see again [66] for a proof; since this time we will be able to by-pass the proof of the alternation theorem and treat it as an axiom Γ, there is no need to give the proof here). Hence there exists an alternant $x_1 < \ldots < x_{n+1}$ in [0,1] for $\psi_1 - f$, i.e. for $j = 0$ or $j = 1$:

$$\bigwedge_{i=1}^{n+1} \left((-1)^{i+j}\big(\psi_1(x_i) - f(x_i)\big) = E_{H,f}\right).$$

Since $\|f - \psi_2\|_\infty \leq E_{H,f}$, it follows that

$$\bigwedge_{i=1}^{n+1} \left((-1)^{i+j}\big(f(x_i) - \psi_2(x_i)\big) \geq -E_{H,f}\right).$$

Hence

$$\bigwedge_{i=1}^{n+1} \Big((-1)^{i+j}\big(\psi_1(x_i) - \psi_2(x_i)\big) \\ = (-1)^{i+j}\big(\psi_1(x_i) - f(x_i)\big) + (-1)^{i+j}\big(f(x_i) - \psi_2(x_i)\big) \geq 0\Big).$$

Using the fact that $(\phi_1, \ldots, \phi_n)$ is a Chebycheff system and $x_1 < \ldots < x_{n+1}$ one concludes that $\psi_1 \equiv \psi_2$ (see lemma 16.38 below). In order to make our metatheorem 15.1 applicable we restrict – as before – H to a compact set. As it will turn out we this time can even take

$$\tilde{K}_{f,H} := \{\psi \in H : \|\psi\|_\infty \leq 2\|f\|_\infty\},$$

i.e. with the bound $2\|f\|_\infty$ instead of $\frac{5}{2}\|f\|_\infty$, and will still be able to extend in the end the modulus we obtain from $\tilde{K}_{f,n}$ to H. Moreover, we modify the proof above in such a way that the alternation theorem is used only in the form

$$(+)\begin{cases} \forall f \in C[0,1] \exists \psi_b \in \tilde{K}_{f,H}, (x_1, \ldots, x_{n+1}) \in [0,1]^{n+1}, j \in \{0,1\} \\ \qquad \Big(\bigwedge_{i=1}^{n} (x_{i+1} \geq x_i) \wedge \bigwedge_{i=1}^{n+1} \big((-1)^{i+j}(\psi_b(x_i) - f(x_i)) = E_{H,f}\big)\Big). \end{cases}$$

$(+)$ follows immediately from the alternation theorem, the existence of a best approximation of f in H and the fact that each best approximation of f must be in $\tilde{K}_{f,H}$. Conversely, $(+)$ implies the alternation theorem if we use already the uniqueness of the best approximation. Since we are just about to prove the uniqueness, the uniqueness proof itself does not 'know' that $(+)$ actually is as strong as the alternation theorem itself and in fact, in the context of this proof, $(+)$ behaves rather different from the alternation theorem in the previous uniqueness proof (analyzed in the preceding section): since $(+)$ has the form Γ (which, as we saw in the previous section, the alternation theorem does not) it can be treated as a 'black box' in the course of the extraction of the modulus of uniqueness.

The uniqueness proof given above can be decomposed into the following main steps:

1. $$\forall \psi_1, \psi_2 \in \tilde{K}_{f,H}, (x_1, \ldots, x_{n+1}) \in [0,1]^{n+1}, j \in \{0,1\}$$
$$\Big(\bigwedge_{i=1}^{n+1} ((-1)^{i+j}(\psi_1(x_i) - f(x_i)) = E_{H,f}) \wedge \bigwedge_{i=1}^{n+1} (|\psi_2(x_i) - f(x_i)| \le E_{H,f}) \\ \to \bigwedge_{i=1}^{n+1} ((-1)^{i+j}(\psi_1(x_i) - \psi_2(x_i)) \ge 0) \Big).$$

2. $$\forall \psi_1, \psi_2 \in \tilde{K}_{f,H}, (x_1, \ldots, x_{n+1}) \in [0,1]^{n+1}, j \in \{0,1\}$$
$$\Big(\bigwedge_{i=1}^{n} (x_{i+1} > x_i) \wedge \bigwedge_{i=1}^{n+1} ((-1)^{i+j}(\psi_1(x_i) - \psi_2(x_i)) \ge 0) \to \|\psi_1 - \psi_2\|_\infty = 0 \Big).$$

3. $$\forall \psi \in \tilde{K}_{f,H}, (x_1, \ldots, x_{n+1}) \in [0,1]^{n+1}, j \in \{0,1\}$$
$$\Big(\bigwedge_{i=1}^{n} (x_{i+1} \ge x_i) \wedge \bigwedge_{i=1}^{n+1} ((-1)^{i+j}(\psi(x_i) - f(x_i)) = E_{H,f} > 0) \to \bigwedge_{i=1}^{n} (x_{i+1} > x_i) \Big).$$

4. $\forall \psi, \psi_1, \psi_2 \in H \big(\|\psi - \psi_1\|_\infty = \|\psi - \psi_2\|_\infty = 0 \to \psi_1 \equiv \psi_2 \big)$.

Together with $(+)$, 1.–4. imply that for all $f \in C[0,1]$:

$$E_{H,f} > 0 \to \forall \psi_1, \psi_2 \in \tilde{K}_{f,H} \big(\|f - \psi_1\|_\infty = E_{H,f} = \|f - \psi_2\|_\infty \to \psi_1 \equiv \psi_2 \big).$$

Assume $E_{H,f} > 0$: By $(+)$ and '3.' there exists ψ_b, $(x_1, \ldots, x_{n+1})$ such that (for, say, $j = 0$)

$$\bigwedge_{i=1}^{n} (x_{i+1} > x_i) \wedge \bigwedge_{i=1}^{n+1} ((-1)^i(\psi_b(x_i) - f(x_i)) = E_{H,f} > 0).$$

Applying '1.' to $\psi_1' := \psi_b, \psi_2' := \psi_1$ as well as to $\psi_1' := \psi_b, \psi_2' := \psi_2$ and $(x_1, \ldots, x_{n+1})$ yields that

$$\bigwedge_{i=1}^{n+1}(-1)^i\big(\psi_b(x_i)-\psi_k(x_i)\big)\geq 0 \quad (k=1,2)$$

and thus, by '2.', $\|\psi_b-\psi_1\|_\infty=0$ and $\|\psi_b-\psi_2\|_\infty=0$. Hence by '4.' $\psi_1\equiv\psi_2$.

Since $(+)$ (as discussed above) is of the form of a sentence Γ in theorem 15.1 its proof is not relevant for the extraction of the modulus of uniqueness as we may simply treat $(+)$ as an axiom (for the course of the extraction).

Rewriting e.g. '$|\ldots|\leq E_{n,f}$' as '$\forall r\in\mathbb{Q}^*_+\,(|\ldots|\leq E_{n,f}+r)$' and using partial prenexation '1.'–'4.' can be transformed into:

$$1^*.\ \forall\psi_1,\psi_2\in\tilde{K}_{f,H},(x_1,\ldots,x_{n+1})\in[0,1]^{n+1},j\in\{0,1\},q\in\mathbb{Q}^*_+\exists r\in\mathbb{Q}^*_+$$
$$\Big(\bigwedge_{i=1}^{n+1}((-1)^{i+j}(\psi_1(x_i)-f(x_i))=E_{H,f})\wedge\bigwedge_{i=1}^{n+1}(|\psi_2(x_i)-f(x_i)|\leq E_{H,f}+r)$$
$$\to\bigwedge_{i=1}^{n+1}((-1)^{i+j}(\psi_1(x_i)-\psi_2(x_i))>-q)\Big).$$

$$2^*.\ \forall\psi_1,\psi_2\in\tilde{K}_{f,H},(x_1,\ldots,x_{n+1})\in[0,1]^{n+1},j\in\{0,1\},q,l\in\mathbb{Q}^*_+\exists r\in\mathbb{Q}^*_+$$
$$\Big(\bigwedge_{i=1}^{n}(x_{i+1}-x_i\geq l)\wedge\bigwedge_{i=1}^{n+1}((-1)^{i+j}(\psi_1(x_i)-\psi_2(x_i))\geq -r)\to\|\psi_1-\psi_2\|_\infty<q\Big).$$

$$3^*.\ \forall\psi\in\tilde{K}_{f,H},(x_1,\ldots,x_{n+1})\in[0,1]^{n+1},j\in\{0,1\},q\in\mathbb{Q}^*_+\exists r\in\mathbb{Q}^*_+$$
$$\Big(\bigwedge_{i=1}^{n}(x_{i+1}\geq x_i)\wedge\bigwedge_{i=1}^{n+1}((-1)^{i+j}(\psi(x_i)-f(x_i))=E_{H,f}\geq q)\to\bigwedge_{i=1}^{n}(x_{i+1}-x_i>r)\Big).$$

The quantitative version of 4. is trivial:

$$4^*.\ \forall\psi,\psi_1,\psi_2\in H,q\in\mathbb{Q}^*_+\Big(\bigwedge_{i=1}^{2}(\|\psi-\psi_i\|_\infty\leq\frac{q}{2})\to\|\psi_1-\psi_2\|_\infty\leq q\Big).$$

'1^*.' and '3^*.' can easily be proved in E-PA$^\omega$. As we will see below, '2^*.' is provable in E-PA$^\omega$+intermediate value theorem (IVT). The intermediate value theorem has the form of another sentence Γ (alternatively, we could just formalize the proof of the intermediate value theorem in E-PA$^\omega$+QF-AC^{0,0}):

$$(\mathrm{IVT}):\forall f\in C[0,1]\exists x_0\in[0,1]\big(f(0)<0\wedge f(1)>0\to f(x_0)=0\big);$$

define $F(f,x):=\min(f(0),0)\cdot\max(f(1),0)\cdot f(x)$. Then

$$F(f,x)=0\leftrightarrow\big(f(0)<0\wedge f(1)>0\to f(x)=0\big).$$

So (IVT) can be written as

$$\forall f \in C[0,1] \exists x_0 \in [0,1] \big(F(f,x_0) =_{\mathbb{R}} 0\big),$$

which is of the form Γ. Hence we can apply theorem 15.1 to extract bounds from below for '$\exists r \in \mathbb{Q}^*_+$' (i.e. '$\exists r \geq_{\mathbb{Q}} \Phi f q$') which depend only on f (together with a modulus of uniform continuity of f) and q (resp. f,q,l). These bounds realize in fact '$\exists r \in \mathbb{Q}^*_+$' since '1*.'–'3*.' are monotone in r. Thus one can obtain effective operations Φ_1, Φ_2, Φ_3 such that $\Phi_1(f,q), \Phi_2(f,l,q), \Phi_3(f,q) \in \mathbb{Q}^*_+$ for all $q,l \in \mathbb{Q}^*_+$ and

1**. $\forall \psi_1, \psi_2 \in \tilde{K}_{f,H}, (x_1,\ldots,x_{n+1}) \in [0,1]^{n+1}, j \in \{0,1\}, q \in \mathbb{Q}^*_+$

$$\Big(\bigwedge_{i=1}^{n+1} \big((-1)^{i+j}(\psi_1(x_i) - f(x_i)) = E_{H,f}\big) \wedge \bigwedge_{i=1}^{n+1} \big(|\psi_2(x_i) - f(x_i)| \leq E_{H,f} + \Phi_1(f,q)\big)$$
$$\to \bigwedge_{i=1}^{n+1} \big((-1)^{i+j}(\psi_1(x_i) - \psi_2(x_i)) > -q\big)\Big).$$

2**. $\forall \psi_1, \psi_2 \in \tilde{K}_{f,H}, (x_1,\ldots,x_{n+1}) \in [0,1]^{n+1}, j \in \{0,1\}, q,l \in \mathbb{Q}^*_+$

$$\Big(\bigwedge_{i=1}^{n} (x_{i+1} - x_i \geq l) \wedge \bigwedge_{i=1}^{n+1} \big((-1)^{i+j}(\psi_1(x_i) - \psi_2(x_i)) \geq -\Phi_2(f,l,q)\big)$$
$$\to \|\psi_1 - \psi_2\|_\infty < q\Big).$$

3**. $\forall \psi \in \tilde{K}_{f,H}, (x_1,\ldots,x_{n+1}) \in [0,1]^{n+1}, j \in \{0,1\}, q \in \mathbb{Q}^*_+$

$$\Big(\bigwedge_{i=1}^{n} (x_{i+1} \geq x_i) \wedge \bigwedge_{i=1}^{n+1} \big((-1)^{i+j}(\psi(x_i) - f(x_i)) = E_{H,f} \geq q\big)$$
$$\to \bigwedge_{i=1}^{n} \big(x_{i+1} - x_i > \Phi_3(f,q)\big)\Big).$$

Define $\Phi(f,l,q) := \Phi_1\big(f, \Phi_2(f, \Phi_3(f,l), \frac{q}{2})\big)$. One easily verifies that for all $f \in C[0,1]$ and $l \in \mathbb{Q}^*_+$ such that $E_{H,f} \geq l$:

$$(++)\ \forall \psi_1, \psi_2 \in \tilde{K}_{f,H}, q \in \mathbb{Q}^*_+ \big(\bigwedge_{i=1}^{2} (\|\psi_i - f\|_\infty \leq E_{H,f} + \Phi(f,l,q)) \to \|\psi_1 - \psi_2\|_\infty \leq q\big).$$

Proof: Assume that $E_{H,f} > 0$ with $E_{H,f} \geq l$ and that

$$\|\psi_1 - f\|_\infty, \|\psi_2 - f\|_\infty \leq E_{H,f} + \Phi(f,l,q).$$

Let $\psi_b \in \tilde{K}_{f,n}$ and $(x_1,\ldots,x_{n+1}) \in [0,1]^{n+1}$ be by $(+)$ such that (formulating for simplicity only the case $j = 0$):

$$\bigwedge_{i=1}^{n} (x_{i+1} \geq x_i) \wedge \bigwedge_{i=1}^{n+1} \big((-1)^i(\psi_b(x_i) - f(x_i)) = E_{H,f}\big).$$

Then '1**.' applied to ψ_b, ψ_1 (for ψ_1, ψ_2) yields that

$$(1)\ \bigwedge_{i=1}^{n+1} \left((-1)^i(\psi_b(x_i) - \psi_1(x_i)) \geq -\Phi_2\big(f, \Phi_3(f,l), \frac{q}{2}\big)\right).$$

Also '3**.' applied to $\psi := \psi_b$ gives

$$(2)\ \bigwedge_{i=1}^{n} \big(x_{i+1} - x_i \geq \Phi_3(f,l)\big).$$

Because of $(1),(2)$ we can use '2**.' applied to ψ_b, ψ_1 (for ψ_1, ψ_2) to conclude that $\|\psi_b - \psi_1\|_\infty \leq \frac{q}{2}$. An analogous reasoning shows that $\|\psi_b - \psi_2\|_\infty \leq \frac{q}{2}$ which yields the claim. □

Only for the **verification** of Φ, i.e. the proof of $(++)$, the assumption $(+)$ is used. The **construction** of Φ does not use (a proof of) $(+)$.
From the proof above it is clear that we need '1**.'–'3**.' only with '$\geq$' and '$\leq$' instead of '$>$' and '$<$' respectively.

We are now going to construct Φ_1, Φ_2, Φ_3 explicitly (in particular for the special case $H := P_{n-1}$):

Extraction of Φ_1: The construction of Φ_1 is trivial. One easily verifies that $\Phi_1(f,q) := q$ fulfills '1**.' with $\psi_1(x_i) - \psi_2(x_i) \geq -q$ instead of $\psi_1(x_i) - \psi_2(x_i) > -q$ for all $\psi_1, \psi_2 \in H$ which – as we just noticed – is sufficient for $(++)$.

Extraction of Φ_2: We first show that for all $\psi \in H, (x_1, \ldots, x_{n+1}) \in [0,1]^{n+1}$

$$(*)\ \bigwedge_{i=1}^{n}(x_{i+1} > x_i) \wedge \bigwedge_{i=1}^{n+1} \left((-1)^i \psi(x_i) \geq 0\right) \to \psi \equiv 0,$$

which immediately gives '2.'

A zero $x^* \in (0,1)$ of ψ is called 'simple' if ψ changes its sign in x^* and 'double' otherwise. From $\bigwedge_{i=1}^{n+1} (-1)^i \psi(x_i) \geq 0$ one can show that ψ has at least n zeroes provided that one, indeed, counts double zeroes twice, which then yields $\psi \equiv 0$. This argument is due to [381] and presented in detail by J.R. Rice in [313] (pp. 61–62). In order to prove $(*)$ from this, one has to show that double zeroes in fact do count twice. This is done e.g. in [313] (p. 57): To every $\psi \in H$ with $\psi \not\equiv 0$ Rice constructs a $\psi_\varepsilon \in H$, with $\psi_\varepsilon \not\equiv 0$, which has the same simple zeroes as ψ but two simple zeroes for each double zero y_i of ψ ($\psi(y_i)$ is disturbed by a sufficiently small ε).

We simplify this proof in that we apply such an ε-perturbation directly to the points $x_1, \ldots, x_{n+1}$ in $(*)$ and reduce $(*)$ to

$$(**)\ \forall \psi \in H, (x_1,\ldots,x_{n+1}) \in [0,1]^{n+1} \neg\Big(\bigwedge_{i=1}^{n}(x_{i+1} > x_i) \wedge \bigwedge_{i=1}^{n+1}\big((-1)^i\psi(x_i) > 0\big)\Big),$$

which follows from the intermediate value theorem and the definition of a Chebycheff system.

Lemma 16.38. *For all $\psi \in H, (x_1,\ldots,x_{n+1}) \in [0,1]^{n+1}$ the following holds*

$$\bigwedge_{i=1}^{n}(x_{i+1} > x_i) \wedge \bigwedge_{i=1}^{n+1}\big((-1)^i\psi(x_i) \geq 0\big) \to \psi \equiv 0.$$

Proof: Suppose that $\bigwedge_{i=1}^{n}(x_{i+1} > x_i) \wedge \bigwedge_{i=1}^{n+1}\big((-1)^i\psi(x_i) \geq 0\big)$ and $(-1)^{i_0}\psi(x_{i_0}) =: \alpha > 0$ for some $i_0 \in \{1,\ldots,n+1\}$. Since H is a Haar subspace there exists a (uniquely determined) $\chi \in H$ such that

$$\chi(x_i) = (-1)^i \text{ for } i = 1,\ldots,i_0-1, i_0+1,\ldots,n+1.$$

Let $\varepsilon > 0$ be so small that $\varepsilon \cdot \|\chi\|_\infty < \alpha$. Then $\bigwedge_{i=1}^{n+1}\big((-1)^i(\psi+\varepsilon\chi)(x_i) > 0\big)$. But this is impossible by $(**)$ (since $\psi+\varepsilon\chi \in H$). Hence $\bigwedge_{i=1}^{n+1}\big((-1)^i\psi(x_i) = 0\big)$ which implies $\psi \equiv 0$. $\square$

$(**)$ is equivalent to a purely universal formula (and hence a–fortiori to a formula having the form $\forall x \in X \exists y \in K(F(x,y) = 0)$). Thus by theorem 15.1 we only have to analyze the proof of the implication '$(**) \to$ lemma 16.38' which can be carried out in E-PA$^\omega$(+QF-AC^{1,0}+ QF-AC^{0,1} + axioms Γ) if $(\phi_1,\ldots,\phi_n)$ is, provably in E-PA$^\omega$(+QF-AC^{1,0}+ QF-AC^{0,1} + axioms Γ), a Chebycheff system. An unwinding of the proof of $(**)$ or a constructivization of the proof of lemma 16.38 which uses only ε-instances of the nonconstructive intermediate value theorem is neither necessary nor would it help in any way to improve the extracted bound.

In the proof of lemma 16.38 we used the norm $\|\chi\|_\infty$ of the interpolation 'polynomial' χ. Hence for a quantitative version of lemma 16.38 (i.e. for the construction of a Φ_2 satisfying '2^{**}.') we have to give an upper estimate for $\|\chi\|_\infty$ which depends on $q \in \mathbb{Q}_+^*$ only, where $\bigwedge_{i=1}^{n}(x_{i+1} - x_i \geq q)$, but not on the points x_i themselves which is achieved by the majorization technique implicit in the proof of theorem 15.1:

Lemma 16.39. *Let $\underline{\phi} = (\phi_1,\ldots,\phi_n)$ be a Chebycheff system over $[0,1]$ where $\phi_1,\ldots,\phi_n \in C[0,1]$ are definable by closed terms of* E-PA$^\omega$ *and $\underline{\phi}$ is provably in* E-PA$^\omega$+QF-AC^{1,0}+QF-AC^{0,1} *plus (possibly) lemmas of the form Γ a Chebycheff system. Then one can extract from such a proof a function $\delta : \mathbb{Q}_+^* \to \mathbb{Q}_+^*$ (given by a closed term of* E-PA$^\omega$*) satisfying*

$$\text{WE-HA}^\omega + \Gamma_\varepsilon \vdash$$
$$\forall \psi \in H, x_1,\ldots,x_n \in [0,1], q,r \in \mathbb{Q}^*_+ \Big(\bigwedge_{i=1}^{n-1} (x_{i+1} - x_i \geq q) \wedge \bigwedge_{i=1}^{n} \big(|\psi(x_i)| \leq \delta(q)\cdot r\big) \rightarrow \|\psi\|_\infty \leq r \Big).$$

(Here $H := \text{Lin}_{\mathbb{R}}(\phi_1,\ldots,\phi_n)$).

Proof: By the assumption that $\underline{\phi}$ is a Chebycheff system we have (provably in E-PA$^\omega$+QF-AC^{1,0}+ QF-AC^{0,1} + Γ)

$$\forall m \in \mathbb{N}, (x_1,\ldots,x_n) \in N_m \exists k \in \mathbb{N}\big(|\det\big(A(\underline{x})\big)| > 2^{-k}\big),$$

where $N_m := \left\{ (x_1,\ldots,x_n) \in [0,1]^n : \bigwedge_{i=1}^{n-1} (x_{i+1} - x_i \geq \min(\frac{1}{n}, 2^{-m})) \right\}$ and $A(\underline{x})$ is the matrix $\big(\phi_i(x_j)\big)_{1\leq i,j\leq n}$. Since N_m is a (uniformly in the parameter m) E-PA$^\omega$-definable (see chapter 4) compact metric space we can apply theorem 15.1 to obtain a primitive recursive (in the sense of E-PA$^\omega$) function $\beta : \mathbb{N} \to \mathbb{N}$ such that

$$\text{WE-HA}^\omega + \Gamma_\varepsilon \vdash \forall m \in \mathbb{N}, (x_1,\ldots,x_n) \in N_m \Big(|\det\big(A(\underline{x})\big)| > 2^{-\beta(m)}\Big).$$

Using β one can compute $\|A^{-1}(\underline{x})\|$ as a function in $\underline{x}$ and therefore also an $\eta : \mathbb{N} \to \mathbb{N}$ such that $\eta(m) \geq \sup_{\underline{x}\in N_m} \|A^{-1}(\underline{x})\|$, where the norm is given by $\|(a_{ij})\| := \max_{i=1,\ldots,n} \sum_{k=1}^{n} |a_{ik}|$. Let $\psi = c_1\phi_1 + \cdots + c_n\phi_n$, $m \in \mathbb{N}$ be given and suppose that

$$\bigwedge_{i=1}^{n} |\psi(x_i)| \leq \frac{r}{\eta(m)}$$

for an $\underline{x} \in N_m$. Then $\|(c_1,\ldots,c_n)\|_{\max} \leq r$. Let $K \in \mathbb{N}$ be such that $K \geq \max_{i=1,\ldots,n} \|\phi_i\|_\infty$. $\|(c_1,\ldots,c_n)\|_{\max} \leq r$ implies that $\|c_1\phi_1 + \cdots + c_n\phi_n\|_\infty \leq r\cdot n\cdot K$. Hence

$$\delta(m) := (\eta(m)\cdot n \cdot K + 1)^{-1},$$

satisfies the claim for $q = 2^{-m}$. δ can easily be extended to $q \in \mathbb{Q}^*_+$. □

Remark 16.40. (to the proof of lemma 16.39): An explicit formula for the computation of δ in terms of β has been given by D. Bridges in [44] (see lemma 16.51.2 below).

In applications of lemma 16.39 to specific Chebycheff systems one, of course, uses known results such as interpolation formulas etc.:

Example 16.41. Let $H_{\underline{\phi}} := P_n (= \text{Lin}_{\mathbb{R}}(1, X, \ldots, X^n))$. The fact that P_n is a Chebycheff system is easily proved using the interpolation formula of Lagrange used already in the previous section. From the extraction of Φ_4 in that section we see that

$$\delta(q) := \frac{\lfloor \frac{n}{2} \rfloor ! \lceil \frac{n}{2} \rceil !}{n+1} \cdot q^n$$

fulfills lemma 16.39.

Using δ from lemma 16.39, we are now able to define Φ_2:
Claim 1: $\Phi_2(f,l,q) := \delta(l)^2 \cdot q$ fulfills '2**.'.

Proof of Claim 1: Let $\psi \in H$ and suppose that

$$(+) \bigwedge_{i=1}^{n}(x_{i+1} - x_i \geq l) \wedge \bigwedge_{i=1}^{n+1}\left((-1)^i \psi(x_i) > -\delta(l)^2 \cdot q\right).$$

We have to show that $\|\psi\|_\infty \leq q$: Assume that $|\psi(x_{i_0})| > \delta(l) \cdot q$ for an $i_0 \in \{1,\ldots,n+1\}$. Then $(-1)^{i_0}\psi(x_{i_0}) > \delta(l) \cdot q$ since (by $(+)$) $(-1)^{i_0}\psi(x_{i_0}) > -\delta(l)^2 \cdot q \geq -\delta(l) \cdot q$ because of $\delta(l) \leq 1$ ($\delta(l) > 1$ would yield a contradiction when applied to χ below).
Define $\psi_q := \psi + \delta(l)^2 \cdot q \cdot \chi$ where χ is the interpolation 'polynomial' from the proof of lemma 16.38. Then for $i \in \{1,\ldots,n+1\} \setminus \{i_0\}$:

$$(-1)^i \psi_q(x_i) = (-1)^i \psi(x_i) + (-1)^i \delta(l)^2 \cdot q \cdot (-1)^i = (-1)^i \psi(x_i) + \delta(l)^2 \cdot q \overset{(+)}{>} 0$$

and for $i = i_0$

$$\begin{aligned}&(-1)^{i_0}\psi_q(x_{i_0}) = (-1)^{i_0}\psi(x_{i_0}) + (-1)^{i_0}\delta(l)^2 \cdot q \cdot \chi(x_{i_0})\\ &\geq (-1)^{i_0}\psi(x_{i_0}) - \delta(l)^2 \cdot q \cdot \|\chi\|_\infty.\end{aligned}$$

By lemma 16.39 and the definition of χ it follows that $\|\chi\|_\infty \leq \frac{1}{\delta(l)}$.
Hence

$$(-1)^{i_0}\psi_q(x_{i_0}) \geq (-1)^{i_0}\psi(x_{i_0}) - \delta(l) \cdot q > 0.$$

In total we have shown that

$$\bigwedge_{i=1}^{n+1}(-1)^i \psi_q(x_i) > 0$$

which is impossible in view of $(**)$. Hence $\bigwedge_{i=1}^{n+1} |\psi(x_i)| \leq \delta(l) \cdot q$ which in turn yields $\|\psi\|_\infty \leq q$ by lemma 16.39. □

By a refinement of the reasoning above we can improve Φ_2:

Claim 2: $\Phi_2(f,l,q) := \delta(l) \cdot q$ fulfills '2**.'.

Proof of claim 2: Let again $\psi \in H$ and assume

$$\bigwedge_{i=1}^{n}(x_{i+1}-x_i \geq l) \wedge \bigwedge_{i=1}^{n+1}\left((-1)^i\psi(x_i) > -\delta(l)\cdot q\right)$$

and the existence of an $x^* \in [0,1]$ such that $|\psi(x^*)| > q$.
Case 1: $\exists i_0 \in \{1,\ldots,n\} : x^* \in [x_{i_0}, x_{i_0+1}]$.

1.1 $(-1)^{i_0}\psi(x^*) > q$. Consider $\tilde{x}_i := \begin{cases} x_i \text{ if } i \neq i_0 \\ x^* \text{ if } i = i_0. \end{cases}$

Let $\chi \in H$ be such that $\chi(\tilde{x}_i) = (-1)^i$ for $i = 1,\ldots,i_0-1, i_0+1,\ldots,n+1$ and $\psi_q := \psi + \delta(l)\cdot q\cdot\chi$. Then for $i \in \{1,\ldots,n+1\}\setminus\{i_0\}$

$$(-1)^i\psi_q(\tilde{x}_i) = (-1)^i\psi_q(x_i) = (-1)^i\psi(x_i) + (-1)^i\delta(l)\cdot q\cdot(-1)^i > 0$$

and for $i = i_0$

$$\begin{aligned}(-1)^{i_0}\psi_q(\tilde{x}_{i_0}) &= (-1)^{i_0}\psi_q(x^*) = (-1)^{i_0}\psi(x^*) + (-1)^{i_0}\delta(l)\cdot q\cdot\chi(x^*) \\ &\geq (-1)^{i_0}\psi(x^*) - \delta(l)\cdot q\cdot\|\chi\|_\infty > 0 \ \ (\text{since } \|\chi\|_\infty \leq \delta(l)^{-1}).\end{aligned}$$

Put together we have $\bigwedge_{i=1}^{n+1}(-1)^i\psi_q(\tilde{x}_i) > 0$ which is impossible by $(**)$.
1.2: $(-1)^{i_0+1}\psi(x^*) > q$: Analogous to 1.1 with i_0+1 instead of i_0.
Case 2: $x^* \in [0,x_1]$. 2.1 $(-1)^1\psi(x^*) = -\psi(x^*) > q$. Consider

$$\tilde{x}_i := \begin{cases} x_i \text{ for } i = 2,\ldots,n+1 \\ x^* \text{ for } i = 1 \end{cases}$$

and let $\chi \in H$ be such that $\chi(\tilde{x}_i) = (-1)^i$ for $i = 2,\ldots,n+1$. Define

$$\psi_q := \psi + \delta(l)\cdot q\cdot\chi.$$

As in case 1 one shows that $\bigwedge_{i=1}^{n+1}(-1)^i\psi_q(\tilde{x}_i) > 0$ which again yields a contradiction.
2.2 $(-1)^0\psi(x^*) = \psi(x^*) > q$. Consider

$$\tilde{x}_i := \begin{cases} x_i \text{ for } i = 1,\ldots,n \\ x^* \text{ for } i = 0 \end{cases}$$

and let $\chi \in H$ be such that $\chi(\tilde{x}_i) = (-1)^i$ for $i = 1,\ldots,n$. Define $\psi_q := \psi + \delta(l)\cdot q\cdot\chi$.
As in case 1 one shows that $\bigwedge_{i=0}^{n}(-1)^i\psi_q(\tilde{x}_i) > 0$ which again yields a contradiction.
Case 3: $x^* \in [x_{n+1}, 1]$ is treated analogously to case 2.
Cases 1–3 together imply that $\|\psi\|_\infty \leq q$. $\square$

Altogether we have proved that for $\Phi_2(f,l,q) := \delta(l)\cdot q$:

$$\forall \psi_1,\psi_2 \in H, (x_1,\ldots,x_{n+1}) \in [0,1]^{n+1}, j \in \{0,1\}, q,l \in \mathbb{Q}^*$$
$$\Big(\bigwedge_{i=1}^{n} (x_{i+1}-x_i \geq l) \wedge \bigwedge_{i=1}^{n+1} ((-1)^{i+j}(\psi_1(x_i)-\psi_2(x_i)) > -\Phi_2(f,l,q))$$
$$\to \|\psi_1-\psi_2\|_\infty \leq q\Big).$$

Clearly this also holds if '$>$' in the premise is weakened to '$\geq$' (our conclusion $\|\psi_1-\psi_2\|_\infty \leq q$ instead of $\|\psi_1-\psi_2\|_\infty < q$ in '2**.' is sufficient for the verification of Φ).

Extraction of Φ_3: For the construction of Φ_3 we need the following explicit version of the fact that $\tilde{K}_{f,H}$ is compact:

Lemma 16.42. *Let $\underline{\phi} := (\phi_1,\ldots,\phi_n)$ be as in lemma 16.39 and $M \in \mathbb{Q}^*_+$. Then one can construct a closed term of* WE-HA$^\omega$ *which provably in* WE-HA$^\omega + \Gamma_\varepsilon$ *defines a common modulus $\omega_{H,M}$ of uniform continuity on $[0,1]$ for all $\psi \in H$ with $\|\psi\|_\infty \leq M$.*

Proof: Let $\underline{x} = (x_1,\ldots,x_n)$ be defined by $x_i := \frac{i}{n}$ for $i = 1,\ldots,n$ and $K \in \mathbb{N}$ such that $K \geq \|A^{-1}(\underline{x})\|$ (where $\|\cdot\|$ and A are defined as in the proof of lemma 16.39). Then for $\psi = c_1\phi_1 + \ldots + c_n\phi_n$ with $\|\psi\|_\infty \leq M$ it follows that $\|(c_1,\ldots,c_n)\|_{\max} \leq K\cdot M$, i.e. $\bigwedge_{i=1}^{n} (|c_i| \leq K\cdot M)$. Let $\omega_{\underline{\phi}}$ be a common modulus of uniform continuity for $\phi_1,\ldots,\phi_n$ (which can be given by a closed term of WE-HA$^\omega$ as the ϕ_i are given by such closed terms (see proposition 4.23) and define $\omega_{H,M}(q) := \omega_{\underline{\phi}}\left(\frac{q}{n\cdot K\cdot M}\right)$ for $q \in \mathbb{Q}^*_+$. For $\psi = c_1\phi_1 + \ldots + c_n\phi_n$ such that $\|\psi\|_\infty \leq M$, we have

$$|x-y| < \omega_{H,M}(q) \to |\psi(x)-\psi(y)| \leq \sum_{i=1}^{n} |c_i|\cdot|\phi_i(x)-\phi_i(y)|$$
$$< \frac{q}{n\cdot K\cdot M}\sum_{i=1}^{n} |c_i| \leq q.$$

□

Remark 16.43. Again an explicit formula for a modulus $\omega_{H,M}$ in terms of the function β from the proof of lemma 16.39 has been given in [44] and will be presented in lemma 16.51.1.

Corollary 16.44. *Let $\underline{\phi}$ and $\omega_{H,M}$ be as in lemma 16.42, $f \in C[0,1]$, ω_f a modulus of uniform continuity for f and $M \in \mathbb{Q}^*_+$ such that $M \geq \|f\|_\infty$. Then $\omega_{f,H}(q) := \min\left(\omega_f(\frac{q}{2}), \omega_{H,2M}(\frac{q}{2})\right)$ is a common modulus of uniform continuity for all $\psi - f$ where $\psi \in \tilde{K}_{f,H}$ (in particular for $\psi_b - f$ where ψ_b is the best approximation of f in H).*

Example 16.45. Let $H := P_n, \tilde{K}_{f,n} := \{p \in P_n : \|p\|_\infty \leq 2\|f\|_\infty\}$. A very explicit proof of the compactness of $\tilde{K}_{f,n}$ results from the Markov inequality (proposition

16.4). As shown already in the previous section, Markov's inequality can be used to prove that

$$\omega_n^M(q) := \begin{cases} \frac{q}{4n^2M} \text{ if } n \geq 1 \\ 1 \text{ if } n = 0 \end{cases}$$

is a common modulus of uniform continuity on all $p \in \tilde{K}_{f,n}$ whenever $\|f\|_\infty \leq M \in \mathbb{Q}_+^*$. Thus

$$\omega_{f,n}^M(q) := \min\left(\omega_n^M(\frac{q}{2}), \omega_f(\frac{q}{2})\right)$$

is a modulus of uniform continuity for all $p - f$ with $p \in \tilde{K}_{f,n}$ if $M \geq \|f\|_\infty$.

Using $\omega_{f,H}$ from corollary 16.44 we are now able to construct Φ_3:

$$\bigwedge_{i=1}^{n+1} ((-1)^{i+j}(\psi(x_i) - f(x_i)) = E_{n,f} \geq q)$$

implies

$$\bigwedge_{i=1}^{n} \left(|(\psi(x_i) - f(x_i)) - (\psi(x_{i+1}) - f(x_{i+1}))| = 2E_{H,f} \geq 2q\right).$$

Using corollary 16.44, $\psi \in \tilde{K}_{f,H}$ and $\bigwedge_{i=1}^{n}(x_{i+1} - x_i \geq 0)$ this yields $\bigwedge_{i=1}^{n}(x_{i+1} - x_i \geq \omega_{f,H}(2q))$. Hence we can define $\Phi_3(f,q) := \omega_{f,H}(2q)$. Φ_3 satisfies '3**.' (with '$\geq \Phi_3(f,q)$' instead of '$> \Phi_3(f,q)$' which is sufficient for (++)).

We are now ready to combine Φ_1, Φ_2 and Φ_3 into a modulus of uniqueness Φ:

$$\Phi(f,l,q) := \Phi_1\left(f, \Phi_2(f, \Phi_3(f,l), \frac{q}{2})\right) = \frac{1}{2}\delta\left(\omega_{f,H}(2l)\right) \cdot q,$$

where $\delta, \omega_{f,H}$ are from lemma 16.39 resp. corollary 16.44.
The restriction to $\psi_1, \psi_2 \in \tilde{K}_{f,H}$ instead of $\psi_1, \psi_2 \in H$ has been used only for the construction of Φ_3. However as the proof of (++) shows, '3**.' is applied only to the best approximation $\psi_b \in \tilde{K}_{f,H}$ which by (+) exists. Hence Φ is not only a modulus of uniqueness on $\tilde{K}_{f,H}$ but on H.

Theorem 16.46. *Let $\underline{\phi} := (\phi_1, \ldots, \phi_n)$ be a Chebycheff system over $[0,1]$ and $\delta : \mathbb{Q}_+^* \to \mathbb{Q}_+^*$ a function such that for all $\psi \in H := Lin_{\mathbb{R}}(\phi_1, \ldots, \phi_n)$ and $x_1, \ldots, x_n \in [0,1]$*

$$\forall l, q \in \mathbb{Q}_+^* \left(\bigwedge_{i=1}^{n-1}(x_{i+1} - x_i \geq l) \land \bigwedge_{i=1}^{n} |\psi(x_i)| \leq \delta(l) \cdot q \to \|\psi\|_\infty \leq q\right).$$

Furthermore, let $\omega_{f,H}$ be a common modulus of uniform continuity for all $\psi - f$ with $\psi \in H, \|\psi\|_\infty \leq 2\|f\|_\infty$, and $E_{H,f} :=$dist(f,H) where $f \in C[0,1]$. Then the following holds:

1) $\forall f \in C[0,1], l \in \mathbb{Q}^*_+ \Big(l \leq E_{H,f} \to \forall \psi_1, \psi_2 \in H, q \in \mathbb{Q}^*_+$
$\big(\bigwedge_{i=1}^{2} (\|\psi_i - f\|_\infty \leq E_{H,f} + \frac{1}{2}\delta\big(\omega_{f,H}(2l)\big) \cdot q) \to \|\psi_1 - \psi_2\|_\infty \leq q) \Big)$.

In particular, $\delta\big(\omega_{f,H}(2l)\big)$ is a constant of strong unicity (and $\frac{2}{\delta(\omega_{f,H}(2l))}$ is a Lipschitz constant for the Chebycheff projection in f) for f such that $l \leq E_{H,f}$.

2) $\tilde{\Phi}(f,q) := \min\Big(\frac{q}{4}, \frac{1}{2}\delta\big(\omega_{f,H}(\frac{q}{2})\big) \cdot q\Big)$ *is a modulus of uniqueness (and also a modulus of pointwise continuity for the Chebycheff projection) for arbitrary $f \in C[0,1]$ which does not depend on a lower estimate $l \leq E_{H,f}$.*

For every E-PA$^\omega$*-definable Chebycheff system $\underline{\phi}$ provably so in*

$$\text{E-PA}^\omega\text{+QF-AC}^{1,0}\text{+QF-AC}^{0,1} + \Gamma\text{-lemmas},$$

δ and $\omega_{f,H}$ are definable by closed terms of E-PA$^\omega$*. Then 1) and 2) above are provable in* WE-HA$^\omega + \Gamma_\varepsilon$*.*
Instead of $\omega_{f,H}$ one may use any modulus of uniform continuity for $\psi_b - f$, where $\psi_b \in H$ is the best approximation of f.

Proof: 1) From the reasoning above it follows that $\frac{1}{2}\delta\big(\omega_{f,H}(2l)\big) \cdot q$ is a modulus of uniqueness for f such that $l \leq E_{H,f}$. Furthermore if we replace ψ_2 by the best approximation ψ_b then the factor $\frac{1}{2}$ can be omitted (since it is used only in 4^*.). Hence by proposition 16.2 $\delta\big(\omega_{f,H}(2l)\big)$ (resp. $\frac{2}{\delta(\omega_{f,H}(2l))}$) is a constant of strong unicity (resp. local Lipschitz constant).
2) Case 1: $\frac{q}{2} < 2E_{H,f}$: The claim follows from 1). Case 2: $\frac{q}{2} \geq 2E_{H,f}$, i.e. $E_{H,f} \leq \frac{q}{4}$:

$$\bigwedge_{i=1}^{2} (\|\psi_i - f\|_\infty \leq E_{H,f} + \tilde{\Phi}(f,q) \leq \frac{q}{4} + \frac{q}{4}) \to \|\psi_1 - \psi_2\|_\infty \leq q.$$

The second part of the claim follows with the proof of 1) and proposition 16.2.
δ and $\omega_{f,H}$ are given as closed terms of E-PA$^\omega$ by lemma 16.39 and corollary 16.44. The provability (in this case) in WE-HA$^\omega + \Gamma_\varepsilon$ follows using theorem 15.1 and the fact that the alternation theorem (even for general E-PA$^\omega$+WKL+QF-AC^{1,0}+QF-AC^{0,1} $+\Gamma$-provable Chebycheff systems) is provable in

$$\text{E-PA}^\omega\text{+WKL+QF-AC}^{1,0}\text{+QF-AC}^{0,1} + \Gamma.$$

□

Corollary to the proof of theorem 16.46: Instead of $\omega_{f,H}(2l)$ one can use also any rational lower bound $0 < \alpha \leq \omega_{f,H}(2l)$.

Corollary 16.47. 1) *Let $A \subset C[0,1]$ be totally bounded with a common modulus of uniform continuity ω_A and $M \geq \|f\|_\infty$ for all $f \in A$. Let $\omega_{H,2M}$ be the modulus from lemma 16.42 and define*

$$\omega_{A,H}(q) := \min\left(\omega_A(\frac{q}{2}), \omega_{H,2M}(\frac{q}{2})\right).$$

Then $\omega_{A,H}$ is a common modulus of uniform continuity for all $\psi - f$ with $f \in A, \psi \in H, \|\psi\|_\infty \leq 2\|f\|_\infty$. Hence $\min\left(\frac{q}{4}, \frac{1}{2}\delta\left(\omega_{A,H}(\frac{q}{2})\right) \cdot q\right)$ is a common modulus of uniqueness (and a common modulus of pointwise continuity for the Chebycheff projection in f) for all $f \in A$.

2) *Let $A \subset C[0,1]$ be compact with $A \cap H = \emptyset$, which implies that $\inf\limits_{f\in A} E_{H,f} > 0$, and let $0 < l \leq \inf\limits_{f\in A} E_{H,f}$. Then theorem 16.46 yields that $\delta\left(\omega_{A,H}(2l)\right)$ (resp. $2/\delta\left(\omega_{A,H}(2l)\right)$) is a common constant of strong unicity (resp. Lipschitz constant for the Chebycheff projection in f) for all $f \in A$.*

Remark 16.48. The 2nd claim in corollary 16.47 is an effective version of a result by [153].

Using example 16.41 and 16.45 we obtain for the special Haar space P_n (observing that for $n = 0$ the trivial modulus of uniqueness $q/2$ applies):

Corollary 16.49. *Let Φ be defined by*

$$\Phi(f,n,l,M) := \frac{\lfloor\frac{n}{2}\rfloor!\lceil\frac{n}{2}\rceil!}{2(n+1)}\left(\omega^M_{f,n}(2l)\right)^n, \text{ where}$$

$$\omega^M_{f,n}(q) := \begin{cases} \min\left(\omega_f(\frac{q}{2}), \frac{q}{8n^2M}\right), & \text{if } n \geq 1 \\ 1, & \text{if } n = 0. \end{cases}$$

Then

1) $\forall f \in C[0,1], l, M \in \mathbb{Q}^*_+ \Big(\|f\|_\infty \leq M \wedge l \leq E_{n,f} \to \forall p_1, p_2 \in P_n, q \in \mathbb{Q}^*_+$

$$\big(\bigwedge_{i=1}^{2} (\|f - p_i\|_\infty \leq E_{n,f} + (\Phi(f,n,l,M)) \cdot q) \to \|p_1 - p_2\|_\infty \leq q)\Big).$$

In particular $2\Phi(f,n,l,M)$ (resp. $\frac{1}{\Phi(f,n,l,M)}$) is a constant of strong unicity (a Lipschitz constant for the projection) for f if $\|f\|_\infty \leq M$ and $l \leq E_{n,f}$.

2) $\tilde{\Phi}(f,n,M,q) := \min\left(\frac{q}{4}, (\Phi(f,n,\frac{q}{4},M)) \cdot q\right)$ *is a modulus of uniqueness (and a modulus of pointwise continuity for the projection) for arbitrary $f \in C[0,1]$ with $\|f\|_\infty \leq M$ (note that $\tilde{\Phi}$ no longer depends on l).*

If f is Lipschitz continuous on $[0,1]$ with Lipschitz constant $\eta > 0$ then $\omega^M_{f,n}$ can be defined as $\omega^M_{f,n}(q) := \frac{q}{\eta + 4n^2M}$.

We now show that even the remaining dependency of $\tilde{\Phi}$ in corollary 16.49 from $f \in C[0,1]$ via ω_f and M can be further reduced to a dependency from ω alone:

Proof of theorem 16.34: For $f \in C[0,1]$ and $p_1, p_2 \in P_n$ define $\tilde{f}(x) := f(x) - f(0)$ and $\tilde{p}_i(x) := p_i(x) - f(0)$. Then

$$\|\tilde{f} - \tilde{p}_i\|_\infty = \|f - p_i\|_\infty \wedge \|\tilde{p}_1 - \tilde{p}_2\|_\infty = \|p_1 - p_2\|_\infty \wedge E_{n,\tilde{f}} = E_{n,f}.$$

Moreover, any modulus of uniform continuity ω for f is also a modulus of uniform continuity for $\tilde{f}$. Hence we can take as moduli of uniqueness for $f \in C[0,1]$ not only $\Phi(f,n,l,M) \cdot q$ and $\tilde{\Phi}(f,n,M,q)$ but also $\Phi(\tilde{f},n,l,M) \cdot q$ and $\tilde{\Phi}(\tilde{f},n,M,q)$ for $M \geq \|\tilde{f}\|_\infty$. However, one easily verifies that $\|\tilde{f}\|_\infty \leq \lceil \frac{1}{\omega(1)} \rceil$ which concludes the proof. □

Remark 16.50. 1) Inspection of the logical proof-analysis above yields the following result which, essentially, is due to Cline [68] (Theorem 5) (combined with e.g. Theorem 3.5 in [313]) whose proof is quite different:
Let $(x_1, \ldots, x_{n+1}) \in [0,1]^{n+1}$ be an alternant of $\psi_b - f$ and, for $i = 1, \ldots, n+1$, χ_i the uniquely determined function in H such that $\chi_i(x_j) = (-1)^j$ for $j \in \{1, \ldots, i-1, i+1, \ldots, n+1\}$. Then $\frac{1}{\max\limits_{i=1,\ldots,n+1} \|\chi_i\|_\infty}$ (resp. $2 \cdot \max\limits_{i=1,\ldots,n+1} \|\chi_i\|_\infty$) is a constant of strong unicity (Lipschitz constant). Note, however, that these estimates use the knowledge of alternation points which in general are not computable at all (since – as we saw in the previous section – already for $H := P_{n-1}$ and $n = 1$ the existence of alternation points implies (A) and hence WKL).

2) The logical analysis of the uniqueness proof by Young/Rice given above has some similarities with the proof of the Lipschitz continuity of the Chebycheff projection by G. Freud in [106] mentioned already in section 16.1. In fact, Freud's proof may be conceived of as a (partial) proof-analysis in our sense. Although Freud himself does not exhibit the numerical content of his proof one can extract the (ineffective) estimate of [68] (mentioned in '2)' above) for the Lipschitz constant from his proof as was observed by Blatt in [37]. Blatt also noticed that a slight modification of Freud's proof yields the corresponding estimate for the constant of strong unicity although the concept of strong unicity is not even mentioned in Freud's paper.

3) Let us compare the estimates in corollary 16.49 with the ones obtained from de La Vallée Poussin's proof in the previous section: If $\gamma_{Y/R}$ (resp. γ_P) denotes the constant of strong unicity obtained from Young/Rice's (resp. de La Vallée Poussin's) proof then $\gamma_{Y/R}$ is roughly $\sqrt[2]{\gamma_P}$ and similarly for the other estimates. So the bounds obtained by logical analysis from the proof by Young/Rice are (when specialized to the polynomial case) significantly better than the ones extracted from de La Vallée Poussin's proof.

4) If the moduli δ and ω_ϕ (used in the definition of $\omega_{H,2M}$) are given as functions $\mathbb{R}^*_+ \to \mathbb{R}^*_+$ (rather than $\mathbb{Q}^*_+ \to \mathbb{Q}^*_+$) then the moduli constructed in the theorem above apply to all $q \in \mathbb{R}^*_+$ with values in $\mathbb{R}^*_+$ as well.

A variant uniqueness proof due to E. Borel [39]:

Even prior to the uniqueness proof given by Young/Rice (and analyzed above) yet another uniqueness proof can be found in Borel [39], though only for the special case $H := P_{n-1}$. This proof proceeds similarly to the one by Young/Rice (in particular the alternation theorem again is used only in the form of an axiom Γ in theorem 15.1) but uses a different argument for lemma 16.38: in fact, lemma 16.38 for P_n can simply be followed from the fundamental theorem of algebra (this form of the uniqueness proof appeared first in the unpublished part of Kirchberger's dissertation [183]). However, in his proof for the continuity of the Chebycheff projection (which can also be used to establish even the local Lipschitz continuity) Borel uses a more explicit argument: the derivative p' of $p \in P_n$ again belongs to a Haar space, namely P_{n-1}. From this, Borel concludes the following (exercise):

$$\begin{cases} \forall p \in P_n, (x_1,\ldots,x_{n+2}) \in [0,1]^{n+2}, q \in \mathbb{Q}^*_+ \Big(\bigwedge_{i=1}^{n+1} (x_{i+1} > x_i) \wedge \bigwedge_{i=1}^{n+2} ((-1)^i p(x_i) > -q) \\ \qquad\qquad \to \exists k \in \{1,\ldots,n+1\} \forall x \in [x_k, x_{k+1}] \big(|p(x)| \le q\big)\Big). \end{cases} \tag{16.1}$$

If we let q tend to 0 this yields lemma 16.38, since $p|_{[x_k,x_{k+1}]} \equiv 0$ implies $p \equiv 0$. The formula above is (modulo the Σ^0_1-collection principle Σ^0_1-CP if n is treated as a variable) equivalent to a statement of the logical form $\Gamma \equiv \forall x \in X \exists y \in K \forall w \in W\big(F(x,y,w) = 0\big)$ from theorem 15.1. Hence theorem 15.1 guarantees that it is possible to extract a bound Φ_2 (satisfying 2**.) without using the proof of 16.1. Indeed, consider

$$x_{k+1} - x_k \ge l \to \exists y_1,\ldots,y_{n+1} \in [x_k, x_{k+1}]\big(y_{i+1} - y_i \ge \frac{l}{n}\big) \text{ for } i = 1,\ldots,n+1.$$

Taken together with the bound δ from example 16.41 one concludes that

$$\Phi_2(f,n,l,q) := \frac{\lfloor \frac{n}{2} \rfloor! \lceil \frac{n}{2} \rceil!}{n+1} \big(\frac{l}{n}\big)^n \cdot q = \frac{\lfloor \frac{n}{2} \rfloor! \lceil \frac{n}{2} \rceil!}{n^n \cdot (n+1)} l^n \cdot q$$

fulfills '2**.' Due to the presence of the factor $\frac{1}{n^n}$ the bound extracted from Borel's argument (for the special case of $H := P_{n-1}$) is less good less than the modulus of uniqueness obtained by logical analysis of our simplification of Young/Rice's proof. However, the bound is better than the one obtained from de La Vallée Poussin's proof in the previous section.

We conclude this section by comparing theorem 16.46 and corollary 16.47 with prior results due to D. Bridges in [44, 45, 46]. It will turn out that our results can be used to significantly improve the bounds obtained by Bridges whose approach is based on a thorough constructivization of large parts of Chebycheff approximation theory:
Let $\underline{\phi} := (\phi_1,\ldots,\phi_n)$ be a Chebycheff system over $[0,1]$, $\underline{\phi}(x) := \big(\phi_1(x),\ldots,\phi_n(x)\big) \in \mathbb{R}^n$, $\|\underline{\phi}\| := \sup_{x\in[0,1]} \|\underline{\phi}(x)\|_2$. Here $\|\cdot\|_2$ denotes the Euclidean norm on $\mathbb{R}^n$.
We now define $\beta, \gamma, \kappa : (0, \frac{1}{n}] \to \mathbb{R}^*_+$ as follows

$$\beta(\alpha) := \begin{cases} \inf\limits_{x\in[0,1]} |\phi_1(x)|, \text{ if } n = 1 \\ \inf\left\{ |\det(\phi_j(x_i))| : 0 \le x_1, \ldots, x_n \le 1, \bigwedge\limits_{i=1}^{n-1} (x_{i+1} - x_i \ge \alpha) \right\}, \text{ if } n > 1 \end{cases}$$

and

$$\gamma(\alpha) := \min\left(\|\underline{\phi}\|, \frac{\beta(\alpha)}{n^{\frac{1}{2}}(n-1)! \prod\limits_{i=1}^{n}(1+\|\phi_i\|_\infty)} \right), \ \kappa(\alpha) := \gamma(\alpha)^{-1} \cdot \|\underline{\phi}\|$$

for $\alpha \in (0, \frac{1}{n}]$. Note that the fact that ϕ is a Chebycheff system implies that $\beta(\alpha) > 0$. Now define $H :=\mathrm{Lin}_{\mathbb{R}}(\phi_1, \ldots, \phi_n)$ and let $\omega_{\underline{\phi}}$ denote a modulus of uniform continuity of $\underline{\phi}$.

Lemma 16.51 (Bridges ([44, 45])).

1) *Suppose that $A \subset C[0,1]$ is totally bounded and equicontinuous with a common modulus of uniform continuity ω_A for all $f \in A$. Let $M > 0$ be a common bound $M \ge \|f\|_\infty$ for all $f \in A$. Then*

$$\omega_{A,H}(\varepsilon) := \min\left(\omega_A(\frac{\varepsilon}{2}), \omega_{\underline{\phi}}\left(\frac{\varepsilon \cdot \beta(\frac{1}{n})}{4Mn^{\frac{3}{2}}(n-1)! \prod\limits_{i=1}^{n}(1+\|\phi_i\|_\infty)} \right) \right)$$

is a common modulus of uniform continuity for all $\psi_b - f$ where $f \in A$ and ψ_b is the best approximation of f in H. In fact, $\omega_{A,H}$ is a common modulus of uniform continuity for all $\psi - f$ with $f \in A$ and $\|\psi\|_\infty \le 2M$.

2) *Suppose that $0 < \alpha \le \frac{1}{n}$ and $\bigwedge\limits_{i=1}^{n-1} (x_{i+1} - x_i \ge \alpha)$ $(x_1, \ldots, x_n \in [0,1])$ for $n \ge 2$. Then*

$$\forall \psi \in H, \varepsilon > 0 \Big(\bigwedge_{i=1}^{n} |\psi(x_i)| \le \frac{\gamma(\alpha)}{n \cdot \|\underline{\phi}\|} \cdot \varepsilon \to \|\psi\|_\infty \le \varepsilon \Big).$$

Proof: 1) Is proved in Bridges [45] (Lemma). 2) is proved in Bridges [44] (4.3).

Theorem 16.46 taken together with corollary 16.47 and lemma 16.51 yields the following moduli of uniqueness and pointwise continuity as well as common constants of strong unicity and Lipschitz constants:

Corollary 16.52. *Let A, γ, κ be as in lemma 16.51 and $E_{H,A} := \inf\limits_{f\in A} E_{H,f}$. With $E_{H,f}$ being replaced by $E_{H,A}$, $\omega_{f,H}$ by $\omega_{A,H}$ and $\delta(\alpha)$ by $\frac{\gamma(\alpha)}{n\cdot\|\underline{\phi}\|}$, theorem 16.46 holds uniformly for all $f \in A$. This, in particular, yields that*

$$\Phi_A(\varepsilon) :=$$
$$\min\left(\frac{\varepsilon}{4}, \frac{1}{2}\frac{\gamma\left(\min\left(\frac{1}{n}, \omega_{A,H}(\frac{\varepsilon}{2})\right)\right)}{n\cdot\|\underline{\phi}\|}\cdot\varepsilon\right) = \min\left(\frac{\varepsilon}{4}, \frac{\varepsilon}{2n\kappa\left(\min\left(\frac{1}{n}, \omega_{A,H}(\frac{\varepsilon}{2})\right)\right)}\right)$$

is a common modulus of uniqueness (and a common modulus of pointwise continuity for the Chebycheff projection in f) for all $f \in A$.
For $l_{H,A} \in \mathbb{Q}^*_+$ such that $l_{H,A} < E_{H,A}$ and $0 < \alpha \leq \min\left(\frac{1}{n}, \omega_{A,H}(2\cdot l_{H,A})\right)$ we obtain that $\frac{\gamma(\alpha)}{n\cdot\|\underline{\phi}\|}$ (resp. $2n\kappa(\alpha)$) is a common (for all $f \in A$) constant of strong unicity (resp. Lipschitz constant).

Using again the transformation $f \mapsto \tilde{f}$, with $\tilde{f}(x) := f(x) - f(0)$, an argument similar to the one used in the proof of theorem theorem 16.34 allows one to conclude:

Theorem 16.53. *Let $(\phi_1,\ldots,\phi_n)$ be a Chebycheff system such that $1 \in H :=Lin_{\mathbb{R}}(\phi_1,\ldots,\phi_n)$ and let $\omega : \mathbb{R}^*_+ \to \mathbb{R}^*_+$ be any function. Then*

$$\Phi_H(\omega,\varepsilon) := \min\left(\frac{\varepsilon}{4}, \frac{\varepsilon}{2n\kappa\left(\min\left(\frac{1}{n}, \omega^H(\frac{\varepsilon}{2})\right)\right)}\right)$$

with

$$\omega^H(\varepsilon) := \min\left(\omega(\frac{\varepsilon}{2}), \omega_{\underline{\phi}}\left(\frac{\varepsilon\cdot\beta(\frac{1}{n})}{4\lceil\frac{1}{\omega(1)}\rceil n^{\frac{3}{2}}(n-1)!\prod_{i=1}^{n}(1+\|\phi_i\|_\infty)}\right)\right)$$

is a common modulus of uniqueness (and a common modulus of pointwise continuity for the Chebycheff projection) for all functions $f \in C[0,1]$ which have ω as a modulus of uniform continuity.

As a corollary we obtain that for arbitrary Haar spaces having the constant function 1 the continuity behavior of the Chebycheff projection is uniform for any class of equicontinuous functions which generalizes a result of [255] for the case of (trigonometric) polynomials.

In [44, 45], D. Bridges obtained the following bounds which are numerically significantly less good than the bounds in corollary 16.52 and theorem 16.53 which were extracted by proof theoretic analysis of the classical uniqueness proof from [381] and [313]: let $0 < l_{H,A} \leq E_{H,A}$ and $0 < \alpha \leq \min\left(\frac{1}{n}, \omega_{A,H}(l_{H,A})\right)$. Bridges shows that $n^{-2}\left(\frac{\gamma(\alpha)}{\|\underline{\phi}\|}\right)^{2n+1}$ resp. $2n\kappa(\alpha)\cdot\left(\sum_{i=1}^{n+1}\kappa(\alpha)^{n+i-1}-1\right)$ is a common constant of strong unicity resp. a Lipschitz constant for all $f \in A$. As a modulus of continuity for the corresponding Chebycheff projection (on A), Bridges obtains

$$\Omega(\varepsilon) := \min\left(\frac{\varepsilon}{8}, \frac{\varepsilon}{2n\sigma(\varepsilon)\cdot\left(\sum\limits_{i=1}^{n+1}\sigma(\varepsilon)^{n+i-1}-1\right)}\right),$$

where $\sigma(\varepsilon) := \kappa\big(\min\big(\frac{1}{n}, \omega_{A,H}(\frac{\varepsilon}{4})\big)\big)$. In [46], Bridges also constructs a full modulus of uniqueness which, however, is even more complex than Ω. Since, in practice, $\frac{\gamma(\alpha)}{\|\phi\|}(\leq 1)$ will be very close to 0 (for the concrete case of P_{n-1} Bridges takes $\beta(\alpha)$ to be $\alpha^{n(n-1)/2}$), it is clear that the moduli and constants in corollary 16.52 provide a significant improvement compared to those obtained by Bridges in the course of his global constructivization of Chebycheff approximation theory. The bounds in corollary 16.52 yield **effective** moduli if β can be estimated by a computable function $\tilde{\beta} : \mathbb{Q}^*_+ \to \mathbb{Q}^*_+$ such that $0 < \tilde{\beta}(\alpha) \leq \beta(\alpha)$ for all $\alpha \in \mathbb{Q}^*_+$. If ϕ is a Chebycheff system as in lemma 16.39 then such a function $\tilde{\beta}$ can be primitive recursively (in the sense of Gödel) constructed (see the proof of lemma 16.39). In practice it will be better to extract the operation δ in lemma 16.39 directly from a proof of the Chebycheff system property instead of extracting first $\tilde{\beta}$ and then constructing γ above via $\tilde{\beta}$ (e.g. for $H := P_n$ our δ is roughly α^n while $\tilde{\beta}$ in this case is already $\alpha^{(n+1)n/2}$).

From [198] (proofs of thm. 4.1 and lemma 4.2) or [199] (proofs of thm. 8.30 and lemma 8.29) one can also exhibit a modulus of uniqueness Φ for the special case P_n namely:
$\Phi(f,n,q) := \frac{1}{8(n+1)^2}\cdot\big(\omega_{f,n}(\frac{q}{2})\big)^{2n}\cdot q$, where $\omega_{f,n}$ denotes a modulus of uniform continuity for all $p-f$ where $\|p\|_\infty \leq 2\|f\|_\infty$ ($f \in C[0,1], p \in P_n$). Ko's proof is similar to the analysis above but based on our first crude analysis of the proof of lemma 16.38 as formulated in Claim 1 (compare with the proof of lemma 4.2 in [198]). Note, however, that Ko erroneously uses simply a modulus ω_n for all $p \in P_n$ such that $\|p\|_\infty \leq 2\|f\|_\infty$ instead of $\omega_{f,n}$. For the latter he takes $\omega_n(q) := q/(4\|f\|_\infty \cdot n^{2n}\cdot n^2)$ compared to our $\omega_n(q) := q/(4\|f\|_\infty \cdot n^2)$ which we used in defining $\omega_{f,n}$. Neither the continuity of the Chebycheff projection nor estimates on strong unicity are explicitly considered in [198, 199] but estimates on those similar to Ko's modulus of uniqueness could be obtained as well.

The numerical estimates given by Bridges are derived as a by-product of his development of Chebycheff approximation theory within the framework constructive analysis in the sense of E. Bishop [32]. Due to this approach even proofs of lemmas which do not contribute to the bounds of the theorem in question are proven in a sometimes more complicated way than necessary classically. Moreover, occasionally, obvious – but partially ineffective – arguments like our proof of '2.' based on the nonconstructive intermediate value theorem (IVT) are not used in favor of constructive arguments based on induction. The latter may well result in less good bounds (e.g. the exponent $2(n+1)$ in Bridges' estimate for the constant of strong unicity is caused by this) while the use of IVT is totally unproblematic from the point of view of 'proof mining' since it has the form of an axiom Γ, i.e.

$$\forall x \in X \exists y \in K(\forall w \in W)\big(F(x,y,w) =_{\mathbb{R}} 0\big),$$

and so does not contribute to extractable bounds at all (see theorem 15.1). In fact, we showed that the whole use of the alternation theorem can be reduced to such an axiom Γ (if one follows the proof due to Young and Rice). As a result of this, only a small part of the corresponding uniqueness proof needed to be analyzed at all. A logical treatment of proofs of lemmas of the form Γ is superfluous as it is to circumvent them by new constructive arguments. This is all the more true as even if one is interested in a fully constructive verification in the end one can just apply our metatheorem 15.1 to get the constructive verification of the final result guaranteed.

16.4 Best L_1-approximation

The material in this section is taken largely from Kohlenbach-Oliva [235]. We analyze the proof of the uniqueness of best polynomial L_1-approximation of $f \in C[0,1]$ given by Cheney [65]. The main result will be the following strong uniqueness theorem:

Theorem 16.54 (Kohlenbach-Oliva [235]). *Let* $dist_1(f,P_n) := \inf_{p\in P_n} \|f-p\|_1$ *and*

$$\begin{aligned}
&\Phi(\omega,n,\varepsilon) := \min\{\tfrac{c_n\varepsilon}{8(n+1)^2}, \tfrac{c_n\varepsilon}{2}\omega_n(\tfrac{c_n\varepsilon}{2})\}, \textit{ where}\\
&c_n := \tfrac{\lfloor n/2\rfloor!\lceil n/2\rceil!}{2^{4n+2}(n+1)^{3n+1}} \textit{ and}\\
&\omega_n(\varepsilon) := \min\{\omega(\tfrac{\varepsilon}{4}), \tfrac{\varepsilon}{40(n+1)^4\lceil\frac{1}{\omega(1)}\rceil}\}.
\end{aligned}$$

The functional Φ is a uniform modulus of uniqueness for the best L_1-approximation of any function f in $C[0,1]$ having modulus of uniform continuity ω from P_n, i.e.

$$\begin{aligned}
&\forall n \in \mathbb{N}, p_1, p_2 \in P_n, \varepsilon \in \mathbb{Q}^*_+\\
&\quad\big(\bigwedge_{i=1}^{2}(\|f-p_i\|_1 - dist_1(f,P_n) < \Phi(\omega,n,\varepsilon)) \to \|p_1-p_2\|_1 \le \varepsilon\big).
\end{aligned}$$

The uniqueness for the best polynomial L_1-approximation of continuous functions ('Jackson's theorem') was first proved in [178]. The proof analyzed in [235] is a simplification of that proof due to [65] which avoids the use of measure theory but relies on WKL. [35, 36] proved the existence of a modulus of uniform continuity of the form $c_{f,n}\varepsilon\omega_f(c_{f,n}\varepsilon)$ with an (unknown) constant $c_{f,n}$ depending on f,n. [254] shows the existence of such constants $c_{\omega_f,n}$ depending only on a modulus of uniform continuity ω_f of f and n and showed that the ε-dependency in the result of [35] is optimal. Our bound above is the first explicit modulus (in all parameters). Moreover, it also depends only on ω and has the optimal ε-dependency.

The remainder of this section is devoted to the proof of theorem 16.54 as obtained in [235] by a logical analysis of the Cheney's ([65]) proof of Jackson's theorem.

16.4.1 Logical preliminaries on Cheney's proof

Cheney's proof is formulated for the best L_1-approximation of $f \in C[0,1]$ by elements of a general Haar space. We will only treat the most important case of polynomial approximation i.e. of the best L_1-approximation of f by polynomials $p \in P_n$ of degree $\leq n$. The most complicated part will be the logical analysis of the following lemma (label 'Lemma 1' in Cheney's proof, [66], p. 219) whose formalization prima facie seems to require a decision functional for $=_{\mathbb{R}}$ and hence the functional $E^2 \leq \lambda f^1.1$ defined by

$$\forall f^1 (E(f) =_0 0 \leftrightarrow \forall x^0 (f(x) =_0 0)).$$

The functional interpretation of this axiom would require the non-majorizable functional μ^2 defined by

$$\mu(f) := \begin{cases} \min x[f(x) =_0 0], \text{ if existent} \\ 0^0, \text{ otherwise.} \end{cases}$$

However, it will turn out that a slight reformulation of this lemma can – in the way it is used in the uniqueness proof – in fact be proved in e.g. E-PA$^\omega$ plus the already mentioned analytical principle (15.1), i.e.

$$(A)\ \forall f \in C[0,1] \exists x_0 \in [0,1] (f(x_0) = \sup_{x \in [0,1]} f(x))$$

and hence in E-PA$^\omega$+WKL (or – much more conveniently – in E-PA$^\omega + \Sigma^0_1$-UB as studied in chapter 12).

Lemma 16.55 (Cheney [66], Lemma 1). *Let $f, h \in C[0,1]$. If f has at most finitely many roots and if $\int_0^1 h\,sgn(f) \neq 0$, then for some $\lambda \in \mathbb{R}$, $\int_0^1 |f - \lambda h| < \int_0^1 |f|$, where*

$$sgn(f)(x) \stackrel{\mathbb{R}}{=} \begin{cases} 1, \textit{if } f(x) >_{\mathbb{R}} 0 \\ 0, \textit{if } f(x) =_{\mathbb{R}} 0 \\ -1, \textit{if } f(x) <_{\mathbb{R}} 0. \end{cases}$$

In the restriction of Cheney's proof to the Haar space P_n (i.e. to Jackson's theorem), $h \in P_n$ is a polynomial of degree $\leq n$. Moreover, it will be used in the form stating that if f (for the particular f at hand) has at most n roots one can construct an h such that $\int_0^1 h\,sgn(f) \neq 0$. This means that the lemma is applied only in the case where f not just has at most finitely many but in fact not more than n roots. At first glance, the existence of $sgn(f)$ seems to presuppose the existence of the characteristic function $\chi_{=_\mathbb{R}}$ which is not available in systems amenable to monotone functional interpretation. The use of sgn can be eliminated, however, in favor of a finite tuple of sign values: if f has at most n roots then there exist points $x_0 < \ldots < x_{n+1}$ in $[0,1]$ (where $x_0 = 0$ and $x_{n+1} = 1$) which contain all the roots of f. By induction on n and the law-of-excluded middle one easily shows in E-PA$^\omega$ the existence of a vector $(\sigma_1,\ldots,\sigma_{n+1}) \in \{-1,1\}^{n+1}$ such that

$$\sigma_i =_0 \begin{cases} 1, \text{ if } f \text{ is positive on } (x_{i-1},x_i), \\ -1, \text{ if } f \text{ is negative on } (x_{i-1},x_i) \end{cases}$$

for $i = 1,\ldots,n+1$. Using this vector $\underline{\sigma}$, the integral $\int_0^1 h\,sgn(f)$ can be written as $\sum_{i=1}^{n+1} \sigma_i \int_{x_{i-1}}^{x_i} h$ (which avoids the use of $sgn(f)$) and the proof of the lemma can be carried out in E-PA$^\omega + (A)$, i.e. in E-PA$^\omega$+WKL (the principle (A) is used in the middle of p. 219 of [66] to conclude that '$\delta > 0$'; see also section 16.4.10 and section 16.4.14).
The logical quantitative analysis of this reformulation of lemma 1 will be the most difficult part in our analysis of Cheney's proof (see 16.4.10). In particular, the monotone functional interpretation of (the negative translation of) this version of Lemma 1 will automatically introduce a key notion for the quantitative analysis of the proof, namely the concept of so-called 'r-clusters of δ-roots'. Moreover, it is this concept, on which the elimination of the use of (A) – i.e. of **WKL** – in Cheney's proof of lemma 1 will be based.

16.4.2 The general logical structure of the proof

As a preparatory step towards the logical analysis of Cheney's proof we now study the overall logical structure of the proof. To this end we make a list the main formulas used in the proof and to show how they are combined into various lemmas. The lemmas will then be analyzed subsequently once the overall logical structure of the proof has become clear. The monotone functional interpretation of the lemmas shows which functionals *can be* can be extracted from the proof of the lemma. As usual not all the functionals need to be presented, since some of them will disappear in the analysis of the proof. This can be seen already in the treatment of the modus ponens in the proof of the soundness of functional interpretation in chapter 8 where the terms $\underline{t}_2$ disappeared in the conclusion. For the individual lemmas we will use monotone functional interpretation mainly in the form of the 'macro' provided by the logical metatheorem 15.1. However, by analyzing the structure of the whole proof by monotone functional interpretation we see which functionals are relevant and *need to be* extracted and the treatment of modus ponens by monotone functional will tell us how to combine them into the final modulus Φ (see section 16.4.13).

For better readability, we below omit the parameters f, n, p_1 and p_2 in our list of propositions A-K. This means that A strictly speaking has to be read as $A(f, n, p_1, p_2)$, where n ranges over $\mathbb{N}$, $f \in C[0,1]$ and $p_1, p_2 \in P_n$ (resp. $K_{f,n}$ further below). Analogously, this applies also to the other propositions.

For the rest of this section we make the following notational conventions: $p(x) := \frac{p_1(x)+p_2(x)}{2}$ and $f_0(x) := f(x) - p(x)$. In the formulas and in the sketch of the proof presented below we use $\overline{x} := x_1, \ldots, x_n$ and $\overline{\sigma} := \sigma_1, \ldots, \sigma_{n+1}$. The following formulas play a key role in Cheney's proof:

$A :\equiv \bigwedge\limits_{i=1}^{2} (\|f - p_i\|_1 - dist_1(f, P_n) = 0)$, i.e.
p_1 and p_2 are best L_1-*approximations* of f from P_n.

$B :\equiv \|f - p\|_1 - dist_1(f, P_n) = 0$, i.e. p is a best L_1-*approximation* of f.

$C :\equiv \|f_0\|_1 = \frac{1}{2}\|f - p_1\|_1 + \frac{1}{2}\|f - p_2\|_1$.

$C_1 :\equiv \forall \varepsilon \in \mathbb{Q}_+^* \exists \delta \in \mathbb{Q}_+^* \forall x, y \in [0,1](|x - y| < \delta \rightarrow |g(x) - g(y)| < \varepsilon)$,
where $g(x) := |f_0(x)| - \frac{1}{2}|f(x) - p_1(x)| - \frac{1}{2}|f(x) - p_2(x)|$.
The formula C_1 states that g is uniformly continuous.

$D :\equiv \forall x \in [0,1](|f_0(x)| = \frac{1}{2}(|f(x) - p_1(x)| + |f(x) - p_2(x)|))$.

$E :\equiv \exists x_0, \ldots, x_n \in [0,1]\big(\bigwedge\limits_{i=0}^{n} f_0(x_i) = 0 \wedge \bigwedge\limits_{i=1}^{n} x_{i-1} < x_i\big)$, i.e.
f_0 has at least $n+1$ distinct roots.

$F :\equiv \exists x_0, \ldots, x_n \in [0,1]\big(\bigwedge\limits_{i=0}^{n} p_1(x_i) = p_2(x_i) \wedge \bigwedge\limits_{i=1}^{n} x_{i-1} < x_i\big)$, i.e.
p_1 and p_2 coincide on at least $n+1$ distinct points.

$G :\equiv \forall x \in [0,1](p_1(x) = p_2(x))$, alternatively, $\|p_1 - p_2\|_1 = 0$ or $p_1 \equiv p_2$.

$H(h) :\equiv \|f_0 - h\|_1 \geq \|f_0\|_1$.

$I(\overline{x}, \overline{\sigma}, h) :\equiv \sum\limits_{i=1}^{n+1} \sigma_i \int_{x_{i-1}}^{x_i} h(x)dx > 0$, where $x_0 := 0$ and $x_{n+1} := 1$.

$J(\overline{x}) :\equiv \exists y \in [0,1](f_0(y) = 0 \wedge \bigwedge\limits_{i=0}^{n+1} x_i \neq y)$, where $x_0 := 0$ and $x_{n+1} := 1$.

$K :\equiv \forall x \in [0,1](f_0(x) = 0 \rightarrow p_1(x) = p_2(x))$.

The first part of the proof (which we call derivation $\mathscr{D}_1$) derives K from the assumption A (later to be discharged) and the lemmas '$A \rightarrow B$', '$A \wedge B \rightarrow C$', 'C_1', '$C \wedge C_1 \rightarrow D$' and '$D \rightarrow K$':

$$\dfrac{\dfrac{\dfrac{\dfrac{\dfrac{[A] \quad \dfrac{[A] \quad A \rightarrow B}{B}}{A \wedge B} \quad A \wedge B \rightarrow C}{C} \quad C_1}{C \wedge C_1} \quad C \wedge C_1 \rightarrow D}{D} \quad D \rightarrow K}{K}$$

The logically most complicated part of the proof is where – using lemma 1 – the existence of (at least) $n+1$ distinct roots of f_0 is established. In the following derivations we use $\overline{\sigma}' := \sigma_1', \ldots, \sigma_{n+1}'$, where $\sigma_i' := sgn(f_0)(\frac{x_{i-1}+x_i}{2})$. Moreover, $\forall \overline{x} := \forall x_1 \leq \ldots \leq x_n$, where $\forall x_1 \leq \ldots \leq x_n\ Q(\overline{x})$ is an abreviation for $\forall x_1, \ldots, x_n \in [0,1](x_1 \leq \ldots \leq x_n \to Q(\overline{x}))$. $h, \tilde{h}$ always are polynomials in P_n. We denote the derivation

$$\cfrac{\forall \overline{x}, \overline{\sigma} \exists \tilde{h}_{\overline{x},\overline{\sigma}} I(\overline{x}, \overline{\sigma}, \tilde{h}_{\overline{x},\overline{\sigma}}) \quad \cfrac{\forall \overline{x}, h\, (\forall \lambda H(\lambda h) \wedge I(\overline{x}, \overline{\sigma}', h) \to J(\overline{x}))}{\forall \overline{x}\, (\forall \lambda H(\lambda \tilde{h}_{\overline{x},\overline{\sigma}'}) \wedge I(\overline{x}, \overline{\sigma}', \tilde{h}_{\overline{x},\overline{\sigma}'}) \to J(\overline{x}))}}{\forall \overline{x}, \lambda\, H(\lambda \tilde{h}_{\overline{x},\overline{\sigma}'}) \to \forall \overline{x} J(\overline{x})}$$

by $\mathscr{D}_2$. Using again $\mathscr{D}_2$, the assumption A and the lemma '$B \to \forall h H(h)$' we obtain a proof that f_0 has $n+1$ distinct roots:

$$\cfrac{\cfrac{\cfrac{[A] \quad A \to B}{B} \quad B \to \forall h\, H(h)}{\forall h\, H(h)} \quad \cfrac{\mathscr{D}_2}{\forall \overline{x}, \lambda\, H(\lambda \tilde{h}_{\overline{x},\overline{\sigma}'}) \to \forall \overline{x} J(\overline{x})}}{\forall \overline{x}\, J(\overline{x})}$$

The above derivation we denote by $\mathscr{D}_3$. With $\mathscr{D}_1, \mathscr{D}_3$ as subderivations we can now write the whole proof in the form of an informal natural deduction derivation using the additional lemma '$\forall \overline{x} J(\overline{x}) \to E$', '$K \wedge E \to F$' and '$F \to G$' and the fact that F trivially follows from K and E (essentially by the modus ponens rule):

$$\cfrac{\cfrac{\cfrac{\cfrac{\mathscr{D}_1}{K} \quad \cfrac{\cfrac{\mathscr{D}_3}{\forall \overline{x}\, J(\overline{x})} \quad \forall \overline{x}\, J(\overline{x}) \to E}{E}}{F} \quad F \to G}{G}}{A \to G}\,[A]$$

As in the previous section on Chebycheff approximation and explained in the introduction to this chapter we need (in general though not always) to restrict P_n to a suitable compact subspace which – for the case of the norm $\|\cdot\|_1$ at hand – is $K_{f,n} := \{p \in P_n : \|p\|_1 \leq \frac{5}{2}\|f\|_1\}$ in order to apply our metatheorem 15.1. Similarly to the trick used already in the proof of theorem 16.30 for Chebycheff aproximation (where then $K_{f,n}$ was taken as $\{p \in P_n : \|p\|_\infty \leq \frac{5}{2}\|f\|_\infty\}$) we will in the end be able to extend our modulus of uniqueness from $K_{f,n}$ to P_n (see the proof of theorem 16.71 below).

16.4.3 Lemma $A \to B$

This lemma expresses the convexity of the set of best approximations:

$$\begin{cases} \forall f \in C[0,1], n \in \mathbb{N}, p_1, p_2 \in K_{f,n} \\ \quad (\bigwedge_{i=1}^{2} \|f - p_i\|_1 = dist_1(f, P_n) \to \|f - p\|_1 = dist_1(f, P_n)). \end{cases}$$

As a first step in the logical analysis we make explicit the quantifiers hidden in the two occurrences of '=' between real numbers:

$$\begin{cases} \forall f \in C[0,1], n \in \mathbb{N}, p_1, p_2 \in K_{f,n} \\ \quad (\forall \delta \in \mathbb{Q}^*_+ (\bigwedge_{i=1}^{2} \|f - p_i\|_1 - dist_1(f, P_n) \le \delta) \to \\ \qquad \forall \varepsilon \in \mathbb{Q}^*_+ (\|f - p\|_1 - dist_1(f, P_n) < \varepsilon)). \end{cases}$$

Next we bring this into $\forall\exists$-form as guided by functional interpretation:

$$\begin{cases} \forall f \in C[0,1], n \in \mathbb{N}, p_1, p_2 \in K_{f,n}, \varepsilon \in \mathbb{Q}^*_+ \exists \delta \in \mathbb{Q}^*_+ \\ \quad (\bigwedge_{i=1}^{2} \|f - p_i\|_1 - dist_1(f, P_n) \le \delta \to \\ \qquad \|f - p\|_1 - dist_1(f, P_n) < \varepsilon). \end{cases} \tag{16.2}$$

The matrix

$$\bigwedge_{i=1}^{2} \|f - p_i\|_1 - dist_1(f, P_n) \le \delta \to \|f - p\|_1 - dist_1(f, P_n) < \varepsilon$$

in (16.2) can be prenexed as a Σ^0_1-formula and so (16.2) has the logical form required in theorem 15.1. Hence (officially representing δ as 2^{-k}, which we avoid for better readability, and using the monotonicity of the formula in k) it is clear that one can extract a functional Φ_1, depending at most on n, f and ε, such that $\Phi_1(f, n, \varepsilon) \in \mathbb{Q}^*_+$ and

$$\begin{cases} \forall f \in C[0,1], n \in \mathbb{N}, p_1, p_2 \in K_{f,n}, \varepsilon \in \mathbb{Q}^*_+ \\ \quad (\bigwedge_{i=1}^{2} (\|f - p_i\|_1 - dist_1(f, P_n) < \Phi_1(f, n, \varepsilon) \to \\ \qquad \|f - p\|_1 - dist_1(f, P_n) < \varepsilon). \end{cases} \tag{16.3}$$

Given the triviality of the proof of the lemma it is obvious how to construct Φ_1 (for better readability we usually drop the arguments the functionals do not depend on):

Claim. 16.56 *The functional $\Phi_1(f, n, \varepsilon) := \Phi_1(\varepsilon) := \varepsilon$ satisfies (16.3).*

Proof: Assume $(i)\ \|f-p_1\|_1 - dist_1(f,P_n) < \varepsilon$ and $(ii)\ \|f-p_2\|_1 - dist_1(f,P_n) < \varepsilon$. Multiplying (i) and (ii) by 1/2 and adding them together yields $1/2(\|f-p_1\|_1 + \|f-p_2\|_1) - dist_1(f,P_n) < \varepsilon$. By the triangle inequality for the L_1-*norm* we get $1/2(\|2f-p_1-p_2\|_1) - dist_1(f,P_n) < \varepsilon$, i.e. $\|f-p\|_1 - dist_1(f,P_n) < \varepsilon$. □

Remark 16.57. Moving from (16.2) to (16.3) we tacitly changed '$\leq$' to '$<$' in the premise of the implication. We wrote $\leq$ first to establish that the matrix in (16.2) is (equivalent to) a Σ_1^0-formula as required in theorem 15.1. Obviously, replacing '$\leq$' by '$<$' results in an equivalent statement (but a simpler Φ_1). We will use similar variations below even without mentioning them.

16.4.4 Lemma $A \wedge B \rightarrow C$

The lemma states,

$$\begin{cases} \forall f \in C[0,1], n \in \mathbb{N}, p_1, p_2 \in K_{f,n} \\ \quad (\bigwedge_{i=1}^{2} (\|f-p_i\|_1 = dist_1(f,P_n)) \rightarrow \\ \qquad\qquad \|f-p\|_1 - 1/2\|f-p_1\|_1 - 1/2\|f-p_2\|_1 = 0). \end{cases}$$

After presenting the hidden quantifiers and performing the functional interpretation we come again to the same logical structure of the formula in Theorem 15.1, and so we know that there must exist a functional Φ_2 depending at most on n, f and ε such that,

$$\begin{cases} \forall f \in C[0,1], n \in \mathbb{N}, p_1, p_2 \in K_{f,n}; \varepsilon \in \mathbb{Q}_+^* \\ \quad (\bigwedge_{i=1}^{2} (\|f-p_i\|_1 - dist_1(f,P_n) < \Phi_2(f,n,\varepsilon)) \rightarrow \\ \qquad\qquad |\ \|f-p\|_1 - 1/2\|f-p_1\|_1 - 1/2\|f-p_2\|_1\ | < \varepsilon). \end{cases} \tag{16.4}$$

Again, the choice of Φ_2 is simple:

Claim. 16.58 *The functional $\Phi_2(f,n,\varepsilon) := \Phi_2(\varepsilon) := \varepsilon$ satisfies (16.4).*

Proof: Assume $(i)\ \|f-p_1\|_1 - dist_1(f,P_n) < \varepsilon$ and $(ii)\ \|f-p_2\|_1 - dist_1(f,P_n) < \varepsilon$. By claim 16.56 (i.e. the logical analysis of the previous lemma) we get $(iii)\ \|f-p\|_1 - dist_1(f,P_n) < \varepsilon$. $\frac{(i)+(ii)}{2}$ yields $(iv)\ 1/2(\|f-p_1\|_1 + \|f-p_2\|_1) - dist_1(f,P_n) < \varepsilon$. Finally, (iii) and (iv) imply $|\ \|f-p\|_1 - 1/2\|f-p_1\|_1 - 1/2\|f-p_2\|_1\ | < \varepsilon$, since if $a \in [0,m)$ and $b \in [0,m)$ then $|a-b| \in [0,m)$. □

16.4.5 Lemma C_1

Let $g(x) := |f_0(x)| - \frac{1}{2}|f(x) - p_1(x)| - \frac{1}{2}|f(x) - p_2(x)|$. Since f and p_1, p_2 are continuous it follows that g is continuous and so, in particular, the extensionality of g can be proved:

$$\begin{cases} \forall f \in C[0,1], n \in \mathbb{N}, p_1, p_2 \in K_{f,n}, \varepsilon \in \mathbb{Q}^*_+, x, y \in [0,1] \exists \delta \in \mathbb{Q}^*_+ \\ \quad (|x-y| \leq \delta \to |g(x) - g(y)| < \varepsilon). \end{cases}$$

Using the compactness of $K_{f,n}$ (and $[0,1]$), monotone functional interpretation in the form of the metatheorem 15.1 yields – given $f \in C[0,1]$ (represented with a modulus of uniform continuity ω_f), an upper bound for $\|f\|_\infty$ (which can be computed in the representation of $f \in C[0,1]$) and n – a function Δ depending only f, n and ε such that $\Delta(f,n,\varepsilon) \in \mathbb{Q}^*_+$ and

$$\begin{cases} \forall f \in C[0,1], n \in \mathbb{N}, p_1, p_2 \in K_{f,n}, \varepsilon \in \mathbb{Q}^*_+, x, y \in [0,1] \\ \quad (|x-y| < \Delta(f,n,\varepsilon) \to |g(x) - g(y)| < \varepsilon), \end{cases}$$

i.e. a modulus of uniform continuity of g. We write $\Delta(f,n,\varepsilon)$ as $\omega_{f,n}(\varepsilon)$. From the representation of the space $\mathcal{K}_{\omega,m}$ given at the end of chapter 4 it, moreover, is clear already that $\omega_{f,n}$ will depend on ω **only** via a modulus of uniform continuity ω_f of f and an upper bound $M \geq \|f\|_\infty$. As discussed in the introduction to this chapter, we can in the end even eliminate the dependency on M by replacing f by $\tilde{f}(x) := f(x) - f(0)$ whose uniform norm $\|\tilde{f}\|_\infty$ can be bounded by e.g. $\lceil \frac{1}{\omega_f(1)} \rceil$, so that the final result only depends on ω_f and n). The modulus of uniform continuity of a function is not unique. Therefore, when we write $\omega_f(\varepsilon) :=$ '...' in the following we mean that '...' can be taken as some modulus of uniform continuity of the function f.

16.4.5.1 Modulus of the sum

Suppose that ω_f and ω_g are moduli of uniform continuity for the functions f and g respectively. We now construct a modulus of uniform continuity ω_{f+g} for $f+g$. From

$$|x-y| < \omega_f(\varepsilon/2) \to |f(x) - f(y)| < \varepsilon/2$$

and

$$|x-y| < \omega_g(\varepsilon/2) \to |g(x) - g(y)| < \varepsilon/2,$$

we obtain that

$$|x-y| < \min\{\omega_f(\varepsilon/2), \omega_g(\varepsilon/2)\} \to |f(x) - f(y)| < \varepsilon/2 \wedge |g(x) - g(y)| < \varepsilon/2$$

and

$$|x-y| < \min\{\omega_f(\varepsilon/2), \omega_g(\varepsilon/2)\} \to |f(x)+g(x)-f(y)-g(y)| < \varepsilon.$$

So we can take $\omega_{f+g}(\varepsilon) := \min\{\omega_f(\varepsilon/2), \omega_g(\varepsilon/2)\}$.
Analogously, one establishes that $\omega_{f-g}(\varepsilon) := \min\{\omega_f(\varepsilon/2), \omega_g(\varepsilon/2)\}$ also is a modulus for $f-g$.

16.4.5.2 Modulus for multiplication by a scalar

For all $a \in \mathbb{Q}_+^*$, if $|x-y| < \omega_f(\frac{\varepsilon}{a})$ then $|f(x)-f(y)| < \frac{\varepsilon}{a}$, and therefore, $|af(x) - af(y)| < \varepsilon$. Hence we can take $\omega_{af}(\varepsilon) := \omega_f(\frac{\varepsilon}{a})$.

16.4.5.3 Moduli of uniform continuity for $p_1, p_2 \in K_{f,n}$

$p_i \in K_{f,n}$ implies that $\|p_i\|_1 \leq \frac{5}{2}\|f\|_1 \leq \frac{5}{2}\|f\|_\infty$. Now let $p_i(x) = a_n x^n + \ldots + a_1 x + a_0$ and define $p_i^*(x) = \frac{a_n x^{n+1}}{n+1} + \ldots + \frac{a_1 x^2}{2} + a_0 x$. Then

$$|p_i^*(x)| = |\int_0^x p_i(x)dx| \leq \int_0^x |p_i(x)|dx \leq \|p_i\|_1 \leq \frac{5}{2}\|f\|_\infty, \quad \forall x \in [0,1]$$

and so $\|p_i^*\|_\infty \leq \|p_i\|_1 \leq \frac{5}{2}\|f\|_\infty$. We now apply the Markov inequality (proposition 16.4), used already in previous sections, and conclude that

$$\|p_i\|_\infty = \|(p_i^*)'\|_\infty \leq 2(n+1)^2\|p_i^*\|_\infty \leq 2(n+1)^2\|p_i\|_1 \leq 2(n+1)^2(\tfrac{5}{2}\|f\|_\infty) = 5(n+1)^2\|f\|_\infty.$$

Another application of the Markov inequality finally yields

$$\|p_i'\|_\infty \leq 2n^2 5(n+1)^2\|f\|_\infty \leq 10(n+1)^4\|f\|_\infty.$$

Together with the mean value theorem this implies that p_i is Lipschitz continuous on $[0,1]$ with Lipschitz constant $10(n+1)^4\|f\|_\infty$, i.e. $\frac{\varepsilon}{10(n+1)^4\|f\|_\infty}$ is a modulus of uniform continuity for p_i on $[0,1]$. Given an upper bound M_f on $\|f\|_\infty$ (which is easily computable in the representation of f e.g. by $M_f := \max\{|f(i \cdot \omega_f(1))| : 0 \leq i \leq \lfloor \frac{1}{\omega_f(1)} \rfloor\} + 1$) we obtain

$$\omega_{p_i}(\varepsilon) := \omega_{n,M_f} := \frac{\varepsilon}{10(n+1)^4 M_f}.$$

Remark 16.59. Let $p := a_n x^n + \dots a_1 x + a_0 \in P_n$. Using Markov's inequality and the inequality $\|p\|_\infty \le 2(n+1)^2 \|p\|_1$ proved above it follows that $|a_i| \le \frac{(2(n+1)^2)^{i+1}}{i!} \|p\|_1$ for $i \le n$.

Proof: Exercise! □

16.4.5.4 The modulus of uniform continuity $\omega_{f,n}$

Putting the above results together we are now able to construct a modulus of uniform continuity $\omega_{f,n}$ for g as a function of ω_f and n (here we use that any modulus of uniform continuity of f is a modulus of uniform continuity for $|f|$ as well):

$$\begin{aligned}\omega_{f,n}(\varepsilon) &= \min\{\omega_{|f-p|}(\varepsilon/2), \omega_{1/2|f-p_1|}(\varepsilon/4), \omega_{1/2|f-p_2|}(\varepsilon/4)\}\\ &= \min\{\omega_{f-p}(\varepsilon/2), \omega_{f-p_1}(\varepsilon/2), \omega_{f-p_2}(\varepsilon/2)\}\\ &= \min\{\omega_f(\varepsilon/4), \omega_{p_1}(\varepsilon/4), \omega_{p_2}(\varepsilon/4)\}\\ &= \min\{\omega_f(\frac{\varepsilon}{4}), \frac{\varepsilon}{40(n+1)^4 M_f}\}.\end{aligned}$$

Note that the construction above made essential use of the fact that we had restricted the space P_n to the compact set $K_{f,n}$.

16.4.6 Lemma $C \wedge C_1 \to D$

Consider again

$$g(x) := |f(x) - p(x)| - 1/2|f(x) - p_1(x)| - 1/2|f(x) - p_2(x)|.$$

Using g the lemma can be stated as

$$\forall f \in C[0,1], n \in \mathbb{N}, p_1, p_2 \in K_{f,n}\Big(\int_0^1 g(x)dx = 0 \to \forall x \in [0,1](g(x) = 0)\Big).$$

After presenting the hidden quantifiers and applying functional interpretation we observe that again we can apply theorem 15.1 to extract a functional $\Phi_3(f,n,\varepsilon)$ such that $\Phi_3(f,n,\varepsilon) \in \mathbb{Q}_+^*$ and

$$\begin{cases}\forall f \in C[0,1], n \in \mathbb{N}, p_1, p_2 \in K_{f,n}, \varepsilon \in \mathbb{Q}_+^*\\ \quad (|\int_0^1 g(x)dx| \le \Phi_3(f,n,\varepsilon) \to \|g\|_\infty \le \varepsilon).\end{cases} \tag{16.5}$$

In the following $\omega_{f,n} : \mathbb{Q}_+^* \to \mathbb{Q}_+^*$ denotes the modulus of uniform continuity of the function $g \in C[0,1]$ constructed in section 16.4.5.

Claim. 16.60 *The functional* $\Phi_3(f,n,\varepsilon) := \Phi_3(\omega_{f,n},\varepsilon) := \frac{\varepsilon}{2}\cdot\min\{\frac{1}{2},\ \omega_{f,n}(\frac{\varepsilon}{2})\}$ *satisfies (16.5).*

Proof: One easily verifies that $\forall x \in [0,1](g(x) \le 0)$. Now assume that $\|g\|_\infty > \varepsilon$. Then we can conclude that $\exists x_0 \in [0,1](g(x_0) \le -\varepsilon)$. By the continuity of g (with modulus $\omega_{f,n}$) this implies

$$\forall x \in [0,1]\big(|x - x_0| < \omega_{f,n}(\varepsilon/2) \to g(x) < -\varepsilon/2\big).$$

Case 1: $x_0 < 1/2$. Then

$$|\int_0^1 g(x)dx| > |\int_{x_0}^{\min\{1,x_0+\omega_{f,n}(\varepsilon/2)\}} -\varepsilon/2\,dx| \ge \frac{\varepsilon}{2}\min\{\frac{1}{2},\omega_{f,n}(\frac{\varepsilon}{2})\}.$$

Case 2: $x_0 \ge 1/2$. One, analogously, has

$$|\int_0^1 g(x)dx| > |\int_{\max\{0,x_0-\omega_{f,n}(\varepsilon/2)\}}^{x_0} -\varepsilon/2\,dx| \ge \frac{\varepsilon}{2}\min\{\frac{1}{2},\omega_{f,n}(\frac{\varepsilon}{2})\}.$$

So in either case

$$|\int_0^1 g(x)dx| > \frac{\varepsilon}{2}\min\{\frac{1}{2},\omega_{f,n}(\frac{\varepsilon}{2})\}.$$

□

16.4.7 Lemma D → K

Define $f_1(x) := 1/2(|f(x) - p_1(x)| + |f(x) - p_2(x)|)$. The lemma in question can be formally written as:

$$\begin{cases} \forall f \in C[0,1], n \in \mathbb{N}, p_1, p_2 \in K_{f,n}, x \in [0,1] \\ \quad (\| |f_0| - f_1 \|_\infty = 0 \to (|f_0(x)| = 0 \to p_1(x) = p_2(x))). \end{cases}$$

Making the quantifiers hidden in the three equalities between real numbers explicit and prenexing the statement into $\forall\exists$-form theorem 15.1 again predicts the existence of functionals $\Phi_4(f,n,\varepsilon) \in \mathbb{Q}_+^*$ and $\Phi_5(f,n,\varepsilon) \in \mathbb{Q}_+^*$ satisfying

$$\begin{cases} \forall f \in C[0,1], n \in \mathbb{N}, p_1, p_2 \in K_{f,n}, x \in [0,1], \varepsilon \in \mathbb{Q}_+^* \\ \quad (\| |f_0| - f_1 \|_\infty \le \Phi_4(f,n,\varepsilon) \to (|f_0(x)| \le \Phi_5(f,n,\varepsilon) \to |p_1(x) - p_2(x)| \le \varepsilon)). \end{cases} \tag{16.6}$$

Claim. 16.61 *The functionals* $\Phi_4(f,n,\varepsilon) := \Phi_4(\varepsilon) := \varepsilon/4$ *and* $\Phi_5(f,n,\varepsilon) := \Phi_5(\varepsilon) := \varepsilon/4$ *satisfy (16.6).*

Proof: Exercise! □

16.4.8 Lemma $F \to G$

This lemma expresses that P_n is a Haar space which we used already in the previous sections on Chebycheff approximation:

$$\begin{cases} \forall f \in C[0,1], n \in \mathbb{N}, p_1, p_2 \in K_{f,n} \forall x_0, \ldots, x_n \in [0,1] \\ \quad \big(\bigwedge_{i=1}^{n} (x_i < x_{i+1}) \wedge \bigwedge_{i=0}^{n} (p_1(x_i) = p_2(x_i)) \to \|p_1 - p_2\|_\infty = 0 \big). \end{cases}$$

Presenting the hidden quantifiers, prenexing and using the metatheorem 15.1 we get a functional $\Phi_6(f,n,r,\varepsilon) \in \mathbb{Q}^*_+$ realizing '$\exists \delta > 0$' in

$$\begin{cases} \forall f \in C[0,1], n \in \mathbb{N}, p_1, p_2 \in K_{f,n}, r, \varepsilon \in \mathbb{Q}^*_+ \forall x_0, \ldots, x_n \in [0,1] \exists \delta \in \mathbb{Q}^*_+ \\ \quad \big(\bigwedge_{i=1}^{n} (x_{i-1} + r \leq x_i) \wedge \bigwedge_{i=0}^{n} (|p_1(x_i) - p_2(x_i)| \leq \delta) \to \|p_1 - p_2\|_\infty \leq \varepsilon \big). \end{cases} \tag{16.7}$$

Claim. 16.62 *The functional* $\Phi_6(f,n,r,\varepsilon) := \Phi_6(n,r,\varepsilon) := \frac{\lfloor n/2 \rfloor ! \lceil n/2 \rceil ! r^n}{(n+1)} \varepsilon$ *produces a* δ *satisfying (16.7).*

Proof: See the construction of the functional Φ_4 after remark 16.20 in section 16.2. □

In fact, the result above applies not only to $p_1, p_2 \in K_{f,n}$ but even to $p_1, p_2 \in P_n$.

16.4.9 Lemma $B \to \forall h\, H(h)$

The lemma

$$\begin{cases} \forall f \in C[0,1], n \in \mathbb{N}, p_1, p_2 \in K_{f,n} \\ \quad \big(\|f_0\|_1 = dist_1(f, P_n) \to \forall h \in P_n (\|f_0 - h\|_1 \geq \|f_0\|_1) \big) \end{cases}$$

is an immediate consequence of the definitions of f_0 and $dist_1$ and the fact that P_n is a vector space. Observing that the statement trivially holds true if $\|h\|_1 \geq 2\|f\|_1$ so that h can be restricted to the compact set $K_{f,n}$ without loss of generality, monotone functional interpretation (in form of the metatheorem 15.1) guarantees a functional $\Phi_7(f,n,\varepsilon)$ satisfying

$$\begin{cases} \forall f \in C[0,1], n \in \mathbb{N}, p_1, p_2 \in K_{f,n}, \varepsilon \in \mathbb{Q}^*_+ \\ \quad \big(\|f_0\|_1 - dist_1(f, P_n) \leq \Phi_7(f,n,\varepsilon) \to \\ \qquad\qquad \forall h \in P_n (\|f_0 - h\|_1 + \varepsilon \geq \|f_0\|_1) \big). \end{cases} \tag{16.8}$$

Again, the task in fact is trivial:

Claim. 16.63 *The functional $\Phi_7(f,n,\varepsilon) := \Phi_7(\varepsilon) := \varepsilon$ satisfies (16.8).*

Proof: Assume (i) $\|f_0\|_1 - dist_1(f,P_n) \le \varepsilon$. By the definition of $dist_1$ we have (ii) $\|f_0 - h\|_1 = \|f-(p+h)\|_1 \ge dist_1(f,P_n)$ for all $h \in P_n$ since with $p,h \in p_n$ also $p+h \in P_n$. (i) and (ii) yield $\|f_0 - h\|_1 + \varepsilon \ge \|f_0\|_1$. □

16.4.10 Lemma $\forall \overline{x}, h\ (\forall \lambda H(\lambda h) \wedge I(\overline{x}, \overline{\sigma}', h) \rightarrow J(\overline{x}))$

This is the most difficult lemma to analyze logically and we now motivate our analysis in some detail. We first consider the related lemma 1 in the form it is used in Cheney's proof in [66].

Lemma 16.64 (Lemma 1). *Let $f \in C[0,1]$, $n \in \mathbb{N}$ and $h, p_1, p_2 \in P_n$. If f_0 has at most n roots then either $\int_0^1 (h(x)\,sgn(f_0)(x))dx = 0$ or there exists a $\lambda \in \mathbb{R}$ such that $\int_0^1 |f_0(x) - \lambda h(x)|dx < \int_0^1 |f_0(x)|dx$.*

Proof: Since f_0 is assumed to have at most n-many roots there are points $0 = x_0 \le x_1 \le \ldots \le x_{n+1} = 1$ which comprise all roots of f_0. We now assume that $\int_0^1 (h(x)\,sgn(f_0)(x))dx \neq 0$. We only treat the case $\int_0^1 (h(x)\,sgn(f_0)(x))dx > 0$ as the other one is analogous. Let $B' := \bigcup_{i=0}^{n+1} (x_i - r, x_i + r)$, $B := B' \bigcap [0,1]$ and define $A := [0,1] \backslash B$. By making r sufficiently small we can ensure that $\int_A (h(x)\,sgn(f_0)(x))dx > \int_B |h(x)|dx$. Since that A is a finite union of compact intervals which contain no roots of f_0 it follows that $\delta :- \min\{|f_0(x)| : x \in A\} > 0$. Consequently, there is a λ such that $0 < \lambda \|h\|_\infty < \delta$ and $sgn(f_0 - \lambda h)(x) = sgn(f_0)(x)$ for all points $x \in A$. As shown in [66] (as well as in the proof of Claim 16.66 below) this implies that $\int_0^1 |f_0(x) - \lambda h(x)|dx < \int_0^1 |f_0(x)|dx$. □

16.4.10.1 Logical analysis of lemma 1

We now indicate that a slight modification of Lemma 1 can be used in Cheney's proof as well which has the benefit that (due to its logical form) we can apply directly the metatheorem 15.1 to its proof. Again, define $B' := \bigcup_{i=0}^{n+1} (x_i - r, x_i + r)$, $B := B' \bigcap [0,1]$ and $A := [0,1] \backslash B$, where $x_0 := 0$ and $x_{n+1} := 1$. It is clear that A coincides up to at most finitely many points with the union of the smaller intervals $A_i := [x_{i-1} + \min\{r, \frac{x_i - x_{i-1}}{2}\}, x_i - \min\{r, \frac{x_i - x_{i-1}}{2}\}]$, for $1 \le i \le n+1$. As a consequence of this, the two terms $\sum \int_{A_i}$ and $\int_A$ coincide. In the following we always use x_0, x_{n+1}, A, B and A_i as defined above and refer explicitly to r only in cases where this might not be clear from the context. The variant of Lemma 1 we consider is the following one: for all $f \in C[0,1]$ and $n \in \mathbb{N}$

$$\begin{cases} \forall x_1 \leq \ldots \leq x_n \in [0,1], h \in P_n, r \in \mathbb{Q}^*_+ \\ \quad (\forall y \in A(f(y) \neq 0) \wedge \int_A h\, sgn(f) > \int_B |h| \to \\ \qquad \exists \lambda \in \mathbb{R}(\|f - \lambda h\|_1 < \|f\|_1)) \end{cases} \tag{16.9}$$

where A, B depend on $x_1 \leq \ldots \leq x_n$ and r.
We start by indicating how (16.9) can be used in Cheney's uniqueness proof. In our applications of (16.9) we always take as f the function f_0 with p_1, p_2 being best approximations (so that $p := (p_1 + p_2)/2$ again is a best approximation). From this it follows (see below) that $\forall \lambda \in \mathbb{R}, h \in P_n(\|f_0 - \lambda h\|_1 \geq \|f_0\|_1)$. Hence for all $f \in C[0,1]$ and $n \in \mathbb{N}$

$$\begin{cases} \forall x_1 \leq \ldots \leq x_n \in [0,1], h \in P_n, r \in \mathbb{Q}^*_+ \\ \quad (\exists y \in A(f_0(y) = 0) \vee \int_A h\, sgn(f_0) \leq \int_B |h|). \end{cases} \tag{16.10}$$

Furthermore, one can show that

$$\begin{cases} \forall f \in C[0,1], n \in \mathbb{N}, x_1 \leq \ldots \leq x_n \in [0,1] \\ \quad \exists h \in P_n, r \in \mathbb{Q}^*_+ (\forall y \in A(f_0(y) \neq 0) \to \int_A h\, sgn(f_0) > \int_B |h|). \end{cases} \tag{16.11}$$

Let the polynomial h from (16.11) be denoted by $\widehat{h}$. Then applying (16.10) to $\widehat{h}$ and r from (16.11) yields

$$\forall x_1 \leq \ldots \leq x_n \in [0,1](\exists y \in A(f_0(y) = 0)) \tag{16.12}$$

from which we can obtain the existence of $n+1$ roots by induction since $y \in A$ implies that y has distance $\geq r$ from all the points $x_0, \ldots, x_{n+1}$ (recall that $x_0 := 0, x_{n+1} := 1$).

As this proof sketch shows, in our analysis of (16.9) it is sufficient to replace f by $f_0 := f - \frac{p_1 + p_2}{2}$ with $p_1, p_2 \in K_{f,n}$ since the lemma will only be used for such f_0. Finally, as described above, we rewrite the integral of $h\, sgn(f_0)$ over A as the sum of the integrals over the smaller intervals A_i. From the assumption that $\forall y \in A(f_0(y) \neq 0)$ it follows that none of these intervals contains any root of f_0. In total, we formulate the following version of (16.9): for all $f \in C[0,1]$ and $n \in \mathbb{N}$

$$\begin{cases} \forall p_1, p_2 \in K_{f,n}, x_1 \leq \ldots \leq x_n \in [0,1], h \in P_n, r \in \mathbb{Q}^*_+ \\ \quad (\forall y \in A(f_0(y) \neq 0) \wedge \sum_{i=1}^{n+1} \sigma_i \int_{A_i} h > \int_B |h| \to \\ \qquad \exists \lambda \in \mathbb{R}(\|f_0 - \lambda h\|_1 < \|f_0\|_1)), \end{cases} \tag{16.13}$$

where $\sigma_i := sgn(f_0)(\frac{x_{i-1} + x_i}{2})$, $x_0 := 0$ and $x_{n+1} := 1$. Presenting the hidden quantifiers and weakening (16.13) by strengthening the premise '$\forall y \in A(f_0(y) \neq 0)$' to

'$\exists \delta \in \mathbb{Q}^*_+ \forall y \in A(|f(y)| \geq \delta)$' we obtain: for all $f \in C[0,1]$ and $n \in \mathbb{N}$

$$\begin{cases} \forall p_1, p_2 \in K_{f,n}, x_1 \leq \ldots \leq x_n \in [0,1], h \in P_n, \delta, r, \eta \in \mathbb{Q}^*_+ \, \exists l \in \mathbb{Q}^*_+ \\ \quad (\forall y \in A(|f_0(y)| \geq \delta) \wedge \sum\limits_{i=1}^{n+1} \sigma_i \int_{A_i} h \geq \int_B |h| + \eta \to \\ \qquad \exists \lambda \in \mathbb{R}(\|f_0 - \lambda h\|_1 + l < \|f_0\|_1)). \end{cases} \tag{16.14}$$

Remark 16.65. In the presence of WKL the premises '$\forall y \in A(f_0(y) \neq 0)$' and '$\exists \delta \in \mathbb{Q}^*_+ \forall y \in A(|f_0(y)| \geq \delta)$' are in fact equivalent (the equivalence is particularly easy if one uses Σ^0_1-UB instead of WKL, see chapter 12). In the course of the quantitative logical analysis of the proof the notion of 'root' gets replaced by that of 'δ-root' so that it suffices to prove (16.14) which (in contrast to (16.13)) does not need WKL anymore. This kind of possibility of the elimination of WKL is guaranteed already by the metatheorem 15.1 (see also the discussion below).

Note that in (16.14) we can take $\eta = 1$ w.l.o.g. since $h/\eta \in P_n$. Hence, get for all $f \in C[0,1]$ and $n \in \mathbb{N}$

$$\begin{cases} \forall p_1, p_2 \in K_{f,n}, x_1 \leq \ldots \leq x_n \in [0,1], h \in P_n, \delta, r \in \mathbb{Q}^*_+ \, \exists l \in \mathbb{Q}^*_+ \\ \quad (\forall y \in A(|f_0(y)| \geq \delta) \wedge \sum\limits_{i=1}^{n+1} \sigma_i \int_{A_i} h \geq \int_B |h| + 1 \to \\ \qquad \exists \lambda \in \mathbb{R}(\|f_0 - \lambda h\|_1 + l < \|f_0\|_1)) \end{cases} \tag{16.15}$$

16.4.10.2 Extraction of a functional realizing Lemma 1

The matrix

$$\forall y \in A(|f_0(y)| \geq \delta) \wedge \sum_{i=1}^{n+1} \sigma_i \int_{A_i} h \geq \int_B |h| + 1 \to \exists \lambda \in \mathbb{R}(\|f_0 - \lambda h\|_1 + l < \|f_0\|_1)$$

of (16.15) can be prenexed into a formula of the form $\exists \underline{l}^{0/1} A_0$ with quantifier-free A_0.[1] Hence we can apply theorem 15.1 to the proof of (16.15) and know a-priori the extractability of a functional $\Phi_8(f, n, \delta, r, h)$, depending at most on the data shown, which bounds – and hence realizes (due to monotonicity) – '$\exists l$' in (16.15). Thus for all $f \in C[0,1]$ and $n \in \mathbb{N}$ the following holds (where for convenience used below we strengthen the two occurrences of $\geq$ in the premise to $>$):

[1] Note that we can treat σ_i as $\forall \sigma_1, \ldots, \sigma_{n+1} \in \{-1, 1\}$ with the purely universal assumption $\bigwedge_{i=1}^{n+1} (\sigma_i = 1 \to sgn(f_0)(\frac{x_{i-1}+x_i}{2}) \geq 0 \wedge \sigma_i = -1 \to sgn(f_0)(\frac{x_{i-1}+x_i}{2}) \leq 0)$,
since the case where $sgn(f_0)(\frac{x_{i-1}+x_i}{2}) = 0$ does not matter.

$$\begin{cases} \forall p_1, p_2 \in K_{f,n}, x_1 \le \ldots \le x_n \in [0,1], h \in P_n, \delta, r \in \mathbb{Q}^*_+ \\ \quad (\forall y \in A(|f_0(y)| > \delta) \wedge \sum_{i=1}^{n+1} \sigma_i \int_{A_i} h > \int_B |h| + 1 \to \\ \qquad \exists \lambda \in \mathbb{R}(\|f_0 - \lambda h\|_1 + \Phi_8(f,n,\delta,r,h) < \|f_0\|_1)). \end{cases} \tag{16.16}$$

Claim. 16.66 *The functional* $\Phi_8(f,n,\delta,r,h) := \Phi_8(n,\delta,h) := \frac{\delta}{\|h\|_\infty}$ *satisfies (16.16).*

Proof: We have to prove that, for all $f \in C[0,1]$ and $n \in \mathbb{N}$

$$\begin{cases} \forall p_1, p_2 \in K_{f,n}, x_1 \le \ldots \le x_n \in [0,1], h \in P_n, \delta, r \in \mathbb{Q}^*_+ \\ \quad (\forall y \in A(|f_0(y)| > \delta) \wedge \sum_{i=1}^{n+1} \sigma_i \int_{A_i} h > \int_B |h| + 1 \to \\ \qquad \exists \lambda \in \mathbb{R}(\|f_0 - \lambda h\|_1 + \frac{\delta}{\|h\|_\infty} < \|f_0\|_1)). \end{cases}$$

Let $f \in C[0,1], n \in \mathbb{N}, p_1, p_2 \in K_{f,n}, h \in P_n, \delta, r \in \mathbb{Q}^*_+$ be fixed. Note that now we not only require f_0 not to have roots in A but not even δ-roots (i.e. $\forall y \in A(|f_0(y)| > \delta)$). As a consequence, any δ-root y has to be 'r-apart' from all x_i. We say that y does not belong to the (x_i, r)-clusters whose union gives B'. Now we follow the original proof. Take n points, $x_1, \ldots, x_n$, such that (i) $0 = x_0 \le x_1 \le \ldots \le x_{n+1} = 1$ and suppose that (ii) all δ-roots of f_0 belong to at least one of the (x_i, r)- clusters. Moreover, suppose that (iii) $\sum_{i=1}^{n+1} \sigma_i \int_{A_i} h > \int_B |h| + 1$, where $\sigma_i = sgn(f_0)(\frac{x_{i-1}+x_i}{2})$. By assumption (ii) we have $\sigma_i = sgn(f_0)(x)$, for $x \in A_i$ and so $\sum_{i=1}^{n+1} \sigma_i \int_{A_i} h(x)\,dx = \int_A (h(x)\, sgn(f_0)(x))\,dx$. Furthermore, we have $|f_0(x)| > \delta$ for all $x \in A$. Therefore, taking $\lambda := \frac{\delta}{\|h\|_\infty}$ we have (iv) $sgn(f_0 - \lambda h)(x) = sgn(f_0)(x)$, for $x \in A$. Using this we can now reason exactly as in Cheney's original proof and obtain the following inequality:

$$\begin{aligned} \|f_0 - \lambda h\|_1 &= \int_A |f_0 - \lambda h| + \int_B |f_0 - \lambda h| \\ &\overset{(iv)}{=} \int_A (f_0 - \lambda h)\, sgn(f_0) + \int_B |f_0 - \lambda h| \\ &= \int_A f_0\, sgn(f_0) - \lambda \int_A h\, sgn(f_0) + \int_B |f_0 - \lambda h| \\ &\le \int_A f_0\, sgn(f_0) - \lambda \int_A h\, sgn(f_0) + \int_B |f_0| + \lambda \int_B |h| \\ &= \int_A |f_0| + \int_B |f_0| + \lambda \int_B |h| - \lambda \int_A h\, sgn(f_0) \\ &= \int_0^1 |f_0| + \lambda \int_B |h| - \lambda \int_A h\, sgn(f_0). \end{aligned}$$

We now add $\frac{\delta}{\|h\|_\infty}$ on both sides of the inequality and choose $\lambda := \frac{\delta}{\|h\|_\infty}$. Then

$$\|f_0 - \lambda h\|_1 + \frac{\delta}{\|h\|_\infty} \leq \|f_0\|_1 + \frac{\delta}{\|h\|_\infty}(1 + \int_B |h| - \int_A h\, sgn(f_0)) \stackrel{(iii)}{<} \|f_0\|_1.$$

□

Remark 16.67. 1) The functional Φ_8 does not depend on r. This can be explained by fact that (as we will see in section 16.4.11) r is taken to be a function of $\|h\|_\infty$, and such a dependency already appears in Φ_8.
2) The functional Φ_8 is not defined for $\|h\|_\infty$ and even when defined will in general not have a value in $\mathbb{Q}_+^*$. Both issues can be resolved by replacing the argument h by an natural number upper bound N on $\|h\|_\infty$ and defining $\Phi_8(n,\delta,N) := \delta/N$. In fact, for the concrete h, to be constructed in the next section, to which we actually apply Φ_8, we can show that $\|h\|_\infty \leq 8(n+1)^2$ so that we in the end can replace Φ_8 by $\tilde{\Phi}_8(n,\delta) := \delta/(8(n+1)^2)$.

16.4.11 Lemma $\forall \overline{x}, \overline{\sigma} \exists h\, I(\overline{x}, \overline{\sigma}, h)$

Let us recall how Lemma 1 is used in Cheney's proof. Assuming that f_0 has less than $n+1$ roots, Cheney concludes from Lemma 1 that $\forall h \in P_n(\int h\, sgn(f_0) = 0)$. In the second part of his proof Cheney shows, however, that $\forall h \in P_n(\int h\, sgn(f_0) = 0)$ can be refuted, i.e. it holds that

$$\exists h \in P_n(|\int h\, sgn(f_0)| > 0)$$

and so (taking if necessary $-h$)

$$\exists h \in P_n(\int h\, sgn(f_0) > 0). \tag{16.17}$$

As a consequence, the assumption that f_0 has at most n roots has been refuted. As mentioned already before (16.17) yields

$$\exists r \in \mathbb{Q}_+^*(\int_A h\, sgn(f_0) > \int_B |h|).$$

We now show that for any given n points $x_1 \leq \ldots \leq x_n$ in the interval $[0,1]$ and for any $\sigma_1, \ldots, \sigma_{n+1} \in \{-1,1\}$ (where σ_i will denote the sign of the function f_0 in the interval A_i) it is possible to find a function $h \in P_n$ and $r \in \mathbb{Q}_+^*$ such that $\sum_{i=1}^{n+1} \sigma_i \int_{A_i} h > \int_B |h|$, where $x_0 = 0$ and $x_{n+1} = 1$. Formally,

$$\begin{cases} \forall n \in \mathbb{N}, x_1 \leq \ldots \leq x_n \in [0,1], \sigma_1, \ldots, \sigma_{n+1} \in \{-1,1\} \exists h \in P_n, r \in \mathbb{Q}_+^* \\ \quad (\sum_{i=1}^{n+1} \sigma_i \int_{A_i} h > \int_B |h|). \end{cases}$$

In the same way as we did in Section 16.4.10.1 we present the hidden quantifier η in the inequality and since $h/\eta \in P_n$ we have,

$$\begin{cases} \forall n \in \mathbb{N}, x_1 \leq \ldots \leq x_n \in [0,1], \sigma_1, \ldots, \sigma_{n+1} \in \{-1,1\} \exists h \in P_n, r \in \mathbb{Q}_+^* \\ \quad \big(\sum_{i=1}^{n+1} \sigma_i \int_{A_i} h > \int_B |h| + 1\big). \end{cases}$$

The sentence above states the existence of an $r \in \mathbb{Q}_+^*$ and a function $h \in P_n$. Therefore, there exists also a $k \in \mathbb{Q}_+^*$ such that $k \geq \|h\|_\infty$. Here we can again apply our metatheorem 15.1 and we are sure to find functions Φ_9 and Φ_{10} depending only on n such that,[2]

$$\begin{cases} \forall n \in \mathbb{N}, x_1 \leq \ldots \leq x_n \in [0,1], \sigma_1, \ldots, \sigma_{n+1} \in \{-1,1\} \exists h \in P_n, r \geq \Phi_9(n) \\ \quad \big(\sum_{i=1}^{n+1} \sigma_i \int_{A_i} h > \int_B |h| + 1 \wedge \Phi_{10}(n) \geq \|h\|_\infty\big), \end{cases} \tag{16.18}$$

where A and B are defined as before.

Claim. 16.68 *The functions* $\Phi_9(n) := \frac{1}{16(n+1)^3}$ *and* $\Phi_{10}(n) := 8(n+1)^2$ *satisfy (16.18).*

Proof: Let $0 = x_0 \leq x_1 \leq \ldots \leq x_{n+1} = 1$ and $\sigma_1, \ldots, \sigma_{n+1} \in \{-1,1\}$ be given. We first drop all the points x_j such that $x_i = x_j$ and $i < j$. We are left with $\tilde{n}+2$ points $0 = x_{a_0} < x_{a_1} < \ldots < x_{a_{\tilde{n}+1}} = 1$ where $a_{i-1} < a_i, a_i \in \{0, \ldots, n+1\}$ and $\tilde{n} \leq n$. Define $\tilde{x}_i := x_{a_i}$ and $\tilde{\sigma}_i := \sigma_{a_i}$. Since we have eliminated only trivial intervals we have for all $h \in P_n$ that $\sum_{i=1}^{n+1} \sigma_i \int_{x_{i-1}}^{x_i} h(x)\,dx = \sum_{i=1}^{\tilde{n}+1} \tilde{\sigma}_i \int_{\tilde{x}_{i-1}}^{\tilde{x}_i} h(x)\,dx$. Among the points $\tilde{x}_1, \ldots, \tilde{x}_{\tilde{n}}$ we only keep the points $\tilde{x}_i$ for which $\tilde{\sigma}_i \neq \tilde{\sigma}_{i+1}$. Finally, we are left with $m+2$ points $0 = \tilde{x}_{b_0} < \tilde{x}_{b_1} < \ldots < \tilde{x}_{b_{m+1}} = 1$ where $b_{i-1} < b_i, b_i \in \{0, \ldots, \tilde{n}+1\}$ and $m \leq \tilde{n}$. Let $y_i := \tilde{x}_{b_i}$ and $\sigma_i^* := \tilde{\sigma}_{b_i}$. Again we have $\sum_{i=1}^{\tilde{n}+1} \tilde{\sigma}_i \int_{\tilde{x}_{i-1}}^{\tilde{x}_i} h(x)dx = \sum_{i=1}^{m+1} \sigma_i^* \int_{y_{i-1}}^{y_i} h(x)dx$, for any $h \in P_n$. Let $\tilde{h} \in P_n$ be defined by $\tilde{h}(x) := (x - y_1) \ldots (x - y_m)$. Finally, we re-norm $\tilde{h}$ to

$$h(x) := \frac{+/-8(n+1)^2}{\|\tilde{h}\|_\infty} \tilde{h}(x)$$

so that $\|h\|_\infty = 8(n+1)^2$. Here we select $+/-$ in such a way that

$$\sum_{i=1}^{m+1} \sigma_i^* \int_{y_{i-1}}^{y_i} h(x)dx = \sum_{i=1}^{m+1} \int_{y_{i-1}}^{y_i} |h(x)|dx$$

holds. Then

[2] Note that Φ_9 and Φ_{10} do not depend on the points $x_1, \ldots, x_n$ nor on $\sigma_1, \ldots, \sigma_{n+1}$ since they are elements from the compact spaces $[0,1]$ and $\{-1,1\}$, respectively, and $\bigwedge_{i=1}^{n-1} x_i \leq x_{i+1}$ is purely universal.

$$\sum_{i=1}^{n+1} \sigma_i \int_{x_{i-1}}^{x_i} h(x)\,dx = \int_0^1 |h(x)|\,dx.$$

As shown in section 16.4.5.3, the Markov inequality (proposition 16.4) implies that

$$\|h\|_\infty \le 2(n+1)^2 \|h\|_1.$$

Hence

$$\int_0^1 |h(x)|\,dx = \|h\|_1 \ge \frac{\|h\|_\infty}{2(n+1)^2} = 4.$$

Let $r := \Phi_9(n)$. The total length of all the intervals B is at most $\frac{1}{8(n+1)^2}$ and so $\int_B |h(x)|\,dx \le \|h\|_\infty \cdot \frac{1}{8(n+1)^2} = 1$. Hence,

$$\sum_{i=1}^{n+1} \sigma_i \int_{A_i} h(x)\,dx = \int_A |h(x)|\,dx > \int_B |h(x)|\,dx + 1.$$

□

Corollary to the proof of claim 16.68: The proof above shows that we actually can take $r = \Phi_9(n)$ whereas the claim itself (as guaranteed by theorem 15.1) only states the existence of an $r \ge \Phi_9(n)$. This can be explained by the fact that h is taken in the proof in such way that $\sum_i \sigma_i \int_{A_i} h = \int_A |h|$ which has the consequence that the matrix of the claim becomes monotone in $\exists r$.

16.4.12 Elimination of the polynomial h in (16.16)

By claim (16.18) and the corollary to its proof we have

$$\begin{cases} \forall x_1 \le \ldots \le x_n \in [0,1], \sigma_1, \ldots, \sigma_{n+1} \in \{-1,1\} \exists h \in P_n \\ \quad \big(\sum_{i=1}^{n+1} \sigma_i \int_{A_i} h > \int_B |h| + 1 \wedge \Phi_{10}(n) \ge \|h\|_\infty\big), \end{cases} \tag{16.19}$$

where A_i and B are defined with r replaced by $\Phi_9(n)$.

Now, let $f \in C[0,1], n \in \mathbb{N}, p_1, p_2 \in K_{f,n}$ and $x_1 \le \ldots \le x_n \in [0,1], \delta \in \mathbb{Q}^*_+$ be fixed, assume (as in (16.16)) that

$$\forall y \in A(|f_0(y)| > \delta$$

with $r = \Phi_9(n)$ as above and let $\widehat{h}$ be the function from (16.19) where σ_i is the sign of $f_0(\frac{x_{i-1}+x_i}{2})$ with $x_0 := 0$ and $x_{n+1} := 1$. Applying claim 16.66 to $\widehat{h}$ and $\Phi_9(n)$ (i.e. taking $h = \widehat{h}$) we get,

$$\exists \lambda \in \mathbb{R}(\|f_0 - \lambda \widehat{h}\|_1 + \Phi_8(n,\delta,\widehat{h}) < \|f_0\|_1).$$

Taking the contraposition we have shown

$$\begin{cases} \forall f \in C[0,1], n \in \mathbb{N}, p_1, p_2 \in K_{f,n}, x_1 \leq \ldots \leq x_n \in [0,1], \delta \in \mathbb{Q}^*_+ \\ \quad \big(\forall \lambda \in \mathbb{R}(\|f_0 - \lambda \widehat{h}\|_1 + \Phi_8(n,\delta,\widehat{h}) \geq \|f_0\|_1) \to \exists y \in A(|f_0(y)| \leq \delta)\big). \end{cases}$$

Note that in the argument above, the value of σ_i only matters in the case where A_i is not a singleton interval $[(x_{i-1}+x_i)/2]$ (since otherwise $\int_{A_i} h = 0$) and in this case it is easily computable as -1 or 1 under the assumption of (16.16) which implies that $|f_0(y)| > \delta$ on such A_i.

Using the fact that Φ_8 is monotone in $\|h\|_\infty$ and the estimate $\|\widehat{h}\|_\infty \leq \Phi_{10}(n) = 8(n+1)^2$ from (16.19) we can eliminate the dependency of Φ_8 from h (and hence also from $x_1,\ldots,x_n$) by defining $\tilde{\Phi}_8(n,\delta) := \frac{\delta}{8(n+1)^2}$. As a result we conclude

$$\begin{cases} \forall f \in C[0,1], n \in \mathbb{N}, p_1, p_2 \in K_{f,n}, x_1 \leq \ldots \leq x_n \in [0,1], \delta \in \mathbb{Q}^*_+ \\ \quad \big(\forall \lambda \in \mathbb{R}(\|f_0 - \lambda \widehat{h}\|_1 + \tilde{\Phi}_8(n,\delta) \geq \|f_0\|_1) \to \exists y \in A(|f_0(y)| \leq \delta)\big). \end{cases}$$

This in turn implies

$$\begin{cases} \forall f \in C[0,1], n \in \mathbb{N}, p_1, p_2 \in K_{f,n}, \delta \in \mathbb{Q}^*_+ \\ \quad \big(\forall h \in P_n(\|f_0 - h\|_1 + \tilde{\Phi}_8(n,\delta) \geq \|f_0\|_1) \to \\ \qquad\qquad \forall x_1 \leq \ldots \leq x_n \in [0,1] \exists y \in A(|f_0(y)| < \delta)\big). \end{cases} \tag{16.20}$$

Remark 16.69. Note that from the proof of the metatheorem 15.1 it follows that Φ_8 is a majorant of a realizing functional and hence only requires a majorant of h as an argument where the latter can easily be computed given an upper bound l on $\|h\|_\infty$ (rather than referring to the proof of theorem 15.1 one can also just take a majorant Φ_8^* of Φ_8). So the construction of a functional doing the job of $\tilde{\Phi}_8$ is always possible.

In fact, we can replace the conclusion of (16.20) above with the actual existence of $n+1$ roots in the following way (lemma $\forall \overline{x} J(\overline{x}) \to E$). Assume

$$\forall x_1 \leq \ldots \leq x_n \in [0,1] \exists y \in A(|f_0(y)| < \delta),$$

i.e.

$$\forall x_1 \leq \ldots \leq x_n \in [0,1] \exists y \in [0,1](|f_0(y)| < \delta \wedge \bigwedge_{i=0}^{n+1} |x_i - y| \geq \Phi_9(n))). \tag{16.21}$$

Suppose that there would not exist $n+1$-many δ-roots of f_0 which are all pairwise apart from each other by at least $\Phi_9(n)$. Then let $m < n+1$ be the biggest number m such that m-many such δ-roots of f_0 exist. Take a (possibly empty in the case of $m=0$) tuple of such δ-roots in increasing order and fill in (in case $m<n$) further points in this order to get n points $x_1 < \ldots < x_n$ which include these m δ-roots.

Then by (16.21) we get a further δ-root which is $\Phi_9(n)$-apart from each of the other m-many δ-roots and so in total a tuple of $m+1$-many δ-roots all pairwise apart by $\Phi_9(n)$. However, this contradicts the maximality of m. Hence (16.20) implies

$$\exists x_0,\ldots,x_n \in [0,1](\bigwedge_{i=0}^{n} |f_0(x_i)| < \delta \wedge \bigwedge_{i=1}^{n}(x_{i-1}+\Phi_9(n) \leq x_i)). \tag{16.22}$$

So we have shown that (16.20) in fact implies

$$\begin{cases} \forall f \in C[0,1], n \in \mathbb{N}, p_1, p_2 \in K_{f,n}, \delta \in \mathbb{Q}_+^* \\ \quad (\forall h \in P_n(\|f_0 - h\|_1 + \tilde{\Phi}_8(n,\delta) \geq \|f_0\|_1) \to \\ \qquad \exists x_0,\ldots,x_n \in [0,1](\bigwedge_{i=0}^{n} |f_0(x_i)| < \delta \wedge \bigwedge_{i=1}^{n} x_{i-1}+\Phi_9(n) \leq x_i)). \end{cases} \tag{16.23}$$

16.4.13 The modulus of uniqueness for L_1-approximation

Having extracted all the relevant functionals from the various lemmas used in the uniqueness proof we can now combine them into the final modulus of uniqueness. Let $f \in C[0,1]$, $n \in \mathbb{N}, p_1, p_2 \in K_{f,n}$ and $\varepsilon \in \mathbb{Q}_+^*$ be fixed. Assume (for $i \in \{1,2\}$),

$$\|f - p_i\|_1 - dist_1(f,P_n) < \begin{cases} \min\{\Phi_1(\Phi_7(\tilde{\Phi}_8(n,\Phi_5(\Phi_6(n,\Phi_9(n),\varepsilon))))), \\ \quad \Phi_2(\Phi_3(\omega_{f,n},\Phi_4(\Phi_6(n,\Phi_9(n),\varepsilon))))\}. \end{cases} \tag{16.24}$$

(16.24) and claim 16.58 yield

$$|\,\|f_0\|_1 - 1/2\|f-p_1\|_1 - 1/2\|f-p_2\|_1\,| < \Phi_3(\omega_{f,n},\Phi_4(\Phi_6(n,\Phi_9(n),\varepsilon))).$$

By claim 16.60 (using that $\int |f_0| - \frac{1}{2}|f-p_1| - \frac{1}{2}|f-p_2| = \|f_0\|_1 - \frac{1}{2}\|f-p_1\|_1 - \frac{1}{2}\|f-p_2\|_1$) this gives

$$\|\,|f_0| - 1/2|f-p_1| - 1/2|f-p_2|\,\|_\infty \leq \Phi_4(\Phi_6(n,\Phi_9(n),\varepsilon)).$$

Hence, by claim 16.61

$$\begin{cases} \forall x \in [0,1](|f_0(x)| \leq \Phi_5(\Phi_6(n,\Phi_9(n),\varepsilon)) \to \\ \quad |p_1(x) - p_2(x)| \leq \Phi_6(n,\Phi_9(n),\varepsilon)). \end{cases} \tag{16.25}$$

By assumption (16.24) and claim 16.56 we also have

$$\|f_0\|_1 - dist_1(f,P_n) < \Phi_7(\tilde{\Phi}_8(n,\Phi_5(\Phi_6(n,\Phi_9(n),\varepsilon)))).$$

By claim 16.63 we have

$$\forall h \in P_n\big(\|f_0 - h\|_1 + \tilde{\Phi}_8(n,\Phi_5(\Phi_6(n,\Phi_9(n),\varepsilon))) \geq \|f_0\|_1\big).$$

Hence, by (16.23) (taking $\delta = \Phi_5(\Phi_6(n,\Phi_9(n),\varepsilon))$)

$$\begin{cases} \exists x_0,\ldots,x_n \in [0,1] \\ \quad (\bigwedge_{i=0}^{n} |f_0(x_i)| < \Phi_5(\Phi_6(n,\Phi_9(n),\varepsilon)) \wedge \bigwedge_{i=1}^{n} x_{i-1} + \Phi_9(n) \leq x_i). \end{cases}$$

Together with (16.25), this gives

$$\begin{cases} \exists x_0,\ldots,x_n \in [0,1] \\ \quad (\bigwedge_{i=0}^{n} |p_1(x_i) - p_2(x_i)| \leq \Phi_6(n,\Phi_9(n),\varepsilon) \wedge \bigwedge_{i=1}^{n} x_{i-1} + \Phi_9(n) \leq x_i). \end{cases}$$

By claim 16.62 (applied to $r = \Phi_9(n)$) this, finally, implies the desired conclusion

$$\|p_1 - p_2\|_\infty \leq \varepsilon. \tag{16.26}$$

If we write out the definitions of the linear functionals, $\Phi_1, \Phi_2, \Phi_4, \Phi_5$ and Φ_7, to make the modulus Φ more transparent, the implication 16.24 $\rightarrow$ 16.26 becomes

$$\begin{cases} \|f - p_i\|_1 - dist_1(f,P_n) < \\ \quad \min\{\tilde{\Phi}_8(n,\frac{\Phi_6(n,\Phi_9(n),\varepsilon)}{4}),\Phi_3(\omega_{f,n},\frac{\Phi_6(n,\Phi_9(n),\varepsilon)}{4})\} \rightarrow \|p_1 - p_2\|_\infty \leq \varepsilon. \end{cases}$$

Next we unpack the definitions of $\tilde{\Phi}_8$ and Φ_9 which leads to

$$\begin{cases} \|f - p_i\|_1 - dist_1(f,P_n) < \\ \quad \min\{\frac{\Phi_6(n,\frac{1}{16(n+1)^3},\varepsilon)}{32(n+1)^2},\Phi_3(\omega_{f,n},\frac{\Phi_6(n,\frac{1}{16(n+1)^3},\varepsilon)}{4})\} \rightarrow \|p_1 - p_2\|_\infty \leq \varepsilon. \end{cases}$$

Inserting the definition of Φ_6 gives

$$\begin{cases} \|f - p_i\|_1 - dist_1(f,P_n) < \\ \quad \min\{\frac{\frac{\lfloor n/2\rfloor!\lceil n/2\rceil!}{2^{4n+2}(n+1)^{3n+1}}\varepsilon}{8(n+1)^2},\Phi_3(\omega_{f,n},\frac{\lfloor n/2\rfloor!\lceil n/2\rceil!}{2^{4n+2}(n+1)^{3n+1}}\varepsilon)\} \rightarrow \|p_1 - p_2\|_\infty \leq \varepsilon. \end{cases}$$

To make the result more readable we use as an abbreviation $c_n := \frac{\lfloor n/2\rfloor!\lceil n/2\rceil!}{2^{4n+2}(n+1)^{3n+1}}$. The above statement then becomes

$$\begin{cases} \|f-p_i\|_1 - dist_1(f,P_n) < \\ \qquad \min\{\frac{c_n\varepsilon}{8(n+1)^2}, \Phi_3(\omega_{f,n}, c_n\varepsilon)\} \to \|p_1-p_2\|_\infty \le \varepsilon. \end{cases}$$

Finally, we unpack the definition of Φ_3 and obtain

$$\begin{cases} \|f-p_i\|_1 - dist_1(f,P_n) < \\ \qquad \min\{\frac{c_n\varepsilon}{8(n+1)^2}, \frac{c_n\varepsilon}{2}\omega_{f,n}(\frac{c_n\varepsilon}{2})\} \to \|p_1-p_2\|_\infty \le \varepsilon. \end{cases}$$

Putting things together, we have shown the following

Proposition. 16.70 *Let $\tilde{\Phi}(f,n,\varepsilon) := \min\{\frac{c_n\varepsilon}{8(n+1)^2}, \frac{c_n\varepsilon}{2}\omega_{f,n}(\frac{c_n\varepsilon}{2})\}$, where*

$$\omega_{f,n} := \min\{\omega_f(\frac{\varepsilon}{4}), \frac{\varepsilon}{40(n+1)^4 M_f}\}$$

and M_f is a bound on $\|f\|_\infty$. Then $\tilde{\Phi}(f,n,\varepsilon)$ is a modulus of uniqueness for the best L_1-approximation of $f \in C[0,1]$ from P_n for $p_1, p_2 \in K_{f,n}$, i.e.

$$\begin{cases} \forall n \in \mathbb{N}, p_1, p_2 \in K_{f,n}, \varepsilon \in \mathbb{Q}^*_+ \\ \quad (\bigwedge_{i=1}^{2} (\|f-p_i\|_1 - dist_1(f,P_n) < \tilde{\Phi}(f,n,\varepsilon)) \to \|p_1-p_2\|_1 \le \|p_1-p_2\|_\infty \le \varepsilon). \end{cases}$$

Proof: The proposition follows from what we just proved together with $\|p_1 - p_2\|_1 \le \|p_1-p_2\|_\infty$. □

We now extend (similarly to the way used already in the previous sections) proposition 16.70 from $K_{f,n}$ to the whole space P_n. At the same time, we eliminate the dependency of $\tilde{\Phi}$ on M_f and obtain a modulus of uniqueness that depends on f only via a modulus of uniform continuity of f. As a result we obtain theorem 16.54 stated at the beginning of this section:

Theorem 16.71 (Kohlenbach-Oliva [235]). *Let*

$$\Phi(\omega,n,\varepsilon) := \min\{\frac{c_n\varepsilon}{8(n+1)^2}, \frac{c_n\varepsilon}{2}\omega_n(\frac{c_n\varepsilon}{2})\},$$

where the constant $c_n := \frac{\lfloor n/2\rfloor!\lceil n/2\rceil!}{2^{4n+2}(n+1)^{3n+1}}$ and $\omega_n(\varepsilon) := \min\{\omega(\frac{\varepsilon}{4}), \frac{\varepsilon}{40(n+1)^4\lceil\frac{1}{\omega(1)}\rceil}\}$.
For all $f \in C[0,1]$ with modulus of uniform continuity ω

$$\begin{cases} \forall n \in \mathbb{N}, p_1, p_2 \in P_n, \varepsilon \in \mathbb{Q}^*_+ \\ \quad (\bigwedge_{i=1}^{2} (\|f-p_i\|_1 - dist_1(f,P_n) < \Phi(\omega,n,\varepsilon)) \to \|p_1-p_2\|_1 \le \varepsilon). \end{cases}$$

Proof: We first argue that proposition 16.70 also holds with P_n instead of $K_{f,n}$ and then show that we may replace $\tilde{\Phi}$ by Φ which no longer depends on M_f. Suppose without loss of generality that $p_1 \in P_n \backslash K_{f,n}$. Then $\|p_1\|_1 > \frac{5}{2}\|f\|_1$ and hence $\|f - p_1\|_1 > \frac{3}{2}\|f\|_1 \geq \frac{3}{2} dist_1(f,P_n)$. Assume that $\|f - p_i\|_1 < dist_1(f,P_n) + \tilde{\Phi}(f,n,\varepsilon) \leq dist_1(f,P_n) + \frac{\varepsilon}{8}$. We conclude that $\frac{\varepsilon}{8} > \frac{1}{2} dist_1(f,P_n)$, i.e. $dist_1(f,P_n) < \frac{\varepsilon}{4}$. So $\|f - p_i\|_1 < dist_1(f,P_n) + \frac{\varepsilon}{8} < \frac{\varepsilon}{2}$ and, therefore, $\|p_1 - p_2\|_1 \leq \varepsilon$. Analogous to the proof of theorem 16.34 (after corollary 16.49) one concludes the proof by showing that the upper bound $M_f \geq \|f\|_\infty$ in $\tilde{\Phi}$ (used to define $\omega_{f,n}$ in proposition 16.70) can be replaced by an upper bound $N_f \geq \sup_{x \in [0,1]} |f(x) - f(0)|$ where the latter can be computed just using a modulus of uniform continuity ω of f but not f itself, e.g. we may take $N_f := \left\lceil \frac{1}{\omega(1)} \right\rceil$. □

As mentioned in remark 16.59, the function $\Psi(n) := \frac{n!}{2^{n+1}(n+1)^{2n+2}}$ relates the L_1-norm of a polynomial $p \in P_n$ to its actual coefficients, i.e.

$$\forall n \in \mathbb{N} \forall p \in P_n \big(\|p\|_1 \leq \Psi(n) \cdot \varepsilon \to \|p\|_{\max} \leq \varepsilon \big),$$

where $\|p\|_{\max}$ denotes the maximum absolute value of the coefficients of p. Therefore, we obtain the following corollary.

Corollary 16.72. *Let $\Phi(\omega,n,\varepsilon)$ be as defined above. For all $f \in C[0,1]$ with modulus of uniform continuity ω*

$$\begin{cases} \forall n \in \mathbb{N}, p_1, p_2 \in P_n, \varepsilon \in \mathbb{Q}_+^* \\ \quad \big(\bigwedge_{i=1}^{2} (\|f - p_i\|_1 - dist_1(f,P_n) < \Phi(\omega,n,\Psi(n) \cdot \varepsilon)) \to \|p_1 - p_2\|_{\max} \leq \varepsilon \big). \end{cases}$$

For special classes of functions the above modulus of uniqueness can be further simplified:

Definition 16.73. $f \in C[0,1]$ is λ/α-Hölder-Lipschitz-continuous with constant $\lambda \in \mathbb{R}_+^*$ and exponent $0 < \alpha \leq 1$, if

$$|f(x) - f(y)| \leq \lambda |x - y|^\alpha \qquad (\forall x, y \in [0,1]).$$

If $f \in C[0,1]$ is λ/α-Hölder-Lipschitz-continuous, then $(\frac{\varepsilon}{\lambda})^{1/\alpha}$ is a modulus of uniform continuity in our sense for f. Moreover, for $f \in C[0,1]$ being λ/α-Hölder-Lipschitz-continuous we have $\sup_{x \in [0,1]} |f(x) - f(0)| \leq \lambda$ and so we can take λ instead of $\lceil \frac{1}{\omega(1)} \rceil$ in theorem 16.71. Hence theorem 16.71 implies the following

Corollary 16.74. *If $f \in C[0,1]$ is Lipschitz-α continuous with constant λ and exponent $0 < \alpha \leq 1$, then the functional*

$$\Phi_{L_\alpha}(\lambda,\alpha,n,\varepsilon) := \min\{\frac{c_n\varepsilon}{8(n+1)^2}, \frac{c_n\varepsilon}{2}(\frac{c_n\varepsilon}{8\lambda})^{1/\alpha}, \frac{c_n^2\varepsilon^2}{160(n+1)^4\lambda}\}$$

is a modulus of uniqueness for f.

Theorem 16.71 together with (the proof of) proposition 16.2.1) implies

Theorem 16.75. *Let $\mathscr{P}(f,n)$ denote the operator which assigns to any given function $f \in C[0,1]$ and any $n \in \mathbb{N}$ the best L_1-approximation of $f \in C[0,1]$ from P_n. Then $\Phi_P(\omega_f,n,\varepsilon) := \frac{\Phi(\omega_f,n,\varepsilon)}{2}$, Φ as defined in Theorem 16.71, is a modulus of pointwise continuity for the operator $\mathscr{P}(f,n)$, i.e.,*

$$\begin{cases} \forall f,\tilde{f} \in C[0,1], n \in \mathbb{N}, \varepsilon \in \mathbb{Q}^*_+ \\ \quad (\|f-\tilde{f}\|_1 < \Phi_P(\omega_f,n,\varepsilon) \to \|\mathscr{P}(f,n) - \mathscr{P}(\tilde{f},n)\|_1 \leq \varepsilon). \end{cases}$$

16.4.14 General logical remarks on the extraction of the modulus of uniqueness

In the previous sections we carried out the extraction of an effective modulus of uniqueness from Cheney's uniqueness proof as a-priori guaranteed to be possible by theorem 15.4 and the fact that Cheney's proof can be formalized in E-PA$^\omega + (A)$ (and hence in E-PA$^\omega$+WKL), where

$$(A)\ \forall f \in C[0,1] \exists x_0 \in [0,1] (f(x_0) = \sup_{x \in [0,1]} f(x)).$$

That theorem not only provides the extractability of such a modulus but also its verification already in WE-HA$^\omega$ as the latter proves the ε-weakening (A_ε) of (A). In our extraction we did not discuss this issue as we were interested in the modulus itself and not in a necessarily constructive verification. However, the reader might have noticed that by our quantitative analysis of Lemma 1 in section 16.4.10 the need of (A) has indeed disappeared: whereas in Cheney's original proof (A) was used to show

$$\forall x \in [x_{i-1}+r, x_i - r](f(x) > 0) \to \inf_{x \in [x_{i-1}+r, x_i - r]} f(x) > 0$$

this has become in the quantitative analysis (replacing 'roots' by 'r-clusters of δ-roots') just

$$\forall x \in [x_{i-1}+r, x_i - r](f(x) > \delta) \to \inf_{x \in [x_{i-1}+r, x_i - r]} f(x) \geq \delta$$

which follows constructively without the use of (A). The use of classical logic can be eliminated either by appealing to theorem 15.4 (in fact just negative translation from chapter 10 is needed) or by making use throughout the proof that e.g. in the quantitative version we no longer have to find σ_i such that

$$\sigma_i =_0 0 \leftrightarrow f(\frac{x_{i-1}+x_i}{2}) \geq_{\mathbb{R}} 0$$

since we now have that

$$f(\frac{x_{i-1}+x_i}{2}) \geq_{\mathbb{R}} \delta \vee f(\frac{x_{i-1}+x_i}{2}) \leq_{\mathbb{R}} -\delta$$

which easily is decidable for $\delta \in \mathbb{Q}_+^*$.

16.4.15 Estimating the computational complexity of best L_1-approximation

As shown in section 16.1, an effective modulus of uniqueness for best L_1-approximation (as the one extracted above) can be used to compute uniformly in $n \in \mathbb{N}$ the coefficients of the unique best L_1-approximation $p_{n,b}(f)$ of $f \in C[0,1]$ in P_n with arbitrary prescribed precision. In particular, the sequence $(p_{n,b}(f))_{n\in\mathbb{N}}$ is computable in the sense of computable analysis ([377]). Moreover, from a concrete subrecursive modulus of uniqueness one can obtain an upper bound on the computational complexity (in the sense of [199]) of $(p_{n,b}(f))_{n\in\mathbb{N}}$. This is achieved by combining the complexity of the computation of approximate best approximations $p_n(f) \in K_{f,n}$, i.e.

$$\|f - p_n(f)\|_1 \leq dist_1(f,P_n) + 2^{-n},$$

and the complexity of the modulus of uniqueness. $p_n(f)$ can be computed by searching through a sufficiently fine ε-net for $K_{f,n}$ or – more precisely – for the compact rectangle

$$\left\{ (a_0,\ldots,a_n) : \bigwedge_{i=0}^{n} (|a_i| \leq \frac{5(2(n+1)^2)^{i+1}}{i!2} \|f\|_1) \right\}$$

in $\mathbb{R}^{n+1}$ (see remark 16.59), where that net consists of $(n+1)$-tuples of rational coefficients.

The complexity upper bound on $(p_{n,b}(f))_{n\in\mathbb{N}}$ resulting from this approach and the modulus of uniqueness constructed in theorem 16.71 has been calculated in [290]:

Theorem 16.76 (Oliva [290]). *For polynomial time computable $f \in C[0,1]$ the sequence $(p_{n,b}(f))_{n\in\mathbb{N}}$ is strongly* **NP** *computable in* $\mathbf{NP}[B_f]$, *where B_f is an oracle deciding left cuts for integration.*

Proof: See [290]. □

16.4.16 Comparison with previously known results in the literature

The uniqueness of the best L_1-approximation of $f \in C[0,1]$ by polynomials in P_n was proved first in 1921 by Jackson [178] making some use of measure theory. The uniqueness proof the analysis given in this section was based upon is due to Cheney [65] in 1965 (reprinted in [66]). Cheney's proof again is ineffective but more elementary than Jackson's original proof in the sense that it uses only the Riemann integral. Moreover, Cheney actually considers arbitrary Haar spaces (in the sense of the previous section). The first result towards the general qualitative form of the rate of strong unicity for the best L_1-approximation of $f \in C[0,1]$ by polynomials in P_n was obtained in 1975 by Björnestål [35]:

Theorem 16.77 (Björnestål [35]). *Let $f \in C[0,1]$ and the modulus Ω_f be defined as*

$$\Omega_f(\varepsilon) := \sup_{|x-y|<\varepsilon} |f(x) - p_b(x) - f(y) + p_b(y)|,$$

where p_b is the best L_1-approximation of f from P_n. Then, for $p \in P_n$, ε sufficiently small and for some constant c depending on f and n,

$$\|p - p_b\|_1 \geq \varepsilon \to \|f - p\|_1 - \|f - p_b\|_1 \geq 2\int_0^{\Omega_f^{-1}(c\varepsilon)} c\varepsilon - \Omega_f(x)\,dx,$$

where $\Omega_f^{-1}(\varepsilon)$ is defined as

$$\Omega_f^{-1}(\varepsilon) := \inf\{\delta : \Omega_f(\delta) = \varepsilon\}.$$

Note that $\Omega_f^{-1}(\varepsilon)$ (for ε small enough so that $\Omega_f^{-1}(\varepsilon)$ is defined) is a special modulus of uniform continuity for $f - p_b$ in our sense.

The constant c, however, is not presented by Björnestål and, moreover, the function Ω_f^{-1} is usually noncomputable.
In 1978, Kroó [254] showed that the constant c in Björnestål's can be chosen not to depend on the function f as such but only on its modulus of continuity. As in Björnestål [35], also Kroó does not present any explicit constant.

We now prove that theorem 16.71 provides an effective version of Kroó's result and so, a-fortiori, of Björnestål's theorem. First we rewrite the latter as a statement about a modulus of uniqueness:

$$\|f - p\|_1 < dist_1(f, P_n) + 2\int_0^{\Omega_f^{-1}(c\varepsilon)} c\varepsilon - \Omega_f(x)\,dx \to \|p - p_b\|_1 < \varepsilon. \quad (16.27)$$

We start by showing that $\int_0^{\Omega_f^{-1}(c\varepsilon)} c\varepsilon - \Omega_f(x)\,dx$ can be written as $c'\varepsilon\Omega_f^{-1}(c'\varepsilon)$, for some constant $\frac{c}{2} \leq c' \leq c$. For that purpose note that

$$\int_0^{\Omega_f^{-1}(c\varepsilon)} c\varepsilon - \Omega_f(x)\,dx \le \int_0^{\Omega_f^{-1}(c\varepsilon)} c\varepsilon\,dx = c\varepsilon\Omega_f^{-1}(c\varepsilon).$$

On the other hand we have

$$\int_0^{\Omega_f^{-1}(c\varepsilon)} c\varepsilon - \Omega_f(x)\,dx \ge \int_0^{\Omega_f^{-1}(\frac{c}{2}\varepsilon)} c\varepsilon - \Omega_f(x)\,dx$$
$$\ge \int_0^{\Omega_f^{-1}(\frac{c}{2}\varepsilon)} \frac{c}{2}\varepsilon\,dx = \frac{c}{2}\varepsilon\Omega_f^{-1}(\frac{c}{2}\varepsilon).$$

So, for some $\frac{c}{2} \le c' \le c$, (16.27) is equivalent to

$$\|f-p\|_1 < dist_1(f,P_n) + 2c'\varepsilon\Omega_f^{-1}(c'\varepsilon) \to \|p-p_b\|_1 < \varepsilon.$$

Taking $p_1 := p_b$ and $p_2 := p$ in the definition of f_0 we can rewrite g from section 16.4.5 as

$$g(x) = |(f-p_b) + \frac{1}{2}(p_b - p)| - \frac{1}{2}|f-p_b| - \frac{1}{2}|(f-p_b)+(p_b-p)|.$$

Now let $\omega_{f\text{-}p_b}$ be any modulus of uniform continuity for $f-p_b$. Since the best approximation p_b is in $K_{f,n}$ we can prove as in section 16.4.5 that

$$\omega_{f\text{-}p_b,n}(\varepsilon) := \min\left\{\omega_{f\text{-}p_b}(\frac{\varepsilon}{4}), \frac{\varepsilon}{40(n+1)^4\lceil\frac{1}{\omega_f(1)}\rceil}\right\}$$

is a modulus of uniform continuity of g for all $p \in K_{f,n}$ where ω_f is a modulus of uniform continuity of f.

Hence we can replace $\omega_{f,n}$ in the modulus of uniqueness by $\omega_{f\text{-}p_b,n}$ and obtain that for all $p \in P_n, \varepsilon > 0$

$$\|f-p\|_1 < dist_1(f,P_n) + \min\{\frac{c_n\varepsilon}{8(n+1)^2}, \frac{c_n\varepsilon}{2}\omega_{f\text{-}p_b,n}(\frac{c_n\varepsilon}{2})\} \to \|p-p_b\|_1 \le \varepsilon.$$

Assuming that $\omega_{f\text{-}p_b}(\varepsilon) \le \varepsilon$ and taking $\tilde{c}_{n,\omega_f} := \frac{c_n}{20(n+1)^4\lceil\frac{1}{\omega_f(1)}\rceil}$ this yields for all $0 < \varepsilon \le 1$ and all $p \in P_n$

$$\|f-p\|_1 < dist_1(f,P_n) + \tilde{c}_{n,\omega_f}\varepsilon\,\omega_{f\text{-}p_b}(\frac{c_n\varepsilon}{8}) \to \|p-p_b\|_1 \le \varepsilon.$$

Hence we got an effective version of Kroó's result since our constant $\tilde{c}_{n,\omega_f}$ indeed only depends on n and ω_f.
Moreover, this result holds for an arbitrary modulus of uniform continuity of $f-p_b$ rather than only for the special one Ω_f^{-1} considered by Björnestål and Kroó and such a modulus can (using ω_{n,M_f} from section 16.4.5) be constructed in terms of a

modulus of uniform continuity ω_f for f and an upper bound for $\|f\|_\infty$ which led to our fully effective modulus of uniqueness $\Phi(\omega_f, n, \varepsilon)$ in theorem 16.71.

16.5 Exercises, historical comments and suggested further reading

Exercises:

1) Prove proposition 16.36.
2) Prove the statement (16.1) in the analysis of the uniqueness proof due to Borel.
3) Prove the claim in remark 16.59.
4) Prove claim 16.61.

Historical comments and suggested further reading:
Most of the material from sections 16.1 and 16.2 is taken from Kohlenbach [204]. The computability (without any subrecursive complexity information though) of solutions for unique existence theorems of the kind considered in 16.1 (uniformly in the data) was first obtained in Kreinovich [239] (see also Kreinovich [240] and – for a modern treatment of a special case of this – Weihrauch [377]). The material of section 16.3 is taken from Kohlenbach [205]. Kohlenbach [204, 205] in turn are based on chapters viii and ix of Kohlenbach [200]. A survey of these results is given in Kohlenbach [206]. The first effective moduli of uniqueness for best Chebycheff approximation were obtained by Bridges in [44, 46] (and – for the polynomial case – also in Ko [198]). The material of section 16.4 is mostly taken from Kohlenbach-Oliva [235]. The fact that Cheney's uniqueness proof can be formalized in E-PA$^\omega$+WKL and hence permits the extraction of a primitive recursive (in the sense of Gödel) modulus of uniqueness was already observed in Kohlenbach [200]. Theorem 16.76 is from Oliva [290]. For general information on best Chebycheff approximation we refer to Cheney [66]. The most comprehensive treatment of best L_1-approximation can be found in Pinkus [301]. A general survey on the relevance of the concept of strong uniqueness can be found in Bartelt-Li [12].

Chapter 17
Applications to analysis: general metatheorems II

17.1 Introduction

In chapter 15 we proved a general metatheorem (theorem 15.1) on the extractability of effective uniform bounds which are independent from parameters in compact metric spaces K but only depend on (representatives) of elements in Polish spaces X. We saw that both the total boundedness as well as the completeness of K were necessary for this result to hold in general. In this chapter we show that if we deal with general classes of abstract metric spaces (rather than individual spaces) we can in certain contexts obtain bounds which are independent even from noncompact but only metrically bounded (sub-)spaces. It will turn out that for this to hold we – in particular – must not use any separability assumptions on the spaces. In the area of metric fixed point theory there are numerous theorems which hold for general classes of spaces such as arbitrary metric, hyperbolic, normed, uniformly convex or inner product spaces. In chapter 18 we will present a number of applications of the general metatheorems proved in this chapter to (proofs of) such theorems and show how to obtain even qualitatively new information concerning the independence of certain convergence results from parameters in bounded metric spaces.

In order to motivate the results in this chapter we start with a simple example which again is taken from approximation theory. Instead of considering the concrete Polish space $(C[0,1], \|\cdot\|_\infty)$ we treat a general class of abstract spaces namely so-called strictly convex spaces:

Definition 17.1. A normed linear space $(X, \|\cdot\|)$ is called strictly convex if

$$\forall x_1, x_2 \in X\big(\|x_1\|, \|x_2\| \leq 1 \wedge \|\frac{1}{2}(x_1 + x_2)\| = 1 \rightarrow x_1 = x_2\big).$$

Proposition 17.2. *Let $(X, \|\cdot\|)$ be a strictly convex space and $C \subseteq X$ be a convex subset. Each element $x \in X$ has at most one element $y_b \in C$ of best approximation in C, i.e. at most one element $y_b \in C$ such that $\|x - y_b\| = d := \inf\{\|x - y\| : y \in C\}$.*

Proof: Let $x \in X$ and $y_1, y_2 \in C$ be such that $\|x - y_1\| = d = \|x - y_2\|$. Then

$$\left\| x - \frac{y_1 + y_2}{2} \right\| \le \frac{1}{2}\|x - y_1\| + \frac{1}{2}\|x - y_2\| = d.$$

Since $(y_1 + y_2)/2 \in C$ we get $\left\| x - \frac{y_1+y_2}{2} \right\| = d$.
Case 1: $d = 0$. Then $y_1 = x = y_2$.
Case 2: $d > 0$. $\frac{x-y_1}{d}, \frac{x-y_2}{d}$ and $\frac{1}{2}\left(\frac{x-y_1}{d} + \frac{x-y_2}{d}\right) = \frac{x-\frac{1}{2}(y_1+y_2)}{d}$ all have norm 1. By the strict convexity of $(X, \|\cdot\|)$ it follows that $(x - y_1)/d = (x - y_2)/d$, i.e. $y_1 = y_2$. □

Remark 17.3. Note that neither $(C[0,1], \|\cdot\|_\infty)$ nor $(C[0,1], \|\cdot\|_1)$ are strictly convex (exercise!) so that this simple uniqueness proof does not apply to best Chebycheff or to best L_1-approximation.

The property of $(X, \|\cdot\|)$ being strictly convex can be written in the following trivially equivalent form: let $B := \{x \in X : \|x\| \le 1\}$. Then

$$\forall x_1, x_2 \in B \forall k \in \mathbb{N} \exists n \in \mathbb{N}\Big(\left\| \frac{1}{2}(x_1 + x_2) \right\| \ge 1 - 2^{-n} \to \|x_1 - x_2\| < 2^{-k}\Big),$$

where '$\left\| \frac{1}{2}(x_1 + x_2) \right\| \ge 1 - 2^{-n} \to \|x_1 - x_2\| < 2^{-k}$' is (logically equivalent to) a Σ^0_1-formula (if we use the representation of real numbers from chapter 4 and take the space X together with the vector space operations and the norm as primitive elements of the language).

Now suppose that $(X, \|\cdot\|)$ would be a (real) separable Banach space which is provably (say in E-PA$^\omega$+WKL+QF-AC^{1,0}) strictly convex with the property that B is compact (which, however, only is the case when X is finite dimensional, i.e. when X is isomorphic to $\mathbb{R}^n$ for some n endowed with a suitable norm). Moreover, suppose that C is a (constructive) representable convex subset. Then the metatheorems from chapter 15 would allow us to extract a (in this case primitive recursive in the sense of Gödel) modulus of uniqueness from the above uniqueness proof. The first step would be to extract from the proof of strict convexity a B-uniform bound $\Phi_0 : \mathbb{N} \to \mathbb{N}$ on '$\exists n$', i.e. a bound (and so by the monotonicity of '$\exists n$' in fact a realizer) that does not depend on $x_1, x_2 \in B$ but only on k:

$$\forall x_1, x_2 \in B \forall k \in \mathbb{N}\Big(\left\| \frac{1}{2}(x_1 + x_2) \right\| \ge 1 - 2^{-\Phi_0(k)} \to \|x_1 - x_2\| < 2^{-k}\Big).$$

Writing this in the more convenient ε/δ-notation we get a $\Phi_1 : \mathbb{Q}^*_+ \to \mathbb{Q}^*_+$ with

$$\forall x_1, x_2 \in B \forall \varepsilon \in \mathbb{Q}^*_+\Big(\left\| \frac{1}{2}(x_1 + x_2) \right\| \ge 1 - \Phi_1(\varepsilon) \to \|x_1 - x_2\| \le \varepsilon\Big)$$

(e.g. take $\Phi_1(\varepsilon) := 2^{-\Phi_0(\min k[2^{-k} \le \varepsilon])}$). We may assume that $\Phi_1(\varepsilon) < 1$.

Such a Φ_1 is called a modulus of uniform convexity. We now show that if we drop all the completeness/separability/compactness assumptions, the provability of the strict convexity and the constructive representability of C and just **assume** that we have such a modulus of uniform convexity, we still can extract a modulus of uniqueness (despite the fact that the y_1, y_2 no longer can be assumed to range over a compact subspace only). Spaces $(X, \|\cdot\|)$ which have such a modulus of uniform convexity are called uniformly convex spaces and by no means have to be finite dimensional. E.g. the L_p-spaces with $1 < p < \infty$ and all Hilbert spaces uniformly convex. So we obtain a vast generalization of the compact case. This is due to the fact that the original uniqueness proof did not use any completeness, separability or compactness conditions. Moreover, the resulting modulus will hold in any Φ_1-uniformly convex normed space and for any convex subset C and will only depend on an upper bound on d.

The main part of this chapter will be devoted to proving general logical metatheorems which guarantee such uniform-bound-extraction results for large classes of abstract structures (metric spaces, hyperbolic spaces, CAT(0)-spaces, normed spaces $(X, \|\cdot\|)$ with convex subsets C, uniformly convex such spaces and inner product spaces) and proofs. However, before we start to embark on this we will continue with the example and prove the claim we made:

A straightforward analysis of the above uniqueness proof yields the following modulus of uniqueness:

Proposition 17.4. *Let $(X, \|\cdot\|)$ be a uniformly convex normed space with modulus of uniform convexity $\Phi_1 : \mathbb{Q}^*_+ \to \mathbb{Q}^*_+ \cap (0,1)$ and $C \subseteq X$ be convex.*
Define

$$\Phi(\varepsilon) := \min\left\{1, \frac{\varepsilon}{4}, \tilde{D} \cdot \frac{\Phi_1(\varepsilon/(D+1))}{1 - \Phi_1(\varepsilon/(D+1))}\right\},$$

where $D, \tilde{D} \in \mathbb{N}$ with $D \geq d := \inf_{y \in C} \|x - y\|$ and $\tilde{D} \leq \max\left\{\frac{\varepsilon}{4}, d\right\}$ (e.g. we may take $\tilde{D} := \varepsilon/4$). Then Φ is a modulus of uniqueness, i.e.

$$\forall y_1, y_2 \in C \forall \varepsilon \in \mathbb{Q}^*_+ \big(\bigwedge_{i=1}^{2} (\|x - y_i\| \leq d + \Phi(\varepsilon)) \to \|y_1 - y_2\| \leq \varepsilon\big).$$

Proof: Let $\varepsilon \in \mathbb{Q}^*_+$.
Case 1: $d \geq \frac{\varepsilon}{4}$. Put

$$\Phi_2(\varepsilon) := \min\left\{1, \tilde{D} \cdot \frac{\Phi_1(\varepsilon)}{1 - \Phi_1(\varepsilon)}\right\}$$

and assume that $y_1, y_2 \in C$ with

$$\bigwedge_{i=1}^{2} (\|x - y_i\| \leq d + \Phi_2(\varepsilon)(\leq d + 1).$$

Then

$$\left\| \frac{x - y_i}{d + \Phi_2(\varepsilon)} \right\| \leq 1$$

and

$$\left\| \frac{1}{2} \left(\frac{x-y_1}{d+\Phi_2(\varepsilon)} + \frac{x-y_2}{d+\Phi_2(\varepsilon)} \right) \right\| = \left\| \frac{x-(y_1+y_2)/2}{d+\Phi_2(\varepsilon)} \right\| \geq \frac{d}{d+\Phi_2(\varepsilon)} \geq 1-\Phi_1(\varepsilon)$$

since

$$\Phi_2(\varepsilon) \leq \tilde{D} \frac{\Phi_1(\varepsilon)}{1-\Phi_1(\varepsilon)} \leq d \frac{\Phi_1(\varepsilon)}{1-\Phi_1(\varepsilon)} = \frac{d}{1-\Phi_1(\varepsilon)} - d.$$

By the properties of Φ_1 we conclude that

$$\left\| \frac{x-y_1}{d+\Phi_2(\varepsilon)} - \frac{x-y_2}{d+\Phi_2(\varepsilon)} \right\| = \frac{1}{d+\Phi_2(\varepsilon)} \|y_1-y_2\| \leq \varepsilon$$

and so

$$\|y_1-y_2\| \leq \varepsilon(D+\Phi_2(\varepsilon)) \leq \varepsilon(D+1).$$

Take $\Phi_3(\varepsilon) := \min\left\{1, \tilde{D} \cdot \frac{\Phi_1(\varepsilon/(D+1))}{1-\Phi_1(\varepsilon/(D+1))}\right\}$. Applying the above results to $\varepsilon' := \varepsilon/(D+1)$ (note that with $d \geq \varepsilon/4$ also $d \geq \varepsilon'/4$) we conclude that

$$\forall y_1, y_2 \in C(\bigwedge_{i=1}^{2} (\|x-y_i\| \leq d+\Phi_3(\varepsilon) \rightarrow \|y_1-y_2\| \leq \varepsilon)$$

for all $\varepsilon \in \mathbb{Q}_+^*$ with $d \geq \varepsilon/4$.

Case 2: $d < \varepsilon/4$. Let $y_1, y_2 \in C$ with $\bigwedge_{i=1}^{2} (\|x-y_i\| \leq d+\varepsilon/4)$. Then

$$\|y_1-y_2\| \leq \|y_1-x\| + \|x-y_2\| \leq 2d + \frac{\varepsilon}{2} < \frac{\varepsilon}{2} + \frac{\varepsilon}{2} = \varepsilon.$$

Case 1 and 2 together yield the claim since $\Phi(\varepsilon) = \min\left\{\frac{\varepsilon}{4}, \Phi_3(\varepsilon)\right\}$. □

In this chapter we will prove general metatheorems which allow us to conclude beforehand that a uniform bound such as Φ is extractable which only depends on upper bounds $N \geq \|x-y_1\|, \|x-y_2\|$ and $M \geq \|x\|$ (see theorem 17.69 and its corollaries below). Note that

$$\bigwedge_{i=1}^{2} (\|x-y_i\| \leq d+2^{-n}) \rightarrow \|y_1-y_2\| < 2^{-k}$$

can be written as

$$\forall y \in C(\bigwedge_{i=1}^{2} (\|x-y_i\| \leq \|x-y\| + 2^{-n})) \rightarrow \|y_1-y_2\| < 2^{-k}$$

and hence as $\exists$-formula without assuming the existence of d.

To have a bound N is – in the example above – in fact equivalent to having an upper bound D on d : clearly $D := N$ works in one direction and in the other direction we can take $N := D+1$ since we only consider y_1, y_2 with $\|x-y_1\|, \|x-y_2\| \leq$

$d + \Phi(\varepsilon) \le D + 1$. The dependence of the extracted bound on some upper bound $M \ge \|x\|$ can (in the setting of normed spaces) in general not be avoided as the following example shows

$$\forall x, y \in X \exists n(n > \|x\| + \|y\|).$$

Clearly any upper bound on n depends on upper bounds on $\|x\|$ and $\|x-y\|$. The reason why the modulus of uniqueness extracted above does not depend on an upper bound M on $\|x\|$ has to do with the fact that the argument can be re-casted in a setting where the whole context X is just a convex subset of a normed space and $\|\cdot\|$ only is applied to measure the distance of two points in that subset rather than a normed space itself (see the beginning of section 17.3 below for how to make this precise). In fact, the proof works in the even more general context of (uniformly convex) hyperbolic spaces (see [128] and [262] for the definition of uniform convexity in this setting) and – as we will prove below – in such contexts (as well as for convex subsets of normed spaces) one can achieve that indeed uniform bounds are guaranteed which only depend on bounds on the relative distances of x and y_i rather than the absolute norms of elements (see theorem 17.52 and its corollaries below).

From uniform uniqueness to existence

In chapter 15 we showed how to extract moduli of uniqueness from ineffective uniqueness proofs in the presence of compactness and how they can be used to convert ineffective compactness-based existence proofs into effective existence proofs. In the absence of compactness, uniform uniqueness (as expressed by the existence of a modulus of uniqueness) can yield existence theorems which even ineffectively might not be able to arrive at without this: e.g. in the example above we get as an immediate corollary the following (well-known) existence theorem which, in contrast to the plain uniqueness proof but just as the proof of uniform uniqueness, does need uniform convexity rather than just strict convexity:

Proposition 17.5. *Let $(X, \|\cdot\|)$ be a uniformly convex Banach space and $C \subseteq X$ a closed convex subset. Then to any given point $x \in X$ there exists a unique point $y \in C$ of best approximation in C.*

Proof: By the definition of d there is a sequence (y_n) in C such that $\|x - y_n\| \le d + 2^{-n}$. With proposition 17.4 it follows that $(\tilde{y}_n)_n$ with $\tilde{y}_n := y_{\lceil \log_2(1/\Phi(2^{-n})) \rceil}$ is a Cauchy sequence (with rate 2^{-n}) which converges in C since C is closed: Let $m \ge n$. Case 1: $\Phi(2^{-n}) \ge \Phi(2^{-m})$. Then $\|x - \tilde{y}_m\|, \|x - \tilde{y}_n\| \le d + \Phi(2^{-n})$. Hence $\|\tilde{y}_m - \tilde{y}_n\| \le 2^{-n}$. Case 2: $\Phi(2^{-n}) < \Phi(2^{-m})$. As in case 1 one shows that $\|\tilde{y}_m - \tilde{y}_n\| \le 2^{-m} \le 2^{-n}$.
By the continuity of the norm it is clear that the limit $\widehat{y}$ of $(\tilde{y})_n$ is a (unique) best approximation: Let $l := \max\{k+2, \lceil \log_2(1/\Phi(2^{-k-2})) \rceil\}$. Then $\|x - y_l\| \le d + 2^{-k-2}$ as well as $\|x - y_l\| \le d + \Phi(2^{-k-2})$. The latter, together with $\|x - \tilde{y}_{k+2}\| \le d + \Phi(2^{-k-2})$, implies that $\|y_l - \tilde{y}_{k+2}\| \le 2^{-k-2}$. Hence $\|x - \tilde{y}_{k+2}\| \le d + 2^{-k-1}$ and so $\|x - \widehat{y}\| \le d + 2^{-k}$ since $\|\widehat{y} - \tilde{y}_{k+2}\| \le 2^{-k-2}$. Since $k \in \mathbb{N}$ was arbitrary, the proposition follows. $\square$

Remark 17.6. Using special properties of the modulus Φ from proposition 17.4 such as $\Phi(\varepsilon) \leq \varepsilon$) the above proof can be much simplified and essentially becomes trivial. However, we wrote the proof in such a way that the argument works for any modulus of uniqueness.

A less trivial application of this methodology which led to a new fixed point theorem by removing a compactness assumption in a previously known theorem was recently given in [50, 51], see theorem 17.122 below.

In order to be able to talk about – say – an arbitrary metric space we axiomatically 'add' abstract structures like general (classes of) metric spaces (X,d) (or normed spaces) to systems $\mathcal{A}^\omega$ like WE-PA$^\omega$ and extensions thereof resulting in theories $\mathcal{A}^\omega[X,d]$ which are based on two ground types $0,X$ rather than only 0. Since the main focus in the applications is on qualitative uniformity results (rather than complexity issues) we aim at making the underlying system as strong as possible and take

$$\mathcal{A}^\omega := \text{WE-PA}^\omega + \text{QF-AC} + \text{DC}$$

as the underlying system, where DC:= $\{\text{DC}^{\underline{\rho}} : \underline{\rho} \in \mathbf{T}\}$ is the axiom schema of dependent choice

$$\text{DC}^{\underline{\rho}} : \forall x^0, \underline{y}^{\underline{\rho}} \exists \underline{z}^{\underline{\rho}} A(x,\underline{y},\underline{z}) \to \exists \underline{f}^{\underline{\rho}(0)} \forall x^0 A(x, \underline{f}(x), \underline{f}(S(x)))$$

which was studied already in chapter 11. Here A is an arbitrary formula and $\underline{\rho}$ a tuple of arbitrary type. $\underline{f}^{\underline{\rho}(0)}$ stands for $f_1^{\rho_1(0)}, \ldots, f_n^{\rho_n(0)}$. We formulate DC here with tuples since for the extension to the new types discussed below we do not have pairing functionals.
As we will also indicate, the main results can also be adapted to fragments (e.g. with WKL instead of DC) where then additional properties concerning the growth of the extractable bounds can be guaranteed.

The extension of $\mathcal{A}^\omega$, which we will introduce now, is based on functionals of all finites types over $0,X$ including, in particular, variables $x^X, y^X, z^X, \ldots$ and quantifiers $\forall x^X, \exists x^X$, where these variables are intended to vary over the elements of the set X. We also add a new constant d_X for the (pseudo-)metric to the system with the usual axioms. In order to do so we rely on the representation of real numbers from chapter 4, i.e. d_X is of type $1(X)(X)$ where the type-1 value is interpreted as a representative of a real number. In interpreting this constant in the full set-theoretic model over the base types 0 and X where X is a metric space with some metric d (to be defined further below) we have to select a canonical representative $(x)_\circ$ for the real number $d(x,y)$ $(x,y \in X)$. This, moreover, has to be done in such a way that the order on the reals $x \leq y$ gets translated into the (strong) majorization relation $(y)_\circ$ $s\text{-}maj_1$ $(x)_\circ$. As mentioned already in chapter 4 such a selection operator necessarily will be discontinuous and hence noncomputable. However, it will be easily majorizable (by a simple effective functional) which is all we need in the proofs of the results in this chapter.
We now define the following (ineffective) construction which selects to a given real

number $x \in [0,\infty)$ a unique representative $(x)_\circ \in \mathbb{N}^\mathbb{N}$ out of all the representatives $f \in \mathbb{N}^\mathbb{N}$ of x such that certain properties are satisfied (here and in the next lemma and definition, $[0,\infty)$ refers to the 'real' space of all positive reals, i.e. not to the sets of representatives, $\leq_1$ is pointwise order on $\mathbb{N}^\mathbb{N}$, and $\leq$ the usual order on $[0,\infty)$):

Definition 17.7. 1) For $x \in [0,\infty)$ define $(x)_\circ \in \mathbb{N}^\mathbb{N}$ by

$$(x)_\circ(n) := j(2k_0, 2^{n+1}-1),$$

where

$$k_0 := \max k \left[\frac{k}{2^{n+1}} \leq x\right]$$

and j is the Cantor pairing function considered already in chapter 3, definition 3.30.

2) $M(b) := \lambda n.j(b2^{n+2}, 2^{n+1}-1)$.

Lemma 17.8. 1) *If $x \in [0,\infty)$, then $(x)_\circ$ is a representative of x in the sense of the representation of real numbers from chapter 4. In particular: $\widehat{(x)_\circ} =_1 (x)_\circ$.*

2) *If $x,y \in [0,\infty)$ and $x \leq y$ (in the sense of $\mathbb{R}$), then $(x)_\circ \leq_\mathbb{R} (y)_\circ$ and also $(x)_\circ \leq_1 (y)_\circ$ (i.e. $\forall n \in \mathbb{N}((x)_\circ(n) \leq_0 (y)_\circ(n))$).*

3) *If $x \in [0,\infty)$, then $(x)_\circ$ is monotone, i.e. $\forall n \in \mathbb{N}((x)_\circ(n) \leq_0 (x)_\circ(n+1))$.*

4) *If $x,y \in [0,\infty)$ and $x \leq y$ (in the sense of $\mathbb{R}$), then $(y)_\circ$ s-maj$_1$ $(x)_\circ$.*

5) *If $b \in \mathbb{N}$ and $x \in [0,b]$, then $(x)_\circ \leq_1 M(b)$.*

6) *$M(b)$ is monotone, i.e. $\forall n \in \mathbb{N}\big((M(b))(n) \leq_0 (M(b))(n+1)\big)$.*

Proof: 1) We only have to verify that $(x)_\circ$ satisfies the condition $(*)$ from the representation of real numbers as carried out in chapter 4 (and hence passes the $(**)$-test), i.e. $\widehat{(x)_\circ} =_1 (x)_\circ$.

2) follows from the definition of $(x)_\circ$ and 1) using the monotonicity of j (in its first argument).

3), 5) and 6) follow from the monotonicity of the Cantor pairing function j in its arguments, while 4) follows from 2) and 3). □

In the applications to metric fixed point theory carried out in this and the following chapter we will work in the context of hyperbolic spaces which generalizes the context of normed linear spaces:

Definition 17.9. (X,d,W) is called a hyperbolic space if (X,d) is a metric space and $W : X \times X \times [0,1] \to X$ a function satisfying

(i) $\forall x,y,z \in X \forall \lambda \in [0,1]\big(d(z,W(x,y,\lambda)) \leq (1-\lambda)d(z,x) + \lambda d(z,y)\big)$,

(ii) $\forall x,y \in X \forall \lambda_1,\lambda_2 \in [0,1]\big(d(W(x,y,\lambda_1),W(x,y,\lambda_2)) = |\lambda_1 - \lambda_2| \cdot d(x,y)\big)$,

(iii) $\forall x,y \in X \forall \lambda \in [0,1]\big(W(x,y,\lambda) = W(y,x,1-\lambda)\big)$,

$$\text{(iv)} \begin{cases} \forall x,y,z,w \in X, \lambda \in [0,1] \\ \big(d(W(x,z,\lambda),W(y,w,\lambda)) \leq (1-\lambda)d(x,y) + \lambda d(z,w)\big). \end{cases}$$

Definition 17.10. Let (X,d,W) be a hyperbolic space. The set

$$seg(x,y) := \{ W(x,y,\lambda) : \lambda \in [0,1] \}$$

is called the metric segment with endpoints x,y.

Remark 17.11. 1) The condition (ii) ensures that $seg(x,y)$ is an isometric image of the real line segment $[0,d(x,y)]$. For this reason we will often write $[x,y]$ instead of $seg(x,y)$.
2) A subset $C \subseteq X$ of a hyperbolic space (X,d,W) is called convex is $W(x,y,\lambda) \in C$, whenever $x,y \in C, \lambda \in [0,1]$. C itself is a hyperbolic space with the restriction of d and W to C.

Condition (i) has been first considered by W. Takahashi in [354] where a triple (X,d,W) satisfying (i) (and with (X,d) being a metric space) is called a convex metric space. Note, however, that our term '$W(x,y,\lambda)$' corresponds to '$W(y,x,\lambda)$' in [354]. (i)–(iii) together are equivalent to (X,d,W) being a space of hyperbolic type in the sense of Goebel and Kirk [126]. The condition (iv) (first considered as 'condition III' in [177]) is used by Reich and Shafrir in [310] to define the class of hyperbolic spaces as well as in [185]. The notion of hyperbolic space in our sense contains all normed linear spaces and convex subsets thereof (where a subset $C \subseteq X$ is called convex if with $x,y \in C$ and $\lambda \in [0,1]$ also $W(x,y,\lambda) \in C$) but also the open unit ball in complex Hilbert space with the hyperbolic metric (or 'Poincaré distance, see below) as well as Hadamard manifolds (see [128, 309, 310, 311]) and CAT(0)-spaces in the sense of Gromov (see definition 17.14 below). In order to achieve this, our definition of 'hyperbolic space' has, in fact, been made slightly more general than the one given in [310] (following [185]): [310] considers a metric space (X,d) together with a family M of metric lines (rather than metric segments). Hence nontrivial hyperbolic spaces in this sense always are unbounded. This, in particular has the consequence, that only those CAT(0) spaces which have the so-called unique geodesic line extension property are hyperbolic spaces in the sense of [310]. Our definition (like the notion of space of hyperbolic type from [126] and Takahashi's notion of convex metric space) is in contrast to this such that every convex subset of a hyperbolic space is itself a hyperbolic space.
Using a set M of segments has the consequence that in general it is not guaranteed (as it is in the case of metric lines) that for $u,v \in seg(x,y)$ with (u,v) different from (x,y), $seg(u,v)$ is a subsegment of $seg(x,y)$ unless M is closed under subsegments (a consequence of this is that, apparently, one cannot derive (iv) from the special case for $\lambda := \frac{1}{2}$ as in the setting of [310] which follows [185] rather than [126] and that is why we formulate (iv) for general $\lambda \in [0,1]$). So all the examples listed above are a-fortiori hyperbolic spaces in our sense. Also the main result of [40] whose proof we will analyze in chapter 18 is valid for our extended notion of hyperbolic space (see [233]) but does not seem to hold for general spaces of hyperbolic type (however some results in [330] do seem to require the existence of metric lines). As we will see below, adding the axiom (iv) to the concept of 'space of hyperbolic type' has the additional benefit that the extensionality of W_X can be derived in a formal context

which only contains a weak quantifier-free extensionality rule. For all these reasons it seems that our definition is the most useful one.

Definition 17.12. Let (X,d) be a metric space. A geodesic in X is a mapping $\gamma:[0,l]\to\mathbb{R}$ satisfying

$$\forall s,t\in[0,l]\big(d(\gamma(s),\gamma(t))=|s-t|\big).$$

A geodesic segment is the image of a geodesic $\gamma:[0,l]\to\mathbb{R}$ in X and the points $x:=\gamma(0)$ and $y:=\gamma(l)$ are called the endpoints of that segment. We say that x and y are joint by this segment. In this case, obviously, $l=d(x,y)$. (X,d) is a (unique) geodesic space if every two points in X are joint by a (unique) geodesic segment.

Note that by remark 17.11 each hyperbolic space is a geodesic space and the metric segment from definition 17.10 always is a geodesic segment, namely $\gamma([0,d(x,y)])$, where $\gamma:[0,d(x,y)]\to\mathbb{R}$ is the geodesic defined by

$$\gamma(\alpha):=W\left(x,y,\frac{\alpha}{d(x,y)}\right).$$

Example 17.13. 1) Every convex subset $C\subseteq X$ of a normed linear space $(X,\|\cdot\|)$ is a hyperbolic space w.r.t. the metric induced by $\|\cdot\|$ and $W(x,y,\lambda):=(1-\lambda)x+\lambda y$.

2) Another important example of a hyperbolic space is the open unit disk Δ in $\mathbb{C}$ w.r.t. the Poincaré metric (also called 'Poincaré distance')

$$d_\Delta(z,w):=\operatorname{arg\,tanh}\left|\frac{z-w}{1-z\overline{w}}\right|=\operatorname{arg\,tanh}(1-\sigma(z,w))^{\frac{1}{2}},$$

where

$$\sigma(z,w):=\frac{(1-|z|^2)(1-|w|^2)}{|1-z\overline{w}|^2},\quad z,w\in\Delta.$$

For each pair $x,y\in\Delta$ there is a unique geodesic $[x,y]$ joining x,y. Now define $W(x,y,\lambda)$ as the unique point $z\in[x,y]$ with $d_\Delta(x,z)=\lambda d_\Delta(x,y)$ and $d_\Delta(z,y)=(1-\lambda)d_\Delta(x,y)$, where $\lambda\in[0,1]$. (Δ,d_Δ,W) is a hyperbolic space.
The example is of importance in metric fixed point theory since holomorphic mappings $f:\Delta\to\Delta$ are nonexpansive w.r.t. d_Δ, i.e.

$$\forall z,w\in\Delta\,\big(d_\Delta(f(z),f(w))\leq d_\Delta(z,w)\big).$$

See Goebel et al. [129] and Goebel-Reich [128] for all this.

3) Example 2) can be extended from $\mathbb{C}$ to general complex Hilbert spaces $(H,\langle\cdot,\cdot\rangle)$: let B_H be the open unit ball in H. Then

$$k_{B_H}(x,y):=\operatorname{arg\,tanh}(1-\sigma(x,y))^{\frac{1}{2}},$$

where

$$\sigma(x,y):=\frac{(1-\|x\|^2)(1-\|y\|^2)}{|1-\langle x,y\rangle|^2},\quad x,y\in B_H,$$

defines a metric on B_H (also called Kobayashi distance). (B_H, k_{B_H}) is also called Hilbert ball. Again, (B_H, k_{B_H}) is a unique geodesic space and becomes a hyperbolic space by defining W via this fact as in example 2. Holomorphic mappings $f : B_H \to B_H$ are k_{B_H}-nonexpansive. See [129, 128, 257] for details.

Definition 17.14. A CAT(0)-space is a geodesic space that satisfies the so-called **CN**-inequality

$$\mathbf{CN}: \begin{cases} \forall x, y_0, y_1, y_2 \in X \big(d(y_0, y_1) = \frac{1}{2}d(y_1, y_2) = d(y_0, y_2) \to \\ \qquad\qquad d\big(x, y_0\big)^2 \leq \frac{1}{2}d(x, y_1)^2 + \frac{1}{2}d(x, y_2)^2 - \frac{1}{4}d(y_1, y_2)^2\big) \end{cases}$$

of Bruhat and Tits ([60], see also [187]).

Remark 17.15. It is easy to show that each CAT(0)-space is a unique geodesic space (exercise).

From [49] (p.163) it follows that this definition of a CAT(0)-space is equivalent to the more usual definition in terms of comparison triangles (see again [49]). Since every hyperbolic space as we noticed already, in particular, is a geodesic space it follows that a hyperbolic space which satisfies **CN** is CAT(0). The converse also holds since a CAT(0)-space (X, d) becomes a hyperbolic space with W defined via the unique geodesic segment in X connecting two points $x, y \in X$, i.e. $W(x, y, \lambda) := \gamma(\lambda \cdot d(x, y))$, where γ is the unique geodesic $\gamma : [0, d(x, y)] \to \mathbb{R}$ with $\gamma(0) = x$ and $\gamma(d(x, y)) = y$ (see Kirk [189]). When referring to a CAT(0)-space as a hyperbolic space (satisfying **CN**) we always refer to this unique convexity structure W (note that if there would exist two such convexity structures W and W' satisfying the axioms (i)–(iv) but being different on some arguments x, y, λ, this would give rise to different geodesic segments joining x, y contradicting that fact that CAT(0)-spaces are unique geodesic spaces). Hence a CAT(0)-space can be defined equivalently as a hyperbolic space that satisfies **CN**.

Example 17.16. 1) In normed spaces whose norm satisfies the parallelogram law (i.e. in inner product or pre-Hilbert spaces, see below) the CN-inequality holds with '$=$' instead of '$\leq$'. So, trivially, any pre-Hilbert space is a CAT(0)-space. Conversely, any real normed linear space which is a CAT(0)-space already is a pre-Hilbert space (see Bridson-Haefliger [49], proposition II.1.14).

2) The Hilbert ball from example 17.13.3) is a CAT(0)-space as follows from Reich-Shafrir [310] (inequality (4.3) and the subsequent remarks).

3) The planar set X defined as the complement of the quadrant

$$\{(x, y) : x > 0, y > 0\}$$

in $\mathbb{R}^2$ endowed with the induced length metric on $X \subset \mathbb{R}^2$ is a CAT(0)-space (see Bridson-Haefliger [49]). The induced length metric between two points in X is defined as the infimum of the length of rectifiable curves in X joining them. Here 'rectifiable curve' is understood w.r.t. the Euclidean metric.

Metric fixed point theory in the context of CAT(0)-spaces has received quite some interest recently, see e.g. [187, 189, 190, 95, 263].

The following basic facts hold in any hyperbolic space (actually in any convex metric space, see the comments after remark 17.11):

Proposition 17.17 (Takahashi [354]).

1) $W(x,y,0) = x$ *and* $W(x,y,1) = y$ *for all* $x,y \in X$.
2) $d(x,W(x,y,\lambda)) = \lambda d(x,y)$ *and* $d(y,W(x,y,\lambda)) = (1-\lambda)d(x,y)$ *for all* $x,y \in X$ *and* $\lambda \in [0,1]$.
3) $W(x,x,\lambda) = x$ *for all* $x \in X, \lambda \in [0,1]$.

Proof: 1) follows immediately from 2).
2) By axiom (i) we have

$$(1)\ d(x,W(x,y,\lambda)) \le (1-\lambda)d(x,x) + \lambda d(x,y) \le \lambda d(x,y)$$

and

$$(2)\ d(y,W(x,y,\lambda)) \le (1-\lambda)d(x,y).$$

Hence

$$\begin{aligned} d(x,y) &\le d(x,W(x,y,\lambda)) + d(W(x,y,\lambda),y) \\ &\le \lambda d(x,y) + (1-\lambda)d(x,y) = d(x,y) \end{aligned}$$

and, therefore,

$$(3)\ d(x,y) = d(x,W(x,y,\lambda)) + d(W(x,y,\lambda),y).$$

(1)–(3) yield the proposition.
3) is an immediate consequence of (i). □

Remark 17.18. We have shown in proposition 17.17.3) above that the axiom (i) implies that

$$(\mathrm{i})'\ \forall x \in X \forall \lambda \in [0,1]\big(W(x,x,\lambda) = x\big).$$

It is easy to show that conversely in the presence of (i)′ one can derive (i) from (iv) (exercise). So one would get an equivalent axiomatization of hyperbolic spaces by replacing (i) by (i)′. We have not chosen this option since our axiomatization allows us to define other important classes of structures such as spaces of hyperbolic type and convex metric spaces by dropping (iv) resp. (ii)–(iv) as discussed above.

By proposition 17.17.2), CN implies (relative to the axioms for hyperbolic spaces)

$$\mathrm{CN}^- :\ \forall x,y_1,y_2 \in X\big(d(x,W(y_1,y_2,\tfrac{1}{2}))^2 \le \frac{1}{2}d(x,y_1)^2 + \frac{1}{2}d(x,y_2)^2 - \frac{1}{4}d(y_1,y_2)^2\big).$$

Conversely, CN^- implies CN since by CN^- any midpoint y_0 of y_1,y_2 (i.e. any point y_0 such that $d(y_0,y_1) = \frac{1}{2}d(y_1,y_2) = d(y_0,y_2)$) must coincide with $W(y_1,y_2,\frac{1}{2})$: Apply CN^- to $x := y_0$. Then

$$d(y_0, W(y_1, y_2, \frac{1}{2}))^2 \le \frac{1}{2}(\frac{1}{4}d(y_1, y_2)^2) + \frac{1}{2}(\frac{1}{4}d(y_1, y_2)^2) - \frac{1}{4}d(y_1, y_2)^2 = 0$$

and hence $y_0 = W(y_1, y_2, \frac{1}{2})$.
So a CAT(0) space can also be defined as a hyperbolic space (X, d, W) that satisfies CN^- which (in contrast to CN) is purely universal (when formalized as below).

Remark 17.19. CN^- even implies a formally stronger quantitative version CN^* of CN (and hence is also equivalent to CN^*):

$$\mathrm{CN}^*: \begin{cases} \forall x, y_1, y_2, z \in X \forall \varepsilon \in \mathbb{Q}^*_+ \big(\max(d(z, y_1), d(z, y_2)) < \frac{1}{2}d(y_1, y_2)(1+\varepsilon) \to \\ \quad d(x,z)^2 \le \frac{1}{2}d(x, y_1)^2 + \frac{1}{2}d(x, y_2)^2 - \frac{1}{4}d(y_1, y_2)^2 \\ \qquad +2d(x, W(y_1, y_2, \frac{1}{2}))\delta_{y_1, y_2}(\varepsilon) + \delta_{y_1, y_2}(\varepsilon)^2 \big), \end{cases}$$

where $\delta_{y_1, y_2}(\varepsilon) := \frac{1}{2}d(y_1, y_2)\sqrt{\varepsilon^2 + 2\varepsilon}$. Just as CN^- also CN^* is purely universal and, therefore, could be used as well for another (equivalent) universal axiomatization of the class of CAT(0)-spaces. In fact, it is used in [226] as axiom. However, CN^- is simpler to state so that we use this below.

CN^* trivially implies CN since the error term tends to zero as ε does (note that in the case where $y_1 = y_2$, CN is trivial). That the converse is also true follows in the following way:
CN^* holds in every CAT(0)-space since it follows from CN^- by the following easy lemma which we leave as an exercise to the reader (see also [49], p. 286):

Lemma 17.20. *Let (X, d, W) be a hyperbolic space satisfying CN^-. Then for all $y_1, y_2, z \in X$ and $\varepsilon \in \mathbb{Q}^*_+$ the following holds*

$$\max(d(z, y_1), d(z, y_2)) \le \frac{1}{2}d(y_1, y_2)(1+\varepsilon) \\ \to d(z, W(y_1, y_2, \tfrac{1}{2})) \le \tfrac{1}{2}d(y_1, y_2)\sqrt{\varepsilon^2 + 2\varepsilon}.$$

Definition 17.21. The set of all finite types $\mathbf{T}^X$ over the ground types $\mathbb{N}, X$ is defined inductively as follows (where, again, the type $\mathbb{N}$ is denoted by 0):

$$0, X \in \mathbf{T}^X, \ \rho, \tau \in \mathbf{T}^X \Rightarrow \tau(\rho) \in \mathbf{T}^X.$$

The theories $\mathcal{A}^\omega[X,d]$, $\mathcal{A}^\omega[X,d,W]$ and $\mathcal{A}^\omega[X,d,W,\mathrm{CAT}(0)]$:

The language $\mathcal{L}(\mathcal{A}^\omega[X,d])$ of $\mathcal{A}^\omega[X,d]$ results from $\mathcal{L}(\mathcal{A}^\omega)$ by extending it to all the types in the set $\mathbf{T}^X$, i.e. we have (countably many) variables and quantifiers for each type in $\mathbf{T}^X$ and the constants $\Pi_{\rho,\tau}, \Sigma_{\delta,\rho,\tau}, \underline{R}_{\underline{\rho}}$ are now included for all types $\delta, \rho, \tau, \underline{\rho} \in \mathbf{T}^X$. Moreover we have additional constants 0_X of type X, b_X of type 0 and a constant d_X of type $1(X)(X)$.

Axioms and rules of $\mathcal{A}^\omega[X,d]$:

1) We have all the axioms and rules of $\mathcal{A}^\omega$ extended to the new set of types $\mathbf{T}^X$. In particular, the defining axioms for $\Pi_{\rho,\tau}, \Sigma_{\delta,\rho,\tau}, \underline{R}_{\underline{\rho}}$ as well as the schemas IA, QF-AC, DC and the rule QF-ER are now formulated for all types in $\mathbf{T}^X$.
2) For the new constants d_X and b_X we have the following axioms:

   (1) $\forall x^X (d_X(x,x) =_{\mathbb{R}} 0_{\mathbb{R}})$,
   (2) $\forall x^X, y^X (d_X(x,y) =_{\mathbb{R}} d_X(y,x))$,
   (3) $\forall x^X, y^X, z^X (d_X(x,z) \leq_{\mathbb{R}} d_X(x,y) +_{\mathbb{R}} d_X(y,z))$,
   (4) $\forall x^X, y^X (d_X(x,y) \leq_{\mathbb{R}} (b_X)_{\mathbb{R}} (:=_1 \lambda k^0.j(2b_X, 0^0)))$.

   In the formulation of these axioms we refer to the representation of real numbers (including the definition of $=_{\mathbb{R}}, \leq_{\mathbb{R}}$) from chapter 4.

Remark 17.22. 1) Since it does not seem to be possible to contract tuples of variables in our system (unless we add a new product type $0 \times X$) we have to take simultaneous primitive recursion as a primitive concept while it was merely a matter of convenience in the case of the types $\mathbf{T}$.
2) Note that the axioms (1)–(3) of d_X imply that $d_X(x,y) \geq_{\mathbb{R}} 0_{\mathbb{R}}$ for all x^X, y^X.

Equality $=_0$ at type 0 is the only primitive equality predicate. $x^X =_X y^X$ is an abbreviation for $d_X(x,y) =_{\mathbb{R}} 0_{\mathbb{R}}$. As in the case of WE-PA$^\omega$, equality for complex types is defined as extensional equality using $=_0$ and $=_X$ for the base cases.

Clearly, $\mathcal{A}^\omega[X,d]$ proves that $=_X$ is an equivalence relation: the reflexivity follows from axiom (1) whereas symmetry and transitivity follow from axioms (2) and (3) respectively (using in the case of transitivity also remark 17.22.2).

$\mathcal{A}^\omega[X,d,W]$ results from $\mathcal{A}^\omega[X,d]$ by adding a new constant W_X of type $X(1)(X)(X)$ to the language together with the axioms (where $\tilde{\lambda}$ refers to definition 4.24 from chapter 4):

(5) $\forall x^X, y^X, z^X \forall \lambda^1 (d_X(z, W_X(x,y,\lambda)) \leq_{\mathbb{R}} (1_{\mathbb{R}} -_{\mathbb{R}} \tilde{\lambda}) \cdot_{\mathbb{R}} d_X(z,x) +_{\mathbb{R}} \tilde{\lambda} \cdot_{\mathbb{R}} d_X(z,y))$,
(6) $\forall x^X, y^X \forall \lambda_1^1, \lambda_2^1 (d_X(W_X(x,y,\lambda_1), W_X(x,y,\lambda_2)) =_{\mathbb{R}} |\tilde{\lambda}_1 -_{\mathbb{R}} \tilde{\lambda}_2|_{\mathbb{R}} \cdot_{\mathbb{R}} d_X(x,y))$,
(7) $\forall x^X, y^X \forall \lambda^1 (W_X(x,y,\lambda) =_X W_X(y,x,(1_{\mathbb{R}} -_{\mathbb{R}} \lambda)))$,
(8) $\begin{cases} \forall x^X, y^X, z^X, w^X, \lambda^1 \\ (d_X(W_X(x,z,\lambda), W_X(y,w,\lambda)) \leq_{\mathbb{R}} (1_{\mathbb{R}} -_{\mathbb{R}} \tilde{\lambda}) \cdot_{\mathbb{R}} d_X(x,y) +_{\mathbb{R}} \tilde{\lambda} \cdot_{\mathbb{R}} d_X(z,w)). \end{cases}$

Remark 17.23. The intended interpretation of W_X in a hyperbolic space (X,d,W) is that $W_X(x,y,\lambda)$ denotes for each $x,y \in X$ the element $W(x,y,r_{\tilde{\lambda}}) \in X$ where $r_{\tilde{\lambda}}$ is the uniquely determined real number in $[0,1]$ which is represented by $\tilde{\lambda}$. The axiomatization above coincides with that from [120] and is obtained from the one used in [226] if one adds the axiom $W_X(x,y,\lambda) =_X W_X(x,y,\tilde{\lambda})$ to the latter.

$\mathcal{A}^\omega[X,d,W,\text{CAT}(0)]$ results from $\mathcal{A}^\omega[X,d,W]$ by adding the formalized form of the CN$^-$-inequality, i.e.

$$\forall x^X, y_1^X, y_2^X \left(d_X(x, W_X(y_1,y_2,\frac{1}{2}))^2 \leq_{\mathbb{R}} \frac{1}{2} d_X(x,y_1)^2 + \frac{1}{2} d_X(x,y_2)^2 - \frac{1}{4} d_X(y_1,y_2)^2\right),$$

as axiom.

Remark 17.24. The new axioms (1)–(4) of $\mathcal{A}^{\omega}[X,d]$ express (via the representation of $\mathbb{R}$ from chapter 4) that d_X represents a pseudo-metric d (on the domain the variables of type X are ranging over) which is bounded by b_X. Thus d_X represents a (b_X-bounded) metric on the set of equivalence classes generated by $=_X$. We do not form these equivalence classes explicitly but talk instead only about representatives x^X, y^X. However, we have to keep in mind that a functional $f^{X \to X}$ represents a function $X \to X$ only if it respects this equivalence relation, i.e.

$$\forall x^X, y^X (x =_X y \to f(x) =_X f(y)).$$

Since our systems only include the weak (quantifier-free) rule of extensionality we cannot prove that each $f^{X \to X}$ represents a function $: X \to X$ but, in general, only can infer $f(s) =_X f(t)$ from a proof of $s =_X t$. This restriction on the availability of extensionality is crucial for our results to hold (see the discussion further below). Fortunately, however, the mathematical properties of the functions we are considering in various applications usually imply the full extensionality of the functions in questions. In particular, our systems are strong enough to establish the extensionality of the new constants d_X and W_X which we will prove next.

Proposition 17.25. 1) $\mathcal{A}^{\omega}[X,d]$ *proves that*

$$\forall x_1^X, x_2^X, y_1^X, y_2^X (x_1 =_X x_2 \wedge y_1 =_X y_2 \to d_X(x_1,y_1) =_{\mathbb{R}} d_X(x_2,y_2)).$$

2) $\mathcal{A}^{\omega}[X,d,W]$ *proves the extensionality of* W_X*, i.e. for all* $x_1^X, x_2^X, y_1^X, y_2^X, \lambda_1^1, \lambda_2^1$

$$x_1 =_X x_2 \wedge y_1 =_X y_2 \wedge \lambda_1 =_{\mathbb{R}} \lambda_2 \to W_X(x_1,y_1,\lambda_1) =_X W_X(x_2,y_2,\lambda_2).$$

Proof: 1) $d_X(x_1,x_2) =_{\mathbb{R}} 0_{\mathbb{R}}$ and $d_X(y_1,y_2) =_{\mathbb{R}} 0_{\mathbb{R}}$ imply (using axioms (2) and (3))

$$d_X(x_1,y_1) \le_{\mathbb{R}} d_X(x_1,x_2) +_{\mathbb{R}} d_X(x_2,y_2) +_{\mathbb{R}} d_X(y_2,y_1) =_{\mathbb{R}} d_X(x_2,y_2).$$

Analogously, it follows that $d_X(x_2,y_2) \le_{\mathbb{R}} d_X(x_1,y_1)$.
2) Assume that $d_X(x_1,x_2) =_{\mathbb{R}} 0_{\mathbb{R}} =_{\mathbb{R}} d_X(y_1,y_2)$ and $\lambda_1 =_{\mathbb{R}} \lambda_2$. From axioms (6) and (8) together with lemma 4.25.6 it follows that

$$\begin{aligned}
&d_X(W_X(x_1,y_1,\lambda_1), W_X(x_2,y_2,\lambda_2)) \le_{\mathbb{R}} \\
&d_X(W_X(x_1,y_1,\lambda_1), W_X(x_2,y_2,\lambda_1)) +_{\mathbb{R}} d_X(W_X(x_2,y_2,\lambda_1), W_X(x_2,y_2,\lambda_2)) \le_{\mathbb{R}} \\
&(1_{\mathbb{R}} -_{\mathbb{R}} \tilde{\lambda}_1) \cdot_{\mathbb{R}} d_X(x_1,x_2) +_{\mathbb{R}} \tilde{\lambda}_1 \cdot_{\mathbb{R}} d(y_1,y_2) +_{\mathbb{R}} |\tilde{\lambda}_1 -_{\mathbb{R}} \tilde{\lambda}_2|_{\mathbb{R}} \cdot_{\mathbb{R}} d_X(x_2,y_2) =_{\mathbb{R}} 0_{\mathbb{R}}.
\end{aligned}$$

□

Hence (5)–(8) in fact express (via the representation of $\mathbb{R}$ and $[0,1]$ from chapter 4) that W_X represents a function $W : X \times X \times [0,1] \to X$ (actually even on $X \times X \times \mathbb{R}$ via the composition with the retract on $[0,1]$ implicit in $\lambda \mapsto \tilde{\lambda}$) which makes the bounded metric space induced by d into a bounded hyperbolic space. We include a constant 0_X of type X in order to make explicit the fact that X is nonempty (which is already implicit in the logical laws) and to have closed terms of each type in our language. Further below, when we treat normed linear spaces, the role of 0_X will be taken by the zero vector. That is why we already here call this (up to now arbitrary) constant '0_X'.

In the proofs of the metatheorems below we will make use of the fact that the axioms (1)–(8) are all purely universal (recall from chapter 4 that $=_X, =_{\mathbb{R}}, \leq_{\mathbb{R}} \in \Pi^0_1$).

Remark 17.26. 1) As in chapter 3 (lemma 3.15) we can define λ-abstraction in $\mathcal{A}^\omega[X,d]$ and $\mathcal{A}^\omega[X,d,W]$ for all types $\rho \in \mathbf{T}^X$.
2) Every type $\rho \in \mathbf{T}^X$ can be written as $\rho = \tau(\rho_k)\ldots(\rho_1)$ where $\tau = 0$ or $\tau = X$. We define $0^\rho := \lambda v_1^{\rho_1},\ldots,v_k^{\rho_k}.0^0$ resp. $0^\rho := \lambda v_1^{\rho_1},\ldots,v_k^{\rho_k}.0_X$.

Notation. 17.27 Following [310] we often write '$(1-\lambda)x \oplus \lambda y$' for '$W(x,y,\lambda)$'.

17.2 Main results in the metric and hyperbolic case

A bounded hyperbolic space is a hyperbolic space (X,d,W) whose underlying metric space (X,d) is a bounded, i.e. for some $C \in \mathbb{N}$: $d(x,y) \leq C$ for all $x,y \in X$.

Definition 17.28. Let X be a nonempty set. The full set-theoretic type structure $\mathcal{S}^{\omega,X} := \langle S_\rho \rangle_{\rho \in \mathbf{T}^X}$ over $\mathbb{N}$ and X is defined by

$$S_0 := \mathbb{N}, \;\; S_X := X, \;\; S_{\rho(\tau)} := S_\rho^{S_\tau}.$$

Here $S_\rho^{S_\tau}$ is the set of all set-theoretic functions $S_\tau \to S_\rho$.

Let (X,d) be a nonempty bounded metric space. $\mathcal{S}^{\omega,X}$ becomes a model of $\mathcal{A}^\omega[X,d]$ by letting the variables of type ρ range over S_ρ if we give the obvious interpretations to $0^0, S^1, \Pi_{\rho,\tau}, \Sigma_{\delta,\rho,\tau}$ and $\underline{R}_{\underline{\rho}}$ for all types $\delta,\rho,\tau,\underline{\rho} \in \mathbf{T}^X$, interpret 0_X by an arbitrary element of X, interpret b_X by some natural number which is an upper bound for d and – finally – interpret $d_X(x,y)$ for $x,y \in X$ by $(d(x,y))_\circ$, where $(\cdot)_\circ$ refers to the construction in definition 17.7.
Note that this model satisfies the quantifier-free rule of extensionality (and even full extensionality) since in a metric space $d(x,y) = 0 \leftrightarrow x = y$ and so

$$x[=_X]_{\mathcal{S}^{\omega,X}}\, y \equiv [d_X(x,y) =_{\mathbb{R}} 0_{\mathbb{R}}]_{\mathcal{S}^{\omega,X}} \leftrightarrow x = y$$

for all $x,y \in X$.
Let (X,d,W) be, moreover, a nonempty bounded hyperbolic space. Then $\mathcal{S}^{\omega,X}$ becomes a model of $\mathcal{A}^\omega[X,d,W]$ if we extend the above interpretation by interpreting (for $x,y \in X, \lambda \in \mathbb{N}^{\mathbb{N}}$) $W_X(x,y,\lambda)$ as $W(x,y,r_{\tilde{\lambda}})$, where $r_{\tilde{\lambda}} \in [0,1]$ is the unique real number represented by $\tilde{\lambda}$ (see definition 4.24).

Definition 17.29. We say that a sentence of $\mathcal{L}(\mathcal{A}^\omega[X,d])$ or of $\mathcal{L}(\mathcal{A}^\omega[X,d,W])$ holds in a nonempty bounded metric space (X,d) or hyperbolic space (X,d,W) if it holds in the models of $\mathcal{A}^\omega[X,d]$ or $\mathcal{A}^\omega[X,d,W]$, respectively, obtained from $\mathcal{S}^{\omega,X}$ as specified above.

Remark 17.30. We use the plural 'models' in definition 17.29 since the interpretations of 0_X and b_X are not uniquely determined.

Notation. 17.31 When we want to express that a sentence A holds in (X,d,W) we usually write '$d(x,y) \le 2^{-k}$' or '$\forall \lambda \in [0,1](W(x,y,\lambda) = \ldots)$' in A instead of '$d_X(x,y) \le_{\mathbb{R}} 2^{-k}$' or '$\forall \lambda^1 (W_X(x,y,\lambda) =_X \ldots)$' etc. in order to improve the readability. Only in as much as the syntactical form of A as a formal sentence of $\mathcal{L}(\mathcal{A}^\omega[X,d,W])$ matters we will spell out the precise formal representation. Similarly for (X,d) and $\mathcal{L}(\mathcal{A}^\omega[X,d])$

Definition 17.32. For $\rho \in \mathbf{T}^X$ we define $\widehat{\rho} \in \mathbf{T}$ inductively as follows:

$$\widehat{0} := 0,\ \widehat{X} := 0,\ \widehat{\tau(\rho)} := \widehat{\tau}(\widehat{\rho}).$$

Definition 17.33. We say that a type $\rho \in \mathbf{T}^X$ has degree $(\le)n \in \mathbb{N}$ if $\rho \in \mathbf{T}$ and $deg(\rho) \le n$, where deg is defined as in section 3.3 of chapter 3.
ρ has degree $(0,X)$ if $\rho = X(0)\ldots(0)$ (including $\rho = X$). A type $\rho \in \mathbf{T}^X$ has degree $(1,X)$ if it has the form $X(\tau_k)\ldots(\tau_1)$ (including $\rho = X$), where τ_i has degree ≤ 1 or $(0,X)$. ρ is of degree n^* if $\widehat{\rho}$ is of degree $\le n \in \mathbb{N}$.

Definition 17.34. A formula F is called $\forall$-formula (resp. $\exists$-formula) if it has the form $F \equiv \forall \underline{a}^{\underline{\sigma}} F_{qf}(\underline{a})$ (resp. $F \equiv \exists \underline{a}^{\underline{\sigma}} F_{qf}(a)$) where F_{qf} is quantifier-free and the types in $\underline{\sigma}$ are of degree 1^* or $(1,X)$.

In the following theorem, for $\rho \in \mathbf{T}$, '$\le_\rho$' is the relation from definition 3.32:

Theorem 17.35. 1) *Let σ,ρ be types of degree ≤ 1 and τ be a type of degree $(1,X)$. Let $s^{\rho(\sigma)}$ be a closed term of $\mathcal{A}^\omega[X,d]$ and $B_\forall(x^\sigma,y^\rho,z^\tau,u^0)$ ($C_\exists(x^\sigma,y^\rho,z^\tau,v^0)$) be a $\forall$-formula containing only x,y,z,u free (resp. a $\exists$-formula containing only x,y,z,v free).*
If

$$\forall x^\sigma \forall y \le_\rho s(x) \forall z^\tau \left(\forall u^0 B_\forall(x,y,z,u) \to \exists v^0 C_\exists(x,y,z,v)\right)$$

is provable in $\mathcal{A}^\omega[X,d]$, then one can extract a computable functional $\Phi : S_\sigma \times \mathbb{N} \to \mathbb{N}$ such that for all $x \in S_\sigma$ and all $b \in \mathbb{N}$

$$\forall y \le_\rho s(x) \forall z^\tau \left[\forall u \le \Phi(x,b) B_\forall(x,y,z,u) \to \exists v \le \Phi(x,b) C_\exists(x,y,z,v)\right]$$

holds in any (nonempty) metric space (X,d) whose metric is bounded by $b \in \mathbb{N}$ (with 'b_X' interpreted by 'b').

2) *If the premise is proved in '$\mathcal{A}^\omega[X,d,W]$ instead of '$\mathcal{A}^\omega[X,d]$', then the conclusion holds in all b-bounded hyperbolic spaces.*

3) *If the premise is proved in '*$\mathscr{A}^\omega[X,d,W,\mathrm{CAT}(0)]$*, instead of '*$\mathscr{A}^\omega[X,d]$*', then the conclusion holds in all b-bounded* CAT(0)*-spaces.*

Instead of single variables x,y,z,u,v *we may also have finite tuples of variables* $\underline{x},\underline{y},\underline{z},\underline{u},\underline{v}$ *as long as the elements of the respective tuples satisfy the same type restrictions as* x,y,z,u,v*. Moreover, instead of a single premise of the form '*$\forall u^0 B_\forall(x,y,z,u)$*' we may have a finite conjunction of such premises.*

Remark 17.36. In practice, the bounds extracted on u and on v will be different but taking their maximum one can always achieve a common bound which makes the general metatheorems easier to state. Numerically, of course, it is not advisable to throw away information by taking the maximum and we will keep the bounds separate in concrete unwindings of proofs.

The theorem will be proved in section 17.4 below.

Arguably, the most remarkable aspect of theorem 17.35 is that the bound $\Phi(x,b)$ not only is independent from y but, moreover, does not depend on z nor does it depend on (X,d) (or W) as long as (X,d) is b-bounded. In the compact case of theorem 15.1 it mainly was the subrecursive complexity of the bound Φ extractable from a given proof which was of interest as the existence of an effective uniform bound could have been achieved by unbounded search and subsequent use of the fact that computable functionals of type 2 on Cantor space are uniformly continuous (with a computable – in additional function parameters – modulus of uniform continuity) and hence bounded. In the absence of compactness, however, it is not clear at all why even an ineffective uniform bound (independent of z) should exist.

Remark 17.37. 1) The proof of theorem 17.35 which we will give below is based on ND from chapter 10 extended to the new types and subsequent majorization (in a novel sense). Hence the proof, actually, provides an extraction algorithm for Φ. The functional Φ is given by a closed term of WE-PA$^\omega$+(BR) where (BR) refers to Spector's ([343]) schema of bar recursion studied in chapter 11, i.e. Φ is a bar recursive functional. However, for concrete proofs usually only small fragments of $\mathscr{A}^\omega[X,d,W]$ (corresponding to fragments of $\mathscr{A}^\omega$ such as WE-PA$^\omega$+QF-AC+WKL) will be needed to formalize the proof. For many such fragments the complexity of the extractable uniform bounds has been calibrated (see chapter 13 and [207, 210] and also the results from other previous chapters). In particular, it follows from the results in chapter 13 and [210] that a single use of sequential compactness (over a sufficiently weak base system such as G$_\infty$A$^\omega$+QF-AC) only gives rise to at most primitive recursive complexity in the sense of Kleene (often only simple exponential complexity) and this corresponds exactly to the complexity of the bounds we will extract in chapter 18 following [220, 232] (see applications 18.12, 18.16 and 18.17 below and [236] for a general discussion). In other cases (e.g. if instead of sequential compactness only Heine-Borel compactness is used relative to weak arithmetic reasoning) even bounds which are polynomial in the input data can be obtained adapting the method of proof from theorem 12.32.

2) It is the interpretation of dependent choice using bar recursion that necessitates the restriction on the types ρ, τ (as well as in the definition of $\forall$- and $\exists$-formulas) since we have to pass through (an extension to the new types of) the model $\mathcal{M}^{\omega}$ of strongly majorizable functionals from chapter 3 to satisfy (BR) (see chapter 11). If a given proof does not use dependent choice, we can allow far more general types in the parameters.

We now show by means of a counterexample that theorem 17.35 would no longer hold if we had included the full extensionality axiom or just the special case

$$(E^X)\ \forall f^{X\to X}\forall x^X, y^X\left(x =_X y \to f(x) =_X f(y)\right)$$

as an axiom: suppose that the theorem would still hold. (E^X) can equivalently be written as

$$(\tilde{E}^X)\ \forall f^{X\to X}\forall x^X, y^X\forall k\in\mathbb{N}\exists n\in\mathbb{N}\big(d_X(x,y)\le_{\mathbb{R}} 2^{-n}\to d_X(f(x),f(y)) <_{\mathbb{R}} 2^{-k}\big),$$

where $d_X(x,y)\le_{\mathbb{R}} 2^{-n}$ (resp. $d_X(f(x),f(y)) <_{\mathbb{R}} 2^{-k}$) is a $\forall$-formula (resp. an $\exists$-formula), see chapter 4.
If (E^X) had been included as an axiom, then the resulting system would prove $(\tilde{E}^X)$. Theorem 17.35 applied this would then yield the existence of a (computable) function $g:\mathbb{N}\times\mathbb{N}\to\mathbb{N}$ such that

$$\forall f^{X\to X}, \forall x^X, y^X\forall k\in\mathbb{N}\big(d_X(x,y)\le_{\mathbb{R}} 2^{-g(k,b)}\to d_X(f(x),f(y)) <_{\mathbb{R}} 2^{-k}\big)$$

holds for any b-bounded metric space (X,d), i.e. we would get that all functions $f:X\to X$ are equicontinuous with a common modulus of uniform continuity which, of course, is false for general b-bounded spaces (X,d). Similarly, if we add a new function constant $F^{X\to X}$ of type $X\to X$ to the system together with the axiom stating that F is extensional: if the resulting system would still satisfy theorem 17.35 then we could apply theorem 17.35 again to infer that F is uniformly continuous on X. If, on the other hand, we add in addition a function symbol ω_F^1 to the system together with the (up to logical equivalence) purely universal axiom

$$(UC)\ \forall k^0, x^X, y^X\,(d_X(x,y) <_{\mathbb{R}} 2^{-\omega_F(k)}\to d_X(F(x),F(y))\le_{\mathbb{R}} 2^{-k})$$

stating that ω_F is a modulus of uniform continuity of F, then theorem 17.35 remains valid (since we can – essentially – treat ω_F as just another parameter x^1 type 1) and yields a bound Φ depending additionally on ω_F. Clearly, (UC) implies the extensionality of F. So the only way to be a able to use full extensionality for objects of type $X\to X$ is by stipulating an appropriate uniform continuity axiom. This precisely is how we proved the full extensionality of d_X, W_X from the d_X, W_X-axioms implying the uniform continuity of d_X, W_X. Similarly, in the case of normed spaces to be treated below we will be able to prove the extensionality of the norm and the vector space operations from their uniform continuity properties. Also in our applications to various classes of functions $f^{X\to X}$ we often will be able to derive their full extensionality from uniform equicontinuity assumptions. E.g. this applies to the important class of nonexpansive functions (see below) or the more general class of L-Lipschitzian functions which **are** equicontinuous with a common modulus of uniform continuity. There seems to be a relation to the notion of 'uniform families of L-structures' which plays in important role in the model theory of Banach spaces (see e.g. [154]). In our proof theoretic approach based on weak extensionality, however, we do not have to make such strong uniform continuity assumptions if the only use of extensionality we make is that provided by QF-ER. This allows us to apply our results not only to classes of functions such as the nonexpansive ones but also to e.g. directionally nonexpansive functions and weakly quasi-nonexpansive functions etc. (see below) which no longer can be proved to be extensional in our setting (due to the lack of continuity). In addition to the effective nature of our results, this is yet another benefit of the proof theoretic approach to functional analysis and there does not seem to be any natural model theoretic counterpart to the weak form of extensionality formalized by QF-ER.

Remark 17.38. Theorem 17.35 holds also for convex metric spaces (resp. spaces of hyperbolic type) if in $\mathcal{A}^{\omega}[X,d,W]$ the W_X-axioms (6)–(8) (resp. (8)) are dropped. However, as discussed above, this has the consequence that the extensionality of W_X is no longer provable so that one has to rely on the weak rule of quantifier-free extensionality instead. If only (8) is dropped one still has full extensionality in λ by (6). In the absence of (6), one naturally would extend the existing rule QF-ER by

$$(+)\ \frac{A_0 \to s^1 =_{\mathbb{R}} t^1}{A_0 \to W_X(x,y,\tilde{s}) =_X W_X(x,y,\tilde{t})} \quad (A_0 \text{ quantifier-free})$$

(which is redundant in the presence of (6)) to have also for the scalar at least weak extensionality of W_X (A_0 is quantifier-free). This is not an instance of QF-ER as formulated so far since the 'official' equality relation for type-1 objects is $=_1$. The proofs of the main results also hold with this extended form of QF-ER.

Definition 17.39. 1) Let (X,d) be a metric space. A function $f : X \to X$ is called nonexpansive (short: 'f n.e.') if

$$\forall x,y \in X\big(d(f(x),f(y)) \le d(x,y)\big).$$

2) f is quasi-nonexpansive if

$$\forall p,x \in X(d(p,f(p)) = 0 \to d(f(x),p) \le d(x,p)).$$

3) f is weakly quasi-nonexpansive if

$$\exists p \in X \forall x \in X\big(d(f(x),p) \le d(x,p))\big).$$

4) f is Lipschitz continuous with Lipschitz constant $L > 0$ if

$$\forall x,y \in X(d(f(x),f(y)) \le L \cdot d(x,y)).$$

5) f is Hölder-Lipschitz continuous with constants $L > 0$ and $0 < \alpha \le 1$ if

$$\forall x,y \in X(d(f(x),f(y)) \leq L \cdot d(x,y)^\alpha).$$

6) f is uniformly continuous with modulus of uniform continuity $\omega : \mathbb{N} \to \mathbb{N}$ if

$$\forall x,y \in X \forall n \in \mathbb{N}\big(d(x,y) < 2^{-\omega(n)} \to d(f(x),f(y)) \leq 2^{-n}\big).$$

7) f is bounded with bounding function $\Omega : \mathbb{N} \to \mathbb{N}$ if

$$\forall x,y \in X \forall n \in \mathbb{N}\big(d(x,y) < n \to d(f(x),f(y)) \leq \Omega(n)\big).$$

8) If (X,d,W) is a hyperbolic space, then $f : X \to X$ is called directionally nonexpansive (short 'f d.n.e') if

$$\forall x \in X \forall y \in seg(x,f(x))\big(d(f(x),f(y)) \leq d(x,y)\big).$$

For normed linear spaces $(X,\|\cdot\|)$ definition 17.39 is understood with respect to the induced metric $d(x,y) := \|x-y\|$.

Remark 17.40. Whereas the concepts 'quasi-nonexpansive', 'weakly quasi-nonexpansive' and 'directionally nonexpansive' only apply to selfmappings $f : X \to X$, the other notions defined above, of course, generalize in the obvious way to functions $f : X \to Y$, where (X,d_X) and (Y,d_Y) are potentially different metric spaces.

Lemma 17.41. *f is weakly quasi-nonexpansive iff*

$$\exists p \in X(d(p,f(p)) = 0 \land \forall x \in X(d(f(x),f(p)) \leq d(x,p))).$$

Proof: '$\to$': Let f be weakly quasi-nonexpansive, i.e.

$$\exists p \in X \forall x \in X\big(d(f(x),p) \leq d(x,p)\big).$$

Taking $x := p$ it follows that $f(p) = p$ and hence also

$$\forall x \in X\big(d(f(x),f(p)) \leq d(x,p)\big).$$

So in total

$$\exists p \in X\big(f(p) = p \land \forall x \in X\big(d(f(x),f(p)) \leq d(x,p)\big)\big).$$

'$\leftarrow$': obvious. □

Quasi-nonexpansive mappings were first considered by Dotson in [90] (based on a related earlier notion due to Diaz and Metcalf [85, 86]). Weakly quasi-nonexpansive mappings were introduced (implicitly) by Kohlenbach and Lambov in [231] in the equivalent formulation of lemma 17.41 (and in an asymptotic version of this notion). In contrast to the notion of 'quasi-nonexpansive functions' the concept of weakly quasi-nonexpansive functions has a nice logical behavior w.r.t. our metatheorems to be proved below. Most proofs of results about quasi-nonexpansive functions immediately generalize also to weakly quasi-nonexpansive functions although the latter is a bigger class of functions (quasi-nonexpansive functions are only considered in contexts where f has at least one fixed point so that it is a stronger requirement to be quasi-nonexpansive than only weakly quasi-nonexpansive): e.g. all selfmappings of a subset C of a normed space which contains the zero vector 0_X that satisfy $\|f(x)\| \leq \|x\|$ obviously are weakly quasi-nonexpansive w.r.t. the fixed point 0, but in general are not quasi-nonexpansive as the following example shows:

Example 17.42. Consider the function $f : [0,1] \to [0,1], x \mapsto x^2$. Clearly, the fixed point set of f is $\{0,1\}$. However, quasi-nonexpansivity fails for 1 since e.g. for $x := 1/2$ we have $|f(x) - 1| = \frac{3}{4} > \frac{1}{2} = |x - 1|$.
Note that the restriction of f to $[0,1)$ **is** a quasi-nonexpansive selfmapping of $[0,1)$.

These examples were communicated to us by Laurenţiu Leuştean. The notion of 'weakly quasi-nonexpansive' functions was recently also considered (independently) under the name of J-type mappings in Garcia-Falset et al. [96]. In that paper the importance of this notion is demonstrated by numerous fixed point results which hold for this class of functions.
The notion of directionally nonexpansive mappings is due to Kirk in [186]. Obviously, any nonexpansive selfmapping of a hyperbolic space is directionally nonexpansive, but the converse fails as directionally nonexpansive mappings not even need to be continuous on the whole space as the example below shows.

Example 17.43. (simplified by Paulo Oliva): Consider the convex subset $[0,1]^2$ of the normed space $(\mathbb{R}^2, \|\cdot\|_{\max})$ and the function

$$f : [0,1]^2 \to [0,1]^2,\ f(x,y) := \begin{cases} (1,y), \text{ if } y > 0 \\ (0,y), \text{ if } y = 0. \end{cases}$$

Clearly, f is directionally nonexpansive (even directionally constant) but discontinuous at $(0,0)$.

All the above properties of f (with the exception of 'quasi-nonexpansive' and 'weakly quasi-nonexpansive') can be written as $\forall$-formulas when formalized in $\mathcal{A}^\omega[X,d]$ resp. $\mathcal{A}^\omega[X,d,W]$ if the data $L, \alpha \in \mathbb{Q}^*_+, \omega$ or Ω are given which can be represented as objects of degree ≤ 1 and hence can be added to the parameter x^1 in theorem 17.35 (in the case of uniformly continuous and of bounded functions f this is due to the display of $<$ and $\leq$ we have chosen). Of course, one can also consider proofs where L, α or Ω are given by fixed terms of the underlying systems as long as these terms only contain free variables of degree ≤ 1. Even the property of f being weakly quasi-nonexpansive becomes an $\forall$-formula if the fixed point p is pulled out as an additional parameter. Because of this we can (see further below) apply our metatheorems to weakly quasi-nonexpansive functions but not to quasi-nonexpansive ones. Note, however, that while the conditions on f to be nonexpansive, (Hölder-)Lipschitz continuous or uniformly continuous imply the extensionality of f, this is not the case for the other notions considered above.

In the following results we use the representation of Polish spaces X and compact metric spaces K from chapter 4. Using this representation a statement of the kind, say,

$$(*)\ \forall x \in P \forall y \in K \big(\forall n \in \mathbb{N} A(x,y,n) \to \exists m \in \mathbb{N} B(x,y,m)\big)$$

has – formalized in $\mathcal{A}^\omega[X,d,W]$ – the form

$$\forall x^1 \forall y \leq_1 M \big(\forall n^0 A(x,y,n) \to \exists m^0 B(x,y,m)\big).$$

We make the convention that when writing $(*)$ we always tacitly assume that $A(x^1,y^1,n^0)$ and $B(x^1,y^1,m^0)$ are (when interpreted in $\mathcal{S}^{\omega,X}$ in the sense of definition 17.28) extensional w.r.t. $=_P, =_K$

$$x_1 =_P x_2 \ \wedge \ y_1 =_K y_2 \ \wedge \ A(x_1,y_1,n) \to A(x_2,y_2,n)$$

(analogously for $B(x,y,m)$) and therefore really express properties about elements in P, K. Actually, it suffices to use the weak form of the representation of compact metric spaces discussed at the end of chapter 4 here as this just adds another purely universal premise.

Notation. 17.44 *Let $f : X \to X$ be a selfmapping of a metric space (X,d). Then the fixed point set of f is defined as $Fix(f) := \{x \in X \mid x = f(x)\}$.*

In the following, we usually write the type $X(X)$ more suggestively as $X \to X$. Also $Fix(f) \neq \emptyset$ denotes the formalized statement

$$\exists p^X \big(f^{X \to X}(p) =_X p\big).$$

Corollary 17.45. 1) *Let P (resp. K) be a $\mathcal{A}^\omega$-definable Polish space (resp. compact metric space) and $B_\forall, C_\exists$ be as before $\forall$- resp. $\exists$-formulas.*
If $\mathcal{A}^\omega[X,d,W]$ proves a sentence

$$\forall x \in P \forall y \in K \forall z^X, f^{X \to X} \big(f \text{ n.e. } \wedge Fix(f) \neq \emptyset \wedge \forall u^0 B_\forall \to \exists v^0 C_\exists \big),$$

then one extract from the proof a computable functional $\Phi : \mathbb{N}^\mathbb{N} \times \mathbb{N} \to \mathbb{N}$ (on representatives $r_x : \mathbb{N} \to \mathbb{N}$ of elements x of P) such that for all $r_x \in \mathbb{N}^\mathbb{N}, b \in \mathbb{N}$

$$\forall y \in K \forall z^X \forall f^{X \to X} \big(f \text{ n.e.} \wedge \forall u \leq \Phi(r_x,b)\, B_\forall \to \exists v \leq \Phi(r_x,b)\, C_\exists \big)$$

holds in any (nonempty) hyperbolic space (X,d,W) whose metric is bounded by $b \in \mathbb{N}$ (where 'b_X' is to be interpreted by 'b').
For $\mathcal{A}^\omega[X,d,W,\mathrm{CAT}(0)]$ instead of $\mathcal{A}^\omega[X,d,W]$ the conclusion holds in any (nonempty) CAT(0)*-space (X,d) whose metric is bounded by b.*
2) *An analogous result holds if 'f n.e.' is replaced by 'f d.n.e'.*
Except for the elimination of the assumption '$Fix(f) \neq \emptyset$', the result holds as well for $\mathcal{A}^\omega[X,d], (X,d)$. However, even in the latter case we still can reduce '$Fix(f) \neq \emptyset$' to '$\forall k^0 \exists p^X (d_X(f(p),p) \leq_\mathbb{R} 2^{-k})$'.
Instead of single Polish and compact metric spaces P, K we have tuples of (potentially different) such spaces and corresponding tuples $\underline{x}, \underline{y}$.

Remark 17.46. Remark 17.37 applies to corollary 17.45 as well.

Proof: 1) The statement provable, by assumption, in $\mathcal{A}^{\omega}[X,d,W]$ can be written as

$$\forall x \in P \forall y \in K \forall z^X, p^X, f^{X \to X}\big(f \text{ n.e.} \wedge f(p) =_X p \wedge \forall u^0 B_\forall \to \exists v^0 C_\exists\big).$$

'$f(p) =_X p$' can be formalized as '$\forall k^0 \big(d_X(p, f(p)) \leq_{\mathbb{R}} 2^{-k}\big)$', where '$d_X(p, f(p)) \leq_{\mathbb{R}} 2^{-k}$' and '$f$ n.e.' are $\forall$-formulas. Moreover, using the representation of P (resp. K) in $\mathcal{A}^{\omega}$ (see chapter 4) quantification over $x \in P$ (resp. $y \in K$) is expressed as quantification over all x^1 (all $y^1 \leq s$ for some closed function term s). Hence by theorem 17.35 there is a functional Φ such that for all $x \in P, b \in \mathbb{N}$, if $r_x \in \mathbb{N}^{\mathbb{N}}$ represents x then

$$\begin{cases} \forall y \in K \forall z^X, p^X \forall f^{X \to X} \\ \quad \big(f \text{ n.e.} \wedge d_X(p, f(p)) \leq_{\mathbb{R}} 2^{-\Phi(r_x, b)} \wedge \forall u \leq_0 \Phi(r_x, b) B_\forall \to \exists v \leq_0 \Phi(r_x, b) C_\exists\big) \end{cases}$$

holds in any b-bounded hyperbolic space (X, d, W), where $\Phi(r_x, b)$ depends on the representative $r_x \in \mathbb{N}^{\mathbb{N}}$ of $x \in P$.
By theorem 1 in [126] (see proposition 18.25 in chapter 18 below) we have (since X is a bounded hyperbolic space),

$$\forall k \in \mathbb{N} \exists p \in X \big(d(p, f(p)) \leq 2^{-k}\big).$$

Hence the corollary follows.
2) follows like 1) observing that 'f directionally nonexpansive' is – formalized in $\mathcal{L}(\mathcal{A}^{\omega}[X,d,W])$ – a $\forall$-formula as well, namely

$$\forall x^X \forall \lambda^1 \big(d_X(f(x), f(W_X(x, f(x), \tilde{\lambda}))) \leq_{\mathbb{R}} d_X(x, W_X(x, f(x), \tilde{\lambda}))\big).$$

□

Remark 17.47. From the proof above it follows that parameters x in $\mathbb{N}$ or $\mathbb{N}^{\mathbb{N}}$ can be treated directly (without having to represent these spaces as Polish spaces). The same is true for $\mathbb{Q}^*_+$ since rational numbers can be encoded by natural numbers.

The reduction of the assumption $Fix(f) \neq \emptyset$ on f to its ε-weakening

$$\forall \varepsilon > 0 \exists p_\varepsilon \in X(d(p_\varepsilon, f(p_\varepsilon)) < \varepsilon)$$

(and subsequent elimination) in the proof of the corollary is reminiscent of the reduction of the axioms Γ to Γ_ε in theorem 15.1 in chapter 15 and the corresponding WKL-elimination (from chapter 10). However, whereas in these cases the benefit of this only was to replace the use of ineffective principles by classically equivalent (using WKL) constructive versions of these principles, in the absence of compactness this allows one to replace assumptions by even classically strictly weaker ones: while nonexpansive selfmappings of bounded hyperbolic spaces always have approximate fixed points they, in general, do not have fixed points (not even for closed bounded convex subsets of Banach spaces such as c_0 (see [340] for a general survey on such fixed point free mappings) unless e.g. X is a bounded convex subset of a uniformly convex Banach space. Even in the latter case this reduction is of relevance as it allows one to remove the use of the corresponding nontrivial fixed point theorem (due to Browder [55], Göhde [137] and Kirk [184]) and the additional completeness and closedness assumptions of this theorem, yielding in many cases fully elementary proofs. In the context of metric fixed point theory there are numerous proofs of theorems having the form required in the corollary which use the assumption that fixed points exist. In [224] it is shown how to achieve the elimination of this assumption in the concrete case of a proof due to Groetsch [144]. The corresponding results in [224] vastly generalize bounds from [191]. An extension of this result to uniformly convex hyperbolic spaces (with a monotone modulus of uniform convexity) is given in [263]. Further applications of this type can be found in [219, 231]. In the latter paper the use of an even more complicated fixed point theorem for so-called asymptotically nonexpansive mappings due to [125] could be eliminated from a proof of a result on the asymptotic behavior of certain iterations of these mappings (the paper even treats asymptotically weakly quasi-nonexpansive mappings and permits error terms in the iteration). Again this has been extended to uniformly convex hyperbolic spaces (with monotone modulus of uniform convexity), see [234].

We will now show that the elimination of the assumption '$Fix(f) \neq \emptyset$' can be carried out for a much more general class of formulas than just $\exists$-formulas (the latter restriction was necessary in the previous corollary only for the extractability of the bound Φ).

Definition 17.48. The class $\mathcal{K}$ of formulas consists of all formulas F that have a prenexation $F' \equiv \exists x_1^{\rho_1} \forall y_1^{\tau_1} \ldots \exists x_n^{\rho_n} \forall y_n^{\tau_n} F_\exists(\underline{x}, \underline{y})$ where $F_\exists$ is an $\exists$-formula, the types ρ_i are 0 and the types τ_i are of degree ≤ 1 or $(1,X)$. If $\tau_i, \ldots, \tau_n$ are of degree $(1,X)$, then ρ_i might even be of degree ≤ 1 or $(0,X)$.

Corollary 17.49. *Let P (resp. K) be a $\mathcal{A}^\omega$-definable Polish space (resp. compact metric space) and let the formula A be in the class $\mathcal{K}$. If $\mathcal{A}^\omega[X,d,W]$ proves a sentence*

$$\forall x \in P\, \forall y \in K\, \forall z^X, f^{X \to X} \big(f \text{ n.e.} \wedge Fix(f) \neq \emptyset \to A\big),$$

then the following holds in all (nonempty) bounded hyperbolic spaces (X,d,W):

$$\forall x \in P\, \forall y \in K\, \forall z^X, f^{X \to X} \big(f \text{ n.e.} \to A\big).$$

For $\mathcal{A}^\omega[X,d,W,\mathrm{CAT}(0)]$ instead of $\mathcal{A}^\omega[X,d,W]$ the conclusion holds in all (nonempty) bounded CAT(0)*-spaces (X,d).*
For $\mathcal{A}^\omega[X,d]$ one still can (as in corollary 17.45) replace $Fix(f) \neq \emptyset$ by the existence of approximate fixed points.

Proof: Let A be in prenex normal form of the form guaranteed by $A \in \mathcal{K}$. Consider the Herbrand normal form

$$A^H :\equiv \forall Y_1,\ldots,Y_n \exists x_1,\ldots,x_n A_\exists(x_1,\ldots,x_n,Y_1x_1,\ldots,Y_nx_1\ldots x_n)$$

of A. Since $A \to A^H$ holds by logic, the assumption implies that

$$\mathcal{A}^\omega[X,d,W] \vdash \forall x \in P\, \forall y \in K\, \forall z^X, f^{X\to X}\,(f \text{ n.e.} \to A^H).$$

The types of $\underline{Y}$ are of degree 1 or $(1,X)$ and so meet the requirements on τ and σ made in theorem 17.35 Moreover, the types of $\underline{x}$ are of degree 1 or $(0,X)$ and hence a-fortiori of degree 1 or $(1,X)$ so that

$$\exists x_1,\ldots,x_n A_\exists(x_1,\ldots,x_n,Y_1x_1,\ldots,Y_nx_1\ldots x_n)$$

is an $\exists$-formula. So we can apply theorem 17.35 (reasoning similarly to the proof of corollary 17.45) to conclude that

$$\forall x \in P\, \forall y \in K\, \forall z^X, f^{X\to X}\,(f \text{ n.e.} \to A^H)$$

holds in any (nonempty) bounded hyperbolic space (X,d) (with 'b_X' being interpreted by an upper bound $b \in \mathbb{N}$ of d). Since

$$\mathcal{S}^{\omega,X} \models A^H \to A$$

(using the axiom of choice on the meta-level), the corollary follows. □

Theorem 17.35 originally was proved in Kohlenbach [226] using the combination of negative translation and functional interpretation ND from chapter 10 extended to the new types using bar recursion (BR) for these types and interpreting the result in an extension of the model of strongly majorizable functionals (see chapter 11) to the situation at hand. For the latter we defined *s-maj*$_X$ on the ground type X in a trivial way as the always true relation which was possible by the boundedness of (X,d). We now – following Gerhardy-Kohlenbach [120] – develop an approach which also works for unbounded spaces and which will yield theorem 17.35 as a special case. This approach is based on a nontrivial extension of the strong majorizability relation which we need first as a syntactic relation and then in the proof also for the definition of the according type structure of strongly majorizable (in this sense) functionals. The main features of the new majorizability relation are the following:

- Majorants of functionals of type $\rho \in \mathbf{T}^X$ are functionals of type $\widehat{\rho} \in \mathbf{T}$ as defined in definition 17.32. Consequently, as all constructions take place on the level of the majorants, we can use the ordinary notions of computability or primitive recursive computability etc. for functionals over $\mathbb{N}$ and do not have to stipulate in any way that the underlying space (X,d) is equipped with a computability structure or anything of this sort.
- The majorization relation is a ternary relation which additionally depends on a reference point $a \in X$.
- While the construction of majorants will depend on the choice of $a \in X$, the domains of all strongly majorizable functionals of a given type will not.

- For types $\rho \in \mathbf{T}$, our relation coincides with the Bezem's original strong majorizability relation $s\text{-}maj_\rho$ and so, in particular, is independent from the choice of $a \in X$ in this case.

The theories $\mathcal{A}^\omega[X,d]_{-b}, \mathcal{A}^\omega[X,d,W]_{-b}$ **and** $\mathcal{A}^\omega[X,d,W,\mathrm{CAT}(0)]_{-b}$: These theories result from $\mathcal{A}^\omega[X,d], \mathcal{A}^\omega[X,d,W]$ and $\mathcal{A}^\omega[X,d,W,\mathrm{CAT}(0)]$ by dropping the axiom $d_X(x,y) \le_{\mathbb{R}} (b_X)_{\mathbb{R}}$ and deleting the constant b_X from the language.
The notions of a model (over $\mathcal{S}^{\omega,X}$) and of validity in this model are the same as before except that the interpretation of b_X is dropped.

We now define the new strong majorizability relation for all types $\rho \in \mathbf{T}^X$ which we denote – following [120] – by $\gtrsim^a_\rho$:

Definition 17.50. We define a ternary relation $\gtrsim^a_\rho$ between objects x, y and a of type $\widehat{\rho}, \rho$ and X respectively as follows:

- $x^0 \gtrsim^a_0 y^0 :\equiv x \ge_0 y$,
- $x^0 \gtrsim^a_X y^X :\equiv (x)_{\mathbb{R}} \ge_{\mathbb{R}} d_X(y,a)$,
- $x \gtrsim^a_{\tau(\rho)} y :\equiv \forall z',z(z' \gtrsim^a_\rho z \to xz' \gtrsim^a_\tau yz) \wedge \forall z',z(z' \gtrsim^a_{\widehat{\rho}} z \to xz' \gtrsim^a_{\widehat{\tau}} xz)$.

In the case of normed linear spaces (to be treated below) we always choose a to be the zero vector 0_X, i.e. $d_X(x,a) =_{\mathbb{R}} \|x\|_X$.

As $\gtrsim^a$ is a relation between objects of different types, the definition of $\gtrsim^a_{\tau(\rho)}$ is slightly more complicated than the corresponding definition of $s\text{-}maj_{\tau(\rho)}$. The first part of the clause ensures that x is a 'majorant' for y, the second part ensures that x also majorizes itself. Since majorants are of type $\widehat{\rho}$, this corresponds to requiring that for all majorants x $s\text{-}maj$ x, and so the definition of $\gtrsim^a_{\tau(\rho)}$ could equivalently be rewritten as:

$$x \gtrsim^a_{\tau(\rho)} y :\equiv \forall z',z(z' \gtrsim^a_\rho z \to xz' \gtrsim^a_\tau yz) \wedge x\ s\text{-}maj_{\widehat{\tau}(\widehat{\rho})}\ x.$$

Remark 17.51. Restricted to the types $\mathbf{T}$, the relation $\gtrsim^a$ is equivalent to Bezem's notion of strong majorizability $s\text{-}maj$.

In the following, majorization relative to the relation $\gtrsim^a$ will be called (strong) 'a-majorization', i.e. if $t_1 \gtrsim^a t_2$ for terms t_1, t_2 we say that t_1 a-majorizes t_2 and we call t_1 an a-majorant. If the term t_1 does not depend on a and a-majorizes t_2 for any a we say that t_1 *uniformly* a-majorizes t_2. We will in general aim for uniform majorants so as to produce uniform bounds.

Below we write again $F_\forall$ (resp. $F_\exists$) for $\forall$-formulas (resp. $\exists$-formulas).

When dealing with the theories $\mathcal{A}^\omega[X,d]_{-b}, \mathcal{A}^\omega[X,d,W]_{-b}$ and $\mathcal{A}^\omega[X,d,W,\mathrm{CAT}(0)]_{-b}$ we usually will assume that the constant 0_X does not occur in the sentences we consider. This is no restriction since 0_X just is an arbitrary Skolem constant which could have been replaced by any new variable of type X which – taking the universal closure – would just add another input that had to be a-majorized. In the case of normed spaces, however, the constant 0_X denotes the zero vector whose use is crucial. Nevertheless, since in this case we always take $a := 0_X$ this constant is trivially a-majorized by 0^0.

Theorem 17.52 (General theorem on proof mining: the abstract metric and hyperbolic case, Gerhardy-Kohlenbach [120]).

1) *Let ρ be of degree $(1,X)$ or ≤ 2 and let $B_\forall(x,u)$, resp. $C_\exists(x,v)$, be $\forall$- resp. $\exists$-formulas that contain only x,u free, resp. x,v free. Assume that 0_X does not occur in $B_\forall, C_\exists$. From a proof in $\mathcal{A}^\omega[X,d]_{-b}$ of*

$$\forall x^\rho (\forall u^0 B_\forall(x,u) \to \exists v^0 C_\exists(x,v)).$$

one can extract a partial functional $\Phi : S_{\widehat{\rho}} \rightharpoonup \mathbb{N}$ whose restriction to the strongly majorizable elements of $S_{\widehat{\rho}}$ is a total computable functional of $\mathcal{M}^\omega$ (in the sense of [195] relativized to $\mathcal{M}^\omega$) such that the following is true: for all (nonempty) metric spaces (X,d) and for all $x \in S_\rho, x^* \in S_{\widehat{\rho}}$ if there exists an $a \in X$ s.t. $x^* \gtrsim^a x$ then

$$\forall u \leq \Phi(x^*) B_\forall(x,u) \to \exists v \leq \Phi(x^*) C_\exists(x,v))$$

holds in (X,d).

In particular, if ρ is of degree 1^*, then $\Phi : S_{\widehat{\rho}} \mapsto \mathbb{N}$ is a total computable functional (in the ordinary sense of type-2 recursion theory).

2) *If the premise of the theorem is provable in $\mathcal{A}^\omega[X,d,W]_{-b}$ (resp. in $\mathcal{A}^\omega[X,d,W,\mathrm{CAT}(0)]_{-b}$) instead of $\mathcal{A}^\omega[X,d]_{-b}$, then the conclusion is valid in any nonempty hyperbolic space (X,d,W) (resp. $\mathrm{CAT}(0)$-space (X,d)).*

Instead of single variables x,u,v and a single premise $\forall u B_\forall(x,u)$ we may have tuples of variables $\underline{x}$ and a finite conjunction of premises, where in the case of $\underline{x} = x_1,\ldots,x_n$ there have to exist componentwise majorants $\underline{x}^* = x_1^*,\ldots,x_n^*$ for a common point $a \in X$.

The theorem will be proved in section 17.4 below.

Remark 17.53. Note that in the above theorem the functional Φ is defined for x^*, since $x^* \gtrsim^a x$ implies that x^* *s-maj* x^*.

In concrete applications specially designed corollaries of the general metatheorem 17.52 will be used. For most of our applications, the following two, more concrete, versions of the metatheorem are sufficient:

Corollary 17.54 (Gerhardy-Kohlenbach [120]).

1) *Let P (resp. K) be a $\mathcal{A}^\omega$-definable Polish space (resp. compact metric space), let τ be of degree 1^* and let $B_\forall(x,y,z,u)$, resp. $C_\exists(x,y,z,v)$, be $\forall$- resp. $\exists$-formulas that contain only x,y,z,u free, resp. x,y,z,v free, where furthermore 0_X does not occur in $B_\forall, C_\exists$. From a proof of a sentence*

$$\forall x \in P \forall y \in K \forall z^\tau (\forall u^0 B_\forall(x,y,z,u) \to \exists v^0 C_\exists(x,y,z,v))$$

in $\mathscr{A}^\omega[X,d]_{-b}$ *one can extract a computable functional* $\Phi : \mathbb{N}^\mathbb{N} \times \mathbb{N}^{(\mathbb{N}\times\ldots\times\mathbb{N})} \to \mathbb{N}$ *s.t. the following holds in every nonempty metric space: for all representatives* $r_x \in \mathbb{N}^\mathbb{N}$ *of* $x \in P$ *and all* $z \in S_\tau, z^* \in \mathbb{N}^{(\mathbb{N}\times\ldots\times\mathbb{N})}$ *if there exists an* $a \in X$ *for which* $z^* \gtrsim^a_\tau z$ *then*

$$\forall y \in K\big(\forall u \leq \Phi(r_x, z^*)B_\forall \to \exists v \leq \Phi(r_x, z^*)C_\exists\big).$$

As in theorem 17.52 we have a tuple of variables $\underline{z}^{\underline{\tau}}$ *as long as all the types in* $\underline{\tau}$ *are of degree* 1^* *and in the conclusion* $z^*_i \gtrsim^a_\tau z_i$ *is assumed for a common* $a \in X$ *for all* τ_i *in* $\underline{\tau} = \tau_1, \ldots, \tau_n$. *Also instead of single Polish and compact metric spaces P,K we may have tuples of (potentially different) such spaces and corresponding tuples* $\underline{x}, \underline{y}$.

2) *If the premise of the theorem is provable in* $\mathscr{A}^\omega[X,d,W]_{-b}$ *(resp. in* $\mathscr{A}^\omega[X,d,W,\mathrm{CAT}(0)]_{-b}$*) instead of* $\mathscr{A}^\omega[X,d]_{-b}$*, then the conclusion is valid in any nonempty hyperbolic space* (X,d,W) *(resp.* CAT(0)*-space* (X,d)*).*

Proof: Using the representation of P and K in $\mathscr{A}^\omega$ from chapter 4 as in the proof of corollary 17.45, quantification over $x \in P$ and $y \in K$ can be expressed as quantification over all x^1, resp. all $y^1 \leq s$ for some closed function term s. Clearly, x^M s-$maj_1 x$ for all x^1 (and so we have $x^M \gtrsim^a x$ for all $a \in X$) and – likewise – $s^M \gtrsim^a y$ for all $y \leq_1 s$, where $x^M(n) := \max\{x(i) : i \leq n\}$.
Since τ has degree 1^*, by Theorem 17.52 we obtain a computable functional $\tilde{\Phi}$ such that $\tilde{\Phi}(r^M_x, s^M, z^*)$ is a bound on u and v, whenever $r^1_x \in \mathbb{N}^\mathbb{N}$ is a representative of $x \in P$. Now define $\Phi(r_x, z^*) := \tilde{\Phi}(r^M_x, s^M, z^*)$. □

Corollary 17.55 (Gerhardy-Kohlenbach [120]).

1) *Let P (resp. K) be a* $\mathscr{A}^\omega$*-definable Polish space (resp. compact metric space). Assume one can prove in* $\mathscr{A}^\omega[X,d,W]_{-b}$ *a sentence:*

$$\forall x \in P \forall y \in K \forall z^X \forall f^{X\to X}(f\,n.e. \wedge \forall u^0 B_\forall(x,y,z,f,u) \to \exists v^0 C_\exists(x,y,z,f,v)),$$

where 0_X *does not occur in* $B_\forall$ *and* $C_\exists$*. Then one can extract a computable functional* $\Phi : \mathbb{N}^\mathbb{N} \times \mathbb{N} \to \mathbb{N}$ *s.t. for all representatives* $r_x \in \mathbb{N}^\mathbb{N}$ *of* $x \in P$ *and all* $b \in \mathbb{N}$

$$\begin{aligned}&\forall y \in K \forall z^X, \forall f^{X\to X}(f\ n.e. \wedge d_X(z, f(z)) \leq_\mathbb{R} (b)_\mathbb{R} \wedge \\ &\qquad \forall u^0 \leq \Phi(r_x, b) B_\forall(x,y,z,f,u) \to \exists v^0 \leq \Phi(r_x, b) C_\exists(x,y,z,f,v))\end{aligned}$$

holds in all (nonempty) hyperbolic spaces (X,d,W).
If the premise of this rule is proved in $\mathscr{A}^\omega[X,d,W,\mathrm{CAT}(0)]_{-b}$*, then the conclusion holds in all (nonempty)* CAT(0)*-spaces* (X,d).

2) *The corollary also holds for an additional parameter* $\forall \tilde{z}^X$ *if one adds the additional premise* $d_X(z, \tilde{z}) \leq_\mathbb{R} (b)_\mathbb{R}$ *in the conclusion, i.e.*

$$\forall y \in K \forall z^X \tilde{z}^X \forall f^{X \to X}(f \text{ n.e. } \wedge d_X(z, f(z)) \leq_{\mathbb{R}} (b)_{\mathbb{R}} \wedge d_X(z, \tilde{z}) \leq_{\mathbb{R}} (b)_{\mathbb{R}}$$
$$\forall u^0 \leq \Phi(r_x, b) B_\forall(x, y, z, \tilde{z}, f, u) \to \exists v^0 \leq \Phi(r_x, b) C_\exists(x, y, z, \tilde{z}, f, v))$$

holds in all (nonempty) hyperbolic spaces.

3) *Furthermore, the corollary holds for an additional parameter* $\forall c^{X(0)}$ *if one adds the additional premise* $\forall n(d_X(z, c(n)) \leq_{\mathbb{R}} (g(n))_{\mathbb{R}})$ *in the conclusion, where the extracted bound then additionally depends on* $g : \mathbb{N} \to \mathbb{N}$.
4) *The results 1), 2) and 3) also hold if we replace 'f n.e.' with* f *being Lipschitz continuous and Hölder-Lipschitz continuous functions (with L, resp.* $L, \alpha \in \mathbb{Q}^*_+$, *where* $\alpha \leq 1$, *as parameters), as well as uniformly continuous functions (with a modulus of uniform continuity* $\omega : \mathbb{N} \to \mathbb{N}$ *taken as parameter). For Lipschitz and Hölder-Lipschitz continuous functions the bound depends on parameters L resp.* L, α. *For uniformly continuous functions the bound depends on a given modulus of uniform continuity* ω^1.
5) *Furthermore, if we replace 'f n.e.' with 'f weakly quasi-nonexpansive (with fixed point p)', then 1), 2) and 3) hold if one adds the premise '*$d_X(z, p) \leq_{\mathbb{R}} (b)_{\mathbb{R}}$*' in the conclusion.*
6) *Moreover, 1), 2) and 3) also hold if we replace 'f n.e.' with* f *satisfying*

$$\forall x^X, y^X, n^0(d_X(x, y) <_{\mathbb{R}} (n)_{\mathbb{R}} \to d_X(f(x), f(y)) \leq_{\mathbb{R}} (\Omega(n))_{\mathbb{R}}),$$

where Ω^1 *is treated as parameter. Then the bound will depend additionally on the function* $\Omega : \mathbb{N} \to \mathbb{N}$.

7) *Finally, 1), 2) and 3) also hold if we replace '*f *n.e.* $\wedge d_X(z, f(z)) \leq_{\mathbb{R}} (b)_{\mathbb{R}}$*' with* f *satisfying*

$$\forall \tilde{z}, n^0(d_X(z, \tilde{z}) <_{\mathbb{R}} (n)_{\mathbb{R}} \to d_X(z, f(\tilde{z})) \leq_{\mathbb{R}} (\tilde{\Omega}(n))_{\mathbb{R}}),$$

which – for monotone $\tilde{\Omega}$ *and with '*$d_X(z, \tilde{z}) \leq_{\mathbb{R}} (n)_{\mathbb{R}}$*' instead of '*$d_X(z, \tilde{z}) <_{\mathbb{R}} (n)_{\mathbb{R}}$ *– just expresses that* $\tilde{\Omega}$ *is a z-majorant of* f.
Then the bound will depend instead of b on the parameter $\tilde{\Omega} : \mathbb{N} \to \mathbb{N}$.

Remark 17.56. 1) The conditions on f in 1)–4) automatically imply the extensionality of f. For 5)–7) this is no longer the case so that in these cases one has to rely on the weak quantifier-free rule of extensionality instead in applications.
2) Even if 'z' does not occur in neither $B_\forall$ nor $C_\exists$ we need in 1)–4) and 6) for the bound Φ a number b which is an upper bound of $d(z, f(z))$ for **some** z as this is used to construct a majorant for f. In 5), we can instead use the fixed point p and as b e.g. 1 since $1 \geq d(p, f(p))$ even for ε-fixed points of f as long as $\varepsilon \leq 1$. In 7) we don't need any b as $\lambda n^0.\tilde{\Omega}^M(n+1)$ is already a z-majorant of f (see the proof below).
3) In '4)' and '6)–7)' above we may even allow that the formulas $B_\forall, C_\exists$ may depend on the additional parameters $L, \alpha, \omega, \Omega, \tilde{\Omega}$. Instead of having $L, \alpha, \omega, \Omega, \tilde{\Omega}$ as free variables these quantities might also be given as terms of the formal systems

in question as long as these terms only contain free variables of degree ≤ 1. Then the bound extractable will depend in general on these variables as additional inputs.

Proof: 1) By the comment after example 17.42 the premise 'f n.e.' is a $\forall$-formula and hence an admissible premise in Theorem 17.52. The quantifiers $\forall x \in P, y \in K$ ranging over the Polish space P, resp, compact metric space K, are treated as in the proof of corollary 17.54. Choose $a := z$, then trivially $0^0 \gtrsim^a z$. Moreover, using $d(z, f(z)) \leq b$ and the nonexpansivity of f we get

$$d(z,\tilde{z}) \leq n \to d(z, f(\tilde{z})) \leq d(z, f(z)) + d(f(z), f(\tilde{z})) \leq b + d(z,\tilde{z}) \leq n + b.$$

Hence $\lambda n^0.(n+b) \gtrsim^a f$.

For better readability we write here and for the rest of this proof simply d and b instead of d_X and $(b)_{\mathbb{R}}$ etc.

For 2) and 3) is suffices to additionally observe that with $a := z$ and $b \geq d(z,\tilde{z})$, trivially $b \gtrsim^a \tilde{z}$ and, with g such that $\forall n(d(z,c(n)) \leq g(n))$, $g^M \gtrsim^a c$.
For 4), 6), we will show that $d(z, f(z)) \leq b$ in conjunction with the requirement that f is Lipschitz continuous (with constant $L \in \mathbb{Q}^*_+$), Hölder-Lipschitz continuous (with constants $L, \alpha \in \mathbb{Q}^*_+, \alpha \leq 1$), uniformly continuous (with modulus $\omega : \mathbb{N} \to \mathbb{N}$) or bounded (with bounding function $\Omega : \mathbb{N} \to \mathbb{N}$) allows one to construct in these data a modulus $\tilde{\Omega}$ as in 7). Similarly, for 5), we show that if f is weakly quasi-nonexpansive and the additional premise $d(z,p) \leq b$ is satisfied, then one can construct in b a modulus $\tilde{\Omega}$ as in 7) for f. As mentioned above, all conditions can be written as $\forall$-formulas and may hence serve as a premise according to our metatheorem. Hence it then remains to prove 7) which, however, is almost trivial: if f satisfies

$$(*)\ \forall \tilde{z} \in X (d(z,\tilde{z}) < n \to d(z, f(\tilde{z})) \leq \tilde{\Omega}(n)),$$

then trivially $\lambda n.\tilde{\Omega}^M(n+1) \gtrsim^a f$ for $a := z$ since

$$d(z,\tilde{z}) \leq n \to d(z,\tilde{z}) < n+1 \overset{(*)}{\to} d(z, f(\tilde{z})) \leq \tilde{\Omega}(n+1).$$

Using the fact that $<_{\mathbb{R}}$ is a Σ^0_1-statement and $\leq_{\mathbb{R}}$ is a Π^0_1-statement we can express $(*)$ as a $\forall$-formula. The results then follow using corollary 17.54 and remark 17.47.

To conclude the proof of 4), let f be Hölder-Lipschitz continuous with constants $L, \alpha \in \mathbb{Q}^*_+$, where $0 < \alpha \leq 1$, and assume that $d(z, f(z)) \leq b \in \mathbb{N}$ and $d(z,\tilde{z}) \leq n$. Then

$$d(z, f(\tilde{z})) \leq d(z, f(z)) + d(f(z), f(\tilde{z})) \leq L \cdot d(z,\tilde{z})^\alpha + b \leq L \cdot n^\alpha + b \leq \lceil L \rceil \cdot n + b \in \mathbb{N}.$$

This means that f satisfies $(*)$ with $\tilde{\Omega}(n) := \lceil L \rceil \cdot n + b$.

Next, let $f : X \to X$ be uniformly continuous with modulus ω^1, i.e.

$$\forall x, y \in X \forall k \in \mathbb{N}(d(x,y) < 2^{-\omega(k)} \to d(f(x), f(y)) \leq 2^{-k}).$$

Then f satisfies $(*)$ with $\tilde{\Omega}(n) := n \cdot 2^{\omega(0)} + 1 + b$. We now use that (X,d,W) is a hyperbolic space: given $z, \tilde{z} \in X$ with $d(z,\tilde{z}) \leq n$ we can (using the fact that $W(x,y,\frac{1}{2})$ is a midpoint of x,y) inductively construct $n \cdot 2^{\omega(0)}$-many points $z_1, \ldots z_{k-1}$ such that

$$d(z,z_1), d(z_1,z_2), \ldots, d(z_{k-1},\tilde{z}) < 2^{-\omega(0)}$$

and hence

$$d(f(z),f(z_1)), d(f(z_1),f(z_2)), \ldots, d(f(z_{k-1}),f(\tilde{z})) \leq 1 (= 2^{-0}).$$

Then by the triangle inequality $d(f(z),f(\tilde{z})) \leq k = n \cdot 2^{\omega(0)} + 1$ and another use of the triangle inequality yields $d(z,f(\tilde{z})) \leq d(z,f(z)) + d(f(z),f(\tilde{z})) \leq n \cdot 2^{\omega(0)} + 1 + b$.

The proof of 5) is concluded as follows: in order to express 'f is weakly quasi-nonexpansive' as a $\forall$-condition, we need to take the fixed point p as an additional parameter. Therefore, for weakly quasi-nonexpansive functions f, we need to add an additional premise: '$d(z,p) \leq b$'. Then for $a = z$ the function f satisfies $(*)$ with $\tilde{\Omega}(n) := n + 2b$, as given $d(z,\tilde{z}) < n$

$$\begin{aligned} d(z,f(\tilde{z})) &\leq d(z,p) + d(f(\tilde{z}),p) \leq d(z,p) + d(\tilde{z},p) \\ &\leq d(z,p) + d(\tilde{z},z) + d(z,p) \leq n + 2b. \end{aligned}$$

Alternatively, choosing $a = p$ (and adjusting the other majorants accordingly) f even satisfies $(*)$ with $\tilde{\Omega}(n) := n$, as given $d(p,\tilde{z}) < n$

$$d(p,f(\tilde{z})) \leq d(p,\tilde{z}) \leq n.$$

Finally, for f, b and Ω as in 6) one easily shows that $\tilde{\Omega}(n) + b$ satisfies the condition in 7) so that this case, which was already treated, applies. □

Remark 17.57. Except for the case of f being uniformly continuous all results also hold for $\mathcal{A}^\omega[X,d]_{-b}$ and general metric spaces (X,d) instead of $\mathcal{A}^\omega[X,d,W]_{-b}$ and hyperbolic spaces (X,d,W). Note, that in metric spaces uniformly continuous functions f need not be bounded in the sense of having a modulus Ω as in 6) and, in general, will not even have a modulus $\tilde{\Omega}$ as in 7). This is due to the fact that for given $x,y \in X$ one in general cannot construct intermediate points (see also [289] for some related discussion).

Consider e.g. $(\mathbb{N}^2, D)$, where

$$D\big((n_1,m_1),(n_2,m_2)\big) := |n_1 - n_2| + \min\{1, |m_1 - m_2|\}.$$

One easily verifies that D is a metric and that any function $f : \mathbb{N}^2 \to \mathbb{N}^2$ is uniformly continuous (e.g. we may take $\omega_f(\varepsilon) := 1/2$ as modulus of uniform continuity). However, $\varphi\big((n,m)\big) := (m,n)$ is not $(0,0)$-majorizable (and so also not a-

majorizable for any other $a \in \mathbb{N}^2$). This follows from the fact that $D\big((0,0),(0,n)\big) \leq 1$ for all $n \in \mathbb{N}$ while $D\big((0,0),\varphi(0,n)\big) = n$.

Definition 17.58. Let $f : X \to X$, then $Fix_\varepsilon(f,y,b) := \{x \in X \mid d(x,f(x)) \leq \varepsilon \wedge d(x,y) \leq b\}$, i.e. $Fix_\varepsilon(f,y,b) \neq \emptyset$ expresses f has an ε-fixed point in a b-neighborhood of y.

As a generalization of corollary 17.45 to the unbounded case (for nonexpansive functions) we prove the following:

Corollary 17.59 (Gerhardy-Kohlenbach [120]).

1) *Let P (resp. K) be a $\mathcal{A}^\omega$-definable Polish space (resp. compact metric space) and let $B_\forall$ and $C_\exists$ be as before. From a proof of a sentence*

$$\forall x \in P\, \forall y \in K\, \forall z^X, f^{X\to X}\big(f \text{ n.e. } \wedge Fix(f) \neq \emptyset \wedge \forall u^0 B_\forall \to \exists v^0 C_\exists\big).$$

in $\mathcal{A}^\omega[X,d]_{-b}$ one can extract a computable functional $\Phi : \mathbb{N}^\mathbb{N} \times \mathbb{N} \to \mathbb{N}$ (on representatives $x : \mathbb{N} \to \mathbb{N}$ of elements of P) s.t. for all $x \in \mathbb{N}^\mathbb{N}, b \in \mathbb{N}$

$$\forall y \in K\, \forall z^X, f^{X\to X}\big(f \text{ n.e. } \wedge \forall \varepsilon > 0 (Fix_\varepsilon(f,z,b) \neq \emptyset)$$
$$\wedge d_X(z,f(z)) \leq_\mathbb{R} (b)_\mathbb{R} \wedge \forall u^0 \leq \Phi(x,b) B_\forall \to \exists v^0 \leq \Phi(x,b) C_\exists\big).$$

holds in any (nonempty) metric space (X,d).
Similarly, if f instead of being nonexpansive satisfies one of the other conditions from the previous corollary (modulo rem. 17.57).

2) *If the premise of the theorem is provable in $\mathcal{A}^\omega[X,d,W]_{-b}$ (resp. in $\mathcal{A}^\omega[X,d,W,\mathrm{CAT}(0)]_{-b}$) instead of $\mathcal{A}^\omega[X,d]_{-b}$, then the conclusion is valid in any nonempty hyperbolic space (X,d,W) (resp.* CAT(0)*-space (X,d)).*

Proof: 1) The statement assumed to be provable can be written as

$$\forall x \in P\, \forall y \in K\, \forall z^X, w^X, f^{X\to X}\big(f \text{ n.e. } \wedge f(w) =_X w \wedge \forall u^0 B_\forall \to \exists v^0 C_\exists\big),$$

where '$f(w) =_X w$' can be written as $\forall k^0 (d(w,f(w)) \leq 2^{-k})$, and both '$d(w,f(w)) \leq 2^{-k}$' and '$f$ n.e. ' are $\forall$-formulas.
By corollary 17.55 and rem. 17.57, we extract a functional Φ s.t. for all $x \in P$ if $r_x \in \mathbb{N}^\mathbb{N}$ represents x then

$$\forall y \in K\, \forall z^X, w^X, f^{X\to X}\big(d(z,w), d(z,f(z) \leq b \wedge f \text{ n.e. } \wedge$$
$$d(f(w),w) \leq 2^{-\Phi(r_x,b)} \wedge \forall u^0 \leq \Phi(r_x,b) B_\forall \to \exists v^0 \leq \Phi(r_x,b) C_\exists\big)$$

holds in all (nonempty) metric spaces (X,d).
The statement $d(z,w) \leq b \wedge d(f(w),w) \leq 2^{-\Phi(r_x,b)}$ expresses that f has $2^{-\Phi(r_x,b)}$-fixed points in a b-neighborhood of z, which, since $2^{-\Phi(r_x,b)}$ does not depend on w, is implied by $Fix_\varepsilon(f,z,b) \neq \emptyset$, so the corollary follows. □
2) is proved analogously.

Similarly to corollary 17.49 also in the above context one can carry out the reduction of the assumption

$$Fix(f) \neq \emptyset$$

to

$$\exists b^0 \forall \varepsilon > 0 \big(Fix_\varepsilon(f,z,b) \neq \emptyset\big)$$

for a rather general class $\mathcal{H}$ of formulas A instead of purely existential ones. However, compared to the class $\mathcal{K}$ in corollary 17.49 we have to restrict the types slightly more in order to ensure the majorizability of the Herbrand index functions in the absence of the assumption on X being bounded. The appropriate class is the following one:

Definition 17.60. The class $\mathcal{H}$ consists of all formulas (in the language of the theory in question) that have a prenex normal form

$$\exists x_1^0 \forall y_1^{\tau_1} \ldots \exists x_n^0 \forall y_n^{\tau_n} F_\exists(\underline{x},\underline{y}),$$

where $F_\exists$ is an $\exists$-formula and the types τ_i are of degree ≤ 1 or $(0,X)$.

The proof of the following result is similar to the one of corollary 17.49 and can be found in [120]:

Corollary 17.61. *Let P (resp. K) be a $\mathcal{A}^\omega$-definable Polish (resp. compact) metric space and let the formula A be in the class $\mathcal{H}$. If $\mathcal{A}^\omega[X,d]_{-b}$ proves a sentence*

$$\forall x \in P\, \forall y \in K\, \forall z^X, f^{X\to X} \big(f \text{ n.e.} \wedge Fix(f) \neq \emptyset \to A\big)$$

then the following holds in every nonempty metric space (X,d):

$$\forall x \in P\, \forall y \in K\, \forall z^X, f^{X\to X}$$
$$\big(f \text{ n.e.} \wedge \exists b^0 \forall \varepsilon > 0 (Fix_\varepsilon(f,z,b) \neq \emptyset) \to A\big).$$

The analogous statement holds for $\mathcal{A}^\omega[X,d,W]_{-b}$ (resp. $\mathcal{A}^\omega[X,d,W,\mathrm{CAT}(0)]_{-b}$) and nonempty hyperbolic spaces (X,d,W) (resp. CAT(0)-spaces (X,d)).

Instead of $Fix(f) \neq \emptyset$ one can also treat different assumptions, e.g.

$$\exists p^X \left(\Psi(x^1, y^1, z^X, f^{X\to X}, p^X) =_{\mathbb{R}} 0_{\mathbb{R}}\right),$$

where Ψ is a functional given by a closed term of type $1(X)(X \to X)(X)(1)(1)$ of the underlying system. Then for the same class of formulas $\mathcal{H}$ as in corollary 17.61 one can reduce this assumption in a proof to the existence of approximate roots within some ball around z, i.e. to

$$\exists b^0 \forall \varepsilon > 0 \exists p^X \left(d_X(z,p) \leq_{\mathbb{R}} (b)_{\mathbb{R}} \wedge |\Psi(x,y,z,f,p)|_{\mathbb{R}} \leq_{\mathbb{R}} \varepsilon\right).$$

We leave it to the reader to carry out the details.

17.3 The case of normed spaces

In the following, we adapt the results obtained above to the setting of (real) normed linear spaces with convex subsets C. As we saw above, our metatheorems for metric and hyperbolic spaces became particularly easy to formulate in the case where these spaces were assumed to be bounded. For normed spaces this is not possible since nontrivial normed spaces always are unbounded, while many theorems using convex subsets C of normed spaces assume C to be bounded. It, therefore, sometimes is convenient to be able to formulate theories dealing with bounded convex subsets of normed spaces without having to formalize the underlying (unbounded) normed space. According to a nice result due to Machado [269], one, in fact, can characterize convex subsets of normed spaces (up to an isometric embedding) in the setting of hyperbolic spaces by adding two further axioms on W. The additional conditions are

- (I) that the convex combinations do not depend on the order in which they are carried out, and
- (II) that the distance is homothetic.

Formally stated, the conditions (I) and (II) read as follows:

$$\begin{aligned}&(\mathrm{I})\ \forall x,y,z \in X \forall \lambda_1,\lambda_2,\lambda_3 \in [0,1] \big(\lambda_1+\lambda_2+\lambda_3 =_{\mathbb{R}} 1 \to \\ &\qquad W(z,W(y,x,\tfrac{\lambda_1}{1-\lambda_3}),1-\lambda_3) = W(x,W(z,y,\tfrac{\lambda_2}{1-\lambda_1}),1-\lambda_1)\big),\\ &(\mathrm{II})\ \forall x,y,z \in X \forall \lambda \in [0,1] \big(d(W(z,x,\lambda),W(z,y,\lambda)) = \lambda \cdot d(x,y)\big).\end{aligned}$$

In order to express axiom (I) as a purely universal sentence we have to avoid the universal premise $\lambda_1+\lambda_2+\lambda_3 =_{\mathbb{R}} 1$ (recall that equality on the reals is a universal statement and hence the axiom itself would no longer be purely universal).
Instead, given λ_1,λ_2 we explicitly define $\bar{\lambda}_1,\bar{\lambda}_2$ and $\bar{\lambda}_3$ so that provably (in $\mathcal{A}^\omega$) $\bar{\lambda}_1,\bar{\lambda}_2,\bar{\lambda}_3$ always represent real numbers in $[0,1]$ with $\bar{\lambda}_1+\bar{\lambda}_2+\bar{\lambda}_3 =_{\mathbb{R}} 1$ and, conversely, for $\lambda_i \in [0,1]$ with $\lambda_1+\lambda_2+\lambda_3 =_{\mathbb{R}} 1$ we have that $\bar{\lambda}_i = \lambda_i$ for $i=1,2,3$. The formal versions of the axioms are then as follows:

$$\begin{aligned}&(\mathrm{I})\ \forall x^X,y^X,z^X \forall \lambda_1^1,\lambda_2^1\\ &\qquad \big(W_X(z,W_X(y,x,\tfrac{\bar{\lambda}_1}{1-\bar{\lambda}_3}),1-\bar{\lambda}_3) = W_X(x,W_X(z,y,\tfrac{\bar{\lambda}_2}{1-\bar{\lambda}_1}),1-\bar{\lambda}_1)\big),\\ &\quad \text{where } \bar{\lambda}_1 =_1 \tilde{\lambda}_1,\ \bar{\lambda}_2 =_1 \min_{\mathbb{R}}(\tilde{\lambda}_2, 1-_{\mathbb{R}}\bar{\lambda}_1) \text{ and } \bar{\lambda}_3 =_1 1-_{\mathbb{R}}(\bar{\lambda}_1+_{\mathbb{R}}\bar{\lambda}_2),\\ &(\mathrm{II})\ \forall x^X,y^X,z^X \forall \lambda^1 \big(d_X(W_X(z,x,\lambda),W_X(z,y,\lambda)) =_{\mathbb{R}} \tilde{\lambda} \cdot_{\mathbb{R}} d_X(x,y)\big),\end{aligned}$$

where $\tilde{\lambda}$ is the construction in Definition 4.24. As discussed for the other (X,d,W) axioms in Remark 17.24, the axiom (II) is formulated with W_X to implicitly satisfy $W_X(x,y,\lambda) =_X W_X(x,y,\tilde{\lambda})$.

From the proofs of theorems 17.35 and 17.52 to be given further below it will be clear that the theorems immediately apply also to any extension of the systems considered so far by new axioms which can be written as $\forall$-formulas (like (I), (II) above), where then the conclusion holds in all spaces (X,d) resp. (X,d,W) which satisfy these additional axioms. Hence just as we treated the case of CAT(0)-spaces by adding the $\forall$-axiom CN$^-$ to the system for hyperbolic spaces, we can treat the case of convex subsets of normed spaces by adding instead (I), (II). However, for many applications it is useful or even necessary to be able to speak also about the underlying (real) normed space directly. For this reason we now develop formal systems for abstract (real) normed spaces as well as normed spaces together with an abstract convex subset C. The latter will be formalized via its characteristic function, where again we will have to consider extensionality issues as we only can stipulate weak extensionality for that characteristic function. Nevertheless, our theorems cover substantial applications to fixed point theory as the ones given in Kohlenbach [224] (providing an effective form of a theorem of [144]) as well as in Kohlenbach-Lambov [231] (dealing with asymptotically weakly quasi-nonexpansive selfmappings of convex subsets of uniformly convex normed spaces).

The theory $\mathscr{A}^\omega[X,\|\cdot\|]$:

$\mathscr{A}^\omega[X,\|\cdot\|]$ is defined as follows:

(i) Just as in the case of $\mathscr{A}^\omega[X,d]$ we first extend $\mathscr{A}^\omega$ to the larger set $\mathbf{T}^X$ of all finite types over the two ground types 0 (representing as before $\mathbb{N}$) and X.

(ii) Next, constants $0_X, 1_X$ of type X are added.

(iii) Finally, instead of b_X^0 and d_X and their corresponding axioms we now include constants $+_X$ of type $X(X)(X)$, $-_X$ of type $X(X)$, $\cdot_X$ of type $X(X)(1)$, $\|\cdot\|_X$ of type $1(X)$ together with the axioms (writing as usual $x+_X y$, $x-_X y$, $\|x\|_X$ and $\alpha\cdot_X x$ (or even αx) for $+_X(x,y)$, $+_X(x,-_X y)$, $\|\cdot\|_X(x)$ and $\cdot_X(\alpha,x)$):

   (0) The (purely universal) vector space axioms for $+_X$, $-_X$, $\cdot_X, 0_X$, formulated with the equality relation $=_X$ between objects of type X as defined below,
   (1) $\forall x^X\,(\|x-_X x\|_X =_{\mathbb{R}} 0_{\mathbb{R}})$,
   (2) $\forall x^X,y^X\,(\|x-_X y\|_X =_{\mathbb{R}} \|y-_X x\|_X)$,
   (3) $\forall x^X,y^X,z^X\,(\|x-_X z\|_X \le_{\mathbb{R}} \|x-_X y\|_X +_{\mathbb{R}} \|y-_X z\|_X)$,
   (4) $\forall \alpha^1,x^X,y^X\,(\|\alpha x-_X \alpha y\|_X =_{\mathbb{R}} |\alpha|_{\mathbb{R}}\cdot_{\mathbb{R}} \|x-_X y\|_X)$,
   (5) $\forall \alpha^1,\beta^1,x^X\,(\|\alpha x-_X \beta x\|_X =_{\mathbb{R}} |\alpha-_{\mathbb{R}}\beta|_{\mathbb{R}}\cdot_{\mathbb{R}} \|x\|_X)$,
   (6) $\begin{cases} \forall x^X,y^X,u^X,v^X \\ \quad (\|(x+_X y)-_X(u+_X v)\|_X \le_{\mathbb{R}} \|x-_X u\|_X +_{\mathbb{R}} \|y-_X v\|_X), \end{cases}$
   (7) $\forall x^X,y^X\,(\|(-_X x)-_X(-_X y)\|_X =_{\mathbb{R}} \|x-_X y\|_X)$,
   (8) $\forall x^X,y^X\,(|\|x\|_X -_{\mathbb{R}} \|y\|_X|_{\mathbb{R}} \le_{\mathbb{R}} \|x-_X y\|_X)$.
   (9) $\|1_X\|_X =_{\mathbb{R}} 1_{\mathbb{R}}$.

As before, the language of $\mathscr{A}^\omega[X,\|\cdot\|]$ only contains equality $=_0$ for objects of type 0 as a primitive predicate. $x^X =_X y^X$ is an abbreviation for $\|x-_X y\|_X =_{\mathbb{R}} 0_{\mathbb{R}}$. Equality for complex types is defined as before as extensional equality using $=_0$ and $=_X$ for the base types.

It is clear that the axioms (0)–(8) all express valid (modulo our representation of $\mathbb{R}$) facts about (real) normed spaces. Conversely, the usual axioms for (real) normed linear spaces can be derived in $\mathscr{A}^\omega[X,\|\cdot\|]$. To see this we have to check that the equality axioms (reflexivity, symmetry and transitivity) hold for our defined equality

$=_X$ and that the vector space operations and the norm function are provably extensional w.r.t. $=_X$ (it then follows that we can carry out the usual proofs of the basic identities in normed spaces which – together with the extensionality of $\|\cdot\|_X$ – imply the usual norm axioms):

The equality axioms follow immediately from (1)–(3).
The extensionality of the scalar product

$$\forall \alpha^1, \beta^1, x^X, y^X \left(\alpha =_{\mathbb{R}} \beta \wedge x =_X y \to \alpha x =_X \beta y\right)$$

follows from (4), (5).
The extensionality of $+_X$ and $-_X$

$$\forall x^X, y^X, u^X, v^X \left(x =_X u \wedge y =_X v \to x +_X y =_X u +_X v\right)$$

and

$$\forall x^X, y^X \left(x =_X y \to -_X x =_X -_X y\right)$$

follows from (6), (7).
Finally, (8) yields the extensionality of $\|\cdot\|_X$

$$\forall x^X, y^X \left(x =_X y \to \|x\|_X =_{\mathbb{R}} \|y\|_X\right).$$

Hence $\|\cdot\|_X$ is a norm on the set of equivalence classes generated by $=_X$ and we can now prove all the usual basic vector space laws and properties of the norm (exercise).

Axiom (9) is a Skolemized form of expressing that the normed space is nontrivial, i.e. contains an element x whose norm is strictly positive. We then can define $1_X := \frac{x}{\|x\|_X}$ for such an x to get an element of norm 1.

The theory $\mathcal{A}^\omega[X, \|\cdot\|, \eta]$ results from $\mathcal{A}^\omega[X, \|\cdot\|]$ by adding a new constant η^1 of type 1 together with the axiom (writing more short $\|\cdot\|$ instead of $\|\cdot\|_X$)

$$(10)\, \forall x^X, y^X \forall k^0 \left(\|x\|, \|y\| <_{\mathbb{R}} 1_{\mathbb{R}} \wedge \left\|\frac{x +_X y}{2}\right\| >_{\mathbb{R}} 1 - 2^{-\eta(k)} \to \|x -_X y\| \leq_{\mathbb{R}} 2^{-k}\right).$$

As mentioned already in the introduction to this chapter, the fact that η is a modulus of uniform convexity is usually defined as follows:

$$(10^*)\, \forall x^X, y^X \forall k^0 \left(\|x\|, \|y\| \leq_{\mathbb{R}} 1_{\mathbb{R}} \wedge \left\|\frac{x +_X y}{2}\right\| \geq_{\mathbb{R}} 1 - 2^{-\eta(k)} \to \|x -_X y\| \leq_{\mathbb{R}} 2^{-k}\right).$$

It is clear that (10^*) implies (10). Conversely, $\mathcal{A}^\omega[X, \|\cdot\|, \eta]$ proves (10^*) with $\tilde{\eta}(k) := \eta(k) + 1$ using the continuity of the norm and the scalar product which can be derived in $\mathcal{A}^\omega[X, \|\cdot\|]$. The reason why we use the formulation (10) as our axiom is that this formulation – in contrast to (10^*) – is logically equivalent to a purely universal statement since $<_{\mathbb{R}} \in \Sigma_1^0$ and $\leq_{\mathbb{R}} \in \Pi_1^0$.

The theory $\mathcal{A}^\omega[X, \langle \cdot, \cdot \rangle]$ is obtained from $\mathcal{A}^\omega[X, \|\cdot\|]$ by the addition of the parallelogram law as a further axiom (11) to $(1)-(9)$:

$$(11)\ \forall x^X, y^X \left(\|x +_X y\|_X^2 +_{\mathbb{R}} \|x -_X y\|_X^2 =_{\mathbb{R}} 2_{\mathbb{R}} \cdot_{\mathbb{R}} (\|x\|_X^2 +_{\mathbb{R}} \|y\|_X^2\right),$$

where $(\cdot)^2$ is a functional of type $1(1)$ which represents on the representations of real numbers the function $x \mapsto x^2$ on $\mathbb{R}$.

Any norm that satisfies (11) gives rise to an inner product function $\langle \cdot, \cdot \rangle : X \times X \to \mathbb{R}$ in the following way: define a functional $\langle \cdot, \cdot \rangle_X$ of type $1(X)(X)$ by (writing $\langle x, y \rangle_X$ for $\langle \cdot, \cdot \rangle_X(x, y)$):

$$(+)\ \langle x^X, y^X \rangle_X :=_1 \left(\langle \frac{1}{4} \rangle\right)_{\mathbb{R}} \cdot_{\mathbb{R}} \left(\|x +_X y\|_X^2 -_{\mathbb{R}} \|x -_X y\|_X^2\right).$$

$\langle \cdot, \cdot \rangle_X$ represents an inner product on the space (of $=_X$-equivalence classes of) X and the norm $\|\cdot\|_X$ can be recovered from $\langle \cdot, \cdot \rangle_X$ in the usual way

$$(++)\ \|x\|_X := sqrt(\langle x, x \rangle_X),$$

where $sqrt^{1\to 1}$ represents the square root function $\mathbb{R}_+ \to \mathbb{R}_+$ on the representation of $\mathbb{R}$ (which can easily be defined by a closed term of $\mathcal{A}^\omega$). Conversely, whenever a norm $\|\cdot\|$ is given by a (real valued) inner product via $(++)$, then the norm satisfies (11) and the inner product can be recovered from that norm by $(+)$. The standard proofs of these facts (see e.g. [375]) can easily be carried out in our formal setting. Thus $\mathcal{A}^\omega[X, \langle \cdot, \cdot \rangle]$ contains a proper axiomatization of the notion of a real inner product space (also called pre-Hilbert space).

Let $(X, \|\cdot\|)$ be a nontrivial real normed linear space. $\mathcal{S}^{\omega,X}$ becomes a model of $\mathcal{A}^\omega[X, \|\cdot\|]$ by letting the variables of type ρ range over S_ρ if

- we give the obvious interpretations to $0^0, S^1, \Pi_{\rho,\tau}, \Sigma_{\delta,\rho,\tau}$ and $\underline{R}_{\underline{\rho}}$ for all types $\delta, \rho, \tau, \underline{\rho} \in \mathbf{T}^X$,
- 0_X is interpreted by the zero vector 0^X of the linear space X, 1_X by some vector $a \in X$ with $\|a\| = 1$, $+_X$ is interpreted as addition in X, $-_X$ is the inverse of x w.r.t. $+$ in X, $\cdot_X$ is interpreted as $\lambda \alpha \in \mathbb{N}^{\mathbb{N}}, x \in X. r_\alpha \cdot x$, where r_α is the unique real number represented by α and '$\cdot$' refers to the scalar multiplication in the $\mathbb{R}$-linear space X. Finally, $\|\cdot\|_X$ is interpreted by $\lambda x \in X.(\|x\|)_\circ$, where $(r)_\circ$ – for $r \in \mathbb{R}_+$ – is the function from definition 17.7.

Similarly for $\mathcal{A}^\omega[X, \|\cdot\|, \eta]$ and $\mathcal{A}^\omega[X, \langle \cdot, \cdot \rangle]$, where than X is has to be a uniformly convex space (and η must be interpreted by some modulus of uniform convexity) or a pre-Hilbert space, respectively.

Definition 17.62. We say that a sentence of $\mathcal{L}(\mathcal{A}^\omega[X, \|\cdot\|])$ holds in a nontrivial (real) normed linear space $(X, \|\cdot\|)$ if it holds in the models of $\mathcal{A}^\omega[X, \|\cdot\|]$ obtained from $\mathcal{S}^{\omega,X}$ as specified above.
Analogously for $\mathcal{A}^\omega[X, \|\cdot\|, \eta]$ and $\mathcal{A}^\omega[X, \langle \cdot, \cdot \rangle]$.

Remark 17.63. Again we strictly speaking have to use the plural 'models' since the interpretation of 1_X is not uniquely determined. What is meant is that the sentence holds no matter how one interprets 1_X as long as it is interpreted by an element having norm 1. In the case of $\mathcal{A}^\omega[X,\|\cdot\|,\eta]$ also the interpretation of η^1 is not uniquely determined but may be interpreted by any modulus of uniform convexity of the underlying space $(X,\|\cdot\|)$.

In $\mathcal{A}^\omega[X,\|\cdot\|]$ we can extend the relation $x \leq_\rho y$ from types $\rho \in \mathbf{T}$ to $\rho \in \mathbf{T}^X$:

Definition 17.64. For functionals x^ρ, y^ρ of type $\rho \in \mathbf{T}^X$ define $x \leq_\rho y$ by

$$x \leq_0 y :\equiv x \leq y,$$

$$x \leq_X y :\equiv \|x\|_X \leq_{\mathbb{R}} \|y\|_X,$$

$$x \leq_{\tau(\rho)} y :\equiv \forall z^\rho (x(z) \leq_\tau y(z)).$$

Lemma 17.65. *The following is provable in* $\mathcal{A}^\omega[X,\|\cdot\|]$

$$\forall x^*, x, x(x^* \gtrsim_\rho^{0_X} x \wedge x \geq_\rho y \to x^* \gtrsim_\rho^{0_X} y).$$

Proof: *Induction on* $\rho \in \mathbf{T}^X$. □

The theory $\mathcal{A}^\omega[X,\|\cdot\|,C]$**:**

$\mathcal{A}^\omega[X,\|\cdot\|,C]$ results from $\mathcal{A}^\omega[X,\|\cdot\|]$ by adding new constants b_X of type 0, c_X of type X and χ_C of type $0(X)$ together with the axioms

(12) $\forall x^X (\chi_C(x) =_0 0 \to \|x\|_X \leq_{\mathbb{R}} (b_X)_{\mathbb{R}},$
(13) $\forall x^X, y^X, \alpha^1 (\chi_C(x) =_0 \chi_C(y) =_0 0 \to \chi_C((1 -_{\mathbb{R}} \tilde{\alpha}) \cdot_X x +_X \tilde{\alpha} \cdot_X y) =_0 0),$
(14) $\chi_C(c_X) =_0 0,$
(15) $\forall x^X (\chi_C(x) \leq_0 1).$

The theories $\mathcal{A}^\omega[X,\|\cdot\|,\eta,C]$ and $\mathcal{A}^\omega[X,\langle\cdot,\cdot\rangle,C]$ are defined analogously
Note that although the vector space operations and $\|\cdot\|_X$ are provably extensional (w.r.t. $=_X, =_{\mathbb{R}}$), the characteristic function χ_C is not. However, by QF-ER we have the following weak form of χ_C-extensionality

$$\frac{A_0 \to s =_X t}{A_0 \to \chi_C(s) =_0 \chi_C(t)} \text{ for quantifier-free } A_0$$

(see also the discussion at the end of this section).
The axioms (12)–(14) express that the set $C := \{x \in X \mid \exists y \in X (x =_X y \wedge \chi_C(y) =_0 0\}$ is a nonempty b-bounded convex subset of X.

The intended interpretation of χ_C is to be the characteristic function of some nonempty convex subset $C \subseteq X$.

In the following '$\forall x^C A(x)$', '$\forall f^{1\to C} A(f)$', '$\forall f^{X\to C} A(f)$' and '$\forall f^{C\to C} A(f)$' abbreviate

$$\forall x^X (\chi_C(x^X) =_0 0 \to A(x)),$$
$$\forall f^{1\to X} (\forall y^1 (\chi_C(f(y)) =_0 0) \to A(f)),$$
$$\forall f^{X\to X} (\forall y^X (\chi_C(f(y)) =_0 0) \to A(f)) \text{ and}$$
$$\forall f^{X\to X} (\forall x^X (\chi_C(x) =_0 0 \to \chi_C(f(x)) =_0 0) \to A(\tilde{f})),$$

where $\tilde{f}(x) := \begin{cases} f(x), \text{ if } \chi_C(x) =_0 0 \\ c_X, \text{ otherwise.} \end{cases}$

Analogously for the corresponding $\exists$-quantifiers with '$\wedge$' instead of '$\to$'. This extends to types of degree $(1,X,C)$ where ρ is of degree $(1,X,C)$ if it has the form $C(\tau_k)\dots(\tau_1)$, where τ_i has degree $1, \tau_i = X$ or $\tau_i = C$.

Remark 17.66. Note that for ρ of degree $(1,X,C)$ a quantifier '$\forall x^\rho$' abbreviates

$$\forall x^{\rho'} (B(x) \to \dots),$$

where ρ' is the type of degree $(1,X)$ resulting from ρ by replacing everywhere 'C' by 'X' and B is (logically equivalent to) a $\forall$-formula.

Officially, the types involving C are not included in our language and a statement like '$\forall x^\rho A(x)$' for ρ of degree $(1,X,C)$ is used an abbreviation for '$\forall x^{\rho'} (B(x) \to A(x))$' for ρ' and B as in remark 17.66.

Remark 17.67. If one defines

$$f^{X\to X} =_{C\to X} g^{X\to X} :\equiv \forall x^X (\chi_C(x) = 0 \to f(x) =_X g(x))$$

then for all $f^{X\to X}, g^{X\to X}$ the following provably holds

$$f =_{C\to X} \tilde{f} \text{ and } \tilde{f} =_{C\to X} \tilde{g} \leftrightarrow \tilde{f} =_{X\to X} \tilde{g}.$$

In the following, when writing '$\forall f^{C\to C} A(f)$' we not only assume that $A(\tilde{f})$ is – when interpreted in $\mathcal{S}^{\omega,X}$ – extensional w.r.t. $=_{C\to X}$ (which in the light of the previous remark always is the case) but that already $A(f)$ is in this sense extensional for functions f satisfying

$$\forall x^X (\chi_C(x) =_0 0 \to \chi_C(f(x)) =_0 0)$$

which automatically is satisfied whenever $A(f)$ results from formalizing a property of functions $f : C \to C$. This guarantees that the meaning of $A(\tilde{f})$ does not depend on the interpretation of the constant c_X used to define $f \mapsto \tilde{f}$.

For $f^{C\to C}$ (i.e. for $f^{X\to X}$ satisfying $\forall x^X (\chi_C(x) =_0 0 \to \chi_C(f(x)) =_0 0)$) '$f$ nonexpansive' is the $\forall$-formula

$$\forall x^X, y^X (\chi_C(x) =_0 0 =_0 \chi_C(y) \to \|f(x) -_X f(y)\|_X \le_{\mathbb{R}} \|x -_X y\|_X).$$

Definition 17.68. We say that a sentence A holds in a nontrivial real normed linear space $(X,\|\cdot\|)$ and a nonempty bounded convex subset $C \subseteq X$ if in addition to the requirements in definition 17.62 we stipulate that χ_C is interpreted as the characteristic function for C, c_X by some arbitrary element in C and b_X by some integer $b \in \mathbb{N}$ with $b \geq \|x\|$ for all $x \in C$.

The theory $\mathcal{A}^\omega[X,\|\cdot\|,C]_{-b}$ results from $\mathcal{A}^\omega[X,\|\cdot\|,C]$ if we delete the constant b_X and drop the axiom expressing that the elements of C are bounded in norm by b_X. The previous definition is then adapted accordingly. Analogously, we define the theories $\mathcal{A}^\omega[X,\|\cdot\|,\eta,C]_{-b}$ and $\mathcal{A}^\omega[X,\langle\cdot,\cdot\rangle,C]_{-b}$

Theorem 17.69 (Gerhardy-Kohlenbach [120]).

1) *Let ρ be of degree $(1,X),(1,X,C)$ or 2 and let $B_\forall(x,u)$, resp. $C_\exists(x,v)$, be $\forall$- resp. $\exists$-formulas that contain only x,u free, resp. x,v free. Assume that*

$$\mathcal{A}^\omega[X,\|\cdot\|,C]_{-b} \vdash \forall x^\rho\,(\forall u^0 B_\forall(x,u) \to \exists v^0 C_\exists(x,v)),$$

Then there exists a partial functional $\Phi : S_{\widehat{\rho}} \times \mathbb{N} \rightharpoonup \mathbb{N}$ s.t. Φ is defined on all strongly majorizable elements of $S_{\widehat{\rho}}$, the restriction to those elements is a (bar recursively) computable functional of $\mathcal{M}^\omega$ and the following holds in all nontrivial (real) normed linear spaces $(X,\|\cdot\|,C)$ and all nonempty convex subsets $C \subseteq X$: for all $x \in S_\rho$, $x^* \in S_{\widehat{\rho}}$ and $n \in \mathbb{N}$ if $x^* \gtrsim^{0_X} x$ and $(n)_\mathbb{R} \geq_\mathbb{R} \|c_X\|_X$ then

$$\forall u \leq \Phi(x^*,n) B_\forall(x,u) \to \exists v \leq \Phi(x^*,n) C_\exists(x,v)).$$

In particular, if ρ is in addition of degree 1^*, then $\Phi : S_{\widehat{\rho}} \times \mathbb{N} \to \mathbb{N}$ is totally computable.

2) *For nontrivial (real) uniformly convex spaces with modulus of uniform convexity η statement 1) holds with $(X,\|\cdot\|,\eta,C)$, $\mathcal{A}^\omega[X,\|\cdot\|,\eta,C]_{-b}$ instead of $(X,\|\cdot\|,C)$, $\mathcal{A}^\omega[X,\|\cdot\|,C]_{-b}$, where the extracted bound Φ additionally depends on η.*
3) *Analogously, for $\mathcal{A}^\omega[x,\langle\cdot,\cdot\rangle,C]_{-b}$ and nontrivial (real) inner product spaces $(X,\langle\cdot,\cdot\rangle)$.*

As in the metric case, instead of single variables x,u,v and single premises $\forall u B_\forall(x,u)$ we may have tuples of variables and finite conjunctions of premises.

Again, this theorem will be proved in section 17.4 below.

In the case of metric and hyperbolic spaces the proofs of the main metatheorems (to be given in section 17.4) use the fact that all the constants (except 0_X) are uniformly a-majorizable and – since we can interpret 0_X arbitrarily (if it does not occur in the theorem to be proved) and so, in particular, by a – this essentially also applies to 0_X. In the normed case, however, this is no longer true as the norm $\|x\|$ measures the distance of x from the zero vector 0_X and so we have to choose the reference point $a \in X$ as 0_X. More precisely, if we would choose a different reference point $a \in X$ then the a-majorant for $\|x\|$ would depend on (an upper bound of) $\|a\|$ and so would not be uniform in a. So to keep a as a variable parameter which can be fixed later (e.g. as one of the input data of the theorem to be proved as we did in the proof of corollary 17.55) without changing the extracted bound does not work out here. As a consequence of this, when dealing with theorems say of the form

$$\forall x^X, y^X \exists n^0 A_\exists(x,y,n),$$

then differently from the metric case it will no longer suffice to have a bound b on the relative distance $\|x-y\|$ in order to obtain a bound on n but we need in addition an absolute norm bound on either $\|x\|$ or $\|y\|$. This is clear e.g. from the trivial example

$$\forall x^X, y^X \exists n^0 (n > \|x\| + \|y\|),$$

which we briefly mentioned already in the introduction to this chapter.

The constant c_X is necessary to witness the nonemptyness of C and since we fix a as 0_X our bounds depend on an upper bound n for the norm of c_X as well. However, if c_X does not occur in the formulas $B_\forall$ and $C_\exists$ and we have another parameter $z \in C$ for which we have a bound on the norm, we do not need a further bound on $\|c_X\|$, since in the model c_X may be interpreted by an arbitrary element of C and we then may interpret c_X by z.

Corollary 17.70. 1) *Let σ be of degree 1 and ρ of degree 1 or $(1,X)$ and let τ be a type of degree $(1,X,C)$. Let s be a closed term of type $\rho(\sigma)$ and $B_\forall, C_\exists$ as before. If a sentence*

$$\forall x^\sigma \forall y \leq_\rho s(x) \forall z^\tau \left(\forall u^0 B_\forall(x,y,z,u) \to \exists v^0 C_\exists(x,y,z,v)\right)$$

is provable in $\mathcal{A}^\omega[X,\|\cdot\|,C]$ then one can extract a computable functional Φ : $S_\sigma \times \mathbb{N} \to \mathbb{N}$ s.t. for all $x \in S_\sigma$

$$\forall y \leq_\rho s(x) \forall z^\tau \left(\forall u^0 \leq \Phi(x,b) B_\forall(x,y,z,u) \to \exists v^0 \leq \Phi(x,b) C_\exists(x,y,z,v)\right)$$

holds in any nontrivial (real) normed linear space $(X,\|\cdot\|)$ and any nonempty b-bounded convex (in the norm) subset $C \subset X$ (with 'b_X' interpreted by 'b').

2) *'1)' holds analogously with $\mathcal{A}^\omega[X,\|\cdot\|,\eta,C]$ and nontrivial (real) uniformly convex spaces $(X,\|\cdot\|,\eta,C)$ and convex subsets C instead of $\mathcal{A}^\omega[X,\|\cdot\|,C]$ and $(X,\|\cdot\|,C)$. This time Φ is a computable functional in x,b and a modulus η of uniform convexity for $(X,\|\cdot\|)$ (which interprets the constant 'η').*

3) *Analogously, for (real) inner-product spaces $(X,\langle\cdot,\cdot\rangle)$.*

Instead of single variables x,y,z,u,v we may also have finite tuples of variables $\underline{x},\underline{y},\underline{z},\underline{u},\underline{v}$ as long as the elements of the respective tuples satisfy the same type restrictions as x,y,z,u,v. Moreover, instead of a single premise of the form '$\forall u^0 B_\forall(x,y,z,u)$' we may have a finite conjunction of such premises.

Proof: 1) W.l.o.g. we may assume that $\sigma = 1$. We, therefore, have $x^M \gtrsim_\sigma^{0_X} x$ and, consequently, $s^*(x^M) \gtrsim^{0_X} sx$, where s^* is some majorant of s (which exists by (the proof of) lemma 17.83 below as a closed term of $\mathcal{A}^\omega$). By lemma 17.65, therefore,

$s(x) \geq_\rho y$ implies that $s^*(x^M) \gtrsim^{0_X} y$.
Next, given a norm bound $b \in \mathbb{N}$ on C, trivially $(b)_\mathbb{R} \geq \|c_X\|$ and writing $\tau = C(\underline{\tau})$, then also $\lambda \underline{x}^{\widehat{\tau_i}}.b \gtrsim_\tau^{0_X} z$.
Hence by theorem 17.69 we can extract a bar recursive functional ϕ such that $\phi(x^M, s^*(x^M), \lambda \underline{x}^{\widehat{\tau_i}}.b, b)$ is a bound on both '$\exists v$' and '$\forall u$', for any nontrivial real normed linear space and any (nonempty) b-bounded convex subset C. Since both the functional $(\cdot)^M$, the 0_X-majorant s^* for s and the 0_X-majorant $\lambda \underline{x}^{\widehat{\tau_i}}.b$ for z are given by closed terms of $\mathcal{A}^\omega$, the functional

$$\Phi :\equiv \lambda x, b.\phi(x^M, s^*(x^M), \lambda \underline{x}^{\widehat{\tau_i}}.b, b)$$

is computable and yields the desired bound.
Note, that in $\mathcal{A}^\omega[X, \|\cdot\|, C]$ we have the boundedness of C as an axiom, while theorem 17.69 only allows one to treat the boundedness as an implicative assumption. Therefore, this corollary strictly speaking does not follow from theorem 17.69 but easily from its proof (to be given below) which works for any extension by purely universal axioms. □
2) and 3) are proved analogously.

Corollary 17.71 (Gerhardy-Kohlenbach [120]).

1) *Let P (resp. K) be a $\mathcal{A}^\omega$-definable Polish space (resp. compact metric space). Suppose that $\mathcal{A}^\omega[X, \|\cdot\|, C]_{-b}$ proves a sentence*

$$\forall x \in P \forall y \in K \forall z^C \forall f^{C \to C}(f \text{ n.e.} \wedge \forall u^0 B_\forall(x,y,z,f,u) \to \exists v^0 C_\exists(x,y,z,f,v)),$$

where c_X does not occur in $B_\forall$ and $C_\exists$. Then one can extract from the proof a computable functional $\Phi : \mathbb{N}^\mathbb{N} \times \mathbb{N} \to \mathbb{N}$ s.t. for all representatives $r_x \in \mathbb{N}^\mathbb{N}$ of $x \in P$ and all $b \in \mathbb{N}$

$$\forall y \in K \forall z^C \forall f^{C \to C}(f \text{ n.e.} \wedge \|z\|_X, \|z - f(z)\|_X \leq_\mathbb{R} (b)_\mathbb{R}$$
$$\wedge \forall u^0 \leq \Phi(r_x, b) B_\forall(x,y,z,f,u) \to \exists v^0 \leq \Phi(r_x, b) C_\exists(x,y,z,f,v))$$

holds in all nontrivial (real) normed linear spaces $(X, \|\cdot\|)$ and nonempty convex subsets C.
Analogously, for uniformly convex spaces $(X, \|\cdot\|, \eta, C)$ and (real) inner product spaces $(X, \langle\cdot,\cdot\rangle)$, where for uniformly convex spaces the bound Φ additionally depends on the modulus of uniform convexity η.
2) *The corollary also holds for an additional parameter $\forall \tilde{z}^C$, if we add the additional premise $\|z - \tilde{z}\|_X \leq_\mathbb{R} (b)_\mathbb{R}$ to the conclusion.*
3) *Furthermore, the corollary holds for an additional parameter $\forall c^{0 \to C}$ if we add the additional premise $\forall n(\|z - c(n)\|_X \leq_\mathbb{R} (g(n))_\mathbb{R})$ to the conclusion, where the bound then additionally depends on $g : \mathbb{N} \to \mathbb{N}$.*
4) *Moreover, 1), 2) and 3) also hold if we replace 'f n.e.' with 'f Lipschitz continuous' and 'f Hölder-Lipschitz continuous (with parameters $L, \alpha \in \mathbb{Q}_+^*$, where $\alpha \leq 1$) or 'f uniformly continuous' (with modulus $\omega : \mathbb{N} \to \mathbb{N}$ as parameter). For Lipschitz and Hölder-Lipschitz continuous functions the bound depends on*

the parameters L resp. L, α and for uniformly continuous functions the bound depends on a given modulus of uniform continuity.

5) *Furthermore, 1), 2) and 3) hold if we replace 'f n.e.' with 'f weakly quasi-nonexpansive'. For weakly quasi-nonexpansive functions (with fixed point p) we need to state the additional premise* $\|p\|_X \leq_{\mathbb{R}} (b)_{\mathbb{R}}$ *in the conclusion.*
6) *1), 2) and 3) also hold if we replace 'f n.e.' in the premise and the conclusion by*

$$\forall n^0, z_1^C, z_2^C(\|z_1 - z_2\|_X <_{\mathbb{R}} (n)_{\mathbb{R}} \to \|f(z_1) - f(z_2)\|_X \leq_{\mathbb{R}} (\Omega_0(n))_{\mathbb{R}}), \ (*)$$

where Ω_0 *is a function* $\mathbb{N} \to \mathbb{N}$ *treated as a parameter and the bound additionally depends on* Ω_0.

7) *Finally, 1), 2) and 3) hold if the previous conditions on f are replaced by*

$$\forall n^0, \tilde{z}^C(\|\tilde{z}\|_X <_{\mathbb{R}} (n)_{\mathbb{R}} \to \|f(\tilde{z})\|_X \leq_{\mathbb{R}} (\Omega(n))_{\mathbb{R}}), \ (**)$$

where Ω *is a function* $\mathbb{N} \to \mathbb{N}$ *is treated as a parameter and the bound additionally depends on* Ω. *In this case we can drop the assumption '*$\|z - f(z)\|_X \leq (b)_{\mathbb{R}}$*' in the conclusion whereas '*$\|z\|_X \leq (b)_{\mathbb{R}}$*' has to remain.*

Note that $(*), (**)$ *are logically equivalent to* $\forall$*-formulas.*

Proof: The corollary follows from theorem 17.69 in a similar way as we proved corollary 17.55 from theorem 17.52, except for two points: (1) as discussed already above we need to fix $a = 0_X$ and hence have to add the premise $\|z\| \leq b$, which implies that $b \gtrsim_X^{0_X} z$, in the conclusion. (2) the 0_X-majorization of f (actually $\tilde{f}$) requires extra care. From the definition of $\tilde{f}$ it is obvious that $n \gtrsim_X^{0_X} f(x)$ for $x \in X \setminus C$ if $n \geq \|c_X\|$. Also note, that since we assume c_X does not occur in $B_\forall$ and $C_\exists$ we may interpret c_X by the parameter z in the model, so that $\|c_X\| \leq b$. Hence, given an a-majorant $\lambda n.f^*(n) \gtrsim_X^{0_X} f$ on the convex subset C, we obtain the 0_X-majorant $\lambda n.max(f^*(n), b)$ for $\tilde{f}$ and thus the extracted bound does not depend on an explicit bound on the norm of c_X. In the following we may, therefore, focus on 0_X-majorants for f on the convex subset C.

For 1), 2) and 3) we have that $b \gtrsim^{0_X} z$, $2b \gtrsim^{0_X} \tilde{z}$ and $\lambda n.g^M(n) + b \gtrsim^{0_X} c$. For $\lambda n^0.n + 3b \gtrsim^{0_X} f$, where f is nonexpansive, we reason as follows: assume $\|\tilde{z}\| \leq n$ then

$$\begin{aligned}
\|f(\tilde{z})\| &= \|f(\tilde{z}) - f(z) + f(z) - z + z\| \\
&\leq \|f(\tilde{z}) - f(z)\| + \|f(z) - z\| + \|z\| \\
&\leq \|\tilde{z} - z\| + b + b \\
&\leq \|\tilde{z}\| + \|z\| + 2b \\
&\leq n + 3b.
\end{aligned}$$

The proofs of 4), 5), 6) and 7) are now very similar to the corresponding clauses of corollary 17.55 and are left to the reader (alternatively, the details can be found in [120]). □

17.4 Proofs of theorems 17.35, 17.52 and 17.69

This section is devoted to the proofs of theorems 17.52 and 17.69. As an immediate corollary to the proof of theorem 17.52 we also obtain theorem 17.35.
We focus on the case of the theory $\mathcal{A}^\omega[X,d,W]_{-b}$. The case of $\mathcal{A}[X,d]_{-b}$ then follows simply by dropping all considerations involving the constant W and its axioms. For $\mathcal{A}^\omega[X,d,W,\mathrm{CAT}(0)]_{-b}$ we only need to observe that lemma 17.72 below immediately extends to the situation where the purely universal axiom $\mathbf{CN}^-$ is added (as such axioms are interpreted by themselves).
The next lemma extends theorem 11.9 due to Spector ([343]) and Howard ([162, 266]), which we proved in chapter 11, from $\mathcal{A}^\omega$ to $\mathcal{A}^\omega[X,d,W]_{-b}$ and states that $\mathcal{A}^\omega[X,d,W]_{-b}$ has (via negative translation) a Gödel functional interpretation in $\mathcal{A}^\omega[X,d,W]_{-b}\setminus\{\text{QF-AC,DC}\}$ (actually even in a quantifier-free fragment of this theory) augmented by the schema (BR) of simultaneous bar recursion from chapter 11 which now is extended to all types of $\mathbf{T}^X$. Here we use the constant-0 functionals 0^ρ extended to all types in $\mathbf{T}^X$ (needed in the formulation of (BR)) by defining

$$0^\rho := \lambda \underline{x}.0^X$$

for types $\rho = X(\rho_k)\ldots(\rho_1)$ and $\underline{x} = x_1^{\rho_1},\ldots,x_k^{\rho_k}$.

Let $\mathcal{A}^\omega[\ldots]^-_{-b} := \mathcal{A}^\omega[\ldots]_{-b}\setminus\{\text{QF-AC,DC}\}$.

Lemma 17.72. *Let $A(\underline{a})$ be an arbitrary formula in the language of $\mathcal{A}^\omega[X,d,W]_{-b}$ containing only the free variables $\underline{a}$. Then the following rule holds:*

$$\begin{cases} \mathcal{A}^\omega[X,d,W]_{-b} \vdash A(\underline{a}) \ \textit{implies that} \\ \mathcal{A}^\omega[X,d,W]^-_{-b} + (\mathrm{BR}) \vdash \forall \underline{y}\,(A')_D(\underline{t}\,\underline{a},\underline{y},\underline{a}), \end{cases}$$

where $\underline{t}$ is a suitable tuple of closed terms of $\mathcal{A}^\omega[X,d,W]^-_{-b}$+(BR) which can be extracted from a given proof of the assumption, A' is the negative translation of A (see definition 10.1) and $(A')^D \equiv \exists\underline{x}\forall\underline{y}(A')_D(\underline{x},\underline{y},\underline{a})$ is the Gödel functional interpretation of $A'(\underline{a})$ (see definition 8.1).

Proof: For $\mathcal{A}^\omega$ the theorem was prove already in chapter 11 (theorem 11.9). Both the negative translation as well as the subsequent functional interpretation generalize without any changes to the extension of the axioms and rules of $\mathcal{A}^\omega$ to the new types $\mathbf{T}^X$ since we extended the simultaneous recursors and bar recursors to these types as well and still the only prime formulas we have are of the form $s =_0 t$ and hence decidable (recall that $=_X$ is a defined notion). Moreover, it is clear that the definability of λ-terms from the combinators Π,Σ extends to the new types using the fact that we extended our combinators accordingly. The new axioms for metric, hyperbolic and CAT(0) spaces are all purely universal and do not contain $\vee$. Hence they are intuitionistically equivalent to their negative interpretation and subsequently interpreted by themselves by the functional interpretation.
The only subtle point to check is that the fact (used in the proof of lemma 11.5) that

the Leibniz identity

$$\forall \Phi^{0\rho}\,(\Phi s =_0 \Phi t)$$

implies the extensional equality $s =_\rho t$ still holds for the extended types. Obviously, the only critical case is $\rho = X$: Assume that

$$\forall \Phi^{0X}\,(\Phi s^X =_0 \Phi t^X)$$

and apply this to $\Phi_n := \lambda x^X.d_X(x,t)(n^0)$. Then

$$\forall n^0 (d_X(s,t)(n) =_0 d_X(t,t)(n)),\ \text{i.e.}\ d_X(s,t) =_1 d_X(t,t)$$

and hence $d_X(s,t) =_{\mathbb{R}} d_X(t,t) =_{\mathbb{R}} 0_{\mathbb{R}}$. So $d_X(s,t) =_{\mathbb{R}} 0_{\mathbb{R}}$, i.e. $s =_X t$. □

Based on $\gtrsim^a$, we define an extension of Bezem's [27] type structure of hereditarily strongly majorizable set-theoretic functionals from chapter 3 to all types $\mathbf{T}^X$. The definition is performed by induction on $|\rho|$, where $|0| := |X| := 0, |\tau(\rho)| := \max\{|\rho|,|\tau|\}+1$:

Definition 17.73. Let (X,d) be a nonempty metric space and let $a \in X$ be given. The extensional type structure $\mathcal{M}^{\omega,X}$ of all hereditarily strongly a-majorizable set-theoretic functionals of type $\rho \in \mathbf{T}^X$ over $\mathbb{N}$ and X is defined as

$$\begin{cases} M_0 := \mathbb{N},\ n \gtrsim^a_0 m :\equiv n \geq m \wedge n,m \in \mathbb{N}, \\ M_X := X,\ n \gtrsim^a_X x :\equiv n \geq d(x,a) \wedge n \in M_0, x \in M_X, \\ \qquad\qquad\qquad\qquad \text{where} \geq \text{ is the usual order on } \mathbb{R}; \\ x^* \gtrsim^a_{\tau(\rho)} x :\equiv x^* \in M_{\widehat{\tau}}^{M_{\widehat{\rho}}} \wedge x \in M_\tau^{M_\rho} \\ \qquad\qquad \wedge \forall y^* \in M_{\widehat{\rho}}, y \in M_\rho\,(y^* \gtrsim^a_\rho y \rightarrow x^*y^* \gtrsim^a_\tau xy) \\ \qquad\qquad \wedge \forall y^*, y \in M_{\widehat{\rho}}\,(y^* \gtrsim^a_{\widehat{\rho}} y \rightarrow x^*y^* \gtrsim^a_{\widehat{\tau}} x^*y), \\ M_{\tau(\rho)} := \left\{ x \in M_\tau^{M_\rho} \;\middle|\; \exists x^* \in M_{\widehat{\tau}}^{M_{\widehat{\rho}}} : x^* \gtrsim^a_{\tau(\rho)} x \right\} \quad (\rho,\tau \in \mathbf{T}^X). \end{cases}$$

Remark 17.74. For the types $\rho \in \mathbf{T}$ over $\mathbb{N}$ only, this type structure is identical to Bezem's type structure $\mathcal{M}^\omega$ of strongly hereditarily majorizable functionals as defined in chapter 3 and $\gtrsim^a_\rho$ coincides with $s\text{-}maj_\rho$ from definition 3.61. This follows by an easy induction on ρ (using that $\widehat{\rho} = \rho$ for $\rho \in \mathbf{T}$). Hence for such types we may write $s\text{-}maj_\rho$ instead of $\gtrsim^a_\rho$ and, in particular, the relation does not depend on the parameter $a \in X$.

Lemma 17.75. $x^* \gtrsim^a_\rho x \rightarrow x^* \gtrsim^a_{\widehat{\rho}} x^* \wedge x^* \in M_{\widehat{\rho}}$ *for all* $\rho \in \mathbf{T}^X$.

Proof: The lemma is immediate for $\rho = 0, X$ and follows using that $x^* \gtrsim^a_{\tau(\rho)} x$ implies $\forall z^*, z(z^* \gtrsim^a_{\widehat{\rho}} z \rightarrow x^*z^* \gtrsim^a_{\widehat{\tau}} x^*z)$ for complex types. □

Remark 17.76.

$$x^* \gtrsim^a_{\tau(\rho)} x \leftrightarrow x^* \in M^{M_{\widehat{\rho}}}_{\widehat{\tau}} \wedge x \in M^{M_\rho}_\tau \wedge$$
$$\forall y^*, \widehat{y} \in M_{\widehat{\rho}}, y \in M_\rho \, (y^* \gtrsim^a_\rho y \wedge y^* \gtrsim^a_{\widehat{\rho}} \widehat{y} \to x^*y^* \gtrsim^a_\tau xy \wedge x^*y^* \gtrsim^a_{\widehat{\tau}} x^*\widehat{y}).$$

Proof: '$\to$' is obvious. '$\leftarrow$' follows by applying the right-hand side twice: first to general y^*, y with defined $\widehat{y} := O^{\widehat{\rho}}$ and again to general $y^*, \widehat{y}$ with defined $y := O^\rho$ if $\rho = 0\rho_k \ldots \rho_1$ and $y := \lambda \underline{v}.a^X$ if $\rho = X\rho_k \ldots \rho_1$, respectively. Here we use lemma 17.75. □

Remark 17.77. The remark 3.62 applies here accordingly using for the case $\rho = X$ that for $n \in \mathbb{N}, x, a \in X$ the interpretation of $(n)_\mathbb{R} \geq_\mathbb{R} d_X(x,a)$ in $\mathcal{M}^{\omega,X}$ gives $(n)_\mathbb{R} \geq_\mathbb{R} (d(x,a))_\circ$ which is equivalent to $n \geq d(x,a)$.

The a-majorization relation is parametrized by an element $a \in X$ (and prima-facie this also applies to the definition of M_ρ). However, the resulting type structure of all hereditarily strongly a-majorizable functionals is in fact independent of the choice of $a \in X$. This is a consequence of the next lemma (by the previous remark this is already clear for types $\widehat{\rho}$ with $\rho \in \mathbf{T}^X$ which we will use implicitly in the formulation and the proof of this lemma):

Lemma 17.78. *For every $\rho \in \mathbf{T}^X$ there is a (primitive recursive) functional Φ_ρ of type $\widehat{\rho}0\widehat{\rho}$ in $M_{\widehat{\rho}0\widehat{\rho}}$ s.t. for all $a, b \in X$, all $x^* \in M_{\widehat{\rho}}$, $x \in M^a_\rho$ and all $n \in \mathbb{N}$ with $d(a,b) \leq n$,*

$$x^* \gtrsim^a_\rho x \to \Phi_\rho(x^*, n) \gtrsim^b_\rho x.$$

In particular, $M^a_\rho = M^b_\rho$ for all $a, b \in X$.
Moreover, for all $x^*, \widehat{x} \in M_{\widehat{\rho}}$ and all $n, m \in \mathbb{N}$ with $n \geq m$

$$x^* \text{ s-maj}_{\widehat{\rho}} \widehat{x} \to \Phi_\rho(x^*, n) \text{ s-maj}_{\widehat{\rho}} \Phi_\rho(\widehat{x}, m),$$

i.e. Φ_ρ s-maj Φ_ρ.

Proof: The lemma is proved using induction on $\rho \in \mathbf{T}^X$ to simultaneously construct Φ_ρ with the properties above and showing that $M^a_\rho = M^b_\rho$.
For $\rho = 0$ define $\Phi_0(x,n) := x$. $M^a_0 = M^b_0$ holds by definition.
For $\rho = X$ we define Φ_X by $\Phi_X(x^*, n) := x^* + n$ for $x^*, n \in \mathbb{N}$. Now let $x^* \in \mathbb{N}$ and $x \in X$ be such that $x^* \gtrsim^a_X x$, i.e. $x^* \geq d(x,a)$. Then the triangle inequality implies $x^* + n \geq d(x,b)$ and hence $x^* + n \gtrsim^b_X x$. The second condition on Φ_X (to majorize itself) is trivial. Again, by definition, we also have $M^a_X = M^b_X$.
For $\rho = \tau(\sigma)$ we need to construct the mapping $\Phi_{\tau(\sigma)}$ and show that $x \in M^a_{\tau(\sigma)}$ implies $x \in M^b_{\tau(\sigma)}$. Assume $x^* \gtrsim^a_{\tau(\sigma)} x$ for $x \in M^a_{\tau(\sigma)}$, and let $y^* \in M_{\widehat{\sigma}}$ and $y \in M^b_\sigma \stackrel{I.H.}{=} M^a_\sigma$ be given such that $y^* \gtrsim^b_\sigma y$. By the induction hypothesis for σ we have constructed Φ_σ such that, using the symmetry in a and b, $\Phi_\sigma(y^*, n) \gtrsim^a_\sigma y$. Next, by the definition of $\gtrsim^a_{\tau(\sigma)}$ we have that $x^*(\Phi_\sigma(y^*, n)) \gtrsim^a_\tau xy$. But then, by the induction hypothesis for τ, we have constructed Φ_τ such that

$$\Phi_\tau(x^*(\Phi_\sigma(y^*, n)), n) \gtrsim^b_\sigma xy.$$

Also for y^* s-$maj_{\widehat{\sigma}}\,\widehat{y}$ and $n \geq m$ we have by I.H. that $\Phi_\sigma(y^*,n)$ s-$maj_{\widehat{\sigma}}$ $\Phi_\sigma(\widehat{y},m)$ and so for x^* s-$maj_{\widehat{\rho}}\,\widehat{x}$ we get $x^*(\Phi_\sigma(y^*,n))$ s-$maj_{\widehat{\tau}}\widehat{x}(\Phi_\sigma(\widehat{y},m))$ which in turn implies that $\Phi_\tau(x^*(\Phi_\sigma(y^*,n)),n)$ s-$maj_{\widehat{\tau}}\Phi_\tau(\widehat{x}(\Phi_\sigma(\widehat{y},m)),m)$.
Since, in particular, x^* s-$maj_{\widehat{\tau}}x^*$ we obtain

$$\Phi_\tau(x^*(\Phi_\sigma(y^*,n)),n)\ s\text{-}maj_{\widehat{\tau}}\ \Phi_\tau(x^*(\Phi_\sigma(\widehat{y},n)),n).$$

So $\Phi_{\tau(\sigma)} := \lambda x^* \in M_{\widehat{\tau}(\widehat{\sigma})}\lambda n \in \mathbb{N}\lambda y^* \in M_{\widehat{\sigma}}.\Phi_\tau(x^*(\Phi_\sigma(y^*,n)),n)$ satisfies the conditions to be proven. In particular, $\Phi_{\tau(\sigma)}(x^*,n)$ is a b-majorant for x and hence $x \in M^b_{\tau(\sigma)}$. □

Remark 17.79. 1) Although the question whether or not a certain functional is a-majorizable is independent from the particular choice of $a \in X$, the complexity and possible uniformities of the majorants may depend crucially on that choice.
2) Φ_ρ in the previous lemma is already definable by a closed term of $\mathrm{G_1A}^\omega$.

We also need to extend lemma 3.63 from chapter 3 to the new types:

Lemma 17.80. *Let $\rho = \tau\rho_k\ldots\rho_1$. Then for $x^* : M_{\widehat{\rho}_1} \to (M_{\widehat{\rho}_2} \to \ldots \to M_{\widehat{\tau}})\ldots)$ and $x : M_{\rho_1} \to (M_{\rho_2} \to \ldots \to M_\tau)\ldots)$ the following holds*
$x^* \gtrsim^a_\rho x$ iff

(I) $\forall y_1^*, y_1, \ldots, y_k^*, y_k\big(\bigwedge\limits_{i=1}^{k}(y_i^* \gtrsim^a_{\rho_i} y_i) \to x^*y_1^*\ldots y_k^* \gtrsim^a_\tau xy_1\ldots y_k\big)$ and

(II) $\forall y_1^*, y_1, \ldots, y_k^*, y_k\big(\bigwedge\limits_{i=1}^{k}(y_i^* \gtrsim^a_{\rho_i} y_i) \to x^*y_1^*\ldots y_k^* \gtrsim^a_{\widehat{\tau}} x^*y_1\ldots y_k\big)$.

Proof: The lemma is proved by induction on k using lemma 17.75. The case $k=1$ follows from the definition of $\gtrsim^a$.
$k = n+1$: Let $\tau_0 = \tau(\rho_{n+1})$. For '⇒', we argue as follows: $x^* \gtrsim^a_\rho x$, in particular, implies that $x^* : M_{\widehat{\rho}_1} \to (M_{\widehat{\rho}_2} \to \ldots \to M_{\widehat{\tau}_0})\ldots)$ and $x : M_{\rho_1} \to (M_{\rho_2} \to \ldots \to M_{\tau_0})\ldots)$.
Hence,by induction hypothesis, we have

$$\forall y_1^*, y_1, \ldots y_n^*, y_n\Big(\bigwedge_{i=1}^{n}(y_i^* \gtrsim^a_{\rho_i} y_i) \to x^*y_1^*\ldots y_n^* \gtrsim^a_{\tau_0} xy_1\ldots y_n\Big).$$

Now assume $y_{n+1}^* \gtrsim^a_{\rho_{n+1}} y_{n+1}$. By definition of $\gtrsim^a_{\tau_0}$ we have

$$x^*y_n^*\ldots y_n^*y_{n+1}^* \gtrsim^a_\tau xy_1\ldots y_ny_{n+1},$$

so (I) follows. (II) can be treated analogously.
For '⇐', assume

$$\forall y_1^*, y_1, \ldots y_{n+1}^*, y_{n+1}\Big(\bigwedge_{i=1}^{n+1}(y_i^* \gtrsim^a_{\rho_i} y_i) \to x^*y_1^*\ldots y_{n+1}^* \gtrsim^a_\tau xy_1\ldots y_{n+1}\Big)$$

and

$$\forall y_1^*, y_1, \dots y_{n+1}^*, y_{n+1}\big(\bigwedge_{i=1}^{n+1} (y_i^* \gtrsim^a_{\hat{\rho}_i} y_i) \to x^* y_1^* \dots y_{n+1}^* \gtrsim^a_{\hat{\tau}} x^* y_1 \dots y_{n+1}\big).$$

We need to show that under these assumptions

$$(1)\ \forall y_1^*, y_1, \dots, y_n^*, y_n(\bigwedge_{i=1}^{n} y_i^* \gtrsim^a_{\rho_i} y_i \to x^* y_1^* \dots y_n^* \gtrsim^a_{\tau_0} x y_1 \dots y_n)$$

and

$$(2)\ \forall y_1^*, y_1, \dots, y_n^*, y_n(\bigwedge_{i=1}^{n} y_i^* \gtrsim^a_{\hat{\rho}_i} y_i \to x^* y_1^* \dots y_n^* \gtrsim^a_{\hat{\tau}_0} x^* y_1 \dots y_n)$$

hold. Then using the induction hypothesis we are done (note that $(1), (2)$ imply, in particular, that $x^* : M_{\hat{\rho}_1} \to (M_{\hat{\rho}_2} \to \dots \to M_{\hat{\tau}_0}) \dots)$ and $x : M_{\rho_1} \to (M_{\rho_2} \to \dots \to M_{\tau_0}) \dots)$).

There are four clauses to verify:

(1a) $\forall y_1^*, y_1, \dots, y_{n+1}^*, y_{n+1}(\bigwedge_{i=1}^{n+1} y_i^* \gtrsim^a_{\rho_i} y_i \to (x^* y_1^* \dots y_n^*) y_{n+1}^* \gtrsim^a_{\tau} (x y_1 \dots y_n) y_{n+1}),$

(1b) $\forall y_1^*, y_1, \dots, y_{n+1}^*, y_{n+1}(\bigwedge_{i=1}^{n} y_i^* \gtrsim^a_{\rho_i} y_i \wedge y_{n+1}^* \gtrsim^a_{\hat{\rho}_{n+1}} y_{n+1} \to (x^* y_1^* \dots y_n^*) y_{n+1}^* \gtrsim^a_{\hat{\tau}} (x^* y_1^* \dots y_n^*) y_{n+1}),$

(2a) $\forall y_1^*, y_1, \dots, y_{n+1}^*, y_{n+1}(\bigwedge_{i=1}^{n+1} y_i^* \gtrsim^a_{\hat{\rho}_i} y_i \to (x^* y_1^* \dots y_n^*) y_{n+1}^* \gtrsim^a_{\hat{\tau}} (x^* y_1 \dots y_n) y_{n+1}),$

(2b) $\forall y_1^*, y_1, \dots, y_{n+1}^*, y_{n+1}(\bigwedge_{i=1}^{n+1} y_i^* \gtrsim^a_{\hat{\rho}_i} y_i \to (x^* y_1^* \dots y_n^*) y_{n+1}^* \gtrsim^a_{\hat{\tau}} (x^* y_1^* \dots y_n^*) y_{n+1}).$

(1a) and (2a) hold by assumption, (1b) and (2b) follow from (2a) using lemma 17.75. □

The following (primitive recursive) functionals, which we only need for the types $\rho \in \mathbf{T}$, were already defined in chapter 3 (definition 3.65). We recall the definition here:

Definition 17.81. For $\rho = 0\rho_k \dots \rho_1 \in \mathbf{T}$ we define $\max_\rho$ by

$$\max{}_\rho(x, y) := \lambda v_1^{\rho_1}, \dots, v_k^{\rho_k}. \max{}_{\mathbb{N}}(x\underline{v}, y\underline{v})$$

For types $\rho 0$ with $\rho = 0\rho_k \dots \rho_1$, we define functionals $(\cdot)^M$ in $M_{\rho 0(\rho 0)}$ by :

$$x^M(y^0) := \lambda \underline{v}^{\underline{\rho}}. \max{}_{\mathbb{N}}\{x(i, \underline{v}) \mid i = 1, \dots, y\}.$$

The next lemma is proved analogously to lemma 3.66 whose extension to the new types it constitutes:

Lemma 17.82. *Let* $\rho \in \mathrm{T}^X$.

1) *Let* $x_1^*, x_2^* \in M_{\widehat{\rho}}$ *and* $x_1, x_2 \in M_\rho$.
   Then $x_1^* \gtrsim^a_\rho x_1 \wedge x_2^* \gtrsim^a_\rho x_2 \to \max_{\widehat{\rho}}(x_1^*, x_2^*) \gtrsim^a_\rho x_1, x_2$.
2) *Let* $x^* \in M_{\widehat{\rho}}^{M_0}, x \in M_\rho^{M_0}$ *and* $\forall n(x^*(n) \gtrsim^a_\rho x(n))$. *Then* $(x^*)^M \gtrsim^a_{\rho 0} x \in M_{\rho 0}$.
   Let $x^*, x \in M_{\widehat{\rho}}^{M_0}$ *and* $\forall n(x^*(n) \gtrsim^a_{\widehat{\rho}} x(n))$. *Then* $(x^*)^M \gtrsim^a_{\widehat{\rho}0} x^M \in M_{\widehat{\rho}0}$.
   In particular, $M_\rho^{M_0} = M_{\rho 0}$ *for each* $\rho \in \mathbf{T}^X$.

Note that $\widehat{\rho} \in \mathbf{T}$ *for* $\rho \in \mathbf{T}^X$ *so that the previous definition applies.*

Lemma 17.83. *Let* (X,d,W) *be a nonempty hyperbolic space. Then* $\mathcal{M}^{\omega,X}$ *is a model of* $\mathcal{A}^\omega[X,d,W]^-_{-b}$ +(BR) *(for a suitable interpretation of the constants of* $\mathcal{A}^\omega[X,d,W]^-_{-b}$ +(BR) *in* $\mathcal{M}^{\omega,X}$*), where we may interpret* 0_X *by an arbitrary element* $a \in X$.
Moreover, for any closed term t *of* $\mathcal{A}^\omega[X,d,W]^-_{-b}$ +(BR) *one can construct a closed term* t^* *of* $\mathcal{A}^\omega$ +(BR) *– so, in particular,* t^* *does not contain the constants* $0_X, d_X$ *and* W_X *– such that*

$$\mathcal{M}^{\omega,X} \models \forall a^X \forall n^0((n)_{\mathbb{R}} \geq d(0_X,a) \to t^*(n) \gtrsim^a t).$$

In particular, if we interpret 0_X *by* $a \in X$*, then it holds in* $\mathcal{M}^{\omega,X}$ *that* $t^*(0^0)$ *is a uniform a-majorant of* t *(note that* $t^*(0^0)$ *does not depend on a).*

Proof: The constants of $\mathcal{A}^\omega$+(BR) – which are characterized by their defining axioms – are interpreted and majorized as in chapter 11, theorem 11.17, except that they are now taken over the extended set of types $\mathbf{T}^X$, where $M_X := X$. Using lemma 17.80 one easily verifies that:

- $0 \gtrsim^a_0 0$,
- $S \gtrsim^a_1 S$,
- $\Pi_{\widehat{\rho},\widehat{\tau}} \gtrsim^a \Pi_{\rho,\tau}$,
- $\Sigma_{\widehat{\sigma},\widehat{\rho},\widehat{\tau}} \gtrsim^a \Sigma_{\sigma,\rho,\tau}$.

In particular, these majorants do not depend on a.
In the construction of majorants for the recursors $\underline{R}$ and the bar-recursors $\underline{B}$ in chapters 3 (proposition 3.69) and 11 (theorem 11.17) we only used the functionals $\max_\rho$ and $x^{\rho 0} \mapsto x^M$ and the properties which we established in lemma 17.82 for the extended types.
As a-majorants for $\underline{R}$ and $\underline{B}$ only involve types in $\mathbf{T}$, we do not need to extend these functionals to the types $\mathbf{T}^X$. We now again suppress the tuple notation.
As in the proof of proposition 3.69 one shows by induction on n that $\forall n(R_{\widehat{\rho}}n \gtrsim^a_\rho R_\rho n)$. Here we use lemma 17.80. Lemma 17.82 now implies that $R_\rho^* := R_{\widehat{\rho}}^M \gtrsim^a_\rho R_\rho$. Again, the majorants do not depend on a.
In the case where we interpret 0_X by a, the a-majorant for $B_{\rho,\tau}$ is defined exactly as in the proof of theorem 11.17, i.e. $B_{\rho,\tau}^* := \lambda y,z,u,n,x.(B_{\widehat{\rho},\widehat{\tau}} y^m z^m u_z)^M nx$, where $y^m(x^{\widehat{\rho}0}) := y(x^M)$, $z^m nx := znx^M$ and $u_z := \lambda v,n,x.\max(znx^M, uvnx^M)$ and B^* does not depend on a. The general case can be reduced to this by lemma 17.78: given a

bound n on $d(a,0_X)$ we may transform this majorant into a majorant for any choice of a. If 0_X does not occur in the theorems whose proofs we are analyzing (see below) then we are free to interpret 0_X by $a \in X$ and the dependency on n disappears. Using remark 17.76 the majorization proof proceeds exactly as in the proof of theorem 11.17.

For the new constants we take in accordance with definition 17.29 (writing simply $\mathcal{M}$ for $\mathcal{M}^{\omega,X}$)

$$[0_X]_{\mathcal{M}} := c \text{ for some } c \in X,$$

$$[d_X]_{\mathcal{M}} := \lambda x, y \in X.(d(x,y))_\circ,$$

$$[W_X]_{\mathcal{M}} := \lambda x, y \in X \lambda \alpha \in \mathbb{N}^{\mathbb{N}}.W(x,y,r_{\tilde{\alpha}}),$$

where $(x)_\circ$ is the construction from definition 17.7 and $r_{\tilde{\alpha}} \in [0,1]$ is the unique real number represented by $\tilde{\alpha}$ (see lemma 4.25). To show that these functionals are in $\mathcal{M}^{\omega,X}$ we have to construct a-majorants as follows:

- $n^0 \gtrsim^a 0_X$ for every n with $(n)_{\mathbb{R}} \geq_{\mathbb{R}} d_X(a,0_X)$. If we interpret 0_X by a, then we may simply take $n := 0$.
- $\lambda x^0, y^0.(x+y)_\circ \gtrsim^a d_X^{1(X)(X)}$, where $(\cdot)_\circ : \mathbb{N} \to \mathbb{N}^{\mathbb{N}}$ here denotes the restriction of the operator from definition 17.7 to $\mathbb{N}$ which can be simply calculated as $(x^0)_\circ(n) = j(x \cdot 2^{n+2}, 2^{n+1}-1)$,
- $\lambda x^0, y^0, z^1.\max_0(x,y) \gtrsim^a W_X^{X(1)(X)(X)}$.

The a-majorant for 0_X is obvious. The a-majorant for d_X follows using the triangle inequality: assume $n_1 \gtrsim^a x$ and $n_2 \gtrsim^a y$ then

$$d(x,y) \leq d(x,a) + d(y,a) \leq n_1 + n_2.$$

In the model $\mathcal{M}^{\omega,X}$ the expression $d_X(x,y)$ is interpreted by $(d(x,y))_\circ$ and by lemma 17.8 $n_1 + n_2 \geq d(x,y)$ implies $(n_1+n_2)_\circ$ $s\text{-}maj_1(d(x,y))_\circ$. Moreover, for $n_1^* \geq_0 n_1$ and $n_2^* \geq_0 n_2$ we have $(n_1^* + n_2^*)_\circ$ $s\text{-}maj_1$ $(n_1+n_2)_\circ$. Hence (using that $s\text{-}maj_1 \equiv \gtrsim_1^a$) the validity of the a-majorant for d_X follows by lemma 17.80.
Finally, the a-majorant for W_X can be justified by the first axiom for hyperbolic spaces:

$$\forall x^X, y^X, z^X \forall \lambda^1 (d_X(z, W_X(x,y,\lambda)) \leq_{\mathbb{R}} (1 -_{\mathbb{R}} \tilde{\lambda}) d_X(z,x) +_{\mathbb{R}} \tilde{\lambda} d_X(z,y)).$$

The construction $\tilde{\lambda}$ turns representatives λ of arbitrary real numbers into representatives $\tilde{\lambda}$ of real numbers in the interval $[0,1]$ (lemma 4.25). Hence it follows that $d_X(a, W_X(x,y,\lambda))$ is less than the maximum of $d_X(a,x)$ and $d_X(a,y)$ and hence less than the maximum of respective upper bounds on $d_X(a,x)$ and $d_X(a,y)$. Since the self-majorizability of $\max_0$ is trivial, the claim follows (using again lemma 17.80).

Note, that the a-majorants for d_X, W_X are uniform, i.e. they do not depend on a. Only the a-majorant for 0_X depends on a. Also note, that the $(\cdot)_\circ$-operator, which is ineffective in general, only is applied to natural numbers, where it is trivially computable (see above).

Since we managed to construct for each constant of $\mathcal{A}^\omega[X,d,W]^-_{-b}$+(BR) a closed term of $\mathcal{A}^\omega$+(BR) majorizing the respective constant we have established that we can construct such an a-majorant t^* for any closed term t of $\mathcal{A}^\omega[X,d,W]^-_{-b}$+(BR) where we furthermore λ-abstract the majorant n for 0_X. So in total (using implicitly remark 17.77) we have shown that

$$\mathcal{M}^{\omega,X} \models \forall a^X \forall n^0((n)_{\mathbb{R}} \geq_{\mathbb{R}} d_X(0_X,a) \rightarrow t^*(n) \gtrsim^a t),$$

where t^* does not contain $0_X, d_X$ and W_X (and $\Pi_{\rho,\tau}, \Sigma_{\delta,\rho,\tau}, \underline{R}_{\rho}, \underline{B}^{\rho,\tau}$ only for types in **T**). Moreover, we may take $n := 0$ if we interpret 0_X by a. □

Lemma 17.83 also covers $\mathcal{A}^\omega[X,d]^-_{-b}$, simply by omitting the parts concerning the W-function, and $\mathcal{A}^\omega[X,d,W,\mathrm{CAT}(0)]^-_{-b}$, as this theory contains no additional constants that need to be majorized.

The next lemma extends proposition 3.71 and remark 3.72

Lemma 17.84. *Let $t^{0\underline{\rho}}$ be a closed term of $\mathcal{A}^\omega[X,d,W]_{-b}$ where $\underline{\rho} = \rho_1,\ldots,\rho_k$ with the ρ_i's being of degree $2, 1^*$ or $(1,X)$. Then*

$$\forall x_1 \in M_{\rho_1},\ldots,x_k \in M_{\rho_k}([t]_{\mathcal{S}^{\omega,X}}(\underline{x}) = [t]_{\mathcal{M}^{\omega,X}}(\underline{x}))$$

(here we tacitly assume that 0_X is interpreted in both models $\mathcal{M}^{\omega,X}$ and $\mathcal{S}^{\omega,X}$ by the same element in X).

Proof: The proof proceeds as in the case of proposition 3.71 and remark 3.72 where we now extend $\approx_\rho$ to the new types by taking

$$x \approx_X y :\equiv x,y \in X \wedge x = y.$$

We only need to observe that $M_{\rho_i} \subseteq S_{\rho_i}$ which follows using lemma 17.82. □

Proof of Theorem 17.52: Assume

$$\mathcal{A}^\omega[X,d,W]_{-b} \vdash \forall x^\rho \left(\forall u^0 B_\forall(x,u) \rightarrow \exists v^0 C_\exists(x,v)\right).$$

The formulas $B_\forall(x,u), C_\exists(x,v)$ have the form $\forall \underline{a} B_{qf}(x,u,\underline{a})$ and $\exists \underline{b} C_{qf}(x,v,\underline{b})$ with quantifier-free B_{qf} and C_{qf} respectively. Hence using classical logic we get

$$\mathcal{A}^\omega[X,d,W]_{-b} \vdash \forall x^\rho \exists u,v,\underline{a},\underline{b}\big(B_{qf}(x,u,\underline{a}) \rightarrow C_{qf}(x,v,\underline{b})\big).$$

By lemma 17.72 we obtain (disregarding the realizers for '$\exists \underline{a},\underline{b}$') closed terms t_U and t_V of $\mathcal{A}^\omega[X,d,W]^-_{-b}$+(BR) such that (pushing back '$\exists \underline{a},\underline{b}$' inside)

$$\mathcal{A}^\omega[X,d,W]^-_{-b} + \text{(BR)} \vdash \forall x^\rho \left(B_\forall(x,t_U(x)) \rightarrow C_\exists(x,t_V(x))\right).$$

Applying in turn lemma 17.72 to t_U, t_V yields closed terms t_{U^*}, t_{V^*} of $\mathcal{A}^\omega$+(BR) (i.e. not involving the new type X and, in particular, do not contain $0_X, d_X, W_X$) such that for all $a \in X$ and $n \in \mathbb{N}$ with $n^0 \geq d(0_X,a)$

$$\mathcal{M}^{\omega,X} \models \begin{cases} t_{U^*}(n) \gtrsim^a t_U \wedge t_{V^*}(n) \gtrsim^a t_V \wedge \\ \qquad \forall x^\rho \left(B_\forall(x, t_U(x)) \to C_\exists(x, t_V(x)) \right), \end{cases}$$

where $\mathcal{M}^{\omega,X}$ is defined in terms of an arbitrarily chosen nonempty hyperbolic space (X,d,W). For notational simplicity we identify 0_X with its interpretation in $\mathcal{M}^{\omega,X}$ and write $t_{U^*}(n)(x)$ more conveniently as $t_{U^*}(n,x)$. Now define the functional $\Phi(x^{\widehat{\rho}}, n) := \max(t_{U^*}(n,x), t_{V^*}(n,x))$. Then

$$(+)\ \mathcal{M}^{\omega,X} \models \forall u \le \Phi(x^*,n) B_\forall(x,u) \to \exists v \le \Phi(x^*,n) C_\exists(x,v)$$

holds for all $n \in \mathbb{N}$, $x \in M_\rho$ and $x^* \in M_{\widehat{\rho}}$ for which there exists an $a \in X$ such that $n \ge d(0_X, a)$ and $x^* \gtrsim^a x$.
For the types γ of degree 1^* or $(1,X)$ of the quantifiers hidden in the definition of $\forall/\exists$-formulas we have at least $M_\gamma \subseteq S_\gamma$, which is sufficient for our purposes. This is because types of that kind have arguments for whose types δ one has – using lemma 17.82 – that $M_\delta = S_\delta$. For parameters x^ρ with ρ of degree 2 or $(1,X)$, we restricted ourselves to those $x \in S_\rho$ which have a-majorants $x^* \in S_{\widehat{\rho}}$. Since functionals of such types ρ only have arguments of types τ for which $M_\tau = S_\tau$ we get from $x^* \gtrsim^a_\rho x$ (which implies that $x^* \gtrsim^a_{\widehat{\rho}} x^*$) that $x^* \in M_{\widehat{\rho}}, x \in M_\rho$. Hence $\Psi(x^*,n) := [\Phi]_{\mathcal{M}^\omega}(x^*,n)$ is defined and $(+)$ together with lemma 17.84 yields

$$(++)\ \mathcal{S}^{\omega,X} \models \forall u \le \Psi(x^*,n) B_\forall(x,u) \to \exists v \le \Psi(x^*,n) C_\exists(x,v)$$

holds for all $n \in \mathbb{N}$, $x \in S_\rho$ and $x^* \in S_{\widehat{\rho}}$ for which there exists an $a \in X$ such that $n^0 \ge d(0_X, a)$ and $x^* \gtrsim^a x$.
Φ is a bar recursively defined and hence denotes a computable (in the sense of Kleene's S1-S9 schemata read over $\mathcal{M}^\omega$) functional $\Psi = [\Phi]_{\mathcal{M}^\omega}$ in $\mathcal{M}^\omega$ which is defined on all majorizable elements of $S_{\widehat{\rho}}$. Note that this functional does not depend on (X,d,W). Moreover, for $\rho \in \mathbf{T}^X$ such that $\widehat{\rho}$ is of degree ≤ 1, Ψ defines a computable functional (in the sense of ordinary type-2 computability theory) in $\mathcal{S}^\omega$ by corollary 11.20.
In the case where 0_X does not occur in either $B_\forall$ or $C_\exists$, we may freely interpret 0_X by $a \in X$ so that $n := 0^0$ can be taken as an $\gtrsim^a$-majorant of 0_X which leads to majorants t_{U^*}, t_{V^*} and a resulting term Φ which no longer depend on a bound n on $d(0_X, a)$ as an extra argument. □

Proof of theorem 17.35: For the proof of theorem 17.35 we first observe that lemma 17.72 also applies to $\mathcal{A}^\omega[X,d,W]$ as this theory only contains one additional purely universal axiom (iv) (4) (expressing the fact that b_X is a bound for d_X) which is interpreted by itself. Lemma 17.83 extends as well with $\mathcal{A}^\omega[X,d,W]^-$ in the conclusion if we interpret b_X in $\mathcal{M}^{\omega,X}$ by any number $b \in \mathbb{N}$ which is a bound on d. We now can reason as in the proof of 17.52 with a taken as the interpretation of 0_X and a suitable constant-b functional $\lambda \underline{v}.b$ as 0_X-majorant of x (instead of x^*). □

Proof of theorem 17.69: we first have to prove a lemma analogous to lemma 17.83

Lemma 17.85. *Let $(X, \|\cdot\|)$ be a nontrivial real normed linear space with a nonempty convex subset C. Then $\mathcal{M}^{\omega,X}$ is a model of $\mathcal{A}^\omega[X,\|\cdot\|,C]^-_{-b}$+(BR) (for a suitable interpretation of the constants of $\mathcal{A}^\omega[X,\|\cdot\|,C]^-_{-b}$ +(BR) in $\mathcal{M}^{\omega,X}$ where we have to interpret 0_X by the zero vector 0^X in X and use $\gtrsim^{0_X}$).*
Moreover, for any closed term t of $\mathcal{A}^\omega[X,\|\cdot\|,C]^-_{-b}$ +(BR) one can construct a closed term t^* of $\mathcal{A}^\omega$ +(BR) such that

$$\mathcal{M}^{\omega,X} \models \forall n^0((n)_\mathbb{R} \geq_\mathbb{R} \|c_X\|_X \to t^*(n) \gtrsim^{0_X} t).$$

Similarly for $\mathcal{A}^\omega[X,\|\cdot\|,\eta,C]^-_{-b}$ +(BR) and $\mathcal{A}^\omega[X,\langle\cdot,\cdot\rangle,C]^-_{-b}$ +(BR).

Proof: The proof proceeds similarly to the one of lemma 17.83. The main difference to the proof of Lemma 17.83 is that we fix $a = 0_X$ (where 0_X now has to be interpreted by the zero vector of X).
The constants of $\mathcal{A}^\omega$+(BR) (extended to the types $\mathbf{T}^X$) are interpreted as before. For the new constants we take (writing simply $\mathcal{M}$ for $\mathcal{M}^{\omega,X}$):

$[0_X]_\mathcal{M} := 0^X$, where 0^X is the zero vector of the linear space X,

$[1_X]_\mathcal{M} := a$ for some $a \in X$ with $\|a\| = 1$ (since X is assumed to be nontrivial there exists a $v \in X$ with $\|v\| > 0$ and hence an $a := \frac{v}{\|v\|}$ with $\|a\| = 1$,)

$[c_X]_\mathcal{M} := c$ for some $c \in C$ (which exists since C is assumed to be nonempty),

$[+_X]_\mathcal{M} :=$ addition in X,

$[-_X]_\mathcal{M} :=$ inverse of x w.r.t. $+$ in X,

$[\cdot_X]_\mathcal{M} := \lambda\alpha \in \mathbb{N}^\mathbb{N}, x \in X.r_\alpha \cdot x$, where r_α is the unique real number represented by α in the sense of chapter 4 and '$\cdot$' refers to the scalar multiplication of the $\mathbb{R}$-linear space X,

$[\|\cdot\|_X]_\mathcal{M} := \lambda x \in X.(\|x\|)_\circ$, where $(r)_\circ$ for $r \in \mathbb{R}_+$ is the construction from 17.7,

$$[\chi_C]_\mathcal{M} := \lambda x \in X. \begin{cases} 0^0, \text{ if } x \in C \\ 1^0, \text{ if } x \notin C. \end{cases}$$

We now have to construct 0_X-majorants for the new constants:

- $0^0 \gtrsim^{0_X} 0_X$,
- $1^0 \gtrsim^{0_X} 1_X$,
- $\lambda x^0.(x)_\circ \gtrsim^{0_X} \|\cdot\|_X^{X\to 1}$ (again with $(\cdot)_\circ$ restricted to $\mathbb{N}$),
- $\lambda x^0, y^0.x + y \gtrsim^{0_X} +_X^{X\to X\to X}$,
- $\lambda x^0.x \gtrsim^{0_X} -_X^{X\to X}$,

- $\lambda \alpha^1, x^0.(\alpha(0)+1)\cdot x \gtrsim^{0_X} \cdot_X^{1\to X\to X}$.

For the convex subset C, we have the characteristic term χ_C for the subset C, which is majorized as follows:

$$\lambda x^0.1 \gtrsim^{0_X} \chi_C^{X\to 0}.$$

For the constant $c_X \in C$ we have, given an $n \geq \|c_X\|$, the 0_X-majorant

$$n^0 \gtrsim_X^{0_X} c_X.$$

For uniformly convex spaces we 0_X-majorize the modulus $\eta : \mathbb{N} \to \mathbb{N}$ of uniform convexity by

$$(\eta)^M \gtrsim_1^{0_X} \eta.$$

The 0_X-majorants for $0_X, 1_X, \chi_C, c_X$ and η are obvious. For $\|\cdot\|_X$ we need to consider the interpretation of $\|\cdot\|_X$ in the model $\mathcal{M}^{\omega,X}$: the norm $\|x\|_X$ of an element $x \in X$ is interpreted by the actual norm using the $()_\circ$-operator, i.e. by $(\|x\|)_\circ$. In order to show that (in the model) $\lambda x.(x)_\circ \gtrsim^a \|\cdot\|_X$ we need to show two things: (1) if $n \gtrsim^{0_X} x$ then $(n)_\circ$ *s-maj*$_1(\|x\|)_\circ$ and (2) if $n \geq m$ then $(n)_\circ$ *s-maj*$_1(m)_\circ$ (recall that *s-maj*$_1$ and $\gtrsim_1^a$ are the same relations). For (1), if $n \gtrsim^{0_X} x$ then by definition $n \geq \|x\|$ and hence $(n)_\circ$ *s-maj*$_1(\|x\|)_\circ$ by lemma 17.8. For (2) the result follows directly from lemma 17.8 (or – even more simple – from the definition of $(n)_\circ$).
For $-_X$, the 0_X-majorant is derived straightforwardly from basic properties of the norm $\|\cdot\|_X$.
For $+_X$, we additionally use the triangle inequality to verify the majorant, i.e. $\|x+y\| \leq \|x\|+\|y\|$: if $n_1 \gtrsim^{0_X} x$ and $n_2 \gtrsim^{0_X} y$ then $n_1+n_2 \geq \|x+y\|$ which shows the correctness of the majorant.
For the verification of the majorant for the scalar multiplication $\cdot_X$ we have to recall the details of how α^1 represents (in the sense of chapter 4) a real number via a Cauchy sequence of rational numbers with fixed rate of convergence. The rational numbers in turn are represented by natural numbers using a monotone coding function (see chapter 4) such that $(\alpha(n))_{\mathbb{Q}} \geq_{\mathbb{Q}} |\alpha(n)|_{\mathbb{Q}}$ for all n, where $(\cdot)_{\mathbb{Q}}$ refers to the embedding of $\mathbb{N}$ into $\mathbb{Q}$ modulo the representation of rational numbers. I.e. the inequality expresses that the natural number $\alpha(n)$ is an upper bound for the absolute value of the rational number encoded by $\alpha(n)$. Since $|\lambda n^0.\alpha(0) -_{\mathbb{R}} \alpha| \leq 1$ the natural number $\alpha(0)+1$ is an upper bound for the real number represented by $|\alpha|_{\mathbb{R}}$. Now let α^* *s-maj*$_1 \alpha$. Then $\alpha^*(0)+1 \geq_0 \alpha(0)+1$. Since $\|\alpha\cdot x\|_X =_{\mathbb{R}} |\alpha|_{\mathbb{R}} \cdot_{\mathbb{R}} \|x\|_X$ we, therefore, have that $\alpha^*(0)+1$ taken as a natural number multiplied with an $n \gtrsim^{0_X} x$ is a 0_X-majorant for $\alpha \cdot_X x$. □

With this lemma in place, the proof of theorem 17.69 now proceeds as in the case of theorem 17.52. □

17.5 Further variations

From the proofs above it is clear that the same results hold true for extensions of the theories treated so far obtained by adding

- new axioms which have the form of $\forall$-sentences.
- new constants (possibly together with further $\forall$-axioms) provided that the latter have sufficiently low types to ensure their a-majorizability (or the majorizability is guaranteed by the defining axioms for these constants as long as they have a type of degree $(1,X)$). In particular this applies to constants of type 1 which will be sufficient in the following.

We made use of this already when we treated the case of CAT(0)-spaces by adding the additional axiom CN^- or the case of uniformly convex normed spaces, where we added a constant η^1 together with the $\forall$-sentence (10) as an axiom.

As observed by Leuştean in [262] this approach can also be used to adapt the above metatheorems to two further classes if structures, namely

- $\mathbb{R}$-trees as introduced by J. Tits ([360]) and
- (δ-)hyperbolic spaces in the sense of M. Gromov ([145]).

$\mathbb{R}$-trees generalize the concept of local Bruhat-Tits buildings.

Definition 17.86 (Tits [360]). (X,d) is an $\mathbb{R}$-tree iff it is a geodesic space containing no homeomorphic image of a circle.

Remark 17.87. It is a well-known fact that $\mathbb{R}$-trees always are **unique** geodesic spaces.

Using a number of results from the literature, L. Leuştean [262] gave the following characterization of $\mathbb{R}$-trees in terms of hyperbolic spaces (X,d,W) :

Proposition 17.88 (Leuştean [262]). *Let (X,d) be a metric space. Then the following are equivalent:*

1) *(X,d) is an $\mathbb{R}$-tree,*
2) *For some $W : X \times X \times [0,1] \to X$, (X,d,W) is a hyperbolic space and the following inequality holds for all $x,y,z,w \in X$*

$$d(x,y)+d(z,w) \leq \max\{d(x,z)+d(y,w), d(x,w)+d(y,z)\}.$$

3) *There is a unique function $W : X \times X \times [0,1] \to X$ w.r.t. which (X,d,W) is a hyperbolic space and for all $x,y,z,w \in X$*

$$d(x,y)+d(z,w) \leq \max\{d(x,z)+d(y,w), d(x,w)+d(y,z)\}.$$

We now can define the theory $\mathcal{A}^{\omega}[X,d,W,\mathbb{R}\text{-tree}]_{(-b)}$ by adding to $\mathcal{A}^{\omega}[X,d,W]_{(-b)}$ the universal axiom

$$\forall x^X, y^X, z^X, w^X \big(d_X(x,y) +_{\mathbb{R}} d_X(z,w) \leq_{\mathbb{R}} \\ \leq_{\mathbb{R}} \max_{\mathbb{R}}\{d_X(x,z) +_{\mathbb{R}} d_X(y,w), d_X(x,w) +_{\mathbb{R}} d_X(y,z)\}\big).$$

The previous results on $\mathcal{A}^\omega[X,d,W]_{(-b)}$ also hold for $\mathcal{A}^\omega[X,d,W,\mathbb{R}\text{-tree}]_{(-b)}$, where then the conclusions of the rules hold in all (bounded in the case of $\mathcal{A}^\omega[X,d,W,\mathbb{R}\text{-tree}]$) nonempty $\mathbb{R}$-trees (X,d).

In order to treat Gromov-hyperbolic spaces we need the notion of Gromov product:

Definition 17.89. Let (X,d) be a metric space. The Gromov product $(x\cdot y)_w$ of $x,y \in X$ with respect to the base point $w \in X$ is defined by

$$(x\cdot y)_w := \frac{1}{2}\big(d(x,w)+d(y,w)-d(x,y)\big).$$

Definition 17.90. Let $\mathbb{R} \ni \delta \geq 0$. A metric space (X,d) is called δ-hyperbolic if for all $x,y,z,w \in X$

$$(x\cdot y)_w \geq \min\{(x\cdot z)_w, (y\cdot z)_w\} - \delta.$$

(X,d) is called Gromov-hyperbolic if it is δ-hyperbolic for some $\delta \geq 0$.

The condition on a metric space (X,d) to be δ-hyperbolic can easily be seen to be equivalent to

$$\forall x,y,z,w \in X\big(d(x,y)+d(z,w) \leq \max\{d(x,z)+d(y,w), d(x,w)+d(y,z)\}+2\delta\big).$$

So we can define a corresponding theory $\mathcal{A}^\omega[X,d,\delta\text{-hyperbolic}]_{(-b)}$ for an abstract δ-hyperbolic space (X,d) by extending $\mathcal{A}^\omega[X,d]_{(-b)}$ in the following way:

- add a constant $\delta^1_{\mathbb{R}}$ of type 1 (representing the nonnegative real number δ),
- add the purely universal axioms $\delta_{\mathbb{R}} \geq_{\mathbb{R}} 0_{\mathbb{R}}$ and

$$\forall x^X, y^X, z^X, w^X \big(d_X(x,y) +_{\mathbb{R}} d_X(z,w) \leq_{\mathbb{R}} \\ \max_{\mathbb{R}}\{d_X(x,z) +_{\mathbb{R}} d_X(y,w), d_X(x,w) +_{\mathbb{R}} d_X(y,z)\} +_{\mathbb{R}} (2)_{\mathbb{R}} \cdot_{\mathbb{R}} \delta_{\mathbb{R}}\big).$$

Extending the interpretation of the constants in the model $\mathcal{S}^{\omega,X}$ over a δ-hyperbolic metric space (X,d) by interpreting $\delta^1_{\mathbb{R}}$ as some representative of $\delta \geq 0$, e.g. as $(\delta)_\circ$, the results for $\mathcal{A}^\omega[X,d]_{(-b)}$ also hold for $\mathcal{A}^\omega[X,d,\delta\text{-hyperbolic}]_{(-b)}$ where then the conclusion of our rules hold in all nonempty δ-hyperbolic spaces (X,d). Note that for $\mathbb{N} \ni k \geq \delta$ we can take $\delta^*_{\mathbb{R}} := \lambda n.j(k\cdot 2^{n+2}, 2^{n+1}-1)$ as majorant of $(\delta)_\circ$.

Remark 17.91. Using the notion of δ-hyperbolic space the concept of $\mathbb{R}$-tree can also be defined as a geodesic metric space (X,d) that is 0-hyperbolic.

We conclude this section by showing how to adapt our metatheorems (similarly to the cases treated above) to **complete** metric, hyperbolic etc. spaces. We axiomatize completeness by adding an operator $C^{X(X0)}$ to the systems whose intended meaning it is to assign to any Cauchy sequence $x^{X0}_{(\cdot)}$ in X a limit (for better readability we sometimes switch to the more common notation (x_n) for sequences once there is no danger of confusing the sequence (x_n) with the element x_n). In order to be able to do this by a purely universal axiom we restrict ourselves to Cauchy sequences with a fixed Cauchy rate (say 2^{-n}) and use a construction similar to the construction $f \mapsto \widehat{f}$ from the representation of general Polish spaces in chapter 4 to even avoid to have to write the assumption that (x_n) is a Cauchy sequence with rate of convergence 2^{-n}. In $\mathcal{A}^{\omega}[X,d]_{(-b)}$ we can define a functional $(x_n) \mapsto (\widehat{x}_n)$ of type $X0(X0)$ such that

$$\widehat{x}_n =_X \begin{cases} x_n \text{ if } \forall k < n\big([d_X(x_k,x_{k+1})](k+1) <_{\mathbb{Q}} 6\cdot 2^{-k-1}\big), \\ x_k \text{ for } \min k < n : [d_X(x_k,x_{k+1})](k+1) \geq_{\mathbb{Q}} 6\cdot 2^{-k-1}, \text{ otherwise.} \end{cases}$$

Note that here d_X denotes the constant from the language of $\mathcal{A}^{\omega}[X,d]$ of type $1(X)(X)$ whereas in the definition of $\widehat{h}$ in chapter 4 it denoted a function of type $1(0)(0)$.

It is clear that $(\widehat{x}_n)$ always satisfies

$$\forall n\big(d_X(\widehat{x}_n,\widehat{x}_{n+1}) <_{\mathbb{R}} 7\cdot 2^{-n-1}\big)$$

and so is a Cauchy sequence with Cauchy rate 2^{-n+3}. Conversely, if (x_n) is a Cauchy sequence with rate 2^{-n}, then $x_n =_X \widehat{x}_n$ for all $n \in \mathbb{N}$.
Now our official completeness axiom $(\mathcal{C})$ is formulated as follows

$$(\mathcal{C})\ \forall x^{X0}_{(\cdot)}, \forall k^0\big(d_X(C(x_{(\cdot)}),\widehat{x}_k) \leq_{\mathbb{R}} 2^{-k+3}\big),$$

which is an $\forall$-formula in the sense of definition 17.34. Let us first verify that (over $\mathcal{A}^{\omega}[X,d]_{(-b)}$) this axiom implies the completeness of the space (X,d) : let (x_n) be an **arbitrary** Cauchy sequence in X, i.e.

$$\forall k^0 \exists n^0 \forall m,\tilde{m} \geq n\big(d_X(x_m,x_{\tilde{m}}) <_{\mathbb{R}} 2^{-k}\big).$$

By $\mathrm{AC}^{0,0}$ (which follows from DC as we showed in chapter 11) – in fact Π^0_1-AC is sufficient here – we obtain

$$(+)\ \exists f^1 \forall k^0 \forall m,\tilde{m} \geq f(k)\big(d_X(x_m,x_{\tilde{m}}) <_{\mathbb{R}} 2^{-k}\big).$$

We now define $y_n := x_{f(n)}$ and show that:

1) (y_n) is a Cauchy sequence with rate 2^{-n}.
2) If (y_n) is convergent with limit y, then (x_n) converges as well and its limit again is y.

Ad 1): Let $m,\tilde{m} \geq n$. W.l.o.g. we may assume that $f(m) \leq f(\tilde{m})$. Then by $(+)$

$$d_X(y_m,\tilde{y}_m) =_{\mathbb{R}} d_X(x_{f(m)},x_{f(\tilde{m})}) <_{\mathbb{R}} 2^{-m} \leq 2^{-n}.$$

Ad 2): Let y be the limit of (y_n). Then, in particular,

$$(a)\ \forall k \exists n \geq k \big(d_X(y, x_{f(n)}) =_{\mathbb{R}} d_X(y, y_n) <_{\mathbb{R}} 2^{-k}\big).$$

Since $n \geq k$, $(+)$ yields that

$$(b)\ \forall l \geq f(n) \Big(d_X(x_l, x_{f(n)}) <_{\mathbb{R}} 2^{-k}\Big).$$

(a) and (b) together give

$$\forall l \geq f(n) \big(d_X(y, x_l) <_{\mathbb{R}} 2^{-k+1}\big).$$

So, in total, we have shown that

$$\forall k \exists m \forall l \geq m (d_X(y, x_l) <_{\mathbb{R}} 2^{-k+1}).$$

By '1)' we have $y_n =_X \widehat{y}_n$ for all $n \in \mathbb{N}$ and so $(\mathscr{C})$ implies that $C((y_n))$ is the limit of (y_n) and hence, using '2)', also of (x_n). In particular, we have shown that $(\mathscr{C})$ implies that (x_n) is convergent.

Conversely, it is clear that $(\mathscr{C})$ can be satisfied in any complete metric space (X, d) if we extend the interpretation given in definition 17.29 by interpreting $C(x_{(\cdot)})$ as the limit of the Cauchy sequence given by $\widehat{x}_{(\cdot)}$ (with d_X interpreted as before, i.e. as $(d(x,y))_\circ$).

It remains to show that the constant C is (in fact uniformly) a-majorizable: let $C^* := \lambda g^1.g(0) + 8$. Then $C^* \gtrsim^a_{X(X0)} C$: assume that $g \gtrsim^a_{X0} x_{(\cdot)}$. Then, in particular,

$$(g(0))_{\mathbb{R}} \geq_{\mathbb{R}} d_X(a, x_0) =_{\mathbb{R}} d_X(a, \hat{x_0}) \geq_{\mathbb{R}} d_X(a, C(x_{(\cdot)})) - 8,$$

where the last inequality holds by $(\mathscr{C})$. Hence

$$(g(0) + 8)_{\mathbb{R}} \geq_{\mathbb{R}} d_X(a, C(x_{(\cdot)})).$$

Clearly, C^* is self-majorizing. Hence the claim follows.

As a corollary to the above treatment we obtain that the main metatheorems proved in this section also apply to proofs which make completeness assumptions on the underlying space if these proofs can be formalized then in $\mathscr{A}^\omega[X, d, \mathscr{C}]_{(-b)}$ (and analogously for the other systems) which results from $\mathscr{A}^\omega[X, d]_{(-b)}$ by adding the constant C and the axiom $(\mathscr{C})$. Then the conclusion holds in all **complete** metric spaces (resp. complete hyperbolic spaces etc.).

17.6 Treatment of several metric or normed spaces $X_1 \ldots, X_n$ simultaneously

In the sections above we only considered extensions of $\mathcal{A}^\omega$ by a single metric, hyperbolic or normed space X and finite types over $\mathbb{N}, X$. This is sufficient for the applications in fixed point theory discussed in the next chapter, where one considers selfmappings $f : X \to X$ of some space X (or of a convex subset $C \subseteq X$ in the normed case). It also provides the framework needed for applications in ergodic theory like the recent results in Avigad et al. [8]. However, our approach easily extends to contexts which involve several spaces $X_1, \ldots, X_n$ simultaneously as might be needed in future applications. We sketch here only some of the changes caused by this. Firstly, we have to extend our set of types:
By $\mathbf{T}^{X_1,\ldots,X_n}$ we denote the set of all finite types ρ over the ground types $0, X_1, \ldots, X_n$. For $\rho \in \mathbf{T}^{X_1,\ldots,X_n}$ the type $\widehat{\rho}$ is the type that results from ρ by replacing all occurrences of X_i, $1 \leq i \leq n$ by 0. For each ground type X_i we choose a reference point a_i of type X_i (i.e. an element of X_i in the intended interpretation). For $\underline{a} = a_1^{X_1}, \ldots a_n^{X_n}$ we then define a relation $\gtrsim^{\underline{a}}$ extending $\gtrsim^a$ as follows:

Definition 17.92. Given an n-tuple $\underline{a} = a_1^{X_1}, \ldots, a_n^{X_n}$ of objects of types $X_1, \ldots, X_n$, respectively, we define a relation $\gtrsim_\rho^{\underline{a}}$ between objects $x^{\widehat{\rho}}, y^\rho$ of type $\widehat{\rho}, \rho$ inductively on $\rho \in \mathbf{T}^{X_1,\ldots,X_n}$ as follows:

$$\begin{cases} x^0 \gtrsim_0^{\underline{a}} y^0 :\equiv x \geq_0 y, \\ x^0 \gtrsim_{X_i}^{\underline{a}} y^{X_i} :\equiv (x)_\mathbb{R} \geq_\mathbb{R} d_{X_i}(y, a_i), \\ x \gtrsim_{\rho\to\tau}^{\underline{a}} y :\equiv \forall z', z(z' \gtrsim_\rho^{\underline{a}} z \to xz' \gtrsim_\tau^{\underline{a}} yz) \wedge \forall z', z(z' \gtrsim_{\widehat{\rho}}^{\underline{a}} z \to xz' \gtrsim_{\widehat{\tau}}^{\underline{a}} xz). \end{cases}$$

For those X_i that represent normed linear spaces we again require $a_i = 0_{X_i}$ so that $d_{X_i}(x, a_i) =_\mathbb{R} \|x\|_{X_i}$.

Remark 17.93. Note that if the type X_i does not occur in ρ, then $\gtrsim_\rho^{\underline{a}}$ does not depend on a_i.

Let us illustrate this definition for the special case of two metric spaces X_1 and X_2 and the type $X_1(X_2)$ (which we write as $X_1 \to X_2$) of functions $f : X_1 \to X_2$: for given $a_1^{X_1}$ and $a_2^{X_2}$ an (a_1, a_2)-majorant for $f^{X_1 \to X_2}$ is a monotone function f^* of type 1 such that

$$\forall n^0, x^{X_1} \big(d_{X_1}(x, a_1) \leq_\mathbb{R} (n)_\mathbb{R} \to d_{X_2}(f(x), a_2) \leq_\mathbb{R} (f^*(n))_\mathbb{R}\big).$$

Since we can now choose a_1 and a_2 separately, some classes of functions have particularly simple majorants: e.g. let f be nonexpansive and $a_2 := f(a_1)$. Then f is (a_1, a_2)-majorized by the identity function $\lambda n.n^0$. Since we cannot 'compare' the elements $x \in X_1$ and $f(x) \in X_2$ (as we could in the case of a common metric space $X_1 := X_2 := X$) we don't need the bound $b \geq d(x, f(x))$ that was required before.

Given two metric spaces X_1, X_2 we can also consider elements of the product of these spaces $X_1 \times X_2$ and functions between such product-elements. Functions involving product types are treated using 'currying' (going back to Schönfinkel [323] and hence also called 'Schönfinkelisation' occasionally) based on the following devices (for better readability we use here the notation $\tau \to \rho$ instead of $\rho(\tau)$):

- a function $f : X_1 \times \ldots \times X_n \to \rho$ is represented by $f : X_1 \to \ldots \to X_n \to \rho$,
- a function $\rho \to X_1 \times \ldots \times X_n$ is represented by an n-tuple of functions $f_i : \rho \to X_i$.

Thus e.g. a function $f : X_1 \times X_2 \to X_1 \times X_2$ will be represented by a pair $f_{1,2} : X_1 \to (X_2 \to X_{1,2})$. A function $g : (X_1 \times X_2 \to X_1 \times X_2) \to X_1 \times X_2$ by a pair $g_{1,2} : (X_1 \to (X_2 \to X_1)) \to ((X_1 \to (X_2 \to X_2)) \to X_{1,2})$ and similar for products of greater arity and functions of more complex types. Clearly, given constants d_{X_1} and d_{X_2} for the respective metrics on X_1 and X_2 one can define the usual product metrics e.g.

$$d_\infty((x_1^{X_1}, x_2^{X_2}), (y_1^{X_1}, y_2^{X_2})) := \max_{\mathbb{R}}(d_{X_1}(x_1, y_1), d_{X_2}(x_2, y_2)),$$

where d_∞ is of type $X_1 \to X_2 \to X_1 \to X_2 \to 1$ and similarly for the p-product metric d_p (with $1 \leq p < \infty$).

17.7 A generalized uniform boundedness principle $\exists$-UB^X

In this section we generalize the Σ_1^0-uniform boundedness principle from chapter 12 to bounded metric, hyperbolic or CAT(0)-spaces. Since we do not have an elimination-of-extensionality procedure for $\mathcal{A}^\omega[X,d]$ and the other theories (e.g. the constant d_X is not provable extensional as a functional of type $1(X)(X)$ but only when considered w.r.t. $=_{\mathbb{R}}$ instead of $=_1$) we proceed differently from the syntactic approach of the proof of theorem 12.8 and use the model-theoretic argument from the proof of theorem 12.14 instead (compare also the proof of theorem 4.9 in [207]).

Definition 17.94. A formula F is called a generalized $\exists$-formula if it has the form $\exists \underline{a}^{\underline{\sigma}} F_{qf}(\underline{a})$ where F_{qf} is quantifier-free and $\underline{\sigma}$ are arbitrary types in $\mathbf{T}^X$.

Definition 17.95. 1) A type $\rho \in \mathbf{T}^X$ is of degree $(\cdot, 0)$ (resp. $(\cdot, X)$) if it has the form $0(\rho_k)\ldots(\rho_1)$ (resp. $X(\rho_k)\ldots(\rho_1)$ with $\rho_1, \ldots, \rho_k \in \mathbf{T}^X$ which includes 0 (resp. X) as special case.
2) Let ρ be a type of degree $(\cdot, 0)$. Then

$$\min_\rho(x^\rho, y^\rho) := \lambda \underline{v}. \min_0(x\underline{v}, y\underline{v}).$$

Definition 17.96. The uniform boundedness schema $\exists$-UB^X for generalized $\exists$-formulas is defined as follows

$$\exists\text{-UB}^X :\equiv \begin{cases} \forall y^{\alpha(0)} (\forall k^0, x^\alpha, \underline{z}^{\underline{\beta}} \exists n^0 A_\exists(y, k, \min_\alpha(x, yk), \underline{z}, n) \to \\ \exists \chi^1 \forall k^0, x^\alpha, \underline{z}^{\underline{\beta}} \exists n \leq_0 \chi k A_\exists(y, k, \min_\alpha(x, yk), \underline{z}, n)), \end{cases}$$

where α is of degree $(\cdot,0)$, $\underline{\beta}=\beta_1,\ldots,\beta_m$ is a tuple of types in $\mathbf{T}^X$ of degree $(\cdot,X)$ and $A_\exists$ is a generalized $\exists$-formula which in addition to the variables indicated may have further parameters of arbitrary types.

Remark 17.97. Just as in the case of $\underline{z}$ which we have formulated for tuples, since it is used this way in our applications, we may also have tuples $\underline{x}^{\underline{\alpha}}$ of variables of types $\underline{\alpha}=\alpha_1,\ldots,\alpha_m$, where α_i is of degree $(\cdot,0)$. However, for notational simplicity we do not formulate this. In fact, one can utilize contractions of tuples of variables of degree $(\cdot,0)$ into a single variable of degree $(\cdot,0)$ to reduce the case with tuples to the case with a single variable x^α.

Remark 17.98. Using the axiom of countable choice (derivable in $\mathcal{A}^\omega[X,d]$ by remark 11.8) we, in fact, can prove $\exists$-UB^X from the prima-facie weaker version

$$\begin{aligned}&\forall y^{\alpha(0)}\big(\forall k^0,x^\alpha,\underline{z}^{\underline{\beta}}\exists n^0 A_\exists(y,k,\min_\alpha(x,yk),\underline{z},n)\to\\&\forall k^0\exists\tilde{n}\forall x^\alpha,\underline{z}^{\underline{\beta}}\exists n\le_0\tilde{n}A_\exists(y,k,\min_\alpha(x,yk),\underline{z},n)\big).\end{aligned}$$

However, this does not seem to be possible for the fragment of $\mathcal{A}^\omega[X,d]$ without DC or weaker fragments. We formulate $\exists$-UB^X as above since our proof of theorem 17.101 below establishes that $\exists$-UB^X does not contribute to the complexity of the extractable bounds even for these fragments whereas DC (and already AC^{0,0}) does (see chapter 11).

In the presence of full extensionality, our formulation of $\exists$-UB^X is equivalent to the following formulation which literally contains Σ^0_1-UB from chapter 12 as a special case:

$$\begin{cases}\forall y^{\alpha(0)}\big(\forall k^0\forall x\le_\alpha yk\forall\underline{z}^{\underline{\beta}}\exists n^0 A_\exists(y,k,x,\underline{z},n)\to\\ \exists\chi^1\forall k^0\forall x\le_\alpha yk\forall\underline{z}^{\underline{\beta}}\exists n\le_0\chi k A_\exists(y,k,x,\underline{z},n)\big),\end{cases}$$

In fact, for this it suffices to have the following instance of extensionality available

$$\forall y,k,\underline{z},n,x_1,x_2(x_1=_\alpha x_2\wedge A_\exists(y,k,x_1,\underline{z},n)\to A_\exists(y,k,x_2,\underline{z},n)).$$

In the absence of full extensionality, however, we have to formulate $\exists$-UB^X with $\min_\alpha(x,yk)$ instead in order to achieve that the corresponding extension of the axiom F has the form of an axiom treated trivially by monotone functional interpretation:

Definition 17.99. Let $\underline{\beta}$ be as before.

$$F^X:\equiv\forall\Phi,y\exists X\le y\exists\underline{Z}\forall k^0,x^\alpha,\underline{z}^{\underline{\beta}}\big(\Phi(k,Xk,\underline{Z}k)\ge_0\Phi(k,\min_\alpha(x,yk),\underline{z})\big).$$

Here X has type $\alpha(0)$, Z_i has type $\beta_i(0)$ and Φ has type $0\beta_m\ldots\beta_1\alpha0$.

The next lemma correspond to proposition 12.6 in chapter 12:

Lemma 17.100.

$$\mathcal{A}^\omega[X,d]+F^X\vdash\exists\text{-UB}^X.$$

Analogously for $\mathcal{A}^\omega[X,d,W]$ and the other extensions we consider.

Proof: Let us assume the premise

$$\forall k^0, x^\alpha, \underline{z}^{\underline{\beta}} \exists n^0 A_\exists(y, k, \min_\alpha(x, yk), \underline{z}, n)$$

of $\exists$-UB^X. Applying the schema of quantifier-free choice QF-AC from $\mathcal{A}^\omega[X,d]$ (recall that QF-AC is formulated for tuples of variables) it follows that there exists a functional Φ such that

$$\forall k^0, x^\alpha, \underline{z}^{\underline{\beta}} A_\exists(y, k, \min_\alpha(x, yk), \underline{z}, \Phi kx\underline{z}).$$

Clearly,

$$\mathcal{A}^\omega[X,d] \vdash \min_\alpha(\min_\alpha(x, yk), yk) =_\alpha \min_\alpha(x, yk)$$

Hence by the quantifier-free extensionality rule QF-ER from $\mathcal{A}^\omega[X,d]$ we obtain

$$\forall k^0, x^\alpha, \underline{z}^{\underline{\beta}} A_\exists(y, k, \min_\alpha(x, yk), \underline{z}, \Phi(k, \min_\alpha(x, yk), \underline{z})).$$

Applying F^X to Φ and y yields the existence of functionals $X(\leq y)$ and $\underline{Z}$ with

$$\forall k^0, x^\alpha, \underline{z}^{\underline{\beta}} \big(\Phi(k, Xk, \underline{Z}k) \geq_0 \Phi(k, \min_\alpha(x, yk), \underline{z})\big).$$

Obviously, $\chi(k) := \Phi(k, Xk, \underline{Z}k)$ satisfies the conclusion of $\exists$-UB^X. □

We now show that theorem 17.35 remains valid if we add $\exists$-UB^X to $\mathcal{A}^\omega[X,d]$:

Theorem 17.101. 1) *Let σ, ρ be types of degree ≤ 1 and τ be a type of degree $(1,X)$. Let $s^{\rho(\sigma)}$ be a closed term of $\mathcal{A}^\omega[X,d]$ and $B_\forall(x^\sigma, y^\rho, z^\tau, u^0)$ ($C_\exists(x^\sigma, y^\rho, z^\tau, v^0)$) be a $\forall$-formula containing only x, y, z, u free (resp. an $\exists$-formula containing only x, y, z, v free).*
If

$$\forall x^\sigma \forall y \leq_\rho s(x) \forall z^\tau \big(\forall u^0 B_\forall(x, y, z, u) \to \exists v^0 C_\exists(x, y, z, v)\big)$$

is provable in $\mathcal{A}^\omega[X,d] + \exists$-UB^X, then one can extract a computable functional $\Phi : S_\sigma \times \mathbb{N} \to \mathbb{N}$ such that for all $x \in S_\sigma$ and all $b \in \mathbb{N}$

$$\forall y \leq_\rho s(x) \forall z^\tau \big[\forall u \leq \Phi(x,b) B_\forall(x, y, z, u) \to \exists v \leq \Phi(x,b) C_\exists(x, y, z, v)\big]$$

holds in any (nonempty) metric space (X,d) whose metric is bounded by $b \in \mathbb{N}$ (with 'b_X' is to be interpreted by 'b').
The computational complexity of Φ can be estimated in terms of the strength of the $\mathcal{A}^\omega$-principle instances actually used in the proof (see remark 17.37 below).

2) *If the premise is proved in '$\mathcal{A}^\omega[X,d,W] + \exists$-UB^X' instead of '$\mathcal{A}^\omega[X,d] + \exists$-UB^X', then the conclusion holds in all b-bounded hyperbolic spaces.*

3) *If the premise is proved in '$\mathcal{A}^\omega[X,d,W,\mathrm{CAT}(0)] + \exists$-UB^X' instead of '$\mathcal{A}^\omega[X,d,W] + \exists$-UB^X', then the conclusion holds in all b-bounded* CAT(0)-*spaces.*

Instead of single variables x, y, z, u, v we may also have finite tuples of variables $\underline{x}, \underline{y}, \underline{z}, \underline{u}, \underline{v}$ as long as the elements of the respective tuples satisfy the same type re-

strictions as x,y,z,u,v. Moreover, instead of a single premise of the form '$\forall u^0 B_\forall(x,y,z,u)$' we may have a finite conjunction of such premises.

Remark 17.102. The proof of theorem 17.101, which we will give below, is based is an extension of the proof of theorems 17.52 and 17.35 above and will again provide an extraction algorithm for Φ. The functional Φ is given by a closed term of WE-PA$^\omega$+BR. Also our remark 17.37 on fragments made in connection with theorem 17.35 still applies as $\exists$-UB^X contributes only trivial terms as majorants.

Remark 17.103. Note that the conclusion is true in all b-bounded metric spaces (X,d), hyperbolic spaces (X,d,W) and CAT(0)-spaces, respectively, although the axiom $\exists$-UB^X even in the absence of x^α (so that the discussion in chapter 12 does not apply) is not. In fact, it is not even true in $\mathscr{M}^{\omega,X}$: e.g. consider the formula

$$(*)\ \forall x^X \exists n^0 \big(\Psi(x) =_0 0 \to d_X(x,0_X) >_{\mathbb{R}} 2^{-n}\big),$$

where Ψ is a parameter of type $0(X)$. Now take $X := [0,1]$, $d(x,y) := |x-y|$ and interpret 0_X by 0 and choose as $\Psi \in M_{0(X)} \subset S_{0(X)}$ the functional

$$\Psi(x) :=_0 \begin{cases} 0, \text{ if } x > 0 \\ 1, \text{ if } x = 0. \end{cases}$$

Then $(*)$ holds true, but, obviously, there is no bound on '$\exists n$' that does not depend on $x \in [0,1]$. This example shows that $\exists$-UB^X is already false for the special case $\underline{z} := z^X$ (and without x^α). In fact, as the next example shows, one does not even have to invoke a parameter Ψ but can consider a closed formula of $\mathscr{L}(\mathscr{A}^\omega[X,d])$ to obtain a counterexample: consider the sentence

$$\forall x^X, y^X \exists n^0 (d_X(x,y) \le_{\mathbb{R}} 2^{-n} \to d_X(x,0_X)(0) =_0 d_X(y,0_X)(0))$$

which is true in $\mathscr{M}^{\omega,[0,1]}$ and $\mathscr{S}^{\omega,[0,1]}$ with the metric above and $0_X := 0$. One easily calculates that $(|x|)_\circ(0) = j(4,1)$, if $x = 1$, but $(|x|)_\circ(0) = j(2,1)$, if $\frac{1}{2} \le x < 1$. Hence for $y := 1 \in [0,1]$ there is no x-uniform bound on '$\exists n$'. Further examples of this kind follow from the applications of $\exists$-UB^X given below.

Proof of theorem 17.101: 1) By the previous lemma, the assumption implies that $\mathscr{A}^\omega[X,d] + F^X$ proves that

$$\forall x^\sigma \forall y \le_\rho s(x) \forall z^\tau \big(\forall u^0 B_\forall(x,y,z,u) \to \exists v^0 C_\exists(x,y,z,v)\big).$$

Let $\mathscr{A}^\omega[X,d,\mathscr{X},\underline{\mathscr{Z}}]$ result from $\mathscr{A}^\omega[X,d]$ by adding new constants $\mathscr{X}$ and $\underline{\mathscr{Z}}$ of type $\alpha 0(\alpha 0)(0\beta_m \dots \beta_1 \alpha 0)$ resp. $\beta_i 0(\alpha 0)(0\beta_m \dots \beta_1 \alpha 0)$ to the language.

As in the proof of lemma 17.72 one shows that $\mathscr{A}^\omega[X,d] + F^X$ has a Gödel functional interpretation in $\mathscr{A}^\omega[X,d,\mathscr{X},\underline{\mathscr{Z}}]^- + \tilde{F}^X$+(BR), where

$$\mathscr{A}^\omega[X,d,\mathscr{X},\underline{\mathscr{Z}}]^- := \mathscr{A}^\omega[X,d,\mathscr{X},\underline{\mathscr{Z}}] \setminus \{\text{ QF-AC,DC }\}$$

and

$$\tilde{F}^X :\equiv \mathscr{X} \leq \lambda \Phi, y.y \wedge \forall \Phi, y, k, x, \underline{z}^{\underline{\beta}} (\Phi(k, \mathscr{X}\Phi yk, \underline{\mathscr{Z}}\Phi yk) \geq_0 \Phi(k, \min_\alpha(x, yk), \underline{z})).$$

In addition to the proof of lemma 17.72 we only have to consider the functional interpretation $((F^X)')^D$ of the negative translation $(F^X)'$ of F^X: clearly $(F^X)'$ is intuitionistically implied by F^X so that it suffices to solve the functional interpretation $(F^X)^D$ of F^X. However, $(F^X)^D$ precisely asks for functionals $\mathscr{X}, \underline{\mathscr{Z}}$ satisfying

$$\forall \Phi, y, k^0, x^\alpha, \underline{z}^{\underline{\beta}} (\mathscr{X}\Phi y \leq y \wedge \Phi(k, \mathscr{X}\Phi yk, \underline{\mathscr{Z}}\Phi yk) \geq_0 \Phi(k, \min_\alpha(x, yk), \underline{z})).$$

But this is just what we provided for in $\mathscr{A}^\omega[X, d, \mathscr{X}, \underline{\mathscr{Z}}] + \tilde{F}^X$.
The next step in the proof of theorem 17.52 (on which the proof of theorem 17.35 above was based) consisted in establishing that the model $\mathscr{M}^{\omega,X}$ of all strongly majorizable functionals over $\mathbb{N}$ and an arbitrary nonempty bounded metric space (X, d) is a model of $\mathscr{A}^\omega[X, d]^-$+(BR) (here b_X^0 is to be interpreted by an integer upper bound on the metric of X) and, moreover, that for any closed term t of $\mathscr{A}^\omega[X, d]^-$+(BR) one can construct a closed term t^* of $\mathscr{A}^\omega$+(BR) (plus b_X) such that

$$\mathscr{M}^{\omega,X} \models t^* \gtrsim^{0_X} t.$$

We now extend this by showing that

$$\mathscr{M}^{\omega,X} \models \mathscr{A}^\omega[X, d, \mathscr{X}, \underline{\mathscr{Z}}]^- + \tilde{F}^X + (\text{BR})$$

for a suitable interpretation of the new constants $\mathscr{X}$ and $\underline{\mathscr{Z}}$ and that for any closed term t of $\mathscr{A}^\omega[X, d, \mathscr{X}, \underline{\mathscr{Z}}]^-$+(BR) we can construct a closed term t^* of $\mathscr{A}^\omega$+(BR) (plus b_X) such that

$$\mathscr{M}^{\omega,X} \models t^* \gtrsim^{0_X} t.$$

We reason in $\mathscr{M}^{\omega,X}$. Let Φ, y, k be in $\mathscr{M}^{\omega,X}$ with types as above and let Φ^*, y^* be $\gtrsim^{0_X}$-majorants for Φ, y in $\mathscr{M}^{\omega,X}$. Since $\min_\alpha(x, yk) \leq_\alpha yk$ and $\underline{\beta}$ are types of degree $(\cdot, X)$ it follows (using the b_X-boundedness of X) that

$$y^*k \gtrsim^{0_X} \min_\alpha(x, yk) \wedge z_i^* := \lambda \underline{v}.b_X \gtrsim^{0_X} z_i$$

for all $k \in \mathbb{N}$ and all $x, \underline{z}$ in $\mathscr{M}^{\omega,X}$ of types α and $\underline{\beta}$ and suitable tuples of variables $\underline{v}$. Hence

$$\forall x \in M_\alpha^{\omega,X}, \underline{z} \in M_{\underline{\beta}}^{\omega,X} (\Phi^*(k, y^*k, \underline{z}^*) \geq_0 \Phi(k, \min_\alpha(x, yk), \underline{z})).$$

Thus

$$Max_{\Phi,y,k} := \max\{\Phi(k, \min_\alpha(x, yk), \underline{z}) : x \in M_\alpha^{\omega,X} \wedge \underline{z} \in M_{\underline{\beta}}^{\omega,X}\}$$

exists (note that $M_\rho^{\omega,X} \neq \emptyset$ for all $\rho \in \mathbf{T}^X$) and hence

$$(+)\ \forall \Phi, y, k \in \mathscr{M}^{\omega,X} \exists x, \underline{z} \in \mathscr{M}^{\omega,X} (x \leq_\alpha yk \wedge \Phi(k, x, \underline{z}) =_0 Max_{\Phi,y,k}).$$

By the axiom of choice applied to $(+)$ we obtain functionals $\Xi \leq \lambda \Phi, y.y$ and $\underline{\Theta}$ such that $x := \Xi \Phi y k$ and $\underline{z} := \underline{\Theta} \Phi y k$ satisfy $(+)$. We now put

$$[\mathscr{X}]_{\mathscr{M}^{\omega,X}} := \Xi \wedge [\underline{\mathscr{Z}}]_{\mathscr{M}^{\omega,X}} := \underline{\Theta}.$$

In order to show that $\Xi, \underline{\Theta} \in \mathscr{M}^{\omega,X}$ we construct closed terms $\mathscr{X}^*$ and $\underline{\mathscr{Z}}^*$ such that

$$\mathscr{M}^{\omega,X} \models \mathscr{X}^* \gtrsim^{0_X} \mathscr{X} \wedge \underline{\mathscr{Z}}^* \gtrsim^{0_X} \underline{\mathscr{Z}}.$$

The terms

$$\mathscr{X}^* := \lambda \Phi, y.y, \quad \mathscr{Z}_i^* := \lambda \underline{v}.b_X$$

do the job (using that $\mathscr{M}^{\omega,X} \models \mathscr{X} \leq \lambda \Phi, y.y$) for a suitable tuple $\underline{v}$ of variables, where the length of the tuple and the types of its components only depend on β_i. It is clear that with this interpretation of $\mathscr{X}, \underline{\mathscr{Z}}$ in $\mathscr{M}^{\omega,X}$ the axiom $\tilde{F}^X$ is satisfied. The construction of t^* from t now proceeds as in the proof of theorem 17.52 with the additional clauses that all occurrences of $\mathscr{X}, \underline{\mathscr{Z}}$ are replaced by $\mathscr{X}^*, \underline{\mathscr{Z}}^*$. The rest of the proof is exactly as in the proof of theorems 17.52 and 17.35 (replacing b_X everywhere by a variable b^0 satisfying that $\forall x^X, y^X ((b)_{\mathbb{R}} \geq_{\mathbb{R}} d_X(x,y)))$.
2) and 3) are proved analogously. □

Corollary 17.104. 1) *Let A be a sentence in $\mathscr{K}$ (as defined in 17.48). If*

$$\mathscr{A}^\omega[X,d] + \exists\text{-UB}^X \vdash A,$$

then A holds in any (nonempty) bounded metric space (X,d) (with 'b_X' being interpreted by some upper bound $b \in \mathbb{N}$ for d).

2) *If the premise is proved in '$\mathscr{A}^\omega[X,d,W] + \exists$-UB^X' instead of '$\mathscr{A}^\omega[X,d] + \exists$-UB^X', then the conclusion holds in all bounded hyperbolic spaces.*
3) *If the premise is proved in '$\mathscr{A}^\omega[X,d,W,\mathrm{CAT}(0)] + \exists$-UB^X' instead of '$\mathscr{A}^\omega[X,d,W] + \exists$-UB^X', then the conclusion holds in all bounded* CAT(0)-*spaces.*

Proof: The corollary follows from theorem 17.101 applying the same reasoning that was used already in the proof of corollary 17.49. □

17.8 Applications of $\exists$-UB^X

As the proofs of theorem 17.35 and theorem 17.52 above clearly show, what mainly is required for a class of structures to satisfy these theorems is that they are axiomatized by axioms which have a monotone functional interpretation in the sense of $\gtrsim$, i.e. which have a sufficiently strong uniformity built in. This becomes particularly obvious in the bounded metric case: e.g. the separability of a bounded space (X,d) would be translated by monotone functional interpretation into its total boundedness and, in fact, if e.g. theorem 17.35 would be valid for the class of bounded separable spaces we could use it to conclude the false statement that every such space would be totally bounded. This puts into a general context the phenomenon eluded to already in the counterexample given after remark 15.2 in chapter 15. Similarly, the extensionality axiom for $f^{X\to X}$ would be translated into the uniform (equi-)continuity of all $f:X\to X$. In the case of strictly normed spaces, uniform convexity would be required by monotone functional interpretation etc. While these facts simply show that our metatheorems are not applicable to certain classes of structures (or properties of functions), the principle $\exists$-UB^X brings this point to the more radical consequence that we now can prove all these incorrect rules even as implications (while at the same time preserving the truth of theorems of the form $A\in\mathcal{K}$ that are proved via $\exists$-UB^X as we saw in corollary 17.104). These implications take the form

$$A\to(A)_U,$$

where $(A)_U$ is the uniform version of A. So the consequences of $\exists$-UB^X are much more serious than those of Σ^0_1-UB given in chapter 12 which, after all, always become classically correct when applied only to contexts in which all function(al)s are assumed to be continuous (where then Σ^0_1-UB holds true by a simple compactness argument).
We formulate our results over $\mathcal{A}^\omega[X,d]$ resp. $\mathcal{A}^\omega[X,d,W]$ base systems but rather weak fragments of these systems would suffice as well.
As before, we usually write in the following '$f^{X\to X}$' instead of '$f^{X(X)}$' for better readability.

17.8.1 Application 1:

Our first application is due to F. Ferreira (private communication): we apply $\exists$-UB^X to '$A:=(X,d)$ is incomplete' to conclude a uniform version $(A)_U$ of incompleteness which, however, is inconsistent, i.e. under $\exists$-UB^X we get

$$\exists\text{-UB}^X\text{: 'incompleteness} \Rightarrow \text{falsity', i.e. 'completeness'.}$$

Proposition 17.105. *$\mathcal{A}^\omega[X,d]+\exists$-UB^X proves that (X,d) is complete.*

Proof: Suppose that (X,d) is not complete. Then there is a Cauchy sequence (x_n) in X such that for any $x\in X$ the sequence does not converge towards x, i.e. $\neg(\lim x_n = x)$. Since $\mathcal{A}^\omega[X,d]$ contains AC^{0,0} we may assume that (x_n) has Cauchy modulus 2^{-k}, i.e.

$$(+)\ \forall k\forall m,n\geq k(d(x_m,x_n)<_{\mathbb{R}}2^{-k}).$$

From $\neg(\lim x_n = x)$ for all x^X we conclude

$$\forall x^X\exists n^0(d_X(x_n,x)>_{\mathbb{R}}2^{-n+1}).$$

By $\exists$-UB^X is follows that

$$\exists n^0\forall x^X\exists\tilde{n}\leq n(d_X(x_{\tilde{n}},x)>_{\mathbb{R}}2^{-\tilde{n}+1}).$$

Using $(+)$ this yields

$$\exists n^0 \forall x^X (d_X(x_n, x) >_{\mathbb{R}} 2^{-n})$$

which leads to a contradiction for $x := x_n$. □

17.8.2 Application 2:

We now give the details of the aforementioned fact that ∃-UB^X 'uniformizes' the assumption of separability of the bounded metric space (X,d) to its total boundedness, i.e.

$$\exists\text{-UB}^X: \text{ 'separability} \Rightarrow \text{total boundedness (with modulus)'.}$$

We first need the following

Definition 17.106. Let (X,d) be a totally bounded metric space. A function $\alpha : \mathbb{N} \to \mathbb{N}$ satisfying

$$\exists (a_n)_n \text{ in } X \forall k \in \mathbb{N} \forall x \in X \exists n \leq \alpha(k)(d(x, a_n) < 2^{-k})$$

is called a modulus to total boundedness.

Remark 17.107. Obviously, using countable choice, every totally bounded metric space posses a modulus of total boundedness.

Proposition 17.108. $\mathcal{A}^\omega[X,d] + \exists\text{-UB}^X$ *proves the following:*
'If (X,d) is separable, then (X,d) is totally bounded and has a modulus of total boundedness α'. Formalized in $\mathcal{L}(\mathcal{A}^\omega[X,d])$, this reads as:

$$\mathcal{A}^\omega[X,d] + \exists\text{-UB}^X \vdash \forall f^{X(0)} \big(\forall k^0, x^X \exists n^0 (d_X(f(n), x) <_{\mathbb{R}} 2^{-k}) \to \\ \exists \alpha^1 \forall k^0, x^X \exists n \leq \alpha(k)(d_X(f(n), x) <_{\mathbb{R}} 2^{-k})\big).$$

Proof: ∃-UB^X applied to

$$\forall k^0, x^X \exists n^0 (d_X(f(n), x) <_{\mathbb{R}} 2^{-k})$$

yields (observing that '$d_X(f(n), x) <_{\mathbb{R}} 2^{-k}$' is an ∃-formula) that

$$\exists \alpha^1 \forall k^0, x^X \exists n \leq \alpha(k)(d_X(f(n), x) <_{\mathbb{R}} 2^{-k}).$$

□

17.8.3 Application 3:

Definition 17.109. Let (X,d,W) be a hyperbolic space.

1) (X,d,W) is called strictly convex if

$$\forall x,y,a \in X \forall r > 0 \big(d(x,a) \le r \wedge d(y,a) \le r \wedge d(\frac{1}{2}x \oplus \frac{1}{2}y,a) = r \to x = y\big).$$

2) ([128, 263]) (X,d,W) is called uniformly convex if

$$\begin{aligned}&\forall r > 0 \forall \varepsilon > 0 \exists \delta > 0 \forall x,y,a \in X \\ &\qquad \big(d(x,a) \le r \wedge d(y,a) \le r \wedge d(\tfrac{1}{2}x \oplus \tfrac{1}{2}y,a) > (1-\delta)r \to d(x,y) < \varepsilon r\big).\end{aligned}$$

We now show that

$$\exists\text{-UB}^X\text{: 'strictly convex} \Rightarrow \text{uniformly convex'.}$$

Proposition 17.110. *$\mathscr{A}^\omega[X,d,W] + \exists\text{-UB}^X$ proves the following: 'If (X,d,W) is strictly convex, then it is uniformly convex'.*

Proof: When formalized, the assumption of strict convexity reads as follows

$$\begin{aligned}&\forall r^1 \forall x^X, y^X, a^X, l^0 \\ &\qquad \big(r \ge_{\mathbb{R}} 2^{-l} \wedge d_X(x,a) \le_{\mathbb{R}} r \wedge d_X(y,a) \le_{\mathbb{R}} r \wedge d_X\big(\tfrac{1}{2}x \oplus \tfrac{1}{2}y,a\big) =_{\mathbb{R}} r \to x =_X y\big)\end{aligned}$$

which - by the first W-axiom – is equivalent to

$$\begin{aligned}&\forall r^1 \forall x^X, y^X, a^X . l^0 \\ &\qquad \big(r \ge_{\mathbb{R}} 2^{-l} \wedge d_X(x,a) \le_{\mathbb{R}} r \wedge d_X(y,a) \le_{\mathbb{R}} r \wedge d_X\big(\tfrac{1}{2}x \oplus \tfrac{1}{2}y,a\big) \ge_{\mathbb{R}} r \to x =_X y\big)\end{aligned}$$

and hence in turn to

$$\begin{aligned}&\forall r^1 \forall x^X, y^X, a^X, l^0, k^0 \exists n^0 \\ &\qquad \big(r \ge_{\mathbb{R}} 2^{-l} \wedge d_X(x,a) \le_{\mathbb{R}} r \wedge d_X(y,a) \le_{\mathbb{R}} r \wedge d_X\big(\tfrac{1}{2}x \oplus \tfrac{1}{2}y,a\big) \ge_{\mathbb{R}} (1-2^{-n})r \\ &\qquad\qquad \to d_X(x,y) <_{\mathbb{R}} 2^{-k} \cdot r\big).\end{aligned}$$

Applying to this $\exists\text{-UB}^X$ yields

$$\begin{aligned}&\forall r^1, l^0, k^0 \exists n^0 \forall x^X, y^X, a^X \\ &\qquad \big(r \ge_{\mathbb{R}} 2^{-l} \wedge d_X(x,a) \le_{\mathbb{R}} r \wedge d_X(y,a) \le_{\mathbb{R}} r \wedge d_X\big(\tfrac{1}{2}x \oplus \tfrac{1}{2}y,a\big) \ge_{\mathbb{R}} (1-2^{-n})r \\ &\qquad\qquad \to d_X(x,y) <_{\mathbb{R}} 2^{-k} \cdot r\big)\end{aligned}$$

which (switching to the more common ε/δ-formulation is just the formalized version of

$$\forall r > 0 \forall \varepsilon > 0 \exists \delta > 0 \forall x, y, a \in X$$
$$\big(d(x,a) \le r \wedge d(y,a) \le r \wedge d\big(\tfrac{1}{2}x \oplus \tfrac{1}{2}y, a\big) > (1-\delta)r \to d(x,y) < \varepsilon r\big).$$

□

Remark 17.111. Restricted to rational $r, \varepsilon \in \mathbb{Q}_+^*$ (which can be encoded by natural numbers as in chapter 4) one even obtains a function $\eta(r,\varepsilon) \in \mathbb{Q}_+^*$ producing a rational $\delta > 0$ satisfying the uniform convexity statement, i.e. a modulus of uniform convexity.

17.8.4 Application 4:

The uniform version of '$f^{X \to X}$ is extensional' is 'f is uniformly continuous'. When the extensionality is taken for all functions $f^{X \to X}$ (or subclass axiomatized by $\forall$-axioms), then the uniform version even becomes 'all functions $f : X \to X$ have a common modulus of uniform continuity', i.e.

∃-UB^X: 'extensionality ⇒ uniform continuity with (common) modulus'.

This corresponds to the counterexample given above to the possibility to add full extensionality in theorem 17.35 which shows: if full extensionality is used for some function $f^{X \to X}$ (resp. for some class of functions $f^{X \to X}$ axiomatized by universal axioms) in a proof it has to follow as a consequence of the existence of a modulus uniform continuity of f (resp. the existence of a common modulus of uniform continuity for the whole class as in the case of the class of nonexpansive functions). Otherwise, only extensionality in the weak form of the quantifier-rule of extensionality may be used.

Proposition 17.112. *Let*

$$Ext(f^{X \to X}) :\equiv \forall x^X, y^X (x =_X y \to f(x) =_X f(y)).$$

1) $\mathcal{A}^\omega[X,d] + \exists$-UB^X *proves that*

$$\forall f^{X \to X} (Ext(f) \to$$
$$\exists \omega^1 \forall k^0, x^X, y^X (d_X(x,y) <_{\mathbb{R}} 2^{-\omega(k)} \to d_X(f(x), f(y)) <_{\mathbb{R}} 2^{-k}).$$

2) $\mathcal{A}^\omega[X,d] + \exists$-UB^X *proves that*

$$\forall f^{X \to X} (Ext(f)) \to$$
$$\exists \omega^1 \forall f^{X \to X}, k^0, x^X, y^X (d_X(x,y) <_{\mathbb{R}} 2^{-\omega(k)} \to d_X(f(x), f(y)) <_{\mathbb{R}} 2^{-k}).$$

Proof: 1) By the definition of $=_X$, the assumption $Ext(f)$ can be written as

$$\forall x^X, y^X (\forall n^0 (d_X(x,y) \le_{\mathbb{R}} 2^{-n}) \to \forall k^0 (d_X(f(x), f(y)) <_{\mathbb{R}} 2^{-k}))$$

and so in turn as

$$(+)\ \forall x^X, y^X \forall k^0 \exists n^0 (d_X(x,y) \leq_{\mathbb{R}} 2^{-n} \to d_X(f(x), f(y)) <_{\mathbb{R}} 2^{-k}).$$

Note that

$$d_X(x,y) \leq_{\mathbb{R}} 2^{-n} \to d_X(f(x), f(y)) <_{\mathbb{R}} 2^{-k}$$

is (logically equivalent to) an $\exists$-formula which, moreover, is monotone w.r.t. '$\exists n$'. Hence $\exists$-UB^X applied to $(+)$ yields

$$\exists \omega^1 \forall k^0, x^X, y^X (d_X(x,y) \leq_{\mathbb{R}} 2^{-\omega(k)} \to d_X(f(x), f(y)) <_{\mathbb{R}} 2^{-k})$$

which establishes the claim.
2) is proved analogously by applying $\exists$-UB^X to

$$(++)\ \forall f^{X \to X} \forall x^X, y^X \forall k^0 \exists n^0 (d_X(x,y) \leq_{\mathbb{R}} 2^{-n} \to d_X(f(x), f(y)) <_{\mathbb{R}} 2^{-k}).$$

Note that the type $X \to X$ is admissible as a type β in $\exists$-UB^X just as well as X is. □

17.8.5 Application 5:

The uniform version of 'there exists no root of, say, $\Phi : X \to \mathbb{R}$ in X' is 'there exist not even approximate roots of Φ'. So, taking the contraposite, this can be paraphrased as

$\exists$-UB^X: 'existence of approximate solutions $\Rightarrow$ existence of a real solution'.

This has the consequence that $\exists$-UB^X extends the usual WKL-applications for compact spaces and continuous functions to bounded spaces and arbitrary functions.

Proposition 17.113. *Let β be of degree $(\cdot, X)$. Then $\mathcal{A}^\omega[X,d] + \exists$-UB^X proves the following*

$$\forall \Phi^{1(\beta)} \big(\forall k^0 \exists y^\beta (|\Phi(y)|_{\mathbb{R}} <_{\mathbb{R}} 2^{-k}) \to \exists y^\beta (\Phi(y) =_{\mathbb{R}} 0) \big).$$

This also holds for tuples of variables $\underline{y}^{\underline{\beta}}$ as long as the types $\underline{\beta}$ are all of degree $(\cdot, X)$.

Proof: Suppose that

$$\forall y^\beta (\Phi(y) \neq_{\mathbb{R}} 0).$$

Then

$$\forall y^\beta \exists k^0 (|\Phi(y)|_{\mathbb{R}} >_{\mathbb{R}} 2^{-k})$$

and hence by $\exists$-UB^X

$$\exists k^0 \forall y^\beta (|\Phi(y)|_{\mathbb{R}} >_{\mathbb{R}} 2^{-k})$$

contradicting the assumption. □

17.8.6 Application 6:

This application is a special case of application 17.8.5: any fixed point problem for $f : X \to X$ can be written equivalently as a root problem '$\exists x \in X\ (d(x, f(x)) = 0)$'. As follows from a theorem due to Ishikawa [176] (for the normed case) and Goebel-Kirk [126] (theorem 1, for the hyperbolic case), to be discussed in detail in chapter 18, nonexpansive selfmappings $f : X \to X$ of bounded hyperbolic spaces (X, d, W) always have approximate fixed points. Moreover, this fact is provable in $\mathscr{A}^\omega[X, d, W]$. Hence by application 17.8.5 we obtain the following result:

Proposition 17.114. $\mathscr{A}^\omega[X, d, W] + \exists$-UB^X *proves the following*

$$\forall f^{X \to X} \big(f \text{ nonexpansive } \to \exists x^X (f(x) =_X x) \big).$$

Proof: Since $f(x) =_X x \leftrightarrow d_X(x, f(x)) =_{\mathbb{R}} 0$ we obtain from application 17.8.5 applied to $\Phi(x) := d_X(x, f(x))$ that it suffices to show (in $\mathscr{A}^\omega[X, d, W]$)

$$\forall k^0 \exists x^X (d_X(x, f(x)) <_{\mathbb{R}} 2^{-k}).$$

This, however, follows by the discussion above. □

The remarkable consequence of proposition 17.114 is that $\exists$-UB^X allows one to make free use of fixed points of nonexpansive mappings in proofs (and still obtain correct $\mathscr{K}$-conclusions) despite the fact that such fixed points in general do not exist (not even for nonexpansive selfmappings of bounded, closed, convex subsets of Banach spaces such as c_0, see the counterexample given in the introduction to chapter 18).

Remark 17.115. The existence of approximate fixed points of nonexpansive mappings between bounded hyperbolic spaces used in the proof above rests strongly on the presence of the hyperbolic structure provided by W and is false for general bounded metric spaces: consider $\mathbb{R}$ equipped with the truncated metric $D(x, y) := \min(|x - y|, 1)$ and $f : \mathbb{R} \to \mathbb{R}$ with $f(x) := x + 1$. f is a nonexpansive (even isometric) selfmapping of the bounded metric space $(\mathbb{R}, D)$ but has no ε-fixed points for $0 < \varepsilon < 1$.

17.8.7 Application 7:

An important line of research in metric fixed point theory is concerned with generalizations of the famous Banach fixed point theorem from contractions to more general classes of functions that are of some generalized form of 'contractive type' (see e.g. [92, 308, 315, 316, 188]). As discussed already in section 15.4, often compactness assumptions are imposed in the respective fixed point theorems in order to ensure certain uniform versions of contractivity. In many cases it has turned out that the assumption of compactness can be replaced by boundedness if the functions are assumed right away to satisfy the uniform contractivity notions (which in the absence of compactness usually is a strictly stronger condition). In many applications of proof mining to such fixed point theorems, the need to uniformize contractivity conditions on f has turned out to be crucial as well (see e.g. [119, 116, 50, 51]). In fact, as shown by Briseid [54], under quite general conditions of 'uniform contractivity' the logical metatheorems proved in this chapter can be utilized to obtain uniform effective rates of convergence of the Picard iteration $f^n(x)$ of f towards the unique fixed point. Interestingly, this is possible despite of the fact that the Cauchy property of $((f^n(x))_n)$ is in Π^0_3 so that the metatheorems cannot be applied directly (see [54] for details). The principle $\exists$-UB^X provides a general tool for producing appropriate uniformizations of contractivity notions for bounded metric spaces. From section 15.4 we recall:

Definition 17.116. Let (X,d) be a metric space and $f : X \to X$ a selfmapping of X.

1) f is called contractive (see [92]) if

$$\forall x,y \in X\,(x \neq y \to d(f(x),f(y)) < d(x,y)).$$

2) f is called uniformly contractive with modulus $\alpha : \mathbb{N} \to \mathbb{N}$ (see [308]) if

$$\forall k \in \mathbb{N}\forall x,y \in X\,(d(x,y) > 2^{-k} \to d(f(x),f(y)) < (1-2^{-\alpha(k)})\,d(x,y)).$$

The next proposition can be summarized as

$$\exists\text{-UB}^X:\ \text{'contractive} \Rightarrow \text{uniformly contractive with modulus'}.$$

Proposition 17.117. *$\mathcal{A}^\omega[X,d] + \exists$-UB^X proves the following: 'every contractive mapping $f : X \to X$ is uniformly contractive with some modulus α'.*

Proof: Assume that

$$\forall x^X, y^X\,(x \neq_X y \to d_X(f(x),f(y)) <_{\mathbb{R}} d_X(x,y)).$$

Then

$$\forall x^X, y^X, k^0 \exists n^0 (d_X(x,y) \geq_{\mathbb{R}} 2^{-k} \to d_X(f(x),f(y)) <_{\mathbb{R}} (1-2^{-n})d_X(x,y)),$$

where

$$d_X(x,y) \geq_{\mathbb{R}} 2^{-k} \to d_X(f(x),f(y)) <_{\mathbb{R}} (1-2^{-n})d_X(x,y)$$

is an $\exists$-formula. Hence $\exists$-UB^X yields (using the monotonicity in '$\exists n$')

$$\exists \alpha^1 \forall k^0, x^X, y^X (d_X(x,y) \geq_{\mathbb{R}} 2^{-k} \to d_X(f(x),f(y)) <_{\mathbb{R}} (1-2^{-\alpha(k)})d_X(x,y)).$$

□

In a similar way, $\exists$-UB^X implies corresponding uniform versions of many other more liberal notions of contractivity including the generalized p-contractive mappings ([315, 316]). Such a uniform version is used by E.M. Briseid in his effective form of a fixed point theorem of [182] thereby eliminating the compactness assumption.

Definition 17.118 (Rhoades [315]). Let (X,d) be metric space and $p \in \mathbb{N}$. A self-mapping $f : X \to X$ is called generalized p-contractive if

$$\forall x,y \in X \big(x \neq y \to d(f^p(x), f^p(y)) < \text{ diam}\{x,y,f^p(x),f^p(y)\}\big).$$

Theorem 17.119 (Kincses-Totik [182]). *Let (K,d) be a compact metric space and $f : K \to K$ be a continuous and generalized p-contractive mapping for some $p \in \mathbb{N}$. Then f has a unique fixed point ξ and for every $x \in K$: $\lim_{n\to\infty} f^n(x) = \xi$.*

Definition 17.120 (Briseid [50, 51]). Let (X,d) be a metric space, $p \in \mathbb{N}$. $f : X \to X$ is called uniformly generalized p-contractive with modulus $\eta : \mathbb{Q}^*_+ \to \mathbb{Q}^*_+$ if for all $x,y \in, \varepsilon \in \mathbb{Q}^*_+$

$$d(x,y) > \varepsilon \to d(f^p(x), f^p(y)) + \eta(\varepsilon) < \text{ diam}\{x,y,f^p(x),f^p(y)\}.$$

As above one easily shows that $\mathscr{A}^\omega[X,d] + \exists$-UB^X proves that every generalized p-contractive mapping $f : X \to X$ is uniformly generalized p-contractive with some modulus η.

Remark 17.121. Instead of an additive modulus η one also can consider a multiplicative version (see the discussion in section 15.4).

The significance of this notion of uniform generalized p-contractivity is witnessed by the following result of E.M. Briseid:

Theorem 17.122 (Briseid [50, 51]). *Let (X,d) be a complete metric space and $p \in \mathbb{N}$. Let $f : X \to X$ be a uniformly continuous and uniformly generalized p-contractive with moduli of uniform continuity ω and uniform generalized p-contractivity η. Let $x_0 \in X$ and assume that $(f^n(x_0))$ is bounded by $b \in \mathbb{Q}^*_+$. Then f has a unique fixed point ξ and $(f^n(x_0))$ converges to ξ with rate of convergence $\Phi : \mathbb{Q}^*_+ \to \mathbb{N}$,*

$$\Phi(\varepsilon) := \begin{cases} p\lceil (b-\varepsilon)/\rho(\varepsilon)\rceil \text{ if } b > \varepsilon, \\ 0, \text{otherwise} \end{cases}$$

with

$$\rho(\varepsilon) := \min\left\{\eta(\varepsilon), \frac{\varepsilon}{2}, \eta(\frac{1}{2}\omega^p(\frac{\varepsilon}{2}))\right\}.$$

17.9 Fragments of $\mathscr{A}^{\omega}[\ldots]$

Instead of $\mathscr{A}^{\omega}[\ldots]_{(-b)}$ one can also consider $\mathscr{T}^{\omega}[\ldots]_{(-b)}$ which for $\mathscr{T} :=$WE-PA$^{\omega}$, $\widehat{\text{WE-PA}}^{\omega}\restriction$,G$_{\infty}A^{\omega}$ is defined as in the case of $\mathscr{A}^{\omega}$ except that DC is dropped and – in the cases of $\mathscr{T} := \widehat{\text{WE-PA}}^{\omega}\restriction$,G$_{\infty}A^{\omega}$ – WE-PA$^{\omega}$ is replaced by $\mathscr{T}$. As indicated already in remark 17.37.1), the proofs of the general metatheorems 17.52 and 17.69 above can be adapted to these systems, where then the extractable bounds Φ are even given by closed terms of $\mathscr{T}$. Moreover, as mentioned in remark 17.37.2), some of the type restrictions can be liberalized that were only used for the passage from $\mathscr{M}^{\omega,X}$ to $\mathscr{S}^{\omega,X}$ which now is superfluous as we do not need (BR) anymore and so can work directly in $\mathscr{S}^{\omega,X}$. For this it suffices to note that all the majorants used for the extra constants of the structures $(X,d,W),(X,\|\cdot\|)$ etc. are even in G$_3$R$^{\omega}$.

Rather than going into any details on this we sketch how the technique of elimination of monotone Skolem functions from chapter 13 can be adapted to G$_{\infty}$A$^{\omega}[X,d,W]_{-b}$ (and similarly for G$_{\infty}$A$^{\omega}[X,\|\cdot\|,C]_{-b}$ etc.).

By the corollary to the proof of theorem 13.7 it is clear that this theorem also applies to G$_{\infty}$A$^{\omega}[X,d,W]_{-b}$ (as long as the bounds Ψ_i still are closed terms of G$_{\infty}$A$^{\omega}$). Moreover, since the majorants for the closed terms of G$_{\infty}$A$^{\omega}[X,d,W]_{-b}$ are in G$_{\infty}$R$^{\omega}$, i.e. closed terms of G$_{\infty}$A$^{\omega}$, in principle also theorem 13.10 and corollary 13.12 can be adapted. However, one technical point to address is that in our metatheorems proved in this chapter the extracted uniform bound is not verified in the formal system at hand but only shown to be true in $\mathscr{S}^{\omega,X}$. So we cannot apply again (monotone) functional interpretation to the result from the first use of (monotone) functional interpretation used to extract the bound as we did in the proof of theorem 13.10. This problem, however, can be solved (in most cases) by not only extracting the bounds $\Psi_i u \underline{h}$ as we did in the proof of theorem 13.10 but also exact witnesses for the remaining existential quantifier '$\exists w A_0^H$' by which the statement in question becomes purely universal so that its proof does not matter. We will not give the most general formulation but only a special case corresponding to the adaption of corollary 13.12 to the context of corollary 17.55:

Proposition 17.123. *Let $A(a) :\equiv \forall x^0 \exists y^0 \forall z^0 A_0(x,y,z,a)$ be a formula in $\mathscr{L}$(G$_{\infty}$A$^{\omega}$) containing only 'a' free, where A_0 is quantifier-free. We assume that A satisfies (provably in G$_{\infty}$A$^{\omega}$) the monotonicity condition $(*)$ in corollary 13.12 and that A is provable in G$_{\infty}$A$^{\omega}$ + Σ^0_1-IA. Let $B_{\exists}(u^X, f^{X\to X}, n^0)$ be an $\exists$-formula in $\mathscr{L}$(G$_{\infty}$A$^{\omega}[X,d,W]_{-b}$) containing only u, f, n free. Neither of A, B contains 0_X and ξ is a closed term of* G$_{\infty}$A$^{\omega}[X,d,W]_{-b}$.
Suppose that G$_{\infty}$A$^{\omega}[X,d,W]_{-b}$ *proves that*

$$\forall u^X, f^{X\to X} \left(f \text{ n.e.} \wedge \exists g \forall x, z A_0(x, g(x), z, \xi(u,f)) \to \exists n^0 B_{\exists}(u,f,n) \right),$$

then one can extract a closed term φ^1 of $\widehat{\text{WE-PA}}^{\omega}\restriction$, i.e. an ordinary primitive recursive function, such that

$$\mathscr{S}^{\omega,X} \models \forall u^X, f^{X\to X}, b^0 \big(f \text{ n.e. } d_X(u,f(u)) \le_{\mathbb{R}} (b)_{\mathbb{R}} \to \exists n \le \varphi(b) B_{\exists}(u,f,n)\big).$$

Proof: $B_{\exists}(u,f,n)$ is of the form $\exists \underline{v}^{\underline{\rho}} B_0(u,f,n,\underline{v})$ with B_0 being quantifier-free. From a proof in $\mathrm{G}_\infty\mathrm{A}^{\omega}[X,d,W]_{-b}$ of

$$\forall u^X, f^{X\to X} \big(f \text{ n.e. } \wedge \exists g \forall x, z A_0(x,g(x),z,\xi(u,f)) \to \exists n^0 B_{\exists}(u,f,n)\big)$$

one extracts (adapting lemma 17.72 to $\mathrm{G}_\infty\mathrm{A}^{\omega}[X,d,W]_{-b}$) closed terms $q,r,t,\underline{s}$ of $\mathrm{G}_\infty\mathrm{A}^{\omega}[X,d,W]_{-b}$ such that

$$\mathscr{S}^{\omega,X} \models \forall u,f,g\big(f \text{ n.e. } \wedge A_0(rufg, g(rufg), qufg, \xi uf) \to B_0(u,f,tufg,\underline{s}ufg)\big).$$

By the proof of lemma 17.83 there exists a closed term r^* of $\mathrm{G}_\infty\mathrm{A}^{\omega}$ such that

$$\mathscr{S}^{\omega,X} \models \forall a^X \, (r^* \gtrsim^a r)$$

(in the absence of (BR) the terms and their majorants always define functionals in $\mathscr{S}^{\omega,X}$ and the majorizing property is valid in $\mathscr{S}^{\omega,X}$).
If f is nonexpansive and $(b)_{\mathbb{R}} \ge_{\mathbb{R}} d_X(u,f(u))$, then $0^0 \gtrsim^u u$ and $\lambda n.n+b \gtrsim^u f$. Hence with $\widehat{r}bg := r^*0(\lambda n.n+b)g^M$ we have

$$(1)\quad \begin{cases} \mathscr{S}^{\omega,X} \models \forall u,f,b\big(f \text{ n.e. } \wedge d_X(u,f(u)) \le_{\mathbb{R}} (b)_{\mathbb{R}} \\ \qquad \to \forall g\,(\forall x \le \widehat{r}bg A_0(x,g(x),qufg,\xi uf) \to B_0(u,f,tufg,\underline{s}ufg))\big). \end{cases}$$

By the corollary to the proof the theorem 13.7 (applied to $\exists x \forall y \exists z, n, \underline{v}\,(A_0 \to B_0)$) and the fact that $\mathrm{G}_\infty\mathrm{A}^{\omega}$ is contained in $\mathrm{G}_\infty\mathrm{A}^{\omega}[X,d,W]_{-b}$ it follows that

$$\mathrm{G}_\infty\mathrm{A}^{\omega}[X,d,W]_{-b} \vdash \\ \forall u,f,b\,\big(\forall g(\forall x \le \widehat{r}bg \forall z A_0 \to \exists n B_{\exists}) \to (\forall x \exists y \forall z A_0 \to \exists n B_{\exists})\big).$$

Hence, a-fortiori,

$$\mathrm{G}_\infty\mathrm{A}^{\omega}[X,d,W]_{-b} \vdash \\ \forall u,f,b\,\big(\forall g(\forall x \le \widehat{r}bg A_0(x,g(x),qufg,\xi uf) \to B_0(u,f,tufg,\underline{s}ufg)) \\ \to (\forall x \exists y \forall z A_0 \to \exists n B_{\exists})\big)$$

and so

$$\mathrm{G}_\infty\mathrm{A}^{\omega}[X,d,W]_{-b} + \Sigma^0_1\text{-IA} \vdash \\ \forall u,f,b\Big(\forall g(\forall x \le \widehat{r}ubg A_0 \to B_0(u,f,tufg,\underline{s}ufg)) \to \exists n B_{\exists}\Big).$$

By functional interpretation (i.e. by adapting lemma 17.72 to $\widehat{\text{WE-PA}}^{\omega}\!\upharpoonright\![X,d,W]_{-b}$) we extract a closed term ψ of $\widehat{\text{WE-PA}}^{\omega}\!\upharpoonright\![X,d,W]_{-b}$ such that

$$(2)\ \mathcal{S}^{\omega,X} \models \forall u,f,b\Big(\forall g(\ldots) \to B_{\exists}(f,u,\psi ufb)\Big).$$

(1) and (2) imply that

$$(3)\ \mathcal{S}^{\omega,X} \models \forall u,f,b\Big(f \text{ n.e. } \wedge d_X(u,f(u)) \leq_{\mathbb{R}} (b)_{\mathbb{R}} \to B_{\exists}(f,u,\psi ufb)\Big).$$

ψ has a majorizing term ψ^* in $\widehat{\text{WE-PA}}^{\omega}\!\upharpoonright$ such that

$$\mathcal{S}^{\omega,X} \models \forall u^X\,(\psi^* \gtrsim^u \psi).$$

Hence with $\varphi(b) := \psi^* 0(\lambda n.n+b)b$ we see that (3) implies

$$\mathcal{S}^{\omega,X} \models \forall u,f,b\Big(f \text{ n.e. } \wedge d_X(u,f(u)) \leq_{\mathbb{R}} (b)_{\mathbb{R}} \to \exists n \leq \varphi(b) B_{\exists}(f,u,n)\Big).$$

□

Remark 17.124. The proposition above, in particular, applies to $A(a)$ being $\text{PCM}_{ar}(a)$, where then $\exists g \forall z, zA_0$ is $\text{PCM}(a)$ (see chapter 13).

17.10 Exercises, historical comments and suggested further reading

Exercises:

1) Prove the claim in remark 17.3.
2) Prove the claim in remark 17.18.
3) Prove the claim in remark 17.15.
4) Prove lemma 17.20
5) Verify the claim in example 17.43.

Historical comments and suggested further reading:

Most of the material from sections 17.1 (except for the material on best approximations in uniformly convex spaces), 17.2, 17.3 and 17.4 is based on Kohlenbach [226] and Gerhardy-Kohlenbach [120] where additional information can be found. In particular the treatment of unbounded metric and hyperbolic spaces is from Gerhardy-Kohlenbach [120]. The treatment of the generalized uniform boundedness principle $\exists$-UB^X in section 17.7 and its applications 17.8.2 and 17.8.4–17.8.7 in section 17.8 are taken from Kohlenbach [228] while the other applications are new. The treatment of δ-hyperbolic spaces and $\mathbb{R}$-trees in section 17.5 is due to Leuştean [262] which also includes an adaptation of the metatheorems to uniformly convex hyperbolic spaces.

Versions the results in a semi-intuitionistic setting (extending the framework from chapter 7 to the setting of the new type X for an abstract metric, hyperbolic or normed space) can be found in Gerhardy-Kohlenbach [119]. For general background information on geodesic spaces, hyperbolic spaces and CAT(0)-spaces we refer to Goebel-Reich [128], Bridson-Haefliger [49] and Papadopoulos [297].

Chapter 18
Case study II: Applications to the fixed point theory of nonexpansive mappings

18.1 General facts

The fixed point theory for selfmappings $f : X \to X$ of complete metric spaces (X,d) with Lipschitz constant < 1 (i.e. contractions) is essentially trivial (even from a computational point of view) because of the well-known Banach fixed point theorem: there always exists a unique fixed point and for any $x \in X$ the Picard iteration $(f^n(x))$ of f starting at x converges to the fixed point with an explicit rate of convergence. Already for the wider class of contractive mappings (mentioned before in sections 15.4 and 17.8.7), satisfying

$$\forall x,y \in X\big(x \neq y \to d(f(x),f(y)) < d(x,y)\big),$$

things are more difficult but still some crucial features of the fixed point theory of contractions prevail, most notably the uniqueness of the fixed point in cases where it exists, e.g. for compact X (see [92]). In this case, again $(f^n(x))$ converges to the fixed point, where the rate of convergence depends on a (either additive or multiplicative) modulus of uniform contractivity of f in the sense of sections 15.4 and 17.8.7. In fact, if f is uniformly contractive even the assumption of the compactness of X can be replaced by just completeness (see [308] and – for explicit uniform rates of convergence – [119, 236]). As mentioned already in section 17.8.7 this has been extended by E. Briseid to the much more general class of (uniformly continuous and) uniformly generalized p-contractive mappings (assuming that $(f^n(x))$ is bounded, [50, 51]). Finally, P. Gerhardy and E. Briseid obtained effective quantitative versions for Kirk's ([188]) so-called asymptotic contractions ([116, 52, 53]). All these results were obtained with the help of proof mining techniques such as the ones developed in chapter 17 (see also the comments at the end of this chapter and the survey [229]).

In contrast to this, the fixed point theory for nonexpansive mappings (as defined in definition 17.39.1)

$$\forall x,y \in X\big(d(f(x),f(y)) \leq d(x,y)\big)$$

is rather different and very intricate. In fact, it has been one of the most active research areas in nonlinear functional analysis from the 50's until today.

Since the identity function id_X on X always is nonexpansive, fixed points are no longer unique in general. Moreover:

1) Whereas in Banach's fixed point theorem no conditions on (X,d) other than completeness are necessary, fixed points of nonexpansive selfmappings of complete metric spaces in general do not exist. E.g. take $X := \mathbb{R}$ and $f(x) := x+1$.
2) Even for closed **bounded** convex subsets of Banach spaces such as c_0 fixed points in general will not exist: e.g. consider in c_0 the closed bounded convex subset

$$C := \{(x_n) \in c_0 : \forall n \in \mathbb{N}(0 \le x_n \le 1)\}$$

and the nonexpansive (and even isometric) selfmapping $f : C \to C$ defined by

$$f((x_n)) := (1, x_1, x_2, \ldots)$$

which obviously is fixed point free in c_0. Many more examples are given in [340].
3) Even when C is compact (and therefore fixed points exist by the fixed point theorems of Brouwer and Schauder) and even in cases where in addition the fixed point is unique, it will in general not be approximated by the Picard iteration $x_{n+1} := f(x_n)$: take $X := \mathbb{R}, C := [0,1], f(x) := 1-x$ and $x_0 := 0$. Then x_n alternates between 0 and 1. In fact, the only starting point for which the Picard iteration (x_n) converges to the unique fixed point $\frac{1}{2}$ is that fixed point itself.

The last example already indicates that having only a metric structure might not be sufficient to set up useful iterations in the context of nonexpansive functions and, indeed, one makes use of some convexity structure (see below) provided in normed spaces but also in hyperbolic spaces. The early history of the fixed point theory of nonexpansive mappings mainly was concerned with mappings on convex subsets of **uniformly convex** Banach spaces. From chapter 17 we recall the following definition:

Definition 18.1 (Clarkson, 1936). A normed space $(X, \|\cdot\|)$ is **uniformly convex** if

$$\forall \varepsilon > 0 \exists \delta > 0 \forall x, y \in X \big(\|x\|, \|y\| \le 1 \wedge \|x-y\| \ge \varepsilon \to \|\frac{1}{2}(x+y)\| \le 1-\delta\big).$$

A function $\eta : (0,2] \to (0,1]$ providing such a $\delta := \eta(\varepsilon) \in (0,1]$ for given $\varepsilon \in (0,2]$ is a modulus of uniform convexity.

The classical fixed point theory for nonexpansive mappings in uniformly convex spaces rests on the following two pillars:

Theorem 18.2 (Browder [55], Göhde [137], Kirk [184]). *Let $(X, \|\cdot\|)$ be a uniformly convex Banach space, $\emptyset \ne C \subseteq X$ convex, closed and bounded and $f : C \to C$ nonexpansive. Then f has a fixed point.*

From the example of a fixed point free nonexpansive mapping in c_0 given above it follows that the condition on X being uniformly convex cannot be dropped in theorem 18.2.

The next theorem shows that under an additional compactness condition one can define an effective iteration schema converging towards a fixed point:

Theorem 18.3 (Krasnoselski's theorem, [238]). *Let K be a convex, closed and bounded set in a uniformly convex Banach space $(X, \|\cdot\|)$, f a nonexpansive mapping of K into a compact subset of K. Then for every $x_0 \in K$, the sequence (called the Krasnoselski iteration of f starting at x_0)*

$$x_{k+1} := \frac{x_k + f(x_k)}{2}$$

converges to a fixed point $p \in K$ of f.

While the existence of an effective iteration converging towards a fixed point is reminiscent of the Banach fixed point theorem, the following counterexample to a (uniformly) effective rate of convergence shows that things, in fact, are quite different (essentially due to the lack of uniqueness of the fixed point):

Theorem 18.4. *There exists a (primitive recursively) computable sequence $(f_l)_{l\in\mathbb{N}}$ of nonexpansive functions $f_l : [0,1] \to [0,1]$ such that for $\lambda_n := \frac{1}{2}$ and $x_0^l := 0$ and the corresponding Krasnoselski iterations (x_n^l) there is no computable function $\delta : \mathbb{N} \to \mathbb{N}$ such that*

$$\forall m \geq \delta(l)\big(|x_m^l - x_{\delta(l)}^l| \leq \frac{1}{2}\big).$$

Proof: Using the primitive recursive Kleene T-predicate (see chapter 2) we define a (primitive recursively) computable function $\alpha : \mathbb{N} \times \mathbb{N} \to \mathbb{N}$ by

$$\alpha(l,n) := \begin{cases} 1, \text{ if } \neg T(l,l,n) \\ 0, \text{ otherwise.} \end{cases}$$

Using α we in turn define an again (primitive recursively) computable sequence $(f_l)_{l\in\mathbb{N}}$ of nonexpansive mappings $f_l : [0,1] \to [0,1]$:

$$f_l(x) := a_l x + 1 - a_l, \text{ where } a_l := \sum_{i=0}^{\infty} \alpha(l,i) 2^{-i-1} \in [0,1].$$

Note that the sequence of partial sums $\sum\limits_{i=0}^{n} \alpha(l,i)2^{-i-1}$ is a Cauchy sequence with Cauchy modulus 2^{-n-1} of rational numbers and so its limit is trivially definable (in fact the sequence itself represents that limit in the sense of chapter 4). So, in particular, (f_l) is a computable sequence (in the sense of computability theory, see e.g. [303] or [377]) of nonexpansive functions.
Suppose now that there would exist a computable function δ satisfying

$$\forall m \geq \delta(l)\big(|x^l_m - x^l_{\delta(l)}| \leq \frac{1}{2}\big).$$

Then for $m := \delta(l)$ we have

$$a_l < 1 \Rightarrow \lim_{n\to\infty} x^l_n = 1 \Rightarrow x^l_m \in [\tfrac{1}{2}, 1] \text{ and}$$
$$a_l = 1 \Rightarrow \forall n(x^l_n = 0) \Rightarrow x^l_m = 0.$$

This implies that using δ one can construct (by computing $x^l_{\delta(l)}$ up to say an error $1/3$) a computable function $\chi : \mathbb{N} \to \mathbb{N}$ such that

$$\forall l \in \mathbb{N}(\chi(l) = 0 \leftrightarrow a_l = 1).$$

Using that $a_l = 1 \leftrightarrow \forall n \in \mathbb{N}(\alpha(l,n) = 1) \leftrightarrow \forall n \in \mathbb{N}\neg T(l,l,n)$ this yields

$$\forall l \in \mathbb{N}(\chi(l) = 0 \leftrightarrow \forall n \in \mathbb{N}\neg T(l,l,n))$$

which contradicts the well-known undecidability of the special halting problem $\{l \in \mathbb{N} : \exists n \in \mathbb{N} T(l,l,n)\}$. □

Logically, this ineffectivity in Krasnoselski's theorem corresponds to the fact that the statement that $(x_k)_{k\in\mathbb{N}}$ is a Cauchy sequence is Π^0_3.
On the other hand if we consider the weaker question of how far we have to go in the iteration to obtain an ε-fixed point, then we notice that the logical form of the statement

$$(+)\ \forall k \in \mathbb{N}\exists n \in \mathbb{N}\big(\|x_n - f(x_n)\| < 2^{-k}\big)$$

is Π^0_2 (assuming that real numbers are represented as Cauchy sequences with fixed rate of convergence as in chapter 4 so that $<_{\mathbb{R}} \in \Sigma^0_1$).
The following crucial monotonicity property holds (see lemma 18.7 below):

$$\|x_{m+1} - f(x_{m+1})\| \leq \|x_m - f(x_m)\|$$

Hence the formula

$$\|x_n - f(x_n)\| < 2^{-k}$$

is equivalent to

$$\forall m \geq n\big(\|x_m - f(x_m)\| < 2^{-k}\big).$$

Thus any bound on $(+)$ provides a rate of convergence for

$$(++)\ \|x_n - f(x_n)\| \stackrel{n\to\infty}{\to} 0,$$

where $(++)$ is called the asymptotic regularity of f. By remark 18.6.3 this coincides with the previously defined notion for f_λ.

We now come to a vast generalization of Krasnoselski's theorem due to Ishikawa ([176]) and – slightly less general – Edelstein and O'Brien ([93]):

- Krasnoselski's theorem holds in **arbitrary** Banach spaces. So in contrast to the Browder-Göhde-Kirk theorem, Krasnoselski's theorem surprisingly does not depend on the geometry of the unit ball.
- Asymptotic regularity holds for arbitrary bounded convex sets C. In fact, it even suffices that the iteration sequence (x_n) is bounded.
- More general, so-called Krasnoselski-Mann iterations [273], can be allowed.

Definition 18.5. Let $(X, \|\cdot\|)$ be a normed space, $C \subseteq X$ convex and (λ_k) be a sequence in $[0,1]$. The general **Krasnoselski-Mann iteration** of a selfmapping $f : C \to C$ starting from $x \in C$ is defined by

$$x_0 := x, \quad x_{k+1} := (1-\lambda_k)x_k + \lambda_k f(x_k).$$

Remark 18.6. 1) If $\lambda_k := 1$ for all $k \in \mathbb{N}$, then (x_k) coincides with the Picard iteration $(f^k(x))$ of f.
2) If $\lambda_k := 0$ for all $k \in \mathbb{N}$, then (x_k) is the constant-x sequence.
3) If $\lambda_k := \lambda$ is a constant sequence, then (x_k) is the Picard iteration $(f^k_\lambda(x))$ of $f_\lambda(x) := (1-\lambda)x + \lambda f(x)$.

Lemma 18.7. *Let $C \subseteq X$ be a convex subset of a normed space, $f : C \to C$ a nonexpansive mapping, $x \in C$ and (λ_k) a sequence in $[0,1]$. Then for the Krasnoselski-Mann iteration of f starting at x the following holds:*

$$\forall k \in \mathbb{N}\,\big(\|x_{k+1} - f(x_{k+1})\| \le \|x_k - f(x_k)\|\big).$$

Proof:

$$\begin{aligned}
&\|x_{k+1} - f(x_{k+1})\| = \|(1-\lambda_k)x_k + \lambda_k f(x_k) - f((1-\lambda_k)x_k + \lambda_k f(x_k))\| = \\
&\|((1-\lambda_k)x_k - (1-\lambda_k)f(x_k)) + (f(x_k) - f((1-\lambda_k)x_k + \lambda_k f(x_k)))\| \le \\
&\|(1-\lambda_k)x_k - (1-\lambda_k)f(x_k)\| + \|f(x_k) - f((1-\lambda_k)x_k + \lambda_k f(x_k))\| \le \\
&\|(1-\lambda_k)x_k - (1-\lambda_k)f(x_k)\| + \|x_k - ((1-\lambda_k)x_k + \lambda_k f(x_k))\| = \\
&(1-\lambda_k)\|x_k - f(x_k)\| + \lambda_k\|x_k - f(x_k)\| = \|x_k - f(x_k)\|.
\end{aligned}$$

□

In the following theorems we assume (following [176]) that $(\lambda_k)_{k\in\mathbb{N}}$ satisfies the following conditions:

- (λ_k) is divergent in sum, i.e.

$$(A)\ \ \forall n, i \in \mathbb{N} \exists k \in \mathbb{N} \left(\sum_{j=i}^{i+k} \lambda_j \ge n \right).$$

- $\limsup_{k\to\infty} \lambda_k < 1$, i.e.

$$(B)\ \exists K, k_0 \in \mathbb{N} \forall k \geq k_0 \big(\lambda_k \leq 1 - \frac{1}{K}\big).$$

Theorem 18.8 (Borwein-Reich-Shafrir theorem [40]). *Let $(X, \|\cdot\|)$ be a normed space, $C \subseteq X$ convex, $f : C \to C$ nonexpansive and $(\lambda_k)_{k\in\mathbb{N}}$ be a sequence in $[0,1]$ which satisfies $(A), (B)$ above. Then for the Krasnoselski-Mann iteration (x_k) of f starting from $x \in C$ one has*

$$\|x_k - f(x_k)\| \stackrel{k\to\infty}{\to} r_C(f),$$

where $r_C(f) := \inf_{x\in C} \|x - f(x)\|$ is the so-called minimal displacement of f.

Theorem 18.9 (Ishikawa's theorem [176]). *Under the same assumptions as in the previous theorem the following holds:*

$$(x_k)_{k\in\mathbb{N}} \text{ bounded } \to \|x_k - f(x_k)\| \stackrel{k\to\infty}{\to} 0.$$

Remark 18.10. As we will see in proposition 18.15 below, if C (or just $f(C)$) is compact, then it follows from theorem 18.9 that (x_k) converges towards a fixed point of f and so (as mentioned already) theorem 18.9 provides a far reaching generalization of Krasnoselski's theorem.

Under the stronger assumption of C being bounded (which trivially implies the boundedness of the sequence $(x_k)_{k\in\mathbb{N}}$) one can easily obtain theorem 18.9 as a corollary to theorem 18.8 using the following simple

Proposition 18.11. *Let $(X, \|\cdot\|)$ be a normed linear space, let $\emptyset \neq C \subseteq X$ be convex with bounded diameter $d(C) < \infty$ and let $f : C \to C$ be nonexpansive. Then f has ε-fixed points in C for every $\varepsilon > 0$, where $x \in C$ is an ε-fixed point of f if $\|x - f(x)\| \leq \varepsilon$.*

Proof: Let $\varepsilon > 0$. Obviously, the lemma is trivial for $\varepsilon > d(C)$. Hence we may assume that $\varepsilon \leq d(C)$. To reduce the situation to the Banach fixed point theorem we use the following well-known construction (see e.g. [137]): Pick a point $c \in C$ and define for $t \in (0,1]$ a selfmapping $f_t : C \to C$ as follows

$$f_t(x) := (1-t)f(x) + tc.$$

$f_t : C \to C$ is a contraction and therefore (the usual proof of) Banach's fixed point theorem applies and yields the existence of approximate fixed points of f_t (since we do not assume X to be complete and C to be closed we will not get a fixed point in general). In particular, for $t := \varepsilon/d(C)$ the function f_t has an ε-fixed point which is a 2ε-fixed point of f. □

The proofs of the Borwein-Reich-Shafrir theorem and of the Ishikawa theorem both use as the main ineffective tool the convergence of the nonincreasing sequence $(\|x_n - f(x_n)\|)$, i.e. the principle PCM discussed already in chapter 13. This principle, however, is easily provable from Π^0_1-AC and Σ^0_1-IA and so in $\mathcal{A}^\omega$ and its extensions discussed in chapter 17. Moreover, as we will see below, the proofs can easily be modified in such a way that only the arithmetic version PCM_{ar}, i.e. the Cauchy property of $(\|x_n - f(x_n)\|)$, is needed (see chapter 13). Note that the proof analysis then crucially involves the treatment of the $\Pi_3^0 \to \Pi_2^0$ modus ponens (when the Π_3^0-Cauchy property is discharged to derive the Π_2^0-asymptotic regularity conclusion) which we discussed already in chapter 2 and then in detail in chapter 10.

18.2 Applications of the metatheorems from chapter 17

Both, the Borwein-Reich-Shafrir theorem 18.8 as well as Ishikawa's theorem 18.9 hold in the more general setting of hyperbolic spaces as introduced in chapter 17. For the former this is proved also in [40] and for the latter by Goebel and Kirk in [126] (Ishikawa's theorem even holds for spaces of hyperbolic type, i.e. hyperbolic spaces with the axiom '(iv)' dropped, see [126]). We call the extension of Ishikawa's theorem to hyperbolic spaces the Ishikawa-Goebel-Kirk theorem. It is not hard to see that the proofs given in [40] and [126] can be formalized in $\mathscr{A}^\omega[X,d,W]_{-b}$ (both proofs will be presented below where the extraction of effective rates of convergence will be carried out) so that the metatheorems from chapter 17 apply which guarantees the following results:

Application. 18.12 Let (X,d,W) be an arbitrary hyperbolic space, $k \in \mathbb{N},\ k \geq 1$ and $(\lambda_n)_{n\in\mathbb{N}}$ a sequence in $[0, 1 - \frac{1}{k}]$ with $\sum\limits_{n=0}^{\infty} \lambda_n = \infty$ and define for $f : X \to X, x \in X$ the Krasnoselski-Mann iteration $(x_n)_n$ starting from x by

$$x_0 := x,\ x_{n+1} := (1-\lambda_n)x_n \oplus \lambda_n f(x_n).$$

In [126](Theorem 1) and [176] (for the normed case) the following is proved

$$\forall x \in X, f : X \to X \big((x_n)_n \text{ bounded and } f \text{ nonexpansive } \to \lim_{n\to\infty} d(x_n, f(x_n)) = 0\big).$$

Corollary 17.55 a-priori guarantees (see the proof below) the extractability of computable bounds $\Phi(k,\alpha,b,\tilde{b},l)$, $\Psi(k,\alpha,b,\tilde{b},l)$ so that in any hyperbolic space (X,d,W), for any $l,b,\tilde{b},k \in \mathbb{N}$, $k \geq 1$, and any $\alpha : \mathbb{N} \to \mathbb{N}$ such that

$$\forall n \in \mathbb{N}(\lambda_n \leq 1 - \frac{1}{k} \wedge n \leq \sum_{i=0}^{\alpha(n)} \lambda_i)$$

the following holds

$$\begin{aligned}\forall x \in X \forall f : X \to X (d(x,f(x)) \leq \tilde{b} \wedge \forall i,j \leq \Psi(k,\alpha,b,\tilde{b},l)\ (d(x_i,x_j) \leq b)\\ \wedge f \text{ nonexpansive } \to \forall m \geq \Phi(k,\alpha,b,\tilde{b},l)\ (d(x_m,f(x_m)) < 2^{-l})).\end{aligned}$$

Proof: As mentioned already, $\mathcal{A}^\omega[X,d,W]_{-b}$ proves the formalized version of Ishikawa's theorem (Theorem 1 in [126]): if $k \geq 1, \lambda^{0\to 1}_{(\cdot)}$ represents an element of the compact metric space $[0,1]^{\mathbb{N}}$ (with the product metric) and $\alpha : \mathbb{N} \to \mathbb{N}$ such that

$$(*)\ \forall n \in \mathbb{N}\big(\lambda_n \leq_{\mathbb{R}} 1 - \frac{1}{k} \wedge n \leq_{\mathbb{R}} \sum_{i=0}^{\alpha(n)} \lambda_i\big),$$

where $\sum_{i=0}^{\alpha(n)} \lambda_i$ represents the corresponding summation of the real numbers in $[0,1]$ represented by λ_i, then for all $l^0, b^0, x^X, f^{X(X)}$

$$\forall i,j(d_X(x_i,x_j) \leq_{\mathbb{R}} b) \wedge f \text{ nonexpansive } \to \exists n^0(d_X(x_n,f(x_n)) <_{\mathbb{R}} 2^{-l})),$$

where '$(*)$' and '$d_X(x_i,x_j) \leq_{\mathbb{R}} b$' are a $\forall$-formulas and '$d_X(x_n,f(x_n)) <_{\mathbb{R}} 2^{-l}$' is an $\exists$-formula.
Now corollary 17.55 yields the existence of computable functionals Φ, Ψ (of course, by taking the maximum, one can make a single functional out of Ψ, Φ as in corollary 17.55, but, numerically, it is better to keep them separate) such that for all $(\lambda_m) \in [0,1]^{\mathbb{N}}, x \in X, f : X \to X$

$$\begin{cases} (*) \wedge d(x,f(x)) \leq \tilde{b} \wedge \forall i,j \leq \Psi(k,\alpha,b,\tilde{b},l)\ (d(x_i,x_j) \leq b) \wedge\ f \text{ n.e.} \\ \quad \to \exists m \leq \Phi(k,\alpha,b,\tilde{b},l)\ (d(x_m,f(x_m)) < 2^{-l}) \end{cases}$$

holds for all $k,\alpha,b,\tilde{b},l$ in any hyperbolic space (X,d,W).
Since $(d(x_n,f(x_n)))_n$ is a nonincreasing sequence (exercise) the conclusion actually implies

$$\forall m \geq \Phi(k,\alpha,b,\tilde{b},l)\ \big(d(x_m,f(x_m)) < 2^{-l}\big).$$

□

Remark 18.13. By a simple trick (which uses the truth of the theorem whose proof we are analyzing) one can see the that the assumption to have a bound $\tilde{b}$ on $d(x,f(x))$ as an input of Ψ, Φ is redundant if we assume that (x_n) is b-bounded: by Ishikawa's theorem we know, in particular, that $d(x_n,f(x_n)) \to 0$ and so a-fortiori

$$\exists n \in \mathbb{N}\big(d(x_n,f(x_n)) \leq b\big).$$

Using $d(x_i,x_j) \leq b$ for all i,j and the nonexpansivity of f yields (note that $x_0 := x$)

$$d(x,f(x)) \leq d(x,x_n) + d(x_n,f(x_n)) + d(f(x_n),f(x)) \leq 3b.$$

As a qualitative consequence of the bound Φ one immediately concludes that for bounded hyperbolic spaces (X,d,W) the convergence $d(x_n,f(x_n)) \to 0$ is uniform in x, f and – except for a bound b on the metric – in (X,d,W). This fact was first proved as theorem 2 (called 'main result') in [126] but – as we just saw – follows directly from theorem 1 of that paper by a general logical metatheorem. For this particular application, corollary 17.45 above already suffices (see [226] for a discussion of this point). However, whereas the proof of theorem 2 in [126] (and similarly the alternative proof, for the case of constant $\lambda_n = \lambda$, from [186]) makes essential use of the fact that the whole space is bounded, the uniform bound obtained from corollary 17.55 (together with the remark above) shows that a bound on just (x_n) is sufficient.

Remark 18.14. For the case of bounded convex subsets of normed spaces and constant $\lambda_n = \lambda \in (0,1)$ the uniformity in x was already shown in [93] and – for $(\lambda_n)_n$ in $[a,b] \subset (0,1)$ and nonincreasing – in [67].

In [232], the extraction of a concrete effective uniform rate of convergence was carried out. In that paper, this result was obtained using a logical analysis of the proof of the Borwein-Reich-Shafrir theorem to be discussed below (rather than the proof of Ishikawa's theorem from [126]). In fact, combining Ishikawa's theorem with the Borwein-Reich-Shafrir theorem it follows that instead of assuming (x_n) to be bounded it suffices to assume that for **some** $x^* \in X$ the Krasnoselski-Mann iteration (x_n^*) starting with x^* is bounded: By Ishikawa's theorem it then follows that $r_X(f) = 0$ and so by the Borwein-Reich-Shafrir theorem we get $d(x_n, f(x_n)) \to 0$. In [120] it is shown how the existence of an effective rate of convergence depending on x, x^*, f and (X,d,W) only via a bound b on (x_n^*) follows from the refined logical metatheorems established in that paper (our corollary 17.55 above).

As we will see below, for the extraction of such a bound it suffices to analyze the proof of the Borwein-Reich-Shafrir theorem and to use the truth (rather a proof) of Ishikawa's theorem. Then, however, no information on how much of the boundedness of (x_n^*) is needed (i.e. no bound Ψ) is obtained. Further below we will carry out all this in detail and extract bounds on the convergence in Ishikawa's theorem both via the logical analysis of the proof of the Borwein-Reich-Shafrir theorem as in done in [232] (see theorem 18.42 below) as well as (for $x^* := x$) via the direct analysis of the proof of Ishikawa's theorem in [126] for hyperbolic spaces (see theorem 18.49).

As a preparation for the next application we need the following

Proposition 18.15 (Ishikawa, Goebel, Kirk [176, 126]). *Let (X,d,W) be a compact hyperbolic space and $(\lambda_n), f, (x_n)$ as in application 18.12. Then $(x_n)_n$ converges towards a fixed point of f.*

Proof: From Ishikawa's theorem it follows that $d(x_n, f(x_n)) \to 0$ since the compactness of X implies that X – and hence $(x_n)_n$ – is bounded. Using again the compactness of X, we know that $(x_n)_n$ has a convergent subsequence $(x_{n_k})_k$ with limit $\widehat{x}$. One easily shows (using the continuity of f) that $\widehat{x}$ is a fixed point of f. The proof is concluded by verifying the easy fact that for any fixed point $\widehat{x}$ of f

$$(*)\ \forall n \in \mathbb{N}(d(x_{n+1}, \widehat{x}) \leq d(x_n, \widehat{x}))$$

which implies that already $(x_n)_n$ converges towards $\widehat{x}$.
Proof of $(*)$:

$$
\begin{aligned}
d_X(x_{n+1},\widehat{x}) &=_{\mathbb{R}} d_X((1-\lambda_n)x_n \oplus \lambda_n f(x_n),\widehat{x})\\
&\leq_{\mathbb{R}} (1-\lambda_n)d_X(x_n,\widehat{x}) +_{\mathbb{R}} \lambda_n d_X(f(x_n),\widehat{x})\\
&=_{\mathbb{R}} (1-\lambda_n)d_X(x_n,\widehat{x}) +_{\mathbb{R}} \lambda_n d_X(f(x_n),f(\widehat{x}))\\
&\leq_{\mathbb{R}} (1-\lambda_n)d_X(x_n,\widehat{x}) +_{\mathbb{R}} \lambda_n d_X(x_n,\widehat{x})\\
&=_{\mathbb{R}} d_X(x_n,\widehat{x}).
\end{aligned}
$$

□

Application. 18.16 Let us consider the proof of the Cauchy property of (x_n) from the asymptotic regularity (i.e. $d(x_n, f(x_n)) \to 0$) (taken as assumption) under the additional assumption of X being compact. The sequential compactness used in the proof follows relative to $\mathcal{A}^{\omega}[X,d,W,\mathcal{C}]_{-b}$ (which already contains the completeness axiom for X) from the total boundedness of X. Using again the fact that the proof of Ishikawa's theorem from [126] can be formalized in $\mathcal{A}^{\omega}[X,d,W]_{-b}$ it follows in total that

$$(+)\ X \text{ tot. bounded} \wedge \lim d(x_n, f(x_n)) = 0 \to \forall k \in \mathbb{N} \exists n \in \mathbb{N} \forall m \geq n (d(x_n, x_m) \leq 2^{-k})$$

is provable in $\mathcal{A}^{\omega}[X,d,W,\mathcal{C}]_{-b}$.
The proof of the implication $(+)$ only uses that $(\lambda_n)_n$ is a sequence in $[0,1]$ but not the other assumptions on $(\lambda_n)_n$ (which are only needed to prove that $d(x_n, f(x_n)) \to 0$).

Before we can see how to apply corollary 17.55 to the proof of $(+)$ we have to make explicit the logical form of the various clauses involved.

- since $(d(x_n, f(x_n)))_n$ is nonincreasing, we can – as before – write the asymptotic regularity equivalently as $\forall l \in \mathbb{N} \exists n \in \mathbb{N} (d(x_n, f(x_n)) \leq 2^{-l})$ which asks for a witnessing rate of asymptotic regularity $\delta : \mathbb{N} \to \mathbb{N}$ such that

$$(1)\ \forall l \in \mathbb{N} (d(x_{\delta(l)}, f(x_{\delta(l)})) \leq 2^{-l}).$$

  If we provide such a δ as input, the remaining formula (1) is a $\forall$-formula.
- the total boundedness of X is expressed by the existence of a sequence $(a_n)_n$ of points in X and a function $\gamma : \mathbb{N} \to \mathbb{N}$ such that

$$(2)\ \forall l \in \mathbb{N}, x \in X \exists n \leq \gamma(l) (d(x, a_n) \leq 2^{-l}).$$

  A function γ such that a sequence $(a_n)_n$ in X satisfying (2) exists is called a modulus of total boundedness for X. So we add γ as yet another input and verify (using Σ^0_1-CP) that (2) is equivalent to a $\forall$-formula.

As discussed already above, i.e. the Cauchy property of (x_n), a Π^0_3-formula, is logically too complicated to be covered by our metatheorems and, in fact, we showed in theorem 18.4 that there simply is no (uniformly) effective Cauchy rate in general. We, therefore, modify the conclusion to its Herbrand normal form (or 'metastable' version to use the terminology from T. Tao [357])

$$(H)\ \forall l^0 \forall g^1 \exists n^0 \forall i,j \in [n;n+g(n)](d(x_i,x_j) < 2^{-l}),$$

where $[n;m]$ denotes the subset $\{n,n+1,\ldots,m-1,m\}$ of $\mathbb{N}$ for $m \geq n$ and – using just Σ^0_1-CP – '$\forall i,j \in [n;n+g(n)](d(x_i,x_j) < 2^{-l})$' is equivalent to an $\exists$-formula.

More precisely, (H) is the Herbrand normal form of a slightly different but trivially equivalent formulation of the Cauchy property.

Classically, (H) is a equivalent to the Cauchy property for $(x_n)_n$ but – since the proof is ineffective – a computable bound on (H) does not yield a computable Cauchy modulus for $(x_n)_n$ (see the discussion of the Herbrand normal form in chapters 2 and 10).

$\mathcal{A}^\omega[X,d,W,\mathcal{C}]_{-b}$ proves that

$$\forall (\lambda_m) \in [0,1]^{\mathbb{N}} \forall x^X \forall f^{X(X)}, (a_n)^{X(0)}, l^0, \gamma^1, \delta^1, g^1$$
$$((1) \wedge (2) \wedge f \text{ n.e. } \rightarrow \exists m \in \mathbb{N} \forall i,j \in [m;m+g(m)](d(x_i,x_j) < 2^{-l})).$$

The total boundedness of X implies that X is bounded and a bound can be computed by $b := \max\{d(a_i,a_j) : i,j \leq \gamma(0)\} + 2$. However, in order to guarantee our result to be independent from $(a_n)_n$ we add a bound b of X as an additional input. Hence by either corollary 17.45 or corollary 17.55 (together with the treatment of the completeness axiom at the end of section 17.5) we obtain a computable bound $\Omega(l,b,\gamma,\delta,g)$ such that for all (λ_n) in $[0,1]$, $x \in X$, (a_n) in X, $f : X \to X$, $l \in \mathbb{N}$ and $\gamma, \delta, g : \mathbb{N} \to \mathbb{N}$:

$$(1) \wedge (2) \wedge f \text{ n.e. } \rightarrow \exists m \leq \Omega(l,b,\gamma,\delta,g) \forall i,j \in [m;m+g(m)](d(x_i,x_j) \leq 2^{-l})$$

holds in any complete, b-bounded, totally bounded (with modulus γ) hyperbolic space (X,d,W) (using a standard completion argument one sees that this conclusion even holds without the completeness assumption).

A concrete bound Ω of this kind has in fact been extracted in [227], where there extraction itself was guided by the algorithm provided by the proof of corollary 17.45 as well as the proof-theoretic study of the Bolzano-Weierstraß principle carried out in [210]. This concrete Ω even is independent from b and is defined as follows

$$\Omega(l,g,\delta,\gamma) := \max_{i \leq \gamma(l+3)} \Psi_0(i,l,g,\delta),$$

where

$$\begin{cases} \Psi_0(0,l,g,\delta) := 0 \\ \Psi_0(n+1,l,g,\delta) := \delta\left(l+2+\lceil \log_2(\max_{i \leq n} g(\Psi_0(i,l,g,\delta))+1)\rceil\right). \end{cases}$$

This bound will be derived in detail further below (see theorem 18.58).

For X being b-bounded and (λ_n) in $[0,1-\frac{1}{k}]$ $(k \geq 1)$ with $n \leq \sum_{i=0}^{\alpha(n)} \lambda_i$ we can take

$\delta(m) := \Phi(k,\alpha,b,\tilde{b},m)$ from application 18.12. Even the fact that the bound does not depend on b can largely be explained based on our metatheorems if one expresses that (X,d) has a modulus of total boundedness γ in the form $\forall l^0, a^{X(0)} \exists i,j(0 \le i < j \le \gamma(l) \wedge d(a(i),a(j)) \le 2^{-l})$ (see also [117]) which avoids the reference to an ε-net.

Discussion on the Herbrand normal form (H) (of the Cauchy property of (x_n)): a bound on the Herbrand normal form (H) of the Cauchy property of (x_n), which classically is equivalent to the convergence of (x_n) (given the completeness of (X,d,W)), also is rather natural from an ordinary mathematical point of view as a generalization of a rate of asymptotic regularity, where with the latter we mean

$$\text{(AR)}\ \ d(x_n,f(x_n)) \overset{n\to\infty}{\to} 0.$$

Let $\lambda_n \in (a,b)$ with $0 < a < b < 1$ for all $n \in \mathbb{N}$, i.e. (λ_n) is bounded away from 1 **and** 0. Because of

$$(+)\ \ d(x_n,x_{n+1}) = \lambda_n \cdot d(x_n,f(x_n))$$

we then have that (AR) holds iff

$$(++)\ \ d(x_n,x_{n+1}) \overset{n\to\infty}{\to} 0$$

which in turn is equivalent to the special case (H_1) of (H) where $g \equiv 1$:
'$\Rightarrow$': Since, in particular, $\lambda_n \le 1$, $(++)$ – and so a-fortiori (H_1) – trivially follows from (AR) using $(+)$.
'$\Leftarrow$': Let $k \in \mathbb{N}$ and $\tilde{k} \in \mathbb{N}$ be such that $2^{-\tilde{k}} \le a \cdot 2^{-k}$. Then by (H_1) there exists an $n \in \mathbb{N}$ such that

$$d(x_n,x_{n+1}) \le 2^{-\tilde{k}} \le a \cdot 2^{-k}.$$

Hence by $(+)$

$$d(x_n,f(x_n)) \le a \cdot 2^{-k}/\lambda_n \le 2^{-k}$$

and so, using that $(d(x_n,f(x_n)))_{n\in\mathbb{N}}$ is nonincreasing, (AR) follows.
This proof also shows that any rate of asymptotic regularity can effectively be converted into a bound on (H_1) and vice versa.

Now whereas (H_1 (i.e. (AR)) holds without any compactness assumption on (X,d,W) (as long as (X,d,W) is bounded), the full Herbrand normal form (H) in general fails for noncompact spaces (X,d,W) : e.g. consider as X the closed bounded convex subset $C := \{(x_n) \in c_0 : \forall n \in \mathbb{N}(0 \le x_n \le 1)\}$ of c_0. As we have seen above, there are fixed point free nonexpansive selfmappings of C. Since (H) is equivalent to the Cauchy property and hence convergence of (x_n) and the limit of (x_n) necessarily would be a fixed point of f (because of (AR)) it is clear that (H) for **general** g fails for C (while it holds for $g :\equiv 1$ and even for all constant functions $g(n) := k$).

Application. 18.17 Let $(X,d,W),k,(\lambda_n),f,x$ and (x_n) be as in application 18.12. In [40], the following result is proved:

$$\forall x \in X, f: X \to X\big(f \text{ nonexpansive } \to \lim_{n\to\infty} d(x_n,f(x_n)) = r_X(f)\big),$$

where $r_X(f) := \inf_{y\in X} d(y, f(y))$ is the so-called minimal displacement of f. As (x_n) is no longer assumed to be bounded, $r_X(f)$ can very well be strictly positive: e.g. for $\mathbb{R}$ (with the natural metric) and $f : \mathbb{R} \to \mathbb{R}, f(x) := x+1$ we have $r_{\mathbb{R}}(f) = 1$ although f is nonexpansive.
The most direct formalization of the above theorem is as follows (using again that $(d(x_n, f(x_n)))$ is nonincreasing):

$$(a)\ \forall l \in \mathbb{N} \forall x \in X, f : X \to X \exists n \in \mathbb{N} \forall x^* \in X (d(x_n, f(x_n)) < d(x^*, f(x^*)) + 2^{-l}).$$

Due to the quantifier '$\forall x^* \in X$', this statement does not have the $\forall\exists$-form required by our metatheorem. Indeed, an effective bound on '$\exists n \in \mathbb{N}$' would allow us to compute (in x_0, f) the infimum $r_X(f)$ which is unlikely to be possible in general. However, the following seemingly weaker formulation is (ineffectively) actually equivalent to (a) :

$$(b)\ \forall l \in \mathbb{N} \forall x, x^* \in X, f : X \to X \exists n \in \mathbb{N} (d(x_n, f(x_n)) < d(x^*, f(x^*)) + 2^{-l}).$$

The proof given in [40] (for both of the formulations (a) and (b) of the theorem) can be formalized in $\mathcal{A}^{\omega}[X,d,W]_{-b}$ and so corollary 17.55 (see the next section) yields (like in the proof of application 18.12 above) an effective bound $\Psi(k, \alpha, b, \tilde{b}, l)$ such that in any hyperbolic space (X,d,W), for any $l, b, \tilde{b}, k \in \mathbb{N}$, $k \geq 1$, and any $\alpha : \mathbb{N} \to \mathbb{N}$, satisfying $\lambda_n \leq 1 - \frac{1}{k}$ and $n \leq \sum_{i=0}^{\alpha(n)} \lambda_i$ for all $n \in \mathbb{N}$, the following holds

$$\begin{aligned} &\forall x, x^* \in X \forall f : X \to X (\ d(x,x^*) \leq b \wedge d(x, f(x)) \leq \tilde{b} \wedge f \text{ nonexpansive} \\ &\qquad \to \exists m \leq \Psi(k, \alpha, b, \tilde{b}, l)\ (d(x_m, f(x_m)) < d(x^*, f(x^*)) + 2^{-l})) \end{aligned}$$

and so (by the fact that $(d(x_n, f(x_n)))_n$ is nonincreasing)

$$\begin{aligned} &\forall x, x^* \in X \forall f : X \to X \big(\ d(x,x^*) \leq b \wedge d(x, f(x)) \leq \tilde{b} \wedge f \text{ nonexpansive} \\ &\qquad \to \forall m \geq \Psi(k, \alpha, b, \tilde{b}, l)\ \big(d(x_m, f(x_m)) < d(x^*, f(x^*)) + 2^{-l}\big)\big). \end{aligned}$$

An explicit such bound Ψ (which is very similar to the bound Φ mentioned in connection with application 18.12) has been extracted in [232] (for the special case of convex subsets of normed spaces this is already due to [220]). We will carry out the extraction in detail in theorem 18.30 and remark 18.31 below.

Instead of replacing (a) by (b) we could consider also the following intermediate version (c) of (a) which, constructively speaking, is stronger than (b) :

$$(c)\ \forall l \in \mathbb{N} \forall x \in X, f : X \to X \forall (y_n) \in X^{\mathbb{N}} \exists n \in \mathbb{N} (d(x_n, f(x_n)) < d(y_n, f(y_n)) + 2^{-l}),$$

i.e. instead of just considering a constant sequence $y_n := x^*$ we allow an arbitrary sequence (y_n) in X (not necessarily a Krasnoselski-Mann iteration!). Now corollary

17.55 yields (as above) an effective bound $\tilde{\Psi}(k,\alpha^1,b^1,l)$ such that in any hyperbolic space (X,d,W), for any $l,k\in\mathbb{N}$, $k\geq 1$, any $b:\mathbb{N}\to\mathbb{N}$ and any $\alpha:\mathbb{N}\to\mathbb{N}$, satisfying $\lambda_n\leq\frac{1}{k}$ and $n\leq\sum_{i=0}^{\alpha(n)}\lambda_i$ for all $n\in\mathbb{N}$, the following holds

$$\forall x\in X\forall f:X\to X\forall(y_n)\in X^{\mathbb{N}}(\ \forall m(d(x,y_m),d(x,f(x))\leq b(m))\wedge f\text{ nonexp.}\to$$
$$\exists n\leq\tilde{\Psi}(k,\alpha^1,b^1,l)\ (d(x_n,f(x_n))<d(y_n,f(y_n))+2^{-l})).$$

We will extract an explicit such $\tilde{\Psi}$ in theorem 18.34 below (which for the normed case is due to [224] and for the hyperbolic case to [229]).

18.3 Logical analysis of the proof of the Borwein-Reich-Shafrir theorem

In the following, let $(\lambda_n)_{n\in\mathbb{N}}\subseteq[0,1)$.
For $i,n\in\mathbb{N}$, we define

$$S_{i,n}:=\sum_{s=i}^{i+n-1}\lambda_s,$$

$$P_{i,n}:=\prod_{s=i}^{i+n-1}\frac{1}{1-\lambda_s}.$$

For $n=0$ we adopt the usual conventions that the empty sum is defined as 0 whereas the empty product is defined to be 1. In the following, (X,d,W) is an arbitrary nonempty hyperbolic space. Let $(x_n)_{n\in\mathbb{N}},(y_n)_{n\in\mathbb{N}}$ be two sequences in X such that for all $n\in\mathbb{N}$,

$$x_{n+1}=(1-\lambda_n)x_n\oplus\lambda_n y_n.$$

The main lemma used in the proof of the Borwein-Reich-Shafrir theorem is an inequality (see proposition 18.20) which in turn is based on the following quite nontrivial result (first was proved in [126]) for spaces of hyperbolic type (a related inequality for the case of normed spaces was already proved in [176]) and so a-fortiori holds for hyperbolic spaces.

Proposition 18.18 (Goebel-Kirk [126]). *Suppose that $(x_n)_{n\in\mathbb{N}},(y_n)_{n\in\mathbb{N}}$ satisfy for all $n\in\mathbb{N}$,*

$$d(y_n,y_{n+1})\leq d(x_n,x_{n+1}).$$

Then the sequence $(d(x_n,y_n))_{n\in\mathbb{N}}$ is nonincreasing and for all $i,n\in\mathbb{N}$,

$$(1+S_{i,n})d(x_i,y_i)\leq d(x_i,y_{i+n})+P_{i,n}[d(x_i,y_i)-d(x_{i+n},y_{i+n})].$$

Remark 18.19. In our applications further below the premise of proposition 18.18 will always be satisfied. Since the conclusion is a $\forall$-formula $A_\forall$ (in the sense of definition 17.34) it can be treated simply as another universal premise in applications of our metatheorems. This is the reason why we will not have to consider the (rather tedious) proof of proposition 18.18.

In [40], the following consequence of the above inequality is derived:

Proposition 18.20 (Borwein-Reich-Shafrir [40]). *Under the assumptions of proposition 18.18,*

$$S_{i,n}d(x_i,y_i) \le d(x_i,x_{i+n}) + P_{i,n}[d(x_i,y_i) - d(x_{i+n},y_{i+n})].$$

Proof: Using that $(d(x_n,y_n))_n$ is nonincreasing it follows that

$$d(x_i,y_{i+n}) - d(x_i,y_i) \le d(x_i,x_{i+n}) + d(x_{i+n},y_{i+n}) - d(x_i,y_i) \le d(x_i,x_{i+n}).$$

Together with proposition 18.18 this proves the claim. □

The following theorem is due to [126] (see, however, also [176] for the normed case):

Theorem 18.21. *Let (X,d,W) be a nonempty hyperbolic space and $(\lambda_n)_{n\in\mathbb{N}}$ a sequence in $[0,1)$. Suppose that $(\lambda_n)_{n\in\mathbb{N}}$ is divergent in sum and* $\limsup\limits_{n\to\infty}\lambda_n < 1$.
Let $(x_n)_{n\in\mathbb{N}}$, $(y_n)_{n\in\mathbb{N}}$ be two sequences in X which satisfy for all $n\in\mathbb{N}$:

$$x_{n+1} = (1-\lambda_n)x_n \oplus \lambda_n y_n \text{ and}$$
$$d(y_n,y_{n+1}) \le d(x_n,x_{n+1}).$$

If $(x_n)_{n\in\mathbb{N}}$ is bounded, then $\lim\limits_{n\to\infty} d(x_n,y_n) = 0$.

Proof: The proof is given in [126]. Modulo the use of proposition 18.18 we will present this proof in the next section. □

18.3.1 Uniform asymptotic regularity for directionally nonexpansive mappings

In the following we not only give a quantitative analysis of the generalization of the theorem of Borwein-Reich-Shafrir to hyperbolic spaces in our general sense but also extend things to the wider class of directionally nonexpansive mappings whose definition given in chapter 17 we recall here:

Definition 18.22 (Kirk [186]). Let (X,d,W) be a nonempty hyperbolic space. A mapping $f: X\to X$ is called directionally nonexpansive if

$$d(f(x),f(y)) \le d(x,y),$$

for all $x\in X$ and $y\in [x,f(x)]$.

In chapter 17 we observed that both conditions 'f is nonexpansive' as well as 'f is directionally nonexpansive' are $\forall$-formulas.
In the following, (X,d,W) will be an arbitrary nonempty hyperbolic space and $f : X \to X$ a directionally nonexpansive mapping. Let us recall the definition of the minimal displacement of f:

$$r_X(f) := \inf\{d(x,f(x)) \mid x \in X\}.$$

Again, we consider the Krasnoselski-Mann iteration starting from $x \in X$

$$x_0 := x, \ \ x_{n+1} := (1-\lambda_n)x_n \oplus \lambda_n f(x_n),$$

where $(\lambda_n)_{n\in\mathbb{N}}$ is a sequence of real numbers in $[0,1)$.

Lemma 18.23. *For all $n \in \mathbb{N}$,*

$$d(f(x_n),f(x_{n+1})) \leq d(x_n,x_{n+1}).$$

Proof: Since $x_{n+1} \in [x_n,f(x_n)]$, we can apply the fact that f is directionally nonexpansive to obtain that $d(f(x_n),f(x_{n+1})) \leq d(x_n,x_{n+1})$. □

Thus, the sequences $(x_n)_{n\in\mathbb{N}}, (f(x_n))_{n\in\mathbb{N}}$ satisfy the hypotheses of proposition 18.18 with $y_n := f(x_n)$. We get immediately the following results.

Proposition 18.24. *The sequence $(d(x_n,f(x_n)))_{n\in\mathbb{N}} \subseteq \mathbb{R}$ is nonincreasing and for all $i,n \in \mathbb{N}$,*

$$S_{i,n}d(x_i,f(x_i)) \leq d(x_i,x_{i+n}) + P_{i,n}[d(x_i,f(x_i)) - d(x_{i+n},f(x_{i+n}))].$$

Proof: Apply lemma 18.23, proposition 18.18 and proposition 18.20. □

For nonexpansive mappings the following proposition (which we call the Ishikawa-Goebel-Kirk theorem) is due to [176] (for normed spaces and – for constant $\lambda_n := \lambda$ – independently also to [93]) and [126] for hyperbolic spaces. Using Lemma 18.23, the proof from [126] extends to directionally nonexpansive mappings:

Proposition 18.25.
Suppose that $(\lambda_n)_{n\in\mathbb{N}}$ is divergent in sum and $\limsup_{n\to\infty} \lambda_n < 1$.

If $(x_n)_{n\in\mathbb{N}}$ is bounded, then $\lim_{n\to\infty} d(x_n,f(x_n)) = 0$.

Proof: By theorem 18.21 and lemma 18.23. □

Corollary 18.26. *Suppose that $(\lambda_n)_{n\in\mathbb{N}}$ is divergent in sum and* $\limsup_{n\to\infty} \lambda_n < 1$.

If X is bounded, then for every $x \in X$, $\lim_{n\to\infty} d(x_n,f(x_n)) = 0$.

Corollary 18.27. *Suppose that $(\lambda_n)_{n\in\mathbb{N}}$ is divergent in sum and* $\limsup_{n\to\infty} \lambda_n < 1$.

If X is bounded or – even weaker – there is $x \in X$ such that $(x_n)_{n\in\mathbb{N}}$ is bounded, then $r_X(f) = 0$.

Let $x^* \in X$ and $(x_n^*)_{n\in\mathbb{N}}$ be the Krasnoselski-Mann iteration starting from x^*.

The next inequality is due to [40]:

Lemma 18.28. *If f is nonexpansive, then for all $n \in \mathbb{N}$,*

$$d(x_{n+1}, x_{n+1}^*) \leq d(x_n, x_n^*).$$

Proof: Applying axiom (iv) of hyperbolic spaces and the definition of a nonexpansive mapping, we get that

$$\begin{aligned} d(x_{n+1}, x_{n+1}^*) &= d((1-\lambda_n)x_n \oplus \lambda_n f(x_n), (1-\lambda_n)x_n^* \oplus \lambda_n f(x_n^*)) \\ &\leq (1-\lambda_n)d(x_n, x_n^*) + \lambda_n d(f(x_n), f(x_n^*)) \\ &\leq (1-\lambda_n)d(x_n, x_n^*) + \lambda_n d(x_n, x_n^*) \\ &= d(x_n, x_n^*). \end{aligned}$$

□

Since in general $x_n^* \not\in [x_n, f(x_n)]$, we cannot prove the inequality

$$d(f(x_n), f(x_n^*)) \leq d(x_n, x_n^*)$$

on which the proof of lemma 18.28 is based for directionally nonexpansive mappings f.

We now present the

Proof of the Borwein-Reich-Shafrir theorem (formulated in the setting of hyperbolic spaces based on [40] with some steps referred in [40] to the literature filled in): Let (X, d, W) be a hyperbolic space, $f : X \to X$ a nonexpansive selfmapping and (λ_n) a sequence in $[0, 1]$ satisfying $(A), (B)$ above with $K, k_0 \in \mathbb{N}$, where we may assume w.l.o.g. that $k_0 = 0$. For $x \in X$ let (x_n) denote the Krasnoselski-Mann iteration of f based on (λ_n) and define $y_n := f(x_n)$. By proposition 18.24 the sequence $(d(x_n, y_n))_n$ is nonincreasing and bounded from below by 0. Hence $r(x) := \lim d(x_n, y_n)$ exists. We first show that $r(x)$ does not depend on x : suppose on the contrary that there are $x, x^* \in X$ with $r(x) > r(x^*)$. Let $\varepsilon > 0$ be small enough so that $r(x) > r(x^*) + \varepsilon$. Choose $i \in \mathbb{N}$ large enough so that

$$(+)\ \forall j \geq i \big(d(x_j, y_j) < r(x) + \varepsilon \wedge d(x_j^*, y_j^*) < r(x^*) + \varepsilon\big).$$

Next (reasoning as in [176]), we have

$$\begin{aligned}
P_{i,n} &= \prod_{s=i}^{i+n-1}(1+\tfrac{\lambda_s}{1-\lambda_s}) = \exp(\ln \prod_{s=i}^{i+n-1}(1+\tfrac{\lambda_s}{1-\lambda_s}))\\
&= \exp(\sum_{s=i}^{i+n-1}\ln(1+\tfrac{\lambda_s}{1-\lambda_s}))\\
&\le \exp(\sum_{s=i}^{i+n-1}\tfrac{\lambda_s}{1-\lambda_s}), \text{ since } \ln(1+x)\le x \text{ for } x\ge 0\\
&\le \exp(K\cdot \sum_{s=i}^{i+n-1}\lambda_s) = \exp(K\cdot S_{i,n}),
\end{aligned}$$

since $\lambda_s \le 1-\frac{1}{K}$ implies $1-\lambda_s \ge \frac{1}{K}$, so $\frac{1}{1-\lambda_s}\le K$ for all $s\in\mathbb{N}$. By proposition 18.24 and $(+)$ we obtain

$$\begin{aligned}
&S_{i,n}\cdot r(x) \le S_{i,n}\cdot d(x_i,y_i) \le d(x_i,x_{i+n}) + \varepsilon\cdot\exp(K\cdot S_{i,n})\\
&\le d(x_i^*,x_{i+n}^*) + d(x_i,x_i^*) + d(x_{i+n},x_{i+n}^*) + \varepsilon\cdot\exp(K\cdot S_{i,n})\\
&\overset{\text{lemma 18.28}}{\le} d(x_i^*,x_{i+n}^*) + 2d(x,x^*) + \varepsilon\exp(K\cdot S_{i,n}).
\end{aligned}$$

Moreover, using the definition of the Krasnoselski-Mann iteration (x_n^*) and the fact that also $(d(x_n^*,y_n^*))_n$ is nonincreasing we have

$$\begin{aligned}
&d(x_i^*,x_{i+n}^*) \le \sum_{s=i}^{n+i-1} d(x_s^*,x_{s+1}^*) = \sum_{s=i}^{n+i-1}\lambda_s d(x_s^*,y_s^*)\\
&\le \sum_{s=i}^{n+i-1}\lambda_s d(x_i^*,y_i^*) = S_{i,n}\cdot d(x_i^*,y_i^*) \overset{(+)}{\le} S_{i,n}(r(x^*)+\varepsilon).
\end{aligned}$$

So put together we have shown that

$$(*)\ S_{i,n}[r(x)-r(x^*)-\varepsilon] \le 2d(x,x^*) + \varepsilon\exp(K\cdot S_{i,n}).$$

Now define

$$M := \frac{1+2d(x,x^*)}{r(x)-r(x^*)}.$$

Using the assumption that (λ_n) is divergent in sum and $\lambda_i < 1$ for all i it is clear that there is an $n\in\mathbb{N}$ such that

$$M \le S_{i,n} \le M+1.$$

Hence

$$(1+2d(x,x^*))\cdot\frac{r(x)-r(x^*)-\varepsilon}{r(x)-r(x^*)} \le 2d(x,x^*) + \varepsilon\cdot\exp(K(M+1)).$$

Note that $x, x^*, K, M, r(x), r(x^*)$ do not depend on ε. So letting tend $\varepsilon\to 0$ we obtain

$$1+2d(x,x^*) \le 2d(x,x^*).$$

This contradiction shows that $r(x) \le r(x^*)$. Hence we have established that $r(x)$ is independent of $x \in X$. Since $r(x) \le d(x, f(x))$ this implies that $r(x) = r_X(f) = \inf\{d(y, f(y)) : y \in X\}$. □

Some logical pre-processing of the proof: We now indicate how to formalize the above proof: firstly as we cannot form $r_X(f)$ in $\mathcal{A}^\omega[X,d,W]_{-b}$ we have to restate the conclusion of the theorem as in application 18.17 either in the form (a) or (the formally weaker form) (b). We first consider (b) and then discuss how to derive (a) from (b). A straightforward formalization of the proof above seemingly requires the use of the principle PCM of convergence of monotone bounded sequences of real numbers which – as discussed before in chapter 13 – amounts to the use of arithmetical comprehension which is formalizable using Π^0_1-AC (and so clearly is available in $\mathcal{A}^\omega[x,d,w]_{-b}$, see chapter 11). However, it is easy to see that if the conclusion is stated as $\exists n \in \mathbb{N}(d(x_n, f(x_n)) < d(x^*, f(x^*)) + \varepsilon)$ as in (b) we do not even have to form explicitly the limits $r(x)$ or $r(x^*)$: to eliminate $r(x^*)$ just note that the result that $r(x)$ is independent of x is only used to conclude that $r(x) \le r(x^*) \le d(x^*, y^*)$. Hence we can run that proof also directly with the assumption $r(x) > r(x^*) + \varepsilon$ (made in the course of the reductio-ad-absurdum) replaced by $r(x) \ge d(x^*, y^*) + \varepsilon$ and, in turn, by $\forall n \in \mathbb{N}(d(x_n, y_n) \ge d(x^*, y^*) + \varepsilon)$ which instead of $(*)$ leads to

$$(**)\ S_{i,n}[d(x_i, y_i) - d(x^*, y^*)] \le 2d(x, x^*) + \tilde{\varepsilon}\exp(K \cdot S_{i,n})$$

if we replace $(+)$ by

$$(++)\ \forall j \ge i\big(d(x_i, y_i) < d(x_j, y_j) + \tilde{\varepsilon}\big)$$

for $\tilde{\varepsilon} > 0$ being arbitrary (the term '$-\varepsilon$' in $(*)$ came from $d(x_i^*, y_i^*) \le r(x^*) + \varepsilon$ which now is replaced by $d(x_i^*, y_i^*) \le d(x^*, y^*)$). The proof then goes through with $M := \frac{1+2d(x,x^*)}{\varepsilon}$. Since in eliminating $r(x)$ we no longer can form the denominator $r(x) - r(x^*)$ resp. $r(x) - d(x^*, y^*)$, we use the lower estimate ε instead and so need to rename the original ε used in $(+)$ into $\tilde{\varepsilon}$ in $(++)$ to led $\tilde{\varepsilon}$ tend to 0 while ε is kept fixed.

So while we do not have to use the ineffective principle of the convergence of $(d(x_n, f(x_n)))_n$ and $(d(x_n^*, f(x_n^*)))_n$ anymore, we still do need the Cauchy property of $(d(x_n, f(x_n)))_n$ to find an i satisfying $(++)$. As discussed already in chapter 2 this can be proved using Σ^0_1-IA and has an ND-interpretation (which in this case coincides with the no-counterexample interpretation) which is solvable using only R_0. The various inequalities used in the above proof are all purely universal and so can be treated in the logical proof analysis just as additional universal assumptions (just like 'f is nonexpansive'). Hence we do not have to consider their proofs. For completeness we note, however, that they all easily follow from proposition 18.24 which in turn is an easy consequence (see [40]) of the deep inequality proved first in [126] (based on prior work in [176]) stated already in proposition 18.18:

$$(1 + S_{i,n})d(x_i, f(x_i)) \le d(x_i, f(x_{i+n})) + P_{i,n}[d(x_i, f(x_i)) - d(x_{i+n}, f(x_{i+n}))].$$

This inequality is proved in [126] by a very complicated (but elementary) induction which can be formalized in $\mathcal{A}^\omega[X,d,W]_{-b}\setminus\{\text{DC}\}$ though (as mentioned above) we don't need this.

We now sketch how to obtain the formulation (a) of the Borwein-Reich-Shafrir theorem from (b) : obviously it suffices to show that

$$\forall k\in\mathbb{N}\exists x^*\in X\forall y\in X(d_X(x^*,f(x^*))\leq_{\mathbb{R}} d_X(y,f(y))+2^{-k})$$

or -equivalently –

$$\forall k\in\mathbb{N}\exists x^*\in X\forall y\in X(d_X(\widehat{x^*,f(x^*)})(k+1)\leq_{\mathbb{Q}} d_X(\widehat{y,f(y)})(k+1)+2^{-k+1}).$$

The latter follows from dependent choice in the form DAC (and hence from DC, see chapter 11; actually only a finite form of DAC is necessary). Since the matrix in this application of DAC is quantifier-free we actually can prove it using QF-AC^{X,X} together with R_X so that the proof can be formalized already in $\mathcal{A}^\omega[X,d,W]_{-b}\setminus\{\text{DC}\}$. By the use of R_X we need R_0 in order to majorize R_X and so (as we will see in theorem 18.34 below) the bound for (c) in application 18.17 resulting from the functional interpretation of the proof of (a) will (in addition to the use of R_0 for (b)) need another use of R_0.

Remark 18.29. The formulation (c) of the Borwein-Reich-Shafrir theorem in application 18.17 can even be proved without QF-AC^{X,X} as follows from the above discussion and functional interpretation. Since (c) implies (a) using only QF-AC^{0,X} it follows that the latter version of QF-AC is sufficient as well to prove (a).

We now first give a detailed quantitative analysis of the proof of the Borwein-Reich-Shafrir theorem (being pre-processed as indicated above and $\tilde{\varepsilon}$ being denoted by δ) formulated in the form (b) which – in general terms – was obtained already in application 18.17 as a consequence of the logical metatheorems developed in chapter 17. The explicit logical analysis actually shows that under an additional assumption (which is redundant for nonexpansive mappings) the result even holds for directionally nonexpansive mappings.

Theorem 18.30 (Kohlenbach-Leuştean [232]). *Let (X,d,W) be a nonempty hyperbolic space and $f:X\to X$ a directionally nonexpansive mapping. Let $(\lambda_n)_{n\in\mathbb{N}}$ be a sequence in $[0,1)$ which is divergent in sum and satisfies*

$$\forall n\in\mathbb{N}(\lambda_n\leq 1-\frac{1}{K})$$

for some $K\in\mathbb{N}$, $K\geq 1$.
Let $\alpha:\mathbb{N}\times\mathbb{N}\to\mathbb{N}$ be such that

$$\forall i,n\in\mathbb{N}((\alpha(i,n)\leq\alpha(i+1,n))\wedge(n\leq\sum_{s=i}^{i+\alpha(i,n)-1}\lambda_s)).$$

Let $x,x^*\in X$ and $b,\tilde{b}>0$ be such that

$$\forall n \in \mathbb{N}\big(d(x_n, x_n^*) \le b\big) \text{ and } d(x, f(x)) \le \tilde{b}$$

where $(x_n)_{n\in\mathbb{N}}$ *and* $(x_n^*)_{n\in\mathbb{N}}$ *are the Krasnoselski-Mann iterations starting from* x *and* x^*.
Then the following holds

$$\forall \varepsilon > 0 \forall n \ge h(\varepsilon, b, \tilde{b}, K, \alpha)(d(x_n, f(x_n)) < d(x^*, f(x^*)) + \varepsilon),$$

where (suing $n \dot{-} 1 = \max(0, n-1)$*)*

$$h(\varepsilon, b, \tilde{b}, K, \alpha) := \widehat{\alpha}(\lceil 2\tilde{b} \cdot \exp(K(M+1)) \rceil \dot{-} 1, M), \text{ with}$$
$$M := \left\lceil \tfrac{1+2b}{\varepsilon} \right\rceil \text{ and}$$
$$\widehat{\alpha}(0,n) := \tilde{\alpha}(0,n),\ \widehat{\alpha}(i+1,n) := \tilde{\alpha}(\widehat{\alpha}(i,n),n) \text{ with}$$
$$\tilde{\alpha}(i,n) := i + \alpha(i,n) \quad (i,n \in \mathbb{N})$$

Proof: Let $\varepsilon > 0$ and define

$$(1)\ \gamma := d(x^*, f(x^*)).$$

Choose $M \in \mathbb{N}$ in such a way that

$$(2)\ M \ge \frac{1+2b}{\varepsilon}.$$

For example, we may take $M := \left\lceil \frac{1+2b}{\varepsilon} \right\rceil$.

Next choose $\delta > 0$ so small that

$$(3)\ \delta \exp(K(M+1)) < 1.$$

This is satisfied e.g. for $\delta := \frac{1}{2\exp(K(M+1))}$.

Let $i, n \in \mathbb{N}$. Then (as shown in the proof above) for all $i, n \in \mathbb{N}$,

$$(4)\ P_{i,n} \le \exp(K \cdot S_{i,n}).$$

We now define $\alpha^* : \mathbb{N} \times \mathbb{N} \to \mathbb{N}$ by

$$(5)\ \alpha^*(i,n) := \min\{m \in \mathbb{N} \mid n \le S_{i,m}\}.$$

Since $(\lambda_n)_{n\in\mathbb{N}}$ is divergent in sum, it follows that for all $i \in \mathbb{N}$, the sequence $(S_{i,m})_{m\in\mathbb{N}}$ is not bounded above, so for all $n \in \mathbb{N}$ the set $A_{i,n} := \{m \in \mathbb{N} \mid n \le S_{i,m}\}$ is nonempty, hence it has a least element. Thus, α^* is well-defined. For $\alpha^*(i,n) > 0$ we also have that $\alpha^*(i,n) - 1 \notin A_{i,n}$ and so $S_{i,\alpha^*(i,n)-1} = S_{i,\alpha^*(i,n)} - \lambda_{i+\alpha^*(i,n)-1} < n$. Hence $S_{i,\alpha^*(i,n)} < n + \lambda_{i+\alpha^*(i,n)-1} < n+1$. Hence, for all $i, n \in \mathbb{N}$

$$(6)\ n \le S_{i,\alpha^*(i,n)} < n+1$$

(observing that this holds trivially in the case where $\alpha^*(i,n) = 0$ and so $S_{i,\alpha^*(i,n)} = n = 0$).
Let $(x_n^*)_{n\in\mathbb{N}}$ be the Krasnoselski-Mann iteration starting from x^* and (x_n) be the Krasnoselski-Mann iteration starting from x. Then

$$d(x_i^*, x_{i+n}^*) \le \sum_{s=i}^{i+n-1} d(x_s^*, x_{s+1}^*) = \sum_{s=i}^{i+n-1} \lambda_s d(x_s^*, f(x_s^*))$$
$$\le (\sum_{s=i}^{i+n-1} \lambda_s) d(x_i^*, f(x_i^*)) = S_{i,n} \cdot d(x_i^*, f(x_i^*)) \le S_{i,n} \cdot d(x^*, f(x^*)),$$

since, by proposition 18.24, $(d(x_n^*, f(x_n^*)))_{n\in\mathbb{N}}$ is nonincreasing. Hence, for all $i,n \in \mathbb{N}$,

$$(7)\ d(x_i^*, x_{i+n}^*) \le S_{i,n} \cdot d(x^*, f(x^*)).$$

Applying again proposition 18.24, we also know that the sequence $(d(x_n, f(x_n)))_{n\in\mathbb{N}}$ is nonincreasing and hence – since it is bounded from below by 0 – a Cauchy sequence. Thus, for $\delta > 0$ there exists an i such that

$$(8)\ d(x_i, f(x_i)) - d(x_{i+\alpha^*(i,M)}, f(x_{i+\alpha^*(i,M)})) \le \delta.$$

Let i now be such that (8) is satisfied.

Applying proposition 18.24 and (8), we get that

$$S_{i,\alpha^*(i,M)} \cdot d(x_i, f(x_i)) \le d(x_i, x_{i+\alpha^*(i,M)}) + \delta \cdot P_{i,\alpha^*(i,M)}$$
$$\le d(x_i, x_i^*) + d(x_i^*, x_{i+\alpha^*(i,M)}^*) + d(x_{i+\alpha^*(i,M)}^*, x_{i+\alpha^*(i,M)}) + \delta \cdot P_{i,\alpha^*(i,M)}$$
$$\le 2b + S_{i,\alpha^*(i,M)} \cdot d(x^*, f(x^*)) + \delta \cdot P_{i,\alpha^*(i,M)}, \text{ by the hypothesis and (7)}$$
$$= 2b + S_{i,\alpha^*(i,M)} \cdot \gamma + \delta \cdot P_{i,\alpha^*(i,M)}, \text{ by (1)}.$$

So in total we have shown that

$$(9)\ S_{i,\alpha^*(i,M)} \cdot d(x_i, f(x_i)) \le 2b + S_{i,\alpha^*(i,M)} \cdot \gamma + \delta \cdot P_{i,\alpha^*(i,M)}.$$

We are now ready to prove that i satisfies the claim, i.e.

$$(10)\ d(x_i, f(x_i)) < \gamma + \varepsilon.$$

Suppose that on the contrary $d(x_i, f(x_i)) \ge \gamma + \varepsilon$. This implies that

$$S_{i,\alpha^*(i,M)}(\gamma + \varepsilon) \le S_{i,\alpha^*(i,M)} \cdot d(x_i, f(x_i))$$

and so – using (9) –

$$S_{i,\alpha^*(i,M)}(\gamma+\varepsilon) \leq 2b + S_{i,\alpha^*(i,M)} \cdot \gamma + \delta \cdot P_{i,\alpha^*(i,M)}.$$

Thus

$$(11)\ S_{i,\alpha^*(i,M)} \cdot \varepsilon \leq 2b + \delta \cdot P_{i,\alpha^*(i,M)}.$$

Putting things together we conclude

$$\begin{aligned}
1+2b &\leq M \cdot \varepsilon && \text{by (2)}\\
&\leq S_{i,\alpha^*(i,M)} \cdot \varepsilon && \text{by (6)}\\
&\leq 2b + \delta \cdot P_{i,\alpha^*(i,M)} && \text{by (11)}\\
&\leq 2b + \delta \cdot \exp(K \cdot S_{i,\alpha^*(i,M)}) && \text{by (4)}\\
&< 2b + \delta \cdot \exp(K(M+1)) && \text{by (6)}\\
&< 2b+1 && \text{by (3).}
\end{aligned}$$

This contradiction concludes the proof of (10).

Summarizing, we have proved that if $i \in \mathbb{N}$ satisfies

$$(8)\ d(x_i, f(x_i)) - d(x_{i+\alpha^*(i,M)}, f(x_{i+\alpha^*(i,M)})) \leq \delta,$$

then

$$(10)\ d(x_i, f(x_i)) < \gamma + \varepsilon.$$

It remains to show that the bound in the theorem is indeed an upper bound below which we can find an i satisfying (8).

Define $\tilde{\alpha}^*$, $\widehat{\alpha^*} : \mathbb{N} \times \mathbb{N} \to \mathbb{N}$ by

$$\tilde{\alpha}^*(k,n) := k + \alpha^*(k,n) \text{ and}$$

$$\widehat{\alpha^*}(0,n) := \tilde{\alpha}^*(0,n) \text{ and } \widehat{\alpha^*}(k+1,n) := \tilde{\alpha}^*(\widehat{\alpha^*}(k,n),n).$$

Since $\widehat{\alpha^*}(k+1,n) = \tilde{\alpha}^*(\widehat{\alpha^*}(k,n),n) = \widehat{\alpha^*}(k,n) + \alpha^*(\widehat{\alpha^*}(k,n),n) \geq \widehat{\alpha^*}(k,n)$, it follows that for all $k,n \in \mathbb{N}$,

$$(12)\ \widehat{\alpha^*}(k,n) \leq \widehat{\alpha^*}(k+1,n).$$

Claim: Let $j := \left\lceil \frac{d(x,f(x))}{\delta} \right\rceil \dot{-} 1$. For all $n \in \mathbb{N}$,

$$(13)\ \exists k \leq j \big(d(x_{\widehat{\alpha^*}(k,n)}, f(x_{\widehat{\alpha^*}(k,n)})) - d(x_{\widehat{\alpha^*}(k+1,n)}, f(x_{\widehat{\alpha^*}(k+1,n)})) \leq \delta\big).$$

Proof of Claim: Let $n \in \mathbb{N}$. For $k \in \mathbb{N}$ we define

$$T_k := d(x_{\widehat{\alpha^*}(k,n)}, f(x_{\widehat{\alpha^*}(k,n)})) - d(x_{\widehat{\alpha^*}(k+1,n)}, f(x_{\widehat{\alpha^*}(k+1,n)})).$$

Now suppose that the claim would be false. Then $T_k > \delta$ for all $k \leq j$, and so

$$\sum_{k=0}^{j} T_k > \delta \cdot (j+1).$$

Hence

$$d(x_{\widehat{\alpha^*}(0,n)}, f(x_{\widehat{\alpha^*}(0,n)})) - d(x_{\widehat{\alpha^*}(j+1,n)}, f(x_{\widehat{\alpha^*}(j+1,n)})) \\ > \delta \cdot (j+1) \geq \delta \cdot \left\lceil \frac{d(x,f(x))}{\delta} \right\rceil \geq d(x,f(x)).$$

This, in particular, implies that

$$d(x_{\widehat{\alpha^*}(0,n)}, f(x_{\widehat{\alpha^*}(0,n)})) > d(x,f(x)),$$

which contradicts the fact that the sequence $(d(x_n, f(x_n)))_{n\in\mathbb{N}}$ is nonincreasing (recall that $x_0 := x$) and so concludes the proof of the claim.

By the claim, now choose a $k \leq j$ satisfying (13) with $n := M$ and let $i := \widehat{\alpha^*}(k,M)$. Then by (13) and the definition of $\widehat{\alpha^*}$, it follows immediately that i satisfies (8) and hence (by the reasoning above) also (10).

Define

$$h(\varepsilon,b,\tilde{b},K,\alpha^*) := \widehat{\alpha^*}(\lceil 2\tilde{b} \cdot \exp(K(M+1)) \rceil \dot{-} 1, M)$$

and fix (as above) $\delta := \frac{1}{2\exp(K(M+1))}$. Then

$$\frac{d(x,f(x))}{\delta} = 2d(x,f(x)) \cdot \exp(K(M+1)) \leq 2\tilde{b} \cdot \exp(K(M+1))$$

and, consequently,

$$k \leq \left\lceil \frac{d(x,f(x))}{\delta} \right\rceil \dot{-} 1 \leq \lceil 2\tilde{b} \cdot \exp(K(M+1)) \rceil \dot{-} 1.$$

Applying (12), it follows that $i \leq h(\varepsilon,b,\tilde{b},K,\alpha^*)$. Since i satisfies (10) and $(d(x_m, f(x_m)))_{m\in\mathbb{N}}$ is nonincreasing we conclude from this that

$$(14)\ \forall n \geq h(\varepsilon,b,\tilde{b},K,\alpha^*)(d(x_n,f(x_n)) < d(x^*,f(x^*)) + \varepsilon).$$

So we have obtained the conclusion of the theorem but with α^* instead of α. We now use a majorization argument to replace α^* by any α satisfying the more liberal requirements from the hypothesis of the theorem, i.e.

$$(15)\ \forall i,n \in \mathbb{N}\big((\alpha(i,n) \leq \alpha(i+1,n)) \wedge (n \leq \sum_{s=i}^{i+\alpha(i,n)-1} \lambda_s)\big).$$

From $n \leq S_{i,\alpha(i,n)}$ and the definition of α^* it follows that for all $i,n \in \mathbb{N}$

$$(16)\ \alpha^*(i,n) \le \alpha(i,n).$$

We now show that for all $i,n \in \mathbb{N}$

$$(17)\ \widehat{\alpha^*}(i,n) \le \widehat{\alpha}(i,n).$$

Let $n \in \mathbb{N}$ be fixed. We proceed by induction on i : the case $i = 0$ is clear since

$$\widehat{\alpha^*}(0,n) = \tilde{\alpha}^*(0,n) = \alpha^*(0,n) \le \alpha(0,n) = \tilde{\alpha}(0,n) = \widehat{\alpha}(0,n).$$

Suppose that $\widehat{\alpha^*}(i,n) \le \widehat{\alpha}(i,n)$. The induction step follows using (16) and the fact that, by assumption, α is nondecreasing in the first argument. Indeed

$$\widehat{\alpha^*}(i+1,n) = \tilde{\alpha}^*(\widehat{\alpha^*}(i,n),n) = \widehat{\alpha^*}(i,n) + \alpha^*(\widehat{\alpha^*}(i,n),n) \le$$
$$\widehat{\alpha}(i,n) + \alpha(\widehat{\alpha^*}(i,n),n) \le \widehat{\alpha}(i,n) + \alpha(\widehat{\alpha}(i,n),n) = \tilde{\alpha}(\widehat{\alpha}(i,n),n) = \widehat{\alpha}(i+1,n).$$

Using (17) we can conclude that

$$\begin{aligned} h(\varepsilon,b,\tilde{b},K,\alpha^*) &= \widehat{\alpha^*}(\lceil 2\tilde{b}\cdot\exp(K(M+1))\rceil \dot{-} 1, M) \\ &\le \widehat{\alpha}(\lceil 2\tilde{b}\cdot\exp(K(M+1))\rceil \dot{-} 1, M) \\ &= h(\varepsilon,b,\tilde{b},K,\alpha). \end{aligned}$$

Hence (14) implies

$$\forall n \ge h(\varepsilon,b,\tilde{b},K,\alpha)(d(x_n,f(x_n)) < d(x^*,f(x^*)) + \varepsilon)$$

which concludes the proof of the theorem. □

Remark 18.31. If f is nonexpansive, applying lemma 18.28, it follows that the sequence $(d(x_n,x_n^*))_{n\in\mathbb{N}}$ is nonincreasing, so letting $b \ge d(x,x^*)$ we get that

$$\forall n \in \mathbb{N}\big(d(x_n,x_n^*) \le b\big).$$

Hence, theorem 18.30 holds with

$$h(\varepsilon,b,\tilde{b},K,\alpha) = \widehat{\alpha}(\lceil 2\tilde{b}\cdot\exp(K(M+1))\rceil \dot{-} 1, M), \text{ where}$$
$$M := \left\lceil \tfrac{1+2b}{\varepsilon} \right\rceil,\ \tilde{b} \ge d(x,f(x)) \text{ and}$$
$$\tilde{\alpha} \text{ and } \widehat{\alpha} \text{ are as above.}$$

It is this bound (together with remark 18.32 below) whose general form was guaranteed already in application 18.17 by our metatheorems from the previous chapter.

Remark 18.32. 1) The condition on α in theorem 18.30 to be monotone in the first argument obviously is not any real restriction: let $\alpha : \mathbb{N} \times \mathbb{N} \to \mathbb{N}$ be such that

$$(*)\ \forall i,n \in \mathbb{N}\big(n \leq \sum_{s=i}^{i+\alpha(i,n)-1} \lambda_s\big),$$

then $\alpha^+ : \mathbb{N} \times \mathbb{N} \to \mathbb{N}$ defined by

$$\alpha^+(i,n) := \max_{j \leq i} \alpha(j,n)$$

still satisfies $(*)$ and is nondecreasing in the first argument.

2) Instead of requiring an α satisfying $(*)$ it also is sufficient to have a function $\beta : \mathbb{N} \to \mathbb{N}$ satisfying the weaker requirement

$$(**)\ \ \forall n(n \leq \sum_{s=0}^{\beta(n)} \lambda_s).$$

If we now define $\beta'(i,n) := \beta(n+i) - i + 1$ and $\alpha(i,n) := \max_{j \leq i} \beta'(j,n)$, then α satisfies the conditions in theorem 18.30 (exercise). However, in practice it usually is better to directly choose an α satisfying these conditions than to apply the above construction.

3) Instead of $\lceil 2\tilde{b} \cdot \exp(K(M+1)) \rceil$ one can use any other natural number upper bound $m \geq 2\tilde{b} \cdot \exp(K(M+1))$ in the bound h as well.

As a corollary to theorem 18.30 we get the following (non-quantitative) generalization of the original Borwein-Reich-Shafrir theorem to directionally nonexpansive mappings:

Corollary 18.33. *Let (X,d,W) be a nonempty hyperbolic space and $f : X \to X$ a directionally nonexpansive mapping. Let $(\lambda_n)_{n\in\mathbb{N}}$ be a sequence in $[0,1)$ which is divergent in sum and satisfies that* $\limsup_{n\to\infty} \lambda_n < 1$. *Then for all $x \in X$ if*

$$\forall \varepsilon > 0 \exists x^* \in X\big(d(x_n, x_n^*) \text{ bounded } \wedge d(x^*, f(x^*)) \leq r_X(f) + \varepsilon\big)$$

then

$$d(x_n, f(x_n)) \overset{n\to\infty}{\to} r_X(f).$$

As predicted at the end of application 18.17 we also get a quantitative version of the following version of the Borwein-Reich-Shafrir theorem for sequences (y_n) of points in X instead of a single point x^* only. Recall that whereas (y_n) is an arbitrary sequence of points in X, (x_n) denotes the Krasnoselski-Mann iteration of f starting from x.

Theorem 18.34. *Under the same assumptions on $(X,d,W), (\lambda_n), K, \alpha$ as in theorem 18.30 the following holds: Let $f : X \to X$ be nonexpansive and (b_n) be a sequence of strictly positive real numbers. Then for all $x \in X$, $(y_n)_{n\in\mathbb{N}} \in X^{\mathbb{N}}$ with*

$$\forall n \in \mathbb{N}(d(x, f(x)), d(x, y_n) \leq b_n)$$

and all $\varepsilon > 0$ there exists an $i \leq j(K, \alpha, (b_n)_{n\in\mathbb{N}}, \varepsilon)$ s.t.

$$d(x_i, f(x_i)) < d(y_i, f(y_i)) + \varepsilon,$$

where (omitting the arguments K, α for better readability)

$$j((b_n)_{n\in\mathbb{N}}, \varepsilon) := \max_{i \le \tilde{h}((b_n)_{n\in\mathbb{N}}, \varepsilon)} h(\varepsilon/2, b_i, b_0, K, \alpha)$$

with

$$\tilde{h}((b_n)_{n\in\mathbb{N}}, \varepsilon) := \max_{i<N} g^i(0), \ g(n) := h(\varepsilon/2, b_n, b_0, K, \alpha), N := \left\lceil \frac{6b_0}{\varepsilon} \right\rceil.$$

Here h is the bound from theorem 18.30 and $g^n(0)$ is defined primitive recursively: $g^0(0) := 0, \ g^{n+1}(0) := g(g^n(0))$.
Instead of N, we can take any integer upper bound for $6b_0/\varepsilon$.

Proof: By theorem 18.30 and remark 18.31 we have that

$$(1) \ \forall n \in \mathbb{N}\big(d(x_{g(n)}, f(x_{g(n)})) < d(y_n, f(y_n)) + \frac{\varepsilon}{2}\big),$$

where $g(n) := h(\varepsilon/2, b_n, b_0, K, \alpha)$. Let $N := \left\lceil \frac{6b_0}{\varepsilon} \right\rceil$ and $l := \max_{i<N} g^i(0)$. Using that

$$(2) \ d(y_0, f(y_0)) \le d(y_0, x) + d(x, f(x)) + d(f(x), f(y_0)) \le 2d(y_0, x) + d(x, f(x)) \le 3b_0$$

we now show that

$$(3) \ \exists i < N\big(d(y_{(g^i(0))}, f(y_{(g^i(0))})) \le d(y_{(g^{i+1}(0))}, f(y_{(g^{i+1}(0))})) + \frac{\varepsilon}{2}\big):$$

Suppose not, then for all $i < N$

$$d(y_{(g^{i+1}(0))}, f(y_{(g^{i+1}(0))})) < d(y_{(g^i(0))}, f(y_{(g^i(0))})) - \frac{\varepsilon}{2}$$

and, therefore,

$$d(y_{(g^N(0))}, f(y_{(g^N(0))})) < d(y_0, f(y_0)) - N\frac{\varepsilon}{2} \overset{(2)}{\le} 3b_0 - N\frac{\varepsilon}{2} \le 0,$$

which is a contradiction and finishes the proof of (3).
Let i be as in (3). Then by (1) we get for $p := g^i(0)$

$$(4) \ \forall n \in \mathbb{N}\big(d(x_{g(p)}, f(x_{g(p)})) < d(y_{g(p)}, f(y_{g(p)})) + \varepsilon\big),$$

where $p \le l$. Hence the theorem is satisfied with $j((b_n)_n, \varepsilon) := \max_{i \le l} g(i)$. □

As a first application of the quantitative version of the Borwein-Reich-Shafrir theorem we derive a fully uniform bound on the asymptotic regularity $d(x_n, f(x_n)) \to 0$ in the case of bounded hyperbolic spaces (X, d, W). 'Fully uniform' here means that

the rate of convergence only depends on the error ε, an upper bound b on the metric d and the quantities K,α on λ_k but not on x, f or any special features of X. For the special case of a bounded convex subsets $C \subset X$ of a normed space $(X, \|\cdot\|)$ uniformity in x for constant $\lambda_k := \lambda$ was first established in [93] and for (λ_n) in $[a,b] \subset (0,1)$ and nonincreasing in [67]. In [126], uniformity in x and f has been proved for in the setting of bounded spaces X of hyperbolic type for general λ_k and b, but no uniformity in X or λ_k. Moreover, no effective bounds were obtained. In ([186], Theorem 1), Kirk established in the context of bounded convex subsets of normed spaces uniformity in x, f for directionally nonexpansive mappings in the case of constant $\lambda_k := \lambda$.
The next result contains all these previous ones as special case:

Corollary 18.35 (Kohlenbach-Leuştean [232]). *Let (X,d,W) be a nonempty and b-bounded hyperbolic space. Let $f : X \to X$ a directionally nonexpansive mapping. Let $(\lambda_n)_{n\in\mathbb{N}}$ be a sequence in $[0,1)$ which is divergent in sum and satisfies*

$$\forall n \in \mathbb{N}\big(\lambda_n \leq 1 - \frac{1}{K}\big)$$

for some $K \in \mathbb{N}$, $K \geq 1$.
Let $\alpha : \mathbb{N}\times\mathbb{N} \to \mathbb{N}$ be such that

$$\forall i,n \in \mathbb{N}\big((\alpha(i,n) \leq \alpha(i+1,n)) \wedge (n \leq \sum_{s=i}^{i+\alpha(i,n)-1} \lambda_s)\big).$$

Then the following holds

$$\forall x \in X \forall \varepsilon > 0 \forall n \geq h(\varepsilon,b,K,\alpha)\big(d(x_n, f(x_n)) \leq \varepsilon\big),$$

where

$$h(\varepsilon,b,K,\alpha) := \widehat{\alpha}(\lceil 2b\cdot\exp(K(M+1))\rceil - 1, M)) \text{ with } M := \left\lceil \tfrac{1+2b}{\varepsilon} \right\rceil \text{ and}$$
$$\widehat{\alpha}(0,n) := \tilde{\alpha}(0,n),\ \widehat{\alpha}(i+1,n) := \tilde{\alpha}(\widehat{\alpha}(i,n),n) \text{ with}$$
$$\tilde{\alpha}(i,n) := i + \alpha(i,n).$$

Proof: Let $x \in X$ and $\varepsilon > 0$. Then for every $x^* \in X$, we have that $d(x_n, x_n^*) \leq b$ and $d(x, f(x)) \leq b$ since X is bounded by b. Hence, for every $x^* \in X$, we can apply Theorem 18.30 to get

$$\forall n \geq h(\varepsilon,b,K,\alpha)(d(x_n, f(x_n)) < d(x^*, f(x^*)) + \varepsilon),$$

where

$$h(\varepsilon,b,K,\alpha) := \widehat{\alpha}(\lceil 2b\cdot\exp(K(M+1))\rceil \dot{-} 1, M), \text{ with}$$
$$M := \left\lceil \tfrac{1+2b}{\varepsilon} \right\rceil \text{ and } \tilde{\alpha}, \widehat{\alpha} \text{ are defined as above.}$$

Let $n \geq h(\varepsilon,b,K,\alpha)$. Since $h(\varepsilon,b,K,\alpha)$ does not depend on x^*, it follows that

$$\forall x^* \in X \big(d(x_n, f(x_n)) < d(x^*, f(x^*)) + \varepsilon\big),$$

hence

$$d(x_n, f(x_n)) \leq \inf\{d(x^*, f(x^*)) \mid x^* \in X\} + \varepsilon,$$

that is

$$d(x_n, f(x_n)) \leq r_X(f) + \varepsilon.$$

Apply now the fact that $r_X(f) = 0$, by corollary 18.27. □

Remark 18.36. Although we work in the more restricted setting of hyperbolic spaces rather than that of spaces of hyperbolic type as used in [126] one actually realizes that the proof above does not use the axiom (*iv*) of hyperbolic spaces since we no longer need lemma 18.28. Hence 18.35 also holds for spaces of hyperbolic type.

Remark 18.37. In corollary 18.35, the bound $h(\varepsilon, b, K, \alpha)$ can be replaced by $h(\varepsilon/b, 1, K, \alpha)$ just by applying the old bound to the modified hyperbolic space, where $d_b(x,y) := \frac{1}{b}d(x,y)$. With (X,d,W) also (X,d_b,W) is a hyperbolic space and (directionally) nonexpansive mappings w.r.t. d again are (directionally) nonexpansive w.r.t. d_b.

Corollary 18.38 (Kohlenbach-Leuştean [232]). *Let $b, \varepsilon > 0$, $K \in \mathbb{N}$, $K \geq 1$, and $\beta : \mathbb{N} \to \mathbb{N}$ be an arbitrary mapping. Then there exists an $N \in \mathbb{N}$ such that for any nonempty b-bounded hyperbolic space (X,d,W), any directionally nonexpansive mapping $f : X \to X$, any sequence $\lambda_n \in [0, 1 - \frac{1}{K}]$ satisfying $n \leq \sum_{s=0}^{\beta(n)} \lambda_s$ (for all $n \in \mathbb{N}$) and any $x \in X$, the following holds*

$$\forall n \geq N \big(d(x_n, f(x_n)) \leq \varepsilon\big).$$

Proof: From $n \leq \sum_{s=0}^{\beta(n)} \lambda_s$ for all $n \in \mathbb{N}$, it follows that $(\lambda_n)_{n\in\mathbb{N}}$ is divergent in sum. Apply remark 18.32 and corollary 18.35. □

Corollary 18.39 (Kohlenbach-Leuştean [232]). *Let (X,d,W) be a nonempty b-bounded hyperbolic space and $f : X \to X$ a directionally nonexpansive mapping. Let $K \in \mathbb{N}, K \geq 2$ and $(\lambda_n)_{n\in\mathbb{N}}$ be a sequence in $[\frac{1}{K}, 1 - \frac{1}{K}]$. Then the following holds:*

$$\forall x \in X \forall \varepsilon > 0 \forall n \geq h(\varepsilon, b, K) \big(d(x_n, f(x_n)) \leq \varepsilon\big),$$

where

$$h(\varepsilon, d, K) := K \cdot M \cdot \lceil 2b \cdot \exp(K(M+1)) \rceil \text{ with } M := \left\lceil \frac{1+2b}{\varepsilon} \right\rceil.$$

Proof: Define $\alpha : \mathbb{N} \times \mathbb{N} \to \mathbb{N}$ by

$$\alpha(i,n) = Kn.$$

Then $\sum\limits_{s=i}^{i+\alpha(i,n)-1} \lambda_s \geq \sum\limits_{s=i}^{i+\alpha(i,n)-1} \frac{1}{K} = \frac{1}{K}\alpha(i,n) = n$ and $\alpha(i,n) = \alpha(i+1,n) = Kn$, so α satisfies the conditions of corollary 18.35.

We also get immediately that

$$\tilde{\alpha}(i,n) = i + \alpha(i,n) = i + Kn \text{ and}$$

$$\widehat{\alpha}(i,n) = K(i+1)n.$$

Applying corollary 18.35, it follows that

$$\forall x \in X \forall \varepsilon > 0 \forall n \geq h(\varepsilon,b,K,\alpha)\big(d(x_n, f(x_n)) \leq \varepsilon\big),$$

where

$$\begin{aligned} h(\varepsilon,b,K,\alpha) &= \widehat{\alpha}(\lceil 2b\cdot\exp(K(M+1))\rceil - 1, M) \\ &= K\cdot M\cdot\lceil 2b\cdot\exp(K(M+1))\rceil \\ &= h(\varepsilon,b,K). \end{aligned}$$

□

Remark 18.40. For the special case of constant $\lambda_n = \lambda \in (0,1)$, normed spaces and nonexpansive functions the exponential bound in corollary 18.39 is not optimal. In fact, [10] establishes – using an extremely complicated proof involving computer aided calculations – an optimal quadratic bound in this special case. Later that computer part (due to P. Paule) was replaced by a classical proof in [300] and a simpler proof for the quadratic bound in the case $\lambda = \frac{1}{2}$ was given in [59]. However, even for normed spaces and nonexpansive mappings the bounds presented above (and first established in [220] using proof mining) are the only effective bounds known at all for non-constant sequences λ_n.

Corollary 18.41. *Let $b,\varepsilon > 0$ and $K \in \mathbb{N}, K \geq 2$. Then there exists an $N \in \mathbb{N}$ such that for any nonempty b-bounded hyperbolic space (X,d,W), any directionally nonexpansive mapping $f : X \to X$, any sequence $(\lambda_n)_{n\in\mathbb{N}}$ in $[\frac{1}{K}, 1-\frac{1}{K}]$ and any $x \in X$, the following holds*

$$\forall n \geq N\big(d(x_n, f(x_n)) \leq \varepsilon\big).$$

Proof: Apply corollary 18.39. □

In the following we extend corollary 18.35 for nonexpansive mappings to the situation where X no longer is required to be bounded but only the existence of a point $x^* \in X$ whose iteration sequence $(x^*_n)_{n\in\mathbb{N}}$ is bounded is assumed. This results in a fully uniform bound on Ishikawa's theorem which only depends on an upper bound b on $d(x,x^*)$ and $d(x^*_n,x^*_m)$ (and ε,K,α). This is of interest since the functional analytic embedding techniques from [126, 186] requires that the whole space X is bounded. The following theorem presents (taken together with remark 18.32) the bound predicted (in its general form) already in application 18.12:

Theorem 18.42 (Kohlenbach-Leuştean [232]). *Let (X,d,W) be a nonempty hyperbolic space and $f : X \to X$ a nonexpansive mapping, $(\lambda_n)_{n\in\mathbb{N}}, \alpha$ and K be as before. Let $b > 0, x, x^* \in X$ be such that*

$$d(x,x^*) \le b \wedge \forall n,m \in \mathbb{N}(d(x_n^*,x_m^*) \le b).$$

Then the following holds

$$\forall \varepsilon > 0 \forall n \ge h(\varepsilon,b,K,\alpha)\big(d(x_n,f(x_n)) \le \varepsilon\big),$$

where

$$h(\varepsilon,b,K,\alpha) := \widehat{\alpha}(\lceil 10b \cdot \exp(K(M+1)) \rceil - 1, M)), \text{ with}$$
$$M := \left\lceil \frac{1+4b}{\varepsilon} \right\rceil \text{ and } \widehat{\alpha} \text{ as before.}$$

Proof: The assumptions on b imply

$$d(x,x_n^*) \le 2b.$$

With the nonexpansiveness of f and using that $x^* = x_0^*$ we obtain

$$d(f(x^*),f(x_n^*)) \le b \text{ and } d(f(x^*),f(x)) \le b.$$

By proposition 18.25 we know that for any $\delta > 0$ there exists an n such that

$$d(x_n^*,f(x_n^*)) \le \delta.$$

Thus

$$d(x,f(x)) \le d(x,x^*) + d(x^*,x_n^*) + d(x_n^*,f(x_n^*)) + d(f(x_n^*),f(x^*)) + d(f(x^*),f(x))$$
$$\le 5b + \delta.$$

Letting δ tend to 0 this yields

$$d(x,f(x)) \le 5b.$$

Let n_δ again be such that

$$d(x_{n_\delta}^*,f(x_{n_\delta}^*)) \le \delta.$$

Applying now theorem 18.30 and remark 18.31 to x and $x_{n_\delta}^*$ with the upper bounds $5b$ and $2b$ on $d(x,f(x))$ and $d(x,x_{n_\delta}^*)$ respectively, we obtain

$$\forall \varepsilon > 0 \forall n \ge h(\varepsilon,b,K,\alpha)\big(d(x_n,f(x_n)) < \delta + \varepsilon\big).$$

Since $\delta > 0$ was arbitrary, the theorem follows. □

If in the proof of theorem 18.42 we add a bound $\tilde{b} \ge d(x,f(x))$ as a new input rather than using $d(x, f(x)) \leq 5b$ and consider only the case $x^* := x$ so that we even have $d(x, x_{n_\delta}) \leq b$ instead of $d(x, x^*_{n_\delta}) \leq 2b$, then we get the following bound:

Theorem 18.43. *Let (X,d,W) be a nonempty hyperbolic space and $f : X \to X$ a nonexpansive mapping, $(\lambda_n)_{n\in\mathbb{N}}, \alpha$ and K be as before. Let $b, \tilde{b} > 0, x \in X$ be such that*

$$d(x, f(x)) \leq \tilde{b} \wedge \forall n, m \in \mathbb{N}(d(x_n, x_m) \leq b).$$

Then the following holds

$$\forall \varepsilon > 0 \forall n \geq h(\varepsilon, b, \tilde{b}, K, \alpha)\big(d(x_n, f(x_n)) \leq \varepsilon\big),$$

where

$$h(\varepsilon, b, \tilde{b}, K, \alpha) := \widehat{\alpha}(\lceil 2\tilde{b} \cdot \exp(K(M+1)) \rceil - 1, M)), \text{ with}$$
$$M := \left\lceil \frac{1+2b}{\varepsilon} \right\rceil \text{ and } \widehat{\alpha} \text{ as before.}$$

In the case of directionally nonexpansive mappings we no longer can derive the estimate $d(x, f(x)) \leq 5b$ used in the proof of theorem 18.42. Even to add an upper bound $\tilde{b} \geq d(x, f(x))$ as an additional input as we did in theorem 18.43 does not help since the proof still relies on remark 18.31 which requires f to be nonexpansive.

In the following we, nevertheless, are able to extend theorem 18.42 to directionally nonexpansive mappings where we will, however, only consider the case where (x_n) itself is bounded (i.e. $x = x^*$) and use an additional assumption which for the case of constant $\lambda_k := \lambda$ though is redundant. Using this we will obtain a different bound on $d(x, f(x))$ which depends on α.

For any $k \in \mathbb{N}$, we define the sequence $((x_k)_m)_{m\in\mathbb{N}}$ by:

$$(x_k)_0 = x_k, \qquad (x_k)_{m+1} = (1 - \lambda_m)(x_k)_m \oplus \lambda_k f((x_k)_m).$$

Hence, for any $k \in \mathbb{N}$, $((x_k)_m)_{m\in\mathbb{N}}$ is the Krasnoselski-Mann iteration starting with x_k.

Remark 18.44. $((x_k)_m)_{m\in\mathbb{N}}$ is not in general a subsequence of $(x_n)_{n\in\mathbb{N}}$. But if $(\lambda_n)_{n\in\mathbb{N}}$ is a constant sequence, $\lambda_n = \lambda$, then $(x_k)_m = x_{k+m}$ for all $m, k \in \mathbb{N}$, hence $((x_k)_m)_{m\in\mathbb{N}}$ is a subsequence of $(x_n)_{n\in\mathbb{N}}$.

Theorem 18.45 (Kohlenbach-Leuştean [232]). *Let (X,d,W) be a nonempty hyperbolic space and $f : X \to X$ a directionally nonexpansive mapping. Let $(\lambda_n)_{n\in\mathbb{N}}$ be a sequence in $[0,1)$ which is divergent in sum and satisfies*

$$\forall n \in \mathbb{N}\big(\lambda_n \leq 1 - \frac{1}{K}\big)$$

for some $K \in \mathbb{N}, K \geq 1$.
Let $\alpha : \mathbb{N} \times \mathbb{N} \to \mathbb{N}$ be such that

$$\forall i,n \in \mathbb{N}\big((\alpha(i,n) \le \alpha(i+1,n)) \wedge (n \le \sum_{s=i}^{i+\alpha(i,n)-1} \lambda_s)\big).$$

Let $b > 0$ and $x \in X$ such that

$$\forall n,k,m \in \mathbb{N}\big(d(x_n,(x_k)_m) \le b\big).$$

Then the following holds

$$\forall \varepsilon > 0 \forall n \ge h(\varepsilon,b,K,\alpha)\big(d(x_n,f(x_n)) \le \varepsilon\big),$$

where

$h(\varepsilon,b,K,\alpha) := \alpha(0,1) + \widehat{\alpha^*}(\lceil 2b \cdot \alpha(0,1) \cdot \exp(K(M+1))\rceil - 1, M)$, *with*
$M := \lceil \frac{1+2b}{\varepsilon} \rceil$ *and* $\widehat{\alpha^*}(0,n) := \tilde{\alpha}^*(0,n)$, $\widehat{\alpha^*}(i+1,n) := \tilde{\alpha}^*(\widehat{\alpha^*}(i,n),n)$ *with*
$\tilde{\alpha}^*(i,n) := i + \alpha^*(i,n)$,
$\alpha^*(i,n) := \alpha(i+\alpha(0,1),n)$ $(i,n \in \mathbb{N})$.

Proof: Since $d(x_n x_m) = d(x_n,(x_0)_m)$ the assumption on b implies that the sequence $(x_n)_{n\in\mathbb{N}}$ is b-bounded, i.e.

$$\forall m,n \in \mathbb{N}\big(d(x_n,x_m) \le b\big).$$

From the second property of α it follows that

$$\sum_{s=0}^{\alpha(0,1)-1} \lambda_s \ge 1.$$

So, clearly, there exists an $N \in \mathbb{N}$, $N \le \alpha(0,1) - 1$ such that

$$\lambda_N \ge \frac{1}{\alpha(0,1)}.$$

Using that $((d(x_n,f(x_n)))_{n\in\mathbb{N}}$ is nonincreasing we conclude that

$$(1)\ d(x_{\alpha(0,1)},f(x_{\alpha(0,1)})) \le d(x_N,f(x_N)) = \frac{1}{\lambda_N} d(x_N,x_{N+1}) \le b \cdot \alpha(0,1).$$

Define $\mu_n := \lambda_{\alpha(0,1)+n}$ for all $n \in \mathbb{N}$. Obviously, $(\mu_n)_{n\in\mathbb{N}}$ still is divergent in sum and $\mu_n \le 1 - \frac{1}{K}$ for all $n \in \mathbb{N}$.
Now consider the Krasnoselski-Mann iteration $(y_n)_{n\in\mathbb{N}}$ of f with starting point $y := x_{\alpha(0,1)}$ associated with $(\mu_n)_{n\in\mathbb{N}}$

$$y_0 := y := x_{\alpha(0,1)}, \qquad y_{n+1} := (1-\mu_n)y_n \oplus \mu_n f(y_n).$$

By an easy induction on n one verifies that

$$y_n = x_{\alpha(0,1)+n}$$

for all $n \in \mathbb{N}$. So

$$\forall m,n \in \mathbb{N}\big(d(y_n,y_m) = d(x_{\alpha(0,1)+n},x_{\alpha(0,1)+m}) \le b\big).$$

Now we are in the position to apply proposition 18.25 to conclude that $\lim\limits_{n\to\infty} d(y_n, f(y_n)) = 0$, i.e.

$$(2)\ \forall \delta > 0 \exists N_\delta \forall n \ge N_\delta \big(d(y_n, f(y_n)) < \delta\big).$$

Let $y^* := y_{N_\delta}$ and $(y_n^*)_{n\in\mathbb{N}}$ be the Krasnoselski-Mann iteration of f starting from y^* based on $(\mu_n)_{n\in\mathbb{N}}$. Then, by the assumption on b,

$$\forall n \in \mathbb{N}\big(d(y_n,y_n^*) = d(x_{\alpha(0,1)+n},(x_{N_\delta+\alpha(0,1)})_n) \le b\big).$$

Define for all $i,n \in \mathbb{N}$,

$$\alpha^*(i,n) := \alpha(i+\alpha(0,1),n).$$

It follows immediately that $\alpha^*(i,n) \le \alpha^*(i+1,n)$ and that

$$\sum_{s=i}^{i+\alpha^*(i,n)-1} \mu_s = \sum_{s=i}^{i+\alpha(i+\alpha(0,1),n)-1} \lambda_{\alpha(0,1)+s} = \sum_{s=i+\alpha(0,1)}^{i+\alpha(0,1)+\alpha(i+\alpha(0,1),n)-1} \lambda_s \ge n.$$

The hypotheses of theorem 18.30 are satisfied with μ_n, α^*, y, y^* instead of $\lambda_n, \alpha, x, x^*$, so we can apply it to get

$$\forall \varepsilon > 0 \forall n \ge h^*(\varepsilon,b,\tilde{b},K,\alpha^*)(d(y_n,f(y_n)) < d(y^*,f(y^*)) + \varepsilon),$$

where

$$h^*(\varepsilon,b,\tilde{b},K,\alpha^*) := \widehat{\alpha^*}(\lceil 2\tilde{b}\cdot\exp(K(M+1))\rceil \dot{-} 1, M), \text{ where}$$
$$M := \left\lceil \tfrac{1+2b}{\varepsilon} \right\rceil \text{ and } \tilde{b} > 0 \text{ is such that } \tilde{b} \ge d(y,f(y)).$$

By (1), we have that

$$d(y,f(y)) = d(x_{\alpha(0,1)}, f(x_{\alpha(0,1)})) \le b\cdot\alpha(0,1)$$

and so we can take $\tilde{b} := b\cdot\alpha(0,1)$.

Now consider

$$h'(\varepsilon,b,K,\alpha) := \widehat{\alpha^*}(\lceil 2b\cdot\alpha(0,1)\cdot\exp(K(M+1))\rceil - 1, M).$$

Applying now (2), it follows that

$$(3)\ \forall \varepsilon > 0 \forall n \geq h'(\varepsilon, b, K, \alpha)(d(y_n, f(y_n)) < \delta + \varepsilon).$$

Since $\delta > 0$ was arbitrary, this yields

$$\forall \varepsilon > 0 \forall n \geq h'(\varepsilon, b, K, \alpha)(d(y_n, f(y_n)) \leq \varepsilon), \text{i.e.}$$

$$\forall \varepsilon > 0 \forall n \geq h'(\varepsilon, b, K, \alpha)(d(x_{\alpha(0,1)+n}, f(x_{\alpha(0,1)+n})) \leq \varepsilon).$$

Finally, letting $h(\varepsilon, b, K, \alpha) := \alpha(0,1) + h'(\varepsilon, b, K, \alpha)$, we get

$$\forall \varepsilon > 0 \forall n \geq h(\varepsilon, b, K, \alpha)(d(x_n, f(x_n)) \leq \varepsilon).$$

□

As mentioned already, the condition

$$\forall n, k, m \in \mathbb{N}(d(x_n, (x_k)_m) \leq b)$$

is equivalent to the boundedness of (x_n) by b

$$\forall n, m \in \mathbb{N}(d(x_n, x_m) \leq b)$$

in the case of constant $\lambda_n = \lambda$. Hence we obtain the following strong uniform version of theorem 2 in [186] (note that theorem 2 in [186] does not state any uniformity of the convergence at all).

Corollary 18.46. *Let (X, d, W) be a nonempty hyperbolic space and $f : X \to X$ a directionally nonexpansive mapping. Let $b > 0$, $K \in \mathbb{N}, K \geq 2$ and $\lambda \in [\frac{1}{K}, 1 - \frac{1}{K}]$. Let $\lambda_n := \lambda$ for all $n \in \mathbb{N}$. Let $x \in X$ such that $d(x_n, x_m) \leq b$ for all $m, n \in \mathbb{N}$. Then the following holds*

$$\forall \varepsilon > 0 \forall n \geq h(\varepsilon, b, K)\big(d(x_n, f(x_n)) \leq \varepsilon\big),$$

where

$$h(\varepsilon, b, K) := K + K \cdot M \cdot \lceil 2b \cdot K \cdot \exp(K(M+1)) \rceil \text{ and } M := \left\lceil \frac{1+2b}{\varepsilon} \right\rceil.$$

Proof: The corollary follows using theorem 18.45 by a reasoning similar to the one used in the proof of corollary 18.39. We leave the details to the reader (alternatively: see [232]). □

The quantitative version of the Ishikawa-Goebel-Kirk theorem provided in theorem 18.43 was obtained from our analysis of the proof of the Borwein-Reich-Shafrir theorem using the truth of the Ishikawa-Goebel-Kirk theorem rather than its proof. We now give a direct logical analysis of the proof of the Ishikawa-Goebel-Kirk theorem yielding another quantitative version of this theorem (as predicted in application 18.12 above but treating as in [126] only the case $x^* := x$). This time we will also be able to weaken the requirement of (x_n) being bounded by not growing to fast (in distance from x). We first present this proof (extended at the same time to directionally nonexpansive mappings).

For the proof of proposition 18.25 we need the following

Lemma 18.47. *Under the hypotheses of proposition 18.18, the following are equivalent*

(i) $(x_n)_{n\in\mathbb{N}}$ *is bounded;*

(ii) $(y_n)_{n\in\mathbb{N}}$ *is bounded;*

(iii) *the set* $\{d(x_n,y_{n+i}) \mid n,i\in\mathbb{N}\}$ *is bounded.*

Proof: Let $n,i\in\mathbb{N}$.
(i)⇒(ii): $d(y_n,y_{n+i})\le d(y_n,x_n)+d(x_n,x_{n+i})+d(x_{n+i},y_{n+i})\le 2d(x_0,y_0)+d(x_n,x_{n+i})$, since $(d(x_n,y_n))_{n\in\mathbb{N}}$ is nonincreasing, by proposition 18.18.
(ii)⇒(iii): $d(x_n,y_{n+i})\le d(x_n,y_n)+d(y_n,y_{n+i})\le d(x_0,y_0)+d(y_n,y_{n+i})$.
(iii)⇒(i): $d(x_n,x_{n+i})\le d(x_n,y_{n+i})+d(y_{n+i},x_{n+i})\le d(x_n,y_{n+i})+d(x_0,y_0)$. □

Proof of proposition 18.25 (Ishikawa-Goebel-Kirk theorem for directionally nonexpansive mappings): Let (X,d,W) be a hyperbolic space, $f:X\to X$ directionally nonexpansive and (λ_n) a sequence in $[0,1-\frac{1}{K}]$ for some K (assuming w.l.o.g. that (λ_n) is bounded away from 1 already from the beginning). Let (λ_n) be divergent in sum and let (x_n) be the Krasnoselski-Mann iteration of f starting from x based on (λ_n). We assume that (x_n) is bounded. By lemma 18.47 there exists a $b\in\mathbb{N}$ such that

$$(0)\ \forall i,n\in\mathbb{N}(b\ge d(x_i,y_{i+n})),$$

where $y_k:=f(x_k)$. By the fact that $(d(x_i,y_i))_n$ is nonincreasing (proposition 18.24) $r:=\lim_{n\to\infty} d(x_n,y_n)$ exists. To show: $r=0$. Assume that $r>0$ and let $\delta>0$ be so small that

$$(1)\ \delta\exp\left(K\left(\frac{b}{r}+1\right)\right)\le r.$$

Choose $i\in\mathbb{N}$ so that

$$(2)\ \forall j\in\mathbb{N}\big(d(x_i,y_i)-d(x_{i+j},y_{i+j})\le\delta\big).$$

Let $n\in\mathbb{N}$ be so that

$$(3)\ r\cdot S_{i,n-1}\le b\le r\cdot S_{i,n}.$$

Then

$$(4)\ r\cdot S_{i,n}<b+r.$$

In the proof of the Borwein-Reich-Shafrir theorem above we already established that

$$(5)\ P_{i,n}\le\exp(K\cdot S_{i,n}).$$

By the inequality of Goebel and Kirk stated in proposition 18.18 and the choices of $\delta>0$ and $i,n\in\mathbb{N}$ we get

$$b+r \overset{(3)}{\leq} (1+S_{i,n})r \leq (1+S_{i,n})d(x_i,y_i)$$
$$\overset{(2),(5),\text{prop.18.18}}{\leq} d(x_i,y_{i+n}) + \delta \cdot \exp(K \cdot S_{i,n})$$
$$\overset{(0),(4)}{<} b+\delta \cdot \exp\left(K\left(\tfrac{b}{r}+1\right)\right) \overset{(1)}{\leq} b+r.$$

This contradiction shows that $r = 0$. □

Remark 18.48. Inspection of the proof above shows that it also establishes theorem 18.21.

Logical analysis of the proof: as in the case of the proof of the Borwein-Reich-Shafrir theorem it is clear that the proof above can be formalized in $\mathcal{A}^\omega[X,d,W]_{-b}$ using PCM (i.e. arithmetical comprehension) to show the existence of r. Again (and much easier than in the case of the proof of the Borwein-Reich-Shafrir theorem) it even follows that instead of PCM we only need the Cauchy property PCM_{ar} of $(d(x_n,y_n))_n$ (to satisfy (2)): this is verified by replacing the assumption that $r > 0$ by

$$\forall n \in \mathbb{N}(d(x_n,y_n) \geq \varepsilon)$$

for some arbitrarily chosen $\varepsilon > 0$. The proof then goes through unchanged with r replaced by ε. If $\tilde{b} \geq d(x_0,y_0)$ and b is a bound on (x_n) then (by the proof of lemma 18.47) $3\tilde{b}+b$ is a bound on $d(x_i,y_{i+n})$ for all $i,n \in \mathbb{N}$. As δ we, therefore, may take

$$\delta := \frac{\varepsilon}{\exp\left(K \cdot \left(\frac{3\tilde{b}+b}{\varepsilon}+1\right)\right)}.$$

Put

$$M := \left\lceil \frac{3\tilde{b}+b}{\varepsilon} \right\rceil$$

and define $n \in \mathbb{N}$ as

$$n := \alpha^*(i,M) := \min k \left[\frac{3\tilde{b}+b}{\varepsilon} \leq S_{i,k}\right].$$

Then $3\tilde{b}+b \leq \varepsilon \cdot S_{i,n}$ and $\varepsilon \cdot S_{i,n-1} \leq 3\tilde{b}+b$ and so also $\varepsilon \cdot S_{i,n} < 3\tilde{b}+b+\varepsilon$. Let $i \in \mathbb{N}$ be such that

$$d(x_i,y_i) - d(x_{i+\alpha^*(i,M)},y_{i+\alpha^*(i,M)}) \leq \delta.$$

As in the proof above one concludes that $d(x_i,y_i) < \varepsilon$. We now can reason exactly as in the second half of the proof of theorem 18.30 (logical analysis of the Borwein-Reich-Shafrir theorem) to conclude that the following function provides a bound on i:

$$h^*(\varepsilon,b,\tilde{b},K,\alpha) := \widehat{\alpha}\left(\left\lceil \frac{\tilde{b}\cdot\exp\left(K\cdot\left(\frac{3\tilde{b}+b}{\varepsilon}+1\right)\right)}{\varepsilon}\right\rceil \dot{-} 1, M\right),$$

where $\widehat{\alpha}$ as before. Thus

$$\forall n \geq h^*(\varepsilon,b,\tilde{b},K,\alpha) \; \big(d(x_n,f(x_n)) \leq \varepsilon)\big).$$

Inspection of the proof shows that instead of the fact that the whole sequence (x_n) is bounded by b we have only used that for $i \leq h^*(\varepsilon,b,\tilde{b},K,\alpha)$

$$d(x_i,x_{i+\alpha^*(i,M)}) \leq b$$

to conclude (using again the proof of lemma 18.47) that

$$d(x_i,y_{i+\alpha^*(i,M)}) \leq 3\tilde{b}+b.$$

For $i \leq h^*(\varepsilon,b,\tilde{b},K,\alpha)$ it follows from (15) and (16) in the proof of theorem 18.30 that

$$\alpha^*(i,M) \leq \alpha(h^*(\varepsilon,b,\tilde{b},K,\alpha),M).$$

So instead of $\forall i,j\; (d(x_i,x_j) \leq b)$, actually the assumption

$$\forall i \leq h^*(\varepsilon,b,\tilde{b},K,\alpha) \forall j \leq \alpha(h^*(\varepsilon,b,\tilde{b},K,\alpha),M) \; \big(d(x_i,x_{i+j}) \leq b\big)$$

suffices for the above conclusion. Thus in total we have shown the following:

Theorem 18.49. *Direct logical analysis of the proof of the Ishikawa-Goebel-Kirk theorem shows the following: Let (X,d,W) be a nonempty hyperbolic space, $f : X \to X$ a directionally nonexpansive mapping, $(\lambda_n), K, \alpha$ as in theorem 18.30, $x \in X$ and (x_n) the Krasnoselski-Mann iteration of f starting from x and $\tilde{b} > 0$ so that $d(x,f(x)) \leq \tilde{b}$. Let h^* be defined as above. Then for every $\varepsilon, b > 0$ the following holds (abbreviating $h^*(\varepsilon,b,\tilde{b},K,\alpha)$ by h^*):*

$$\forall i \leq h^* \forall j \leq \alpha(h^*,M) \; (d(x_i,x_{i+j}) \leq b) \to \forall n \geq h^* \; \big(d(x_n,f(x_n)) < \varepsilon\big).$$

For the case of sequences (λ_n) in $[a,b]$ for $0 < a < b < 1$ we obtain from theorem 18.49 the following qualitative improvement of the Ishikawa-Goebel-Kirk theorem concerning the requirement of (x_n) being bounded (which for the case of constant $\lambda_n := \lambda \in (0,1)$ and convex subsets of normed spaces was first observed in [11] (theorem 2.1)):

Theorem 18.50. *Let (X,d,W) be a hyperbolic space and $f : X \to X$ nonexpansive. For $x \in X$ and (λ_n) in $[a,b]$, where $0 < a < b < 1$, let (x_n) be the corresponding Krasnoselski-Mann iteration of f starting from x. Let*

$$c(n) := \max\{d(x,x_j) : j \leq n\}.$$

Then

$$\lim_{n\to\infty} \frac{c(n)}{n} \to 0$$

implies that

$$\lim_{n\to\infty} d(x_n, f(x_n)) = 0.$$

Proof: Let $K \in \mathbb{N}$, $K \geq 2$, so that $\lambda_n \in [\frac{1}{K}, 1 - \frac{1}{K}]$ for all $n \in \mathbb{N}$ and $\varepsilon > 0$. As in the proof of corollary 18.39 we can use in this case in the previous result $\alpha(i,M) := K \cdot M = K \cdot \left\lceil \frac{3\tilde{b}+b}{\varepsilon} \right\rceil$. Let $\tilde{b} \geq 1$ such that $d(x, f(x)) \leq \tilde{b}$. By the assumption that $\lim \frac{c(n)}{n} \overset{n\to\infty}{\to} 0$ it follows that

$$\exists N \forall n \geq N \big(c(n) \leq \frac{\varepsilon}{3K\tilde{b}} \cdot n\big).$$

Let $b \geq 2$ be such that $\mathbb{N} \ni \frac{K(3\tilde{b}+b)}{\varepsilon} \geq N$. Then

$$c\left(\frac{K(3\tilde{b}+b)}{\varepsilon}\right) \leq \frac{K(3\tilde{b}+b)}{\varepsilon} \cdot \frac{\varepsilon}{3K\tilde{b}} \leq b.$$

Hence

$$\forall j \leq K \cdot M \ (d(x, x_j) \leq b).$$

Since $x_{i+n} = (x_n)_i$, where $((x_n)_i)_i$ denotes the Krasnoselski-Mann iteration starting from $x^* := x_n$, we can apply lemma 18.28 to conclude that

$$\forall i, n \in \mathbb{N} \ (d(x_i, x_{i+n}) \leq d(x, x_n)).$$

So put together we have that

$$\forall i \in \mathbb{N} \forall j \leq K \cdot M \ (d(x_i, x_{i+j}) \leq b).$$

Hence by the previous result we obtain $d(x_n, f(x_n)) < \varepsilon$ for all $n \geq h^*(\cdot)$.
Since $\varepsilon > 0$ was arbitrary, the conclusion follows. □

Remark 18.51. The previous result shows that $d(x_n, f(x_n)) \to 0$ provided (x_n) grows with a lower than linear (in n) rate. This is optimal in the sense that linear growth does not suffice as follows from the following simple example: $X := \mathbb{R}, f(x) := x+1$ and $\lambda := \frac{1}{2}$. For the starting point $x_0 := 0$ we have for the Krasnoselski-Mann iteration (x_n) that $x_n = \frac{n}{2}$, but $d(x_n, f(x_n)) = 1$ for all $n \in \mathbb{N}$.

Discussion: The bound obtained in theorem 18.49 is very similar to the one we got combining our quantitative version of the Borwein-Reich-Shafrir theorem with the Ishikawa-Goebel-Kirk theorem in theorem 18.43. However, while the latter applied to general x^* and not only $x^* := x$ the former also extended to directionally nonexpansive mappings without any extra assumption as we needed in theorem 18.45. The main difference, however, is that in the above direct analysis we could also investigate how much of the boundedness assumptions of (x_n) is actually needed for the proof of the Ishikawa-Goebel-Reich theorem, whereas in the approach via the Borwein-Reich-Shafrir theorem this was blocked since we used only the truth of the former (rather than its proof).

We continue our case study of proof mining in fixed point theory by establishing the bound Ω predicted already in application 18.16 as an instance of the metatheorems derived in chapter 17.

We recall the definition of the concept of modulus of total boundedness in a slightly different from:

Definition 18.52. Let (M,d) be a totally bounded metric space. We call $\gamma : \mathbb{N} \to \mathbb{N}$ a modulus of total boundedness for M if for any $k \in \mathbb{N}$ there exist elements $a_0, \ldots, a_{\gamma(k)} \in M$ such that

$$\forall x \in M \exists i \leq \gamma(k) \big(d(x, a_i) \leq 2^{-k}\big).$$

Remark 18.53. Every modulus of total boundedness in the sense of application 18.16 is also one in the sense of the previous definition. Conversely, if γ is a modulus in the sense of the previous definition then $k \cdot \delta(k)$ is one in the sense of application 18.16.

Definition 18.54. Let (M,d) be a metric space, $f : M \to M$ a selfmapping of M and (x_n) a sequence in M. A function $\delta : \mathbb{N} \to \mathbb{N}$ is called an approximate fixed point bound for (x_n) if

$$\forall k \in \mathbb{N} \exists m \leq \delta(k) \big(d(x_m, f(x_m)) \leq 2^{-k}\big).$$

Of course, an approximate fixed point bound only exists is (x_n) contains arbitrarily good approximate fixed points.

In the following, (X,d,W) is a nonempty hyperbolic space, $f : X \to X$ is a nonexpansive selfmapping of X and (λ_n) a sequence in $[0,1]$. (x_n) denotes the corresponding Krasnoselski-Mann iteration starting from $x_0 \in X$.

Remark 18.55. As we have proved already above, the sequence $(d(x_n, f(x_n))$ is always nonincreasing. Hence any approximate fixed point bound Φ for (x_n) is in fact a rate of convergence for $d(x_n, f(x_n))$.

Lemma 18.56. *Let $\varepsilon > 0$ and $u \in X$ be an ε-fixed point of f, i.e. $d(u, f(u)) \leq \varepsilon$. Then*

$$\forall n, m \in \mathbb{N} (d(x_{n+m}, u) \leq d(x_n, u) + m \cdot \varepsilon).$$

Proof: Let $n \in \mathbb{N}$ be fixed. We proceed by induction on $\mathbb{N}$:
For $m = 0$, the lemma trivially is true.
$m \mapsto m+1$:

$$\begin{aligned}
&d(x_{n+m+1}, u) = d((1-\lambda_{n+m})x_{n+m} \oplus \lambda_{n+m} f(x_{n+m}), u) \\
&\overset{W\text{-axiom (i)}}{\leq} (1-\lambda_{n+m}) d(x_{n+m}, u) + \lambda_{n+m} d(f(x_{n+m}), u) \\
&\leq (1-\lambda_{n+m}) d(x_{n+m}, u) + \lambda_{n+m} d(f(x_{n+m}), f(u)) + \lambda_{n+m} d(f(u), u) \\
&\overset{f \text{ n.e.}}{\leq} (1-\lambda_{n+m}) d(x_{n+m}, u) + \lambda_{n+m} d(x_{n+m}, u) + \varepsilon \overset{\text{I.H.}}{\leq} d(x_n, u) + (m+1)\varepsilon.
\end{aligned}$$

$\square$

Notation. 18.57 *For $n,m \in \mathbb{N}$ with $m \geq n$, we use $[n;m]$ to denote the set $\{n,n+1,\ldots,m\} \subset \mathbb{N}$.*

We now assume that X is totally bounded.

Theorem 18.58. *Let $k \in \mathbb{N}, g : \mathbb{N} \to \mathbb{N}, \delta : \mathbb{N} \to \mathbb{N}$ and $\gamma : \mathbb{N} \to \mathbb{N}$. We define a function $\Omega(k,g,\delta,\gamma)$ (primitive) recursively as follows:*

$$\Omega(k,g,\delta,\gamma) := \max_{i \leq \gamma(k+3)} \Psi_0(i,k,g,\delta),$$

where

$$\begin{cases} \Psi_0(0,k,g,\delta) := 0 \\ \Psi_0(n+1,k,g,\delta) := \delta\left(k+2+\lceil \log_2(\max_{l \leq n} g(\Psi_0(l,k,g,\delta))+1)\rceil\right). \end{cases}$$

If δ is an approximate fixed point bound for (x_n) and γ a modulus of total boundedness for X, then

$$\forall k \in \mathbb{N} \forall g : \mathbb{N} \to \mathbb{N} \exists n \leq \Omega(k,g,\delta,\gamma) \forall i,j \in [n;n+g(n)] \big(d(x_i,x_j) \leq 2^{-k}\big).$$

Proof: We first note that by remark 18.55 $x_{\delta(k)}$ is a 2^{-k}-fixed point for any $k \in \mathbb{N}$. Define $n_i := \Psi_0(i,k,g,\delta)$.
Claim: $\exists i,j \leq \gamma(k+3)+1, i \neq j(d(x_{n_i},x_{n_j}) \leq 2^{-k-2})$.
Proof of claim: By the assumption on γ it follows that there exist points $a_0,\ldots,a_{\gamma(k+3)} \in X$ such that for at least two of the $(\gamma(k+3)+2)$-many indices $0 \leq i \leq \gamma(k+3)+1$ the corresponding x_{n_i}'s must be in a common 2^{-k-3}-ball around some a_l with $l \leq \gamma(k+3)$, i.e.

$$\begin{aligned} \exists i,j \leq \gamma(k+3)+1, i \neq j, \exists l \leq \gamma(k+3) : \\ d(a_l,x_{n_i}) \leq 2^{-k-3} \wedge d(a_l,x_{n_j}) \leq 2^{-k-3} \end{aligned}$$

and hence $d(x_{n_i},x_{n_j}) \leq 2^{-k-2}$.
End of the proof of the claim.
By the claim, let $i < j \leq \gamma(k+3)+1$ be such that

$$d(x_{n_i},x_{n_j}) \leq 2^{-k-2}.$$

By construction and $j > 0$, x_{n_j} is a $(2^{-k-2-\lceil \log_2(\max_{l<j} g(\Psi_0(l,k,g,\delta))+1)\rceil})$-fixed point of f and hence a-fortiori a $(2^{-k-2-\lceil \log_2(g(\Psi_0(i,k,g,\delta))+1)\rceil}) = (2^{-k-2-\lceil \log_2(g(n_i)+1)\rceil})$-fixed point of f. By the lemma above we, therefore, obtain for all $l \leq g(n_i)$:

$$\begin{aligned} d(x_{n_i+l}, x_{n_j}) &\leq d(x_{n_i}, x_{n_j}) + l \cdot 2^{-k-2-\lceil \log_2(g(n_i)+1) \rceil} \\ &\leq d(x_{n_i}, x_{n_j}) + 2^{-k-2} \leq 2^{-k-1}. \end{aligned}$$

Thus

$$\forall j_1, j_2 \in [n_i; n_i + g(n_i)](d(x_{j_1}, x_{j_2}) \leq 2^{-k}),$$

where $i \leq \gamma(k+3)$. Since $\Omega(k, g, \delta, \gamma) = \max\{n_i : i \leq \gamma(k+3)\}$, the theorem follows. □

Remark 18.59. Note that Ψ is a primitive recursive functional in the sense of Kleene [194] which corresponds to proposition 13.27.

If (λ_n) is in $[0, 1 - \frac{1}{K}]$ (for some $K \in \mathbb{N}, K \geq 1$) and divergent in sum with $\forall n(n \leq \sum_{i=0}^{\alpha(n)} \lambda_i)$ and X is b-bounded then we can take the approximate fixed point bound δ as $\delta(k) := h(2^{-k}, b, K, \alpha)$ where h is the bound from corollary 18.35.

18.4 Asymptotically nonexpansive mappings

As discussed already earlier in this chapter, asymptotic regularity results prior to Ishikawa's theorem [176] were proved only in the special context of uniformly convex normed spaces (see e.g. Krasnoselski [238] and Browder-Petryshyn [56]). While these results to a large extent are superseded by Ishikawa's theorem the particular proof technique used first in Krasnoselski's pioneering paper [238] is still of interest for the following reasons (among others):

- The technique can be used to show asymptotic regularity in uniformly convex normed spaces for Krasnoselski-Mann iterations based on sequences of scalars (λ_n) in $[0, 1]$ that only have to satisfy the condition

  $$\sum_{i=0}^{\infty} \lambda_i(1 - \lambda_i) = \infty.$$

  This was shown in Groetsch [144] for the normed case and generalized in Leuştean [263] to the hyperbolic case.
  The above condition is more general than the condition in Ishikawa's theorem and, in fact, is optimal even in the special case of Hilbert spaces.
- In uniformly convex spaces one can generalize many results to the class of asymptotically nonexpansive mappings (see below) while for the general normed case almost nothing is known for this class. However, one now has to require that $\lambda_n \in [\frac{1}{L}, 1 - \frac{1}{L}]$ for some $L \geq 2$.
- The bounds extracted from such proofs that are based on the uniform convexity usually are very simple constructions in the modulus of uniform convexity and, in fact, often yield quadratic bounds in the case of Hilbert spaces (see

[219, 224, 231]). Since the proofs usually generalize to uniformly convex hyperbolic spaces one even gets such quadratic bounds in the case of CAT(0)-spaces (see [263, 234]). Only for the special case of constant $\lambda_n := \lambda \in (0,1)$ and the normed setting quadratic bounds could be obtained so far without assuming uniform convexity by the deep work of Baillon-Bruck [10].

Typical for all these asymptotic regularity proofs relying on uniform convexity is that they assume the existence of a fixed point of the function in question. In the bounded case this assumption usually can be shown to be true using appropriate fixed point theorems. At the time of Krasnoselski's paper [238] only Schauder's fixed point theorem was available and hence he had to restrict himself to the compact case. Browder and Petryshyn [56] could rely on the Browder-Göhde-Kirk fixed point theorem that was established shortly before and so replaced compactness by 'closed and bounded'. For asymptotically nonexpansive mappings a fixed point theorem corresponding to the Browder-Göhde-Kirk theorem was established in Goebel-Kirk [125] which introduced this class of functions. By the corollary 17.59 (that can be adapted to the uniformly convex hyperbolic case [262] and which also holds for the uniformly convex normed case [120]) it follows that in the course of an asymptotic regularity proofs the assumption of the existence of a fixed point can be replaced by the existence of approximate fixed points in some fixed neighborhood. In the bounded case, the latter condition usually can easily be established (and e.g. for the normed case and nonexpansive mappings trivially follows, see proposition 18.11) without the use of completeness/closedness-assumptions. In particular, one obtains by logical analysis of Krasnoselski's original proof not only an explicit rate of asymptotic regularity but even a totally elementary proof of asymptotic regularity (at the same time generalized from the compact to the bounded case) that neither needs Schauder's fixed point theorem nor the Browder-Göhde-Kirk fixed point theorem and that could have been conceived of already by Banach (see Kohlenbach [219]).

In the following we present (without proof) the maybe most advanced result of this type so far achieved by the logical metatheorems proved in chapter 17. This result is concerned with the aforementioned asymptotically nonexpansive mappings:

Definition 18.60. Let (X,d) be a metric space. A function $f : X \to X$ is called asymptotically nonexpansive if for some sequence (k_n) in $[0,\infty)$ with $\lim_{n\to\infty} k_n = 0$ one has (with f^n denoting the n-th iteration of f)

$$d(f^n x, f^n y) \le (1+k_n)d(x,y), \quad \forall n \in \mathbb{N}, \forall x,y \in X.$$

In the case of asymptotically nonexpansive mappings one considers the following version of the Krasnoselski-Mann iteration:

$$x_0 := x,\ x_{n+1} := (1-\lambda_n)x_n \oplus \lambda_n f^n(x_n).$$

We recall the definition of uniformly convex hyperbolic spaces from definition 17.109 in chapter 17:

Definition 18.61 ([128, 262]). A hyperbolic space (X,d,W) is uniformly convex if for any $r > 0$ and any $\varepsilon \in (0,2]$ there exists $\delta \in (0,1]$ such that for all $a,x,y \in X$,

$$\left.\begin{array}{l} d(x,a) \leq r \\ d(y,a) \leq r \\ d(x,y) \geq \varepsilon r \end{array}\right\} \quad \Rightarrow \quad d\left(\frac{1}{2}x \oplus \frac{1}{2}y, a\right) \leq (1-\delta)r. \tag{18.1}$$

A mapping $\eta : (0,\infty) \times (0,2] \to (0,1]$ providing such a $\delta := \eta(r,\varepsilon)$ for given $r > 0$ and $\varepsilon \in (0,2]$ is called a *modulus of uniform convexity*.
We say that η is monotone if it decreases with r (for any fixed ε).

Using an appropriate version of the metatheorems from chapter 17 the following result was recently extracted from an asymptotic regularity proof that – for the normed case – essentially was implicit in the literature (see Kohlenbach-Lambov [231] for details) and which could be generalized to the hyperbolic case. For the uniformly convex normed case the theorem below was first established in Kohlenbach-Lambov [231].

Theorem 18.62 (Kohlenbach-Leuştean [234]). *Let (X,d,W) be a uniformly convex hyperbolic space with a monotone modulus of uniform convexity η and $f : X \to X$ be asymptotically nonexpansive with sequence (k_n).*
Assume that $K \geq 0$ is such that $\sum_{n=0}^{\infty} k_n \leq K$ and that $L \in \mathbb{N}, L \geq 2$ is such that $\frac{1}{L} \leq \lambda_n \leq 1 - \frac{1}{L}$ for all $n \in \mathbb{N}$.
Let $x \in X$ and $b > 0$ be such that for any $\delta > 0$ there is $p \in X$ with

$$d(x,p) \leq b \wedge d(f(p),p) \leq \delta.$$

Then for all $\varepsilon \in (0,1]$ and for all $g : \mathbb{N} \to \mathbb{N}$,

$$\exists N \leq \Phi(K,L,b,\eta,\varepsilon,g) \forall m \in [N, N+g(N)] \, (d(x_m, f(x_m)) < \varepsilon),$$

where

$$\begin{aligned}
&\Phi(K,L,b,\eta,\varepsilon,g) := h^{(M)}(0), \; h(n) := g(n+1)+n+2, \\
&M := \left\lceil \frac{3\left(5KD+D+\frac{11}{2}\right)}{\delta} \right\rceil, \; D := e^K(b+2), \\
&\delta := \frac{\varepsilon}{L^2 F(K)} \cdot \eta\left((1+K)D+1, \frac{\varepsilon}{F(K)((1+K)D+1)}\right), \\
&F(K) := 2(1+(1+K)^2(2+K)).
\end{aligned}$$

Moreover, $N = h^{(i)}(0)+1$ for some $i < M$.
If $\eta(r,\varepsilon)$ can be written as $\varepsilon \cdot \tilde{\eta}(r,\varepsilon)$, where, for any fixed r, the function $\tilde{\eta}(r,\varepsilon)$ decreases with ε, then even

$$\exists N \leq \Phi(K,L,b,\tilde{\eta},\varepsilon,g) \forall m \in [N, N+g(N)] \, (d(x_m, f(x_m)) < \varepsilon)$$

with Φ as above.

Proof: See Kohlenbach-Leuştean [234]. □

In the case where (X,d,W) is bounded one can show that asymptotically nonexpansive mappings have approximate fixed points and – if (X,d) is complete – even fixed points (see [234]) so that the condition on x made in the previous theorem always is satisfied with b_X being any upper bound on the metric d.

For CAT(0)-spaces (X,d) one can take as modulus of uniform convexity $\eta(\varepsilon) := \eta(r,\varepsilon) := \frac{\varepsilon^2}{8}$. Since $\eta(\varepsilon) = \varepsilon \cdot \tilde{\eta}(\varepsilon)$ with $\tilde{\eta}(\varepsilon) := \varepsilon/8$ being monotone, the last part in the theorem above applies. Moreover, if instead of metastability one only is interested in a bound on some N such that $d(x_N, f(x_N)) < \varepsilon$ (which is the special case where $g \equiv 0$), then this yields the following quadratic bound:

Corollary 18.63 (Kohlenbach-Leuştean [234]). *Let (X,d) be a bounded* CAT(0)-*space with diameter b_X and $f : X \to X$ be asymptotically nonexpansive with sequence (k_n).*
Assume that $K \geq 0$ is such that $\sum_{n=0}^{\infty} k_n \leq K$ and that $L \in \mathbb{N}, L \geq 2$ is such that $\frac{1}{L} \leq \lambda_n \leq 1 - \frac{1}{L}$ for all $n \in \mathbb{N}$.
Then the following holds for all $x \in X$:

$$\forall \varepsilon \in (0,1] \exists N \leq \Phi(K,L,b_X,\varepsilon)\,(d(x_N, f(x_N)) < \varepsilon),$$

where

$$\begin{aligned} &\Phi(K,L,b_X,\varepsilon) := 2M, \\ &M := \left\lceil \frac{1}{\varepsilon^2} \cdot 24L^2 \left(5KD + D + \tfrac{11}{2}\right) (F(K))^3((1+K)D+1)^2 \right\rceil, \\ &D := e^K (b_X + 2), \quad F(K) := 2(1 + (1+K)^2(2+K)). \end{aligned}$$

18.5 Applications of proof mining in ergodic theory

The field of ergodic theory has close connections to metric fixed point theory as nonexpansive mappings f as well as isometries again feature prominently. In recent years, ergodic theory has become a powerful tool in combinatorics and number theory. This line of research has its origin in Furstenberg's proof of the famous Szemerédi Theorem using ergodic theory and has given rise to ergodic theoretic proofs of combinatorial results including van der Waerden's theorem (see Furstenberg [110]). Most recently, this development has culminated in the spectacular Green-Tao theorem on the existence of arbitrarily long progressions in the set of prime numbers (see Green-Tao [142]). This interplay between (often ineffective) infinitary ergodic theoretic methods and their use in finite combinatorics makes the application of proof interpretations (that unwind the combinatorial skeleton of the ergodic theoretic arguments used) particularly promising.

An early use of proof mining in this context was Girard's ([122], pp. 237–251, 483–496) logical analysis of two versions of the proof of van der Waerden's theorem given by Furstenberg and Weiss (based on topological dynamics). This work led to interesting logical insights (e.g. on the phenomenon that small changes in a proof can result in big differences w.r.t. the growth of extractable bounds; see also [117]) but not to new information on van der Waerden's theorem. Recently, however, new quantitative results in ergodic theory have been achieved (based on the metatheorems from chapter 17) in connection with the so-called von Neumann mean ergodic theorem:

Ergodic theory, in particular, studies the asymptotic behavior of the averaging operator defined by

$$A_n(x) := \frac{1}{n} S_n(x), \text{ where } S_n(x) := \sum_{i=0}^{n-1} f^i(x).$$

The context, typically, is that of Hilbert spaces (or, more generally, uniformly convex Banach spaces).
A classical result is the following

Theorem 18.64 (von Neumann mean ergodic theorem). *Let X be a Hilbert space and $f : X \to X$ a nonexpansive linear operator. Then for any point $x \in X$ the sequence $(A_n(x))_n$ defined above converges (in the Hilbert space norm).*

Based on (the extraction algorithm from the proofs of) the metatheorems from chapter 17 as well as the methodology of eliminating fixed instances of PCM from chapter 13, Avigad-Gerhardy-Towsner [8] give a thorough logical analysis of a standard (ineffective) proof of the mean ergodic theorem. Since, as the authors show, there is (in general) no computable bound on the convergence itself, one instead extracts (as in theorem 18.58 above) a bound on the no-counterexample version of the Cauchy property (i.e. metastability in the sense of Tao):

$$\forall g : \mathbb{N} \to \mathbb{N}\, \forall \varepsilon > 0\, \forall x \in X\, \exists n \forall i, j \in [n; n+g(n)]\, (\|A_i(x) - A_j(x)\| < \varepsilon).$$

From corollary 17.71.1) proved in chapter 17 (together with the treatment of the completeness assumption in section 17.5) it can be inferred (after some preprocessing of the proof) that one can extract a computable bound $\Phi(d, \varepsilon, g)$ on $\exists n$ that only depends on a norm upper bound $d \geq \|x\|, \varepsilon$ and g (note that since f is linear and nonexpansive, one has $\|f(x)\| \leq \|x\|$ so that $\|f(x) - x\| \leq 2\|x\|$).
Among many other results, the following explicit bound Φ is extracted in [8]:

Theorem 18.65 (Avigad-Gerhardy-Towsner [8]). *Let X and f be as in theorem 18.64. Then*

$$\forall g : \mathbb{N} \to \mathbb{N}\, \forall \varepsilon > 0\, \forall x \in X\, \exists n \leq \Phi\, \forall i, j \in [n; n+g(n)]\, (\|A_i(x) - A_j(x)\| \leq \varepsilon),$$

where $\Phi = h^{(k)}(0)$ with $\rho := \lceil \frac{\|x\|}{\varepsilon} \rceil$, $k := 2^9 \rho^2$, $h(n) := n + 2^{13} \rho^4 \tilde{g}((n+1)\tilde{g}(2n\rho)\rho^2)$ and $\tilde{g}(n) := \max\{i + g(i) : i \leq n\}$.

18.6 Exercises, historical comments and suggested further reading

Exercises:

1) Show that lemma 18.7 also holds for hyperbolic spaces (X,d,W) instead of C.
2) Prove the statements in remark 18.32.
3) Fill in the details in the proof of corollary 18.46.

Historical comments and suggested further reading: For general information on metric fixed point theory see Goebel-Kirk [127] and Kirk-Sims [192]. Ishikawa's theorem is due to Ishikawa [176] and was generalized to spaces of hyperbolic type in Goebel-Kirk [126]. The Borwein-Reich-Shafrir theorem was established in [40]. The results from section 18.2 are essentially due to Gerhardy-Kohlenbach [120]. Many of the results from section 18.3 are taken from Kohlenbach-Leuştean [232] (the case of convex subsets in normed spaces was first treated in Kohlenbach [220] with some additions in Kohlenbach [224]). That paper was based on the somewhat more restricted notion of hyperbolic space as defined in Kirk [185] and Reich-Shafrir [310]. However, is was stressed that most results remain correct in the setting of hyperbolic spaces which is even more general as our notion of hyperbolic space. The few other results, where the more restricted concept of hyperbolic space in the sense of [310] was needed, easily extend to our notion as we showed above. Since for our concept of hyperbolic space we have that a convex subset of a hyperbolic space is hyperbolic itself, we did not have to consider convex subsets in this chapter. Theorems 18.49 and 18.50 are proved here for the first time. Theorem 18.34 was proved in the normed case first in Kohlenbach [224] and was generalized to hyperbolic spaces in Kohlenbach [229]. Theorem 18.58 is taken from Kohlenbach [227] which also contains a generalization to asymptotically nonexpansive functions as well as the noncomputability theorem 18.4.
The uniform bounds on the Borwein-Reich-Shafrir result as well as on Ishikawa's theorem presented in this chapter are used by Kohlenbach and Leuştean in [233] to prove new qualitative results on the asymptotic fixed point property of product spaces.
Many other applications of proof mining and the general metatheorems from chapter 17 are concerned with the case of convex subsets of uniformly convex normed spaces: In Kohlenbach [224] a full quantitative version of a theorem of Groetsch is given. This is generalized in Leuştean [263] to uniformly convex hyperbolic spaces yielding, in particular, a quadratic rate of asymptotic regularity for CAT(0)-spaces. Kohlenbach-Lambov [231] gives effective approximate fixed point bounds for Krasnoselski-Mann iterations of asymptotically weakly quasi-nonexpansive mappings in uniformly convex spaces. For asymptotically nonexpansive mappings these results are generalized to uniformly convex hyperbolic spaces in Kohlenbach-Leuştean [234] (see section 18.4). Lambov [260] uses proof mining to obtain a quantitative version of Hillam's theorem (a generalization of Krasnoselski's theorem to Lipschitz continuous functions on the real line). In Leuştean [264] effective uniform rates on the asymptotic regularity of so-called Halpern iterations of nonex-

pansive mappings are extracted. Gerhardy [116] provides effective bounds on Kirk's fixed point theorem for asymptotic contractions (Kirk [188]) which has been further improved in Briseid [52]. Using results from [52] it is shown in Briseid [53] that Picard iterations of asymptotic contraction are always bounded so that the corresponding assumption in Kirk's theorem is redundant. Briseid [50] gives a thorough logical analysis (based on the metatheorems developed in chapter 17) of the proof of a fixed point theorem due to Kincses and Totik [182] for so-called 'generalized p-contractive mappings' which is a particularly general class of mappings of contractive type (see Rhoades [315, 316]). In fact, Briseid [50] (and the subsequent Briseid [51]) obtains the first effective quantitative version of this theorem. A survey on all these results can be found in Kohlenbach [229].
The result discussed in section 18.5 is due to Avigad et al. [8] which contains many further interesting applications of proof mining in the context of ergodic theory.

Chapter 19
Final comments

In this book we have studied various proof interpretations and demonstrated their use as tools for extracting new (both qualitative and quantitative) information from given proofs. In particular, we presented extended case studies with applications of these techniques to concrete proofs in the areas of approximation theory and metric fixed point theory.

A common feature of proof interpretations is that they translate a system $\mathcal{T}$ into another system $\mathcal{S}$ by assigning to every formula A of the former a formula A^I of the latter such that the implication

$$\mathcal{T} \vdash A \Rightarrow \mathcal{S} \vdash A^I$$

holds. Moreover the proof of A^I in $\mathcal{S}$ can be obtained by a simple recursion over a given proof of A in $\mathcal{T}$ since the interpretations respect the logical deduction rules (**locality** or **modularity** of proof interpretations).

As a consequence of this such proof interpretations preserve to a certain extent the structure of the original proof and the resulting $\mathcal{S}$-proof of A^I will not be much longer than the original proof of A in $\mathcal{T}$ (see [159]). This is in sharp contrast to structural proof transformations like cut-elimination or normalization which in general cause a non-elementary recursive blow-up of the original proof ([344, 296, 305]). Of course, at a few places (proposition 10.55, corollary 10.59) we had to normalize the term extracted by the proof interpretation which again is of non-elementary complexity. However, as we have seen one can also make substantial use of terms involving higher types by exploiting the mathematical structure of the functionals denoted by these terms without having to normalize them (see e.g. theorem 6.8, corollary 6.10, proposition 9.10 and – if we allow Φ to be an arbitrary closed term of $\widehat{\text{WE-HA}}^{\omega}\!\upharpoonright$ – also theorem 10.58)! So proof interpretations of the sort we investigated in this book allow one to separate those aspects of unwinding proofs which can be carried out locally by recursion over the proof from those which involve a global rebuilding of a proof or a term like normalization.

The metatheorems obtained in chapter 17 even guarantee a-priori (without any compactness assumptions) new qualitative uniformity results (i.e. the existence of strongly uniform bounds) without having to carry out any actual extraction (which, however, is possible as well by the algorithm provided by the proof of the respective metatheorem).

Another important consequence of the modularity of proof interpretations is that they can easily be extended to systems $\tilde{\mathscr{T}} \supseteq \mathscr{T}$ obtained by adding further non-logical axioms Γ to $\mathscr{T}$. If the interpretation Γ^I of Γ is provable in $\mathscr{S}$ (resp. in some extension $\tilde{\mathscr{S}}$ of $\mathscr{S}$), then the given interpretation immediately extends to an interpretation of $\tilde{\mathscr{T}}$ in $\mathscr{S}$ (resp. in $\tilde{\mathscr{S}}$). So it suffices to consider the new axioms.

As a simple example for such an extension we recall e.g. the following: both functional interpretation and negative translation are trivial for purely universal sentences $\mathscr{P} :\equiv \forall \underline{x} A_0(\underline{x})$. Because of this the proofs of e.g. theorem 10.7 and theorem 10.8 immediately applied to WE-PA$^\omega + \mathscr{P}$, WE-HA$^\omega + \mathscr{P}$ and not just to WE-PA$^\omega$, WE-HA$^\omega$ (for $\mathscr{P}$ in the language of WE-PA$^\omega$).
As a corollary we obtained that the addition of ($\mathscr{S}^\omega$-true) universal axioms to WE-PA$^\omega$+ QF-AC doesn't change the provably recursive functionals of the system. This observation – which has been stressed in the context of first order arithmetic by G. Kreisel – can be extended also to more general classes of formulas. In the context of our general metatheorem on proof mining 15.1 from chapter 15 it turned out that we may add arbitrary lemmas Γ of the form

$$\forall x \in X \exists y \in K (F(x,y) =_{\mathbb{R}} 0)$$

as axioms where X, K are (constructively representable) Polish spaces, K is compact and $F : X \times K \to \mathbb{R}$ is a constructive and hence continuous function. The fact that the proofs of such lemmas need not to be analyzed in applications of monotone functional interpretation to the extraction of uniform bounds was extensively used in the applications presented in chapters 16 and 18. The general metatheorems obtained in chapter 17 make it possible to add even certain principles of the form above where K does no longer need to be compact but only metrically bounded.

Finally, proof interpretations can easily be combined with each other: e.g. in chapter 10 we used a combination of three different interpretations: elimination of extensionality, negative translation and (monotone) functional interpretation. These techniques, moreover, can be combined with the method of elimination of monotone Skolem functions as we showed in chapter 13.

In the applications to proofs in mathematics presented in this book we focussed on approximation theory and nonlinear analysis. However, the applicability of the proof-theoretic methods developed in this book by no means is restricted to these areas. We conclude the book with some wild speculations on rather ambitious future projects for proof mining that seem to be approachable by the techniques developed in this book (or suitable extensions thereof):

1) W.r.t. to applications of the metatheorems for the case of concrete Polish and compact metric spaces presented in chapter 15 an interesting candidate for proof mining is A. Bressan's celebrated uniqueness (and stability result) for entropy-weak solutions of hyperbolic systems of conservation laws (see Bressan [41] and Bianchini-Bressan [30]). Here the existence of global solutions is proved by applying a sequential compactness argument (Helly's theorem) to a sequence of approximate solutions. This example (communicated to us by A. Yoshikawa) looks very promising from our point of view since logical metatheorems suggest the possibility to

   (a) extract a rate Φ of convergence towards approximate solutions,
   (b) extract of a rate of strong unicity Ψ from the uniqueness proof,
   (c) and to combine Φ and Ψ (as we did in chapter 15) to provide an algorithm computing the unique exact solution.

2) A proof mining project in the context of algebraic number theory that recently has been suggested to the author by G. Kreisel is the following one: As conjectured by G. Frey and proved by K. Ribet, the modularity of a certain class of elliptic curves (called 'Frey curves') implies Fermat's last theorem: let $a^n + b^n = c^n$ be a counterexample to Fermat's last theorem, then the corresponding Frey curve is defined by

$$y^2 = x(x-a^n)(x+b^n).$$

The Taniyama-Shimura conjecture and now theorem (also called Taniyama-Weil conjecture T/W) states that all elliptic curves over $\mathbb{Q}$ are modular. Wiles and Taylor [379, 359] proved this conjecture for a large class of elliptic curves that includes all Frey curves (let us denote the corresponding special case of T/W by $(\text{T/W})_F$) thereby establishing Fermat's last theorem. Meanwhile the full T/W-conjecture has been proved by Breuil-Conrad-Diamond-Taylor [42].
$(\text{T/W})_F$ can be expressed in (maybe a suitable extensions of) the language of Peano arithmetic PA as a Π^0_1-sentence as can – trivially – Fermat's last theorem (FLT). So

$$(\text{T/W})_F \to \text{FLT}$$

has the form $\Pi^0_1 \to \Pi^0_1$. As recently outlined by A. Macintyre, the proof of T/W apparently can be reformulated in such a way that it becomes formalizable in PA and it also seems to follow that

$$\text{PA} \vdash (\text{T/W})_F \to \text{FLT}.$$

(Monotone) functional interpretation would allow one to extract from such a proof (but also from proofs in theories stronger than PA that are closer to the original proof) of this implication an explicit function that tells one how much of $(\text{T/W})_F$ actually is needed to prove a particular instance (or family of instances of) Fermat's last theorem which could shed new light on the relation between

T/W and FLT and could be of interest in connection with number fields other than $\mathbb{Q}$ (or even other structures).

3) Concerning further applications of the metatheorems developed in chapter 17 we mentioned already the area of ergodic theory (in particular when related to combinatorics). The success of Avigad-Gerhardy-Towsner [8] discussed briefly at the end of chapter 18 makes it very promising to analyze proofs of more recent results in ergodic theory in a similar way. Although usually proofs based on general ergodic theory may only yield numerically poor bounds, this approach might lead to new uniformity results as well as generalized parametric version of the theorems in question. Also already known results may receive a new understanding as instances of general logical metatheorems.

   Based on the so-called Gowers uniformity norms, much better bounds have been achieved in a number of cases, notably van der Waerden's theorem and Szemerédi's theorem (see e.g. Gowers [141]). An interesting question seems to be to investigate whether Gowers' technique can be viewed as a sophisticated form of majorizability that could be incorporated into a specially designed monotone functional interpretation.

4) The metatheorems proved in chapter 17, in particular, apply to CAT(0)-spaces and $\mathbb{R}$-trees. These structure have played an important role e.g. in M. Gromov's (see e.g. [145, 146]) work on geometric aspects of group representations. There are many interesting problems in this context (e.g. related to Gromov's theorem that finitely generated groups have polynomial growth iff they have a nilpotent subgroup of finite index) involving both quantitative as well as qualitative uniformity aspects which should be amenable to apply our proof-theoretic machinery to.

We hope that the material presented in this book has succeeded to convince the reader about the theoretical importance of proof interpretations as well as their applicability to different parts of mathematics.

References

The numbers at the end of each item refer to the pages on which the respective paper is cited.

1. Ackermann, W., Zum Aufbau der rellen Zahlen. Math. Ann. **99**, pp. 118–133 (1928). (54)
2. Aigner, M., Ziegler, G.M., Proof from THE BOOK, 3rd edn. Springer, Berlin (2003). viii+239 pp. (17)
3. Akama, Y., Berardi, S., Hayashi, S., Kohlenbach, U., An arithmetical hierarchy of the law of excluded middle and related principles. Proc. of the 19th Annual IEEE Symposium on Logic in Computer Science (LICS'04), pp. 192–201, IEEE Press (2004). (124)
4. Artemov, S., Explicit provability and constructive semantics. Bull. Symb. Log. **7**, pp. 1–36 (2001). (43)
5. Avigad, J., Interpreting classical theories in constructive ones. J. Symb. Log. **65**, pp. 1785–1812 (2000). (274, 277)
6. Avigad, J., Weak theories of nonstandard arithmetic and analysis. In: Simpson, S. (ed.), Reverse Mathematics. Lecture Notes in Logic **21**, pp. 19–46. A K Peters (2005). (242)
7. Avigad, J., Feferman, S., Gödel's functional ('Dialectica') interpretation. In: [63], pp. 337–405 (1998). (vii, 53, 139, 149, 196)
8. Avigad, J., Gerhardy, P., Towsner, H., Local stability of ergodic averages. arXiv:0706.1512v1 [math.DS] (2007). (196, 272, 435, 500, 502, 506)
9. Babaev, A.A., Solovjov, S.V., A coherence theorem for canonical maps in cartesian closed categories. Zap. Naucn. Semin. Leningr. Otd. Ordena Lenina Mat. Inst. Im. V.A. Steklova Akad. Nauk SSSR (LOMI) **88**, pp. 3–29 (1979). (Russian with English summary) Translation in J. Sov. Math. **20**, pp. 2263–2279 (1982). (40)
10. Baillon, J., Bruck, R.E., The rate of asymptotic regularity is $0(\frac{1}{\sqrt{n}})$. Theory and Applications of Nonlinear Operators of Accretive and Monotone Type. Lecture Notes in Pure and Appl. Math. **178**, pp. 51–81. Dekker, New York (1996). (484, 497)
11. Baillon, J.B., Bruck, R.E., Reich, S., On the asymptotic behavior of nonexpansive mappings and semigroups in Banach spaces. Houst. J. Math. **4**, pp. 1–9 (1978). (492)
12. Bartelt, M., Li, W., Error estimates and Lipschitz constants for best approximation in continuous function spaces. Comput. Math. Appl. **30**, pp. 255–268 (1995). (376)
13. Beeson, M., A type-free Gödel interpretation. J. Symb. Log. **43**, pp. 213–227 (1978). (140)
14. Beeson, M., Goodman's theorem and beyond. Pac. J. Math. **84**, pp. 1–16 (1979). (102)
15. Beeson, M., Foundations of Constructive Mathematics. Springer, Berlin (1985). (ix, 78, 95, 114)

16. Bellin, G., Ramsey interpreted: a parametric version of Ramsey's theorem. In: Logic and Computation (Pittsburgh, PA, 1987). Contemp. Math. **106**, pp. 17–37. Am. Math. Soc., Providence (1990). (39, 40)
17. Berardi, S., Bezem, M., Coquand, T., On the computational content of the axiom of choice. J. Symb. Log. **63**, pp. 600–622 (1998). (196, 221)
18. Berger, U., Program extraction from normalization proofs. In: Bezem, M. et al. (eds.), TLCA'93. LNCS **664**, pp. 91–106. Springer, New York (1993). (107)
19. Berger, U., Uniform Heyting arithmetic. Ann. Pure Appl. Log. **133**, pp. 125–148 (2005). (139)
20. Berger, U., Berghofer, S., Letouzey, P., Schwichtenberg, H., Program extraction from normalization proofs. Stud. Log. **82**, pp. 25–49 (2006). (ix, 107)
21. Berger, U., Buchholz, W., Schwichtenberg, H., Refined program extraction from classical proofs. Ann. Pure Appl. Log. **114**, pp. 3–25 (2002). (ix, 40, 276, 277)
22. Berger, U., Oliva, P., Modified bar recursion and classical dependent choice. In: Baaz, M., Friedman, S.-D., Krajicek, J. (eds.), Logic Colloquium '01. Lecture Notes in Logic **20**, pp. 89–107, Association for Symbolic Logic. A K Peters, Ltd., Wellesley (2005) (76, 196, 221, 277)
23. Berger, U., Oliva, P., Modified bar recursion. Math. Struct. Comput. Sci. **16**, pp. 163–183 (2006). (221, 277)
24. Berger, U., Schwichtenberg, H., An inverse of the evaluation functional for typed λ-calculus. In: Vemuri, R. (ed.), Proc. of the Sixth Annual IEEE Symposium on Logic in Computer Science, pp. 203–211. IEEE Computer Society, Los Alamitos (1991). (107)
25. Berger, U., Schwichtenberg, H., Program extraction from classical proofs. In: Leivant, D. (ed.), Logic and Computational Complexity Workshop LCC'94. LNCS **960**, pp. 77–97. Springer, New York (1995). (107, 277)
26. Berger, U., Schwichtenberg, H., Seisenberger, M., The Warshall algorithm and Dickson's lemma: Two examples of realistic program extraction. J. Autom. Reason. **26**, pp. 205–221 (2001). (ix, 276, 277)
27. Bezem, M., Strongly majorizable functionals of finite type: a model for bar recursion containing discontinuous functionals. J. Symb. Log. **50**, pp. 652–660 (1985). (57, 66, 69, 71, 76, 215, 221, 421)
28. Bezem, M., Strong normalization of barrecursive terms without using infinite terms. Arch. Math. Log. Grundlagenforsch. **25**, pp. 175–181 (1985). (221)
29. Bezem, M., Equivalence of bar recursors in the theory of functionals of finite type. Arch. Math. Log. **27**, pp. 149–160 (1988). (221)
30. Bianchini, S., Bressan, A., Vanashing viscosity solutions of nonlinear hyperbolic system. Ann. Math. **161**, pp. 223–342 (2005). (505)
31. Biering, B., Cartesian closed Dialectica categories. Preprint (2007). (140)
32. Bishop, E., Foundations of Constructive Analysis. McGraw-Hill, New York (1967). (78, 161, 347)
33. Bishop, E., Mathematics as a numerical language. In: Kino, Myhill, Vesley (eds.), Intuitionism and Proof Theory, pp. 53–71. North-Holland, Amsterdam (1970). (140, 161)
34. Bishop, E., Bridges, D., Constructive Analysis. Springer, Berlin (1985). xii+477 pp. (161)
35. Björnestal, B.O., Continuity of the metric projection operator I-III. The preprint series of Department of Mathematics. Royal Institute of Technology. Stockholm, TRITA-MAT **17** (1974), **20** (1974), **12** (1975). (302, 348, 374)
36. Björnestal, B.O., Local Lipschitz continuity of the metric projection operator. Approximation theory. In: Papers, VIth Semester, Stefan Banach Internat. Math. Center, Warsaw, 1975, pp. 43–53. Banach Center Publ. **4**. PWN, Warsaw (1979). (348)
37. Blatt, H.-P., Lipschitz continuity and strong unicity in G. Freud's work. J. Approx. Theory **46**, pp. 25–31 (1986). (302, 343)
38. Bombieri, E., van der Poorten, A.J., Some quantitative results related to Roth's theorem. J. Aust. Math. Soc. Ser. A **45**, pp. 233–248 (1988). (21)

39. Borel, E., Lecons sur les Fonctions de Variables Réelles. Gauthier-Villars, Paris (1905). (301, 343, 344)
40. Borwein, J., Reich, S., Shafrir, I., Krasnoselski-Mann iterations in normed spaces. Can. Math. Bull. **35**, pp. 21–28 (1992). (384, 460, 461, 466, 467, 469, 471, 473, 501)
41. Bressan, A., Hyperbolic Systems of Conservation Laws. The One-dimensional Cauchy Problem. Oxford Lecture Series in Mathematics and its Applications, **20**. Oxford University Press, Oxford (2000). xii+250 pp. (505)
42. Breuil, C., Conrad, B., Diamond, F., Taylor, R., On the modularity of elliptic curves over $\mathbb{Q}$: wild 3-adic exercises. J. Am. Math. Soc. **14**, pp. 843–939 (2001). (505)
43. Bridges, D.S., On the foundation of best approximation theory. J. Approx. Theory **28**, pp. 273–292 (1980). (315, 329)
44. Bridges, D.S., A constructive development of Chebychev approximation theory. J. Approx. Theory **30**, pp. 99–120 (1980). (302, 336, 339, 344, 345, 346, 376)
45. Bridges, D.S., Lipschitz constants and moduli of continuity for the Chebychev projection. Proc. Am. Math. Soc. **85**, pp. 557–561 (1982). (302, 329, 344, 345, 346)
46. Bridges, D.S., Recent progress in constructive approximation theory. In: Troelstra, A.S., van Dalen, D. (eds.), The L.E.J. Brouwer Centenary Symposium, pp. 41–50. North-Holland, Amsterdam (1982). (302, 305, 344, 347, 376)
47. Bridges, D.S., Richman, F., Varieties of Constructive Mathematics. London Mathematical Society LNS **97**. Cambridge University Press, Cambridge (1987). (120, 156)
48. Bridges, D.S., Richman, F., Julian, W.H., Mines, R., Extensions and fixed points of contractive maps in R^n. J. Math. Anal. Appl. **165**, pp. 438–456 (1992). (293)
49. Bridson, M.R., Haefliger, A., Metric Spaces of Non-positive Curvature. Springer, Berlin (1999). (386, 388, 453)
50. Briseid, E.M., Proof mining applied to fixed point theorems for mappings of contractive type. Master Thesis, Oslo, 70 pp. (2005). (196, 382, 448, 449, 455, 502)
51. Briseid, E.M., Fixed points of generalized contractive mappings. To appear in: J. Nonlinear Convex Anal. (196, 293, 382, 448, 449, 455, 502)
52. Briseid, E.M., A rate of convergence for asymptotic contractions. J. Math. Anal. Appl. **330**, pp. 364–376 (2007). (196, 293, 455, 502)
53. Briseid, E.M., Some results on Kirk's asymptotic contractions. Fixed Point Theory **8**, pp. 17–27 (2007). (196, 455, 502)
54. Briseid, E.M., Logical aspects of rates of convergence in metric spaces. In preparation. (448)
55. Browder, F.E., Nonexpansive nonlinear operators in a Banach space. Proc. Natl. Acad. Sci. U.S.A. **54**, pp. 1041–1044 (1965). (400, 456)
56. Browder, F.E., Petryshyn, W.V., The solution by iteration of nonlinear functional equations in Banach spaces. Bull. Am. Math. Soc. **72**, pp. 571–575 (1966). (496, 497)
57. Brown, D.K., Simpson, S.G., Which set existence axioms are needed to prove the separable Hahn-Banach theorem? Ann. Pure Appl. Log. **31**, pp. 123–144 (1986). (78, 149)
58. Brown, D., Giusto, M., Simpson, S., Vitali's theorem and WWKL. Arch. Math. Log. **41**, pp. 191–206 (2002). (149)
59. Bruck, R.E., A simple proof that the rate of asymptotic regularity of $(I+T)/2$ is $0(1/\sqrt{n})$. In: Recent Advances on Metric Fixed Point Theory (Seville, 1995). Ciencias **48**, pp. 11–18, Univ. Sevilla, Seville (1996). (484)
60. Bruhat, F., Tits, J., Groupes réductifs sur un corps local. I. Données radicielles valuées. Inst. Hautes Études Sci. Publ. Math. **41**, pp. 5–251 (1972). (386)
61. Burr, W., Functional interpretation of Aczel's constructive set theory. Ann. Pure Appl. Log. **104**, pp. 31–73 (2000). (140)
62. Buss, S.R., On Herbrand's Theorem. In: Leivant, D. (ed.), Logic and Computational Complexity. LNCS **960**, pp. 195–209. Springer, New York (1995). (39)
63. Buss, S.R. (editor), Handbook of Proof Theory. Studies in Logic and the Foundations of Mathematics, Vol. 137, Elsevier, Amsterdam (1998). vii+811 pp. (507, 522)

64. Chebycheff, P.L., Sur les questions de minima qui se rattachent a la représentation approximative des fonctions. Oeuvres I, pp. 273–378 (1859). (303, 305)
65. Cheney, E.W., An elementary proof of Jackson's theorem on mean-approximation. Math. Mag. **38**, pp. 189–191 (1965). (301, 302, 348, 349, 374)
66. Cheney, E.W., Approximation Theory. AMS Chelsea Publishing (1966). (299, 300, 330, 349, 350, 360, 374, 376)
67. Chidume, C.E., On the approximation of fixed points of nonexpansive mappings. Houst. J. Math. **7**, no. 3, pp. 345–355 (1981). (463, 482)
68. Cline, A.K., Lipschitz conditions on uniform approximation operators. J. Approx. Theory **8**, pp. 160–172 (1973). (343)
69. Collaço, P., Silva, J.C., A complete comparison of 25 contraction conditions. Nonlinear Anal. **30**, pp. 471–476 (1998). (293)
70. Collatz, L., Krabs, W., Approximationstheorie. Teubner Studienbücher Mathematik. B.G. Teubner, Stuttgart (1973). 208 pp. (329)
71. Cook, S., Urquhart, A., Functional interpretations of feasibly constructive arithmetic. Ann. Pure Appl. Log. **63**, pp. 103–200 (1993). (107, 138, 140)
72. Coquand, T., A semantics of evidence for classical arithmetic. J. Symb. Log. **60**, pp. 325–337 (1995). (40)
73. Coquand, T., Computational content of classical logic. In: Semantics and Logics of Computation (Cambridge, 1995). Publ. Newton Inst. **14**, pp. 33–78. Cambridge Univ. Press, Cambridge (1997). (40)
74. Coquand, Th., Sur un théorème de Kronecker concernant les variétés algébriques. C.R. Acad. Sci. Paris, Ser. I **338**, pp. 291–294 (2004). (vii)
75. Coquand, T., Hofmann, M., A new method for establishing conservativity of classical systems over their intuitionistic version. Lambda-calculus and logic. Math. Struct. Comput. Sci. **9**, pp. 323–333 (1999). (274, 277)
76. Coquand, Th., Lombardi, H., Quitte, C., Generating non-Noetherian modules constructively. Manuscr. Math. **115**, pp. 513–520 (2004). (vii)
77. Coste, M., Lombardi, H., Roy, M.F., Dynamical methods in algebra: effective Nullstellensätze. Ann. Pure Appl. Log. **111**, pp. 203–256 (2001). (vii)
78. de Paiva, V.C.V., The Dialectica categories. In: Categories in Computer Science and Logic (Boulder, CO, 1987). Contemporary Math. **92**, pp. 47–62. Am. Math. Soc., Providence (1989). (140)
79. Delzell, C.N., Continuous sums of squares of forms. In: Proc. L.E.J. Brouwer Centenary Symposium. Noordwijkerhout, pp. 65–75 (1981). (40)
80. Delzell, C.N., Case distinctions are necessary for representing polynomials as sums of squares. Proc. Herbrand Symposium, pp. 87–103 (1981). (40)
81. Delzell, C.N., A finiteness theorem for open semi-algebraic sets, with applications to Hilbert's 17th problem. Contemporary Math. **8**, pp. 79–97. Am. Math. Soc., Providence (1982). (40)
82. Delzell, C.N., A continuous, constructive solution to Hilbert's 17th problem. Invent. Math. **76**, pp. 365–384 (1984). (40)
83. Delzell, C.N., Continuous, piecewise-polynomial functions which solve Hilbert's 17th problem. J. Reine Angew. Math. **440**, pp. 157–173 (1993). (40)
84. Delzell, C.N., Kreisel's unwinding of Artin's proof-Part I. In: Odifreddi, P. (ed.), Kreiseliana, pp. 113–246. A K Peters, Wellesley (1996). (vii, 39, 40)
85. Diaz, J.B., Metcalf, F.T., On the structure of the set of subsequential limit points of successive approximations. Bull. Am. Math. Soc. **73**, pp. 516–519 (1967). (396)
86. Diaz, J.B., Metcalf, F.T., On the set of subsequential limit points of successive approximations. Trans. Am. Math. Soc. **135**, pp. 459–485 (1969). (396)
87. Diller, J., Logical problems of functional interpretations. Ann. Pure Appl. Log. **114**, pp. 27–42 (2002). (139)
88. Diller, J., Nahm, W., Eine Variante zur Dialectica-Interpretation der Heyting-Arithmetik endlicher Typen. Arch. Math. Log. Grundlagenforsch. **16**, pp. 49–66 (1974). (139)

89. Diller, J., Vogel, H., Intensionale Funktionalinterpretation der Analysis. In: Diller, J., Müller (eds.), Proof Theory Symposium, Kiel 1974. LNM **500**, pp. 56–72. Springer, New York (1975). (220, 221)
90. Dotson, W.G. Jr., On the Mann iterative process. Trans. Am. Math. Soc. **149**, pp. 65–73 (1970). (396)
91. Dragalin, A.G., New kinds of realizability and the Markov rule. Dokl. Akad. Nauk SSSR **251**, pp. 534–537 (1980) (Russian). English translation: Sov. Math. Dokl. **21**, pp. 461–464 (1980). (274, 277)
92. Edelstein, M., On fixed and periodic points under contractive mappings. J. Lond. Math. Soc. **37**, pp. 74–79 (1962). (292, 293, 447, 448, 455)
93. Edelstein, M., O'Brien, R.C., Nonexpansive mappings, asymptotic regularity and successive approximations. J. Lond. Math. Soc. **17**, pp. 547–554 (1978). (458, 463, 470, 482)
94. Esnault, H., Viehweg, E., Dyson's lemma for polynomials in several variables (and the theorem of Roth). Invent. Math. **78**, pp. 445–490 (1984). (21)
95. Espínola, R., Kirk, W.A., Fixed point theorems in $\mathbb{R}$-trees with applications to graph theory. Topol. Appl. **153**, Issue 7, pp. 1046–1055 (2006). (387)
96. Garcia-Falset, J., Llorens-Fuster, E., Prus, S., The fixed point property for mappings admitting a center. Nonlinear Anal. **66**, pp. 1257–1274 (2007). (397)
97. Faltings, G., Endlichkeitssätze für abelsche Varietäten über Zahlkörpern. Invent. Math. **73**, pp. 349–366 (1983). (21)
98. Feferman, S., Theories of finite type related to mathematical practice. In: Barwise, J., (ed.), Handbook of Mathematical Logic, pp. 913–972. North-Holland, Amsterdam (1977). (53, 76, 189, 190, 194, 196, 221)
99. Feferman, S., Kreisel's 'Unwinding Program'. In: Odifreddi, P. (ed.), Kreiseliana: about and around Georg Kreisel, pp. 247–273. A.K. Peters, Wellesley (1996). (39)
100. Feferman, S., In the Light of Logic. Oxford University Press, New York (1998). x+340 pp. (210)
101. Felscher, W., Berechenbarkeit: Rekursive und programmierbare Funktionen. Springer–Lehrbuch, Berlin (1993). (60)
102. Fernandes, A.M., Ferreira, F., Groundwork for weak analysis. J. Symb. Log. **67**, pp. 557–578 (2002). (161)
103. Ferreira, F., Nunes, A., Bounded modified realizability. J. Symb. Log. **71**, pp. 329–346 (2006). (107)
104. Ferreira, F., Oliva, P., Bounded functional interpretation. Ann. Pure Appl. Log. **135**, pp. 73–112 (2005). (140, 161, 232, 242)
105. Ferreira, F., Oliva, P., Bounded functional interpretation and feasible analysis. Ann. Pure Appl. Log. **145**, pp. 115–129 (2007). (161, 196)
106. Freud, G., Eine Ungleichung für Tschebyscheffsche Approximationspolynome. Acta Sci. Math. (Szeged) **19**, pp. 162–164 (1958). (302, 343)
107. Friedman, H., Systems of second-order arithmetic with restricted induction (abstract). J. Symb. Log. **41**, pp. 558–559 (1976). (149)
108. Friedman, H., Classical and intuitionistically provably recursive functions. In: Müller, G.H., Scott, D.S. (eds.), Higher Set Theory. LNM **669**, pp. 21–27. Springer, New York (1978). (273, 274, 277)
109. Friedrich, W., Gödelsche Funktionalinterpretation für eine Erweiterung der klassischen Analysis. Z. Math. Log. Grundl. Math. **31**, pp. 3–29 (1985). (196, 204, 220)
110. Furstenberg, H., Recurrence in Ergodic Theory and Combinatorial Number Theory. Princeton University Press, Princeton (1981). (499)
111. Gandy, R.O., Hyland, J.M.E., Computable and recursively countable functionals of higher type. In: Gandy, R.O., Hyland, J.M.E. (eds.), Logic Colloquium 1976, pp. 407–438. North-Holland, Amsterdam (1977). (75)
112. Gehlen, W., On a conjecture concerning strong unicity constants. J. Approx. Theory **101**, pp. 221–239 (1999). (327)

113. Gehlen, W., Unboundedness of the Lipschitz constants of best polynomial approximation. J. Approx. Theory **106**, pp. 110–142 (2000). (327)
114. Gerhardy, P., Refined complexity analysis of cut elimination. In: Baaz, M., Makowsky, J.A. (eds.), Proc. of 17th International Workshop CSL 2003. LNCS **2803**, pp. 212–225. Springer, New York (2003). (39)
115. Gerhardy, P., The role of quantifier alternations in cut elimination. Notre Dame J. Form. Log. **46**, pp. 165–171 (2005). (39)
116. Gerhardy, P., A quantitative version of Kirk's fixed point theorem for asymptotic contractions. J. Math. Anal. Appl. **316**, pp. 339–345 (2006). (196, 293, 448, 455, 502)
117. Gerhardy, P., Proof mining in topological dynamics. Preprint 2007. (466, 500)
118. Gerhardy, P., Kohlenbach, U., Extracting Herbrand disjunctions by functional interpretation. Arch. Math. Log. **44**, pp. 633–644 (2005). (3, 39)
119. Gerhardy, P., Kohlenbach, U., Strongly uniform bounds from semi-constructive proofs. Ann. Pure Appl. Log. **141**, pp. 89–107 (2006). (114, 293, 448, 453, 455, 516)
120. Gerhardy, P., Kohlenbach, U., General logical metatheorems for functional analysis. Trans. Am. Math. Soc. **360**, pp. 2615–2660 (2008) (10, 220, 389, 401, 402, 403, 404, 408, 409, 416, 418, 419, 452, 463, 497, 501)
121. Girard, J.-Y., Une extension de l'interpretation de Gödel à l'analyse, et son application à l' élimination des coupures dans l'analyse et dans la théorie des types. In: Fenstad, J.E. (ed.), Proc. of the Second Scandinavian Logic Symposium, pp. 63–92. North-Holland, Amsterdam (1971). (196, 221)
122. Girard, J.-Y., Proof Theory and Logical Complexity Vol. I. Studies in Proof Theory. Bibliopolis, Napoli and Elsevier Science Publishers, Amsterdam (1987). (vii, 17, 39, 40, 499)
123. Girard, J.-Y., Linear Logic. Theor. Comput. Sci. **50**, pp. 1–101 (1987). (140)
124. Glivenko, V.I., Sur quelques points de la logique de M. Brouwer. Bull. Soc. Math. Belg. **15**, pp. 183–188 (1929). (163)
125. Goebel, K., Kirk, W.A., A fixed point theorem for asymptotically nonexpansive mappings. Proc. Am. Math. Soc. **35**, pp. 171–174 (1972). (400, 497)
126. Goebel, K., Kirk, W.A., Iteration processes for nonexpansive mappings. In: Singh, S.P., Thomeier, S., Watson, B. (eds.), Topological Methods in Nonlinear Functional Analysis. Contemporary Math. **21**, pp. 115–123. AMS, Providence (1983). (384, 399, 447, 461, 462, 463, 464, 468, 469, 470, 473, 474, 482, 483, 484, 489, 501)
127. Goebel, K., Kirk, W.A., Topics in Metric Fixed Point Theory. Cambridge Studies in Advanced Mathematics **28**. Cambridge University Press, Cambridge (1990). (501)
128. Goebel, K., Reich, S., Uniform Convexity, Hyperbolic Geometry, and Nonexpansive Mappings. Monographs and Textbooks in Pure and Applied Mathematics **83**. Dekker, New York (1984). ix+170 pp. (381, 384, 385, 386, 444, 453, 498)
129. Goebel, K., Sękowski, T., Stachura, A., Uniform convexity of the hyperbolic metric and fixed points of holomorphic mappings in the Hilbert ball. Nonlinear Anal. **4**, pp. 1011–1021 (1980). (385, 386)
130. Gödel, K., Zur intuitionistischen Arithmetik und Zahlentheorie. Ergebnisse eines Mathematischen Kolloquiums, Vol. 4, pp. 34–38 (1933). (126, 163)
131. Gödel, K., Vortrag bei Zilsel (1938). First published in: [136], pp. 86–113 (1995). (139)
132. Gödel, K., In what sense is intuitionistic logic constructive. Lecture at Yale (1941). First published in: [136], pp. 189–200 (1995). (139)
133. Gödel, K., Über eine bisher noch nicht benützte Erweiterung des finiten Standpunktes. Dialectica **12**, pp. 280–287 (1958). (27, 32, 34, 41, 51, 76, 125, 129, 130, 139, 212)
134. Gödel, K., On an extension of finitary mathematics which has not yet been used (1972). First published in: [135], pp. 271–280 (1990). (139)
135. Gödel, K., Collected Work, Vol. 2, S. Feferman et al. (eds.). Oxford University Press, New York (1990). (139, 512)
136. Gödel, K., Collected Work, Vol. 3, S. Feferman et al. (eds.). Oxford University Press, New York (1995). (139, 512)

137. Göhde, D., Zum Prinzip der kontraktiven Abbildung. Math. Nachr. **30**, pp. 251–258 (1965). (400, 456, 460)
138. Goodman, N., The faithfulness of the interpretation of arithmetic in the theory of constructions. J. Symb. Log. **38**, pp. 453–459 (1973). (102)
139. Goodman, N., The theory of the Gödel functionals. J. Symb. Log. **41**, pp. 574–582 (1976). (102)
140. Goodman, N., Relativized realizability in intuitionistic arithmetic of all finite types. J. Symb. Log. **43**, pp. 23–44 (1978). (102)
141. Gowers, T., A new proof of Szemerédi's theorem. Geom. Funct. Anal. **11**, pp. 465–588 (2001). (506)
142. Green, B., Tao, T., The primes contain arbitrarily long arithmetic progressions. To appear in: Ann. Math. (499)
143. Grilliot, T.J., On effectively discontinuous type-2 objects. J. Symb. Log. **36**, pp. 245–248 (1971). (191)
144. Groetsch, C.W., A note on segmenting Mann iterates. J. Math. Anal. Appl. **40**, pp. 369–372 (1972). (400, 411, 496)
145. Gromov, M., Hyperbolic groups. In: Gersten, S.M. (ed.), Essays in Group Theory. MSRI Publ. **8**, pp. 75–263. Springer, New York (1987). (431, 506)
146. Gromov, M., Metric Structures for Riemannian and Non-Riemannian Spaces (with appendices by Katz, M., Pansu, P., Semmes, S.), edited by LaFontaine, L., Pansu, P. Modern Birkhäuser Classics, 2nd printing. Birkhäuser, Boston (2001). xix+585 pp. (506)
147. Grzegorczyk, A., Some classes of recursive functions. Rozprawny Matematyczne, 46 pp. Warsaw (1953). (54)
148. Hacks, J., Einige Anwendungen der Funktion $\lfloor x \rfloor$. Acta Math. **14**, pp. 329–336 (1890). (38)
149. Hardy, G.H., Wright, E.M., An Introduction to the Theory of Numbers, 5th edn., Oxford Science Publications, Oxford (1979). (17, 22)
150. Hayashi, S., Nakano, H., PX: A Computational Logic. MIT Press, Cambridge (1988). xiv+200 pp. (107)
151. Hayashi, S., Nakata, M., Towards limit computable mathematics. In: Callaghan, P. (ed.), TYPES 2000. LNCS **2277**, pp. 125–144. Springer, New York (2002). (123)
152. Henry, M.S., Roulier, J.A., Lipschitz and strong unicity constants for changing dimensions. J. Approx. Theory **22**, pp. 86–94 (1978). (327)
153. Henry, M.S., Schmidt, D., Continuity theorems for the product approximation operator. In: Law, A.G., Sahney, B.N. (eds.), Theory of Approximation with Applications, Alberta 1975, pp. 24–42. Academic, New York (1976). (301, 327, 329, 342)
154. Henson, C.W., Iovino, J., Ultraproducts in analysis. In: Analysis and Logic. London Mathematical Society LNS **262**, pp. 1–112. Cambridge University Press, Cambridge (2002). (395)
155. Hernest, M.-D., Light functional interpretation. An optimization of Gödel's technique towards the extraction of (more) efficient programs from (classical) proofs. In: Ong, L. (ed.), CSL 2005. LNCS **3634**, pp. 477–492. Springer, New York (2005). (139)
156. Hernest, M.-D., The MinLog proof-system for Dialectica program-extraction. Free software available at http://www.brics.dk/∼danher/MinLogForDialectica. (135)
157. Hernest, M.-D., Light Dialectica program exraction from a classical Fibonacci proof. Electron. Notes Theor. Comput. Sci. **171**, pp. 43–53 (2007). (139)
158. Hernest, M.-D., Synthesis of moduli of uniform continuity by Monotone Dialectica Interpretation in the proof-system MinLog. Electron. Notes Theor. Comput. Sci. **174**, pp. 141–149 (2007). (ix, 161)
159. Hernest, M.-D., Kohlenbach, U., A complexity analysis of functional interpretations. Theor. Comput. Sci. **338**, pp. 200–246 (2005). (3, 140, 141, 503)
160. Hertz, A., A constructive version of the Hilbert basis theorem. Master Thesis, Carnegie Mellon University (2004). (196)
161. Hilbert, D., Über das Unendliche. Math. Ann. **95**, pp. 161–190 (1926). (4, 51, 73, 76, 139)

162. Howard, W.A., Functional interpretation of bar induction by bar recursion. Compos. Math. **20**, pp. 107–124 (1968). (196, 206, 207, 220, 420)
163. Howard, W.A., Hereditarily majorizable functionals of finite type. In: Troelstra (ed.), Metamathematical Investigation of Intuitionistic Arithmetic and Analysis. LNM **344**, pp. 454–461. Springer, New York (1973). (61, 66, 75, 109, 110, 127, 135)
164. Howard, W.A., Ordinal analysis of simple cases of bar recursion. J. Symb. Log. **46**, pp. 17–30 (1981). (212, 220)
165. Howard, W.A., Ordinal analysis of bar recursion of type zero. Compos. Math. **42**, pp. 105–119 (1981). (220)
166. Howard, W.A., The formulae-as-types notion of construction. In: Hindley, J.R., Seldin, J.P. (eds.), To H.B. Curry, pp. 479–490. Academic, London (1980). (52)
167. Howard, W.A., Kreisel, G., Transfinite induction and bar induction of types zero and one, and the role of continuity in intuitionistic analysis. J. Symb. Log. **31**, pp. 325–358 (1966). (206)
168. Humphreys, A.J., Simpson, S., Separation and weak König's lemma. Preprint (1997). (149)
169. Hyland, J.M.E., Filter spaces and continuous functionals. Ann. Math. Log. **16**, pp. 101–143 (1979). (68, 76)
170. Hyland, J.M.E., Proof theory in the abstract. Ann. Pure Appl. Log. **114**, pp. 43–78 (2002). (140)
171. Ishihara, H., An omniscience principle, the König lemma and the Hahn-Banach theorem. Z. Math. Log. Grundl. Math. **36**, pp. 237–240 (1990). (118)
172. Ishihara, H., Continuity and nondiscontinuity in constructive mathematics. J. Symb. Log. **56**, pp. 1349–1354 (1991). (120)
173. Ishihara, H., Continuity properties in constructive mathematics. J. Symb. Log. **57**, pp. 557–565 (1992). (120)
174. Ishihara, H., Markov's principle, Church's thesis and Lindelöf's theorem. Idag. Math. **4**, pp. 321–325 (1993). (120)
175. Ishihara, H., Weak König's lemma implies Brouwer's Fan theorem: a direct proof. Notre Dame J. Form. Log. **47**, pp. 249–252 (2006). (118)
176. Ishikawa, S., Fixed points and iterations of a nonexpansive mapping in a Banach space. Proc. Am. Math. Soc. **59**, pp. 65–71 (1976). (447, 458, 459, 460, 461, 463, 468, 469, 470, 471, 473, 496, 501)
177. Itoh, S., Some fixed point theorems in metric spaces. Fund. Math. **102**, pp. 109–117 (1979). (384)
178. Jackson, D., Note on a class of polynomials of approximation. Trans. Am. Math. Soc. **22**, pp. 320–326 (1921). (301, 348, 374)
179. Jørgensen, K.F., Finite type arithmetic: computable existence analysed by modified realizability and functional interpretation. Master Thesis, University of Roskilde (2001). viii+121 pp. (127, 135, 139, 196)
180. Jørgensen, K.F., Functional interpretation and the existence property. Math. Log. Q. **50**, pp. 573–576 (2004). (139)
181. Ketonen, J., Solovay, R., Rapidly growing Ramsey functions. Ann. Math. **113**, pp. 267–314 (1981). (40)
182. Kincses, J., Totik, V., Theorems and counterexamples on contractive mappings. Math. Balk., New Series **4**, pp. 69–90 (1990). (293, 449, 502)
183. Kirchberger, P., Über Tschebychefsche Annäherungsmethoden. Dissertation, Göttingen (1902). (303, 305, 344)
184. Kirk, W.A., A fixed point theorem for mappings which do not increase distances. Am. Math. Mon. **72**, pp. 1004–1006 (1965). (400, 456)
185. Kirk, W.A., Krasnosel'skii iteration process in hyperbolic spaces. Numer. Funct. Anal. Optim. **4**, pp. 371–381 (1982). (384, 501)
186. Kirk, W.A., Nonexpansive mappings and asymptotic regularity. Nonlinear Anal. **40**, pp. 323–332 (2000). (397, 463, 469, 482, 484, 489)

187. Kirk, W.A., Geodesic geometry and fixed point theory. In: Seminar of Mathematical Analysis (Malaga/Seville, 2002/2003). Colecc. Abierta **64**, pp. 195–225. Univ. Seville Secr. Publ., Seville (2003). (386, 387)
188. Kirk, W.A., Fixed points of asymptotic contractions. J. Math. Anal. Appl. **277**, pp. 645–650 (2003). (293, 447, 455, 502)
189. Kirk, W.A., Geodesic geometry and fixed point theory II. In: G-Falset, J., L-Fuster, E., Sims, B. (eds.), Proc. International Conference on Fixed Point Theory, Valencia 2003, pp. 113–142. Yokohama Press (2004). (386, 387)
190. Kirk, W.A., Fixed point theorems in CAT(0) spaces and $\mathbb{R}$-trees. Fixed Point Theory Appl. **2004**(4), pp. 309–316 (2004). (387)
191. Kirk, W.A., Martinez-Yanez, C., Approximate fixed points for nonexpansive mappings in uniformly convex spaces. Ann. Pol. Math. **51**, pp. 189–193 (1990). (400)
192. Kirk, W.A., Sims, B. (eds.), Handbook of Metric Fixed Point Theory. Kluwer Academic, Dordrecht (2001). xi+703 pp. (501, 518, 521)
193. Kleene, S.C., On the interpretation of intuitionistic number theory. J. Symb. Log. **10**, pp. 109–124 (1945). (43, 139)
194. Kleene, S.C., Introduction to Metamathematics. North-Holland, Amsterdam, Noordhoff, Groningen, Van Nostrand, New-York (1952). (15, 18, 79, 496)
195. Kleene, S.C., Recursive functionals and quantifiers of finite types, I. Trans. Am. Math. Soc. **91**, pp. 1–52 (1959). (52, 68, 75, 76, 212, 403)
196. Kleene, S.C., Countable functionals. In: Heyting, A. (ed.), Constructivity in Mathematics, pp. 81–100. North-Holland, Amsterdam (1959). (68, 76)
197. Kleene, S.C., Formalized Recursive Functionals and Formalizes Realizability. Memoirs of the American Mathematical Society **89**. AMS, Providence (1969). (68)
198. Ko, K.-I., On the computational complexity of best Chebycheff approximation. J. Complex. **2**, pp. 95–120 (1986). (302, 329, 347, 376)
199. Ko, K.-I., Complexity Theory of Real Functions. Birkhäuser, Boston (1991). x+309 pp. (78, 329, 347, 373)
200. Kohlenbach, U., Theorie der majorisierbaren und stetigen Funktionale und ihre Anwendung bei der Extraktion von Schranken aus inkonstruktiven Beweisen: Effektive Eindeutigkeitsmodule bei besten Approximationen aus ineffektiven Beweisen. PhD Thesis, Frankfurt am Main (1990). xxii+278 pp. (88, 160, 221, 301, 376)
201. Kohlenbach, U., Pointwise hereditary majorization and some applications. Arch. Math. Log. **31**, pp. 227–241 (1992). (75, 113, 114, 160)
202. Kohlenbach, U., Remarks on Herbrand normal forms and Herbrand realizations. Arch. Math. Log. **31**, pp. 305–317 (1992). (26, 39)
203. Kohlenbach, U., Effective bounds from ineffective proofs in analysis: an application of functional interpretation and majorization. J. Symb. Log. **57**, pp. 1239–1273 (1992). [For an erratum see the end of [204]]. (149, 160, 161, 177, 196)
204. Kohlenbach, U., Effective moduli from ineffective uniqueness proofs. An unwinding of de La Vallée Poussin's proof for Chebycheff approximation. Ann. Pure Appl. Log. **64**, pp. 27–94 (1993). (17, 77, 95, 196, 288, 295, 300, 301, 303, 376, 515)
205. Kohlenbach, U., New effective moduli of uniqueness and uniform a-priori estimates for constants of strong unicity by logical analysis of known proofs in best approximation theory. Numer. Funct. Anal. Optim. **14**, pp. 581–606 (1993). (17, 196, 301, 328, 376)
206. Kohlenbach, U., Analysing proofs in analysis. In: Hodges, W., Hyland, M., Steinhorn, C., Truss, J. (eds.), Logic: From Foundations to Applications. European Logic Colloquium (Keele, 1993), pp. 225–260. Oxford University Press, Oxford (1996). (39, 140, 160, 196, 242, 376)
207. Kohlenbach, U., Mathematically strong subsystems of analysis with low rate of growth of provably recursive functionals. Arch. Math. Log. **36**, pp. 31–71 (1996). (61, 75, 76, 79, 107, 113, 138, 189, 196, 242, 393, 436)
208. Kohlenbach, U., Elimination of Skolem functions for monotone formulas in analysis. Arch. Math. Log. **37**, pp. 363–390 (1998). (196, 213, 246, 269, 272)

209. Kohlenbach, U., Proof theory and computational analysis. Electron. Notes Theor. Comput. Sci. **13**, Elsevier (1998). (http://www.elsevier.nl/locate/entcs/volume13.html), 34 pages. (78, 94, 95, 293, 295)
210. Kohlenbach, U., Arithmetizing proofs in analysis. In: Larrazabal, J.M., Lascar, D., Mints, G. (eds.), Logic Colloquium '96. Springer Lecture Notes in Logic **12**, pp. 115–158 (1998). (39, 196, 213, 242, 262, 271, 272, 294, 295, 393, 465)
211. Kohlenbach, U., On the arithmetical content of restricted forms of comprehension, choice and general uniform boundedness. Ann. Pure Appl. Log. **95**, pp. 257–285 (1998). (35, 269, 272)
212. Kohlenbach, U., Relative constructivity. J. Symb. Log. **63**, pp. 1218–1238 (1998). (107, 113, 114, 123, 160, 242)
213. Kohlenbach, U., The use of a logical principle of uniform boundedness in analysis. In: Cantini, A., Casari, E., Minari, P. (eds.), Logic and Foundations of Mathematics. Synthese Library **280**, pp. 93–106. Kluwer Academic, Dordrecht (1999). (196, 239, 242)
214. Kohlenbach, U., A note on Goodman's theorem. Stud. Log. **63**, pp. 1–5 (1999). (102)
215. Kohlenbach, U., On the no-counterexample interpretation. J. Symb. Log. **64**, pp. 1491–1511 (1999). (34, 39, 212, 213, 221)
216. Kohlenbach, U., Things that can and things that cannot be done in PRA. Ann. Pure Appl. Log. **102**, pp. 223–245 (2000). (221, 263, 272, 295)
217. Kohlenbach, U., A note on Spector's quantifier-free rule of extensionality. Arch. Math. Log. **40**, pp. 89–92 (2001). (161)
218. Kohlenbach, U., Intuitionistic choice and restricted classical logic. Math. Log. Q. **47**, pp. 455–460 (2001). (156)
219. Kohlenbach, U., On the computational content of the Krasnoselski and Ishikawa fixed point theorems. In: Blanck, J., Brattka, V., Hertling, P. (eds.), Proceedings of the Fourth Workshop on Computability and Complexity in Analysis. LNCS **2064**, pp. 119–145. Springer, New York (2001). (39, 196, 400, 497)
220. Kohlenbach, U., A quantitative version of a theorem due to Borwein-Reich-Shafrir. Numer. Funct. Anal. Optim. **22**, pp. 641–656 (2001). (196, 393, 467, 484, 501)
221. Kohlenbach, U., On uniform weak König's lemma. Ann. Pure Appl. Log. **114**, pp. 103–116 (2002). (180, 196)
222. Kohlenbach, U., On weak Markov's principle. Math. Log. Q. **48**, Suppl. 1, pp. 59–65 (2002). (120, 124)
223. Kohlenbach, U., Foundational and mathematical uses of higher types. In: Sieg, W., Sommer, R., Talcott, C. (eds.), Reflections on the Foundations of Mathematics. Essays in Honor of Solomon Feferman. Lecture Notes in Logic **15**, pp. 92–120. A.K. Peters, Wellesley (2002). (37, 83, 242)
224. Kohlenbach, U., Uniform asymptotic regularity for Mann iterates. J. Math. Anal. Appl. **279**, pp. 531–544 (2003). (196, 400, 411, 468, 497, 501)
225. Kohlenbach, U., Higher order reverse mathematics. In: Simpson, S. (ed.), Reverse Mathematics. Lecture Notes in Logic **21**, pp. 281–295. A.K. Peters, Wellesley (2005). (193)
226. Kohlenbach, U., Some logical metatheorems with applications in functional analysis. Trans. Am. Math. Soc. **357**, no. 1, pp. 89–128 (2005). [Some minor errata are corrected at the end of [119]]. (39, 220, 388, 389, 401, 452, 463)
227. Kohlenbach, U., Some computational aspects of metric fixed point theory. Nonlinear Anal. **61**, pp. 823–837 (2005). (196, 272, 465, 501)
228. Kohlenbach, U., A logical uniform boundedness principle for abstract metric and hyperbolic spaces. In: Proceedings of WoLLIC 2006. Electron. Notes Theor. Comput. Sci. **165**, pp. 81–93 (2006). (242, 452)
229. Kohlenbach, U., Effective uniform bounds from proofs in abstract functional analysis. In: Cooper, B., Loewe, B., Sorbi, A. (eds.), New Computational Paradigms: Changing Conceptions of What is Computable, pp. 223–258. Springer, New York (2008). (39, 455, 468, 501, 502)

230. Kohlenbach, U., Gödel's functional interpretation and its use in current mathematics. To appear in: Horizons of Truth, Gödel Centenary. Cambridge University Press. (139)
231. Kohlenbach, U., Lambov, B., Bounds on iterations of asymptotically quasi-nonexpansive mappings. In: G-Falset, J., L-Fuster, E., Sims, B. (eds.), Proc. International Conference on Fixed Point Theory, Valencia 2003, pp. 143–172. Yokohama Press (2004). (38, 196, 396, 400, 411, 497, 498, 501)
232. Kohlenbach, U., Leuştean, L., Mann iterates of directionally nonexpansive mappings in hyperbolic spaces. Abstr. Appl. Anal. **2003**, no. 8, pp. 449–477 (2003). (196, 393, 463, 467, 474, 482, 483, 485, 486, 489, 501)
233. Kohlenbach, U., Leuştean, L., The approximate fixed point property in product spaces. Nonlinear Anal. **66**, pp. 806–818 (2007). (196, 384, 501)
234. Kohlenbach, U., Leuştean, L., Asymptotically nonexpansive mappings in uniformly convex hyperbolic spaces. arXiv:0707.1626 [math.LO], submitted (2007). (196, 400, 497, 498, 499, 501)
235. Kohlenbach, U., Oliva, P., Proof mining in L_1-approximation. Ann. Pure Appl. Log. **121**, pp. 1–38 (2003). (196, 301, 302, 348, 349, 370, 376)
236. Kohlenbach, U., Oliva, P., Proof mining: a systematic way of analysing proofs in mathematics. Proc. Steklov Inst. Math. **242**, pp. 136–164 (2003). (39, 161, 293, 295, 393, 455)
237. Kolmogorov, A.N., On the principle of the excluded middle. Mat. Sb. **32**, pp. 646–667 (1925) (Russian). (163)
238. Krasnoselski, M.A., Two remarks on the method of successive approximation. Usp. Math. Nauk (N.S.) **10**, pp. 123–127 (1955) (Russian). (457, 496, 497)
239. Kreinovich, V.Ja., Categories of space-time models (Russian). Candidate Thesis, Novosibirsk, Gos. Univ., Novosibirsk (1979). (376)
240. Kreinovich, V.Ja., Review of 'Bridges, DS: Constructive functional analysis'. Zentralbl. Math. **401**, p. 03027 (1982). (376)
241. Kreisel, G., On the interpretation of non-finitist proofs, part I. J. Symb. Log. **16**, pp. 241–267 (1951). (8, 26, 39)
242. Kreisel, G., On the interpretation of non-finitist proofs, part II: Interpretation of number theory, applications. J. Symb. Log. **17**, pp. 43–58 (1952). (26, 39)
243. Kreisel, G., Mathematical significance of consistency proofs. J. Symb. Log. **23**, pp. 155–182 (1958). (39)
244. Kreisel, G., Interpretation of analysis by means of constructive functionals of finite types. In: Heyting, A. (ed.), Constructivity in Mathematics, pp. 101–128. North-Holland, Amsterdam (1959). (43, 68, 76, 97, 107, 168)
245. Kreisel, G., Sums of squares. In: Summaries of talks presented at the Summer Institute for Symbolic Logic, Cornell University, 1957, Communications Research Division, Institute for Defense Analyses, Princeton, NJ, pp. 313–320 (1960). (40)
246. Kreisel, G., On weak completeness of intuitionistic predicate logic. J. Symb. Log. **27**, pp. 139–158 (1962). (43, 97, 107)
247. Kreisel, G., Foundations of intuitionistic logic. In: Nagel, E., Suppes, P., Tarski, A. (eds.), Proc. Logic Methodology and Philosophy of Science, pp. 198–210. Stanford University Press, Stanford (1962). (43)
248. Kreisel, G., Functions, ordinals, species. In: Proc. 3rd Int. Congr. Amsterdam, pp. 145–159 (1968). (220)
249. Kreisel, G., Finiteness theorems in arithmetic: an application of Herbrand's theorem for Σ_2-formulas. In: Proc. of the Herbrand Symposium (Marseille, 1981), pp. 39–55. North-Holland, Amsterdam (1982). (vii, 21, 25, 39)
250. Kreisel, G., Proof theory and synthesis of programs: potentials and limitations. In: Eurocal '85 (Linz 1985). LNCS **203**, pp. 136–150. Springer, New York (1986). (39)
251. Kreisel, G., Logical aspects of computation: contributions and distractions. In: Odifreddi, P. (ed.), Logic and Computer Science, pp. 205–278. Academic, London (1990). (39, 172)

252. Kreisel, G., Macintyre, A., Constructive logic versus algebraization I. In: Troelstra, A.S., van Dalen, D. (eds.), Proc. L.E.J. Brouwer Centenary Symposium (Noordwijkerhout 1981), pp. 217–260. North-Holland, Amsterdam (1982). (vii, 39, 172)
253. Krivine, J.-L., Dependent choice, 'quote' and the clock. Theor. Comput. Sci. **308**, pp. 259–276 (2003). (221, 276)
254. Kroó, A., On the continuity of best approximations in the space of integrable functions. Acta Math. Acad. Sci. Hung. **32**, pp. 331–348 (1978). (302, 348, 374)
255. Kroó, A., On the uniform modulus of continuity of the operator of best approximation in the space of periodic functions. Acta Math. Acad. Sci. Hung. **34**, no. 1–2, pp. 185–203 (1979). (346)
256. Kroó, A., On strong unicity of L_1-approximation. Proc. Am. Math. Soc. **83**, pp. 725–729 (1981). (302)
257. Kuczumow, T., Reich, S., Shoikhet, D., Fixed points of holomorphic mappings: a metric approach. In: [192], pp. 437–515 (2001). (386)
258. Kuratowski, C., Topologie, Vol. I. Warsawa (1952). (67)
259. Kuroda, S., Intuitionistische Untersuchungen der formalistischen Logik. Nagoya Math. **3**, pp. 35–47 (1951). (163)
260. Lambov, B., Rates of convergence of recursively defined sequences. Electron. Notes Theor. Comput. Sci. **120**, pp. 125–133 (2005). (286, 501)
261. Leivant, D., Syntactic translations and provably recursive functions. J. Symb. Log. **50**, pp. 252–258 (1985). (277)
262. Leuştean, L., Proof mining in $\mathbb{R}$-trees and hyperbolic spaces. In: Proceedings of WoLLIC 2006. Electron. Notes Theor. Comput. Sci. **165**, pp. 95–106 (2006). (381, 431, 452, 497, 498)
263. Leuştean, L., A quadratic rate of asymptotic regularity for CAT(0)-spaces. J. Math. Anal. Appl. **325**, pp. 386–399 (2007). (196, 387, 400, 444, 496, 497, 501)
264. Leuştean, L., Rates of asymptotic regularity for Halpern iterations of nonexpansive mappings. In: Calude, C.S., Stefanescu, G., Zimand, M. (eds.), Combinatorics and Related Areas. A Collection of Papers in Honour of the 65th Birthday of Ioan Tomescu. J. Univ. Comput. Sci. Vol. 13, No. 11, pp. 1680–1691 (2007). (501)
265. Lorenzen, P., Differential und Integral. Eine konstruktive Einführung in die klassische Analysis. Akademische Verlagsgesellschaft, Frankfurt a.M. (1965). v+293 pp. (210)
266. Luckhardt, H., Extensional Gödel Functional Interpretation. Springer Lecture Notes in Mathematics **306** (1973). (vii, 139, 180, 182, 196, 204, 206, 207, 209, 215, 219, 220, 221, 420)
267. Luckhardt, H., Herbrand-Analysen zweier Beweise des Satzes von Roth: Polynomiale Anzahlschranken. J. Symb. Log. **54**, pp. 234–263 (1989). (vii, 17, 20, 21, 25, 39)
268. Luckhardt, H., Bounds extracted by Kreisel from ineffective proofs. In: Odifreddi, P. (ed.), Kreiseliana, pp. 289–300. A.K. Peters, Wellesley (1996). (vii, 39, 40)
269. Machado, H.V., A characterization of convex subsets of normed spaces. Kodai Math. Sem. Rep. **25**, pp. 307–320 (1973). (410)
270. Macintyre, A., The mathematical significance of proof theory. Philos. Trans. R. Soc. A **363**, pp. 2419–2435 (2005). (39)
271. Mandelkern, M., Constructive Continuity. Mem. Am. Math. Soc. **277**. AMS, Providence (1983). (120)
272. Mandelkern, M., Constructive complete finite sets. Z. Math. Log. Grundl. Math. **34**, pp. 97–103 (1988). (120)
273. Mann, W.R., Mean value methods in iteration. Proc. Am. Math. Soc. **4**, pp. 506–510 (1953). (459)
274. Markov, A.A., Sur une question posee par Mendeleieff. Izv. Akad. Nauk SSSR **62**, pp. 1–24 (1889). (301)
275. Meszáros, J., A comparison of various definitions of contractive type mappings. Bull. Calcutta Math. Soc. **84**, pp. 167–194 (1992). (293)

276. Mints, G.E., Closed categories and the theory of proofs. Zap. Naucn. Semin. Leningr. Otd. Ordena Lenina Mat. Inst. Im. V.A. Steklova Akad. Nauk SSSR (LOMI) **68**, pp. 83–114 (1977) (Russian). Translation in J. Sov. Math. **15**, pp. 45–62 (1981) and also revised translation in [280], pp. 183–212. (40)
277. Mints, G.E., Finite investigations of transfinite derivations. J. Sov. Math. **10**, pp. 548–596 (1978). Reprinted in [280] (pp. 17–71). (102)
278. Mints, G.E., A simple proof of the coherence theorem for cartesian closed categories. In: [280] (pp. 213–220). (40)
279. Mints, G.E., Proof theory and category theory. In: [280] (pp. 157–182). (40)
280. Mints, G.E., Selected Papers in Proof Theory. Studies in Proof Theory, Vol. 3. Bibliopolis, Napoli, North-Holland, Amsterdam (1992). 294 pp. (519)
281. Mints, G.E., Unwinding a non-effective cut elimination proof. In: Grigoriev, D. Harrison, J. Hirsch, E.A. (eds.), Computer Science – Theory and Applications, First International Computer Science Symposium in Russia, CSR 2006, Proceedings, St. Petersburg, Russia, June 8–12, 2006. LNCS **3967**, pp. 259–269. Springer, New York (2006). (197)
282. Moschovakis, Y.N., Descriptive Set Theory. Studies in Logic and the Foundations of Mathematics. North-Holland, Amsterdam (1980). xii+637 pp. (95)
283. Murthy, C., Extracting constructive content from classical proofs. PhD thesis, Cornell University (1990). (277)
284. Natanson, I.P., Konstruktive Funktionentheorie. Akademie-Verlag, Berlin (German translations of the original Russian edition 1949). (301, 303, 304, 305, 310, 325)
285. Nesterenko, Yu.V., Modular functions and transcendence questions. Sb. Math. **187**, no. 9, pp. 1319–1348 (1996). (22)
286. Newman, D.J., Shapiro, H.S., Some theorems on Cebysev approximation. Duke Math. J. **30**, pp. 673–682 (1963). (298, 301, 309, 327, 329)
287. Normann, D., Recursion on the Countable Functionals. Springer Lecture Notes in Mathematics **811** (1980). v+190 p. (76, 193)
288. Nürnberger, G., Strong unicity constants for spline functions. Numer. Funct. Anal. Optim. **5**, pp. 319–347 (1983). (303)
289. O'Farrell, A.G., When uniformly-continuous implies bounded. Irish Math. Soc. Bull. **53**, pp. 53–56 (2004). (407)
290. Oliva, P., On the computational complexity of best L_1-approximation. Math. Log. Q. **48**, suppl. I, pp. 66–77 (2002). (303, 373, 376)
291. Oliva, P., Polynomial-time algorithms from ineffective proofs. In: Proc. of the Eighteenth Annual IEEE Symposium on Logic in Computer Science LICS'03, pp. 128–137 (2003). (221)
292. Oliva, P., Unifying functional interpretations. Notre Dame J. Form. Log. **47**, pp. 263–290 (2006). (139, 161, 196)
293. Oliva, P., Understanding and using Spector's bar recursive interpretation of classical analysis. In: Proceedings of CiE 2006. LNCS **3988**, pp. 423–434. Springer, New York (2006). (36, 174, 213, 220, 238)
294. Oliva, P., An analysis of Gödel's Dialectica interpretation via linear logic. To appear in: Dialectica. (140)
295. Oliva, P., Streicher, T., On Krivine's realizability interpretation of classical second-order arithmetic. To appear in: Fund. Math. (221)
296. Orevkov, V.P., Complexity of Proofs and Their Transformations in Axiomatic Theories. Translations of Mathematical Monographs **128**. American Mathematical Society, Providence (1993). (3, 503)
297. Papadopoulos, A., Metric Spaces, Convexity and Nonpositive Curvature. IRMA Lectures in Mathematics and Theoretical Physics **6**. European Mathematical Society, Zürich (2005). (453)
298. Parsons, C., Proof-theoretic analysis of restricted induction schemata (abstract). J. Symb. Log. **36**, p. 361 (1971). (246)
299. Parsons, C., On n-quantifier induction. J. Symb. Log. **37**, pp. 466–482 (1972). (27, 51, 53, 196, 246, 272, 274)

300. Paule, P., A classical hypergeometric proof of an important transformation formula found by J.-B. Baillon and R.E. Bruck. In: Theory and Applications of Nonlinear Operators of Accretive and Monotone Type. Lecture Notes in Pure and Appl. Math. **178**, pp. 241–242. Dekker, New York (1996). (484)
301. Pinkus, A., On L_1-Approximation. Cambridge Tracts in Mathematics **93**. Cambridge University Press, Cambridge (1989). (303, 376)
302. Poreda, S.J., Counterexamples in best approximation. Proc. Am. Math. Soc. **56**, pp. 167–171 (1976). (327)
303. Pour-El, M.B., Richards, J.I., Computability in Analysis and Physics. Springer, Berlin (1989). (281, 457)
304. de La Vallée Poussin, C.J., Lecons sur l'Approximation des Fonctions d'une Variable Réelle. Gauthier-Villars, Paris (1919). (301, 302, 303, 304, 310, 325)
305. Pudlak, P., The length of proofs. In: Buss, S. (ed.), Handbook of Proof Theory, pp. 548–637. Elsevier Science B.V., Amsterdam (1998). (3, 503)
306. Qihou, L., Iteration sequences for asymptotically quasi-nonexpansive mappings with error member. J. Math. Anal. Appl. **259**, pp. 18–24 (2001). (38)
307. Raffalli, C., Getting results from programs extracted from classical proofs. Theor. Comput. Sci. **323**, pp. 49–70 (2004). (276)
308. Rakotch, E., A note on contractive mappings. Proc. Am. Math. Soc. **13**, pp. 459–465 (1962). (293, 447, 448, 455)
309. Reich, S., The alternating algorithm of von Neumann in the Hilbert ball. Dyn. Syst. Appl. **2**, pp. 21–25 (1993). (384)
310. Reich, S., Shafrir, I., Nonexpansive iterations in hyperbolic spaces. Nonlinear Anal. Theory Methods Appl. **15**, pp. 537–558 (1990). (384, 386, 391, 501)
311. Reich, S., Zaslavski, A.J., Generic aspects of metric fixed point theory. In: Kirk, W.A., Sims, B. (eds.), Handbook of Metric Fixed Point Theory, pp. 557–576. Kluwer Academic, Dordrecht (2001). (384)
312. Renardel de Lavalette, G.R., Extended bar induction in applicative theories. Ann. Pure Appl. Log. **50**, pp. 139–189 (1990). (102)
313. Rice, J.R., The Approximation of Functions, Vol. 1. Addison-Wesley, Reading (1964). (301, 302, 328, 334, 343, 346)
314. Ritchie, R.W., Classes of recursive functions based on Ackermann's function. Pac. J. Math. **15**, pp. 1027–1044 (1965). (54)
315. Rhoades, B.E., A comparision of various definitions of contractive mappings. Trans. Am. Math. Soc. **226**, pp. 257–290 (1977) (293, 447, 449, 502)
316. Rhoades, B.E., Contractive definitions. In: Rassias, Th.M. (ed.), Nonlinear Analysis, pp. 513–526. World Sci. Publishing, Singapore (1987). (293, 447, 449, 502)
317. Roth, K.F., Rational approximations to algebraic numbers. Mathematika **2**, pp. 1–20 (1955). (21)
318. Ruitenberg, W., Basic logic and Fregean set theory. In: Barendregt, Bezem, Klop (eds.), Dirk van Dalen Festschrift. Quaestiones Infinitae, Vol. V. Utrecht University, Utrecht (1993). (42)
319. Safarik, P., On the interpretation of the Bolzano-Weierstraß principle using bar recursion. Master Thesis TU Darmstadt (under preparation). (212)
320. Sakamoto, N., Yamazaki, T., Uniform versions of some axioms of second order arithmetic. Math. Log. Q. **50**, pp. 587–593 (2004). (193)
321. Scarpellini, B., A model for bar recursion of higher types. Compos. Math. **23**, pp. 123–153 (1971). (67, 68, 76, 214, 215, 221)
322. Scarpellini, B., A formally constructive model for barrecursion of higher types. Z. Math. Log. Grundl. Math. **18**, pp. 321–383 (1972). (76, 221)
323. Schönfinkel, M., Über die Bausteine der mathematischen Logik. Math. Ann. **92**, pp. 305–316 (1924). (51, 436)
324. Schütte, K., Proof Theory. Grundlehren der Mathematischen Wissenschaften **225**. Springer, Berlin (1977). xii+302 pp. (139)

325. Schwichtenberg, H., Einige Anwendungen von unendlichen Termen und Wertfunktionalen. Münster, Habilitationsschrift (1973). (161)
326. Schwichtenberg, H., Proof theory: some aspects of cut-elimination. In: Barwise, J. (ed.), The Handbook of Mathematical Logic, pp. 867–895. North-Holland, Amsterdam (1977). (39)
327. Schwichtenberg, H., On bar recursion of types 0 and 1. J. Symb. Log. **44**, pp. 325–329 (1979). (213, 221)
328. Schwichtenberg, H., Inverting monotone continuous functions in constructive analysis. In: Logical Approaches to Computational Barriers Second Conference on Computability in Europe, CiE 2006, Swansea, UK, June 30–July 5, 2006. LNCS **3988**, pp. 490–504. Springer, New York (2006). (237)
329. Schwichtenberg, H., Dialectica interpretation of well-founded induction. To appear in: Math. Log. Q. (140)
330. Shafrir, I., The approximate fixed point property in Banach and hyperbolic spaces. Isr. J. Math. **71**, pp. 211–223 (1990). (384)
331. Shioji, N., Tanaka, K., Fixed point theory in weak second-order arithmetic. Ann. Pure Appl. Log. **47**, pp. 167–188 (1990). (149)
332. Shoenfield, J.S., Mathematical Logic. Addison-Wesley, Reading (1967). (25, 39, 195, 196, 244)
333. Shoenfield, J.S., Recursion Theory. Lecture Notes in Logic **1**. ASL and A.K. Peters (2001). 96 pp. (ix)
334. Sieg, W., Fragments of arithmetic. Ann. Pure Appl. Log. **28**, pp. 33–71 (1985). (149)
335. Simpson, S.G., Which set existence axioms are needed to prove the Cauchy/Peano theorem for ordinary differential equations. J. Symb. Log. **49**, pp. 783–801 (1984). (149)
336. Simpson, S.G., Reverse mathematics. In: Proc. Symp. Pure Math. **42**, pp. 461–471. AMS, Providence (1985). (149)
337. Simpson, S.G., Ordinal numbers and the Hilbert basis theorem. J. Symb. Log. **53**, pp. 961–974 (1988). (197)
338. Simpson, S.G., Subsystems of Second Order Arithmetic. Perspectives in Mathematical Logic. Springer, New York (1999). xiv+445 pp. (vii, 7, 10, 78, 83, 88, 95, 149, 196, 199, 210, 212, 242, 263, 294)
339. Simpson, S.G., Tanaka, K., Yamazaki, T., Some conservation results on weak König's lemma. Ann. Pure Appl. Log. **118**, pp. 87–114 (2002). (196)
340. Sims, B., Examples of fixed point free mappings. In: [192], pp. 35–48 (2001). (400, 456)
341. Smorynski, C., Logical Number Theory I. Universitext. Springer, Berlin (1991). (46, 54, 60)
342. Specker, E., Nicht konstruktiv beweisbare Sätze der Analysis. J. Symb. Log. **14**, pp. 145–158 (1949). (2, 31)
343. Spector, C., Provably recursive functionals of analysis: a consistency proof of analysis by an extension of principles formulated in current intuitionistic mathematics. In: Dekker, J.C.E. (ed.), Recursive Function Theory, Proceedings of Symposia in Pure Mathematics, Vol. 5, pp. 1–27. AMS, Providence (1962). (33, 196, 199, 202, 204, 205, 220, 393, 420)
344. Statman, R., Lower bounds on Herbrand's theorem. Proc. Am. Math. Soc. **75**, pp. 104–107 (1979). (3, 503)
345. Stein, M., Eine Hybrid-Interpretation der Heyting-Arithmetik. Master Thesis. Universität Münster (1974). (139)
346. Stein, M., Interpretationen der Heyting-Arithmetik endlicher Typen. Arch. Math. Log. Grundlagenforsch. **19**, pp. 175–189 (1978). (139)
347. Streicher, T., Kohlenbach, U., Shoenfield is Gödel after Krivine. Math. Log. Q. **53**, pp. 176–179 (2007). (195)
348. Streicher, T., Reus, B., Classical logic: continuation semantics and abstract machines. J. Funct. Program. **8**(6), pp. 543–572 (1998). (195)

349. Sørensen, M.H., Urzyczyn, P., Lectures on the Curry-Howard Isomorphism. Studies in Logic and the Foundations of Mathematic **149**. Elsevier Science, Amsterdam (2006). 456 pp. (ix, 47, 52, 62, 76)
350. Tait, W.W., The substitution method. J. Symb. Log. **30**, pp. 175–192 (1965). (39)
351. Tait, W.W., Normal derivability in classical logic. In: Barwise, J. (ed.), The Syntax and Semantics of Infinitary Languages, pp. 204–236. Springer, Berlin (1968). (39)
352. Tait, W.W., Normal form theorem for bar recursive functions of finite type. In: Proc. 2nd Scandinavian Logic Sympos. 1970. Stud. Log. Found. Math. **63**, 353–367 (1971). (221)
353. Tait, W.W., Gödel's reformulation of Gentzen's first consistency proof for arithmetic: the no-counterexample interpretation. Bull. Symb. Log. **11**, pp. 225–238 (2005). (39)
354. Takahashi, W., A convexity in metric space and nonexpansive mappings, I. Kodai Math. Sem. Rep. **22**, pp. 142–149 (1970). (384, 387)
355. Takeuti, G., Proof Theory. Studies in Logic and the Foundations of Mathematic **81**. North-Holland, Amsterdam (1975). vii+372 pp. Second Edition: x+490 pp., Elsevier (1987). (246)
356. Takeuti, G., A conservative extension of Peano arithmetic. Part II of 'Two applications of logic to mathematics'. Publ. Math. Soc. Jpn. **13** (1978). (221)
357. Tao, T., Soft analysis, hard analysis, and the finite convergence principle. Essay posted May 23, 2007. Available at: http://terrytao.wordpress.com/2007/05/23/soft-analysis-hard-analysis-and-the-finite-convergence-principle/ (32, 37, 464)
358. Tao, T., Norm convergence of multiple ergodic averages for commuting transformations. arXiv:0707.1117v1 [math.DS] (2007). To appear in: Ergodic Theory Dyn. Syst. (4, 32)
359. Taylor, R., Wiles, A., Ring-theoretic properties of certain Hecke algebras. Ann. Math. **141**, pp. 553–572 (1995). (505)
360. Tits, J., A theorem of Lie-Kolchin for trees. In: Bass, H., Cassidy, P., Kovacic, J. (eds.), Contributions to Algebra: A Collection of Papers Dedicated to Ellis Kolchin, pp. 377–388. Academic, New York (1977). (431)
361. Toftdal, M., A calibration of ineffective mathematical theorems by a hierarchy of semi-classical logical principles over HA. In: Diaz, J., et al. (eds.), Proc. ICALP'2004. LNCS **3142**, pp. 1188–1200. Springer, New York (2004). (124)
362. Tonelli, L., I polinomi d'approssimazione di Tchebychev. Ann. Mat. Pura Appl. **15**, pp. 47–119 (1908). (307)
363. Tourlakis, G., Lectures in Logic and Set Theory. Volume I: Mathematical Logic. Cambridge Studies in Advanced Mathematics **82**. Cambridge University Press, Cambridge (2003). xi+328 pp. (56)
364. Troelstra, A.S., Intuitionistic general topology. PhD Thesis. Universiteit van Amsterdam (1966). (95)
365. Troelstra, A.S., Principles of Intuitionism. Lecture Notes in Mathematics **95**. Springer, Berlin (1969). (76)
366. Troelstra, A.S. (ed.), Metamathematical Investigation of Intuitionistic Arithmetic and Analysis. Lecture Notes in Mathematics **344**. Springer, Berlin (1973). (vii, 4, 41, 50, 59, 68, 69, 76, 88, 98, 100, 101, 102, 107, 121, 130, 136, 138, 139, 146, 148, 149, 159, 164, 168, 215, 221, 225)
367. Troelstra, A.S., Note on the fan theorem. J. Symb. Log. **39**, pp. 584–596 (1974). (149)
368. Troelstra, A.S., Some models for intuitionistic finite type arithmetic with fan functional. J. Symb. Log. **42**, pp. 194–202 (1977). (111, 114)
369. Troelstra, A.S., Realizability. In: [63], pp. 407–473 (1998). (vii, 102, 107)
370. Troelstra, A.S., Schwichtenberg, H., Basic Proof Theory, 2nd edn. Cambridge Tracts in Theoretical Computer Science **43**. Cambridge University Press, Cambridge (2000). xii+417 pp. (62)
371. Troelstra, A.S., van Dalen, D., Constructivism in Mathematics, Vol. I and II. North-Holland, Amsterdam (1988). (ix, 15, 19, 45, 76, 78, 95, 114, 276)

372. van Dalen, D., Logic and Structure, 4th edn. Universitext. Springer, Berlin (2004). 263pp. (ix, 76)
373. Vogel, H., Ein starker Normalisationssatz für die bar-rekursiven Funktionale. Arch. Math. Log. Grundlagenforsch. **18**, pp. 81–84 (1976). (221)
374. Vogel, H., Über die mit dem Bar-Rekursor vom Typ 0 definierbaren Ordinalzahlen. Arch. Math. Log. Grundlagenforsch. **19**, pp. 165–173 (1978/79). (220)
375. Weidmann, J., Linear operators in Hilbert Spaces. Graduate Texts in Mathematics **68**. Springer, Berlin (1980). xii+402 pp. (413)
376. Weiermann, A., Analytic combinatorics, proof-theoretic ordinals, and phase transitions for independence results. Ann. Pure Appl. Log. **136**, pp. 189–218 (2005). (40)
377. Weihrauch, K., Computable Analysis. Springer, Berlin (2000). (78, 95, 373, 376, 457)
378. Weyl, H., Das Kontinuum. Kritische Untersuchungen über die Grundlagen der Analysis. Veit, Leipzig (1918). English translation: Weyl, H., The Continuum: A Critical Examination of the Foundation of Analysis. Dover, New York (1994). (210)
379. Wiles, A., Modular elliptic curves and Fermat's Last Theorem. Ann. Math. **141**, pp. 443–551 (1995). (505)
380. Yasugi, M., Intuitionistic analysis and Gödel's interpretation. J. Math. Soc. Jpn. **15**, pp. 101–112 (1963). (130, 136)
381. Young, J.W., General theory of approximation by functions involving a given number of arbitrary parameters. Trans. Am. Math. Soc. **8**, pp. 331–344 (1907). (301, 302, 328, 330, 334, 346)
382. Zeilfelder, F., Strong unicity of best uniform approximations from periodic spline spaces. J. Approx. Theory **99**, pp. 1–29 (1999). (303)
383. Zucker, J.I., Iterated inductive definitions, trees and ordinals. In: Troelstra (ed.), Metamathematical Investigation of Intuitionistic Arithmetic and Analysis. LNM **344**, pp. 392–453. Springer, Berlin (1973). (160)

List of formal systems and term classes

List of axioms and rules

Index

Springer Monographs in Mathematics

This series publishes advanced monographs giving well-written presentations of the "state-of-the-art" in fields of mathematical research that have acquired the maturity needed for such a treatment. They are sufficiently self-contained to be accessible to more than just the intimate specialists of the subject, and sufficiently comprehensive to remain valuable references for many years. Besides the current state of knowledge in its field, an SMM volume should also describe its relevance to and interaction with neighbouring fields of mathematics, and give pointers to future directions of research.

Abhyankar, S.S. **Resolution of Singularities of Embedded Algebraic Surfaces** 2nd enlarged ed. 1998
Alexandrov, A.D. **Convex Polyhedra** 2005
Andrievskii, V.V.; Blatt, H.-P. **Discrepancy of Signed Measures and Polynomial Approximation** 2002
Angell, T.S.; Kirsch, A. **Optimization Methods in Electromagnetic Radiation** 2004
Ara, P.; Mathieu, M. **Local Multipliers of C^*-Algebras** 2003
Armitage, D.H.; Gardiner, S.J. **Classical Potential Theory** 2001
Arnold, L. **Random Dynamical Systems** corr. 2nd printing 2003 (1st ed. 1998)
Arveson, W. **Noncommutative Dynamics and E-Semigroups** 2003
Aubin, T. **Some Nonlinear Problems in Riemannian Geometry** 1998
Auslender, A.; Teboulle, M. **Asymptotic Cones and Functions in Optimization and Variational Inequalities** 2003
Banagl, M. **Topological Invariants of Stratified Spaces** 2006
Banasiak, J.; Arlotti, L. **Perturbations of Positive Semigroups with Applications** 2006
Bang-Jensen, J.; Gutin, G. **Digraphs** 2001
Baues, H.-J. **Combinatorial Foundation of Homology and Homotopy** 1999
Bonfiglioli, A.; Lanconelli, E.; Uguzzoni, F. **Stratified Lie Groups and Potential Theory for Their Sub-Laplacians** 2007
Böttcher, A.; Silbermann, B. **Analysis of Toeplitz Operators** 2006
Brown, K.S. **Buildings** 3rd printing 2000 (1st ed. 1998)
Chang, K.-C. **Methods in Nonlinear Analysis** 2005
Cherry, W.; Ye, Z. **Nevanlinna's Theory of Value Distribution** 2001
Ching, W.K. **Iterative Methods for Queuing and Manufacturing Systems** 2001
Chudinovich, I. **Variational and Potential Methods for a Class of Linear Hyperbolic Evolutionary Processes** 2005
Coates, J.; Sujatha, R. **Cyclotomic Fields and Zeta Values** 2006
Crabb, M.C.; James, I.M. **Fibrewise Homotopy Theory** 1998
Dineen, S. **Complex Analysis on Infinite Dimensional Spaces** 1999
Dugundji, J.; Granas, A. **Fixed Point Theory** 2003
Ebbinghaus, H.-D.; Flum, J. **Finite Model Theory** 2006
Edmunds, D.E.; Evans, W.D. **Hardy Operators, Function Spaces and Embeddings** 2004
Elstrodt, J.; Grunewald, F.; Mennicke, J. **Groups Acting on Hyperbolic Space** 1998
Engler, A.J.; Prestel, A. **Valued Fields** 2005
Fadell, E.R.; Husseini, S.Y. **Geometry and Topology of Configuration Spaces** 2001
Fedorov, Yu. N.; Kozlov, V.V. **A Memoir on Integrable Systems** 2001
Flenner, H.; O'Carroll, L.; Vogel, W. **Joins and Intersections** 1999
Gelfand, S.I.; Manin, Y.I. **Methods of Homological Algebra** 2nd ed. 2003 (1st ed. 1996)
Griess, R.L. Jr. **Twelve Sporadic Groups** 1998
Gras, G. **Class Field Theory** corr. 2nd printing 2005
Greuel, G.-M.; Lossen, C.; Shustin, E. **Introduction to Singularities and Deformations** 2007
Hida, H. ***p*-Adic Automorphic Forms on Shimura Varieties** 2004
Ischebeck, E; Rao, R.A. **Ideals and Reality** 2005
Ivrii, V. **Microlocal Analysis and Precise Spectral Asymptotics** 1998
Jakimovski, A.; Sharma, A.; Szabados, J. **Walsh Equiconvergence of Complex Interpolating Polynomials** 2006
Jech, T. **Set Theory** (3rd revised edition 2002)
Jorgenson, J.; Lang, S. **Spherical Inversion on SLn (R)** 2001
Kanamori, A. **The Higher Infinite** corr. 2nd printing 2005 (2nd ed. 2003)
Kanovei, V. **Nonstandard Analysis, Axiomatically** 2005

Khoshnevisan, D. **Multiparameter Processes** 2002
Koch, H. **Galois Theory of p-Extensions** 2002
Komornik, V. **Fourier Series in Control Theory** 2005
Kozlov, V.; Maz'ya, V. **Differential Equations with Operator Coefficients** 1999
Lam, T.Y. **Serre's Problem on Projective Modules** 2006
Landsman, N.P. **Mathematical Topics between Classical & Quantum Mechanics** 1998
Lau, D. **Function Algebras on Finite Sets: Basic Course on Many-Valued Logic and Clone Theory** 2006
Leach, J.A.; Needham, D.J. **Matched Asymptotic Expansions in Reaction-Diffusion Theory** 2004
Lebedev, L.P.; Vorovich, I.I. **Functional Analysis in Mechanics** 2002
Lemmermeyer, F. **Reciprocity Laws: From Euler to Eisenstein** 2000
Levine, M.; Morel, F. **Algebraic Cobordism** 2007
Malle, G.; Matzat, B.H. **Inverse Galois Theory** 1999
Mardesic, S. **Strong Shape and Homology** 2000
Margulis, G.A. **On Some Aspects of the Theory of Anosov Systems** 2004
Miyake, T. **Modular Forms** 2006
Murdock, J. **Normal Forms and Unfoldings for Local Dynamical Systems** 2002
Narkiewicz, W. **Elementary and Analytic Theory of Algebraic Numbers** 3rd ed. 2004
Narkiewicz, W. **The Development of Prime Number Theory** 2000
Neittaanmaki, P.; Sprekels, J.; Tiba, D. **Optimization of Elliptic Systems. Theory and Applications** 2006
Onishchik, A.L. **Projective and Cayley–Klein Geometries** 2006
Parker, C.; Rowley, P. **Symplectic Amalgams** 2002
Peller, V. (Ed.) **Hankel Operators and Their Applications** 2003
Prestel, A.; Delzell, C.N. **Positive Polynomials** 2001
Puig, L. **Blocks of Finite Groups** 2002
Ranicki, A. **High-dimensional Knot Theory** 1998
Ribenboim, P. **The Theory of Classical Valuations** 1999
Rowe, E.G.P. **Geometrical Physics in Minkowski Spacetime** 2001
Rudyak, Y.B. **On Thorn Spectra, Orientability and Cobordism** 1998
Ryan, R.A. **Introduction to Tensor Products of Banach Spaces** 2002
Saranen, J.; Vainikko, G. **Periodic Integral and Pseudodifferential Equations with Numerical Approximation** 2002
Schneider, P. **Nonarchimedean Functional Analysis** 2002
Serre, J-P. **Complex Semisimple Lie Algebras** 2001 (reprint of first ed. 1987)
Serre, J-P. **Galois Cohomology** corr. 2nd printing 2002 (1st ed. 1997)
Serre, J-P. **Local Algebra** 2000
Serre, J-P. **Trees** corr. 2nd printing 2003 (1st ed. 1980)
Smirnov, E. **Hausdorff Spectra in Functional Analysis** 2002
Springer, T.A.; Veldkamp, F.D. **Octonions, Jordan Algebras, and Exceptional Groups** 2000
Székelyhidi, L. **Discrete Spectral Synthesis and Its Applications** 2006
Sznitman, A.-S. **Brownian Motion, Obstacles and Random Media** 1998
Taira, K. **Semigroups, Boundary Value Problems and Markov Processes** 2003
Talagrand, M. **The Generic Chaining** 2005
Tauvel, P.; Yu, R.W.T. **Lie Algebras and Algebraic Groups** 2005
Tits, J.; Weiss, R.M. **Moufang Polygons** 2002
Uchiyama, A. **Hardy Spaces on the Euclidean Space** 2001
Üstünel, A.-S.; Zakai, M. **Transformation of Measure on Wiener Space** 2000
Vasconcelos, W. **Integral Closure. Rees Algebras, Multiplicities, Algorithms** 2005
Yang, Y. **Solitons in Field Theory and Nonlinear Analysis** 2001
Zieschang, P.-H. **Theory of Association Schemes** 2005